Finite Mathematics and Calculus

Under the Editorship of the Late Carl B. Allendoerfer

Finite Mathematics and Calculus

Applications in Business and the Social and Life Sciences

Hugh G. Campbell
Virginia Polytechnic Institute and State University

Robert E. Spencer
Virginia Polytechnic Institute and State University

Macmillan Publishing Co., Inc.
New York
Collier Macmillan Publishers
London

Copyright © 1977, Hugh G. Campbell and Robert E. Spencer

PRINTED IN THE UNITED STATES OF AMERICA

All rights reserved. No part of this book may be reproduced or transmitted in any form or by any means, electronic or mechanical, including photocopying, recording, or any information storage and retrieval system, without permission in writing from the Publisher.

Based on the authors' *Finite Mathematics*, copyright © 1974 by Hugh G. Campbell and Robert E. Spencer, and *A Short Course in Calculus with Applications*, copyright © 1975 by Hugh G. Campbell and Robert E. Spencer.

MACMILLAN PUBLISHING CO., INC.
866 Third Avenue, New York, New York 10022

COLLIER MACMILLAN CANADA, LTD.

Library of Congress Cataloging in Publication Data

Campbell, Hugh G
  Finite mathematics and calculus.

  Based on the authors' Finite mathematics, published in 1974 and on their A short course in calculus with applications, published in 1975.
    Bibliography: p.
    Includes index.
    1. Mathematics—1961–   2. Calculus. I. Spencer, Robert E., joint author. II. Title.
QA39.2.C34   510   75-45286
ISBN 0-02-318600-3

Printing: 1 2 3 4 5 6 7 8   Year: 7 8 9 0 1 2 3

To Carole Spencer and Greg, Ginger, and Allen Campbell

Preface

For some time many colleges and universities have recognized the need for a sequence of mathematics courses specifically designed for freshman-sophomore students who major in business or in the social or life sciences. Nevertheless, the length of the sequence and the choice of its contents continue to vary widely. In recognition of these facts, this book was written to accommodate a diversity of needs and preferences of instructors who bear responsibility for these sequences.

A conscious effort was made to organize the book in such a way that the instructor can easily alter the emphasis or length of a course to suit the needs of the class. Many of the sections are independent of other sections and are so advertised at the outset of each chapter; all the applications are optional; and the format of the text is such that the theory or applications can be emphasized or avoided as desired.

The material is organized into fifteen chapters with a total of ninety-eight sections; each section is designed to constitute one lesson, but two lessons will be needed in some cases if the optional applications are studied in detail. Thus the material can be used for periods of one to five quarters or for periods of one to three semesters. At the beginning of each chapter, its prerequisites are stated, except for a few that are assumed in the optional subsections on applications. These statements of prerequisites make it clear what previous sections in the book are needed before studying each new section.

One distinguishing feature of this book is its preoccupation with motivation—a vital part of the learning process. The authors have devoted considerable energy toward the goal of furnishing the reader with reasons for studying the mathematics included in this book. Almost every section includes a subsection entitled "Applications"; these subsections, however, are not designed as comprehensive lists of real-world applications, nor is any claim made concerning the realism of many examples. Rather, these "indications of usefulness" are designed to point out in a limited and elementary way that the material is relevant to a broad spectrum of disciplines and also that it has considerable potential for future applications. Many of the applications are documented to permit further investigation by the reader.

A second distinguishing feature of this book is its emphasis on calculus and the applications of calculus in probability and statistics. The presentation of calculus in this book is more comprehensive than found in many books written for the same audience. Also, there are many examples that show how calculus is related to a study of certain topics of probability and statistics; these examples are labeled clearly and occur in Sections 9.3, 10.2, 11.4, 13.2, 13.3, 13.5, 13.6, and 14.7.

Above all, the guiding principle in the construction of this book has been consideration for the needs of the student. The sincerity of the authors in this regard is illustrated by the attempt to motivate through the subsections on applications, the list of new vocabulary at the end of each chapter, the numerous exercises, and the many examples illustrating the concepts.

The authors express sincere appreciation to the many people who have assisted us in the production of this book. Special gratitude is expressed for the help of Everett Smethurst, William F. Tyndall, H. Earl Spencer, and the late Carl Allendoerfer, all of whom have made many valuable suggestions. Our respective wives, Allen and Carole, also deserve considerable thanks for their encouragement and typing during the development of the book.

Blacksburg, Virginia

H. G. C.
R. E. S.

Contents

Chapter 1 Sets 1
1.1 Basic Definitions 1
1.2 Set Operations 7
1.3 Properties of Set Operations (Optional) 15
1.4 The Number of Elements in a Set 20
1.5 Cartesian Product 26

Chapter 2 Logic 31
2.1 Statement Forms and Connectives 31
2.2 Evaluation of Statement Forms 37
2.3 Equivalent Statement Forms 41
2.4 Properties of the Algebra of Components 45
2.5 Arguments 50
2.6 Series and Parallel Systems 55
2.7 Evaluation of Compound Components 61
2.8 Compound Components and the Computer 69

Chapter 3 Review of Algebra; Functions and Graphs 73
3.1 Exponents 73
3.2 Factoring, Special Products, and Fractions 77
3.3 Solving Equations 83
3.4 Formulas from Geometry 90
3.5 Inequalities and Absolute Values 94
3.6 An Introduction to Analytic Geometry 101
3.7 Functions 107
3.8 Straight Lines 117
3.9 Some Special Types of Functions 126
3.10 Rectangular Coordinate System in Three Dimensions 134

Chapter 4 Probability 141
4.1 Basic Terminology 141
4.2 Probabilities of Complements, Unions, and Intersections of Events 150
4.3 Expected Value 156
4.4 Conditional Probability 162
4.5 Independent Events 171
4.6 Stochastic Experiments 178
4.7 Counting Techniques: Permutations 185
4.8 Counting Techniques: Combinations 195
4.9 Counting Techniques: Partitions (Optional) 203
4.10 Binomial Experiments 209

Chapter 5 The Algebra of Matrices   219
5.1 Matrices   219
5.2 Vectors   227
5.3 Matrix Multiplication   235
5.4 Special Matrices   245
5.5 Systems of Linear Equations Having a Unique Solution   256
5.6 The Inverse Matrix   264
5.7 Systems of Linear Equations Not Having a Unique Solution   272
5.8 The Gauss-Jordan Method   278
5.9 The Determinant of a Matrix (Optional)   282
5.10 The Characteristic Value Problem (Optional)   288

Chapter 6 Linear Programming   295
6.1 The Feasible Set of a Linear Programming Problem   295
6.2 A Geometric Method of Solution   298
6.3 An Algebraic Method of Solution   305
6.4 An Algebraic Method of Solution (Continued)   309
6.5 Matrix Notation   313
6.6 An Introduction to the Simplex Method   315
6.7 The Simplex Method   318

Chapter 7 Game Theory   328
7.1 Games and Strategies   328
7.2 Pure Strategies   335
7.3 An Optimal Strategy for the Row Player   342
7.4 An Optimal Strategy for the Column Player   347

Chapter 8 Statistics   354
8.1 Introduction   354
8.2 Grouped Data and Summation Notation   359
8.3 Measures of Central Tendency   365
8.4 Measures of Variation   371

Chapter 9 Limits and Continuity   379
9.1 Introduction to Calculus   379
9.2 Limits   382
9.3 Limits (Continued)   390
9.4 Continuity   398

Chapter 10 Differentiation   407
10.1 Instantaneous Rates of Change and Tangent Lines   407
10.2 The Derivative and Its Interpretations   416
10.3 Basic Differentiation Formulas   426
10.4 Basic Differentiation Formulas (Continued)   432
10.5 Higher Derivatives   438

Contents

10.6 Differentials   442
10.7 Implicit Differentiation (Optional)   448

Chapter 11 Maximum and Minimum Problems   455
11.1 Introduction   455
11.2 Increasing and Decreasing Functions   461
11.3 First Derivative Test for Maximum and Minimum Points   467
11.4 Applied Maximum and Minimum Problems   476
11.5 Concavity and Second Derivative Test   485
11.6 Curve Sketching   492
11.7 Mean Value Theorem for Derivatives (Optional)   501

Chapter 12 Antiderivatives and Mathematical Models   507
12.1 Introduction   507
12.2 An Important Integration Formula   511
12.3 Differential Equations   516
12.4 Mathematical Models (Optional)   523
12.5 Mathematical Models (Continued) (Optional)   526

Chapter 13 The Definite Integral   532
13.1 Definition   532
13.2 Fundamental Theorem of Integral Calculus   539
13.3 Properties of Definite Integrals   548
13.4 Area Between Curves   555
13.5 More About Calculation of Definite Integrals (Optional)   562
13.6 Discussion of Proof of Fundamental Theorem (Optional)   568

Chapter 14 Exponential and Logarithmic Functions   577
14.1 Exponential Functions   577
14.2 Logarithmic Functions   584
14.3 Differentiation Formulas   591
14.4 Integration Formulas   597
14.5 Exponential Growth and Decay   606
14.6 Integration Using Table of Integration Formulas   612
14.7 Integrals with Infinite Limits   617

Chapter 15 Functions of More Than One Variable   627
15.1 Introduction   627
15.2 Partial Derivatives   631
15.3 Maximum and Minimum Points   637
15.4 Lagrange Multipliers   644

References   652
Table of Integration Formulas   656
Answers to Odd-numbered Exercises   658
Index   703

Finite Mathematics and Calculus

1 Sets

Over the years the development of mathematics has closely paralleled the need for quantification of natural and other phenomena. From the time man invented symbols that correspond to our numerals 1, 2, ..., there has been a relentless march toward the use of mathematics in ever more diverse ways. Certainly, the rate of this advance has not been constant and often has been related to the intellectual and political climates of the age as well as the maturity and influence of sister disciplines. Early interest in astronomy, for example, provided considerable impetus for the development of related mathematical concepts and techniques. Over a period of years, many of these concepts and techniques have led to a comparatively sophisticated development of the engineering sciences; the development of the computer and space travel give ample evidence of this fact. The computer and other engineering developments, in turn, are now having a tremendous effect upon modern mathematics, as well as having the effect of opening vast new areas for quantification in the social, management, and life sciences.

In this book we have chosen a few mathematical topics that have shown considerable promise in the social, management, and life sciences. Extensive quantification in these areas is a relatively new thing, and there is not yet universal acceptance of the usefulness of mathematical models in these disciplines. Although many criticisms are probably valid and useful, we should not lose sight of the fact that similar battles were fought over the usefulness of calculus in the natural sciences back in the 17th and 18th centuries. It should be emphasized that the art of application of the topics of this book to the behavioral, management, and life sciences is in its infancy, and while there is considerable reason to believe that mathematics can contribute significantly to our understanding of these areas of knowledge, much work remains to be done.

In this first chapter we begin with a brief study of sets—a topic that is fundamental to much of modern mathematics.

Prerequisites: high school algebra.
Suggested Sections for Moderate Emphasis: 1.1, 1.2, 1.4, 1.5.
Suggested Sections for Minimum Emphasis: 1.1, 1.2, 1.4.

1.1 Basic Definitions

Although it is customary in mathematics to treat the words **set** and **element** as undefined terms, it is useful to have intuitive interpretations of these

terms when undertaking an introductory study of set theory. Think of a well-defined collection of distinguishable objects; by a well-defined collection, we mean that it is clear whether an arbitrary object does or does not belong to the collection, and by distinguishable, we mean that we can distinguish one object from another. For example, we may think of a collection of postage stamps, or a collection of current U.S. Senators, or a collection of real numbers that satisfies the equation $x^2 = 1$. Such well-defined collections of distinguishable objects serve as intuitive interpretations of the word *set*, while the objects themselves are examples of the *elements* of a set.

We usually denote sets by capital letters $(A, B, C, \ldots)$ and the elements of a set by lower-case letters. The statement "b is an element of B" is expressed as $b \in B$.

There are several ways that a specific set can be described or defined. First, a defining statement can be used: "A is the set of the first three U.S. Presidents" defines a set A. Second, the same set can be defined by listing the elements within braces; for example,

$$A = \{\text{Washington, Adams, Jefferson}\}.$$

Third, using a colon to represent the expression "such that," we can use the set-builder notation; for example,

$$A = \{x : x \text{ is one of the first three U.S. Presidents}\},$$

which is read "the set of all x such that x is one of the first three U.S. Presidents." The letter x in the set-builder notation is simply a placeholder, and other symbols could have been used.

Next, the concept of a **subset** of a set is introduced by means of an example. Consider the set C of the first two U.S. Presidents; that is,

$$C = \{\text{Washington, Adams}\}.$$

Each element of C is also an element of the set

$$A = \{\text{Washington, Adams, Jefferson}\},$$

and we choose to express this important relationship between A and C by saying that C is a subset of A.

▶ **Definition 1**

Set R is a **subset** of a set S if every element of R is also an element of S; "R is a subset of S" will be denoted

$$R \subseteq S.$$

Example 1

If P represents the set of all living people in the world and F represents the set of living people that are female, then we can say that

$$F \subseteq P.$$

which is read "F is a subset of P." If there is at least one living nonfemale person, then we say that F is a **proper subset** of P, and that concept is denoted as

$$F \subset P.$$

In other words, a proper subset does not include all the elements of the parent set; a subset may or may not include all the elements of the parent set. ∎

It is possible to describe a set in which there are no elements. For example, suppose we say: Let S be the set of U.S. Presidents who were born in Mongolia. Because there are no elements of this set, we say that the set is the **empty set** or the **null set** and denote such a set with the symbol ∅ (Scandinavian modified "oh"). When working with sets, experience has shown that it is useful to identify a set U such that all sets in the discussion are subsets of U; such a set is called the **universal set.** If, for example, a discussion revolves around various sets of nations in the world, then we could choose the set of all nations of the world as the universal set, that is, the universe of the discussion.

Often it is helpful to draw graphical representations of sets. First, we draw a large rectangle that represents the universal set, as shown in Figure 1.1. Then we can represent a subset A of U by a circle (or other closed curve) within the rectangle, as shown in Figure 1.2. We may think of points within the circle as representing elements x of U for which $x \in A$ becomes a true statement and points outside of the circle as representing elements x of U for which $x \in A$ becomes a false statement. Next, we can represent a subset of A, which we will call B, by another closed curve within A, as shown in Figure 1.3; such diagrams are often called **Venn diagrams.**

Next, we wish to use an illustration to make a distinction between "a subset of a set" and "an element of a set." Consider a set A of all states in

Figure 1.1. *Representation of universal set.*

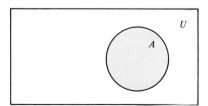

Figure 1.2. *Representation of set A.*

Sec. 1.1] Basic Definitions

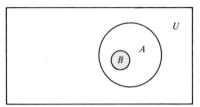

Figure 1.3. *Representation of a subset of A.*

the United States. Consider a second set E of states that lie entirely east of the Mississippi River. We say that E is a subset of A; it is wrong to say that E is an element of A. Now consider the state of Maine, which we shall call m. We say that m is an element of A, that is, $m \in A$; but it is wrong to say that m is a subset of A. We can say, however, that the set consisting of the single state of Maine is a subset of A; that is, we can say that $\{m\} \subseteq A$. Suppose that we let w represent the state of Wyoming; by the symbol $w \notin E$ we mean that Wyoming does *not* belong to set E. By the symbol $\{w\} \subseteq E$ we mean that the set consisting of the state of Wyoming is *not* a subset of E.

Example 2
Let $A = \{a, b, c\}$. The subsets of A are $\{a, b, c\}$, $\{a, b\}$, $\{a, c\}$, $\{b, c\}$, $\{a\}$, $\{b\}$, $\{c\}$, and $\emptyset$; notice that both the empty set and the whole set are included in the list of subsets. The elements of A are a, b, and c. The following are *correct* statements:

$$\{a\} \subset A, \qquad a \in A,$$
$$\{a, b\} \subset A, \qquad \{a, b\} \subseteq A,$$
$$\{a, b, c\} \subseteq A, \qquad \emptyset \subset A.$$

The following are *incorrect* statements:

$$\{b\} \in A, \qquad b \subset A, \qquad \{a, b, c\} \subset A.$$

The first one is incorrect because $\{b\}$ represents a subset and not an element; the second is incorrect because b represents an element and not a subset; the third statement is incorrect because $\{a, b, c\}$ is *not* a *proper* subset of A. ∎

Many of the sets encountered in this book will be subsets of the set of real numbers. Rather than attempt to state all the definitions and postulates needed for a rigorous development of the properties of the real number system, we assume some familiarity with real numbers from high school algebra and proceed with facts that the reader will need.

The real numbers can be represented geometrically by the **real line,** shown in Figure 1.4. We assume that for each real number there is a corresponding point on the real line, and vice versa. The point O is called the **origin,** and the number corresponding to any point P is called the

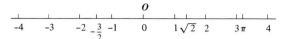

Figure 1.4. *Real line.*

coordinate of *P*. Included among the real numbers are the **positive integers** or **natural numbers,**

$$1, 2, 3, 4, \ldots,$$

the **negative integers,**

$$-1, -2, -3, -4, \ldots,$$

and the integer 0 (**zero**), which is neither positive nor negative. A **rational number** is a real number that can be expressed as the ratio a/b of two integers a and b, where $b \neq 0$. The rational numbers include the integers (any integer a can be expressed as $a/1$) and fractions such as $-\frac{3}{2}, \frac{1}{3}$, and $\frac{22}{7}$. Any real number that is not a rational number is called an **irrational number.** It can be shown that numbers such as $\sqrt{2}, \sqrt[3]{5}$, and π are irrational numbers. Every real number is known to have a decimal representation, and it may be helpful to know that terminating or repeating decimal representations always correspond to rational numbers: for example,

$$3.125 = \tfrac{25}{8}$$

and

$$.124242424\ldots = \tfrac{41}{330}.$$

Long division of 41 by 330 will verify the second equality. An irrational number has a decimal representation that is nonterminating and nonrepeating.

We describe a **finite set** as a set having n elements, where either n is a positive integer or $n = 0$. The set $\{3, 4, 5, 6\}$ is a finite set because it contains $n = 4$ elements. If a set is not a finite set, it is called an **infinite set.** The set of all real numbers is an infinite set; the set of all positive integers, the set of all rational numbers, and the set of all irrational numbers are also infinite sets.

EXERCISES

Suggested minimum assignment: Exercises 1, 3, 6, 7, 9, 13, 15, 17, 18, 19, 23, and 27.

In each of Exercises 1–8 express the given set in two additional ways.

1. *A* is the set of all states in the region of the United States that is called New England.
2. *B* is the set of living men who have been President of the United States.
3. *C* is the set of integers that satisfy the equation $x^2 - 1 = 0$.

4. D is the set of rational numbers that satisfy the equation
$$(2x + 3)(x - 2) = 0.$$
5. $E = \{x : (x - 3)(x + 2) = 0,$ and x is a positive integer$\}$.
6. $F = \{x : x$ is the capital city of New York or New Jersey or South Carolina$\}$.
7. $G = \{1, 2, 3\}$.
8. $H = \{$Ford, General Motors, Chrysler, American Motors$\}$.

In each of Exercises 9–12 write a proper subset of the given set in symbolic notation. Also express an element of the given set using symbolic notation.

9. $A = \{3, 4, 5\}$.
10. $B = \{x, y, z\}$.
11. C is the set of all men who currently are President or Vice-President of the United States.
12. $D = \{x : 1 \leq x \leq 3, x$ is an integer$\}$.
13. If the universal set of discussion is the set of cities $S = \{$Chicago, Cleveland, Pittsburgh$\}$, list all the subsets of S.
14. If the universal set of discussion is the set of stocks $B = \{$AT&T, GE, IBM, GM$\}$, list all the subsets of B.

In each of Exercises 15–22 state whether the given statement is true or false. If the statement is false, change it so that it becomes a true statement. Let
$$U = \{2, 3, 4, 5\}.$$

15. $2 \in U$.
16. $\emptyset \subset U$.
17. $\{2, 3\} \in U$.
18. $\{2, 3, 4, 5\} \subseteq U$.
19. $\emptyset \in U$.
20. $3 \subset U$.
21. $6 \notin U$.
22. $\{4, 5\} \subseteq U$.

23. If the universal set of discussion is the U.S. Senate, draw a Venn diagram showing the subset of all Democrats and the subset of all Republicans. List one member of each subset. Draw a second diagram in which another subset of all liberals (defined according to some index) is superimposed on a duplicate of the first diagram.
24. Use set notation to express the relationship between the set of all Republicans and the set of all liberal Republicans in the U.S. Senate.
25. Tell whether each of the following real numbers is rational or irrational:
$$\tfrac{2}{3}, \quad 0, \quad \pi, \quad 2\pi, \quad 5, \quad \sqrt{2}, \quad .44444\ldots, \quad -\tfrac{9}{8}.$$
26. Tell whether each of the following real numbers is rational or irrational:
$$\sqrt{9}, \quad \sqrt[3]{9}, \quad -3, \quad 1 + \pi, \quad \tfrac{3}{4}, \quad .7239, \quad \tfrac{22}{7}, \quad \tfrac{-3}{2}.$$

27. Let A be the set of all real numbers, B the set of all rational numbers, C the set of all irrational numbers, and D the set of all integers. Tell whether each of the following is true or false.

(a) $B \subset A$. (b) $C \subset A$. (c) $D \subseteq C$.
(d) $A \subseteq B$. (e) $D \subseteq B$. (f) $0 \in B$.

1.2 Set Operations

In this section we shall introduce various operations whereby subsets of a universal set U can be combined to produce a unique subset of U. We shall also present the concept of the complement of a single subset of U.

First, however, we establish what is meant by the **equality** of two sets.

▶ **Definition 2**
Two sets A and B are said to be **equal** if their elements are the same.

The fact that the elements of two sets A and B are the same if $A \subseteq B$ and $B \subseteq A$ is frequently used to establish the equality of two sets.

Example 1
Let $A = \{$Ford, Chrysler, General Motors$\}$. Let B be the set of corporations that are the three largest manufacturers of automobiles in the United States. We can show that $A \subseteq B$ and we can show that $B \subseteq A$; therefore, $A = B$; that is, the elements of A and B are identical. ∎

We now turn our attention to ways of combining sets.

▶ **Definition 3**
Let A and B be arbitrary subsets of a universal set U. The set of elements that belong to both A and B is called the **intersection** of A and B and will be designated by

$$A \cap B.$$

Example 2
Let $A = \{a, b, c, d\}$ and let $B = \{c, d, e\}$. Then $A \cap B = \{c, d\}$. ∎

Example 3
Let F be the set of all Ford automobiles in this country. Let R be the set of all red automobiles in this country. Then $F \cap R$ is the set of all red Fords in this country. ∎

We now relate the intersection of two sets to Venn diagrams. For an arbitrary element x of U, the intersection, $A \cap B$, can be thought of as the set of elements of U which when substituted for x in

"$x \in A$ and $x \in B$"

produces a true statement. That is, for $x \in U$,
$$A \cap B = \{x : (x \in A) \text{ and } (x \in B)\}.$$

From Table 1.1 we obtain the Venn diagram representation of $A \cap B$, as shown by the colored region in Figure 1.5. In Figure 1.5 the number of each region corresponds to the number of the row of Table 1.1. The reader should realize that any of the numbered subsets shown in Figure 1.5 can be empty.

Table 1.1. INTERSECTION OF A AND B.

	$x \in A$	$x \in B$	$x \in A \cap B$
(1)	True	True	True
(2)	True	False	False
(3)	False	True	False
(4)	False	False	False

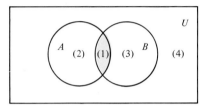

Figure 1.5. *Intersection of A and B.*

If the intersection of two sets is the empty set, then the sets are said to be **disjoint**. Next we define a second set operation known as the **union** of two sets.

▶ **Definition 4**

Let A and B be arbitrary subsets of a universal set U. The set of elements that belong to A or B (including those that belong to both) is called the **union** of A and B and will be designated by
$$A \cup B.$$

Example 4
Let $A = \{a, b, c, d\}$ and let $B = \{c, d, e\}$. Then $A \cup B = \{a, b, c, d, e\}$. ∎

Example 5
By using the same sets that were described in Example 3, $A \cup B$ represents the set of all automobiles in this country that are red or Fords or both. ∎

We now relate the union of two sets to Venn diagrams. For an arbitrary element x of U, the union, $A \cup B$, can be thought of as the set of elements of U which when substituted for x in
$$\text{``} x \in A \text{ or } x \in B \text{''}$$

produces a true statement. That is, for $x \in U$,
$$A \cup B = \{x : (x \in A) \text{ or } (x \in B)\}.$$
From Table 1.2 we obtain the Venn diagram representation of $A \cup B$ as shown by the colored region in Figure 1.6. In Figure 1.6 the number of each region corresponds to the number of the row of Table 1.2. Again, it is pointed out that any of the numbered subsets shown in Figure 1.6 can be empty.

Table 1.2. UNION OF A AND B.

	$x \in A$	$x \in B$	$x \in A \cup B$
(1)	True	True	True
(2)	True	False	True
(3)	False	True	True
(4)	False	False	False

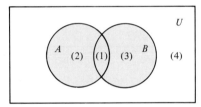

Figure 1.6. *Union of A and B.*

Finally, we define a third set operation, known as the **complementation** of a set.

▶ **Definition 5**
Let A be an arbitrary subset of a given universal set U. The set of all elements that belong to U but do not belong to A is called the **complement** of A and is denoted A'.

Example 6
Let U be the English alphabet, and let A be the set of consonants. Then A' is the set of vowels. ∎

Example 7
Let U be the set of U.S. Senators, and let R be the set of Republican U.S. Senators. Then R' represents the set of all non-Republican U.S. Senators. ∎

We now relate the complement of a set to Venn diagrams. For an arbitrary element x of U, the complement of A can be thought of as the set of elements of U which when substituted for x in

"it is not true that $x \in A$"

produces a true statement. That is, for $x \in U$,
$$A' = \{x : \text{it is not true that } x \in A\}.$$

From Table 1.3 we obtain the Venn diagram representation of A', as shown by the colored region in Figure 1.7. In this figure the number of each region corresponds to the number of the row of Table 1.3.

Table 1.3. COMPLEMENT OF A.

	$x \in A$	It is not the case that $x \in A$
(1)	True	False
(2)	False	True

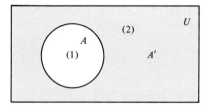

Figure 1.7. *Complement of A.*

Definitions 3 and 4 can be extended to define the intersection and union of three or more sets. Let $A_1, \ldots, A_n$ be subsets of a universal set U. By $A_1 \cap \cdots \cap A_n$ we mean the set of all elements of U that belong to all of the sets $A_1, \ldots, A_n$. By $A_1 \cup \cdots \cup A_n$ we mean the set of all elements of U which belong to at least one of the sets $A_1, \ldots, A_n$.

APPLICATIONS

The applications given at the end of each section are certainly not meant to be an exhaustive listing of the applications of that particular section, nor is any claim made concerning the realism of many of the examples. Rather, these "evidences of usefulness" are supplied primarily in the interest of motivating the reader and with the hope that occasionally one will provide a spark that may arouse a much deeper, productive study in some areas of interest. **All applications are optional in that they are not prerequisite to later material in the main part of the text.**

Example 8

Prior to a blood transfusion certain tests are executed on both the recipient and the prospective donor in order to classify the blood of each and hence determine the suitability of the prospective donor's blood. Blood classification is accomplished by determining whether or not a person reacts to each of three tests. These tests and the corresponding classification are given in Table 1.4. If we let set A represent the set of all people who react to test A,

Table 1.4. SET REPRESENTATIONS OF BLOOD CLASSIFICATIONS.

Blood classification	A test	B test	Rh test	Set notation
AB Rh positive	reacts	reacts	reacts	$A \cap B \cap R$
A Rh positive	reacts	no	reacts	$A \cap B' \cap R$
B Rh positive	no	reacts	reacts	$A' \cap B \cap R$
O Rh positive	no	no	reacts	$A' \cap B' \cap R$
AB Rh negative	reacts	reacts	no	$A \cap B \cap R'$
A Rh negative	reacts	no	no	$A \cap B' \cap R'$
B Rh negative	no	reacts	no	$A' \cap B \cap R'$
O Rh negative	no	no	no	$A' \cap B' \cap R'$

let set B represent the set of all people who react to test B, and let set R represent the set of all people who react to the Rh test, then each of the classifications can be expressed using set notation, as shown in the last column of Table 1.4.

Now construct for each *donor* a set D of tests to which he *did* react, and for each *recipient* a set C of tests to which he *did* react. In order for a transfusion to be safe, $D \subseteq C$, and from this relationship, donors for particular recipients can be identified. For example, a donor in the set

$$A' \cap B \cap R,$$

for which $D = B \cap R$, can give blood only to those in the sets

$$A' \cap B \cap R \quad \text{or} \quad A \cap B \cap R.$$

On the other hand, a person in the set $A' \cap B' \cap R'$ can give blood to anyone but can receive blood only from someone else in the same set. ∎

Example 9

The purpose of this example is to illustrate the use of sets and set operations in the analysis of legal documents. The Tenth Amendment to the U.S. Constitution states: "The powers not delegated to the United States by the constitution, nor prohibited by it to the states, are reserved to the states respectively, or to the people." Let U represent the universe of powers referred to by the Tenth Amendment. Under the assumption that these powers are well defined, let

D = set of all powers delegated to the United States by the Constitution;
P = set of all powers prohibited to the states by the Constitution;
S = set of powers of the states;
T = set of powers of the people.

Under the assumptions that $S \cap P = \emptyset$, $D \cup D' = U$, and $P \cup P' = U$, we have the Venn diagram of Figure 1.8.

The various disjoint subsets defined by intersections of two or more of sets D, P, S, and T are numbered in Figure 1.8 for reference.

Sec. 1.2] Set Operations

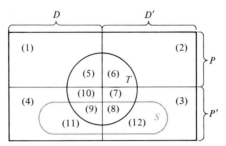

Figure 1.8. *Representation of the Tenth Amendment.*

The Tenth Amendment, which says that

$$(D' \cap P') \subseteq (S \cup T),$$

requires that subset (3) be empty. Various arguments can be made that other of the numbered subsets of Figure 1.8 are empty. One point of view is that subsets (5), (7), (8), (9), (10), and (11) are all empty. Another point of view is that (2), (4), (5), and (6) are empty; many of these disagreements arise because of a violation of our first assumption that the powers are well defined. In other words, there is disagreement over membership in the sets D, P, S, and T, particularly the latter two. The interested reader may wish to list various powers and establish set membership of these powers according to predetermined criteria. Such an analysis using sets may be quite helpful in legal interpretations. ∎

Example 10

In our society we frequently encounter sets of people who have authority to make decisions: for example, boards of directors of corporations, state legislatures, and appeals courts. These sets of people usually operate under by-laws that specify the manner in which decisions can be accomplished. Often a simple majority of those voting is all that is required to effect a decision. In some cases, however, a majority of the members in the set, regardless of the number present, is required. There are also cases where a $\frac{2}{3}$ or $\frac{3}{4}$ majority is required. In the case of the stockholders of corporations, ordinarily a person's vote is weighted according to the number of shares of stock he owns.

Let the universal set U consist of the set of people charged with making certain decisions; a subset W of U is called a *winning coalition* if the voting power of W is sufficient to pass an issue. In such a case the subset W' is called a *losing coalition*. In some situations it is possible for a subset B of members to have sufficient voting strength to keep another coalition from passing a certain measure but not have sufficient voting strength for its own view to prevail; such a subset B is called a *blocking coalition*. An illustration of these concepts may be found in the 15-nation Security Council of the United Nations, where each of the Big Five nations (United States, United Kingdom, U.S.S.R., France, and China) has veto power. Since it takes 9 votes

for a measure to pass, any winning coalition W must consist of all the Big Five plus 4 or more additional nations. The complement of W would constitute a losing coalition. A blocking coalition would be a subset that included from 1 to 5 of the Big Five or a subset that included 7 or more of the "other nations." ∎

EXERCISES

Suggested minimum assignment: Exercises 1, 3, 6, 7, 8, 9, 11, and 13.

1. Let $U = \{1, 2, 3, 4, 5, 6\}$, let $A = \{1, 2, 3, 4\}$, and let $B = \{4, 5, 6\}$. Find:
   (a) $A \cup B$. (b) $A \cap B$. (c) A'. (d) $A \cup (A' \cap B)$.
   (e) $A \cup \emptyset$. (f) $B \cap B'$. (g) $U \cup A$. (h) $\emptyset \cap B$.

2. Let $U = \{a, b, c, d, e\}$, let $A = \{a, b, c\}$, and let $B = \{b, c, d\}$. Find:
   (a) $A \cup B$. (b) $A \cap B$. (c) A'. (d) $B \cup (A \cap B')$.
   (e) $A \cap \emptyset$. (f) $B \cup B'$. (g) $U \cap A$. (h) $\emptyset \cup B$.

3. Let U be the set of all U.S. Senators, let P be the set of U.S. Senators from those states that are adjacent to the Pacific Ocean, and let R be the set of all Republican Senators. Use set-builder notation to represent
   (a) $P \cup R$. (b) $P \cap R$. (c) R'.
   Draw Venn diagrams to represent these sets.

4. Let U be the set of all Justices on the U.S. Supreme Court, let A represent those who were appointed by a Democratic President, and let E be those who resided in states east of the Mississippi at the time of their appointment. Use set-builder notation to represent
   (a) $A \cup E$. (b) $A \cap E$. (c) A'.
   Draw Venn diagrams to represent these sets.

5. For a subset $A = \{1, 2\}$ of a universal set $U = \{1, 2, 3, 4\}$, write a set that is equal to each of the following.
   (a) $A \cap U'$. (b) $A \cup \emptyset'$. (c) $\emptyset \cup U$. (d) $A \cup A$.

6. For a subset $A = \{x, y\}$ of a universe $U = \{u, v, x, y, z\}$, write a set that is equal to each of the following.
   (a) $A \cup U'$. (b) $A \cap \emptyset'$. (c) $\emptyset \cap U$. (d) $A \cap A$.

In Exercises 7–12 use the following set designations, assuming that none are empty:

Let U be the set of all members of a certain legislature.
Let D be the set of all Democrats.
Let $R = D'$ be the set of all Republicans.
Let C be the set of all conservatives.
Let L be the set of all liberals.
Let N be the set of all members that are neither conservatives nor liberals.

Use set notation to identify:

7. The set of liberal Democrats.
8. A conservative–Republican coalition.

9. The relation between liberal Democrats and all Democrats.
10. What is the intersection of N and L (using one term)?
11. What is the union of N and L (using one term)?
12. Draw a Venn diagram showing all sets. Draw this diagram in such a way that no null (empty) set is showing in the diagram.

13. Let U be the set of real numbers.
    Let K be the set of rational numbers.
    Let Z be the set of irrational numbers.
    Let I be the set of integers.
    Let N be the set of natural numbers.

    Use set notation to identify the following.
    (a) K'.   (b) Z'.   (c) $I \cup N$.   (d) $I \cap N$.
    (e) $Z \cup K$.   (f) $Z \cap I$.   (g) $I \cup K$.   (h) $I' \cap Z$.

14. For subsets A and B of a universal set U, we define another operation by
$$A - B = A \cap B',$$
which is sometimes read "A minus B" or "the **relative complement of B in A**"; from the colored region of the Venn diagram shown in Figure 1.9 we see that $A - B$ represents the set of all elements in A that are not in B. For the sets defined in Exercises 1 and 2, find $A - B$ and $B - A$.

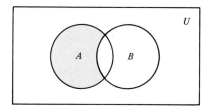

Figure 1.9. *Relative complement of B in A.*

15. For subsets A and B of a universal U, we define another operation by
$$A \triangle B = (A \cup B) \cap (A \cap B)',$$
which is sometimes called the **symmetric difference** of A and B because it can be proved that
$$A \triangle B = (A - B) \cup (B - A).$$

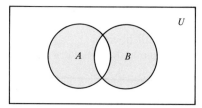

Figure 1.10. *Symmetric difference of A and B.*

See Figure 1.10 for the Venn diagram representation of $A \triangle B$. For the sets defined in Exercises 1 and 2, find $A \triangle B$.

16. Write an expression involving some or all of the symbols $\cap$, $\cup$, and $'$ for the colored region of the Venn diagram shown in Figure 1.11.

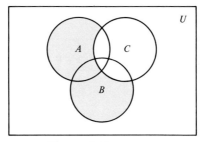

Figure 1.11

17. Write an expression involving some or all of the symbols $\cap$, $\cup$, and $'$ for the colored region of the Venn diagram shown in Figure 1.12.

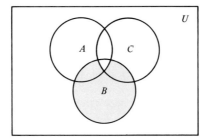

Figure 1.12

1.3 Properties of Set Operations (Optional)

We have laid the foundation for a new mathematical structure called the algebra of sets. The elements of the structure are the subsets of a designated universal set U. Equality is a relation between the subsets of U, and the operations $\cup$, $\cap$, and $'$ are defined over the elements of the structure; we also have the relations $\subseteq$ and $\subset$. We now build on this foundation. The resulting superstructure, presented as theorems and fastened into place by means of the proofs, provides the reader with an increased capability to make use of set theory in realistic applications.

▶ **Theorem 1**
(Commutative Properties) If A and B are arbitrary subsets of a universal set U, then

(i) $A \cup B = B \cup A$.
(ii) $A \cap B = B \cap A$.

Proof of (i): By the definition of the union of two sets we obtain the third and fourth columns of Table 1.5. Because the third and fourth columns are identical, the definition of set equality is satisfied and the conclusion follows.

The proof of (ii) is left as Exercise 20. ◀

Table 1.5

(1) $x \in A$	(2) $x \in B$	(3) $x \in A \cup B$	(4) $x \in B \cup A$
True	True	True	True
True	False	True	True
False	True	True	True
False	False	False	False

▶ **Theorem 2**
(Distributive Properties) If A, B, and C are arbitrary subsets of a universal set U, then

(i) $A \cap (B \cup C) = (A \cap B) \cup (A \cap C)$.
(ii) $A \cup (B \cap C) = (A \cup B) \cap (A \cup C)$.

Proof of (i): By the definitions of the union and intersection of sets we obtain the fourth and fifth columns of Table 1.6. Because the fourth and fifth columns are identical, the definition of set equality is satisfied and the conclusion follows.

The proof of (ii) is left as Exercise 21. ◀

Table 1.6

(1) $x \in A$	(2) $x \in B$	(3) $x \in C$	(4) $x \in A \cap (B \cup C)$	(5) $x \in (A \cap B) \cup (A \cap C)$
True	True	True	True	True
True	True	False	True	True
True	False	True	True	True
True	False	False	False	False
False	True	True	False	False
False	True	False	False	False
False	False	True	False	False
False	False	False	False	False

▶ **Theorem 3**
(Identity Properties) If A is an arbitrary subset of a universal set U, and if $\emptyset$ represents the empty set, then for every A,

(i) $A \cup \emptyset = A$.
(ii) $A \cap U = A$.

Proof of (i): By the definition of the union of two sets we obtain the third column of Table 1.7. Because the first and third columns are identical, the definition of set equality is satisfied and the conclusion follows.

The proof of (ii) is left as Exercise 22. ◀

Table 1.7

(1) $x \in A$	(2) $x \in \emptyset$	(3) $x \in A \cup \emptyset$
True	False	True
False	False	False

▶ **Theorem 4**
(Complement Properties) If A is an arbitrary subset of a universal set U, and if $\emptyset$ represents the empty set, then for every A,

(i) $A \cap A' = \emptyset$.
(ii) $A \cup A' = U$.

Proof of (i): By the definition of the intersection of two sets we obtain the third column of Table 1.8. Because $x \in A \cap A'$ is never true, then $A \cap A'$ must be the empty set.

The proof of (ii) is left as Exercise 23. ◀

Table 1.8

(1) $x \in A$	(2) $x \in A'$	(3) $x \in A \cap A'$
True	False	False
False	True	False

The first four theorems have been proved using only the basic definitions. These four theorems were chosen carefully, however, so that the later theory can be based entirely on these four theorems. In other words, later theorems may be proved by using only preceding theorems, and tables are no longer required. Of course, one may choose to use a table for a proof if convenient. To illustrate these alternative methods of proof, we provide two proofs of the following theorem.

▶ **Theorem 5**
If A is an arbitrary subset of a universal set U, and if $\emptyset$ represents the empty set, then for every A,

(i) $A \cup U = U$.
(ii) $A \cap \emptyset = \emptyset$.

Proof of (i)

	Statement	Reason
1.	$A \cup U = (A \cup U) \cap U$	1. Theorem 3(ii).
2.	$= (A \cup U) \cap (A \cup A')$	2. Theorem 4(ii).
3.	$= A \cup (U \cap A')$	3. Theorem 2(ii).
4.	$= A \cup (A' \cap U)$	4. Theorem 1(ii).
5.	$= A \cup A'$	5. Theorem 3(ii).
6.	$= U$	6. Theorem 4(ii).

Sec. 1.3] Properties of Set Operations

Alternative Proof of (i): By the definition of the union of two sets we obtain the third column of Table 1.9. Because $x \in A \cup U$ is always true, $A \cup U$ must be the universal set.

The proof of (ii) is left as Exercise 5. ◀

Table 1.9

(1) $x \in A$	(2) $x \in U$	(3) $x \in A \cup U$
True	True	True
False	True	True

Further properties of set operations are listed in Exercises 6–17.

APPLICATIONS

Example 1

Consider the following passage from *Alice in Wonderland*, which was written by the mathematician Charles Dodgson (Lewis Carroll) in the 19th century.

"'Now let me see,' mused Alice, 'There are only Red Knights and White Knights, and not many White ones now, since the Duchess beheaded all the mounted White Knights for riding across the croquet lawn. That was the morning when the King confiscated the horses of any Knight who couldn't sing "Humpty-Dumpty" and that was a silly thing to do, because everybody knows that no Red Knight could ever sing a note.'

She looked up, and exclaimed, 'Look! There's a man on a horse! Now I wonder whether he's a Knight or not.'"

We can analyze the passage by the use of set notation as follows. Suppose that we let

$$U = \text{set of all knights,}$$
$$R = \text{set of all red knights,}$$
$$W = \text{set of all white knights,}$$
$$M = \text{set of all mounted knights,}$$
$$S = \text{set of all singing knights.}$$

From the quotation we make the following observations (in order):
1. $U = R \cup W$, $R \cap W = \emptyset$.
2. $M \cap W = \emptyset$.
3. $S' \cap M = \emptyset$.
4. $R \subseteq S'$.

If the intersection of two sets is the empty set (that is, the sets are disjoint), then each of the sets is a subset of the complement of the other. In other words,

$$(A \cap B = \emptyset) \Rightarrow (A \subseteq B').$$

Using this implication, we can say that

and
$$(2.\ M \cap W = \emptyset) \Rightarrow (M \subseteq W')$$
$$(3.\ S' \cap M = \emptyset) \Rightarrow (S' \subseteq M').$$

From observation 1 we have $R = W'$; hence 4 becomes $W' \subseteq S'$. Using these three inclusive relations, we have

$$M \subseteq W' \subseteq S' \subseteq M'.$$

Since M is a subset of its complement, then M must be an empty set. Thus we can conclude that there are no mounted knights, and hence if there is a mounted man he is *not* a knight. ∎

Example 2
A student should have some familiarity with the notation and basic concepts of set theory because of the extent to which the notation is currently being used in the literature of the social, management, and life sciences. Articles and books abound in which the vocabulary of set theory is a prerequisite; for example, in political science, the book entitled *The Theory of Coalitions*, by Riker [55], depends heavily upon the basic ideas and notation of set theory. The student of political science should at least glance through this book to observe the relevance of set theory to the subject discussed in that book. ∎

We shall observe in the remainder of this book that the basic concepts of set theory are also fundamental in the study of probability, linear programming, game theory, and calculus.

EXERCISES

Suggested minimum assignment: Exercises 1, 5, 7, 11, and 18.

In each of Exercises 1–4 illustrate the given theorem with Venn diagrams.
1. Theorem 2(i).
2. Theorem 2(ii).
3. Theorem 3.
4. Theorem 4.
5. Prove Theorem 5(ii).

Twelve theorems are stated in Exercises 6–17; the reader is asked to prove these theorems by the use of previous theorems in this chapter or by tables. Let A, B, and C be arbitrary subsets of a universal set U, and let $\emptyset$ represent the empty set.

6. For every subset A, $A \cup A = A$.
7. For every subset A, $A \cap A = A$.
8. If X is a subset of U such that $A \cup X = A$ for every A, then $X = \emptyset$.
9. If X is a subset of U such that $A \cap X = A$ for every A, then $X = U$.
10. $A \cup (A \cap B) = A$ (Absorption Property).
11. $A \cap (A \cup B) = A$ (Absorption Property).
12. $A \cap B = A$ if and only if $A \cup B = B$.
13. If $A \cup B = A \cup C$ and if $A \cap B = A \cap C$, then $B = C$.

14. $A \cup (B \cup C) = (A \cup B) \cup C$ (Associative Property).
15. $A \cap (B \cap C) = (A \cap B) \cap C$ (Associative Property).
16. $(A \cup B)' = A' \cap B'$ (DeMorgan's Property).
17. $(A \cap B)' = A' \cup B'$ (DeMorgan's Property).

18. Illustrate the properties of Exercises 10, 14, and 16 with Venn diagrams.
19. Illustrate the properties of Exercises 11, 15, and 17 with Venn diagrams.

In Exercises 20–23 prove the indicated theorems using only the material that preceded the statement of the theorem.

20. Prove Theorem 1(*ii*).
21. Prove Theorem 2(*ii*).
22. Prove Theorem 3(*ii*).
23. Prove Theorem 4(*ii*).

1.4 The Number of Elements in a Set

Associated with a finite set A is a natural number $n(A)$ which simply represents the number of elements in the set. In this section we shall develop a few formulas concerning the number of elements of sets.

First we consider the union of two sets. It is easy to see that if two finite sets A and B are disjoint, then $n(A \cup B) = n(A) + n(B)$. Now consider the case where A and B are not disjoint. Suppose that A is the set of red automobiles on a certain parking lot and B is the set of all Fords on the same parking lot. Suppose that we let

$$n(A) = x + y,$$

where x is the number of red nonFords and y is the number of red Fords (see Figure 1.13). Suppose that we let

$$n(B) = z + y,$$

where z is the number of nonred Fords and y is the number of red Fords (see Figure 1.13). Then, since the sets A and B overlap by the amount

$$n(A \cap B) = y,$$

we have

$$n(A \cup B) = x + z + y$$
$$= (x + y) + (z + y) - y$$
$$= n(A) + n(B) - n(A \cap B).$$

This type of reasoning can be used to prove the following theorem.

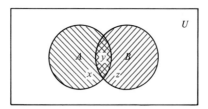

Figure 1.13

▶ **Theorem 6**
If A and B are finite sets, then
$$n(A \cup B) = n(A) + n(B) - n(A \cap B).$$

Outline of Proof: Let $X = A \cap B'$, $Y = A \cap B$, and $Z = A' \cap B$.
1. It can be shown that X, Y, and Z are disjoint and that
$$A \cup B = (X \cup Y) \cup Z,$$
$$A = X \cup Y,$$
$$B = Y \cup Z.$$

2. Because of part 1,
$$n(A \cup B) = n(X) + n(Y) + n(Z),$$
$$n(A) = n(X) + n(Y),$$
$$n(B) = n(Y) + n(Z).$$

3. The last two equations of part 2 can be solved for $n(X)$ and $n(Z)$ and substituted into the first equation of part 2 to produce the desired conclusion. ◀

Example 1
Suppose that in a set of 100 patients, 20 exhibit symptom α, 30 exhibit symptom β, and 5 exhibit both symptoms. How many have symptom α or symptom β?
Let
$$A = \text{set of patients with symptom } \alpha,$$
$$B = \text{set of patients with symptom } \beta,$$
$$A \cap B = \text{set of patients having both symptoms}.$$

Then $A \cup B$ = set of patients having one or the other or both symptoms, and we are asking for $n(A \cup B)$. By Theorem 6,
$$n(A \cup B) = n(A) + n(B) - n(A \cap B)$$
$$= 20 + 30 - 5$$
$$= 45.$$

This problem can be illustrated in Figure 1.14, where the set $A \cup B$ is thought of as being partitioned into three subsets, namely, $(A \cap B')$, $(A' \cap B)$, and $(A \cap B)$, having 15, 25, and 5 elements, respectively. The sum of these three numbers then represents the number of elements in $(A \cup B)$. ∎

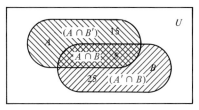

Figure 1.14. $A \cup B$ partitioned into three disjoint subsets.

In Exercise 22 the reader is asked to find a formula for $n(A \cap B')$, which in Example 1 would answer the question of how many patients have symptom α but not symptom β.

Next, suppose that in Example 1 we wanted to know the number of patients who did *not* have symptoms α or β. We could reason then that we are looking for

$$n[(A \cup B)'],$$

which is simply the total number of patients minus those who have symptoms α or β. That is,

$$n[(A \cup B)'] = n(U) - n(A \cup B).$$

We have here an illustration of a rather obvious theorem.

▶ **Theorem 7**
If A is a subset of a finite universal set U, then

$$n(A') = n(U) - n(A).$$

The proof of Theorem 7 is left as Exercise 21. ◀

Now suppose that we have three subsets A, B, and C of a universal set as shown in Figure 1.15, where

$$n(A) = x + r + s + u,$$
$$n(B) = z + s + t + u,$$
$$n(C) = y + r + t + u.$$

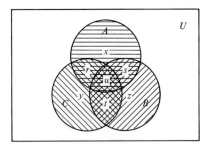

Figure 1.15. $A \cup (B \cup C)$ partitioned into eight disjoint subsets.

We see that

$$n[A \cup (B \cup C)] = x + y + z + r + s + t + u,$$

which, after considerable addition and subtraction of the natural numbers r, s, t, and u, agrees with the conclusion of the following theorem.

▶ **Theorem 8**
If A, B, and C are finite sets, then

Chap. 1] Sets

$$n[A \cup (B \cup C)] = n(A) + n(B) + n(C) - n(A \cap B)$$
$$- n(B \cap C) - n(A \cap C) + n[A \cap (B \cap C)].$$

A proof of Theorem 8 is similar to that outlined for Theorem 6. ◀

Example 2

Given: $n(A) = 30$, $n(B) = 40$, $n(C) = 20$, $n(A \cap B) = 7$, $n(A \cap C) = 5$, $n(B \cap C) = 10$, $n[A \cap (B \cap C)] = 2$, and $n(U) = 100$. Find (i) $n(A \cup B)$, (ii) $n(A')$, (iii) $n[A \cup (B \cup C)]$, and (iv) $n[(A' \cap C) \cup B]$.

Solutions

(i) To find $n(A \cup B)$, we can use Theorem 6.
$$n(A \cup B) = n(A) + n(B) - n(A \cap B)$$
$$= 30 + 40 - 7 = 63.$$

(ii) To find $n(A')$, use Theorem 7.
$$n(A') = n(U) - n(A)$$
$$= 100 - 30 = 70.$$

(iii) To find $n[A \cup (B \cup C)]$, use Theorem 8.
$$n[A \cup (B \cup C)] = n(A) + n(B) + n(C) - n(A \cap B) - n(A \cap C)$$
$$- n(B \cap C) + n[A \cap (B \cap C)]$$
$$= 30 + 40 + 20 - 7 - 5 - 10 + 2$$
$$= 70.$$

(iv) To find $n[(A' \cap C) \cup B]$, construct a Venn diagram like that shown in Figure 1.15 and determine the number of elements in the various disjoint subsets, namely, u, r, s, t, x, y, and z, in that order.

$$u = n[A \cap (B \cap C)] = 2$$
$$r = n(A \cap C) - u = 5 - 2 = 3$$
$$s = n(A \cap B) - u = 7 - 2 = 5$$
$$t = n(B \cap C) - u = 10 - 2 = 8$$
$$x = n(A) - r - s - u = 30 - 3 - 5 - 2 = 20$$
$$y = n(C) - r - t - u = 20 - 3 - 8 - 2 = 7$$
$$z = n(B) - s - t - u = 40 - 5 - 8 - 2 = 25.$$

Therefore, from Figure 1.16, where these numbers are recorded and

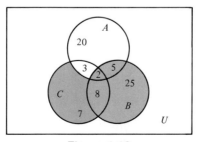

Figure 1.16

where the region $(A' \cap C) \cup B$ is colored, we see that
$$n[(A' \cap C) \cup B] = y + t + u + s + z$$
$$= 7 + 8 + 2 + 5 + 25$$
$$= 47. \blacksquare$$

APPLICATIONS

Example 3
Suppose that the following information is published in a survey.

68 countries receive foreign aid from United States,
40 countries receive foreign aid from U.S.S.R.,
20 countries receive foreign aid from France,
25 countries receive foreign aid from United States and U.S.S.R.,
10 countries receive foreign aid from United States and France,
 4 countries receive foreign aid from U.S.S.R. and France,
 2 countries receive foreign aid from United States, U.S.S.R., and France.

The survey goes on to say that there are 91 countries that obtain foreign aid from the U.S.S.R. or France or the United States and that there are 70 countries that receive aid from the United States or France but not from all three countries. The question arises: Are these last two assertions consistent with the original data? To analyze the problem, we let

S = set of countries receiving aid from United States,
R = set of countries receiving aid from U.S.S.R.,
F = set of countries receiving aid from France.

We are trying to verify that
$$n[S \cup (R \cup F)] = 91$$
and
$$n(S \cup F) - n[S \cap (F \cap R)] = 70.$$

From the information given at the outset of the example, we can write

$n(S) = 68,\quad n(R) = 40,\quad n(F) = 20,\quad n(S \cap R) = 25,$
$n(S \cap F) = 10,\quad n(R \cap F) = 4,\quad n[S \cap (R \cap F)] = 2.$

By Theorem 8,
$$n[S \cup (R \cup F)] = 68 + 40 + 20 - 25 - 10 - 4 + 2 = 91,$$

which is consistent with the report of the survey; but the second assertion of the report is not consistent with the data because

$n(S \cup F) - n[S \cap (F \cap R)] = n(S) + n(F) - n(S \cap F) - n[S \cap (F \cap R)]$
$$= 68 + 20 - 10 - 2$$
$$= 76$$
$$\neq 70.$$

See Figure 1.17 for a Venn diagram with much of the above information shown thereon. ∎

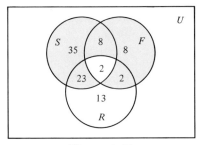

Figure 1.17

Example 4
In addition to the analysis of situations more complex but similar to Example 3, applications of the material of this section are indirectly obtained through the use of probability theory, as we shall demonstrate in Chapter 4. ∎

EXERCISES

Suggested minimum assignment: Exercises 1, 5, 8, 9, 11, 13, 19, and 22(a).

1. Suppose that 600 of 900 employees of a certain company elect to participate in a group life insurance plan, 700 elect to participate in a group health insurance plan, and 500 elect to participate in both plans.
   (a) How many employees are participating in one or both group plans?
   (b) How many are not participating in either plan?
2. Suppose that 900 students participate in intramural football, 800 participate in intramural basketball, and 1100 students are known to have participated in at least one of the sports.
   (a) How many students participated in both sports?
   (b) Of those who did not participate in football, how many did participate in basketball?

In each of Exercises 3–12 find the number of elements in the given sets if we let

$n(A) = 12$, $n(B) = 15$, $n(C) = 30$, $n(A \cap B) = 5$, $n(A \cap C) = 6$, $n(B \cap C) = 4$, $n[A \cap (B \cap C)] = 2$, $n(U) = 50$.

(*Hint:* Draw a Venn diagram similar to Figure 1.15, and first calculate the number of elements in each of the indicated disjoint subsets.)

3. $n(A \cup B)$.
4. $n(A \cup C)$.
5. $n(B \cup C)$.
6. $n[(A \cup B)']$.
7. $n[(A \cup C)']$.
8. $n[(B \cup C)']$.
9. $n[A \cup (B \cup C)]$.
10. $n[A \cup (B \cap C')]$.
11. $n[A \cup (C \cap B')]$.
12. $n[A' \cup (B' \cup C')]$.

Sec. 1.4] The Number of Elements in a Set

In a certain survey of 100 people it was found that 10 people buy *only* brand A, 30 people buy *only* brand B, 15 people buy *only* brand C, 8 people buy both brands A and B, 5 people buy brands A and C, 6 people buy brands B and C, and 4 people buy all three brands. In Exercises 13–18 answer the given questions based on these data.

13. How many people bought at least one of the three brands?
14. How many people bought none of the three brands?
15. How many people bought both brands A and B but not C?
16. How many people bought brand B or C?
17. How many people bought brand A?
18. How many people bought brand B?

19. If there are 55 Democrats in the Senate, 50 liberals, and 30 liberal Democrats, what is the voting strength of a liberal–Democrat coalition?
20. Provide the details for the proof of Theorem 6.
21. Prove Theorem 7. (*Hint:* Use Theorem 4 of Section 1.3.)
22. (a) Develop a formula for finding $n(A \cap B')$ in terms of $n(A)$ and $n(A \cap B)$. [*Hint:* $A = (A \cap B) \cup (A \cap B')$.]
    (b) Develop a formula for finding $n(A \cup B')$ in terms of $n(U)$, $n(B)$, and $n(A \cap B)$.

1.5 Cartesian Product

Prior to the national political conventions it is customary for various national pollsters to pit the leading Democratic candidates against the leading Republican candidates in hypothetical 2-man races. Suppose that the leading Democratic candidates are represented by the set $D = \{d_1, d_2, d_3\}$ and the leading Republican candidates are represented by the set $R = \{r_1, r_2\}$. The 2-man races that can be developed from these two sets can be represented by the set of ordered pairs

$$\{(d_1, r_1), (d_1, r_2), (d_2, r_1), (d_2, r_2), (d_3, r_1), (d_3, r_2)\},$$

where the first entry of each pair is an element of D and the second entry of each pair is an element of R; this set of ordered pairs is an example of a **Cartesian product of D and R** and is designated $D \times R$. It may be helpful to the reader to visualize the ordered pairs of $D \times R$ as entries of the rectangular array

	r_1	r_2
d_1	(d_1, r_1)	(d_1, r_2)
d_2	(d_2, r_1)	(d_2, r_2)
d_3	(d_3, r_1)	(d_3, r_2)

▶ **Definition 6**
If a is an element of set A and if b is an element of set B, then the set of *all* ordered pairs (a, b) that can be constructed from A and B is called the **Cartesian product of A and B.**

Example 1
Let R be the set of all real numbers. The Cartesian product $R \times R$ is simply the set of all ordered pairs of real numbers. The elements (x, y) of $R \times R$ are used to identify points in the Cartesian plane. (The Cartesian coordinate system will be discussed in detail in Section 3.6.) For example, the element $(3, 2)$ graphed in Figure 1.18 is one member of the set $R \times R$; other points in the plane of Figure 1.18 may be thought of as a geometrical representation of other pairs belonging to $R \times R$. From this usage of the Cartesian plane, we get the name Cartesian product. ∎

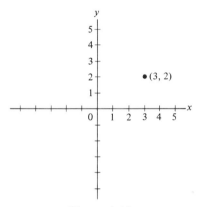

Figure 1.18

Reconsider the sets $D = \{d_1, d_2, d_3\}$, $R = \{r_1, r_2\}$, and $D \times R = \{(d_1, r_1), (d_1, r_2), (d_2, r_1), (d_2, r_2), (d_3, r_1), (d_3, r_2)\}$. Notice that associated with each of the 3 elements of D there are 2 elements of R, thus giving us $3 \cdot 2$, or 6, elements of $D \times R$. Generalizing this idea, we ascertain that for finite sets A and B,

$$n(A \times B) = n(A) \cdot n(B).$$

Example 2
If M is a set of 5 male rabbits and if F is a set of 6 female rabbits, then there are $5 \cdot 6$, or 30, different ways that the rabbits can be paired so that one is male and the other is female. The set of pairs can be represented by $M \times F$ and

$$n(M \times F) = n(M) \cdot n(F) = 5 \cdot 6 = 30. \ \blacksquare$$

The idea of a Cartesian product can be extended to apply to more than two sets, as the next example illustrates.

Example 3
A certain set of astronauts all have secondary training in some field. Suppose that there are 3 astronomers represented by the subset
$$A = \{a_1, a_2, a_3\},$$
4 meteorologists represented by the subset
$$M = \{m_1, m_2, m_3, m_4\},$$
and 2 engineers represented by the subset
$$E = \{e_1, e_2\}.$$
A space team is to be selected consisting of one astronaut from each of these subsets. One possible team is the triple
$$(a_1, m_1, e_1).$$
Another team is the triple
$$(a_1, m_1, e_2),$$
and of course there are many more that could be listed. The set of all such triples (potential teams) illustrates the concept of the Cartesian product $A \times M \times E$ of the three subsets. The number of elements (teams) in the Cartesian product can be determined by arranging the triples in a three-dimensional array shown in Figure 1.19. Thus we have the formula
$$n(A \times M \times E) = n(A) \cdot n(M) \cdot n(E). \blacksquare$$

The ideas presented in Example 3 are now formalized.

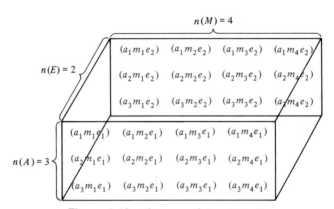

Figure 1.19. *Elements of $A \times M \times E$.*

▶ **Definition 7**
If a_i is an element of set A_i, where $i = 1, 2, \ldots, k$, then the set of all ordered k-tuples $(a_1, a_2, \ldots, a_k)$ is called the **Cartesian product**
$$A_1 \times A_2 \times \cdots \times A_k.$$

We state without proof that the number of elements in the Cartesian product $A_1 \times A_2 \times \cdots \times A_k$ of finite sets is given by the formula

$$n(A_1 \times A_2 \times \cdots \times A_k) = n(A_1)n(A_2) \cdots n(A_k).$$

APPLICATIONS

The concept of the Cartesian product of two or more sets will be used in Chapter 4.

EXERCISES

Suggested minimum assignment: Exercises 1, 3, 5, and 8.

1. If $A = \{a, b, c\}$ and $B = \{x, y, z\}$, write $A \times B$.
2. For the sets in Exercise 1, write $B \times A$.
3. Let $A = \{1, 2, 3\}$, and $B = \{1, 2, 3, 4\}$. Plot the elements of $(A \times B)$ with respect to the Cartesian coordinate system.
4. Repeat Exercise 3 for $B \times A$ rather than $A \times B$. Is $B \times A = A \times B$?
5. Use the appropriate set notation to represent the number of different matings of 7 females and 9 males that are possible in a certain set of like animals. What is the number of matings?
6. In Example 3 suppose that an additional member is to be added to the team, thus making it a 4-man team. Further, suppose that this fourth person is to be selected from a set D of 3 medical doctors. State the set of teams as a Cartesian product and find the number of potential teams.
7. If there are 3 candidates for the Democratic nomination, 4 candidates for the Republican nomination, and 2 potential independent candidates, then for how many 3-man races in the general election must a poll taker prepare? Express the set of potential 3-man races as a Cartesian product.
8. An ordinary die is a cube with a different number on each of the 6 faces of the cube. Let F be the set of numbers that can appear on the upward face of a die lying on a flat surface.
   (a) Use set notation to express the set of pairs of numbers that can be obtained from 2 dice. How many pairs are there?
   (b) Use set notation to express the set of triples of numbers from 3 dice. How many triples are there?

NEW VOCABULARY

set 1.1
subset 1.1
proper subset 1.1
empty set (null set) 1.1
universal set 1.1
Venn diagrams 1.1

real line 1.1
origin 1.1
coordinate 1.1
positive integers 1.1
natural numbers 1.1
negative integers 1.1

zero 1.1
rational number 1.1
irrational number 1.1
finite set 1.1
infinite set 1.1
equal sets 1.2

intersection of sets 1.2
disjoint sets 1.2
union of sets 1.2
complement of a set 1.2
Cartesian product 1.5

2 Logic

The primary purpose of this chapter is best described in the words of Charles Dodgson, a 19th-century mathematician who was better known as Lewis Carroll, the author of Alice in Wonderland. In his book Symbolic Logic, Dodgson made the following ageless observation: "It [symbolic logic] will give you clearness of thought—the ability to see your way through a puzzle—the habit of arranging your ideas in an orderly and get-at-able form—and, more valuable than all, the power to detect fallacies, and to tear to pieces the flimsy illogical arguments which you will so continually encounter in books, in newspapers, in speeches ... and which so easily delude those who have never taken the trouble to master this fascinating art." In the process of accomplishing this primary purpose, we shall introduce a mathematical structure in which elements other than numbers are combined by various operations.

Prerequisites: high school algebra.
Suggested Sections for Moderate Emphasis: 2.1–2.5, or 2.1–2.3, 2.6, 2.7.
Suggested Sections for Minimum Emphasis: 2.1, 2.2.

2.1 Statement Forms and Connectives

In the ordinary usage of language there are certain declarative expressions for which it is meaningful to say that they are either true or false (but not both); in the study of mathematical logic such expressions often are called **statements**.†

▶ **Definition 1**
A **statement** is a declarative sentence that can be assigned exactly one of the values True or False.

Example 1
The following expressions are specific examples of statements.
1. The Consumer Price Index rose last month.
2. Both Senators from Florida are now Republicans.
3. There is a man by the name of William Howard Palm who is a professional engineer.

† The reader should be aware that the terms applied to some of the concepts of mathematical logic vary from one textbook to another.

4. There is a swinging footbridge in Hardwick, Vermont.
5. $x^2 - 4 = (x - 2)(x + 2)$ for all real numbers x.
6. $1 + 3 = 9$. ∎

Example 2
The following expressions are not statements because they are *not* declarative assertions for which it is meaningful to say that they are either true or false.
1. Run home.
2. $x^2 + 1$.
3. How high is the sky? ∎

Hereafter we shall be concerned primarily with symbolic representations of statements rather than with specific statements. We immediately encounter a problem in terminology, however, because the unspecified letters p, q, r, and so on, that we will use do not in themselves constitute a statement according to Definition 1; that definition requires that exactly *one* value, true or false, be assigned to a statement, and, without knowing what p, q, r, and so on, represent, this cannot be done. To escape this dilemma we introduce a new term; the unspecified letters p, q, r, and so on, used in symbolic representations of statements will be called **statement forms.** These letters are not statements but are simply placeholders for statements.

▶ **Definition 2**
A symbolic representation p that becomes a statement when the placeholder p is specified is called a **statement form.**

When p is specified, the resulting statement is called an **instance** of the statement form. The advantage of studying statement forms rather than statements is that we can develop properties for the statement forms that will be valid regardless of specific interpretations that might be given. These properties furnish us with rules of logic that are very useful in such areas as deductive reasoning, legal analysis (see Kort [42]), and computer science.

Compound statement forms will be formed by using the words "or," "and," "if ... then," "if and only if," and "not." Symbols that will be used to represent these words are given in Table 2.1. Therefore, if we are given statement forms p, q, the compound form using the connective "**and**" will be

Table 2.1. Symbols of specified connectives.

or	$\vee$
and	$\wedge$
if ... then	$\rightarrow$
if and only if	$\leftrightarrow$
not	$\sim$

designated by $p \wedge q$. If we let p represent the statement

"John Jones is mayor"

and if we let q represent the statement

"Taxes have increased,"

then the statement

"John Jones is mayor *and* taxes have increased"

is one instance of the statement form $p \wedge q$.

In using the connective "**or**," denoted by the symbol $\vee$, we must be careful because in ordinary usage two interpretations are common. For example, when we say

"John Jones is mayor or taxes have increased,"

do we mean to include or exclude the possibility of both? In mathematical logic it is customary to adopt the inclusive interpretation unless specified otherwise. In other words, $p \vee q$ means either p or q or both. The exclusive "or" will be discussed in Exercise 11.

The statement

"John Jones is mayor *or* taxes have increased"

is one instance of the statement form $p \vee q$.

Statement forms that do not include any of the connectives of Table 2.1 are known as **simple statement forms**. When two or more simple statement forms are connected by connectives or when the symbol $\sim$ is used, the resulting statement form is called a **compound statement form**. Hereafter the phrase *statement form* will mean either a simple or a compound statement form. **Simple** and **compound statements** are defined in like manner. Any simple statement form or connection of two or more simple statement forms that make up a part or all of a compound statement form is a **component** of the compound statement form. Large and complicated compound statements are used frequently in normal communications such as legal documents. The analysis, and perhaps simplification, of compound statements through symbolic representation and the rules of logic can be very helpful in some cases.

Other connectives, "if ... then," "if and only if," and "not," as well as the two already discussed, are presented and illustrated in Table 2.2; this table should be studied carefully.

Example 3

Consider the compound statement: "If income exceeds $40,000 and expenses do not exceed 50 per cent of income, then an extra salesman can be hired." Let "income exceeds $40,000" be an instance of p; let "expenses exceed 50 per cent of income" be an instance of q; let "an extra salesman can be hired" be an instance of r. The original compound statement then is

Table 2.2. SUMMARY OF FIVE CONNECTIVES.

Connective	Symbol	Statement form	Name	Instance
and	$\wedge$	$p \wedge q$	conjunction	John Jones is mayor **and** taxes have increased.
or	$\vee$	$p \vee q$	disjunction	John Jones is mayor **or** taxes have increased.
if ... then	$\rightarrow$	$p \rightarrow q$	conditional	**If** John Jones is mayor, **then** taxes have increased.
if and only if	$\leftrightarrow$	$p \leftrightarrow q$	biconditional	John Jones is mayor **if and only if** taxes have increased.
not	$\sim$	$\sim p$	negation	It is **not** the case that John Jones is mayor.

an instance of the compound statement form

$$[p \wedge (\sim q)] \rightarrow r;$$

the statement forms p, q, r, $\sim q$, $[p \wedge (\sim q)]$, $\{[p \wedge (\sim q)] \rightarrow r\}$ are all components of this compound statement form. The first three components are simple and the last three are compound. ∎

To construct compound statement forms such as those in Example 3, care must be taken in the use of parentheses. The first four connectives listed in Table 2.2 must be used to connect exactly two statement forms (simple or compound); hence it is insufficient in Example 3 to write

$$p \wedge \sim q \rightarrow r$$

because of the ambiguity of whether $[p \wedge (\sim q)] \rightarrow r$ or $p \wedge [(\sim q) \rightarrow r]$ or $p \wedge \sim (q \rightarrow r)$ is the correct interpretation. As is the case in Example 3, sentence punctuation usually reveals the proper location of the parentheses.

In summary, we have begun the construction of a structure called the Algebra of Components. The elements or building blocks of this structure are the components p, q, r, and so on. Over this set of components we have a set of operations that we have identified as the connectives $\wedge$, $\vee$, $\rightarrow$, $\leftrightarrow$, and $\sim$. We continue our construction in the remaining sections of this chapter.

APPLICATIONS

Example 4

In recent years, scholars, both in and out of the legal profession, have begun to explore the relationship between modern mathematics and law. The use of mathematical logic in this study has proved particularly promising. Allen and Caldwell [2] provide a discussion of the use of logic in legal analysis as

well as many references on the subject. Part II, beginning on p. 234, is particularly relevant to the material of this chapter, although the interested reader is warned that the notation and diagrams differ somewhat from this text. As Allen and Caldwell point out, many assertions in legal writings—whether by design or accident—are syntactically ambiguous. They then illustrate this fact with a portion of the 1938 Federal Food and Drug Act:

"Sec. 343. Misbranded Food

"A food shall be deemed to be misbranded . . .

"(c) If it is an imitation of another food, unless its label bears, in type of uniform size and prominence, the word 'imitation' and, immediately thereafter, the name of the food imitated."

The question is whether such an assertion is appropriately interpreted as

(1) If a food is an imitation of another food, then it shall be deemed to be misbranded *if* its label *does not* bear, in type of uniform size and prominence, the word "imitation" and, immediately thereafter, the name of the food imitated;

or whether it is more appropriately interpreted as

(2) If a food is an imitation of another food, then it shall be deemed to be misbranded *if and only if* its label *does not* bear, in type of uniform size and prominence, the word "imitation" and, immediately thereafter, the name of the food imitated.

Allen and Caldwell then use some of the ideas of logic to describe and illustrate the method for representing and detecting alternative interpretations of syntactically ambiguous legal assertions. ∎

EXERCISES

Suggested minimum assignment: Exercises 1, 4, 5, 7, 8, 9, and 11.

In each of Exercises 1–4 state which sentences are statements and which are not.

1. (a) Fix the car.
   (b) The moon is visible now.
   (c) The Republicans hold exactly one of the U.S. Senate seats from Florida.
2. (a) Where is John?
   (b) Eufaula is a town in Alabama.
   (c) Ford Motor Company is one of the three largest manufacturers of automobiles in the United States.
3. (a) George Washington lived in California for 10 years.
   (b) $6 > 2$.
   (c) The painting is good.
4. (a) $3 + 4 = 13$.
   (b) Solve the equation $7x + 6 = 8$.
   (c) For some rational values of x, $x^2 + 2 = 6$.

In each of Exercises 5–8 write the statement form for which the given statement is an instance. Also state the primary connective and the name of the compound statement form.

5. (a) If the object glitters, then it is gold.
   (b) If the vote is favorable, then the labor dispute will be settled and the workers will return to work.
   (c) The following is not the case: "Inflation will continue, and unemployment will increase."
6. (a) The productivity of the people is lower, or there has not been a greater efficiency.
   (b) It is not the case that if we institute an austerity program, then profits will increase.
   (c) If the European common market includes England, then U.S. exports will be affected and the balance of payments may change.
7. (a) The cash reserves of the First State Bank of Eustis, Florida, have increased, and it is not true that their dividend will decrease this year.
   (b) If wishes were horses, then we would all take a ride.
   (c) The real estate prices in Blacksburg, Virginia, will increase next year if and only if the population increases and the University expands.
8. (a) Federal taxes increased, and money became tighter.
   (b) The school bond issue will be approved by the voters, or there will be no new schools next year.
   (c) Greater resources can be allocated to education if and only if the voters support the educational institutions in the referendum and the money is available.

In each of Exercises 9 and 10 give instances of p, q, and write statements that are instances of the given statement forms. State the connective and the name of the compound statement form.

9. (a) $p \vee q$.   (b) $p \rightarrow q$.   (c) $\sim p$.
10. (a) $p \wedge q$.   (b) $p \leftrightarrow q$.   (c) $(\sim p) \rightarrow q$.

11. As mentioned in the text, it is customary in mathematical logic to use the connective "or" in the inclusive sense; legal documents often use the symbol and/or to denote this inclusive intent. The exclusive "or" will be denoted by the symbol $\underline{\vee}$, and the compound statement form $p \underline{\vee} q$ means p or q but not both; the arrangement "either ... or ..." is often used to denote this exclusive intent. Let "The computer is available" be an instance of p and "I will get my project finished" be an instance of q; write an instance of $p \underline{\vee} (\sim q)$.

2.2 Evaluation of Statement Forms

We shall assign values to statements according to the truth or falsity of the statement. We assign the value T to a true statement and a value F to a false statement. We could have assigned values of 1 and 0 rather than T and F; some authors do, and it is permissible for the reader to read a T as a 1 and an F as a 0 if he prefers. The value of a simple statement form may be T or F but not both simultaneously. The values for compound statement forms are defined next.

▶ **Definition 3**
For statement forms p, q, the values of the compound statement forms $p \wedge q$, $p \vee q$, $p \rightarrow q$, $p \leftrightarrow q$, and $\sim p$ are defined according to Tables 2.3–2.7.

The definitions given in Tables 2.3–2.7 are compatible with common usage of the English language; it is important, however, that the reader convince himself of this fact. The only definition likely to cause much difficulty is the conditional (Table 2.5), which we discuss in the next example.

Table 2.3. DEFINITION OF VALUES OF CONJUNCTION.

p	q	$p \wedge q$
T	T	T
T	F	F
F	T	F
F	F	F

Table 2.4. DEFINITION OF VALUES OF DISJUNCTION.

p	q	$p \vee q$
T	T	T
T	F	T
F	T	T
F	F	F

Table 2.5. DEFINITION OF VALUES OF CONDITIONAL.

p	q	$p \rightarrow q$
T	T	T
T	F	F
F	T	T
F	F	T

Table 2.6. DEFINITION OF VALUES OF BICONDITIONAL.

p	q	$p \leftrightarrow q$
T	T	T
T	F	F
F	T	F
F	F	T

Table 2.7
DEFINITION OF VALUES OF NEGATION.

p	$\sim p$
T	F
F	T

Example 1

The purpose of this example is to attempt the sometimes difficult task of convincing the reader that the values assigned to the conditional $p \rightarrow q$ are plausible. Consider the following specific statement made by a candidate for public office. "If I am elected, then the expressway will be completed." Under what circumstances will the candidate's statement be a lie and under what circumstances will the candidate not be lying (that is, telling the truth in a two-value logic)? The alternatives are listed in Table 2.8; notice that Table 2.8 is consistent with Table 2.5. The difficulty that readers have in accepting the plausibility of the definition of the values of $p \rightarrow q$ usually centers on the third and fourth lines of Tables 2.5 and 2.8. "How can $p \rightarrow q$ be true if p is false?" Notice in the illustration that if the candidate is not elected, that is, if p is false, then there is no way that his statement can be untrue regardless of what q represents or whether q is true or false. If his statement is not untrue, then it must be true in a two-value logic. ∎

Table 2.8

Candidate elected	Expressway completed	Candidate told truth (that is, did not lie)
yes	yes	yes
yes	no	no
no	yes	yes
no	no	yes

We now turn our attention to a method of evaluation of more complicated compound statement forms. Such an evaluation can be made by use of evaluation tables and repeated use of Definition 3.

Example 2

Evaluate the compound statement form

$$[(\sim p) \vee q] \wedge (q \rightarrow p).$$

Table 2.9 illustrates an evaluation procedure. We conclude from Table 2.9 that the given compound statement form is true when both p and q are true or when both p and q are false. ∎

Table 2.9. EVALUATION OF $[(\sim p) \vee q] \wedge (q \rightarrow p)$.

(1) p	(2) q	(3) $\sim p$	(4) $(\sim p) \vee q$	(5) $q \rightarrow p$	(4) $\wedge$ (5) $[(\sim p) \vee q] \wedge (q \rightarrow p)$
T	T	F	T	T	T
T	F	F	F	T	F
F	T	T	T	F	F
F	F	T	T	T	T

Example 3
We now illustrate a compound statement form that is always true, namely, $(p \to q) \leftrightarrow [(\sim p) \vee q]$. (See Table 2.10.)

Statement forms that are always true are very important and will be discussed further in the next section. ∎

Table 2.10. EVALUATION OF $(p \to q) \leftrightarrow [(\sim p) \vee q]$.

(1) p	(2) q	(3) $\sim p$	(4) $p \to q$	(5) $(\sim p) \vee q$	(4) ↔ (5) $(p \to q) \leftrightarrow [(\sim p) \vee q]$
T	T	F	T	T	T
T	F	F	F	F	T
F	T	T	T	T	T
F	F	T	T	T	T

In summary, we have defined values of the results of each of the five connectives in a manner that is consistent with common word usage. Moreover, we have shown how to construct evaluation tables to determine values of compound statement forms.

APPLICATIONS

Example 4
An analysis of the First Amendment of the U.S. Constitution can be made as follows: "Congress shall make *no* law {[(respecting an establishment of religion), *or* (prohibiting the free exercise thereof)]; *or* [(abridging the freedom of speech) *or* (of the press)]; *or* [(of the right of the people to peaceably assemble) *and* (to petition the Government for a redress of grievances)]}."
(with labels: p = respecting an establishment of religion; q = prohibiting the free exercise thereof; r = abridging the freedom of speech; s = of the press; t = of the right of the people to peaceably assemble; u = to petition the Government for a redress of grievances)

As a statement form, the First Amendment can be expressed as

$$\sim[(p \vee q) \vee (r \vee s) \vee (t \wedge u)].$$

A violation of the First Amendment occurs when the statement form is false. An evaluation table is one way of determining when this happens.

Later in this chapter we shall find that the statement form above has the same truth values as

$$(\sim p) \wedge (\sim q) \wedge (\sim r) \wedge (\sim s) \wedge [(\sim t) \vee (\sim u)],$$

and can be diagramed as shown in Figure 2.1. From the last compound statement form it is easy to see that a violation of the First Amendment will occur if any one of the components $\sim p$, $\sim q$, $\sim r$, $\sim s$, or $(\sim t) \vee (\sim u)$ is false. For example, a violation occurs if $\sim s$ has value F, which means that s

Figure 2.1. *Diagram of the First Amendment.*

has value T, which means that Congress made a law that abridged the freedom of the press.

An interesting observation can be made by noticing the effect of the connective "and" in the statement of the amendment. A violation of $(\sim t) \vee (\sim u)$ occurs only if both t and u have value T. This analysis shows an interpretation that may not have been intended by its authors and that has *not* been consistent with interpretations made by the courts. Thus we have here an excellent example showing how mathematical analysis can expose flaws in legal draftsmanship or in legal interpretation. ∎

EXERCISES

Suggested minimum assignment: Exercises 1, 3, 6, 9, and 11.

In each of Exercises 1–8 evaluate the given compound statement forms for all possible arrangements of values of the placeholders (or variables) p, q, r:

1. $[p \to (\sim q)] \wedge p$.
2. $(p \to q) \wedge \sim q$.
3. $[(\sim p) \vee (\sim q)] \leftrightarrow \sim (p \wedge q)$.
4. $[(\sim p) \wedge (\sim q)] \leftrightarrow \sim (p \vee q)$.
5. $[(\sim r) \vee q] \vee (p \wedge r)$.
6. $[p \wedge (\sim q)] \vee (r \wedge q)$.
7. $[(p \vee r) \vee q] \to (p \wedge q)$.
8. $[(p \wedge r) \wedge q] \to (p \vee r)$.

In each of Exercises 9 and 10 state a compound statement form that has the given evaluation table:

9. **Table 2.11**

p	q	?
T	T	T
T	F	T
F	T	F
F	F	T

10. **Table 2.12**

p	q	?
T	T	F
T	F	F
F	T	T
F	F	F

11. If we have a compound statement form involving four placeholders (or variables) p, q, r, and s, how many lines will be required in an evaluation table?
12. Refer to Exercise 11 of Section 2.1 and write an evaluation table defining the truth values of the exclusive "or" that are consistent with common usage.
13. The Eighth Amendment to the U.S. Constitution reads: "Excessive bail shall not be required, nor excessive fines imposed, nor cruel and unusual punishments inflicted."

(a) Identify, and label with the letters p, q, r, and s, the four positive simple statements that are included in the amendment.
(b) Construct, using p, q, r, and s, a compound statement form for which the amendment is an instance. (The word "nor" means "and not" or "and no.")

2.3 Equivalent Statement Forms

Consider the compound statement form $p \wedge (p \vee q)$ and the simple statement form p. From Table 2.13 it can be observed that the truth values are identical. In other words, as far as the truth values are concerned, the form $p \wedge (p \vee q)$ and the form p are the same.

Table 2.13. $p \wedge (p \vee q)$ AND p HAVE THE SAME EVALUATION.

p	q	$p \vee q$	$p \wedge (p \vee q)$
T	T	T	T
T	F	T	T
F	T	T	F
F	F	F	F

▶ **Definition 4**

Two statement forms are **equivalent** when they have the same value for all possible arrangements of values of the simple component statement forms. Equivalence will be denoted by the symbol "$\equiv$."

Example 1

Table 2.14 shows that $(p \rightarrow q) \equiv [(\sim q) \rightarrow (\sim p)]$ because the values in column (3) are *identical* with those of column (6). This particular equivalence is used often enough to warrant a special name. The form $(\sim q) \rightarrow (\sim p)$ is called the **contrapositive** of $p \rightarrow q$. The statement "If penicillin is administered, then the patient will recover" is an instance of $(p \rightarrow q)$ and hence by Table 2.14 is equivalent to the statement "If the patient did not recover, then penicillin was not administered." ∎

Table 2.14. $p \rightarrow q$ AND $(\sim q) \rightarrow (\sim p)$ HAVE THE SAME EVALUATION.

(1) p	(2) q	(3) $p \rightarrow q$	(4) $\sim q$	(5) $\sim p$	(6) $(\sim q) \rightarrow (\sim p)$
T	T	T	F	F	T
T	F	F	T	F	F
F	T	T	F	T	T
F	F	T	T	T	T

There are certain compound statement forms that are always true, regardless of the values of the simple components; such statement forms are also important enough to be given a special name.

▶ **Definition 5**
A compound statement form is a **tautology** if the value of the compound statement form is always T, regardless of the values of its simple components. A tautology will be designated by the symbol U.

Example 2
The compound statement form $p \vee (\sim p)$ is a tautology because column (3) of Table 2.15 consists of all T's. ∎

Table 2.15. EXAMPLE OF A TAUTOLOGY.

(1) p	(2) $\sim p$	(3) $p \vee (\sim p)$
T	F	T
F	T	T

Example 3 of Section 2.2 illustrates another tautology.

▶ **Definition 6**
A conditional $p \to q$ that is a tautology is called an **implication**† and is denoted $p \Rightarrow q$.

Example 3
The conditional $[(p \to q) \wedge p] \to q$ is a tautology because column (5) of Table 2.16 consists of all T's. Hence the given conditional is an implication and can be designated

$$[(p \to q) \wedge p] \Rightarrow q. \quad \blacksquare$$

Table 2.16. EXAMPLE OF A TAUTOLOGY.

(1) p	(2) q	(3) $p \to q$	(4) (3) $\wedge$ (1) $(p \to q) \wedge p$	(5) (4) $\to$ (2) $[(p \to q) \wedge p] \to q$
T	T	T	T	T
T	F	F	F	T
F	T	T	F	T
F	F	T	F	T

As might be expected, there are also compound statement forms that have false values, regardless of the values of the simple components.

† The reader is reminded of the previous warning that the terminology used in the study of logic is not uniform. The term "implication" is an example.

▶ **Definition 7**

A compound statement form is a **contradiction** if the value of the compound statement form is always F, regardless of the values of the simple components. A contradiction will be designated by the symbol C.

Example 4

The compound statement form $p \wedge (\sim p)$ is a contradiction, as verified in Table 2.17. [Column (3) has all F's.] ■

Table 2.17. EXAMPLE OF A CONTRADICTION.

(1)	(2)	(3)
p	$\sim p$	$p \wedge (\sim p)$
T	F	F
F	T	F

We have seen that when a conditional $p \rightarrow q$ is a tautology, then the conditional is called an implication and is denoted by $p \Rightarrow q$. When a biconditional $p \leftrightarrow q$ is a tautology, then p and q must have the same values, and hence p and q are *equivalent*. Thus a biconditional $p \leftrightarrow q$ that is a tautology is an equivalence and often is designated $p \Leftrightarrow q$ because $p \Rightarrow q$ and $q \Rightarrow p$. Thus we have two symbols, $\equiv$ and $\Leftrightarrow$, that are commonly used to represent the very important concept of equivalence.

Example 5

In the evaluation table of Example 3 in Section 2.2 we showed that $(p \rightarrow q) \leftrightarrow [(\sim p) \vee q]$ always had a value T. Hence we have a biconditional that is a tautology, and we write

$$(p \rightarrow q) \Leftrightarrow [(\sim p) \vee q]$$

or

$$(p \rightarrow q) \equiv [(\sim p) \vee q],$$

both of which are read "$(p \rightarrow q)$ is equivalent to $[(\sim p) \vee q]$." ■

The significance of the result of Example 5 should be emphasized. Because $(p \rightarrow q)$ is equivalent to $[(\sim p) \vee q]$, the form $(p \rightarrow q)$ can be replaced by a statement form involving the connectives "or" and "not." We can also show that the connective $\leftrightarrow$ can be expressed in terms of the connectives $\vee$, $\wedge$, and $\sim$. Thus the five connectives introduced in Section 2.1 could be reduced to three connectives; it is more convenient, however, to keep all five.

In summary, a pair of statement forms can be shown to be logically equivalent by showing that their truth values are the same for all possible arrangements of values of the simple components; this is the same as showing that the biconditional of the two expressions is a tautology. Also, one statement form can be shown to imply another by showing that the associated conditional is a tautology.

APPLICATIONS

Example 6
An illustration using the algebra of components in a problem of production management may be found in an article by Kattsoff and Simone [35]. The problem is concerned with the production of an item with m components that have been produced from n different materials. ∎

Example 7
Applications of this material for the solution of logical-puzzle-type problems may be found in Arnold [4], pp. 133–137. A simple illustration of such a problem follows. Suppose that a window has been broken and four boys are questioned. They make the following statements: Hippler: "Myers did it." Myers: "Hippler is lying." Palm: "I didn't do it." Stebbins: "Hippler did it." If exactly three of the suspects are lying, then who did it? If exactly one of the suspects is lying, then who did it? Let h, m, p, and s represent the following statements:

h: Hippler did it.
m: Myers did it.
p: Palm did it.
s: Stebbins did it.

The four boys made the following assertions: m, $\sim m$, $\sim p$, h. If three of them are lying, then one of the following statements must be true:

$$m \wedge [\sim(\sim m)] \wedge [\sim(\sim p)] \wedge (\sim h);$$
$$(\sim m) \wedge (\sim m) \wedge [\sim(\sim p)] \wedge (\sim h);$$
$$(\sim m) \wedge [\sim(\sim m)] \wedge (\sim p) \wedge (\sim h);$$
$$(\sim m) \wedge [\sim(\sim m)] \wedge [\sim(\sim p)] \wedge h.$$

By evaluation tables or by the theorems of the next section, the latter four statements can be shown to be equivalent to

$$m \wedge p \wedge (\sim h);$$
$$(\sim m) \wedge p \wedge (\sim h);$$
$$C;$$
$$C.$$

Since exactly one of these must have value T, we can say that

$$[m \wedge p \wedge (\sim h)] \vee (\sim m) \wedge p \wedge (\sim h) \vee C \vee C$$

must have value T. The latter statement is equivalent to $[p \wedge (\sim h)]$, and for this statement to be true, p must be true; hence Palm did it. The solution of the second part (when only one boy is lying) is left as Exercise 28 of Section 2.4. ∎

EXERCISES

Suggested minimum assignment: Exercises 1, 3, 6, 7, 9, 11, and 13.

In each of Exercises 1–8 determine which of the statement forms are tautologies, which are contradictions, and which are neither. Also identify any implications and equivalences.

1. $(p \wedge q) \to (\sim q)$.
2. $\sim(p \vee q) \to p$.
3. $[p \vee (p \wedge q)] \leftrightarrow p$.
4. $[p \wedge (p \vee q)] \leftrightarrow p$.
5. $[(\sim p) \vee q] \leftrightarrow (q \wedge p)$.
6. $[q \wedge (\sim p)] \leftrightarrow [(\sim q) \vee p]$.
7. $[(p \vee q) \vee (\sim r)] \to p$.
8. $p \to [(p \wedge q) \wedge (\sim r)]$.

In each of Exercises 9–12 prove that the given statement forms are equivalent.

9. $[\sim(p \vee q)]$, $[(\sim p) \wedge (\sim q)]$.
10. $[\sim(p \wedge q)]$, $[(\sim p) \vee (\sim q)]$.
11. $[(p \to q) \wedge (q \to p)]$, $[p \leftrightarrow q]$.
12. $[(p \vee q) \vee r]$, $[p \vee (q \vee r)]$.

13. The exclusive "or" has values defined by Table 2.18. Does $p \veebar (p \wedge q)$ imply $\sim q$? Justify your answer.

Table 2.18

p	q	$p \veebar q$
T	T	F
T	F	T
F	T	T
F	F	F

14. Does $[q \wedge (\sim p)] \vee q$ imply $p \vee q$? Justify your answer.
15. Article III of the U.S. Constitution includes the sentence: "No person shall be convicted of treason unless on the testimony of two witnesses to the same overt act, or on confession in open Court."
    (a) Identify the three simple statements and represent them by the letters p, q, and r, respectively.
    (b) Construct, using p, q, and r, a compound statement form for which the given sentence is an instance. (*Hint:* The word "unless" is defined by Webster to mean "if not.")
    (c) We have seen that $(a \to b) \equiv [(\sim a) \vee b]$. Use this equivalence to express the answer of part (b) without the conditional.

2.4 Properties of the Algebra of Components

We have laid the foundation for a new mathematical structure which we call the Algebra of Components. The elements of this structure are statement forms p, q, r, and so on (simple or compound). We have two relations between statement forms called equivalence and implication, and we have

the operations of $\wedge$, $\vee$, $\rightarrow$, $\leftrightarrow$, and $\sim$ that are defined over the elements. Once the foundation is laid, it is natural to begin the superstructure; that is, formulate conjectures involving the elements, operations, and relations, and then try to prove or disprove the conjectures. The proved conjectures are called theorems and become very useful laws of logic. In this section we list some of the more important and basic theorems.

▶ **Theorem 1**
(Commutative Properties) If p and q are arbitrary statement forms, then
(i) $(p \vee q) \equiv (q \vee p)$.
(ii) $(p \wedge q) \equiv (q \wedge p)$.

The proof of Theorem 1 is easily accomplished by the use of evaluation tables and is left as Exercises 18 and 19. ◀

Example 1
The statement "The assets exceed the liabilities or the firm will show a profit this year" is an instance of the form $p \vee q$. Also, "The firm will show a profit this year or the assets exceed the liabilities" is an instance of the form $q \vee p$. According to Theorem 1, these two statements are equivalent; we say that the connective $\vee$ is commutative. The connective $\wedge$ is also commutative. ∎

▶ **Theorem 2**
(Distributive Properties) If p, q, and r are arbitrary statement forms, then
(i) $[p \wedge (q \vee r)] \equiv [(p \wedge q) \vee (p \wedge r)]$.
(ii) $[p \vee (q \wedge r)] \equiv [(p \vee q) \wedge (p \vee r)]$.

The proof of Theorem 2 is left as Exercises 20 and 21. ◀

Example 2
The statement "Sales will increase, and the sales manager or the vice president will be promoted" is an instance of $[p \wedge (q \vee r)]$; by Theorem 2(i) this statement is equivalent to the statement "Sales will increase and the sales manager will be promoted, or sales will increase and the vice president will be promoted." ∎

▶ **Theorem 3**
(Identity Properties) If C, U, and p are statement forms, where C is a contradiction and U is a tautology, then for every p,
(i) $(p \vee C) \equiv p$.
(ii) $(p \wedge U) \equiv p$.

The proof of Theorem 3 is left as Exercises 22 and 23. ◀

Example 3
The statement "If A is elected, then taxes will increase; A is elected; therefore, taxes will increase" is an instance of the tautology
$$[(p \to q) \land p] \to q.$$
According to Theorem 3(ii), any statement connected by $\land$ with this or any other tautology is equivalent to the statement itself. ∎

▶ **Theorem 4**
(Complement Properties) If C, U, and p are statement forms, where C is a contradiction and U is a tautology, then for every p,

(i) $[p \lor (\sim p)] \equiv U.$
(ii) $[p \land (\sim p)] \equiv C.$

The proof of Theorem 4 is left as Exercises 24 and 25. ◀

Example 4
The statement "Candidate A will win or candidate A will lose" is an instance of the form $p \lor (\sim p)$ and according to Theorem 4(i) is a tautology; that is, the given statement is always true. Theorem 4(i) is often referred to as the law of the excluded middle because $p \lor (\sim p)$ must be true. ∎

Theorem 5, which follows, can be proved by using Theorems 1–4 or by using evaluation tables.

▶ **Theorem 5**
If C, U, and p are statement forms, where C is a contradiction and U is a tautology, then for every p,

(i) $(p \lor U) \equiv U.$
(ii) $(p \land C) \equiv C.$

The proof of Theorem 5 is left as Exercises 26 and 27. ◀

Although Theorems 1–4 were proved using evaluation tables, the following theorems may be proved using only those theorems that precede the one in question; therefore, we may think of Theorems 1–4 as a basis or a foundation upon which Theorems 5–17 can be built. Of course, Theorems 5–17 also may be proved by using evaluation tables, if it is more convenient to do so.

▶ **Theorems 6–17**
If C, U, p, q, and r are statement forms, where C is a contradiction and U is a tautology, then
    6. For every p, $(p \lor p) \equiv p$.
    7. For every p, $(p \land p) \equiv p$.
    8. If x is a statement form such that $(p \lor x) \equiv p$ for every p, then $x \equiv C$.

9. If x is a statement form such that $(p \wedge x) \equiv p$ for every p, then $x \equiv U$.
10. $[p \vee (p \wedge q)] \equiv p$ (Absorption Property).
11. $[p \wedge (p \vee q)] \equiv p$ (Absorption Property).
12. $[(p \wedge q) \equiv p]$ if and only if $[(p \vee q) \equiv q]$.
13. If $(p \vee q) \equiv (p \vee r)$ and if $(p \wedge q) \equiv (p \wedge r)$, then $q \equiv r$.
14. $[p \vee (q \vee r)] \equiv [(p \vee q) \vee r]$ (Associative Property).
15. $[p \wedge (q \wedge r)] \equiv [(p \wedge q) \wedge r]$ (Associative Property).
16. $[\sim(p \vee q)] \equiv [(\sim p) \wedge (\sim q)]$ (DeMorgan's Property).
17. $[\sim(p \wedge q)] \equiv [(\sim p) \vee (\sim q)]$ (DeMorgan's Property). ◀

APPLICATIONS

Example 5
Complicated compound statements also can be simplified by using the properties of the algebra of components developed in this section.

For example, consider the following quotation:
"In case of death:
The interest paid to the party of the first part must exceed $6000;
AND
the party of the second part will sell its interest in the company, or the interest paid to the party of the first part must not exceed $6000."

The quotation can be expressed symbolically as

$$p \wedge [q \vee (\sim p)],$$

which by Theorems 2, 3, and 4 is logically equivalent to

$$(p \wedge q).$$

Therefore, the quotation written above has the same meaning as
"In case of death:
The interest paid to the party of the first part must exceed $6000;
AND
the party of the second part will sell its interest in the company." ∎

Example 6
To illustrate the analysis of conflict by means of material presented so far, we consider the following hypothetical situation. (Other situations involving conflict, such as labor–management negotiations, could have been chosen.)

Suppose there is general agreement among certain powers at a peace conference that there is a need for a peacekeeping force consisting of forces from more than two nations. The conferees have agreed that such a force must come from among five nations, but they have imposed certain conditions upon the makeup of the peacekeeping force. These conditions are
1. Nations Q and R cannot serve together.
2. If nation X is excluded, then nation P cannot serve.
3. Nations Q and S cannot serve together.

4. Nations X and R cannot serve together.
5. Nation P or nation Q must serve.
6. Nation R or nation S must serve.

Is the peace conference hopelessly deadlocked or is there a solution that will satisfy all six conditions?

Let p represent the statement "Nation P serves"; let q represent the statement "Nation Q serves"; and so on. Conditions 1 through 6 can be restated as:

1. $(q \wedge r) \equiv C$.
2. $[(\sim x) \wedge p] \equiv C$ [derived from $(\sim x) \Rightarrow (\sim p)$].
3. $(q \wedge s) \equiv C$.
4. $(x \wedge r) \equiv C$.
5. $(p \vee q) \equiv U$.
6. $(r \vee s) \equiv U$.

Using the definitions and theorems stated so far, we can make the following deductions: From 1 and 3 we have

$$[(q \wedge r) \vee (q \wedge s)] \equiv C,$$

or

$$[q \wedge (r \vee s)] \equiv C.$$

From condition 6 the latter expression becomes

$$(q \wedge U) \equiv C.$$

Therefore, $q \equiv C$. From 5 we can now say that $p \equiv U$. Now, from 2, $(\sim x) \equiv C$ and hence $x \equiv U$. From 4, $r \equiv C$ and, from 6, $s \equiv U$. Therefore, nations P, S, and X can serve under the imposed conditions; Q and R cannot serve under the imposed conditions. ∎

EXERCISES

Suggested minimum assignment: Exercises 1, 3, 11, 14, and 17.

In each of Exercises 1–5 simplify the given compound statement form by using Theorems 1–5 of this section.

1. $(\sim p) \wedge (q \vee p)$.
2. $(\sim p) \vee (q \wedge p)$.
3. $[p \wedge (\sim p)] \vee [(q \vee \sim q) \wedge p]$.
4. $p \vee \{q \vee [p \wedge (\sim p)]\}$.
5. $[(\sim p) \wedge (p \vee q)] \vee \{(q \vee p) \wedge [(\sim q) \vee p]\}$.

In each of Exercises 6–17 prove the indicated theorem using evaluation tables or any preceding theorems.

6. Theorem 6.
7. Theorem 7.
8. Theorem 8.
9. Theorem 9.
10. Theorem 10.
11. Theorem 11.
12. Theorem 12.
13. Theorem 13.

14. Theorem 14. (*Hint:* Use evaluation tables—very difficult otherwise.)
15. Theorem 15. (*Hint:* Use evaluation tables—very difficult otherwise.)
16. Theorem 16.
17. Theorem 17.

In each of Exercises 18–25 use evaluation tables to prove the theorem.

18. Prove Theorem 1(*i*).
19. Prove Theorem 1(*ii*).
20. Prove Theorem 2(*i*).
21. Prove Theorem 2(*ii*).
22. Prove Theorem 3(*i*).
23. Prove Theorem 3(*ii*).
24. Prove Theorem 4(*i*).
25. Prove Theorem 4(*ii*).
26. Prove Theorem 5(*i*) by using Theorems 1–4.
27. Prove Theorem 5(*ii*) by using Theorems 1–4.
28. Solve the second part of Example 7 of Section 2.3.
29. The first section of the proposed Equal Rights Amendment to the U.S. Constitution states: "Equality of rights under the law shall not be denied or abridged by the United States or by any state on account of sex."
    (*a*) Identify, and label with the letters p, q, r, and s, the four positive simple statements that are included in the amendment.
    (*b*) Construct, using p, q, r, and s, a compound statement form for which the amendment is an instance.
    (*c*) Write a statement form that is equivalent to your answer to part (*b*) such that only negations of simple (rather than compound) statement forms appear.

2.5 Arguments

Most of us in our normal lives are subjected to a steady stream of arguments from various sources, such as advertisements, candidates for office, public officials, parents, children, teachers, students, peers, revolutionaries, and various other dispensers of rhetoric.

The purpose of this section is to apply the rules of logic (properties of our structure) to analyze the forms of arguments, to classify them as valid or invalid, and to distinguish arguments from nonarguments.

We shall begin with the following definition of an **argument form**, which is simply a special type of statement form.

▶ **Definition 8**

An **argument form** is a conditional of the form

$$(p_1 \wedge p_2 \wedge \cdots \wedge p_n) \to q.$$

The statement forms $p_1, \ldots, p_n$ are called **premises** and q is called the **conclusion**. An argument form is said to be **valid** if it is a tautology; otherwise the argument form is a **fallacy**.

In other words, if the argument form (or conditional) is an implication, then the argument form is said to be valid; if it is not an implication, it is said to be a fallacy.

Example 1
The conditional
$$[(p \to q) \land p] \to q$$
has premise $(p \to q)$ and premise p. The conclusion is q. The conditional is a valid argument form because the conditional is a tautology as verified in Example 3 of Section 2.3 and hence an implication. On the other hand, the conditional
$$[(p \to q) \land q] \to p,$$
with premise $(p \to q)$, premise q, and conclusion p, is a fallacy because the conditional is not a tautology, as can be demonstrated in an evaluation table (Table 2.19). ∎

Table 2.19. EXAMPLE OF A FALLACY.

(1) p	(2) q	(3) $p \to q$	(4) $(p \to q) \land q$	(4) → (1)
T	T	T	T	T
T	F	F	F	T
F	T	T	T	$\boxed{F}$
F	F	T	F	T

For reference purposes we give a list of basic theorems that are particularly useful in establishing the validity or fallacy of complicated argument forms.

▶ **Theorems 18–24**
Let p, q, and r be statement forms.
 18. (Double Negation Property) $[\sim(\sim p)] \Leftrightarrow p$.
 19. (Conjunction Property) $(p \land q) \Rightarrow p$.
 20. (Disjunction Property) $p \Rightarrow (p \lor q)$.
 21. (Detachment Property) $[(p \to q) \land p] \Rightarrow q$.
 22. (Contrapositive Property) $[(p \to q) \land (\sim q)] \Rightarrow (\sim p)$.
 23. (Disjunction Detachment Property) $[(p \lor q) \land (\sim p)] \Rightarrow q$.
 24. (Chain Property) $[(p \to q) \land (q \to r)] \Rightarrow (p \to r)$.

The proofs of Theorems 18–24 are left as Exercises 15–21. These proofs can be accomplished by means of evaluation tables or by means of the theorems of the last section. If the latter approach is used, one must recognize that each proof will involve showing that a conditional is a tautology and hence an implication; one can show this by remembering that a conditional $r \to s$ is equivalent to $[(\sim r) \lor s]$ and then by showing, if possible, that the latter expression is equivalent to U. ◀

In Example 1, we showed how argument forms could be shown to be valid or fallacious by use of evaluation tables. We next illustrate how argument forms may be determined to be valid by use of the theorems that have been stated so far.

Example 2
Show whether or not the argument form
$$\{[(p \vee q) \wedge (\sim p)] \wedge [(\sim p) \to (\sim q)]\} \to p$$
is valid.

$\{[(p \vee q) \wedge (\sim p)] \wedge [(\sim p) \to (\sim q)]\}$
$\Rightarrow \{q \wedge [(\sim p) \to (\sim q)]\}$ by Theorem 23,
$\Rightarrow \{[\sim(\sim q)] \wedge [(\sim p) \to (\sim q)]\}$ by Theorem 18,
$\Rightarrow \sim(\sim p)$ by Theorems 1 and 22,
$\Rightarrow p$ by Theorem 18. ∎

Example 3
The argument forms
1. $(q \to p) \to (p \to q)$,
2. $[(q \to p) \wedge (\sim q)] \to (\sim p)$,
3. $[(p \to q) \wedge (q \to r)] \to (r \to p)$

are common argument forms that are not valid. Unfortunately, they often are treated as valid, and grandiose false arguments are thereby perpetrated on the unwary. The first of the three forms deserves special attention: $(p \to q)$ is called the **converse** of $(q \to p)$, but the truth of $(q \to p)$ does not ensure the truth of $(p \to q)$. Argument forms such as these can be shown to be fallacious by means of evaluation tables. Illustrations of the fallacious forms presented above are given in Example 6 of the Applications. ∎

We now turn our attention to specific arguments. Because an argument form is a special type of statement form, we might expect an argument to be a special type of statement.

▶ **Definition 9**
An argument form in which each placeholder p, q, and so on, is a specified statement is known as an **argument**.

Example 4
Arguments are often written in the following manner:
1. If candidate A wins, then taxes will increase.
2. Candidate A wins.
   C. Therefore, taxes will increase.

The argument above is an instance of the following form (argument form):

$$\begin{bmatrix} p \to q \\ p \\ \hline \text{Therefore, } q. \end{bmatrix} \quad \text{or} \quad [(p \to q) \wedge p] \to q,$$

which is valid according to Theorem 21. ∎

In Example 4, the fact that the form of the argument is valid does not necessarily mean that the conclusion is true. A valid form simply means that

the rules of reasoning are correct and says nothing about the truth or falsity of the premises and the conclusion. For example, if either one of the two premises in Example 4 is false, then the conclusion may or may not be true. If, however, the premises of an argument are true and the corresponding argument form is valid, then the conclusion must be true.

Statements that are not conditionals are not arguments.

APPLICATIONS

Example 5

An argument having a valid form and true premises is called a **sound argument**. An argument having a fallacious form or at least one false premise is called an **unsound argument**. Consider the following argument:
1. If the Florida Power Corporation issued rights in 1971, then stockholders with at least 10 shares had an option to buy more stock at $45.25 per share.
2. The Florida Power Corporation issued rights in 1971.
C. Therefore, stockholders with at least 10 shares had an option to buy more stock at $45.25 per share.

The argument has the valid form $[(p \to q) \land p] \Rightarrow q$ (Theorem 21), and the premises are true; hence the argument is sound.

Consider another argument:
1. If Mexico City is the capital of Mexico, then the moon had trees.
2. If the moon had trees, wood could be brought to earth from the moon.
C. Therefore, if Mexico City is the capital of Mexico, then wood could be brought to the earth from the moon.

The argument has the valid form $[(p \to q) \land (q \to r)] \Rightarrow (p \to r)$, but the first premise is not true; hence the argument is unsound.

Consider a third argument:
1. If candidate A gets a majority of the votes in the general election, then he is elected to office.
2. Candidate A does not get a majority of the votes.
C. Therefore, candidate A is not elected to office.

The form, however, is the fallacious form $[(p \to q) \land (\sim p)] \to (\sim q)$, and hence the argument is unsound, regardless of the truth or falsity of the premises. In other words, the conclusion does not follow logically from the premises even though these premises may be true. Evaluation tables can be used to prove that a form is fallacious.

Finally, we point out that there are some arguments that cannot be labeled as sound or unsound because the truth or falsity of the premises is impossible to ascertain. Consider the assertion:
1. If there are bacteria on Saturn, then there is life on Saturn.
2. There are bacteria on Saturn.
C. Therefore, there is life on Saturn.

Although this argument has a valid form, we cannot say whether the second sentence is true or false, although we know that it is one or the other. Thus we cannot say that the argument is sound or unsound. ∎

Example 6
In everyday life the ordinary citizen is subject to a never-ending barrage of arguments from a variety of sources. Some of these arguments are *not* sound because their premises are not true. Others, however, are not sound simply because their forms are not valid. Instances of the fallacious argument forms of Example 3 follow. We emphasize that in these instances, the truth or falsity of the premises is not considered; rather it is the form that is faulty.

  1. If candidate A is elected, there will be peace for 20 years. To have peace for 20 years it is necessary that candidate A is elected.
  2. If a system provides the means for peaceful change, then no cause justifies violence in the name of change. No means for peaceful change are provided; therefore, violence is justifiable.
  3. If brand X is used, event A will take place, and if event A takes place, happiness will prevail. Happiness prevails; therefore, brand X is used. ∎

EXERCISES

Suggested minimum assignment: Exercises 1, 3, 5, 8, 9, 11, 13, and 17.

In each of Exercises 1–8 determine whether the given statement forms are valid argument forms, fallacious argument forms, or nonargument forms. Give reasons for your answer. Identify the conclusion and each premise.

1. $[(p \vee q) \wedge (\sim p)] \to q$.
2. $[(\sim q) \wedge (p \to q)] \to p$.
3. $[(p \vee q) \wedge q] \to p$.
4. $[q \wedge (q \to p)] \to (\sim p)$.
5. $[q \wedge (p \vee q)]$.
6. $[q \vee (p \wedge \sim q)]$.
7. $\{[p \vee (\sim q)] \wedge [(p \vee q) \wedge q]\} \to (p \wedge q)$.
8. $\{[p \wedge (\sim q)] \wedge [(p \to q) \wedge p]\} \to p$.

In each of Exercises 9–14 determine whether the form (argument form) of the given argument is valid or fallacious.

9. If the experimental mouse is hungry, he will enter the food box on the next trial. If the mouse enters the food box on the next trial, then it is not the case that the hunger response was less on the previous trial. The hunger response on the previous trial was less. Therefore, the mouse is not hungry.
10. If the factory completes production in one day, the factory will receive an extra allocation of funds. The factory gets an extra allocation of funds, or the workers will not get a bonus. The factory completes the production in one day. Therefore, the workers get a bonus.
11. If a candidate running for office is poor, then he cannot spend much money on television. Moreover, if the candidate cannot spend much

money on television, his chances for election are not good. But his chances for election are good; therefore, he is poor.
12. If interest rates decrease, consumers borrow more money; when consumers borrow more, more houses are built. Therefore, interest rates decrease if more houses are built.
13. If the population of China increased, then the Mississippi River flows north. China had increased population. Therefore, the Mississippi River flows north.
14. If the state of Vermont triples its population, she will get at least one extra seat in the House of Representatives. Vermont tripled her population or Mt. Mansfield erupted. Mt. Mansfield did not erupt. Therefore, Vermont will get at least one extra House seat.

15. Prove Theorem 18.
16. Prove Theorem 19.
17. Prove Theorem 20.
18. Prove Theorem 21.
19. Prove Theorem 22.
20. Prove Theorem 23.
21. Prove Theorem 24.
22. Determine whether the argument of Exercise 9 is sound or unsound or neither.
23. Determine whether the arguments of Exercises 11 and 13 are sound or unsound or neither.

2.6 Series and Parallel Systems†

The word **system** is quite common in our current language. We frequently hear about such diverse and important systems as computer systems, digestive systems, business systems, spacecraft systems, political systems, transportation systems, and many others. It is important in various disciplines that participants have some idea of what is meant by a system and the various ways that mathematics can serve in analyzing, modifying, or improving identifiable systems; often this service can be achieved by means of constructing **mathematical models** of physical, natural, or social systems. Mathematical models will be considered in some detail in Sections 12.4 and 12.5.

Webster defines the word **system** to be "a set or arrangement of things (components) so related or connected as to form a unity or organic whole." So far in this chapter we have been concerned with verbal systems in which the components were statements or statement forms, and these components were connected with the word connectives "and," "or," and so on. In this section we shall consider different but analogous systems, where components have different interpretations.

Consider a specific system having n parts or components that are interrelated in a certain way. If the components of a system are interrelated in such a way that the entire system fails when any one of the components fails, then the system is called a **series system**. On the other hand, a **parallel system**

† Sections 2.4 and 2.5 are not prerequisite to this section.

is one that will fail only if all its components fail. Naturally, we also can have systems with both parallel and series systems as components. We now present some specific illustrations of systems and their components.

Example 1

Think of an aircraft in flight as a system; let the components of the system be the pilot, the copilot, and the aircraft. Assume that the pilot and the copilot are trained to perform the same duties; together, they form a parallel system because a failure in their assignment will occur only if both fail to perform. Ordinarily, such a parallel system is diagrammed as shown in Figure 2.2. On the other hand, the crew (pilot and copilot) and the aircraft form a series system because if either the crew or the aircraft fails, the system fails. Such a series system is diagrammed as shown in Figure 2.3. The entire system can be diagrammed as shown in Figure 2.4.

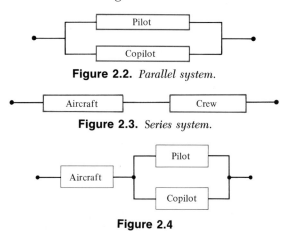

Figure 2.2. *Parallel system.*

Figure 2.3. *Series system.*

Figure 2.4

The value of a component depends upon whether the component performs satisfactorily or not according to some well-defined standards. If a component performs satisfactorily, a value of 1 is assigned; otherwise, a value of 0 is assigned. ∎

Example 2

Consider a set of "on–off" switches similar to a common light switch found near the door of most rooms or a set of switches found in the interior of a digital computer. We shall use the letters p, q, r, s, t, and so on, to represent these switches. An ordinary light switch is either closed or open (that is, on or off); to reflect these two states, a switch p is assigned the value 1 if closed (on), and 0 if open (off). By a closed switch we mean that electrical current will flow through the switch from terminal A to terminal B (Figure 2.5), and

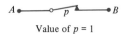

Value of $p = 1$

Figure 2.5. *Closed switch; electricity will flow from A to B.*

by an open switch we mean that electrical current will not flow through the switch from terminal A to terminal B (Figure 2.6). Of course, one need not be confined to the flow of electricity; one can just as well consider the flow of something else, such as information, money, or influence. Switches can be connected in parallel as shown in Figure 2.7, or they can be connected in series as shown in Figure 2.8. ∎

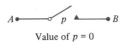

Value of $p = 0$

Figure 2.6. *Open switch; electricity will not flow from A to B.*

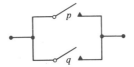

Figure 2.7. *Parallel connection of switches.*

Figure 2.8. *Series connection of switches.*

We shall designate a parallel connection of components by the symbol $\vee$, and hence the parallel connections shown in Figures 2.2 and 2.7 can be expressed symbolically as $p \vee q$. The symbol $\vee$ can be read "is connected in parallel with." The series connections shown in Figures 2.3 and 2.8 can be expressed as $r \wedge s$. The symbol $\wedge$ can be read "is connected in series with." If a component t always has a value opposite to that of p, then component t is said to be the **negation** of component p and is designated $\sim p$. For example, if in Example 2 a switch t is wired in such a way that it opens when switch p is closed and closes when switch p is opened, then switch t is said to be the negation of p, and switch t is designated with the symbol $\sim p$. The symbol $\sim$ can be read "the negation of."

When two or more single components are connected in parallel or series (or both) or when the negation of a component occurs, we call such a system a **compound component**.

Example 3

To diagram a compound component such as $[q \vee (r \wedge q)] \wedge p$, start from within, namely, $(r \wedge q)$, and sketch that as shown in Figure 2.9. In order to emphasize the fact that the value of each component is 0 or 1, we shall adopt the convention of using a switch representation for each component in the diagram of a system. Then connect compound component $(r \wedge q)$ with q in parallel as shown in Figure 2.10. Finally, connect compound component $q \vee (r \wedge q)$ in series with p as shown in Figure 2.11.

Figure 2.9. $r \wedge q$.

Sec. 2.6] Series and Parallel Systems

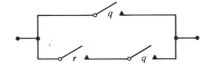

Figure 2.10. $q \vee (r \wedge q)$.

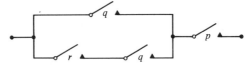

Figure 2.11. $[q \vee (r \wedge q)] \wedge p$.

In Figure 2.11 notice that component q appears twice. Within a given system it is not unusual for a component to appear more than once; for example, two or more switches can be constructed in such a way that they are "off" (or "on") at the same time. ∎

APPLICATIONS

Example 4

Let the **reliability** of a system be defined as the probability that the system will perform satisfactorily for a specified time interval under specified environmental conditions. In Chapter 4 after a study of probability is made, we can verify that the reliability R_s of a series system consisting of n independent components is the product of the reliabilities $R_1, R_2, \ldots, R_n$ of the respective components. That is, $R_s = (R_1)(R_2) \cdots (R_n)$. For example, if a series system had 3 components with reliabilities of $\frac{1}{3}, \frac{3}{4}$, and $\frac{5}{6}$, then the reliability R_s of the system would be

$$R_s = (\tfrac{1}{3})(\tfrac{3}{4})(\tfrac{5}{6}) = \tfrac{15}{72} \approx 21 \text{ per cent.}$$

We can also verify later that the reliability R_p of a parallel system with n independent components is given by the formula

$$R_p = 1 - [(1 - R_1)(1 - R_2) \cdots (1 - R_n)].$$

For example, if a parallel system had components with reliabilities of $\frac{1}{3}, \frac{3}{4}$, and $\frac{5}{6}$, then the reliability of the system would be

$$R_p = 1 - [(1 - \tfrac{1}{3})(1 - \tfrac{3}{4})(1 - \tfrac{5}{6})]$$
$$= 1 - [(\tfrac{2}{3})(\tfrac{1}{4})(\tfrac{1}{6})]$$
$$= 1 - \tfrac{1}{36}$$
$$= \tfrac{35}{36} \approx 97 \text{ per cent.}$$

The reliability of a system can be improved by the appropriate inclusion of parallel components. In space travel such parallel subsystem components are often referred to as "backup systems." Further discussion of reliability may be found in Exercise 10 of this section, in Example 6 of Section 4.5, and in Miller and Freund [49], pp. 362–364. ∎

Example 5

The purpose of this application is to illustrate the use of the material of this chapter in legal analysis. Consider Section 6731(e) of the Business and Professions Code of California: "Nothing in this chapter shall prohibit the preparation of plans, drawings, specifications, estimates, or instruments of service for single or multiple dwellings not more than two stories and basement in height; garages or other structures appurtenant to such dwellings; farm or ranch buildings; or any other buildings, except steel frame and concrete buildings, not over one story in height, where the span between bearing walls does not exceed twenty-five (25) feet."

On the basis of the statutory provision this question arises: What kind of papers pertaining to buildings are not prohibited? To answer the question and to analyze Section 6731(e), the code can be considered as a system of words consisting of component phrases with the connectives "or," "and," and "not." The connective "or" will be considered as a parallel connection of the component phrases; the word "and" will be considered as a series connection of phrases, and the word "not" will call for the negation.

On the basis of the following identification of component phrases, Section 6731(e) can be diagrammed as shown in Figure 2.12: "Nothing in this chapter shall prohibit the preparation of $\overbrace{\text{plans}}^{a}$, $\overbrace{\text{drawings}}^{b}$, $\overbrace{\text{specifications}}^{c}$, $\overbrace{\text{estimates}}^{d}$, or $\overbrace{\text{instruments of service}}^{e}$ for $\overbrace{\text{single}}^{f}$ or $\overbrace{\text{multiple}}^{g}$ dwellings $\overbrace{\text{not more than two stories and basement in height}}^{h}$; $\overbrace{\text{garages}}^{i}$ or $\overbrace{\text{other structures}}^{j}$ $\overbrace{\text{appurtenant to such dwellings}}^{(f \vee g) \wedge \sim h \;\; k}$; $\overbrace{\text{farm or ranch buildings}}^{l}$; or $\overbrace{\text{any other buildings}}^{m}$, except $\overbrace{\text{steel frame}}^{n}$ and $\overbrace{\text{concrete}}^{o}$ buildings, $\overbrace{\text{not over one story in height}}^{p}$, $\overbrace{\text{where the span between the bearing walls does not exceed twenty-five (25) feet}}^{q}$."

Figure 2.12 provides a diagram of the code for purposes of analysis. Such a diagram may be particularly appropriate in many problems involving statutory interpretation, since it makes visible to whom or to what cumbersome statutory provisions apply. In this particular example, any paper pertaining to buildings will not be prohibited, provided a closed path may be found from A to B in Figure 2.12. Section 6731(e) is analyzed in a slightly different way by Allen and Caldwell [2]. This reference, which deals with logic and judicial decision making, was discussed in Section 2.1. ∎

Sec. 2.6] Series and Parallel Systems

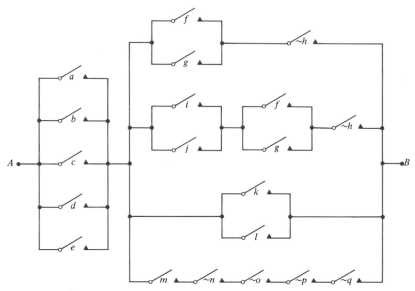

Figure 2.12. *Section of Business and Professions Code.*

EXERCISES

Suggested minimum assignment: Exercises 1, 4, 5, 8, and 9.

In Exercises 1–4 sketch the diagrams of the following systems (compound components).

1. $(p \vee q) \wedge (r \vee s)$.
2. $(p \wedge q) \vee (r \wedge s)$.
3. $[p \vee (p \wedge q)] \vee (p \wedge r)$.
4. $[p \wedge (p \vee q)] \wedge (p \vee r)$.

5–8. For Exercises 5–8 write the symbolic expressions for the systems (compound components) shown in Figures 2.13–2.16.

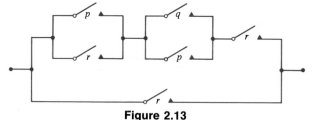

Figure 2.13

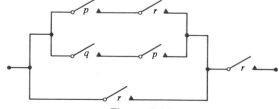

Figure 2.14

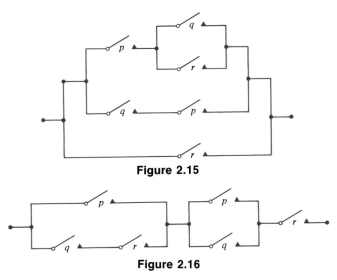

Figure 2.15

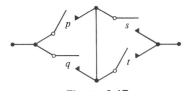

Figure 2.16

9. Using p, q, s, t, $\vee$, and $\wedge$, write an expression that represents the arrangement of switches shown in Figure 2.17. Draw another diagram corresponding to your answer.

Figure 2.17

10. (a) Suppose that we model an oral communication from person A to person B by a series system of three switches. The first switch represents the verbalization of an idea by person A. The second switch represents the hearing of person B, and the third switch represents the interpretation by person B of what was heard. If the reliabilities of these three switches are $\frac{9}{10}$, $\frac{7}{10}$, and $\frac{8}{10}$, respectively, use Example 4 to determine the reliability of this particular communication system (that is, the series system).
    (b) Suppose that the communication is repeated, thus creating a parallel system. What is the reliability of the repetitive communication (that is, the parallel system)?

2.7 Evaluation of Compound Components

Values of the compound components $p \vee q$, $p \wedge q$, and $\sim p$ are defined in Tables 2.20 and 2.21. Of course, the definitions are consistent with the evaluations of the conjunction, disjunction, and negation of statement forms where 1 corresponds to T and 0 corresponds to F. Also, the definitions are

consistent with the performance of the systems in Examples 1 and 2 of Section 2.6. For example, the compound switch $p \vee q$ has value 1 if p has value 1, regardless of the value of q, because current will flow from terminal A to terminal B through the switch p as shown in Figure 2.18.

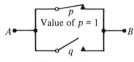

Figure 2.18

▶ **Definition 10**

For arbitrary components p and q, the values of $p \vee q$ and $p \wedge q$ and $\sim p$ are defined according to Tables 2.20 and 2.21.

Table 2.20. DEFINITION OF VALUES OF PARALLEL AND SERIES CONNECTIONS.

p	q	$p \vee q$	$p \wedge q$
1	1	1	1
1	0	1	0
0	1	1	0
0	0	0	0

Table 2.21 DEFINITION OF VALUES OF $\sim p$.

p	$\sim p$
1	0
0	1

We are now in a position to evaluate more complicated compound components by using the definition just given.

Example 1

Evaluate the system (compound component) $[p \wedge (p \vee q)] \vee q$, shown in Figure 2.19, for all possible arrangements of the values of p and q.

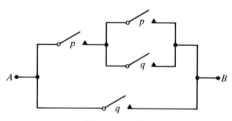

Figure 2.19

The evaluation is made by means of a table similar to that given in Definition 10; see Table 2.22. ∎

One of the purposes of this study is to point the way toward the analysis of systems with a view of simplifying them. For example, consider the compound switch $p \wedge (p \vee q)$ shown in Figure 2.20 and the switch p shown in Figure 2.21. It turns out that when p is closed, $p \wedge (p \vee q)$ is closed, and when p is open, $p \wedge (p \vee q)$ is open. In other words, as far as the flow of

Table 2.22. EVALUATION OF $[p \wedge (p \vee q)] \vee q$.

(1)	(2)	(3) (1) $\vee$ (2) $p \vee q$	(4) (1) $\wedge$ (3) $p \wedge (p \vee q)$	(5) (4) $\vee$ (2) $[p \wedge (p \vee q)] \vee q$
p	q			
1	1	1	1	1
1	0	1	1	1
0	1	1	0	1
0	0	0	0	0

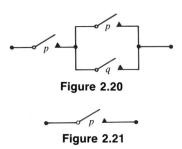

Figure 2.20

Figure 2.21

electricity (or information, or money) is concerned, $p \wedge (p \vee q)$ and p are the same.

▶ **Definition 11**
Two compound components are **equivalent** when they have the same value for all possible arrangements of values of the single components. Equivalence will be denoted by the symbol $\equiv$.

Example 2
The compound component $[p \wedge (p \vee q)] \vee q$ shown in Figure 2.19 is equivalent to the compound component $p \vee q$. This can be observed from the evaluation table of Example 1; notice that the values in columns (3) and (5) are the same. ▮

APPLICATIONS

Example 3
In this example we illustrate the construction of a voting device that will cause a light to flash if a majority of people vote "yes" simultaneously on a yes–no option. For simplicity this illustration deals with a device for three voters. The possible votes with an indication of majority outcomes are given in Table 2.23.

Let p, q, and r be the yes–no switches operated by P, Q, and R, respectively. Electrical current will flow from terminal A to terminal B if all three switches shown in Figure 2.22 have value 1; this compound switch corresponds to the first row of Table 2.23. Electrical current will flow from terminal A to terminal B if all three switches shown in Figure 2.23 have value 1; this situation corresponds to the second row of Table 2.23. Two

Sec. 2.7] Evaluation of Compound Components

Table 2.23

	Voters		
P	Q	R	Majority of yes votes
yes	yes	yes	yes
yes	yes	no	yes
yes	no	yes	yes
no	yes	yes	yes
yes	no	no	no
no	yes	no	no
no	no	yes	no
no	no	no	no

Figure 2.22. *Switch corresponding to the first row of Table 2.23.*

Figure 2.23. *Switch corresponding to the second row of Table 2.23.*

other cases in which electrical current will flow from A to B are shown in Figure 2.24. In both cases all three switches must have value 1; the latter cases correspond to the third and fourth rows of Table 2.23. There will be a majority of yes votes and current will flow from A to B in all four of the situations that have just been described.

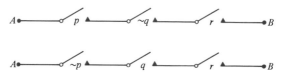

Figure 2.24. *Switches corresponding to the third and fourth rows of Table 2.23.*

If the four compound switches shown in Figures 2.22, 2.23, and 2.24 are connected in parallel, then electrical current will flow from A' to B' as shown in Figure 2.25 if and only if there is a majority of yes votes cast by voters P, Q, and R. The compound switch shown in Figure 2.25 can be represented by s, where we let s be

$$[(p \wedge q) \wedge r] \vee [(p \wedge q) \wedge (\sim r)] \vee \{[p \wedge (\sim q)] \wedge r\} \vee \{[(\sim p) \wedge q] \wedge r\}.$$

It can be shown that the compound switch s can be simplified; that is, it can be shown that the flow of electricity from A' to B' through s is the same as the flow from A' to B' through the simpler compound switch

$$[p \wedge (q \vee r)] \vee (q \wedge r)$$

shown in Figure 2.26. ∎

Chap. 2] Logic

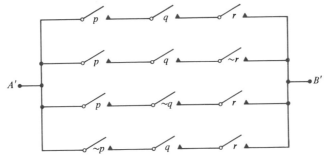

Figure 2.25. Switch representing a majority or nonmajority of yes votes in Table 2.23.

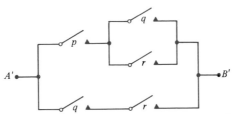

Figure 2.26. Switch representing a majority or nonmajority of yes votes in Table 2.23.

Example 4

As civilization has developed, men have found ways to quantify certain events, situations, or arrangements of matter of the real world. Such mathematical representations are often called **mathematical models**. The compound switch

$$[(p \wedge q) \wedge r] \vee [(p \wedge q) \wedge (\sim r)] \vee \{[(p \wedge (\sim q)] \wedge r\} \vee \{[(\sim p) \wedge q] \wedge r\}$$

that was constructed in Example 3 is an illustration of a simple mathematical model of a small but real voting situation. Example 5 of Section 2.6 provides an example of a mathematical model of a legal statute. Mathematical modeling, which will be discussed further in Sections 12.4 and 12.5, enables a scientist to study the properties of the physical, biological, social, or commercial phenomena by studying the analogous mathematical models. From this study better theoretical understanding of the real situation is gained, and perhaps through this understanding, new theories can be justified. Example 3 further illustrated the concept of modeling in that the reader also was instructed in the construction of a mechanical model of the voting situation.

Mathematical modeling often provides insights into useful application or generalization of a single model in other ways. To show how a model that is constructed to simulate one real situation can be used to originate other applications, we shall generalize Table 2.23 of Example 3 to read "factors" rather than "voters" and "attainment of a goal" rather than "majority of yes votes"; see Table 2.24.

Sec. 2.7] Evaluation of Compound Components

Table 2.24

	Factors		Attainment of a goal
P	Q	R	
yes	yes	yes	yes
yes	yes	no	yes
yes	no	yes	yes
no	yes	yes	yes
yes	no	no	no
no	yes	no	no
no	no	yes	no
no	no	no	no

With this generalization one can give a variety of interpretations to the "factors" and the "goals," with each situation having the same basic mathematical model. For example, the "factors" may be environmental factors involved in the "goal" of maintaining the life of a certain organism.

The factors may be indicators (such as the price–earnings ratio) by which stocks are judged; if specific levels of two of the three indicators are satisfied, then a purchase of the stock is indicated. The factors may represent certain desirable conditions for drilling an oil well or building a hotel. If two of the three conditions are met, then a decision to act is in order. Larger versions of the basic model undoubtedly would be needed in realistic applications, such as meaningful environmental problems, stock purchases, or business ventures.

Moreover, the model can be altered to reflect weights of factors such as might be encountered in the election of a presidential candidate by the electoral college.

Further discussion of the construction of mathematical models may be found later in this book and in Freund [22], pp. 1–6. ∎

Example 5

As mentioned before, the concepts developed in this chapter need not be confined to the flow of electricity. To illustrate, consider an open or closed communication system between two terminals (perhaps nations or people) by way of *other* terminals. Consider a hypothetical situation in which a citizen wishes to communicate a message to the president of a large corporation. Of course, one way is by direct communication. Another alternative is through a friend, who can communicate directly or through one of two influential people who have the ear of the president. A diagram of this communication flow is given in Figure 2.27. The algebraic expression corresponding to Figure 2.27 is

$$p \vee (q \wedge \{r \vee [(u \wedge v) \vee (s \wedge t)]\}).$$ ∎

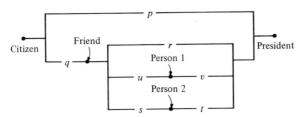

Figure 2.27. *Influence diagram.*

EXERCISES

Suggested minimum assignment: Exercises 1, 4, 5, 9, 11, and 13.

In each of Exercises 1–8 evaluate the given compound component for all possible arrangements of the values of the single components.

1. $p \wedge (p \vee q)$.
2. $p \vee [(\sim p) \wedge q]$.
3. $[(p \vee (\sim q)] \vee [q \vee (\sim p)]$.
4. $(q \wedge p) \wedge [q \vee (\sim p)]$.
5. $(p \vee r) \wedge (\sim q)$. (Eight rows will be needed in the evaluation table.)
6. $(r \vee q) \wedge [(\sim p) \vee (\sim q)]$. (Eight rows will be needed in the evaluation table.)
7. $(p \wedge q) \vee [r \wedge (\sim q)]$.
8. $p \vee [(\sim p) \wedge (r \vee q)]$.

In each of Exercises 9–12 determine whether the two given components are equivalent.

9. $p \vee (p \wedge q)$, p.
10. $\sim(p \vee q)$, $[(\sim p) \wedge (\sim q)]$.
11. $(p \wedge q) \vee [p \wedge (\sim q)]$, $(p \vee q)$.
12. $[p \vee (\sim q)] \wedge [(\sim p) \vee q]$, $[p \wedge (\sim q)]$.
13. (a) Construct a compound component that will have value 1 when components p and q have the values shown in the first row of Table 2.25 and which has value 0 otherwise. (*Hint:* Consider a series connection of some two single components.)

    Table 2.25

p	q
1	1
0	1

    (b) Construct a compound component that will have value 1 when components p and q have the values shown in the second row of Table 2.25 and which has value 0 otherwise.
    (c) Construct and sketch a diagram of a compound switch that will have value 1 when components p and q have the values shown in either row of Table 2.25 and which has value 0 otherwise. [*Hint:* Consider a parallel connection of your answers to parts (a) and (b).]

(d) Determine whether your answer to part (c) is equivalent to the component q.

14. (a) Construct a compound component that will have value 1 when components p, q, and r have the values shown in the first row of Table 2.26 and which has value 0 otherwise. [*Hint:* Consider a series connection of some three single components.]

    Table 2.26

p	q	r
0	0	1
1	0	1

    (b) Construct a compound component that will have value 1 when components p, q, and r have the values shown in the second row of Table 2.26 and which has value 0 otherwise.
    (c) Construct and sketch a diagram of a compound component that will have value 1 when switches p, q, and r have the values shown in either row of Table 2.26 and which has value 0 otherwise. [*Hint:* Consider the parallel connection of your answers to parts (a) and (b).]
    (d) Determine whether your answer to part (c) is equivalent to the compound component shown in Figure 2.28.

    Figure 2.28

15. The Eighth Amendment to the U.S. Constitution reads: "Excessive bail shall not be required, nor excessive fines imposed, nor cruel and unusual punishments inflicted."
    (a) Identify, and label with the letters p, q, r, and s the four positive simple statements that are included in the amendment.
    (b) Construct, using p, q, r, and s, a compound statement form for which the amendment is an instance.
    (c) Write a statement form that is equivalent to your answer to part (b) such that only negations of simple (rather than compound) statement forms appear and diagram your answer. The diagram provides a visual representation of what does *not* violate this amendment.
    (d) Write the negation of the answer to part (c) and express the result such that only negations of simple statement forms appear, and diagram your answer. The diagram provides a visual representation of what *does* constitute a violation of this amendment.
    (e) List any significant or unusual observations that become apparent from this analysis.

16. The first section of the proposed Equal Rights Amendment to the U.S. Constitution states: "Equality of rights under the law shall not be denied

or abridged by the United States or by any state on account of sex." Repeat the first four parts of Exercise 15.

17. Article III of the U.S. Constitution includes the sentence: "No person shall be convicted of treason unless on the testimony of two witnesses to the same overt act, or on confession in open Court."
    (a) Identify the three simple statements and represent them by the letters p, q, and r, respectively.
    (b) Construct using p, q, and r, a compound statement form for which the given sentence is an instance. (*Hint:* The word "unless" is defined by Webster to mean "if not.")
    (c) It can be proved that $(a \rightarrow b) \equiv [(\sim a) \vee b]$. Use this equivalence to express the answer of part (b) without the conditional.
    (d) Diagram your answer to part (c). The diagram gives a visual representation of what does *not* violate this part of the constitution.
    (e) Write the negation of the expression given as an answer to part (d) and diagram your answer. The diagram gives a visual representation of what *does* violate this part of the constitution.

2.8 Compound Components and the Computer

The concepts presented so far in this chapter play a very important part in the design of electronic computers, as well as in certain communications and control systems. An introduction to some ideas in this regard is presented by Hohn [29]; a more elementary but less comprehensive discussion may be found in [33]. The latter reference is a programmed self-instructional manual. A very short elementary introduction to the use of the algebra of switches to computer design may be found in Arnold [4], pp. 128–131.

To give the reader an intuitive idea of the connection between compound switches and computer design we will design a pair of compound switches that can be used to add a certain pair of integers. First we must be able to express integers as binary numerals involving only the digits 0 and 1; this means to write the integers using the base 2 rather than the base 10. For those not familiar with numerals using the base 2, the binary numerals shown in Table 2.27 may be thought of as a "code" of the corresponding

Table 2.27

Integer	Corresponding binary numerals
0	000
1	001
2	010
3	011
4	100
5	101
6	110
7	111

integers. (A more comprehensive presentation of numerals using base 2 may be found in Freund [22], pp. 49–54.)

To perform the addition

$$\begin{array}{r} 1 \\ +5 \\ \hline 6 \end{array}$$

we must add the binary equivalents (found in Table 2.27),

$$\begin{array}{r} 001 \\ +101 \\ \hline 110. \end{array}$$

Although it appears that we are adding pairs of digits in each column, in essence we are adding three digits in each column (except the right) because of the "carry" aspect of addition. In other words, we can think of the sum

$$\begin{array}{r} 001 \\ +101 \\ \hline 110 \end{array}$$

as the succession of sums (reading from right to left)

$$\begin{array}{ccc} 0 & 1 & 0 \\ 0 & 0 & 1 \\ +1 & +0 & +1 \\ \hline 1 & 1 & 0 \end{array}$$

where the top digit in each sum is the carried digit from the previous sum. (The binary addition table is given in Table 2.28.)

Table 2.28
BINARY ADDITION.

+	0	1
0	0	1
1	1	10

When adding three binary digits there are eight possibilities, as shown in Table 2.29.

The next step is to construct† a compound switch whose evaluation will be that of the fourth column of Table 2.29; the switch shown in Figure 2.29

---

† One way of constructing such a switch is to consider only those rows of Table 2.29 where the desired evaluation in the fourth column is 1 (that is, rows 1, 4, 6, and 7). Within the first column, replace each 1 by c and each 0 by $\sim c$. Within the second column, replace each 1 by p and each 0 by $\sim p$. Within the third column, replace each 1 by q and each 0 by $\sim q$. Then treat each row as a series connection of the three component switches and connect the four rows in parallel.

Chap. 2] Logic

Table 2.29. EIGHT POSSIBILITIES WHEN ADDING THREE BINARY DIGITS.

	Digits to be added			Recorded digit	Digit to be carried
	c	p	q		
(1)	1	1	1	1	1
(2)	1	1	0	0	1
(3)	1	0	1	0	1
(4)	1	0	0	1	0
(5)	0	1	1	0	1
(6)	0	1	0	1	0
(7)	0	0	1	1	0
(8)	0	0	0	0	0

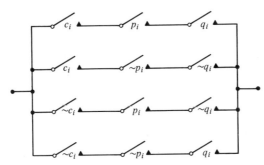

Figure 2.29. *Switch representing recorded digit of Table* 2.29.

is such a switch. Then we construct a compound switch whose evaluation will be that of the fifth column of Table 2.29; Figure 2.30 illustrates such a switch.

The subscript i in c_i, p_i, and q_i represents the column number starting from the right in a sum such as

$$\begin{array}{r} 0\ 0\ 1 \\ 1\ 0\ 1 \\ \hline \end{array}.$$

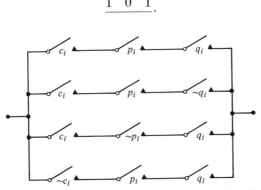

Figure 2.30. *Switch representing carried digit of Table* 2.29.

Sec. 2.8] Compound Components and the Computer

In the right column of this particular sum both p_1 and q_1 are 1, while c_1 (the carried digit) is 0 at the outset; hence the compound switch of Figure 2.29 has value 0, and the compound switch of Figure 2.30 has value 1. Therefore, the recorded digit is 0, and the carried digit is 1. We now have

$$\begin{array}{r} 1 \\ 0\ 0\ 1 \\ +1\ 0\ 1 \\ \hline 0 \end{array},$$

where now in the second column c_2 is 1 and p_2 and q_2 are 0. The process is repeated until terminated. The sum is then decoded (changed back to base 10).

NEW VOCABULARY

statement 2.1
statement form 2.1
instance of a statement form 2.1
connectives of statement forms 2.1
simple statement forms 2.1
compound statement forms 2.1
simple statements 2.1
compound statements 2.1
component of a statement form 2.1
conjunction 2.1
disjunction 2.1
conditional 2.1
biconditional 2.1
algebra of statement forms 2.1
negation 2.1
value of a statement form 2.2
equivalent statement forms 2.3
contrapositive of $p \to q$ 2.3
tautology 2.3
implication 2.3

contradiction 2.3
argument form 2.5
premise 2.5
conclusion 2.5
valid argument form 2.5
fallacious argument form 2.5
converse 2.5
argument 2.5
system 2.6
series system 2.6
parallel system 2.6
series connection of switches 2.6
parallel connection of switches 2.6
negation of a switch 2.6
compound component 2.6
evaluation of compound components 2.7
equivalent components 2.7

3 Review of Algebra; Functions and Graphs

Before one can begin a successful study of many of the subjects considered later in this book, certain preliminary topics must be mastered. This chapter is a collection of such topics.

The first part of the chapter consists of a review of certain algebraic topics that experience has shown to be troublesome to students. Much of the last part of the chapter is classified as analytic geometry. Analytic geometry has been described as "a subject generated by the marriage of algebra and geometry"; its development in the first half of the 17th century by René Descartes (1596–1650) and Pierre de Fermat (1601–1665) provided a necessary foundation for the development of much of modern science and mathematics, including calculus.

Applications of the material of this chapter are to a great extent indirect and will become apparent in later chapters.

Prerequisites: 1.1, 1.2.
Suggested Sections for Moderate Emphasis: 3.1–3.3, 3.5–3.9.
Suggested Sections for Minimum Emphasis: 3.5–3.9; those topics needed as prerequisites to material to be covered later.

3.1 Exponents

We begin our review of algebra with a study of the rules of **exponents**. Rules 1, 2, and 3 below define the use of positive integer exponents, negative integer exponents, and the exponent zero, respectively. Assume that a is a real number and that n is a positive integer.

1. $a^n = \underbrace{(a)(a) \cdots (a)}_{n \text{ times}}$.

2. $a^{-n} = 1/a^n$, provided that $a \neq 0$.
3. $a^0 = 1$, provided that $a \neq 0$.

Example 1
By rule 1,
$$4^3 = (4)(4)(4) = 64,$$
$$\left(-\frac{1}{2}\right)^4 = \left(-\frac{1}{2}\right)\left(-\frac{1}{2}\right)\left(-\frac{1}{2}\right)\left(-\frac{1}{2}\right) = \frac{1}{16},$$
$$0^2 = (0)(0) = 0,$$
and
$$3^1 = 3.$$

By rule 2,
$$2^{-3} = \frac{1}{2^3} = \frac{1}{8},$$
$$\left(\frac{2}{3}\right)^{-1} = \frac{1}{\left(\frac{2}{3}\right)^1} = \frac{3}{2},$$
and
$$(-3)^{-2} = \frac{1}{(-3)^2} = \frac{1}{9}.$$

By rule 3,
$$2^0 = 1,$$
$$0^0 \text{ is undefined,}$$
and
$$(-4)^0 = 1. \blacksquare$$

Now that the use of integer exponents has been defined, we turn our attention to fractional exponents. If a is a positive number and n is a positive integer larger than 1, then $a^{1/n}$ is defined to equal the *positive* number b by saying that
$$a^{1/n} = b \quad \text{means} \quad b^n = a.$$

Example 2
$$4^{1/2} = 2, \quad \text{since } 2^2 = 4;$$
$$(125)^{1/3} = 5, \quad \text{since } 5^3 = 125. \blacksquare$$

The number $a^{1/n}$ is often called the **nth root** of a and is written as $\sqrt[n]{a}$; that is,
$$a^{1/n} = \sqrt[n]{a}.$$
When $n = 2$, $\sqrt[2]{a}$ is usually replaced by $\sqrt{a}$. We call $\sqrt{a}$ the **square root** of a, and $\sqrt[3]{a}$ is commonly called the **cube root** of a.

Chap. 3] Review of Algebra; Functions and Graphs

Example 3

$$9^{1/2} = \sqrt{9} = 3,$$
$$\sqrt[3]{64} = 4,$$
$$\sqrt[4]{16} = 2.$$

Notice that it is *wrong* to say $9^{1/2} = \pm 3$ and $\sqrt{9} = \pm 3$, since we agreed originally that the answer would be positive; hence $\sqrt{9}$ equals just one number and is not ambiguous. ∎

Next, it is logical to define $a^{m/n}$, where a is positive and m and n are integers other than zero (we presuppose that the fraction m/n has been reduced to lowest terms). Notice that n can always be assumed to be positive because a rational number such as $-\frac{4}{3}$ can be written

$$-\frac{4}{3} = \frac{-4}{3}.$$

With this agreement that n is positive we define

$$a^{m/n} = (a^{1/n})^m,$$

or equivalently

$$a^{m/n} = (\sqrt[n]{a})^m.$$

Example 4

$$8^{2/3} = (8^{1/3})^2 = 2^2 = 4,$$
$$8^{-4/3} = (8^{1/3})^{-4} = 2^{-4} = \frac{1}{2^4} = \frac{1}{16},$$
$$(16)^{-3/4} = (\sqrt[4]{16})^{-3} = 2^{-3} = \frac{1}{2^3} = \frac{1}{8},$$
$$1^{6/7} = (\sqrt[7]{1})^6 = 1^6 = 1. \quad \blacksquare$$

If a is negative, $a^{m/n}$ is defined as above, provided that we require n to be an *odd* positive integer; in this case $a^{1/n}$ is negative. If a is negative and n is an *even* positive integer, then $a^{1/n}$ is not a real number b; the reason is that for any real number b and any even positive integer n, the number b^n is not negative.

Example 5

$$(-8)^{1/3} = -2, \quad \text{since } (-2)^3 = -8;$$
$$(-8)^{-2/3} = (\sqrt[3]{-8})^{-2} = (-2)^{-2} = \frac{1}{(-2)^2} = \frac{1}{4},$$
$$\left(-\frac{1}{32}\right)^{-3/5} = \left[\left(-\frac{1}{32}\right)^{1/5}\right]^{-3} = \left(-\frac{1}{2}\right)^{-3} = \frac{1}{(-\frac{1}{2})^3} = \frac{1}{-\frac{1}{8}} = -8,$$
$$(-16)^{1/2} \text{ is not a real number.} \quad \blacksquare$$

The following five rules of exponents can be shown to be valid in all cases in which the indicated quantities have been defined previously:
1. $a^r a^s = a^{r+s}$.
2. $(a^r)^s = a^{rs}$.
3. $(ab)^r = a^r b^r$.
4. $a^{-r} = \dfrac{1}{a^r}$, $a \neq 0$.
5. $\dfrac{a^r}{a^s} = a^{r-s}$, $a \neq 0$.

Example 6

$$2^3 2^4 = 2^7,$$
$$(2^4)^{3/2} = 2^6,$$
$$(ab)^3 = a^3 b^3,$$
$$\sqrt{18} = \sqrt{(9)(2)} = \sqrt{9}\sqrt{2} = 3\sqrt{2},$$
$$(24)^{1/3} = \sqrt[3]{(8)(3)} = 2\sqrt[3]{3},$$
$$(9)^{-1/2} = \frac{1}{\sqrt{9}} = \frac{1}{3},$$
$$\frac{2^8}{2^2} = 2^6. \blacksquare$$

EXERCISES

Suggested minimum assignment: Exercises 1, 4, 5(a), 6(a), and 8(a)(c).

1. Calculate each of the following.
   (a) 3^4. (b) 3^{-2}. (c) 3^0. (d) $(16)^{1/2}$. (e) $\sqrt{25}$.
   (f) $\sqrt[3]{8}$. (g) $(27)^{2/3}$. (h) $9^{-3/2}$. (i) $(-64)^{1/3}$.
2. Calculate each of the following.
   (a) 2^5. (b) 2^{-3}. (c) 8^0. (d) $8^{1/3}$.
   (e) $\sqrt{36}$. (f) $4^{3/2}$. (g) $4^{-3/2}$. (h) $\sqrt[3]{-27}$.
3. Simplify each of the following.
   (a) $(\sqrt{25})(25)^{-3/2}$. (b) $\dfrac{(2^4)(3^3)}{(2^6)(3^{-1})}$. (c) $\sqrt{12}\sqrt{3}$.
   (d) $\left(\dfrac{2}{3^{-2}}\right)^2$. (e) $a^4 b^3 (ab)^2$. (f) $(a^3)^4 (a^{-5})\left(\dfrac{1}{a^2}\right)$.
   (g) $\dfrac{\sqrt{100}}{\sqrt{4}}$. (h) $\dfrac{a^4 b^5}{a^{-3} b^{-4}}$.
4. Simplify each of the following.
   (a) $(3^{-4})(3^{5/2})(3^{-1/2})$. (b) $\left(\dfrac{1}{125}\right)^{1/3}(-8)^{-1/3}$. (c) $\dfrac{\sqrt{72}}{\sqrt{6}}$.

(d) $\dfrac{(x^3)^2 x^4}{x^7}$.   (e) $(49)^{1/2}(5^0)(1)^{2/3}(2)^{-2}$.   (f) $\dfrac{(ab)^4 a^{-3}}{b^{-1}}$.

(g) $\sqrt[3]{1}\sqrt[4]{256}$.   (h) $\sqrt[3]{2^2}\sqrt[3]{2^5}$.

5. Simplify and express each answer without a denominator and without using the symbol $\sqrt[n]{\ }$. Specify the conditions under which your answer is valid.

(a) $\sqrt[3]{\dfrac{x^2 y}{x^5 y^2}}$.   (b) $\dfrac{\sqrt{x^5 y^7}}{\sqrt[3]{x^8 y^0}}$.   (c) $\dfrac{1}{\sqrt{a^3 b^2}\sqrt[3]{a^{-4} b^6}}$.

6. Simplify and express each answer without using negative or fractional exponents. Specify the conditions under which your answer is valid.

(a) $(x^5 y)^{-1/2}(x^3 y^0)^{1/3}$.   (b) $\dfrac{(x^4 y^3)^{1/4}}{(x^3 y^2)^{1/2}}$.   (c) $\dfrac{(x^5 y^7)^0}{(x^3 y^4)^{-1/2}}$.

7. Suppose that r is an integer.
   (a) For what real numbers a and b is it true that $(a/b)^r$ and a^r/b^r are each equal to real numbers?
   (b) Use rules of exponents stated in this section to show that $(a/b)^r = a^r/b^r$ whenever $(a/b)^r$ and a^r/b^r are each equal to real numbers.

8. Simplify each of the following. Assume that x, y, and z are positive real numbers.

(a) $\dfrac{x^2 y^3 z^{-2}}{xy^{-2} z^3}$.   (b) $\sqrt[5]{x^{10} y^{25}}$.

(c) $\dfrac{x^{5/2} z^{-3}}{\sqrt{xy^2 z^{-5}}}$.   (d) $\sqrt[8]{\dfrac{x^{32} y^{-3}}{y^5}}$.

(e) $\dfrac{x^{n+2} x^n}{x}$, where n is a positive integer.

3.2 Factoring, Special Products, and Fractions

In this section we shall review several basic algebraic manipulations. The emphasis will be on factoring, calculating certain special products, adding fractions, and simplifying fractions.

Factoring is the expression of a sum (or difference) of terms as a product. The distributive axiom of algebra states that if a, b, and c are real numbers, then the sum $ab + ac$ can be written as the product $a(b + c)$; thus common factors can be factored out of a sum of terms.

Example 1

$$x^2 - 3x = x(x - 3),$$
$$5x + 4x = x(5 + 4) = 9x,$$
$$x^3 y^4 + x^2 y^3 + x^4 y^5 = x^2 y^3(xy + 1 + x^2 y^2),$$
$$x(x - 2) + 7(x - 2) = (x - 2)(x + 7). \blacksquare$$

The distributive axiom can be used to calculate the important product $(x - y)(x + y)$ as follows:

$$(x - y)(x + y) = (x - y)x + (x - y)y$$
$$= x^2 - yx + xy - y^2$$
$$= x^2 - y^2.$$

Such a "special product" may also be viewed as a formula for factoring;

$$x^2 - y^2 = (x - y)(x + y)$$

is used to factor the difference of two squares.

Example 2
Calculate the product $(2a - 3b)(2a + 3b)$.
The answer is

$$(2a - 3b)(2a + 3b) = 4a^2 - 9b^2. \blacksquare$$

Example 3
Factor:
(a) $4^2 - x^2$.   (b) $9x^2 - 16y^2$.
(c) $x^4y - 16y$. (d) $x^2 - (y + z)^2$.
The answers are
(a) $4^2 - x^2 = (4 - x)(4 + x)$.
(b) $9x^2 - 16y^2 = (3x)^2 - (4y)^2 = (3x - 4y)(3x + 4y)$.
(c) $x^4y - 16y = y(x^4 - 16)$
$\qquad = y(x^2 - 4)(x^2 + 4)$
$\qquad = y(x - 2)(x + 2)(x^2 + 4).$
(d) $x^2 - (y + z)^2 = [x - (y + z)][x + (y + z)]$
$\qquad = (x - y - z)(x + y + z). \blacksquare$

Notice that the answers to factoring problems can be checked by multiplying the factors together. The reader may check the following formulas for factoring the sum or difference of two cubes:

$$a^3 + b^3 = (a + b)(a^2 - ab + b^2),$$
$$a^3 - b^3 = (a - b)(a^2 + ab + b^2).$$

Example 4

$$x^3 + 8 = x^3 + 2^3 = (x + 2)(x^2 - 2x + 4),$$
$$x^3 - 27y^3 = x^3 - (3y)^3 = (x - 3y)(x^2 + 3xy + 9y^2). \blacksquare$$

A common type of expression to factor is a trinomial such as $8x^2 - 2x - 15$. Apparently, the answer is of the form $(ax + b)(cx + d)$; by trial and error one finds that

$$8x^2 - 2x - 15 = (2x - 3)(4x + 5).$$

Chap. 3] Review of Algebra; Functions and Graphs

We have used the fact that
$$(ax + b)(cx + d) = acx^2 + (ad + bc)x + bd.$$

Example 5
$$x^2 - 5x + 6 = (x - 2)(x - 3),$$
$$2x^2 + 5x - 12 = (2x - 3)(x + 4),$$
$$4x^3 - 16x^2 + 16x = 4x(x^2 - 4x + 4)$$
$$= 4x(x - 2)(x - 2)$$
$$= 4x(x - 2)^2. \blacksquare$$

A special type of product that occurs frequently is
$$(a + b)^n,$$
where n is a positive integer. It can be verified that
$$(a + b)^1 = a + b,$$
$$(a + b)^2 = (a + b)(a + b) = a^2 + 2ab + b^2,$$
$$(a + b)^3 = a^3 + 3a^2b + 3ab^2 + b^3,$$
$$(a + b)^4 = a^4 + 4a^3b + 6a^2b^2 + 4ab^3 + b^4.$$

The coefficients in these *binomial expansions* can be found using the following array of numbers:

$$\begin{array}{ccccccc} & & & 1 & & 1 & \\ & & 1 & & 2 & & 1 \\ & 1 & & 3 & & 3 & & 1 \\ 1 & & 4 & & 6 & & 4 & & 1 \end{array}$$

With 1's down each side, this array could be continued indefinitely; and any number in the middle will always be the sum of the two numbers just above it. The determination of the coefficients in this manner and the determination of the correct powers of a and b in the expansion of $(a + b)^n$ are a result of a theorem in algebra known as the **binomial theorem.** Binomial expansions have many applications in statistics, and the array begun above is a time-saving device for a binomial expansion such as that shown in Example 6 below.

Example 6
Calculate $(2x - y)^4$.
$$(2x - y)^4 = [(2x) + (-y)]^4$$
$$= (2x)^4 + 4(2x)^3(-y) + 6(2x)^2(-y)^2 + 4(2x)(-y)^3 + (-y)^4$$
$$= 16x^4 - 32x^3y + 24x^2y^2 - 8xy^3 + y^4. \blacksquare$$

In conclusion we review fractions. An ability to manipulate fractions correctly will be important later, especially in our study of calculus. Two

basic rules are

$$\frac{a}{b} = \frac{ac}{bc}, \quad \text{if } b \neq 0 \text{ and } c \neq 0,$$

and

$$\frac{a}{b} + \frac{c}{b} = \frac{a+c}{b}, \quad \text{if } b \neq 0.$$

Example 7

$$\frac{4}{\sqrt{2}} = \frac{4}{\sqrt{2}} \frac{\sqrt{2}}{\sqrt{2}} = \frac{4\sqrt{2}}{2} = 2\sqrt{2}. \blacksquare$$

Example 8

$$\frac{x}{x^2-4} - \frac{3}{x-2} = \frac{x}{x^2-4} - \frac{3(x+2)}{(x-2)(x+2)}$$
$$= \frac{x - 3(x+2)}{x^2-4}$$
$$= \frac{-2x-6}{x^2-4}, \quad \text{if } x \neq -2 \text{ and } x \neq 2. \blacksquare$$

In the next example we illustrate the important formula

$$\frac{a}{b} + \frac{c}{d} = \frac{ad+bc}{bd}, \quad \text{if } b \neq 0 \text{ and } d \neq 0.$$

Example 9

$$\frac{2}{5} + \frac{3}{4} = \frac{8+15}{20} = \frac{23}{20};$$

also

$$\frac{1}{x} - \frac{1}{x-1} = \frac{(x-1)-x}{x(x-1)} = \frac{-1}{x(x-1)}, \quad \text{if } x \neq 0 \text{ and } x \neq 1. \blacksquare$$

In Examples 10 and 11 miscellaneous problems involving several algebraic manipulations are solved. It is assumed that quantities in the denominator of a fraction are not zero.

Example 10
Simplify

$$\frac{a^{-2}-b^{-2}}{a^{-1}+b^{-1}}.$$

$$\frac{a^{-2}-b^{-2}}{a^{-1}+b^{-1}} = \frac{\dfrac{1}{a^2}-\dfrac{1}{b^2}}{\dfrac{1}{a}+\dfrac{1}{b}}$$

$$= \frac{\dfrac{b^2-a^2}{a^2b^2}}{\dfrac{b+a}{ab}} = \frac{(b-a)(b+a)}{a^2b^2}\cdot\frac{ab}{b+a}$$

$$= \frac{b-a}{ab}. \blacksquare$$

Example 11
Simplify

$$\frac{\sqrt{4x+1}-(x)(\tfrac{1}{2})(4x+1)^{-1/2}(4)}{(\sqrt{4x+1})^2}.$$

We obtain

$$\frac{\sqrt{4x+1}-\dfrac{2x}{\sqrt{4x+1}}}{4x+1}$$

$$= \frac{\dfrac{(4x+1)-2x}{\sqrt{4x+1}}}{4x+1}$$

$$= \frac{2x+1}{\sqrt{4x+1}}\cdot\frac{1}{4x+1}$$

$$= \frac{2x+1}{(4x+1)^{3/2}}. \blacksquare$$

EXERCISES

Suggested minimum assignment: Exercises 1, 4, 5, 7, 11, 14, 15, 19, 20, and 23.

1. Factor each of the following.
   (a) x^2+xy.
   (b) $x(x-4)+3(x-4)$.
   (c) x^2-9.
   (d) $4y-x^2y$.
2. Factor each of the following.
   (a) $tx+ty+2tz$.
   (b) $25x^2-36y^2$.
   (c) $x(y+3)-4(y+3)$.
   (d) $x^2y^2-25y^2$.
3. Factor each of the following.
   (a) x^2+x-12.
   (b) x^3-3x^2+2x.
   (c) $2x^2y+7xy-4y$.
   (d) x^3-8.

Sec. 3.2] Factoring, Special Products, and Fractions

4. Factor each of the following.
   (a) $6t^2 + 11t - 10$.
   (b) $x^4 + 6x^3 + 9x^2$.
   (c) $x(x - y) + (x + 1)(x - y)$.
   (d) $x^3 + 27$.
5. Factor each of the following.
   (a) $x^4 - x$.
   (b) $x^4 - 9$.
   (c) $4a^2 + 12ab + 9b^2$.
   (d) $x^6 - 64$.
6. Factor each of the following.
   (a) $10x^2 + 19x + 6$.
   (b) $10x^2 + 16x + 6$.
   (c) $9x^5 - 4xy^6$.
   (d) $x^4 - (2y + 1)^2$.
7. Calculate the following products.
   (a) $(x - 2y)(x + 2y)$.
   (b) $x(x - 2)(x - 3)$.
   (c) $(2x - 3)(7x + 8)$.
   (d) $(4a + 5b)^2$.
8. Calculate the following products.
   (a) $(2a - 3b)(4a - 7b)$.
   (b) $(2x^2 - 3y^2)(x^2 + 5y^2)$.
   (c) $(x + 2)(2x - 3y + 3)$.
   (d) $(x + 5)(x^2 - 5x + 25)$.
9. Calculate $(x + 2y)^4$.
10. Calculate $(x - y)^4$.
11. Calculate $(3x - y)^3$.
12. Calculate $(a + b)^5$.
13. Rewrite in simplest form without square roots in the denominator.
    (a) $\dfrac{1}{\sqrt{3}}$.
    (b) $\sqrt{8} + \dfrac{6}{\sqrt{2}}$.
14. Rewrite in simplest form without square roots in the denominator.
    (a) $\dfrac{5}{2\sqrt{2}}$.
    (b) $\dfrac{1}{\sqrt{5}} - \sqrt{125} + \dfrac{4}{5}\sqrt{5}$.

In Exercises 15 and 16 combine the given fractions (assume that all denominators are not equal to zero).

15. (a) $\dfrac{3}{x^2 + 2} + \dfrac{x}{x^2 + 2}$.
    (b) $\dfrac{7}{x} + \dfrac{8}{x^2}$.
    (c) $\dfrac{9}{8} - \dfrac{7}{12}$.
    (d) $\dfrac{5}{x - 4} + \dfrac{6}{x + 4}$.

16. (a) $\dfrac{3x + 1}{x + 1} + \dfrac{x - 2}{x + 1}$.
    (b) $\dfrac{4}{x + 2} + \dfrac{5}{(x + 2)^3}$.
    (c) $\dfrac{1}{2} - \dfrac{2}{3} + \dfrac{3}{4}$.
    (d) $\dfrac{1}{x} + \dfrac{2}{x - 1} + \dfrac{3}{x - 2}$.

Simplify as much as possible in Exercises 17–22 (assume that all denominators are not equal to zero).

17. $\dfrac{5}{x^2 - 6x + 5} - \dfrac{5}{x^2 - 1}$.

18. $\dfrac{\dfrac{1}{x - 2} + \dfrac{1}{x + 2}}{1 + \dfrac{2}{x - 2}}$.

19. $\dfrac{a^{-1} - b^{-1}}{a^{-2} - b^{-2}}$.

20. $\dfrac{(x)(-\frac{1}{2})(1+x^2)^{-3/2}(2x) - (1+x^2)^{-1/2}}{x^2}$.

21. $\dfrac{\dfrac{x+2y}{2x-3y} + \dfrac{3x+y}{2x+y}}{\dfrac{2x-5y}{2x-3y}}$.

22. $\dfrac{(x+2)^{-1} - (x-2)^{-1}}{(x+2)^{-1} + (x-2)^{-1}}$.

23. Which of the following are correct?

(a) $\sqrt{x+y} = \sqrt{x} + \sqrt{y}$.   (b) $\dfrac{a+b}{a+c} = \dfrac{b}{c}$.

(c) $\dfrac{a}{b+c} = \dfrac{a}{b} + \dfrac{a}{c}$.   (d) $\dfrac{1}{a^{-1}+b^{-1}} = a+b$.

3.3 Solving Equations

One of the fundamental problems in mathematics is that of solving simple equations. The examples of this section will illustrate the basic techniques that the reader will need for solving equations.

The number $x = -3$ is a **solution** of the equation $2x + 6 = 0$ because $2(-3) + 6 = 0$; since $2(-3) + 6 = 0$, we say that the number $x = -3$ *satisfies the equation* or we say that the number $x = -3$ *checks* when it is substituted in the equation. To **solve an equation** means to find all solutions of the equation.

A **linear equation** in one unknown x is an equation that can be written in the form

$$ax + b = 0,$$

where a and b are real numbers and $a \neq 0$. To solve such an equation, one needs to know that the same expression may be added (or subtracted) to each side of an equation, and that each side of an equation may be multiplied (or divided) by the same nonzero expression.

Example 1

Solve the linear equation $7x - 3 = 5x + 13$.

The addition of $3 - 5x$ to each side of the equation yields

$$2x = 16.$$

Then the multiplication of each side by $\frac{1}{2}$ gives the only solution

$$x = 8,$$

which can be shown to satisfy the original equation. We can use set language and say that the **solution set** (that is, the set of all solutions) is $\{8\}$. ∎

Example 2
Solve for x: $ax + 3 + b(x - 5) = \frac{9}{2}$.
  We proceed as follows:
$$2ax + 6 + 2b(x - 5) = 9,$$
$$(2a + 2b)x - 10b = 3,$$
$$2(a + b)x = 3 + 10b,$$
$$x = \frac{3 + 10b}{2(a + b)}, \qquad \text{provided that } a + b \neq 0. \blacksquare$$

A useful property of real numbers for solving equations is that if m and n are real numbers, then
$$mn = 0 \quad \text{if and only if} \quad m = 0 \text{ or } n = 0.$$
This property is often helpful in solving a **quadratic equation,** which is an equation of the form
$$ax^2 + bx + c = 0, \qquad \text{where } a \neq 0.$$
If $ax^2 + bx + c$ can be factored into two linear factors, then each linear factor can be equated to zero in order to find the two solutions of the quadratic equation (occasionally each linear factor leads to the same solution, in which case the quadratic equation really has only one solution).

Example 3
Solve $12x^2 - 5x - 2 = 0$ by factoring.
$$(3x - 2)(4x + 1) = 0,$$
and therefore
$$3x - 2 = 0, \quad \text{or} \quad 4x + 1 = 0,$$
$$x = \tfrac{2}{3}, \quad \text{or} \quad x = -\tfrac{1}{4}. \blacksquare$$

The algebraic procedure called **completing the square** is important in several situations, one of which is in solving quadratic equations. The procedure is based upon the equality
$$x^2 + bx + \left(\frac{b}{2}\right)^2 = \left(x + \frac{b}{2}\right)^2;$$
for the expression $x^2 + bx$, we can add $(b/2)^2$ in order to complete the square—that is, in order to make it a perfect square.

Example 4
Solve $2x^2 - 3x - 1 = 0$ by completing the square.

$$2\left(x^2 - \frac{3}{2}x\right) = 1,$$

$$2\left(x^2 - \frac{3}{2}x + \frac{9}{16}\right) = 1 + 2\left(\frac{9}{16}\right),$$

$$2\left(x - \frac{3}{4}\right)^2 = \frac{17}{8},$$

$$\left(x - \frac{3}{4}\right)^2 = \frac{17}{16},$$

$$x - \frac{3}{4} = \pm\frac{\sqrt{17}}{4},$$

$$x = \frac{3}{4} \pm \frac{\sqrt{17}}{4}.$$

In the type of calculation above, always factor out the coefficient of x^2 first. The number $\frac{9}{16}$, which is added to make the quantity in parentheses a perfect square, is obtained by squaring one half of the coefficient of x; that is,

$$[\tfrac{1}{2}(-\tfrac{3}{2})]^2 = \tfrac{9}{16}. \blacksquare$$

The method of completing the square can be applied to the general quadratic equation $ax^2 + bx + c = 0$, where $a \neq 0$, in order to obtain the well-known **quadratic formula**.

$$ax^2 + bx + c = 0,$$

$$a\left(x^2 + \frac{b}{a}x\right) = -c,$$

$$a\left[x^2 + \frac{b}{a}x + \left(\frac{1}{2}\frac{b}{a}\right)^2\right] = -c + a\left(\frac{1}{2}\frac{b}{a}\right)^2,$$

$$a\left(x^2 + \frac{b}{a}x + \frac{b^2}{4a^2}\right) = -c + \frac{b^2}{4a},$$

$$a\left(x + \frac{b}{2a}\right)^2 = \frac{-4ac + b^2}{4a},$$

$$\left(x + \frac{b}{2a}\right)^2 = \frac{b^2 - 4ac}{4a^2},$$

$$x + \frac{b}{2a} = \pm\frac{\sqrt{b^2 - 4ac}}{2a},$$

$$x = \frac{-b \pm \sqrt{b^2 - 4ac}}{2a},$$

which is the quadratic formula that can be used to find the solutions (or roots) of a quadratic equation.

Example 5
If $2x^2 - 3x - 1 = 0$, solve for x using the quadratic formula. (The equation is the same as in Example 4.)
Since $a = 2$, $b = -3$, and $c = -1$, we find that
$$x = \frac{3 \pm \sqrt{9 + 8}}{4},$$
and hence the two roots are
$$x = \frac{3}{4} + \frac{\sqrt{17}}{4}$$
and
$$x = \frac{3}{4} - \frac{\sqrt{17}}{4}. \blacksquare$$

If the two roots in the quadratic formula are called r_1 and r_2, then it can be shown that
$$ax^2 + bx + c = a(x - r_1)(x - r_2).$$
This fact enables the quantity $2x^2 - 3x - 1$ in Example 5 to be factored as follows:
$$2x^2 - 3x - 1 = 2\left[x - \left(\frac{3}{4} + \frac{\sqrt{17}}{4}\right)\right]\left[x - \left(\frac{3}{4} - \frac{\sqrt{17}}{4}\right)\right].$$

In the examples above it is possible to check the answers by substitution in the original equations, but such a check is not mandatory. In the next example a check is required because extraneous roots may be introduced by performing steps that are not reversible.

Example 6
Solve the equation $\sqrt{x + 1} = 3x - 7$.
Squaring both sides, we obtain
$$x + 1 = 9x^2 - 42x + 49,$$
$$9x^2 - 43x + 48 = 0,$$
$$(x - 3)(9x - 16) = 0,$$
and apparently the roots are
$$x = 3 \quad \text{and} \quad x = \tfrac{16}{9}.$$
However, the operation of squaring both sides may lead to numbers that do not check. For $x = 3$, the two expressions $\sqrt{x + 1}$ and $3x - 7$ each equal 2 and the original equation is satisfied. The number $x = \tfrac{16}{9}$ does not check because
$$\sqrt{\tfrac{16}{9} + 1} = \tfrac{5}{3} \neq 3(\tfrac{16}{9}) - 7 = -\tfrac{5}{3}.$$

Chap. 3] Review of Algebra; Functions and Graphs

Therefore, the only solution is
$$x = 3. \blacksquare$$

Another useful property of real numbers for solving equations is that if m and n are real numbers, then

$$\frac{m}{n} = 0 \quad \text{if and only if} \quad m = 0 \text{ and } n \neq 0.$$

This property can be used when the unknown in an equation appears in the denominator—just transpose all terms to the left side of the equation and combine the terms to get an equation of the form $\frac{m}{n} = 0$.

Example 7

Solve the equation $\dfrac{x^3 - 9x}{x^3 - 3x^2 - 3x + 9} = 0.$

In order for a fraction to equal 0, it is essential that the numerator equal 0 and thus

$$x^3 - 9x = 0,$$
$$x(x^2 - 9) = 0,$$
$$x(x - 3)(x + 3) = 0,$$
$$x = 0, \quad x = 3, \quad x = -3.$$

The number $x = 3$ does not check in the original equation because the denominator (as well as the numerator) is 0 if $x = 3$. The numbers $x = 0$ and $x = -3$ do check, so the only solutions are

$$x = 0 \quad \text{and} \quad x = -3. \blacksquare$$

In the next example we find values of x and y that satisfy two equations simultaneously.

Example 8

Solve the system $\begin{cases} 2x + y = 4, \\ 2y = x^2 + 2x - 19. \end{cases}$

We substitute $y = 4 - 2x$ from the first equation into the second equation:

$$2(4 - 2x) = x^2 + 2x - 19,$$
$$x^2 + 6x - 27 = 0,$$
$$(x - 3)(x + 9) = 0,$$
$$x = 3, \quad x = -9.$$

If $x = 3$,
$$y = 4 - 2(3) = -2.$$

Sec. 3.3] Solving Equations

If $x = -9$,
$$y = 4 - 2(-9) = 22.$$
Thus one solution is
$$x = 3, \quad y = -2$$
and a second solution is
$$x = -9, \quad y = 22.$$
Each solution checks in each original equation. ∎

We shall not review systems of linear equations here, since they will be considered in detail in Chapter 5.

Let us return briefly to quadratic equations. The quantity $b^2 - 4ac$ in the quadratic formula is called the **discriminant**. If the discriminant is negative, the original quadratic equation has no real solutions, since $\sqrt{b^2 - 4ac}$ is not real when $b^2 - 4ac$ is negative. By the introduction of $i = \sqrt{-1}$ we can allow every quadratic equation to have two solutions, which can be obtained by the quadratic formula. If k is a positive real number, then $-k$ is negative and we define $\sqrt{-k} = \sqrt{k}\sqrt{-1} = \sqrt{k}\,i$. The equation $x^2 = -4$ has solutions $x = \pm\sqrt{-4} = \pm\sqrt{4}\sqrt{-1} = \pm 2i$. Observe that $i^2 = -1$. A **complex number** is a number of the form $a + bi$, where a and b are real numbers. The set of real numbers is a subset of the set of complex numbers because every real number is of the form $a + bi$ with $b = 0$. If $b \neq 0$, the complex number $a + bi$ is also called an **imaginary number**. *In future sections of this book the only numbers under consideration, unless stated otherwise, are the real numbers.*

E X E R C I S E S

Suggested minimum assignment: Exercises 1, 3, 8(b), 9(b), 10(a), 12(a), 13(b), 15(b), 17(a), and 18(a).

1. Solve for x.
    (a) $3x - 7 = 5$.   (b) $4x - 5 = 6x + 2$.
    (c) $3ax - 11 = a(x + 1)$, where $a \neq 0$.
2. Solve for x.
    (a) $2 - 3x = \frac{1}{2} + x$.   (b) $\frac{x}{2} - \frac{x}{3} = 4 - \frac{x}{4}$.
    (c) $a(x - 2) = \frac{bx}{2} + 1$, where $2a - b \neq 0$.
3. Solve for y' in terms of x and y.
$$\frac{y - xy'}{y^2} + 2yy' = 3.$$

4. Solve for z in terms of x and y.

$$xz - \frac{2z}{x-1} = \frac{x+1}{y}.$$

5. A track star can run 1 mile in 4 minutes. What is his average speed in miles per hour? (*Hint:* Solve $d = rt$ for r.)
6. A test driver drove a car at exactly 40 miles per hour for 220 miles and used 8 gallons of gas. How many hours did he travel and how many miles per gallon did he average?
7. Solve for x.
   (a) $(x-3)(x-4)(x-5) = 0$. (b) $x^2 - 9x + 20 = 0$.
   (c) $2x^2 - 9x - 18 = 0$. (d) $(x-3)(x-4) = 6$.
8. Solve by completing the square.
   (a) $3x^2 + 9x + 5 = 0$. (b) $2x^2 + 8x - 1 = 0$.
   (c) $x^2 - 3 = 0$.
9. Use the quadratic formula to solve the following equations.
   (a) $x^2 + 7x + 3 = 0$. (b) $2x^2 - 2x - 3 = 0$.
10. Factor each of the following.
    (a) $2x^2 - 2x - 3$. (b) $2x^2 + x - 4$.
11. Use the quadratic formula to find $\{x : x^2 - 9x + 3 = 0\}$.
12. Solve and check.
    (a) $\sqrt{2x+1} = 7 - x$. (b) $1 + \sqrt{x+2} = \frac{3}{2}x$.
    (c) $\sqrt{x} = -4$.
13. Solve and check.
    (a) $\dfrac{(x-2)(x-3)}{x-4} = 0$. (b) $\dfrac{2x^2 - 23x - 12}{x^2 - 144} = 0$.
14. Solve for x.

$$\frac{2}{x-1} + \frac{3}{x-2} = -1 - \frac{2}{x^2 - 3x + 2}.$$

15. Solve each of the following systems of equations.
    (a) $\begin{cases} y = x^2 + 4, \\ y = x + 6. \end{cases}$ (b) $\begin{cases} 2x + 3y = 5, \\ x = y^2. \end{cases}$
16. Let r_1 and r_2 be the solutions (either real or imaginary) of the quadratic equation $ax^2 + bx + c = 0$.
    (a) Find $r_1 + r_2$. (b) Find $r_1 r_2$.
17. Determine whether or not the solutions of the following equations are real by calculating the discriminant.
    (a) $x^2 + 2x + 7 = 0$. (b) $5x^2 + 8x + 3.1 = 0$.
    (c) $5x^2 + 8x + 3.2 = 0$.
18. Solve by using the quadratic formula.
    (a) $x^2 + 2x + 5 = 0$. (b) $2x^2 + 7x + 7 = 0$.
19. Write in $a + bi$ form.
    (a) $\sqrt{-36} + 2$. (b) $(2-i)(3-2i)$.

Sec. 3.3] Solving Equations

3.4 Formulas from Geometry

In this section we shall recall some facts and formulas about certain geometric figures. These formulas will be useful later in this book when we study optimization problems. We start with two-dimensional figures. The area A of a **rectangle** of length l and width w is given by

$$A = lw.$$

For a **circle** of radius r, the area A and the circumference C are given by

$$A = \pi r^2$$

and

$$C = 2\pi r.$$

For a **semicircle** the area and circumference are, of course, just half as much.

A **trapezoid** has four sides, two of which are parallel. Let l_1 and l_2 be the lengths of the parallel sides. Let the distance between the parallel sides be denoted by h (this number h is called the altitude of the trapezoid). Then the area A is given by

$$A = \tfrac{1}{2}h(l_1 + l_2).$$

Several facts about **triangles** are of importance. The sum of the angles of any triangle is 180°. If one of the angles is 90° (a right angle) the triangle is called a **right triangle**; the side opposite the right angle is called the **hypotenuse** and the other two sides are called the **legs** of the triangle. If the length of the hypotenuse is c and the lengths of the legs are a and b, then by the **Pythagorean theorem**,

$$c^2 = a^2 + b^2.$$

This formula applies only to right triangles.

Getting back to arbitrary triangles (rather than necessarily a right triangle), we recall that a **median** is a segment from a vertex to the midpoint of the opposite side. An **altitude** is a segment from a vertex drawn perpendicular to the opposite side to the point where it reaches the opposite side (or the extension of the opposite side). Let an altitude of length h be drawn perpendicular to a side (the base) of length b; then the area of the triangle is given by

$$A = \tfrac{1}{2}bh.$$

If two sides of a triangle are equal in length, the triangle is called an **isosceles triangle**. Two triangles are **similar** if their corresponding angles are equal (in other words, the triangles have the same shape). An important fact is that corresponding sides of similar triangles are proportional.

Example 1
A racetrack consists of a rectangle of length x and width y with a semicircle at each end (see Figure 3.1). Find the total area inside the racetrack and the total perimeter of the track.

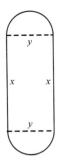

Figure 3.1. *Racetrack.*

The radius of each semicircle is $y/2$. The total area inside the track is

$$A = \frac{1}{2}\pi\left(\frac{y}{2}\right)^2 + xy + \frac{1}{2}\pi\left(\frac{y}{2}\right)^2$$

$$= \frac{1}{4}\pi y^2 + xy.$$

The total perimeter (length around the outside) is

$$P = x + \frac{1}{2}\left(2\pi\frac{y}{2}\right) + x + \frac{1}{2}\left(2\pi\frac{y}{2}\right)$$

$$= 2x + \pi y. \ \blacksquare$$

Example 2
Find the area of the trapezoid in Figure 3.2.

$$A = \tfrac{1}{2}h(l_1 + l_2) = \tfrac{1}{2}(2)(4 + 5) = 9. \ \blacksquare$$

Figure 3.2. *Trapezoid.*

Example 3
Refer to Figure 3.3.
  (a) Find the area of the large triangle.
  (b) Find the hypotenuse z of the large triangle.
  (c) Find the relationship between x and y.

Sec. 3.4] Formulas from Geometry

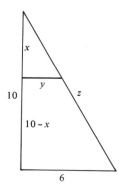

Figure 3.3. *Similar triangles.*

Solutions:
(a) The large triangle has base 6 and altitude 10, so its area is
$$A = \tfrac{1}{2}(6)(10) = 30.$$

(b) By the Pythagorean theorem,
$$\begin{aligned}z^2 &= (10)^2 + (6)^2\\ &= 136,\\ z &= \sqrt{136} = 2\sqrt{34}.\end{aligned}$$

(c) The large and small triangles are similar, and since corresponding sides of similar triangles are proportional

$$\frac{10}{6} = \frac{x}{y},$$

or

$$y = \frac{3}{5}x. \quad\blacksquare$$

We shall occasionally need to know some volume and surface area formulas for common three-dimensional figures. These are given in the following paragraphs, and some straightforward problems on their use are given in the exercises.

For a **sphere** of radius r, the volume is given by
$$V = \tfrac{4}{3}\pi r^3,$$

and the surface area is given by
$$S = 4\pi r^2.$$

Let r be the radius of a **right circular cylinder** (shaped like a tin can) and let h be the height (altitude). The formula for the volume is
$$V = \pi r^2 h,$$

and the total surface area is
$$S = 2\pi rh + 2\pi r^2.$$

Chap. 3] Review of Algebra; Functions and Graphs

In the surface area formula the term $2\pi r^2$ represents the area of the circular top and bottom; if the right circular cylinder had no top on it, πr^2 would replace $2\pi r^2$ in the formula. Sometimes we refer to a right circular cylinder with no top as an **open** right circular cylinder. In contrast, a **closed** right circular cylinder is understood to have the top on it.

The volume of a **right circular cone** with radius r and height h can be calculated from

$$V = \tfrac{1}{3}\pi r^2 h.$$

Consider a box (technically, a right parallelepiped) of length l, width w, and height h. The formula for its volume is

$$V = lwh,$$

and the total surface area of its six sides is

$$S = 2lw + 2lh + 2wh.$$

If it is an open box (instead of a closed box), this means that the top is missing and hence lw replaces $2lw$ in the surface area formula; of course, the volume remains the same.

EXERCISES

Suggested minimum assignment: Exercises 1, 3, 5, 7, 8, 10, 11, 13, 15, and 17.

1. A circle has radius 5 inches. Find the area of the circle, and also find its circumference.
2. A semicircle has area 18π square feet. What is the radius of the semicircle?
3. Draw a trapezoid like the one in Figure 3.2 except that the parallel sides have lengths 10 and 14 and the distance between the parallel sides is 6. Find the area of the trapezoid. Check by dividing the trapezoid into a rectangle and a triangle and adding their areas.
4. The legs of a right triangle have lengths 3 and 4. Find the length of the hypotenuse.
5. The hypotenuse of a right triangle has length 13 and one leg has length 5. Find the length of the other leg of the triangle.
6. A right triangle has sides of lengths 8, 15, and 17. Find the area of the triangle.
7. Two right triangles are similar. The smaller triangle has sides of lengths 6, 8, and 10. The legs of the larger triangle have lengths 9 and 12. How long is the hypotenuse of the larger triangle?
8. A triangle has sides of lengths 6, 6, and 8.
   (a) Is it a right triangle?
   (b) Is it an isosceles triangle?
9. A window is in the shape of a square surmounted by a semicircle. One side of the square is 4 feet long. What is the total area of the window?

Sec. 3.4] Formulas from Geometry

10. Draw a triangle whose angles are approximately 20°, 40°, and 120°. Draw and label all medians and altitudes of the triangle.
11. The radius of a sphere is 4 inches.
    (a) Find the volume of the sphere.
    (b) Find the surface area of the sphere.
12. The volume of a sphere is 36π cubic feet. What is the radius of the sphere?
13. An open right circular cylinder has radius 3 inches and height 6 inches.
    (a) Find its volume.
    (b) Find the total surface area.
14. Repeat Exercise 13 for a closed right circular cylinder.
15. A right circular cone has volume $16\pi/3$ cubic feet. Its height is twice its radius. Find the radius and height.
16. An open box has length 8 feet, width 2 feet, and height 3 feet.
    (a) Find its volume.
    (b) Find its total surface area.
17. A closed box has length 1 foot, width 8 inches, and height 6 inches.
    (a) Find its volume in cubic inches.
    (b) If the material used for the top, bottom, and sides costs 10 cents per square inch, what is the total cost of the material?

3.5 Inequalities and Absolute Values

Thus far in this chapter we have reviewed a few topics from algebra and geometry that will be particularly useful to us. We turn our attention in this section to further discussion of the real number system, a topic that was introduced in Section 1.1.

We say that real number a is less than real number b if the point corresponding to a lies to the left of the point corresponding to b on the real line, and we denote this relationship by $a < b$. For example,

$$\sqrt{2} < 3 \quad \text{and} \quad -7 < -6.$$

Another way to explain the meaning of $a < b$ is to say that $b - a$ is positive. To indicate that a is either equal to or less than b, we write $a \leq b$. For example,

$$4 \leq 5 \quad \text{and} \quad 5 \leq 5$$

are both correct. "Real number b is greater than real number a" is written $b > a$ and means exactly the same thing as $a < b$. Similarly, $b \geq a$ denotes that b is equal to or greater than a and means the same thing as $a \leq b$. The symbols $<$, $>$, $\leq$, and $\geq$ are called **inequality symbols,** and any mathematical statement involving any of these symbols is called an **inequality.** Notice that $a > 0$ means that a is positive, $a < 0$ means that a is negative, $a \geq 0$ means that a is nonnegative, and $a \leq 0$ means that a is nonpositive. If we write

$$2 < a < 3,$$

we mean that $2 < a$ and also $a < 3$. Therefore, to say that $\frac{5}{2}$ is between 2 and 3 we could write $2 < \frac{5}{2} < 3$.

Example 1
A broker recommends to a client that a particular common stock be purchased if the price–earnings ratio of the stock is less than 10. To express this statement as an inequality, let E be the earnings per share for the latest 12 months and let P be the current price per share. The broker's recommendation (assume that $E > 0$) is to buy the stock, provided that

$$\frac{P}{E} < 10. \blacksquare$$

We shall need to be able to refer to subsets of real numbers, and hence it shall be convenient to use set notation. For example,

$$\{x : 2 < x < 5\}$$

denotes the set of numbers that are greater than two and also less than five. If a and b are real numbers and $a < b$, then the set

$$\{x : a < x < b\}$$

is called an **open interval** and is also designated by (a, b). The set

$$\{x : a \leq x \leq b\}$$

is called a **closed interval** and often is denoted by $[a, b]$. The sets

$$\{x : a \leq x < b\} \quad \text{and} \quad \{x : a < x \leq b\},$$

where $a < b$, are called **half-open intervals** and are denoted by $[a, b)$ and $(a, b]$, respectively. In each case a and b are called the **endpoints** of the interval, and each number in (a, b) is called an **interior point** of the interval. See Figure 3.4 for a geometric illustration of the types of intervals discussed above.

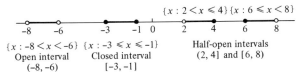

Figure 3.4. *Types of intervals.*

We now turn to an important new concept. If the price of a stock changes from \$120 to \$130 over a period of time, we say that the price changed by \$10 in a positive direction. If, over the next time interval, the stock price changes from \$130 to \$120, we say that the change was $-\$10$. In some situations, only the amount of change, and not the direction of change, is desired. In such cases the concept of **absolute value** of a real number is important.

▶ **Definition 1**
The **absolute value** of a real number a, denoted by $|a|$, is given by
$$|a| = \begin{cases} a & \text{if } a \geq 0, \\ -a & \text{if } a < 0. \end{cases}$$

Example 2
$$|5| = 5,$$
$$|0| = 0,$$
$$|-5| = -(-5) = 5.$$

Notice that the absolute value of a negative number is the negative of that negative number, which gives a positive result. ∎

The number $|a - b|$ is the number of units between a and b on the real line and is called the **distance** between a and b. If $a = -4$ and $b = 8$, then the distance between a and b (see Figure 3.5) is given by
$$|a - b| = |-4 - 8| = |-12| = 12.$$

Observe also that
$$|b - a| = |8 - (-4)| = |12| = 12.$$

In general,
$$|a - b| = |b - a|,$$

which can be interpreted as meaning the distance between a and b without regard to direction.

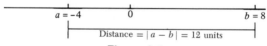

Figure 3.5

For any real number a the rules
$$|a|^2 = a^2$$
and
$$\sqrt{a^2} = |a|$$
are valid.

Example 3
If $a = -5$, then
$$|a|^2 = |(-5)|^2 = 5^2 = 25$$
and
$$a^2 = (-5)^2 = 25,$$
which illustrates that $|a|^2 = a^2$ when $a = -5$. Also
$$\sqrt{a^2} = \sqrt{(-5)^2} = \sqrt{25} = 5$$

Chap. 3] Review of Algebra; Functions and Graphs

and
$$|a| = |-5| = 5,$$
which illustrates that $\sqrt{a^2} = |a|$ when $a = -5$. ∎

An inequality of the form
$$ax + b < cx + d,$$
where a, b, c, and d are real numbers and $a \neq c$ is a **linear inequality** ($<$ may be replaced by $>$, $\leq$, or $\geq$). A **solution** of any inequality in one unknown x is a number that when substituted for x yields a true statement. To **solve an inequality** means to find all solutions of the inequality.

Whereas we found in Section 3.3 that a linear equation in one unknown has exactly one solution, we shall find that a linear inequality in one unknown has an infinite number of solutions. Nevertheless, the methods of solving linear equations and linear inequalities are similar except for one major difference. We will solve linear inequalities by treating both sides with the same expression, through addition, subtraction, multiplication, or division (of course, we will not multiply or divide by zero), but the major difference from solving linear equations occurs when multiplying or dividing by a *negative* number. It can be shown that

$$\text{If } a < b \text{ and } c < 0, \quad \text{then } ac > bc. \tag{3.1}$$

For example, if both sides of
$$3 < 4$$
are multiplied by the *negative* number $c = -2$, then we must **reverse the sense of the inequality** and write
$$-6 > -8.$$

Similar statements to (3.1) can be written if we begin with $a \leq b$, $a > b$, or $a \geq b$, but in each case one must reverse the sense of the inequality when $c < 0$. Given the true statement
$$-10 \geq -20 \geq -30,$$
we obtain by dividing by -10 another true statement,
$$1 \leq 2 \leq 3.$$

Example 4
Solve the linear inequality
$$4x - 5 > 8 + 6x.$$

Add 5 to each side, to obtain
$$4x > 13 + 6x,$$

and then subtract $6x$ from each side, to get
$$-2x > 13.$$
Then divide each side by -2 and reverse the sense of the inequality, to find that any number
$$x < -\tfrac{13}{2}$$
is a solution. There are an infinite number of solutions since there are an infinite number of real numbers less than $-\tfrac{13}{2}$, and any of these solutions will check if substituted into the original inequality. ∎

We conclude this section by solving some equations and inequalities that involve absolute values. Our methods will be based on the facts that if a is a positive real number, then

$$|x| = a \quad \text{if and only if} \quad x = a \text{ or } x = -a, \qquad (3.2)$$
$$|x| < a \quad \text{if and only if} \quad -a < x < a, \qquad (3.3)$$

and

$$|x| > a \quad \text{if and only if} \quad x > a \text{ or } x < -a. \qquad (3.4)$$

Example 5
Solve for x.
(a) $|2x - 13| = 14$.
(b) $|4 - 3x| < 10$.
(c) $|x + 3| \geq 8$.
Solutions:
(a) By (3.2) we obtain two solutions:
$$2x - 13 = 14,$$
$$2x = 27,$$
$$x = \tfrac{27}{2},$$
or
$$2x - 13 = -14,$$
$$2x = -1,$$
$$x = -\tfrac{1}{2}.$$

(b) By (3.3)
$$-10 < 4 - 3x < 10,$$
and hence
$$-14 < -3x < 6,$$
$$\tfrac{14}{3} > x > -2.$$

Using set language, we say that the solution set is the open interval $(-2, \tfrac{14}{3})$.

(c) By (3.4)
$$x + 3 \geq 8,$$
$$x \geq 5,$$

Chap. 3] Review of Algebra; Functions and Graphs

or
$$x + 3 \leq -8,$$
$$x \leq -11.$$

The solution set is
$$\{x : x \geq 5\} \cup \{x : x \leq -11\}. \blacksquare$$

APPLICATIONS

Example 6
Many important decisions are based upon statements that can be expressed mathematically as inequalities. In Exercise 16 the reader is asked to translate each of the following into an inequality:
  (a) The plane can take off as soon as the visibility is at least 1 mile.
  (b) Town council voted to borrow money to build the new school if the interest rate is less than $5\frac{1}{2}$ per cent.
  (c) A doctor advised his patient to come back whenever the patient's normal pulse rate exceeds 85 beats per minute.
  (d) A client instructs his stockbroker to buy 10 shares of stock A if the price of the stock reaches the amount necessary to make the annual dividend exceed 5 per cent of the price of the stock.
  (e) The client in part (d) tells the stockbroker to place a stop-loss order for stock B at $50 per share.
  (f) A basketball coach tells his team to freeze the ball whenever the team has at least a 10-point lead.
  (g) The trailing presidential candidate will campaign in a particular state if the latest opinion poll in that state shows that he is within 12 percentage points of the leading candidate. $\blacksquare$

EXERCISES

Suggested minimum assignment: Exercises 1, 3, 5, 7, 8, 10, 11(a), 13, and 15.

1. Tell which of the following real numbers are nonnegative.
$$\tfrac{2}{3}, \quad 0, \quad \pi, \quad -\tfrac{9}{8}.$$

2. Tell which of the following real numbers are nonpositive.
$$\sqrt{9}, \quad -3, \quad 1 + \pi, \quad 0.$$

3. State whether each of the following is true or false.
   (a) $5 > \pi$.   (b) $-8 > -7$.   (c) $-5 \leq 0$.   (d) $0 \leq 0$.

4. State whether each of the following is true or false.
   (a) $-9 < -8$.   (b) $-8 > -9$.   (c) $-3 \leq 2$.   (d) $4 \geq 4$.

5. Change from set notation to interval notation and illustrate the set geometrically. Also tell in each case whether the interval is open, closed, or half-open.
   (a) $\{x : -1 < x < 6\}$.
   (b) $\{x : -2 \leq x < 1\}$.
   (c) $\{x : -8 \leq x \leq -6\}$.
   (d) $\{x : 4 < x \leq 7\}$.
6. Repeat Exercise 5 for each of the following.
   (a) $\{x : 9 \leq x \leq 10\}$.
   (b) $\{x : 6 \leq x < 7\}$.
   (c) $\{x : -5 < x \leq 0\}$.
   (d) $\{x : -1 < x < 2\}$.
7. Change each of the following intervals to set notation.
   (a) $(-3, 8)$.   (b) $[\pi, 5]$.   (c) $[-19, -5]$.   (d) $(-3, 3]$.
8. Evaluate each of the following.
   (a) $|-8|$.   (b) $|8|$.   (c) $|5 - 5|$.   (d) $|3 - 9|$.
   (e) $|5 - 7 - 1|$.   (f) $|-4|^2$.   (g) $|(-3)(4)|$.
9. Evaluate each of the following.
   (a) $|-4|$.   (b) $|0|$.   (c) $|5 - 3|$.   (d) $\left|\dfrac{-4}{5}\right|$.
   (e) $\dfrac{|7|}{|3-4|}$.   (f) $|-5| + |3|$.   (g) $|-5 + 3|$.
10. If $a = -3$ and $b = 7$, illustrate that
    (a) $|a - b| = |b - a|$.   (b) $|a|^2 = a^2$.
    (c) $\sqrt{a^2} = |a|$.   (d) $|ab| = |a||b|$ (this property is valid for any real numbers a and b).
11. Find the distance between a and b. (In each case your answer should be a positive real number.)
    (a) $a = 5$, $b = -2$.   (b) $a = -10$, $b = -3$.
    (c) $a = 0$, $b = 7$.   (d) $a = 8$, $b = -8$.
12. (a) For what values of x does $|2x + 7| = 2x + 7$?
    (b) For what values of x does $|2x + 7| = -2x - 7$? (*Hint:* The equation $|a| = -a$ is valid for $a = 0$ as well as for $a < 0$.)
13. In order to be considered, suppose that a petition must have at least 100 signatures. Suppose, furthermore, that at least 30 signatures must be women's signatures. Express these conditions as inequalities.
14. Solve for x.
    (a) $9x - 6 > 3x + 12$.   (b) $|2x + 3| = 8$.
    (c) $|2x + 3| < 8$.   (d) $|2x + 3| \geq 8$.
15. Solve for x.
    (a) $4x + 6 < 7x - 5$.   (b) $|5 - 2x| = 8$.
    (c) $|1 - 3x| \leq 9$.   (d) $\left|\dfrac{x}{2} + 4\right| > 5$.
16. Translate each of the statements of Example 6 into an inequality.

3.6 An Introduction to Analytic Geometry

Analytic geometry was first developed in the 17th century as a way of relating geometric and algebraic concepts. This was an important accomplishment because, through the use of analytic geometry, many well-known geometric figures can be represented by means of algebraic equations, and vice versa. Such representations are fundamental in the development of calculus. Necessary to analytic geometry is the concept of a coordinate system. In this section we shall introduce the **rectangular coordinate system,** which is also called the **Cartesian coordinate system** in recognition of René Descartes (1596–1650). Descartes is credited by many as being the founder of analytic geometry.

Let two real lines, one horizontal and one vertical, be drawn in a plane so that the point of intersection is the origin of each real line. The horizontal line is called the ***x*-axis** and its positive direction is chosen to the right; the vertical line is called the ***y*-axis** and its positive direction is chosen upward. The two real lines are called **coordinate axes,** and their point of intersection is called the **origin.** These lines divide a plane into four **quadrants,** which are numbered as shown in Figure 3.6.

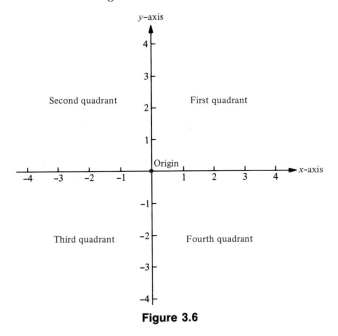

Figure 3.6

Select any point P that is *not* on either coordinate axis and draw perpendiculars from P to each coordinate axis. Let a and b be, respectively, the coordinates on the x and y axes of the points where these perpendiculars meet the axes. See Figure 3.7. The point P will be designated by (a, b), and although the same notation is used for an open interval, it should be clear in context whether (a, b) designates a point or an open interval. A point on the

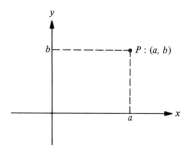

Figure 3.7

x-axis with coordinate a on the horizontal real line will be denoted by $(a, 0)$, and a point on the y-axis with coordinate b on the vertical real line will be denoted by $(0, b)$. The number a is called the **x-coordinate** or **abscissa** of the point (a, b), and b is called the **y-coordinate** or **ordinate**. When we designate a point by (a, b), we agree that the first number a is the one that corresponds to the horizontal axis. A point in any quadrant or on either coordinate axis has associated with it a unique designation (a, b), and conversely, if (a, b) is given, then a unique point in the plane is determined. In Figure 3.8 several points have been plotted and their coordinates are shown.

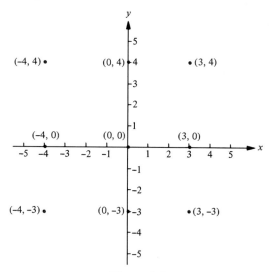

Figure 3.8

Next consider the problem of finding the length d of the line segment between two given points A and B in a plane. Assume that $A : (x_1, y_1)$ and $B : (x_2, y_2)$ are not on the same vertical or horizontal line (see Figure 3.9). Let us think of starting at A and changing the x-coordinate by $x_2 - x_1$ to get $C : (x_2, y_1)$, and then changing the y-coordinate of C by $y_2 - y_1$ in order to obtain B. These changes in x and y are called **increments** and are denoted by

$$\Delta x = x_2 - x_1$$

and
$$\Delta y = y_2 - y_1.$$

(The increments are read "delta x" and "delta y.") In Figure 3.9 the increments Δx and Δy are positive, but one or both of them can be *negative*, so that, in general, the legs of right triangle ABC have lengths $|\Delta x|$ and $|\Delta y|$. The desired length d is found by the Pythagorean theorem:

$$\begin{aligned} d^2 &= |\Delta x|^2 + |\Delta y|^2 \\ &= (\Delta x)^2 + (\Delta y)^2 \\ &= (x_2 - x_1)^2 + (y_2 - y_1)^2, \end{aligned}$$

and therefore

$$d = \sqrt{(x_2 - x_1)^2 + (y_2 - y_1)^2}. \tag{3.5}$$

The number d is called the **distance** between A and B, and (3.5) is known as the **distance formula**. It can also be verified that (3.5) gives the correct length between A and B even if these points are on the same vertical or horizontal line (Exercise 14).

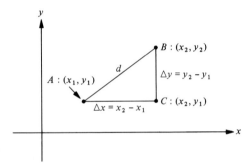

Figure 3.9

Example 1

A map of a region that contains valuable timber has been marked off using a rectangular coordinate system in such a way that one unit on each axis represents 1 mile. By sighting smoke from two different fire towers, it is determined that a fire is breaking out at approximately the location $(-6, -5)$. If a helicopter located at $(-2, 3)$ is ordered to fly directly to the fire, how far must the helicopter travel?

The answer to the problem is the distance d between $(-6, -5)$ and $(-2, 3)$. Let $x_1 = -6$, $y_1 = -5$, $x_2 = -2$, and $y_2 = 3$; then, by (3.5),

$$d = \sqrt{[-2 - (-6)]^2 + [3 - (-5)]^2} = \sqrt{16 + 64}$$
$$= \sqrt{80} = 4\sqrt{5} \text{ miles.} \blacksquare$$

In a problem such as the one in Example 1, it does not matter which of the two points is chosen to be labeled (x_1, y_1); the answer is the same either way because $(x_2 - x_1)^2 = (x_1 - x_2)^2$ and $(y_2 - y_1)^2 = (y_1 - y_2)^2$.

An important problem in analytic geometry is that of sketching graphs (or geometric representations) of algebraic equations. This can be done after plotting a sufficient number of points whose coordinates satisfy a given equation.

Example 2
Sketch the graph of the equation
$$y = 4 - x^2.$$

By giving various real values to x and then calculating the corresponding values of y, each of the following points can be shown to belong to the graph of the given equation: $(-4, -12), (-3, -5), (-2, 0), (-1, 3), (0, 4), (1, 3), (2, 0), (3, -5),$ and $(4, -12)$. Then these points are joined together, as shown in Figure 3.10. ∎

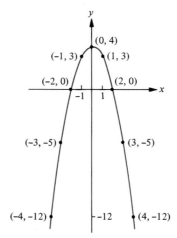

Figure 3.10. $y = 4 - x^2$.

The concept of the graph of an equation, as illustrated in the previous example, can be formalized by the following definition.

▶ **Definition 2**
The **graph of an equation** that involves x and y is the set of all points (x, y) whose coordinates satisfy the equation.

Later we shall frequently use the word **curve** when we are referring to the graph of a set of points (x, y); in this general usage, even a straight line may be referred to as a curve.

Perhaps the most important points on a graph are those, if any, which are on the coordinate axes. In Example 2, if we set $x = 0$ and solve to obtain $y = 4$, we find the y-coordinate of the only point that the graph has in

common with the y-axis—in this case the number 4 is called a **y-intercept**. In general the y-intercepts of the graph of an equation are the ordinates of the points that are on the graph *and* on the y-axis. In Example 2, if we set $y = 0$, we find that $x = 2$ and $x = -2$; these are abscissas of the points at which the graph intersects the x-axis and are called the **x-intercepts** of the graph of the equation.

Another important problem in analytic geometry is that of finding algebraic equations that represent geometric figures.

Example 3

Find an equation for the circle with center $(2, 3)$ and radius 5.

The desired curve consists of that set of points (x, y) which are at a distance of 5 units from the point $(2, 3)$. Hence by (3.5) a point (x, y) is on the circle (see Figure 3.11) if and only if

$$\sqrt{(x-2)^2 + (y-3)^2} = 5$$

or

$$(x-2)^2 + (y-3)^2 = 25,$$

which is the desired equation. ∎

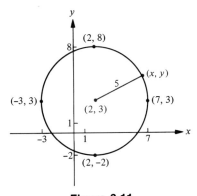

Figure 3.11

Example 3 can be generalized to show that an equation for a circle with center (h, k) and radius r is

$$(x - h)^2 + (y - k)^2 = r^2. \tag{3.6}$$

Sometimes the completing the square technique (Section 3.3) is useful for putting the equation of a circle in the form (3.6).

Example 4

Find the center and radius of the circle

$$x^2 + y^2 + 12x - 6y - 4 = 0.$$

Sec. 3.6] An Introduction to Analytic Geometry

We proceed as follows:

$$(x^2 + 12x + 36) + (y^2 - 6y + 9) = 4 + 36 + 9,$$
$$(x + 6)^2 + (y - 3)^2 = 49,$$
$$[x - (-6)]^2 + (y - 3)^2 = 7^2,$$

which is in the form (3.6) with $h = -6$, $k = 3$, and $r = 7$. Therefore, the circle has center $(-6, 3)$ and radius 7. ∎

APPLICATIONS

Example 5
Consider the hypothetical situation of three councilmen who allocate money for two items. The allocations advocated by each councilman can be thought of as a point (x_i, y_i) in our two-dimensional Cartesian coordinate system; that is, x_i represents the amount that councilman i would allocate to item 1 in the budget and y_i represents the amount that councilman i would allocate to item 2 in the budget. Once the issue is decided, the amounts actually budgeted for the two items can be represented by the point (a, b). The distances between the points (x_i, y_i) and the point (a, b) will provide a measure of the compromise made by the councilmen. Moreover, the distance formula can be generalized to apply to dimensions greater than two [see (3.8) of Section 3.10 for the three-dimensional case]; this will permit an analysis of the allocations for a large number of items by a large group of arbitrators. Such an analysis can be used to rank the degree of compromise by the arbitrators and thereby, over a period of time, to study such things as leadership, influence, and middle-of-the-road positions. ∎

EXERCISES

Suggested minimum assignment: Exercises 1, 3(b), 5, 7, 8(b) (c), 10(b) (d), 12, and 13.

1. Plot each of the following points and tell in which quadrant each point is located: $(-3, 5)$, $(4, -1)$, $(2, 7)$, $(-5, -4)$.
2. Plot each of the following points and tell in which quadrant each point is located: $(4, -7)$, $(0, 5)$, $(-3, -1)$, $(-5, \pi)$.
3. Find the distance between each of the following pairs of points:
   (a) $(1, 4)$, $(5, 7)$.   (b) $(-1, -5)$, $(6, -7)$.   (c) $(-5, -3)$, $(4, -3)$.
4. Find the distance between each of the following pairs of points (let the first point be the initial point and identify Δx and Δy in each part):
   (a) $(3, -6)$, $(2, -5)$.   (b) $(\sqrt{2}, 0)$, $(2\sqrt{2}, -1)$.   (c) $(\frac{1}{2}, 5)$, $(\frac{-7}{2}, 5)$.
5. A circle passes through $(-4, 7)$. Its center is $(3, -1)$. What is the radius of the circle?
6. Use the distance formula and the Pythagorean theorem to determine whether or not the triangle with vertices $(-5, -1)$, $(-1, 4)$, and $(5, -9)$ is a right triangle. Sketch.

7. Given points $A : (-3, 6)$ and $B : (5, -1)$.
   (a) Starting at A and ending at B, find the increments Δx and Δy.
   (b) Starting at B and ending at A, find the increments Δx and Δy.
8. Sketch the graph of each of the following equations:
   (a) $y = x^2 - 9$.   (b) $x = y^2$.
   (c) $y = |x|$.   (d) $x^2 + y^2 = 4$.
   (e) $x^2 + 16y^2 = 16$.   (f) $2x + 3y = 12$.
9. Is the point $(1, 2)$ on the graph of the equation $x^4 + y^3 - 8x - 1 = 0$? Why?
10. Find all x-intercepts and all y-intercepts.
    (a) $y = x^4 - x^2$.   (b) $y = \dfrac{x^2 - 4}{x^2 - 9}$.   (c) $y^2 = \dfrac{x^2 - 4}{x^2 - 9}$.
    (d) $x^2 - y^2 = 16$.   (e) $x^2 + y^2 = 16$.
    (f) $x = (y - 1)(y - 2)(y - 3)$.
11. Find an equation of the set of points (x, y) that are at a distance of 3 from the point $(-1, -2)$.
12. Find an equation of the set of points (x, y) that are equidistant (the same distance) from $(4, 3)$ and $(-3, -2)$.
13. Find the center and radius of each of the following circles.
    (a) $(x - 7)^2 + (y + 10)^2 = 4$.
    (b) $x^2 + y^2 = 29$.
    (c) $x^2 + 2x + y^2 - 8y = 8$.
    (d) $x^2 + y^2 = \frac{5}{2} + 5x - y$.
14. Show that the distance formula gives the correct length between two points if the points are on either the same vertical line or the same horizontal line.
15. A walkie-talkie has a maximum transmitting range of approximately 14 miles. Can the transmission be received by a person, appropriately equipped, who has walked 5 miles east and then 12 miles south of the sender?

3.7 Functions

Expressions of relationships between two or more entities are very common and important in the real world. These expressions may vary from simple statements such as "the demand for a commodity increases as the price decreases" to very complex mathematical formulas. The purpose of this section is to gain some precision in our understanding and expression of special types of relationships known as **functions**.

Example 1

Let $D = \{a, b, c, d\}$ be a set of four small countries. Table 3.1 defines a function f by associating with each country of set D, the gross national product (GNP) of that country in a particular year (to the nearest billion dollars). We say that the **image** of each country is the corresponding GNP.

Table 3.1. Gross National Product of Four Countries.

Country	Gross National Product
a	55,000,000,000
b	37,000,000,000
c	34,000,000,000
d	55,000,000,000

Table 3.1 displays a set of four assignments—each country is assigned its GNP for the year as its image. The set D of countries is called the **domain** of the function f, and the set R of images that appear in the second column of Table 3.1 is called the **range** of f. Hence

$$R = \{34{,}000{,}000{,}000, \quad 37{,}000{,}000{,}000, \quad 55{,}000{,}000{,}000\}.$$

The fact that the GNP of b is 37,000,000,000 is indicated by writing

$$f(b) = 37{,}000{,}000{,}000,$$

which is read "the value of the function f at b is 37,000,000,000." ∎

Figure 3.12 is presented to help the reader understand the concept of a function. An element of the domain can have only one image in the range; however, it is possible for an element of the range to be the image of more than one element of the domain (consider the element 55 billion in Example 1). The sets D and R are different in general, but they may be the same. Next we give a definition of the terminology introduced above.

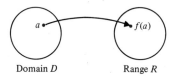

Domain D Range R
Figure 3.12. *Function concept.*

▶ **Definition 3**
Let D be a given nonempty set. A **function** f is a set of assignments whereby to each element a of D there is assigned exactly one element of a set R. The element assigned to a is denoted by $f(a)$ and is called the **image** of a. [The symbol $f(a)$ is read "f of a" or "the value of f at a."] Set D is called the **domain** of the function f, and the set R of all images of elements of D is called the **range** of f.†

---

† As an alternative definition, a **function** f is a set of ordered pairs (x, y), no two of which have the same first term. The **domain** of f is the set of all possible values of x, and the **range** of f is the set of all possible values of y.

In the next example a function will be defined by an equation rather than by a table. Observe also in Example 2 that the domain and range each contain an infinite number of elements rather than a finite number of elements.

Example 2
Corresponding to any real number x, let exactly one real number y be assigned by the equation

$$y = 4 - x^2$$

of Example 2 of Section 3.6. This equation defines a function f whose domain is the set of all real numbers. Figure 3.10 reveals that y can be any real number not greater than 4, and hence the range R of f is

$$R = \{y : y \leq 4\}. \blacksquare$$

Since x and y in the last example may vary throughout our discussion, they are called **variables**. When y is written in terms of x as in Example 2, it is customary to call x the **independent variable** and y the **dependent variable**. If x is replaced by a number in the domain of f, then the associated value of y can be computed. In this manner y is dependent upon x. We shall say that "y is a function of x" and write

$$y = f(x) = 4 - x^2.$$

Many people, particularly those who have seen sophisticated hand calculators, enhance their understanding of the concept of a function by relating it to a "machine." An element of the domain of a function is entered into the machine, which performs its assignment and produces as output *exactly one* corresponding element of the range of the function. The machine will not process an element that does not belong to the domain of the function. The machine is constructed in such a way that it will take an element (the input) of the domain of a function and assign to it the correct image (the output) in the range of the function. For example, if the number 81 is entered in a hand calculator and the $\sqrt{x}$ button is pushed, then the number 9 is displayed; thus the calculator determines that if $f(x) = \sqrt{x}$, then $f(81) = \sqrt{81} = 9$. The set of all such assignments from the domain constitutes the square root function.

Suppose that a function f is said to be defined by $y = f(x)$ and that no mention is made of the domain. *If the domain of f is not specified, we shall assume in this text that it consists of all real numbers which can be substituted for x in order to determine a unique real number y.* For example, if

$$y = f(x) = \frac{x}{x-2},$$

then the domain D of f, unless otherwise indicated, is implied to be

$$D = \{x : x \text{ is a real number and } x \neq 2\}.$$

Since division by zero is undefined in mathematics, the real number 2 is excluded from the domain. The range of the function in this example is not easy to identify from the original equation; one way to find the range is to first solve for x in terms of y as follows:

$$xy - 2y = x,$$
$$x(y - 1) = 2y,$$
$$x = \frac{2y}{y - 1}.$$

Then it is possible to observe that y can take on any real value except 1, and hence the range R is

$$R = \{y : y \text{ is a real number and } y \neq 1\}.$$

If a function is said to be defined by

$$y = g(x) = \sqrt{x - 4},$$

then, in order to assure only real values for y, the domain of g is understood to be

$$\{x : x \geq 4\}.$$

If, however, a certain set of values of the independent variable is specified, such as

$$y = h(x) = \sqrt{x - 4}, \quad \text{where } 7 \leq x \leq 8,$$

then the domain of h is the closed interval $[7, 8]$. Functions g and h above are *not* the same because the domains are different.

Example 3
Consider the function f defined by $f(x) = x^2 + 3x + 1$. The domain of f is assumed to be (as explained above) the set of all real numbers, so any real number can be substituted for x; in other words,

$$f(\ ) = (\ )^2 + 3(\ ) + 1,$$

where any one real number can be inserted in all three parentheses. For example,

(a) $f(-2) = (-2)^2 + 3(-2) + 1 = -1.$
(b) $f(0) + f(3) = [(0)^2 + 3(0) + 1] + [(3)^2 + 3(3) + 1] = 1 + 19 = 20.$
(c) $f(a) = a^2 + 3a + 1.$
(d) $f(2 + \Delta x) - f(2) = [(2 + \Delta x)^2 + 3(2 + \Delta x) + 1] - [(2)^2 + 3(2) + 1]$
$\qquad = 4 + 4\,\Delta x + (\Delta x)^2 + 6 + 3\,\Delta x + 1 - 11$
$\qquad = (\Delta x)^2 + 7\,\Delta x.$
(e) $f(2x) - 2f(x) = [(2x)^2 + 3(2x) + 1] - [2(x^2 + 3x + 1)]$
$\qquad = 4x^2 + 6x + 1 - 2x^2 - 6x - 2$
$\qquad = 2x^2 - 1.$ ∎

An understanding of the notation used, and the calculations made, in Example 3 will be very helpful in preparing for a study of calculus in this text. A geometric understanding of the concept of a function is important also. The graph of the function f of Example 3 is shown in Figure 3.13 [by the **graph of f** we mean the graph of the equation $y = f(x)$]. Graphically, $f(3)$ is the y-coordinate or "height" of the curve at $x = 3$. The y-intercept is $f(0)$. A line drawn parallel to the axis of the dependent variable (the y-axis) cannot intersect the graph of f at more than one point; this is because the basic idea of a function is that each element of the domain has exactly *one* image in the range.

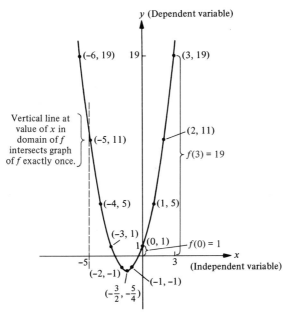

Figure 3.13. $y = f(x) = x^2 + 3x + 1$.

Example 4
Revenue functions, cost functions, and profit functions are the names of some important functions in economics. These functions are introduced here and will be used in future examples.

The revenue (perhaps in dollars) received by a manufacturer for the sale of a number x of a certain item might be given by

$$R(x) = 3x + \tfrac{1}{50}x^2,$$

where R is called a **revenue function** and expresses a relationship between the number of items sold and the amount of money received for these items. Some sort of revenue function may be assumed for planning purposes; often it is possible from experience to make a rather accurate determination of such a function. We can calculate

$$R(20) = 3(20) + \tfrac{1}{50}(20)^2 = 68,$$

which means that the company expects to receive $68 if it can sell 20 items of this particular type.

The cost (perhaps in dollars) of producing a number x of a certain item might be estimated to be

$$C(x) = 10 + 2x,$$

where C is called a **cost function**. Observe that

$$C(0) = 10,$$

which means that there is a **fixed cost** of $10 even if no items are produced—this may represent overhead. The cost of 20 items is given by

$$C(20) = 10 + 2(20) = 50.$$

If the revenue and cost functions above are valid for the same item, we can calculate

$$R(20) - C(20) = 68 - 50 = 18,$$

which gives the profit on the sale of 20 items. In general, the **profit function** P is given by

$$P(x) = R(x) - C(x).$$

Although in real situations, the number of items sold or produced would usually have to be a nonnegative integer, it is customary to let the domain of a revenue, cost, or profit function (which serves as a model of the real situation) be an *interval* of nonnegative real numbers.

Cost accountants often plot both the revenue and cost functions on the same graph. They call this graph a **break-even chart**. The point, if it exists, at which the curves intersect is called the **break-even point**. For the revenue and cost functions of this example the break-even point occurs between

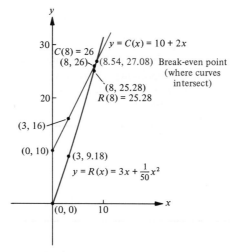

Figure 3.14. *Break-even chart.*

$x = 8$ and $x = 9$ (see Figure 3.14 for the approximate coordinates of the break-even point). Hence, if 8 or fewer items are sold, there is a loss, but if at least 9 items are sold, there is a profit. ∎

APPLICATIONS

*Example 5 Weighting Functions and Probability

This example is the first of a sequence of examples that apply the calculus and preliminary topics to a study of probability. The probability examples, which occur throughout the remainder of this book, often depend upon the content of previous examples in the sequence and will be designated by a ★.

In a study of probability one often begins with the idea of an experiment in which any one of a set of possible outcomes can occur. For instance, consider an experiment of choosing one card from an ordinary deck of 52 cards. There are 52 possible outcomes, and the experiment consists of the selection of one of these outcomes. The nonempty set U of possible outcomes of an experiment is known as a **sample space,** and any subset of U is called an **event.** An event is said to occur if one or more of its outcomes occur.† Several illustrations of sample spaces and events are given in Table 3.2.

Table 3.2. ILLUSTRATIONS OF SAMPLE SPACES AND EVENTS.

Experiment	Sample space	An event
Flip a coin on a flat surface and observe whether a head or a tail appears.	{head, tail}	{head}
Select a sample product from an assembly line and observe whether the product is passable, repairable, or nonrepairable.	{passable, repairable, nonrepairable}	{passable, nonrepairable}
Count the number of days it takes to breed a female offspring in a set of laboratory animals.	$\{1, 2, 3, \ldots\}$	$\{3, 4, 5\}$
Measure the time in hours, or portion thereof, between accidents on a given stretch of highway.	The set of nonnegative real numbers	The set of nonnegative real numbers less than or equal to 4

In the remainder of this example we shall be concerned only with sample spaces having a finite number of outcomes. Prior to the occurrence of an experiment it is common to speculate about the likelihood or probability of achieving a single outcome, or a set of outcomes, of the sample space.

† It is customary to say that the event which contains no outcomes does not occur, and it is called the **impossible event.**

Precision may be gained by thinking of the likelihood or probability of a single outcome in terms of a function w whose domain is the sample space $U = \{u_1, u_2, \ldots, u_n\}$; such a function, for which each $w(u_i) \geq 0$ and for which $w(u_1) + \cdots + w(u_n) = 1$, is often called a **weighting function.** For example, suppose that prior to the experiment of casting a pair of dice, one speculates about the likelihood that a given total will appear. The sample space is the set $U = \{2, 3, 4, 5, 6, 7, 8, 9, 10, 11, 12\}$; the weighting function w for this illustration is given in Table 3.3. The weight of each outcome can be determined by observing that since there are six faces on a single die, there are 6^2 or 36 equally likely ways for a pair of dice to land on a flat surface, and by observing the number of ways each number of the sample space can occur. Notice that the sum of the weights of all possible outcomes is 1. The set of numbers in the first row of Table 3.3 is the domain of the function w, and the set of numbers in the second row is the range of w. The graph of the weighting function w in this example is the set of eleven points shown in Figure 3.15.

Table 3.3. Weights for the outcomes of a pair of dice.

u_i	2	3	4	5	6	7	8	9	10	11	12
$w(u_i)$	$\frac{1}{36}$	$\frac{2}{36}$	$\frac{3}{36}$	$\frac{4}{36}$	$\frac{5}{36}$	$\frac{6}{36}$	$\frac{5}{36}$	$\frac{4}{36}$	$\frac{3}{36}$	$\frac{2}{36}$	$\frac{1}{36}$

For sample spaces that have a finite number of outcomes, it is customary to define $P(E)$, the **probability of an event** E, in terms of the weighting function as follows: If

$$E = \{e_1, e_2, \ldots, e_k\}$$

is a subset of a sample space U, then

$$P(E) = w(e_1) + w(e_2) + \cdots + w(e_k).$$

If E is the empty set, then $P(E)$ is defined to equal zero.

Suppose that it is desired to find the probability of achieving a sum of either seven or eleven for one throw of a pair of dice. In this case the event E is the subset $E = \{7, 11\}$ of the sample space U. Therefore,

$$P(E) = w(7) + w(11) = \tfrac{6}{36} + \tfrac{2}{36} = \tfrac{2}{9}. \ \blacksquare$$

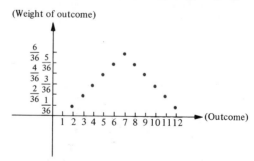

Figure 3.15. *Probabilities of various totals for a pair of dice.*

*Example 6 Random Variable or Measurement

Consider an experiment in which a certain hospital determines a man's blood type in preparation for a possible transfusion. The set of outcomes in this particular test is $U = \{A+, A-, B+, B-, AB+, AB-, O+, O-\}$. The hospital has to buy blood for transfusions from a commercial blood bank. Associated with each of the outcomes of U is a cost in dollars of obtaining a pint of the particular type of blood. This information, given in Table 3.4, defines a function X. The domain and range of X are given in the respective rows of Table 3.4.

Table 3.4. COST OF PINT OF BLOOD.

Outcomes, u	A+	A−	B+	B−	AB+	AB−	O+	O−
Cost, $X(u)$	55	55	60	60	60	65	50	55

A function X that assigns a real number to each outcome u of a sample space U (finite or infinite) is called a **random variable**. In our illustration the values of the function X (the values of the random variable) are the real numbers 50, 55, 60, and 65.

Because a random variable simply assigns a real number to each outcome of a sample space, a random variable may be thought of as a **measurement** of the various outcomes. The term "measurement" seems to promote a better intuitive grasp of the basic concept than does the term "random variable," and for this reason "measurement" will be used more often in the remainder of this book.

Other illustrations of a measurement follow:

1. Let U consist of the set of grades $\{A, B, C, D, F\}$. Grade averages are often computed by the use of a random variable, or measurement, X as defined by Table 3.5.

Table 3.5

u	A	B	C	D	F
$X(u)$	4	3	2	1	0

2. Let $U = \{1, 2, 3, 4, 5, 6\}$ be the set of outcomes of a single throw of a single die. Consider a game in which the thrower will collect the amount in pennies shown on the die. Since the outcomes of U are real numbers, a measurement X can be defined that simply assigns to every outcome the same real number. In other words, $X(u) = u$. On the other hand, suppose that the thrower must *pay* the amount shown on the die if the number is *odd* and will continue to *collect* if the amount is *even*. In this case we have defined a different measurement Z, as shown in Table 3.6.

Table 3.6

u	1	2	3	4	5	6
$Z(u)$	−1	2	−3	4	−5	6

3. Consider an experiment in which a coin is flipped until a head appears. Here U has an infinite number of elements, and $U = \{H, TH, TTH, \ldots\}$. Let a random variable, or measurement, X be defined in such a way that $X(u)$ is the number of flips that it takes to obtain a head.

In all these illustrations it is important to note that a random variable, or measurement, is a *function* whose domain is a sample space U and whose range is a subset of the set of real numbers. ∎

EXERCISES

Suggested minimum assignment: Exercises 3, 5, 6, 9, 13, 15, and 17.

1. Suppose that the domain of a function f is $D = \{1, 3, 5, 7\}$ and that the image of each number of set D is 10 times the number. Find $f(1)$, $f(3)$, $f(5)$, $f(7)$, and the range R of f.
2. Make up an example of a function whose domain is the set of cities in the United States whose population is over 1,000,000.
3. A function f whose domain is the set of all real numbers is defined by $y = f(x) = x^2 + 4$. What is the image of the number 3? Draw a graph of f and state the range of f.
4. Describe a function whose domain is the set of all stocks listed on the New York Stock Exchange.
5. Suppose that a function f is defined by $f(1) = 4$, $f(2) = 5$, and $f(3) = 6$. State the domain and range of f. If $f(2) = 7$ were added above, why would we no longer have a function defined?
6. Suppose that s is a function of t as defined by $s = f(t) = t^4 + 1$, provided that $t \geq 0$. Which variable is the dependent variable and which variable is the independent variable? Find $f(2)$.
7. Suppose that a function g is defined by $y = g(x) = 1/x$, provided that $x \neq 0$. Which variable is the dependent variable and which variable is the independent variable? Find $g(5)$.
8. If $f(x) = 1 - x - 3x^2$, find $f(0)$ and $f(1)$.
9. State the implied domain of f if f is said to be defined by each of the following equations.

    (a) $f(x) = \dfrac{2x}{x + 2}$.   (b) $f(x) = 1 + 2x$.

    (c) $f(x) = \dfrac{1}{x^2 - 1}$.   (d) $f(x) = 1 + \sqrt{x}$.

10. Let $y = f(x)$, where $f(x)$ is given in Exercise 9(a). Solve for x in terms of y and state the range of f.
11. If $f(x) = \dfrac{1}{2x + 3}$, find $f(-3)$ and $f(a + b)$.
12. If $f(x) = 3 - (x/2)$, sketch the graph of f. Indicate $f(-4)$ and $f(10)$ on your sketch.

13. Suppose that $f(x) = x^2$. Calculate each of the following.
    (a) $f(3)$.   (b) $f(3) - f(2)$.   (c) $f(3-2)$.   (d) $f(x+3)$.
    (e) $f(x+3) - f(x)$.   (f) $\dfrac{f(x+h) - f(x)}{h}$, where $h \neq 0$.

14. If $f(x) = 3x - 2x^2$, calculate each of the following.
    (a) $f(4)$.   (b) $f(2)$.   (c) $f(x_1)$.   (d) $f(x_1 + \Delta x)$.
    (e) $f(x_1 + \Delta x) - f(x_1)$.   (f) $\dfrac{f(x + \Delta x) - f(x)}{\Delta x}$, where $\Delta x \neq 0$.

15. Let a cost function C be defined by $C(x) = 5 + 8x - \frac{1}{25}x^2$, where $0 \leq x \leq 10$. What is the domain of C? Find the cost of producing five items.

16. Solve algebraically the system of equations
$$\begin{cases} y = 3x + \frac{1}{50}x^2, \\ y = 10 + 2x, \end{cases}$$
in order to find the break-even point in Example 4.

17. The revenue and cost functions for an item are given by $R(x) = 10x$ and $C(x) = 40 + 8x$, provided that $0 \leq x \leq 100$. Find $P(x)$, and find the profit if 50 items are sold. What is the fixed cost? How many items must be sold to reach the break-even point?

18. An experiment consists of choosing one card from an ordinary deck of 52 cards and observing whether it is a spade, heart, diamond, or club.
    (a) What is the sample space U in this experiment?
    (b) Assign a weight to each outcome in a manner that satisfies the definition of a weighting function; let the outcomes that are equally likely have equal weights. State the domain and range of the weighting function w.
    (c) Find $P(E)$ if the event E is the subset {spade, club} of U.

19. (a) If U is a sample space and w is a weighting function with domain U, is it always correct to call w a random variable (or measurement)? Give reasons for your answer.
    (b) If U is a sample space and X is a measurement with domain U, is it always correct to call X a weighting function? Give reasons for your answer.

20. In illustration 3 of Example 6 a measurement X is defined. What is the image of the element TTH? What is the range of X?

3.8 Straight Lines

The owners of a business spend $15,000 for equipment and decide for income tax purposes to depreciate the equipment over a 10-year period. If the straight-line method of depreciation is used, the value of the equipment is assumed to decrease by $1500 each year for 10 years. In Figure 3.16 the value y of the equipment in thousands of dollars is shown after x years,

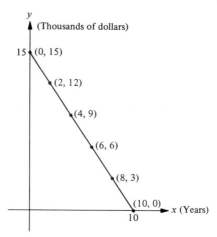

Figure 3.16. *Straight-line depreciation.*

where $0 \leq x \leq 10$. The graph is a portion of a straight line. A straight line is one of the simplest, and yet one of the most important, types of figures encountered in applications. In this section we shall define the useful concept of slope of a straight line; then we shall introduce various forms of the equation of a straight line.

First, let us agree not to distinguish between the terms "line" and "straight line." Also it is understood that if we refer to two points on a line, then we mean two different points. A vertical line is understood to be a line parallel to the y-axis, and a horizontal line is parallel to the x-axis. Assume that two points (x_1, y_1) and (x_2, y_2) determine a unique line l passing through the two points, and that any two points on a given line l can be used to determine l in this manner.

▶ **Definition 4**

If (x_1, y_1) and (x_2, y_2) are points on a nonvertical line l, the **slope** m of l is given by

$$m = \frac{\Delta y}{\Delta x} = \frac{y_2 - y_1}{x_2 - x_1}.$$

The slope of a vertical line is not defined.

The slope of a nonvertical line is a real number that is a measure of the direction that the line has in relation to the coordinate axes. In Figure 3.17 a line with slope $\frac{7}{3}$ is shown. Start at any point on the line and change x by 3 units ($\Delta x = 3$); the change in y is 7 units ($\Delta y = 7$), and the slope is

$$m = \frac{\Delta y}{\Delta x} = \frac{7}{3}.$$

Any line that goes up as one traces it from left to right, as in Figure 3.17, has a positive slope. On the other hand, a line that goes down as one traces it

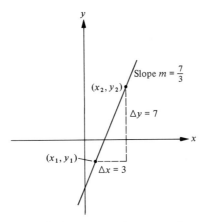

Figure 3.17. *Line with positive slope.*

from left to right has a negative slope; for example, in Figure 3.18 each line goes down 3 units ($\Delta y = -3$) for every 4 units of movement to the right ($\Delta x = 4$), and the slope is

$$m = \frac{\Delta y}{\Delta x} = \frac{-3}{4} = -\frac{3}{4}.$$

As Figure 3.18 suggests, nonvertical parallel lines have the same slope and, conversely, lines with the same slope are parallel.† *The slope of a horizontal line is zero* because the change Δy between any two points on the line is 0, and hence $\Delta y / \Delta x = 0$.

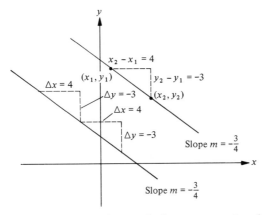

Figure 3.18. *Parallel lines with the same negative slope.*

---

† The reader who has studied trigonometry may observe that the slope of a line equals $\tan \alpha$, where α is the smallest nonnegative angle whose initial side is the positive direction of the x-axis and whose terminal side is the given line.

Sec. 3.8] Straight Lines

Example 1
Find the slope of the line that passes through $(-3, 8)$ and $(6, -4)$.
  This problem can be solved without a graph, although the line is shown in Figure 3.20. Let $(x_1, y_1) = (-3, 8)$ and $(x_2, y_2) = (6, -4)$. Then

$$m = \frac{\Delta y}{\Delta x} = \frac{y_2 - y_1}{x_2 - x_1} = \frac{-4 - 8}{6 - (-3)} = \frac{-12}{9} = -\frac{4}{3}.$$

Since

$$\frac{y_2 - y_1}{x_2 - x_1} = \frac{y_1 - y_2}{x_1 - x_2},$$

it does not matter which of the two points above is labeled (x_1, y_1); the other is then (x_2, y_2). ∎

  We now begin a discussion of *equations* of lines by considering vertical and horizontal lines. A vertical line has an equation of the form

$$x = a,$$

and a horizontal line has an equation of the form

$$y = b.$$

For example, the equation $x = -5$ (which may be written $x + 0y = -5$) is satisfied by points such as $(-5, -2)$, $(-5, 0)$, $(-5, 2)$, and $(-5, y)$ for any real number y; its graph is a vertical line 5 units to the left of the y-axis. In Figure 3.19 some horizontal and vertical lines, together with their equations, are shown. Notice that the graph of $x = 0$ is the y-axis and the graph of $y = 0$ is the x-axis.

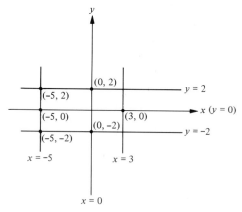

Figure 3.19. *Equations of vertical and horizontal lines.*

  Next we examine the form of equations of other lines. A unique non-vertical line l is determined by a point (x_1, y_1) on the line and the slope m of the line. A point $(x, y) \neq (x_1, y_1)$ is on l if and only if

$$m = \frac{y - y_1}{x - x_1}$$

or

$$y - y_1 = m(x - x_1).$$

This last equation is also satisfied if $(x, y) = (x_1, y_1)$. Therefore, a point (x, y) is on l if and only if

$$y - y_1 = m(x - x_1),$$

which is called the **point-slope form** of the equation of a line.

Example 2

Find an equation of the line that passes through $(-3, 8)$ and $(6, -4)$.

In this type of problem, first find the slope of the line and then substitute in the point-slope form; in Example 1 we found that $m = -\frac{4}{3}$. Either of the given points can be substituted for (x_1, y_1) in the point-slope form; we select $(x_1, y_1) = (-3, 8)$ and substitute to obtain

$$y - 8 = -\tfrac{4}{3}[x - (-3)]$$

or

$$y - 8 = -\tfrac{4}{3}(x + 3). \blacksquare$$

We shall use the letters a and b, respectively, for the x- and y-intercepts of a line. In Example 2 we can set $y = 0$ to find that the x-intercept is

$$a = 3,$$

and we can set $x = 0$ to find that the y-intercept is

$$b = 4.$$

See Figure 3.20.

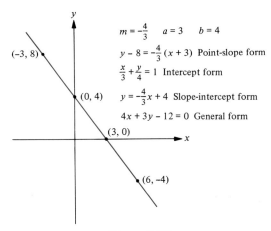

Figure 3.20

There are other useful forms of an equation of a line beside the point-slope form. Three other such forms are shown in Figure 3.20 for the line of Examples 1 and 2. These can be obtained from the point-slope form by simple algebra, and they may be compared with the information about the various forms of an equation of a line in Table 3.7.

Table 3.7. FORMS OF EQUATION OF LINE.

Equation of line	Name	Comments
$y - y_1 = m(x - x_1)$	Point-slope form	Passes through (x_1, y_1); slope m.
$\dfrac{x}{a} + \dfrac{y}{b} = 1$	Intercept form	Passes through $(a, 0)$ and $(0, b)$. ($a \neq 0$, $b \neq 0$.)
$y = mx + b$	Slope-intercept form	Slope m; passes through $(0, b)$. Special case of point-slope form, obtained by setting $(x_1, y_1) = (0, b)$. May be obtained from any other form by simply solving for y.
$Ax + By + C = 0$	General form	Slope is $-A/B$ and y-intercept is $-C/B$, provided that $B \neq 0$. (A, B, and C are any real numbers, except that not both A and B are zero.)

Example 3

Suppose that the graph of an unknown cost function is known to be a line. Find an equation of the line if the fixed cost is $10 and if each item costs an additional $2 above the fixed cost.

The line has y-intercept $b = 10$ because the fixed cost is known to be $10. The line has slope $m = 2$ because it is known that each item costs $2 above the fixed cost. Since

$$b = 10$$

and

$$m = 2,$$

the slope-intercept form can be used immediately, and the equation is

$$y = 2x + 10.$$

(The cost function in this example is the same one that appeared in Example 4 of Section 3.7.) ∎

Example 4

Find the slope of the line whose equation is

$$3x - 2y + 12 = 0$$

and draw its graph.

An excellent procedure for finding the slope is to solve the equation for y first:
$$2y = 3x + 12,$$
$$y = \tfrac{3}{2}x + 6.$$

This puts the equation in slope-intercept form $y = mx + b$, from which we can observe that the coefficient of x is $m = \tfrac{3}{2}$. [An alternative procedure is to observe that according to Table 3.7 the line is given in general form, and its slope is $m = -A/B = -(3/-2) = \tfrac{3}{2}$.] The line can be graphed quickly by plotting any two points, since two points determine a line. One way to find two points on the line is to put the equation in intercept form and read off the x- and y-intercepts:
$$3x - 2y = -12,$$
$$\frac{x}{-4} + \frac{y}{6} = 1.$$

Hence the x-intercept is -4, the y-intercept is 6, and the line passes through $(-4, 0)$ and $(0, 6)$. See Figure 3.21. By observation of the figure we can also verify that the slope of the line is $\tfrac{3}{2}$. ∎

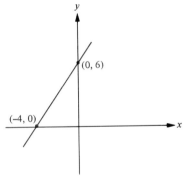

Figure 3.21. $3x - 2y + 12 = 0$.

APPLICATIONS

Example 5
Studies of some biological environments require knowledge of the two customary scales of measurement of temperature. The graph of the equation that relates the temperature y in degrees Celsius (centigrade) to the temperature x in degrees Fahrenheit is known to be a line. The line passes through $(32, 0)$ because water freezes at 32 degrees Fahrenheit and 0 degrees Celsius. It also passes through $(212, 100)$ because water boils at 212 degrees Fahrenheit (at standard pressure) and 100 degrees Celsius. Find the slope of the line, an equation of the line, and sketch. Also find the Celsius temperature that corresponds to 50 degrees Fahrenheit.

Let $(x_1, y_1) = (32, 0)$ and $(x_2, y_2) = (212, 100)$. Then

$$m = \frac{y_2 - y_1}{x_2 - x_1} = \frac{100 - 0}{212 - 32} = \frac{100}{180} = \frac{5}{9}.$$

The equation of the line in point-slope form is

$$y - 0 = \tfrac{5}{9}(x - 32)$$

or

$$y = \tfrac{5}{9}(x - 32).$$

The graph is shown in Figure 3.22. The slope $\tfrac{5}{9}$ may be interpreted to mean that for every 9-degree increase in the Fahrenheit temperature, there is an increase of 5 degrees Celsius. The Celsius temperature corresponding to 50 degrees Fahrenheit is found by substituting $x = 50$ in the equation to obtain

$$y = \tfrac{5}{9}(50 - 32) = \tfrac{5}{9}(18) = 10 \text{ degrees Celsius.} \blacksquare$$

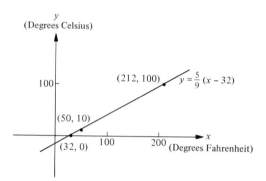

Figure 3.22

EXERCISES

Suggested minimum assignment: Exercises 1, 2(a) (b), 4, 5(b), 8(a), 9(a), 12(a), and 13.

1. Find the slope, if possible, of the line passing through the given pair of points.
   (a) $(4, 8)$ and $(3, 1)$.   (b) $(-5, -6)$ and $(-7, 4)$.
   (c) $(-4, 3)$ and $(3, 3)$.   (d) $(-1, 3)$ and $(-1, -4)$.
2. Sketch the line that passes through the given point and has given slope m.
   (a) $(4, 3)$, $m = \tfrac{5}{4}$.   (b) $(4, 3)$, $m = -\tfrac{5}{4}$.
   (c) $(-1, -2)$, $m = 4$.   (d) $(3, -4)$, $m = -3$.
   (e) $(0, 4)$, $m = \tfrac{1}{3}$.

3. A line l passes through $(-5, -4)$ and $(6, 3)$. What is the slope of any line parallel to l?
4. Graph each of the following lines.
   (a) $x = 6$.   (b) $x = -7$.   (c) $y = 8$.   (d) $y = -9$.
5. Find the point-slope form of the equation of the line through each pair of points.
   (a) $(5, 3)$ and $(6, 5)$.
   (b) $(-2, -3)$ and $(1, 2)$.
   (c) $(3, 1)$ and $(-2, 8)$.
6. Find the general form of the equation of each line in Exercise 5.
7. Find the point-slope form of the equation of the line that passes through the given point and has given slope m.
   (a) $(6, -1)$, $m = \frac{2}{3}$.
   (b) $(-4, 2)$, $m = -5$.
8. Find the intercept form of the equation of the line whose x-intercept a and y-intercept b are given. Graph the line.
   (a) $a = 5, b = 2$.
   (b) $a = -6, b = \frac{1}{2}$.
9. Find the slope-intercept form of the equation of the line whose slope m and y-intercept b are given. Graph the line.
   (a) $m = 3, b = 2$.
   (b) $m = \frac{2}{3}, b = -1$.
   (c) $m = -7, b = 0$.
10. Find the general form of the equation of the line that passes through $(7, 8)$ and that has x-intercept -2.
11. A line passes through $(-7, -5)$ and $(-3, 1)$. Find its equation in
    (a) Point-slope form.   (b) Slope-intercept form.
    (c) General form.   (d) Intercept form.
12. Sketch each of the following lines. Find the x-intercept, the y-intercept, and the slope.
    (a) $5x - 2y + 10 = 0$.   (b) $y = 4x - 7$.
    (c) $x + y = 3$.   (d) $\dfrac{x}{2} - \dfrac{y}{9} = 1$.
    (e) $\dfrac{x}{2} - \dfrac{y}{9} = 2$.   (f) $y - 4 = 2(x - 3)$.
13. Find an equation of the line that passes through $(4, 2)$ and that is parallel to the line $5x + 2y + 7 = 0$. Leave your answer in general form.
14. Explain why the slope of a line, as defined in Definition 4, does not depend upon which two specific points on the line are used.
15. Let nonvertical lines l_1 and l_2 have slopes m_1 and m_2, respectively. It can be shown that l_1 and l_2 are perpendicular if and only if $m_2 = -1/m_1$. What is the slope of any line that is perpendicular to the line $2x - 3y + 5 = 0$?
16. A revenue function is given by $y = R(x) = 3x$, provided that $0 \leq x \leq 10$. Sketch its graph.

17. (a) Find the Celsius temperature corresponding to 140 degrees Fahrenheit.
    (b) Find the Fahrenheit temperature corresponding to 30 degrees Celsius.
    (c) Find the Fahrenheit temperature corresponding to -40 degrees Celsius.

3.9 Some Special Types of Functions

Suppose that a large corporation deposits $10 million in a savings account for 1 year at 6 per cent interest compounded semiannually. After 6 months the savings account contains 103 per cent of the initial investment, which is

$$(10,000,000)(1.03) = \$10,300,000.$$

After 1 year the amount in the account is 103 per cent of the amount held after the first 6 months, namely,

$$(10,300,000)(1.03) = (10,000,000)(1.03)^2 = \$10,609,000.$$

Had the corporation deposited the money at 6 per cent at a place where the money was compounded quarterly, the amount after 1 year would have been

$$(10,000,000)(1.015)^4 = \$10,613,635.51.$$

If the money is compounded n times a year at 6 per cent, where n is a positive integer, the amount in the account after 1 year is a function of n as given by

$$f(n) = (10,000,000)\left(1 + \frac{.06}{n}\right)^n.$$

The domain of the function f above is the set of positive integers, since the number of times a year that interest is compounded is assumed to be a positive integer. The domain of many important functions is the set of positive integers 1, 2, 3, 4, ... ; according to Definition 5, this special type of function is given a name.

▶ **Definition 5**
A **sequence** is a function whose domain is the set of positive integers.

Example 1
Let f be the sequence defined by $f(n) = (n + 1)/(2n + 1)$. Since f is a sequence, we know that the domain of f is the set of positive integers 1, 2, 3, ... The positive integers (in order) can be substituted to obtain the elements of the range, which are called the **terms** of the sequence. Hence the **first term** is

$$f(1) = \frac{1 + 1}{2 + 1} = \frac{2}{3}.$$

The **second term** of the sequence is

$$f(2) = \frac{2+1}{4+1} = \frac{3}{5}.$$

$f(n) = (n+1)/(2n+1)$ is called the **nth term** or **general term** of the sequence because any term can be generated from it by substituting the appropriate positive integer. Since the domain of any sequence is the same, the sequence above can be accurately described by listing the terms of the sequence in order:

$$\frac{2}{3}, \frac{3}{5}, \frac{4}{7}, \frac{5}{9}, \ldots, \frac{n+1}{2n+1}, \ldots \blacksquare$$

Instead of using

$$f(1), f(2), \ldots, f(n), \ldots$$

as the general representation of the terms of a sequence, we find that it is customary to use

$$a_1, a_2, \ldots, a_n, \ldots,$$

where each subscript represents an element of the domain and each term represents an element of the range. In Example 1, $a_n = (n+1)/(2n+1)$. It is also customary to describe a sequence by just putting the nth term in braces—we can refer to the sequence in Example 1 by writing

$$\left\{\frac{n+1}{2n+1}\right\}.$$

We now consider another special type of function. Functions defined by

$$f(x) = \tfrac{3}{2}x^3 - 2x^3 + 5x + \sqrt{2}$$

and

$$g(x) = x^6 - 4x^2 + 7$$

are examples of polynomial functions; this type of function will be a favorite in this text because of its frequency of use in business, the social sciences, and the life sciences.

▶ **Definition 6**
A function f defined by

$$f(x) = b_n x^n + b_{n-1} x^{n-1} + \cdots + b_1 x + b_0,$$

where n is nonnegative integer, $b_n \neq 0$, and $b_n, b_{n-1}, \ldots, b_1, b_0$ are real numbers, is called a **polynomial function of degree n**.

The two functions f and g immediately preceding Definition 6 are polynomial functions of degrees 3 and 6, respectively. The domain of any polynomial function, unless specified otherwise, is the set of all real numbers.

Three special cases of Definition 6 are now mentioned briefly:
1. A polynomial function of degree 0, as given by $f(x) = b_0$ with $b_0 \neq 0$, is called a **constant function.** Its graph is a horizontal line (Section 3.8). The function given by $f(x) = 0$ is also called a constant function.
2. A polynomial function of degree 1, as given by $f(x) = b_1 x + b_0$ with $b_1 \neq 0$, is called a **linear function.** Its graph is a line with slope b_1 and y-intercept b_0 (Section 3.8).
3. A polynomial function of degree 2, as given by $f(x) = b_2 x^2 + b_1 x + b_0$ with $b_2 \neq 0$, is called a **quadratic function.** Its graph is called a **parabola** and always has a characteristic appearance; if $b_2 > 0$, the parabola opens upward (see Figure 3.13), whereas if $b_2 < 0$, the parabola opens downward (see Figure 3.10).

A third special type of function is a **rational function,** which can be defined as a quotient of polynomial functions. For example, the function f defined by

$$f(x) = \frac{x^3 + 8x + 5}{2x^2 - 8}$$

is a rational function whose domain is the set of all real numbers except 2 and -2 (these numbers make the denominator zero).

A fourth special type of function is one that is defined by more than one equation; we illustrate such a function in Example 2.

Example 2

One item sold by a certain dealer in home and building supplies is storm windows. Most customers pay the dealer an extra fee for installing the windows instead of doing the installation themselves. The dealer charges y dollars per window for the installation, where y is a function of the number x of miles from the store to the customer's home—see Figure 3.23. A customer who lives less than 10 miles from the store is within the immediate city area, and the charge is $6 per window for installation [the arrow at $(10, 6)$ indicates that the $6 fee applies up to but *not* including 10 miles]. A customer from a nearby town $(10 \leq x < 40)$ pays $8 per window, and a

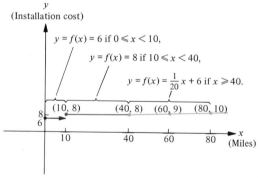

Figure 3.23. *Installation cost per window.*

customer 40 or more miles away pays a little more ($\frac{1}{20}x + 6$), depending upon the mileage. The function f is graphed in Figure 3.23, and it can be defined by three equations as follows:

$$f(x) = \begin{cases} 6 & \text{if } 0 \le x < 10, \\ 8 & \text{if } 10 \le x < 40, \\ \frac{1}{20}x + 6 & \text{if } x \ge 40. \end{cases}$$

It is important to realize in Example 2 that only *one* function f has been defined (not three functions). We are trying to point out that some functions can best be described by separate equations for different parts of the domain.

Example 3

Graph the function f defined by

$$f(x) = \begin{cases} 2 + |x| & \text{if } x < 2, \\ x^2 & \text{if } x \ge 2. \end{cases}$$

The domain of f is the set of all real numbers, but the manner in which $f(x)$ is calculated depends upon whether or not $x < 2$. Table 3.8 gives a table of values, and Figure 3.24 is a sketch of the curve that passes through

Table 3.8

x	−3	−1	0	1	1.9	2	3	4
$f(x)$	5	3	2	3	3.9	4	9	16

Use $f(x) = 2 + |x|$ if $x < 2$.   Use $f(x) = x^2$ if $x \ge 2$.

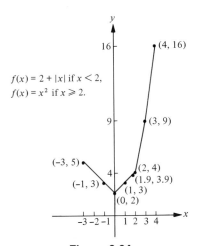

Figure 3.24

Sec. 3.9] Some Special Types of Functions

the points $(x, y) = (x, f(x))$. Using the definition of absolute value (Definition 1), we can define the function f as follows:

$$f(x) = \begin{cases} 2 - x & \text{if } x < 0, \\ 2 + x & \text{if } 0 \leq x < 2, \\ x^2 & \text{if } x \geq 2. \end{cases}$$

Thus the graph of f in Figure 3.24 consists of portions of the graphs of linear and quadratic functions. ∎

APPLICATIONS

Example 4
In the discussion at the beginning of this section, we showed that $10 million deposited at 6 per cent compounded quarterly would yield after 1 year the amount $10,613,635.51. Suppose, instead, that one wonders how much money would have to be deposited now at 6 per cent compounded quarterly in order to yield $10 million 1 year from now. In other words, $10 million 1 year from now has what **present value** at 6 per cent compounded quarterly?
Let P be the required present value. Then

$$P(1.015)^4 = 10{,}000{,}000$$

and

$$P = \frac{10{,}000{,}000}{(1.015)^4} = \frac{10{,}000{,}000}{1.061363550625}$$
$$= \$9{,}421{,}842.30.$$

The type of calculations made in this example and at the beginning of this section are not restricted to problems in which the time period is 1 year. Suppose that a principal P is deposited at an annual interest rate r for t years. If interest is compounded n times a year, then the amount A after t years is given by

$$A = P\left(1 + \frac{r}{n}\right)^{nt}. \blacksquare$$

Example 5
Suppose that a person living in the United States wants to mail a domestic first-class letter, and suppose that the weight of the letter does not exceed 13 ounces ($\frac{13}{16}$ pound). The cost y, in cents, of mailing the letter is a function f of the weight x of the letter in ounces. As of this writing the cost of mailing a letter is 13 cents for the first ounce and 11 cents for each additional ounce, provided that $0 < x \leq 13$. Therefore, the cost is given as follows:

$$f(x) = \begin{cases} 13 & \text{if } 0 < x \le 1, \\ 24 & \text{if } 1 < x \le 2, \\ 35 & \text{if } 2 < x \le 3, \\ \cdot & \cdot \\ \cdot & \cdot \\ \cdot & \cdot \\ 145 & \text{if } 12 < x \le 13. \end{cases}$$

In Exercise 12 the reader is asked to graph $y = f(x)$. ∎

*Example 6 Discrete and Continuous Measurements
The effect of defining a measurement (random variable) on a sample space U is to create a new, and corresponding, sample space S consisting of the set of values of the measurement. Such a set S has the advantage of having elements that are always real numbers. If the set S is finite or countably infinite,† then the measurement is said to be **discrete**. If, however, the set S consists of an interval of real numbers, then the measurement is said to be **continuous**.

The study of probability for discrete measurements presupposes information normally found in a course in finite mathematics, whereas the study of probability for continuous measurements requires some knowledge of calculus; both will be pursued in later sections.

As an illustration of the discrete case, consider the experiment of flipping a coin until a head appears. The sample space U is

$$\{H, TH, TTH, TTTH, \ldots\}.$$

A measurement X is defined that associates with each outcome the number of times the coin is flipped until a head appears. The set of values of the measurement (the range of X) is the set $S = \{1, 2, 3, \ldots\}$ of all positive integers; since S is a countable set, X is a *discrete* measurement. A weighting function w, with domain S, will be defined below such that each weight $w(s) \ge 0$, and the sum of all weights is 1. Since, on one throw, the probability of throwing a head is $\frac{1}{2}$, on two throws the probability of throwing a tail followed by a head is $\frac{1}{4}$, and so on, then w is defined by

$$w(s) = \begin{cases} \dfrac{1}{2} & \text{if } s = 1, \\ \dfrac{1}{4} & \text{if } s = 2, \\ \vdots & \\ \dfrac{1}{2^n} & \text{if } s = n, \\ \vdots & \end{cases}$$

† A set S is said to be **countably infinite** (or **countable**) if there is a one-to-one correspondence between S and the set of positive integers.

Sec. 3.9] Some Special Types of Functions

Notice that w is a sequence in this illustration and could be represented by $w(n) = 1/2^n$ or by $\{w_n\} = \{1/2^n\}$.

For the discrete case, the probability of an event $E = \{s_1, s_2, \ldots, s_k\}$, where E is a finite subset of S, can be defined as follows: If w, with domain S, is the weighting function, then

$$P(E) = w(s_1) + \cdots + w(s_k).$$

In the previous illustration let $E = \{1, 2, 3, 4\}$; hence E is the subset of measurement values that are less than or equal to 4. Here $P(E)$ represents the probability that a head will appear (at least once) if a coin is flipped four times. We calculate $P(E)$ as follows:

$$P(E) = w(1) + w(2) + w(3) + w(4)$$
$$= \tfrac{1}{2} + \tfrac{1}{4} + \tfrac{1}{8} + \tfrac{1}{16}$$
$$= \tfrac{15}{16}.$$

We assumed in the illustration of this example that the sum

$$\frac{1}{2} + \frac{1}{4} + \frac{1}{8} + \frac{1}{16} + \cdots + \frac{1}{2^n} + \cdots \tag{3.7}$$

of weights was equal to 1. An expression, such as (3.7), in which addition of an infinite number of real numbers is indicated, will be discussed in Example 7 of Section 9.2; in that example we will prove that the sum (3.7) of weights is 1. The idea of the sum of an infinite number of real numbers also arises in defining $P(E)$ for an infinite subset $E = \{s_1, s_2, \ldots, s_n, \ldots\}$ of a countable sample space S corresponding to a discrete measurement X. (For example, if $E = \{2, 4, 6, \ldots, 2n, \ldots\}$ in our previous illustration, then $P(E)$ represents the probability that a head will appear first on an even-numbered flip of a coin.) If w, with domain S, is the weighting function, then $P(E)$ is defined by

$$P(E) = w(s_1) + w(s_2) + \cdots + w(s_n) + \cdots.$$

Example 7 of Section 9.2 should also aid in understanding this definition. ∎

EXERCISES

Suggested minimum assignment: Exercises 2(c), 3(a), 4, 5(a), 6, 8, and 11.
1. If $8000 is deposited in a savings account at 5 per cent compounded semiannually, what is the amount in the account after 1 year?
2. Write the first five terms of each of the following sequences.
   (a) $\{4n - 1\}$.   (b) $\left\{\dfrac{10}{n}\right\}$.
   (c) $\left\{\dfrac{n+2}{n+3}\right\}$.   (d) $\left\{\dfrac{(-1)^n}{n+3}\right\}$.

3. Write the nth term of each of the following sequences.
   (a) 10, 20, 30, 40, 50, ...   (b) $\frac{1}{2}, \frac{1}{5}, \frac{1}{8}, \frac{1}{11}, \frac{1}{14}, \ldots$
   [In part (a) assume that the terms continue to increase by 10; in (b) assume that the numerators are always 1 and that the denominators continue to increase by 3.]
4. What is the twenty-fifth term of the sequence $\left\{\dfrac{n+7}{3n}\right\}$?
5. State the degree of each of the following polynomial functions. Also state the implied domain of each function.
   (a) $f(x) = 3 + 7x + x^5$.   (b) $g(x) = (x^4)^2 + x^4 - 5$.
6. Tell whether each of the following polynomial functions is a constant function, a linear function, or a quadratic function.
   (a) $f(x) = 7x - 5$.   (b) $g(x) = x^2 + 4x + 5$.   (c) $h(x) = 8$.
7. Sketch the parabola defined by each of the following.
   (a) $y = f(x) = 16 - x^2$.   (b) $y = g(x) = x^2 + 6x$.
8. State the implied domain of each of the following rational functions.
   (a) $f(x) = \dfrac{x}{(x-1)(x-2)(x-3)}$.   (b) $g(x) = \dfrac{x+5}{x^2-9}$.
9. In Example 2 what is the cost of installation per window for a customer who lives 50 miles from the store?
10. Graph each of the following functions.
    (a) $f(x) = \begin{cases} 2x & \text{if } x \leq 3, \\ 3 & \text{if } x > 3. \end{cases}$ (b) $g(x) = \begin{cases} |x| & \text{if } x < 4, \\ x - 4 & \text{if } x \geq 4. \end{cases}$
    (c) $h(x) = \begin{cases} x & \text{if } 0 \leq x < 2, \\ 3 & \text{if } 2 \leq x \leq 3, \\ 4 & \text{if } x > 3. \end{cases}$
11. Graph the cost function C defined by
    $$C(x) = \begin{cases} x^2 & \text{if } 0 \leq x \leq 2, \\ x + 2 & \text{if } x > 2. \end{cases}$$
12. Graph the function f defined in Example 5.
13. How much money would have to be deposited now at 8 per cent compounded semiannually in order to yield \$2000 one year from now?
14. If \$1000 is deposited at 5 per cent compounded quarterly, set up the formula (but do not calculate) for the amount in the account after 10 years.
15. In Theodore's book *Applied Mathematics: An Introduction* [62], several applications of linear and quadratic functions in business and economics are given on pp. 374–407. Solve the following problem, which is like one of Theodore's simpler examples. A retailer sells an item costing 60 cents for 78 cents and another item costing 70 cents for 91 cents. If these two examples represent the general markup policy for determining selling price, find the linear function that determines the selling price of an item.

Sec. 3.9] Some Special Types of Functions

3.10 Rectangular Coordinate System in Three Dimensions

In Section 3.6 we introduced the two-dimensional rectangular coordinate system. Later in the text we will need the **three-dimensional rectangular coordinate system,** which is also called the **three-dimensional Cartesian coordinate system.**

In Figure 3.25 the x-axis, y-axis, and z-axis of a three-dimensional rectangular coordinate system are shown. Each axis is perpendicular to each other axis, although this may not appear to be the case, since we are attempting to draw a three-dimensional figure on a two-dimensional piece of paper. The x-axis can be thought of as being perpendicular to the plane of the paper, and the units on the x-axis are marked about 70 per cent as far apart as on the other axes in order to give the best perspective. A few points (x, y, z) are also plotted in Figure 3.25; notice that the x-coordinate is given first, then the y-coordinate, and finally the z-coordinate.

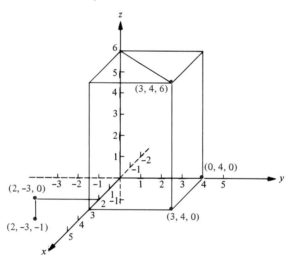

Figure 3.25. *Three-dimensional rectangular coordinate system.*

The plane determined by the x and y axes is called the **xy-plane;** every point in the xy-plane has a z-coordinate of zero, so an equation of the xy-plane is

$$z = 0.$$

In Figure 3.26 the xy-plane, the xz-plane, the yz-plane, and some planes parallel to these **coordinate planes** are shown along with their equations.

Recall that the Pythagorean theorem was used to derive the distance formula (3.5), which is

$$d = \sqrt{(x_2 - x_1)^2 + (y_2 - y_1)^2}.$$

It can be used to find the distance d between points $A : (x_1, y_1)$ and $B : (x_2, y_2)$

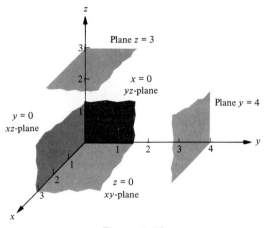

Figure 3.26

in two dimensions. There is an analogous formula in three dimensions that can be derived by two applications of the Pythagorean theorem. Let $A : (x_1, y_1, z_1)$ and $B : (x_2, y_2, z_2)$ be two points in three-dimensional space and let us obtain a formula for the distance d between A and B. See Figure 3.27. First, we shall write a formula for the distance q between A and $C : (x_2, y_2, z_1)$. Points A and C lie in the plane $z = z_1$, which is parallel to the xy-plane; therefore, by (3.5),

$$q = \sqrt{(x_2 - x_1)^2 + (y_2 - y_1)^2}.$$

Triangle ABC is a right triangle with right angle at C. The leg BC is a vertical segment with length $|z_2 - z_1|$. By the Pythagorean theorem

$$d^2 = q^2 + |z_2 - z_1|^2.$$

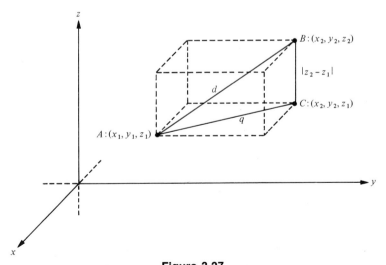

Figure 3.27

Sec. 3.10] Rectangular Coordinate System in Three Dimensions

Therefore,
$$d^2 = (x_2 - x_1)^2 + (y_2 - y_1)^2 + |z_2 - z_1|^2.$$
Since $|z_2 - z_1|^2 = (z_2 - z_1)^2$, we find that the nonnegative distance d is given by
$$d = \sqrt{(x_2 - x_1)^2 + (y_2 - y_1)^2 + (z_2 - z_1)^2}. \tag{3.8}$$
Equation (3.8) is the **distance formula** in three dimensions. In Figure 3.27 we assumed that $x_1 \neq x_2$, $y_1 \neq y_2$, and $z_1 \neq z_2$, but it can be verified that (3.8) is valid if these restrictions are removed.

Example 1
Find the distance between $A : (4, 7, -1)$ and $B : (2, -1, 3)$.
By the distance formula (3.8),
$$d = \sqrt{(2-4)^2 + (-1-7)^2 + [3-(-1)]^2}$$
$$= \sqrt{84} = 2\sqrt{21}. \ \blacksquare$$

In three dimensions the set of points equidistant from two given points clearly is a plane. In fact, the plane is the perpendicular bisector of the line segment joining the given points.

Example 2
Find an equation of the plane that is equidistant from the points $(2, -1, 3)$ and $(1, 2, 0)$.
A point (x, y, z) is in the required plane if and only if its distance to $(2, -1, 3)$ equals its distance to $(1, 2, 0)$; that is,
$$\sqrt{(x-2)^2 + (y+1)^2 + (z-3)^2} = \sqrt{(x-1)^2 + (y-2)^2 + z^2}.$$
It follows that
$$x^2 - 4x + 4 + y^2 + 2y + 1 + z^2 - 6z + 9$$
$$= x^2 - 2x + 1 + y^2 - 4y + 4 + z^2$$
or
$$-4x + 2y - 6z + 14 = -2x - 4y + 5.$$
Therefore, an equation of the plane is
$$2x - 6y + 6z - 9 = 0. \ \blacksquare$$

The method of Example 2 can be used to show that any plane in three-dimensional space has an equation of the form
$$Ax + By + Cz + D = 0, \tag{3.9}$$
where A, B, C, and D are real numbers and A, B, and C are not all zero. Conversely, any equation of this form is an equation of a plane, although we

omit the proof of this statement. Observe that even the coordinate planes such as $z = 0$ are of the form (3.9).

By the graph of an equation that involves x, y, and z we mean the set of all points (x, y, z) whose coordinates satisfy the equation. (This is similar to Definition 2 of Section 3.6.)

Example 3
Sketch the graph of the equation
$$3x + 2y + z - 6 = 0.$$

The graph is a plane since the equation is of the form (3.9). A good method of sketching a plane is based upon finding the intercepts. If y and z are 0, then $x = 2$, and we say that the x-intercept is 2. Similarly, the y-intercept is 3 and the z-intercept is 6. The points $(2, 0, 0)$, $(0, 3, 0)$, and $(0, 0, 6)$ can then be plotted and joined together by line segments as shown in Figure 3.28. The plane extends indefinitely far, but one can visualize the whole plane by drawing just the portion shown in Figure 3.28. ∎

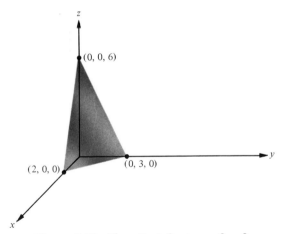

Figure 3.28. Plane $3x + 2y + z - 6 = 0$.

Example 4
Sketch the graph of the equation
$$-y + 2z = 8.$$

The plane passes through $(0, -8, 0)$ and $(0, 0, 4)$. There is no x-intercept and hence the plane is parallel to the x-axis. See Figure 3.29. ∎

In general, if a letter is missing in the equation of a plane as in Example 4, then the plane is parallel to the axis of the missing variable. The plane $y = 4$ is parallel to both the x-axis and the z-axis (see Figure 3.26).

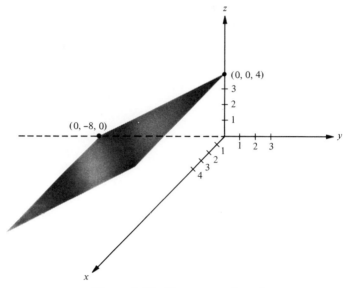

Figure 3.29. Plane $-y + 2z = 8$.

APPLICATIONS

Applications of the material in this section will appear in Chapters 5 and 15.

EXERCISES

Suggested minimum assignment: Exercises 1, 3, 6(a), 7, 14(b), and 14(c).

1. Plot all the given points (x, y, z) on the same drawing and label them clearly:

$$(5, 1, 0), (1, 4, -1), (0, 0, 6).$$

2. Repeat Exercise 1 for the following points:

$$(3, 0, 0), (3, 6, 0), (3, 6, 5), (1, 10, 3).$$

3. (a) What is an equation of the xy-plane?
   (b) What is an equation of the xz-plane?
   (c) What is an equation of the yz-plane?
   (d) What is an equation of the plane containing the points $(0, 0, 3)$, $(2, 5, 3)$, and $(1, -4, 3)$?

4. Sketch the graphs of each of the following equations using the three-dimensional rectangular coordinate system.
   (a) $x = 3$.
   (b) $y = 2$.
   (c) $z = 5$.

5. Find the distance between each of the following pairs of points.
   (a) (4, 7, 1) and (0, 5, 4).
   (b) (−2, 1, 5) and (4, 3, 2).
6. Find the distance between each of the following pairs of points.
   (a) (−3, −2, 7) and (−6, 4, 5).
   (b) (1, 8, −3) and (2, 0, −1).
7. Find an equation of the plane that is equidistant from the points (0, 3, 2) and (−1, −2, 1).
8. Repeat Exercise 7 for the points (1, 2, −4) and (−1, 1, 3).
9. (a) Find a formula for the distance from the origin (0, 0, 0) to a point (x, y, z).
   (b) What is the distance from the origin to (3, 6, −2)?
10. Plot the points (1, 4, −1) and (1, 4, 6) and tell by inspection the distance between these points. Also calculate the distance between these points by the distance formula.
11. A **sphere** with radius $r > 0$ and center (h, k, l) is the set of points whose distance from (h, k, l) is equal to r. Use the distance formula to derive an equation of a sphere with radius r and center (h, k, l).
12. Find an equation of the sphere with center (1, 5, −2) and radius 4. (See Exercise 11.) Also find the volume of the sphere.
13. A sphere has center (4, 5, 3) and passes through (0, 1, 0). Find the radius of the sphere. (See Exercise 11.)
14. Sketch the graphs of each of the following equations using the three-dimensional rectangular coordinate system.
    (a) $x + y + z = 4$.
    (b) $3x + 4y + 2z - 12 = 0$. Also state the intercepts.
    (c) $2x + z - 6 = 0$.
    (d) $5x + 2y - 10 = 0$.
    (e) $y - 6z = 6$.
    (f) $2x + y - z - 6 = 0$.
    (g) $x^2 + y^2 + z^2 = 25$. (See Exercise 11.)
15. Find an equation of the plane that is parallel to the y-axis if the x-intercept is 4 and the z-intercept is 5.

NEW VOCABULARY

exponents 3.1
nth root 3.1
square root 3.1
cube root 3.1
factoring 3.2
binomial expansion 3.2
solution (root) of an equation 3.3
linear equation 3.3
solution set of an equation 3.3
quadratic equation 3.3

completing the square 3.3
quadratic formula 3.3
discriminant 3.3
complex number 3.3
imaginary number 3.3
rectangle 3.4
circle 3.4
semicircle 3.4
trapezoid 3.4
right triangle 3.4

hypotenuse of right triangle 3.4
legs of right triangle 3.4
Pythagorean theorem 3.4
median of a triangle 3.4
altitude of a triangle 3.4
isosceles triangle 3.4
similar triangles 3.4
sphere 3.4, 3.10
right circular cylinder 3.4
open right circular cylinder 3.4
closed right circular cylinder 3.4
right circular cone 3.4
inequality 3.5
open interval 3.5
closed interval 3.5
half-open interval 3.5
endpoints of an interval 3.5
interior points of an interval 3.5
absolute value 3.5
distance 3.5, 3.6, 3.10
linear inequality 3.5
solution of an inequality 3.5
reverse the sense of an inequality 3.5
rectangular (Cartesian) coordinate system in two dimensions 3.6
abscissa 3.6
ordinate 3.6
increment 3.6
distance formula 3.6, 3.10
graph of an equation 3.6, 3.10
curve 3.6

intercepts 3.6, 3.10
function 3.7
image 3.7
domain 3.7
range 3.7
independent variable 3.7
dependent variable 3.7
graph of function 3.7
revenue function 3.7
cost function 3.7
fixed cost 3.7
profit function 3.7
break-even chart 3.7
break-even point 3.7
slope 3.8
point-slope form 3.8
intercept form 3.8
slope-intercept form 3.8
general form 3.8
sequence 3.9
terms of sequence 3.9
polynomial function of degree n 3.9
constant function 3.9
linear function 3.9
quadratic function 3.9
parabola 3.9
rational function 3.9
rectangular (Cartesian) coordinate system in three dimensions 3.10
coordinate planes 3.10
equation of plane 3.10

4 Probability

The actions and behavior of a person often are results of conscious or subconscious decisions by the person. Many of these decisions are based on measurements of the likelihood of certain events. For example, people often make decisions based on a likelihood of bad weather at a future point in time. Of course, the significance of such a decision may range from the significance of scheduling a picnic to that of scheduling a space trip; and the methods of measurement of the likelihood of bad weather may vary from outright guesses to those based on sophisticated scientific analysis. As another illustration, consider a business decision based on the likelihood that the reader will be alive 10 years from now. Measurements of the likelihood of that event may vary from those that are based on little more than outright guesses to those that have been carefully obtained by the insurance industry.

Nevertheless, the decisions of persons, however significant, and however they are based, do shape the course of events. It seems reasonable, therefore, that we concern ourselves with those measurements upon which decisions often can be based.

Prerequisites: 1.1, 1.2, 1.4; also 1.5 is prerequisite to 4.7–4.10.
Suggested Sections for Moderate Emphasis: 4.1–4.6.
Suggested Sections for Minimum Emphasis: 4.1–4.3.

4.1 Basic Terminology†

The word **probability** is usually associated with the measurement of the likelihood of an event; this association is illustrated by the television announcer who says, "the probability of rain tomorrow is 30 per cent." To gain some precision in the meaning of the word "probability," we begin with a nonempty universal set U, which we shall call the **sample space**. We then think of U as the set of outcomes of certain experiments or activities, such as flipping coins, or throwing dice, or swinging at a baseball, or buying stock, or trying to sell a piece of merchandise. Subsets of U are called **events**, or **event spaces**; in other words, an event or an event space is a subset of the set of outcomes.

---

† An alternative presentation of much of the material in this section may be found in Example 5 of Section 3.7. Additional material on probability may be found in Example 6 of Section 3.7 and Example 6 of Section 3.9. The reader should read this material at some point in the study of this chapter.

Example 1

Consider a single die whose six faces appear as shown in Figure 4.1. We may think of casting the die on a level surface and reading the upper face as an experiment; each of the six faces can be considered as an outcome (result) of the experiment. Hence the sample space is the set shown in Figure 4.1. Of course, there are many events that can occur; the subset of odd numbers constitutes one event that we shall call E_1 (see Figure 4.2). The subset consisting of the outcomes shown in Figure 4.3 is another event, which we shall call E_2. If we form the ratio

$$\frac{n(E_1)}{n(U)} = \frac{3}{6} = .5,$$

we have a measure of the likelihood that an odd digit will appear on the upward face when the die is cast on a level surface; we call this ratio the probability of event E_1. Likewise, the probability of event E_2 is

$$\frac{n(E_2)}{n(U)} = \frac{2}{6} \approx .33. \ (\approx \text{ means "is approximately equal to."}) \ \blacksquare$$

Figure 4.1. *Faces of a single die.*

Figure 4.2. *Odd-numbered faces of a die.*

Figure 4.3. *Elements of E_2.*

In Example 1 we implicitly assumed that any one face was as likely to appear as any other face; such outcomes are said to be **equally likely**. Equally likely outcomes are not always the case, however. For example, it might *not* be the case that a magnetized die thrown on a metal table will produce equally likely outcomes.

▶ **Definition 1**
Let U be a finite set of equally likely outcomes of an experiment, and let E be a subset of U. The **probability $P(E)$ of event E** is defined as follows:

$$P(E) = \frac{n(E)}{n(U)}.$$

Example 2
Consider the probability of exactly two heads being thrown if a coin is flipped 3 times. First, we identify the sample space to be the set of outcomes

$$U = \{HHH, HHT, HTH, THH, HTT, THT, TTH, TTT\}.$$

Second, we identify the event space to be the subset of outcomes consisting of exactly two heads,

$$E = \{HHT, HTH, THH\}.$$

Hence, since the outcomes are equally likely,

$$P(E) = \frac{n(E)}{n(U)} = \frac{3}{8}. \blacksquare$$

Example 3
Suppose that an investor intends to buy stock in two companies. He has narrowed his interest to 3 industrials, which we shall label I_1, I_2, and I_3, and 2 utilities, which we shall label U_1 and U_2. If the investor chooses 2 of these 5 stocks at random, what is the probability of choosing 2 industrials?
Solution: Here the sample space of equally likely outcomes is

$$U = \{I_1U_1, I_1U_2, I_1I_2, I_1I_3, I_2U_1, I_2U_2, I_2I_3, I_3U_1, I_3U_2, U_1U_2\},$$

where $n(U) = 10$, and the event space is

$$E = \{I_1I_2, I_1I_3, I_2I_3\},$$

where $n(E) = 3$. Therefore,

$$P(E) = \frac{n(E)}{n(U)} = \frac{3}{10}. \blacksquare$$

Now consider the case of finding the probability of an event where the outcomes are *not* equally likely. Recall that in Definition 1, $P(E)$ was defined to be $n(E)/n(U)$, where E was a subset of the set U of equally likely outcomes. If we examine the probability of each outcome e_i we find that each probability $P(\{e_i\})$ is $1/n(U)$; hence we can think of $P(E)$ as the sum of the probabilities of all the outcomes of the event space. That is,

$$P(E) = \frac{n(E)}{n(U)} = \underbrace{\frac{1}{n(U)} + \frac{1}{n(U)} + \cdots + \frac{1}{n(U)}}_{n(E) \text{ terms}}.$$

Sec. 4.1] Basic Terminology

This idea is extended to sample spaces of outcomes that are not equally likely. We shall redefine the probability of an event to be the sum of the probabilities of obtaining the outcomes (equally likely or not) that comprise the event. The probabilities of obtaining the respective outcomes are often referred to as **weights** of the outcomes.

For example, consider the simple case of a jar containing 10 capsules, of which 3 are red, 6 are white, and 1 is blue. The experiment consists of drawing a capsule from the jar; the outcomes are red, white, and blue capsules, but these outcomes obviously are not equally likely. Therefore we assign the respective weights $\frac{3}{10}$, $\frac{6}{10}$, and $\frac{1}{10}$ to the outcomes. Each weight is simply the probability of the event of obtaining each respective outcome. The probability of an event will be defined to be the sum of the weights of the respective outcomes that belong to the event; for example, the probability of drawing a nonred capsule is the sum of the weights of the nonred outcomes, namely,

$$\frac{6}{10} + \frac{1}{10} = \frac{7}{10}.$$

In Chapter 1 the reader was reminded of the conceptual difference between the element u_i of a set U and the subset $\{u_i\}$ of a set U; it is because of that distinction that we speak of the weight of an outcome (element) u_i, whereas we speak of the probability of an event (subset) $\{u_i\}$. We now state a generalization of Definition 1 to include nonequally likely outcomes.

▶ **Definition 2**
Let U be a finite set of n outcomes $u_1, u_2, \ldots, u_n$ with nonnegative weights $w(u_1), w(u_2), \ldots, w(u_n)$, such that the sum of *all* the weights is 1. If an event E consists of k outcomes $e_1, e_2, \ldots, e_k$, then the **probability of E** is

$$P(E) = w(e_1) + w(e_2) + \cdots + w(e_k).$$

Example 4†

Consider the experiment of throwing two dice and adding the amounts shown on the two upward faces. What is the probability that a 7 or an 11 will occur?

Solution: In Figure 4.4 we illustrate all 36 ways that the dice can appear. We can reason that there are 11 outcomes, namely, the sums 2 through 12, and using Figure 4.4, we can assign weights as shown in Table 4.1.

Table 4.1. WEIGHTS FOR THE OUTCOMES OF A PAIR OF DICE.

Outcome	2	3	4	5	6	7	8	9	10	11	12
Weight	$\frac{1}{36}$	$\frac{2}{36}$	$\frac{3}{36}$	$\frac{4}{36}$	$\frac{5}{36}$	$\frac{6}{36}$	$\frac{5}{36}$	$\frac{4}{36}$	$\frac{3}{36}$	$\frac{2}{36}$	$\frac{1}{36}$

† Historically, the first recorded problem in probability is similar to this problem. See Example 6 of this section.

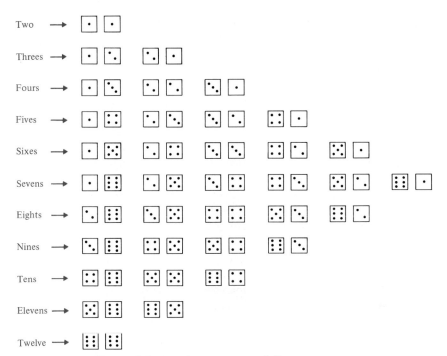

Figure 4.4. *Set of ways a pair of dice can appear.*

Hence by Definition 2 the probability of throwing a 7 or an 11 is the sum

$$P(E) = \tfrac{6}{36} + \tfrac{2}{36} = \tfrac{8}{36} \approx .22.$$

An alternative way of solving this problem is to think of each arrangement in Figure 4.4 as a separate outcome, thus producing a different sample space of 36 equally likely outcomes rather than a sample space of 11 nonequally likely outcomes. Then, using Definition 1, we have

$$P(E) = \frac{n(E)}{n(U)} = \frac{8}{36} \approx .22. \quad \blacksquare$$

It is easy to show that the probability of an event E of a sample space U, as defined in Definition 2, satisfies the properties stated in the following theorem. Remember that sample spaces and events have been defined as sets.

▶ **Theorem 1**

Let E be an event of a sample space U, which has a finite number of outcomes.
   (i) $0 \leq P(E) \leq 1.$
   (ii) $P(U) = 1.$
   (iii) $P(\emptyset) = 0.$
A proof of Theorem 1 is left as Exercises 15–17. ◀

Sec. 4.1] Basic Terminology

At this point we wish to make a distinction between those probabilities that are obtained by a nonexperimental calculation and those that are extrapolated from some type of experiment. The flipping of a fair coin or the casting of a fair die are examples of **a priori probabilities** that can be calculated without experimentation. In the cases of the coin and die the calculations are based on the shape of the objects being thrown. On the other hand, the probability that a certain basketball player will make a free throw or the probability that the reader will die within 10 years are examples of **a posteriori probabilities**—probabilities obtained by past experimentation. The probability that a basketball player will make a free throw can be extrapolated from consideration of his past performance, while the probability that a person will die within 10 years can be extrapolated from a study of the experience of others of the same age. In any case, however, the reader should clearly recognize that a probability measurement is only the *measurement of the likelihood* of an event and in no way affects the event itself. For example, the probability of the event E of not flipping a head 6 times in succession on a fair coin is $\frac{63}{64}$, but if one tries the experiment, the large probability is no guarantee whatsoever that an outcome of the specified event E will occur.

Apparently the study of probability began through an interest in games of chance. It is recorded that Galileo (1564–1642) solved an isolated dice problem for the Grand Duke of Tuscany (see Example 6 of this section), but a serious study of probability did not begin until about 1654. This study was begun by Blaise Pascal (1623–1662) and Pierre de Fermat (1601–1665), and much of it is recorded in the correspondence between the two men. A French nobleman, Chevalier DeMéré, who had considerable interest in gambling, apparently initiated Pascal's interest by posing problems connected with games of chance.

APPLICATIONS

Example 5
In a certain genetic problem the assumption is made that there are two different alleles† A and a at a certain locus. Suppose that genotypes Aa and Aa are mated according to Figure 4.5, under the genetic rule that AA, Aa, aA, and aa are all equiprobable and that Aa and aA are biologically indistinguishable; then the outcomes AA, Aa, and aa are assigned weights

$$w(AA) = \tfrac{1}{4}, \qquad w(Aa) = \tfrac{1}{2}, \qquad w(aa) = \tfrac{1}{4}. \blacksquare$$

Example 6
As with many subjects there is some difference of opinion concerning the origins of mathematical probability. There is, however, considerable agreement that these origins were a direct result of questions posed by

† An *allele* is one of two or more alternative hereditary units or genes at identical loci of homologous chromosomes, giving rise to contrasting Mendelian characters.

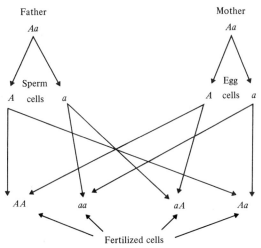

Figure 4.5. *Genotype matings.*

gamblers. According to some, the first recorded result was due to Galileo (1564–1642) in response to a question by the Grand Duke of Tuscany. When 3 dice were thrown, it appeared to the Grand Duke that the sum of 10 occurred more often than did the sum of 9. This seemed strange to the Grand Duke because there were six ways of obtaining both the sum of 10 and the sum of 9, as shown in Figure 4.6. What the Grand Duke overlooked was that the outcomes shown in Figure 4.6 were not equally likely. For example, there are six ways that the sum $6 + 3 + 1$ can be obtained:

$$6 + 3 + 1,$$
$$6 + 1 + 3,$$
$$3 + 1 + 6,$$
$$3 + 6 + 1,$$
$$1 + 3 + 6,$$
$$1 + 6 + 3.$$

Yet there are only three ways that the sum $6 + 2 + 2$ can be obtained, namely,

$$6 + 2 + 2,$$
$$2 + 6 + 2,$$
$$2 + 2 + 6.$$

Galileo found that the probability of obtaining a sum of 10 is $\frac{27}{216}$, whereas the probability of obtaining a sum of 9 is $\frac{25}{216}$. In Exercise 18 the reader is asked to verify Galileo's results by assigning the proper weights to each of the outcomes of Figure 4.6. ∎

Figure 4.6

Sec. 4.1] Basic Terminology

Example 7

In order to determine life insurance premiums and annuity payments, it is necessary for life insurance companies to be able to calculate the probability that a given person will be alive after a certain time interval. These calculations are based on mortality tables which give various types of information concerning deaths of humans. These tables are constantly revised by the insurance industry in order to reflect various influencing factors such as medical advances and living standards. According to one source, the probability of death of United States citizens at different ages is given in Table 4.2. If we let x_i be the outcome of death during the ith decade of life, then the second column of Table 4.2 can be thought of as a set of weights for the corresponding sample space of outcomes. Let E be the event of death during the first 3 decades after birth. Then

$$E = \{x_1, x_2, x_3\},$$

where from Table 4.2 we have

$$w(x_1) = .0323, \qquad w(x_2) = .0065, \qquad w(x_3) = .0121.$$

Therefore, according to Definition 2,

$$\begin{aligned} P(E) &= w(x_1) + w(x_2) + w(x_3) \\ &= .0323 + .0065 + .0121 \\ &= .0509. \end{aligned}$$

Table 4.2. PROBABILITY OF DEATH AT DIFFERENT AGES IN THE UNITED STATES.

Age (years)	Probability of death
0–10	.0323
10–20	.0065
20–30	.0121
30–40	.0184
40–50	.0431
50–60	.0969
60–70	.1821
70–80	.2728
80 and over	.3358
Total	1.0000

SOURCE: *United States Life Tables by Causes of Death: 1959–1961*, Vol. 1, No. 6, 1967, p. 15, National Health Center for Health Statistics, Public Health Service, Washington, D.C.

EXERCISES

Suggested minimum assignment: Exercises 1, 3, 6, 7, 9, 11, and 13.

In solving probability problems, the reader may find it helpful to start by identifying both the sample space U and the event space E. If the elements of these sets are not too numerous, they can be listed; otherwise, the set can be described, or a sufficient number of elements can be listed, to give the reader a clear idea of the spaces with which he is working.

1. What is the probability that a telephone number picked at random will have a last digit which is divisible by 3?
2. What is the probability of drawing a face card (include aces) from an ordinary deck of 52 cards?

In Exercises 3–6 suppose that a pair of ordinary dice are cast.

3. What is the probability that an odd sum is cast?
4. What is the probability that an even sum is cast?
5. What is the probability that a 6 or 7 or 8 is cast?
6. What is the probability that a 2 or 3 or 4 or 10 or 11 or 12 is cast?
7. In a family of 3 children, what is the probability that exactly 2 of them are boys? Assume equally likely outcomes and state sample and event spaces.
8. A pair of animals are inoculated against a certain disease. There is an equally likely chance that the inoculation will take or will not take. What is the probability that the inoculation will take on at least one animal? State the sample space and the event space.
9. Let $U = \{u_1, u_2, u_3, u_4\}$ be a sample space of 4 outcomes. Suppose that the weights of these outcomes are $w(u_1) = \frac{1}{4}$, $w(u_2) = \frac{1}{2}$, $w(u_3) = \frac{1}{6}$, and $w(u_4) = \frac{1}{12}$. What is the probability of $E_1 = \{u_1, u_3\}$?
10. In Exercise 9 suppose that $w(u_1) = \frac{1}{3}$, $w(u_2) = \frac{1}{4}$, and $w(u_3) = \frac{1}{12}$. What must $w(u_4)$ be? Then calculate the probability of $E = \{u_3, u_4\}$.
11. Suppose that a jar contains 6 white capsules, 3 yellow capsules, and 2 orange capsules. What is the probability of picking a nonwhite capsule? List the sample space, the event space, and the weights of the outcomes.
12. Suppose that the probability for rain is 10 per cent, the probability for sleet is 20 per cent, and the probability for snow is 30 per cent; what is the probability that it will rain or snow? Assume that precipitation is classified in exactly one of the categories at any given time. List the sample space, the event space, and weights of the outcomes. Are the outcomes equally likely?
13. A pair of regular tetrahedrons (four-sided objects) are thrown on a flat surface in the same manner that one would throw a pair of dice. The sides of the tetrahedrons are numbered from 1 to 4 (like dice). Let the sum of the numbers on the downward faces of the two objects constitute an outcome of an experiment. Find the probability that the sum on one throw is 5.

14. In Exercise 13 find the probability that the sum shown on one throw is 7.
15. Prove Theorem 1(*i*).
16. Prove Theorem 1(*ii*).
17. Prove Theorem 1(*iii*).
18. In Example 6 verify the probabilities found by Galileo.

4.2 Probabilities of Complements, Unions, and Intersections of Events

Probability problems are often such that the event whose probability is sought can be treated as a union, or an intersection, or a complement of other events.

Example 1
Consider the experiment of drawing a single card from an ordinary deck of 52 cards. Suppose that we are requested to find the probabilities of 3 events: (1) the card drawn is a king *or* a diamond, (2) the card drawn is a king *and* a diamond, and (3) the card drawn is *not* a diamond. Let E_d represent the set of all diamonds. Let E_k represent the set of all kings; let the universal set be the sample space of all 52 cards. The three indicated probabilities that we seek are (1) $P(E_k \cup E_d)$, (2) $P(E_k \cap E_d)$, and (3) $P(E_d')$. Thus we see that event (1) can be thought of as a union; event (2) can be thought of as an intersection; and event (3) can be thought of as a complement. ∎

We now develop theorems that will give us formulas for solving Example 1 and similar examples. In the last section we defined a specific probability function (Definition 2), and we shall now prove that it possesses the following property.

▶ **Theorem 2**†

Let E_1 and E_2 be events of a sample space U, which has a finite number of outcomes. If $E_1 \cap E_2 = \emptyset$, then $P(E_1 \cup E_2) = P(E_1) + P(E_2)$.

Proof: Let $E_1 = \{e_1, \ldots, e_j\}$ and let $E_2 = \{e_k, \ldots, e_m\}$.

Statement	Reason
1. $E_1 \cup E_2$ $= \{e_1, \ldots, e_j, e_k, \ldots, e_m\}$.	1. Definition of $\cup$ and $E_1 \cap E_2 = \emptyset$.
2. $P(E_1 \cup E_2)$ $= w(e_1) + \cdots + w(e_j) + w(e_k)$ $+ \cdots + w(e_m)$	2. Definition 2.
3. $= [w(e_1) + \cdots + w(e_j)]$ $+ [w(e_k) + \cdots + w(e_m)]$	3. Grouping terms.
4. $= P(E_1) + P(E_2)$.	4. Definition 2. ◀

† In an axiomatic approach to probability theory, Theorems 1 and 2 can be considered as axioms from which later theorems follow.

Example 2
Consider the same experiment that we did in Example 1, that of drawing a single card from an ordinary deck of 52 cards. Let E_d be the event of drawing a diamond, and let E_s be the event of drawing a spade. Obviously, these events are disjoint; no card is both a spade and a diamond. Thus the probability of drawing a spade or a diamond can be found by the formula of Theorem 2:
$$P(E_d \cup E_s) = P(E_d) + P(E_s).$$
Because there are 52 equally likely outcomes, 13 of which are diamonds and 13 of which are spades, then
$$P(E_d) + P(E_s) = \tfrac{13}{52} + \tfrac{13}{52} = \tfrac{1}{2}. \blacksquare$$

When two events are disjoint (that is, their intersection is the empty set) it is customary to say that the events are **mutually exclusive**. Events E_d and E_s of Example 2 are mutually exclusive.

A general formula for finding the probability of a union of two events (mutually exclusive or not) will be stated next.

▶ **Theorem 3**

Let E_1 and E_2 be events of a sample space U, which has a finite number of outcomes. Then
$$P(E_1 \cup E_2) = P(E_1) + P(E_2) - P(E_1 \cap E_2).$$

Proof: Let
$$E_1 = \{e_1, \ldots, e_j, e_k, \ldots, e_m\} \quad \text{and} \quad E_2 = \{e_k, \ldots, e_m, e_n, \ldots, e_p\},$$
where $E_1 \cap E_2 = \{e_k, \ldots, e_m\}$.

Statement	Reason
1. $E_1 \cup E_2$ $= \{e_1, \ldots, e_j, e_k, \ldots, e_m, e_n, \ldots, e_p\}.$	1. Definition of the union of two sets.
2. $P(E_1 \cup E_2)$ $= w(e_1) + \cdots + w(e_j)$ $+ w(e_k) + \cdots + w(e_m)$ $+ w(e_n) + \cdots + w(e_p)$	2. Definition 2.
3. $= [w(e_1) + \cdots + w(e_j)$ $+ w(e_k) + \cdots + w(e_m)]$ $+ [w(e_k) + \cdots + w(e_m)$ $+ w(e_n) + \cdots + w(e_p)]$ $- [w(e_k) + \cdots + w(e_m)]$	3. Add and subtract $[w(e_k) + \cdots + w(e_m)]$ and regroup terms.
4. $= P(E_1) + P(E_2) - P(E_1 \cap E_2).$	4. Definition 2. ◀

Example 3
Again returning to Example 1 with its 52 equally likely outcomes, 13 diamonds, 4 kings, and one card that is both a king and a diamond, we have
$$P(E_d) = \tfrac{13}{52}, \quad P(E_k) = \tfrac{4}{52}, \quad \text{and} \quad P(E_d \cap E_k) = \tfrac{1}{52}.$$

Sec. 4.2] Complements, Unions, and Intersections of Events

Therefore, the probability of selecting a card that is a king *or* a diamond is given by the formula of Theorem 3.

$$P(E_d \cup E_k) = P(E_d) + P(E_k) - P(E_d \cap E_k)$$
$$= \tfrac{13}{52} + \tfrac{4}{52} - \tfrac{1}{52}$$
$$= \tfrac{4}{13}. \blacksquare$$

Notice that in the formula of Theorem 3 we need to know how to find $P(E_1 \cap E_2)$. Formulas for $P(E_1 \cap E_2)$ (which follow immediately from Definitions 1 and 2 and Theorem 1) are stated formally for future reference.

▶ **Theorem 4**

Let E_1 and E_2 be events (mutually exclusive or not) of a sample space U, which has a finite number of outcomes.

(i) If $E_1 \cap E_2 = \{e_1, \cdots, e_k\}$ with weights $w(e_1), \cdots, w(e_k)$, then

$$P(E_1 \cap E_2) = w(e_1) + \cdots + w(e_k).$$

If $E_1 \cap E_2 = \emptyset$, then $P(E_1 \cap E_2) = 0$.

(ii) If U consists only of equally likely outcomes, then

$$P(E_1 \cap E_2) = \frac{n(E_1 \cap E_2)}{n(U)}. \quad \blacktriangleleft$$

Additional formulas for $P(E_1 \cap E_2)$ will be given in Section 4.5.

Example 4

Consider the sample space of outcomes consisting of the sums of two dice when cast. The sample space together with their respective weights are given in Table 4.3.

Table 4.3. Weights for the outcomes of a pair of dice.

u	2	3	4	5	6	7	8	9	10	11	12
$w(u)$	$\tfrac{1}{36}$	$\tfrac{2}{36}$	$\tfrac{3}{36}$	$\tfrac{4}{36}$	$\tfrac{5}{36}$	$\tfrac{6}{36}$	$\tfrac{5}{36}$	$\tfrac{4}{36}$	$\tfrac{3}{36}$	$\tfrac{2}{36}$	$\tfrac{1}{36}$

Let E_1 be the set of all sums less than 7; that is,

$$E_1 = \{2, 3, 4, 5, 6\}.$$

Let E_2 be the set of even sums; that is,

$$E_2 = \{2, 4, 6, 8, 10, 12\}.$$

Suppose that one wants the probability of the intersection

$$E_1 \cap E_2 = \{2, 4, 6\}.$$

By Theorem 4(i)

$$P(E_1 \cap E_2) = w(2) + w(4) + w(6)$$
$$= \tfrac{1}{36} + \tfrac{3}{36} + \tfrac{5}{36}$$
$$= \tfrac{9}{36} = 0.25. \blacksquare$$

Finally, we consider the probability of the complement of an event.

▶ **Theorem 5**

Let E be an event of a sample space U, which has a finite number of outcomes, and let E' represent the complement of E. Then

$$P(E') = 1 - P(E).$$

Proof

Statement	Reason
1. $P(E \cup E') = P(U)$.	1. Given that $E \cup E' = U$.
2. $P(E) + P(E') = P(U)$.	2. Theorem 2.
3. $P(E) + P(E') = 1$.	3. Theorem 1.
4. $P(E') = 1 - P(E)$.	4. Subtracting $P(E)$ from both sides. ◀

Example 5

Again, from Example 1, an ordinary deck of 52 cards has 13 diamonds, and if E_d represents the event of selecting a diamond, then $P(E_d) = \tfrac{13}{52}$. Hence the event of *not* selecting a diamond is

$$P(E'_d) = 1 - P(E_d)$$
$$= 1 - \tfrac{13}{52}$$
$$= \tfrac{39}{52} = .75. \blacksquare$$

EXERCISES

Suggested minimum assignment: Exercises 1, 4, 5, 7, and 11.

1. Consider the experiment in which one die is cast. Let E_1 be the event of casting an even number. Let E_2 be the event of casting a number greater than 2.
    (a) List the set of outcomes (sample space).
    (b) List the elements of E_1, E_2, $E_1 \cap E_2$, $E_1 \cup E_2$, and E'_2.
    (c) Calculate $P(E_1)$ and $P(E_2)$.
    (d) Calculate $P(E_1 \cap E_2)$ by Theorem 4.
    (e) Calculate $P(E_1 \cup E_2)$ by Theorem 3.
    (f) Calculate $P(E'_2)$ by Theorem 5.

2. Consider the experiment of flipping a coin 3 times. Let E_1 be the event of achieving a head on the first flip, and let E_2 be the event of achieving exactly two heads.
   (a) List the set of outcomes (sample space).
   (b) List the elements of E_1, E_2, $E_1 \cap E_2$, $E_1 \cup E_2$, and E_2'.
   (c) Calculate $P(E_1)$ and $P(E_2)$.
   (d) Calculate $P(E_1 \cap E_2)$ by Theorem 4.
   (e) Calculate $P(E_1 \cup E_2)$ by Theorem 3.
   (f) Calculate $P(E_2')$ by Theorem 5.
3. A pair of newlyweds plan a family of 3 children. Let E_1 be the event that the first child is a boy, and let E_2 by the event that all 3 children are the same sex. Assume that in a single birth a boy and a girl are equally likely.
   (a) List the set of outcomes (sample space).
   (b) List the elements of E_1, E_2, $E_1 \cap E_2$, $E_1 \cup E_2$, and E_2'.
   (c) Calculate $P(E_1)$ and $P(E_2)$.
   (d) Calculate $P(E_1 \cap E_2)$ by Theorem 4.
   (e) Calculate $P(E_1 \cup E_2)$ by Theorem 3.
   (f) Calculate $P(E_2')$ by Theorem 5.
4. Consider a deck of cards in which all the cards except the kings, queens, and jacks have been removed. Let E_1 be the event of drawing a queen from the deck. Let E_2 be the event of drawing a diamond or a heart from the deck.
   (a) List the set of outcomes (sample space).
   (b) List the elements of E_1, E_2, $E_1 \cap E_2$, $E_1 \cup E_2$, and E_2'.
   (c) Calculate $P(E_1)$ and $P(E_2)$.
   (d) Calculate $P(E_1 \cap E_2)$ by Theorem 4.
   (e) Calculate $P(E_1 \cup E_2)$ by Theorem 3.
   (f) Calculate $P(E_2')$ by Theorem 5.
5. Let the outcomes of an experiment be called O_1, O_2, and O_3, where $w(O_1) = \frac{1}{4}$, $w(O_2) = \frac{1}{3}$, and $w(O_3) = \frac{5}{12}$. Let E_1 consist of outcomes O_1 and O_2, and let E_2 consist of outcomes O_2 and O_3.
   (a) List the set of outcomes (sample space).
   (b) List the elements of E_1, E_2, $E_1 \cap E_2$, $E_1 \cup E_2$, and E_2'.
   (c) Calculate $P(E_1)$ and $P(E_2)$.
   (d) Calculate $P(E_1 \cap E_2)$.
   (e) Calculate $P(E_1 \cup E_2)$.
   (f) Calculate $P(E_2')$.
6. Let the outcomes of an experiment be called O_1, O_2, O_3, and O_4, where $w(O_1) = \frac{1}{4}$, $w(O_2) = \frac{1}{2}$, $w(O_3) = \frac{1}{6}$, and $w(O_4) = \frac{1}{12}$. Let E_1 consist of outcomes O_1 and O_2, and let E_2 consist of outcomes O_3 and O_4.
   (a) List the set of outcomes (sample space).
   (b) List the elements of E_1, E_2, $E_1 \cap E_2$, $E_1 \cup E_2$, and E_2'.
   (c) Calculate $P(E_1)$ and $P(E_2)$.
   (d) Calculate $P(E_1 \cap E_2)$.

(e) Calculate $P(E_1 \cup E_2)$.
(f) Calculate $P(E_2')$.

7. Let U be the sample space of outcomes consisting of the sum shown when two dice are cast. The nonequally likely outcomes are 2, 3, 4, 5, 6, 7, 8, 9, 10, 11, 12, and the weights (or probabilities of occurrence) of these outcomes are given in Table 4.3 (see p. 152). Let E_1 be the event of achieving an even sum, and let E_2 be the event of achieving a sum of 7 or 11. Calculate $P(E_1 \cap E_2)$, $P(E_1 \cup E_2)$, and $P(E_2')$.

8. Suppose that an enclosure contains 4 types of animals. There is 1 animal that has only disease A; there are 3 animals that have only disease B; and there are 4 animals that have only disease C. The enclosure also contains 2 animals that have all 3 diseases. Let E_1 be the event of selecting an animal with disease A or B (this includes those animals having all 3 diseases), and let E_2 be the event of selecting an animal that has disease C. Calculate $P(E_1 \cap E_2)$, $P(E_1 \cup E_2)$, and $P(E_2')$.

9. A certain automobile manufacturer has information that leads the executives to believe that in a certain model, 10 per cent of the autos have a defective starter, 5 per cent of the autos have a defective part in the transmission, and 2 per cent have both defective parts.
   (a) If the executives order a recall of the models, what is the probability that a specific auto selected at random will be free of any defect (under the assumption that the information listed above is accurate)?
   (b) What is the probability that the starter will be defective and the transmission part will be nondefective?

10. In the Senate of a certain state there are 50 members; 30 of these members are Democrats and 20 are Republicans. A certain survey determines that 28 of the Senators favor a certain issue; of these 28, there are 15 that are Democrats.
    (a) Find the probability that a Senator chosen at random will be a Democrat or will favor the issue.
    (b) Find the probability that a Senator chosen at random will be a Republican who does not favor the issue.

11. Let A, B, and C be events of a sample space U, and suppose that

    $n(A) = 10$, $n(B) = 20$, $n(C) = 15$, $n(A \cap B) = 7$,
    $n(B \cap C) = 10$, $n(A \cap C) = 5$, $n(A \cap B \cap C) = 2$,
    $n(U) = 100$.

    Moreover, assume that all outcomes of the sample space are equally likely.
    (a) Draw a Venn diagram showing the number of elements in each of the various identifiable disjoint subsets of U.
    (b) Find the probability of $(A \cup B)'$.
    (c) Find $P(A \cup B \cup C)$.
    (d) What is the probability that an element of U, selected at random, will belong to the set $(A \cap C) \cup (B' \cap C)$?

12. A pair of dice is thrown.
    (a) What is the probability that an even sum is thrown or that a sum larger than seven is thrown or that a sum less than four is thrown?
    (b) What is the probability that an even sum and a sum larger than seven is *not* thrown?

4.3 Expected Value

Before we make a formal definition of **expected value,** we present an intuitive idea of the concept by means of several examples.

Example 1

One of the simplest gambling games is that of roulette. A wheel with 37 slots (some have 38) is spun, and a ball comes to rest in one of the slots numbered 0 through 36. The player may bet $1 on one or more numbers; if he wins on one number, he will be paid $35, and he will get his $1 stake back for the winning number.

If a player bet $1 on some number, then he has a $\frac{1}{37}$ probability of winning $35 and a $\frac{36}{37}$ probability of losing $1. If we designate the $1 loss by $(-\$1)$, we can calculate his expected winnings per game to be

$$\frac{1}{37}(\$35) + \frac{36}{37}(-\$1) = \$\frac{-1}{37}.$$

In other words, in the long run, each player can expect to *lose* at the rate of $\$\frac{1}{37}$ each time he bets $1. In *one* game, however, it is impossible to lose $\$\frac{1}{37}$; rather, the player will win $35 or lose $1. In the long run the owner of the game can expect to *win* at the rate of $\$\frac{1}{37}$ from each player on every spin. ∎

Example 2

A highway engineer knows from experience that his men and equipment can resurface a highway at the rate of 3 miles per day in rainy weather and 5 miles per day in nonrainy weather. The contractor has agreed to take on a large job in an area in which the Weather Bureau predicts a 20 per cent probability of rainfall for that time of year. Based on this information, the engineer can estimate the number of days that it will take him to do the job by calculating the expected mileage per day. There is a $\frac{1}{5}$ probability that in one day he will lay 3 miles of surface and a $\frac{4}{5}$ probability that in one day he will lay 5 miles of surface. The expected amount of surface to be laid per day is

$$\tfrac{1}{5}(3 \text{ miles}) + \tfrac{4}{5}(5 \text{ miles}) = \tfrac{23}{5} \text{ miles.}$$

It should be emphasized that $\frac{23}{5}$ miles per day is a long run expectation and not a literal expectation for one day. ∎

Betting $1 on a number of the roulette wheel in Example 1 may be viewed as an experiment with two possible outcomes, namely, that of

winning $35 and losing $1. These two outcomes had respective weights of $\frac{1}{37}$ and $\frac{36}{37}$. In Example 2 we could think of laying highway surface as an experiment with the two outcomes of 3 miles and 5 miles; the two outcomes had respective weights of $\frac{1}{5}$ and $\frac{4}{5}$. Notice that each of the outcomes was represented by a real number; in Example 1 the real number measured monetary value, in Example 2 the real number measured the mileage of road surface laid per day. In each case we have multiplied the real number associated with each outcome by the probability of obtaining the outcome and then added the products. We now formalize this concept with the following definition.

▶ **Definition 3**
If the outcomes of an experiment are represented by the real numbers $a_1, a_2, \ldots, a_n$, and if the probabilities of obtaining the respective outcomes are $p_1, p_2, \ldots, p_n$, then the **expected value** V for the experiment is

$$V = p_1 a_1 + p_2 a_2 + \cdots + p_n a_n.$$

Example 3
A hotel chain is considering the construction of a resort hotel at a certain beach. On clear sunny days in the summer season, the hotel can expect to fill its anticipated 200-room facility. On rainy days during the summer season, the experience of existing hotels shows that a 50 per cent capacity can be expected, while on threatening days a 70 per cent capacity can be expected. The Weather Bureau records indicate that during the summer season at that location, there is a 60 per cent probability of clear sunny weather, a 10 per cent probability of rainy days, and a 30 per cent probability of threatening days. Thus there is a $\frac{6}{10}$ probability of renting 200 rooms, there is a $\frac{1}{10}$ probability of renting 100 rooms, and there is a $\frac{3}{10}$ probability of renting 140 rooms. By Definition 3, the expected value for the experiment is

$$V = \tfrac{6}{10} (200 \text{ rooms}) + \tfrac{1}{10} (100 \text{ rooms}) + \tfrac{3}{10} (140 \text{ rooms})$$
$$= 172 \text{ rooms}.$$

As far as weather conditions are concerned, the hotel chain can expect in the long run to rent at the rate of 172 rooms per night during the summer season. ■

Example 4
One of the most common ways of separating money from the uninformed (and in many cases the informed) is the "numbers game." There are many variations, but one version is that of choosing a 3-digit number, which, with $1 is delivered to the nearest "numbers runner." According to some previously announced method, a 3-digit number is selected by the operators of the game, and payments of some amount, say $800, are awarded to the winners. From a player's point of view the outcomes of this experiment are ($799) and ($-1). The respective probabilities of these outcomes are $\frac{1}{1000}$

and $\frac{999}{1000}$ (assuming that numbers like 001 or 062 are 3-digit numbers). Therefore, the expected value for the experiment to a player is

$$V = \frac{1}{1000}(\$799) + \frac{999}{1000}(-\$1) = \$\frac{-1}{5}.$$

In other words, in the long run the player can expect to lose at the rate of $.20 per game. Of course, in any *one* game he will *not* lose exactly $.20. ∎

The real numbers $a_1, a_2, \ldots, a_n$ of Definition 3 are often generated by functions called **random variables** or **measurements**; such functions were discussed in Example 6 of Section 3.7. The reader is encouraged to read that example.

For an experiment with n equally likely real number outcomes $a_1, a_2, \ldots, a_n$, the expected value is

$$V = \frac{1}{n}a_1 + \frac{1}{n}a_2 + \cdots + \frac{1}{n}a_n$$
$$= \frac{a_1 + a_2 + \cdots + a_n}{n},$$

which is simply the *average* of the real numbers $a_1, a_2, \ldots, a_n$.

The reader has probably heard the expression "the odds in *favor* of a victory are 3 to 2." The statement simply means that the probability of winning is $\frac{3}{3+2}$ while the probability of not winning is $\frac{2}{3+2}$.

▶ **Definition 4**

If the probability of event E is $\frac{a}{a+b}$, then the **odds in favor of E** are a to b and the **odds against E** are b to a.

Example 5

In horse racing and dog racing one often hears the expression "the odds on a horse are 6 to 1." This means the odds are 6 to 1 against his winning; that is, there is a $\frac{6}{7}$ probability that he will *not* win. ∎

Example 6

Suppose that the final odds on a horse as determined by the bettors are 2 to 1 and that the racetrack operators are authorized by state rules to earn $.10 per $2 bet. How much will they pay if the horse wins? (Show and Place are not considered here.)

Solution: Let P represent the amount the operators will pay per $2 ticket if the specified horse wins. The outcomes for a bettor are $-\$2$ and $+\$(P-2)$, with respective probabilities of $\frac{2}{3}$ and $\frac{1}{3}$; the expected value to the bettor is $-\$.10$; hence we have

$$\tfrac{2}{3}(-2) + \tfrac{1}{3}(P-2) = -.10,$$

$$-\tfrac{4}{3} + \tfrac{P}{3} - \tfrac{2}{3} = -.10,$$

$$\tfrac{P}{3} = -\tfrac{1}{10} + \tfrac{6}{3},$$

$$P = -\tfrac{3}{10} + 6$$

$$= \$5.70. \blacksquare$$

APPLICATIONS

Example 7

The concept of expected value can be very useful in certain types of decision making. For example, suppose that a group of actors is considering two different locations for their summer production. In one city there is an outdoor theater that will seat 500 people and an indoor theater that will serve 100 people. In the second city there is an outdoor theater that will serve 450 people and a 200-seat indoor theater. The probabilities of bad weather (sufficient to require an indoor performance) in the respective cities are 30 and 20 per cent. To which location should the drama group go for the summer if the expected audience size is the sole criteria for the decision? (Assume that the group will perform to a capacity audience each night.) A performance at each of the cities can be considered as an experiment. In the first experiment there is a $\tfrac{7}{10}$ probability of achieving an outcome of 500 people and a $\tfrac{3}{10}$ probability of achieving an outcome of 100 people; hence the expected value for the experiment is

$$V_1 = \tfrac{7}{10}(500 \text{ people}) + \tfrac{3}{10}(100 \text{ people})$$
$$= 380 \text{ people}.$$

In the second experiment there is a $\tfrac{8}{10}$ probability of achieving an outcome of 450 people and a $\tfrac{2}{10}$ probability of achieving an outcome of 200 people; hence the expected value for the experiment is

$$V_2 = \tfrac{8}{10}(450 \text{ people}) + \tfrac{2}{10}(200 \text{ people})$$
$$= 400 \text{ people}.$$

Thus on the basis of the calculated expected values of audience size, the drama group should select the second city. $\blacksquare$

Example 8

A fruit broker must decide how many trucks of watermelons to order for the summer season. Based on past seasons and other considerations, he estimates the probabilities of selling various numbers of truckloads as shown in Table 4.4. Each truckload costs $3000 and he can make a profit of $4000 on each truckload that he can sell. Moreover, we assume that if potential sales

Table 4.4

Number of truckloads	1	2	3	4
Probability of selling	$\frac{1}{6}$	$\frac{1}{4}$	$\frac{1}{3}$	$\frac{1}{4}$

exceed the amount he stocked, then the difference represents a loss for the broker. Because watermelons are perishable, he also loses whatever he overstocks. For the sake of simplicity we do not consider the sale of fractions of truckloads. Table 4.5 provides the net profit for various numbers of truckloads that he might buy and sell. (Each entry of the table is calculated by the formula [entry] = [sales profit] − [loss from overstocking or understocking].)

Table 4.5

Potential purchases (truckloads)	Potential sales (truckloads)			
	1	2	3	4
1	4,000	0	−4,000	−8,000
2	1,000	8,000	4,000	0
3	−2,000	5,000	12,000	8,000
4	−5,000	2,000	9,000	16,000

From Tables 4.4 and 4.5 we calculate the expected value for each *purchase* possibility. We shall call the respective expected values E_1, E_2, E_3, and E_4.

$E_1 = \frac{1}{6}(4{,}000) + \frac{1}{4}(0) + \frac{1}{3}(-4{,}000) + \frac{1}{4}(-8{,}000) = -2{,}667,$

$E_2 = \frac{1}{6}(1{,}000) + \frac{1}{4}(8{,}000) + \frac{1}{3}(4{,}000) + \frac{1}{4}(0) = 3{,}500,$

$E_3 = \frac{1}{6}(-2{,}000) + \frac{1}{4}(5{,}000) + \frac{1}{3}(12{,}000) + \frac{1}{4}(8{,}000) = 6{,}917,$

$E_4 = \frac{1}{6}(-5{,}000) + \frac{1}{4}(2{,}000) + \frac{1}{3}(9{,}000) + \frac{1}{4}(16{,}000) = 6{,}667.$

Based on the assumptions made, the broker has a greatest expected value if he orders 3 truckloads of watermelons. If no information can be determined about the probabilities of sales, the problem can be approached through the use of game theory, which is to be studied in Chapter 7. ∎

Example 9

A very important application of the concept of expected value occurs in the study of game theory. Game theory is a relatively new branch of mathematics that permits the construction of models of adversary relationships; consequently, game theory has considerable potential in the social and managerial sciences and will be introduced in Chapter 7. ∎

EXERCISES

Suggested minimum assignment: Exercises 1, 3, 5, 7, 9, 10, 12, and 13.

1. In Example 1 what is the expected value for a player making a $1 bet on a number of a roulette wheel with 38 slots numbered 00, 0, 1,

2, ..., 36? The "house" will pay $35 for a winner plus the return of the $1 stake.

2. On a roulette wheel as described in Example 1 there are 18 black slots, 18 red slots, and 1 green slot (the number 0). A player may bet on the colors black or red rather than a number, and the "house" will pay $1 for a winner plus a return of the $1 stake. What is the expected value for the experiment of betting $1 on "red"?

3. A game is devised in which a single die is cast and a person agrees to pay in dollars the amount shown on the die if the number is odd and to collect the amount shown on the die if the number is even. What is the expected value of the game to the person?

4. A game is devised in which a single die is cast and a person agrees to collect $10 if a 1 or a 2 is cast and to pay the amount (in dollars) that is equal to the number shown on the die otherwise. What is the expected value of the game to the person?

5. In Exercise 3 suppose that the person has to pay $1 to play the game. What is the expected value of the game then?

6. In Exercise 4 suppose that the person has to pay $1 to play the game. What is the expected value of the game then?

7. A common money-raising device is that of the lottery. Suppose that 10,000 $1 tickets are sold in a lottery that promises a first prize of $1000 and 10 second prizes of $500 apiece. What is the expected value of 1 ticket? What are the odds against winning a prize?

8. A merchant promises a free chance on an automobile to every person who purchases a special television model from him. If the television set cost the consumer $40 more than the same set sold by a competitor, if the automobile would cost the consumer $3000 (although it may not cost the merchant that much), and if the merchant sells 70 sets, what is the expected value of a purchase to the consumer? What are the odds against winning?

9. A certain part takes 2 machine-hours to manufacture. All parts are tested and the defective parts are returned and repaired at an expense of 1 additional machine-hour per part. If there is a 20 per cent probability of manufacturing a defective part initially, what is the expected machine-hour expenditure per part?

10. A certain manufacturer produces a product that is graded according to the following categories: nondefective, defective but usable, and defective unusable; the respective values to the manufacturer are $30, $20, and $0 per item. What is the expected value of each item to the manufacturer if experience has shown that the probabilities of attaining items in these categories are $\frac{7}{10}$, $\frac{2}{10}$, and $\frac{1}{10}$, respectively?

11. Suppose that a person enters a game in which a pair of dice are thrown and a prize of $4 is paid if the numbers showing on the dice total 7. The game costs $1 to play. Find the odds against his winning and the expected value of the game for the player.

12. Three fair dimes are tossed simultaneously and a prize of $4 is paid if

all three faces turn up to be the same; it costs $1 per game to play. Find the odds against winning and the expected value of the game.

13. On the day before election eve a candidate is faced with the decision of whether to go to city *A* or city *B* for an election-eve rally which is planned in each city. The candidate's staff has decided to base the decision on the expected value of the number of people the candidate will address. In city *A* the rally has been scheduled for the 2000-seat city coliseum in case of rain and in the 5000-seat city stadium otherwise. In city *B* the rally has been scheduled for a 1000-seat auditorium in case of rain and in the 10,000-seat baseball park otherwise. If there is a 20 per cent probability of rain in city *A* and a 40 per cent probability of rain in city *B*, and assuming that the candidate's appearance will guarantee a capacity crowd in any case, to which city should the candidate go?

14. Suppose that on the eve of a political rally the campaign manager of a certain candidate is faced with a decision as to whether to hold a rally for his candidate in the local outdoor stadium or the indoor auditorium. If he holds the rally in the outdoor stadium and it rains, he can expect a very small crowd of approximately 100 people, but if it does not rain he expects a crowd of 3000 people. If, however, he holds the rally indoors, he can expect the maximum capacity of 1000 people regardless of whether it rains or not. The latest weather report gives a 40 per cent probability for rain at the time of the rally; at which location should the manager schedule the rally if he bases his decision only on the expected value of the audience size?

15. The final odds on a certain racehorse are 5 to 2. If the racetrack operators are authorized to earn $.20 on each $2 bet, how much will they pay for each $2 bet if the horse wins? (Do not consider Show and Place in this exercise.)

4.4 Conditional Probability

In this section we shall be concerned with the probability of an event if it is known that another event has occurred. In other words, we shall revise our probability measure as additional information becomes known.

Example 1
Again consider the experiment of flipping a coin 3 times in succession. The sample space of *equally likely outcomes* is

$$U = \{\text{HHH, HHT, HTH, THH, HTT, THT, TTH, TTT}\}.$$

In Example 2 of Section 4.1 we found that the probability of obtaining exactly 2 heads was $\frac{3}{8}$. Recall that

$$E = \{\text{HHT, HTH, THH}\}.$$

Suppose that we now are given the condition that the first flip produces a

head. Given this condition, now what is the probability of obtaining exactly 2 heads?

Solution: We shall answer this question by posing a brand-new problem. If we consider the fact that a head has been thrown the first time, then the reduced sample space is

$$F = \{HHH, HHT, HTH, HTT\},$$

and the reduced event space is

$$E \cap F = \{HHT, HTH\};$$

hence the probability that we seek is

$$\frac{n(E \cap F)}{n(F)} = \frac{2}{4} = \frac{1}{2}. \qquad (4.1)$$

This new probability, subject to the condition of flipping a head on the first try, is called a **conditional probability**; notice that it differs from the probability of $\frac{3}{8}$ that was calculated without the condition. Since both numerator and denominator of (4.1) can be divided by $n(U)$, the conditional probability of E, given F, can be written as

$$\frac{n(E \cap F)/n(U)}{n(F)/n(U)}$$

or

$$\frac{P(E \cap F)}{P(F)}.$$

Visual assistance in understanding conditional probability may be obtained from the **tree diagram** of Figure 4.7; each complete sequence (from left to right) of connecting line segments in the tree diagram is called a **path** and represents an outcome of the experiment. In Figure 4.7 there are 3 of

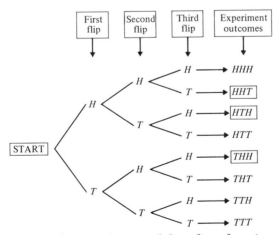

Figure 4.7. *Tree diagram of three flips of a coin.*

Sec. 4.4] Conditional Probability

8 equiprobable paths of the tree that produce 2 H's, hence the probability of $\frac{3}{8}$. However, once we are given that the first flip produces a head, then the tree diagram narrows to only the upper half of the tree; there we see that there are 2 of 4 equiprobable paths that produce 2 H's, hence the probability of $\frac{2}{4}$. ∎

Although the last example illustrated a sample space of equally likely outcomes, the following definition is made for sample spaces with or without equally likely outcomes.

▶ **Definition 5**
If E_1 and E_2 are two events of a finite sample space and if $P(E_1) \neq 0$, then the **conditional probability** of E_2, given E_1, is designated $P(E_2|E_1)$ and defined by

$$P(E_2|E_1) = \frac{P(E_2 \cap E_1)}{P(E_1)}.$$

Example 2
A biologist expects to obtain 2 offspring from a pair of animals. What is the probability that both will be male? Is this probability changed after the first born is a male? The sample space is

$$U = \{mm, mf, fm, ff\}.$$

Suppose that these outcomes are not equally likely and it is known that the weights of the respective outcomes are

$$w(mm) = .36, \quad w(mf) = .24, \quad w(fm) = .24, \quad w(ff) = .16.$$

Solution: To answer the first question, we observe that before any birth the probability that both offspring are male is $P(\{mm\}) = w(mm) = .36$. To answer the second question, we use the formula of Definition 5. Let E_1 be the event of the first born being male; then

$$E_1 = \{mm, mf\}.$$

Let E_2 be the event of both being male; then

$$E_2 = \{mm\}.$$

Therefore,

$$E_2 \cap E_1 = \{mm\}.$$

To use the formula of Definition 5, we need

$$P(E_1) = w(mm) + w(mf) = .36 + .24 = .6,$$

and

$$P(E_2 \cap E_1) = w(mm) = .36.$$

Hence, by Definition 5,

$$P(E_2|E_1) = \frac{P(E_2 \cap E_1)}{P(E_1)} = \frac{.36}{.6} = .6.$$

Again, the formula just used can be illustrated by means of a tree diagram as shown in Figure 4.8. Before any births, the desired probability was the weight associated with the desired path relative to all 4 paths of the tree, but after knowing that the first birth was male, we simply consider only the weight associated with the desired path relative to the top 2 paths. In other words, the weight of the desired outcome is divided by the sum of the weights of the outcomes that are still possible after the condition is known. ∎

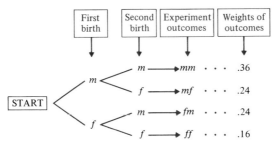

Figure 4.8. *Tree diagram of two births.*

Example 3

Suppose that 55 per cent of the voters of a certain precinct voted for the Republican candidate in the last election, and suppose that 20 per cent of the voters voted for the Republican candidate *and* for the local bond issue. What is the probability that a voter who is selected at random voted for the bond issue if he tells us that he voted for the Republican candidate?

Solution: Let U be the set of all voters in the precinct; let R be the subset of U who voted Republican; let B be the subset of U who voted for the bond issue. We are given that $P(R) = .55$ and that $P(B \cap R) = .20$; therefore,

$$P(B|R) = \frac{P(B \cap R)}{P(R)} = \frac{.20}{.55} \approx .36.$$

Again we seek a visualization of conditional probability by means of a tree diagram; in Figure 4.9 the symbol $\sim r$ indicates that the voter did not vote Republican and the symbol $\sim b$ indicates that the voter did not vote for the bond issue. This time we shall view the tree diagram (Figure 4.9) in two ways. First, as we have done in the preceding two examples, we view the given condition that the voter votes Republican as simply narrowing the sample space to the upper half of the tree and asking what new weight should be assigned to rb with respect to the new sample space $\{rb, r(\sim b)\}$.

Sec. 4.4] Conditional Probability

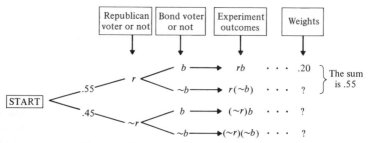

Figure 4.9. *Tree diagram of voter action.*

That is, to find $P(B|R)$ we must calculate the ratio

$$\frac{w(rb)}{w(rb) + w[r(\sim b)]} = \frac{P(R \cap B)}{P(R)} = \frac{.20}{.55}.$$

Second, we view the tree diagram as a systematic way of writing the experiment outcomes rb, $r(\sim b)$, $(\sim r)b$, $(\sim r)(\sim b)$. We are given that $w(rb) = P(R \cap B) = .20$, and this number is associated with the path consisting of the two line segments "start to r" and "r to b." We are also given that $P(R) = .55$, and this number is associated with the line segment "start to r." The unknown number $P(B|R)$ is then associated with the line segment "r to b." The probability $P(R \cap B)$ [or weight $w(rb)$] associated with the *path* is simply the product of the probabilities associated with the respective line segments that make up that path. Therefore,

$$(.55)(P(B|R)) = .20$$

and

$$P(B|R) = \frac{.20}{.55}. \blacksquare$$

In Exercise 21 the reader will be asked to prove, using Definitions 1 and 5, that if E_1 and E_2 are events of a *finite sample space of equally likely outcomes* and if $n(E_1) \neq 0$, then

$$P(E_2|E_1) = \frac{n(E_2 \cap E_1)}{n(E_1)}.$$

APPLICATIONS

Example 4

The probability of death of U.S. citizens at different ages was given in Table 4.2 of Example 7 of Section 4.1. From that table, the probability of death during the fourth decade was listed as .0184. Suppose, however, that the probability of dying during the fourth decade is recalculated under the condition that the person survives the first three decades. Let A be the event that the person survives the first three decades. From Example 7 of Section 4.1, we found that the probability of dying during the first three decades was

.0509; therefore, the probability of living three decades is $P(A) = 1 - .0509$ or $P(A) = .9491$. Let B be the event that the person dies during the fourth decade. The conditional probability of death during the fourth decade, given that death does not occur during the first three decades, is

$$P(B|A) = \frac{P(B \cap A)}{P(A)}$$
$$= \frac{P(\{x_4\})}{P(A)}$$
$$= \frac{.0184}{.9491}$$
$$\approx .0194.$$

Conditional probability can be useful in calculating premiums for life insurance policies; for example, suppose a 30-year-old person wanted to take out a $10,000 term life insurance policy for 10 years and wanted to pay the premium in one lump sum at the outset. Excluding overhead costs, commissions, and so on, the premium, based on the information of this problem, would be

$$(.0194)(\$10,000) = \$194.$$

In reality, methods of calculating insurance premiums are more sophisticated, but this example does illustrate the basic idea. ∎

Example 5
A variety of applied problems can be solved in a manner illustrated by this example. Suppose that a company manufactures a product that includes a timer. The timer is purchased from several different subcontractors, including subcontractor A. The service department estimates that 5 per cent of all products have defective timers. They also estimate that, of the products serviced by them, 20 per cent of the defective timers and 60 per cent of the nondefective timers are purchased from A. What is the probability that a timer purchased from A is defective?

Solution: Let

A = event that a timer was purchased from A,
D = event of a defective timer.

We have been given

$$P(D) = .05, \quad P(A|D) = .20, \quad \text{and} \quad P(A|D') = .60.$$

We wish to find $P(D|A)$:

$$P(D|A) = \frac{P(D \cap A)}{P(A)}.$$

Since $D \cap D' = \emptyset$ and $D \cup D' = U$,

$$P(D|A) = \frac{P(D \cap A)}{P(A \cap D) + P(A \cap D')}$$

and

$$P(D|A) = \frac{P(A|D)P(D)}{P(A|D)P(D) + P(A|D')P(D')}$$

$$= \frac{(.20)(.05)}{(.20)(.05) + (.60)(.95)}$$

$$\approx .017.$$

A visualization of the formula just derived can be obtained from a tree diagram in which D and D' are listed first and A and A' are listed second (Exercise 22). To illustrate how the method can be applied in other contexts, let

A = event that a person passed a certain test,
D = event that a person graduates from college.

The question would be to find the probability that a person would graduate from college if he passed the test. In psychology or medicine the model could be used to find the probability that a person actually had some physical or mental disorder, given that a certain test registered negative. Further discussion of the model presented in this example may be found in other books under the title of Bayes' Rule or Bayes' Theorem (for example, Freund [22], pp. 480–484). ∎

EXERCISES

Suggested minimum assignment: Exercises 1, 3, 5, 9, 11, 12, and 14.

1. Let the experiment consist of choosing a card from a deck in which all cards except the jacks, queens, and kings have been removed. Let E_1 be the event of choosing a king, and let E_2 be the event of choosing a diamond.
   (a) List the set of outcomes (sample space).
   (b) List the elements of E_1, E_2, and $(E_1 \cap E_2)$.
   (c) Find $P(E_1)$, $P(E_2)$, and $P(E_1 \cap E_2)$.
   (d) Find $P(E_1|E_2)$ and $P(E_2|E_1)$.
2. Let the experiment consist of casting one ordinary die. Let E_1 be the event of obtaining an odd number, and let E_2 be the event of obtaining a number less than 5.
   (a) List the set of outcomes (sample space).
   (b) List the elements of E_1, E_2, and $(E_1 \cap E_2)$.
   (c) Find $P(E_1)$, $P(E_2)$, and $P(E_1 \cap E_2)$.
   (d) Find $P(E_1|E_2)$ and $P(E_2|E_1)$.
3. In a certain corporation the wages earned by union men vary directly with seniority. Suppose that 5 per cent of the union men have more

than 10 years seniority and make more than $4 per hour. If 10 per cent of the union men make more than $4 per hour, what is the conditional probability that a man who makes more than $4 per hour has more than 10 years of seniority? (*Hint:* Start by identifying the sample and event spaces.)

4. Suppose that 20 per cent of a seed mixture consists of seeds of red flowers and 5 per cent of the seed mixture consists of seeds of flowers that are both red and have diameters of less than 2 inches. What is the conditional probability that a plant selected at random has a flower with a diameter less than 2 inches if we know that the flower is red? (*Hint:* Start by identifying the sample and event spaces.)

5. Twenty per cent of the population of a certain area are both female and not eligible to vote (under 18). If 60 per cent of the population is female, what is the probability that a female who is selected at random is ineligible to vote? Draw a tree diagram and visualize the conditional probability in two ways as illustrated in Example 3.

6. Suppose that 30 per cent of the people in a certain precinct are over 18 and have registered to vote. If we know that 75 per cent of the people in the precinct are over 18, what is the probability that a person over 18 is registered? Draw a tree diagram and visualize the conditional probability in two ways as illustrated in Example 3.

7. Suppose that there is a probability of $\frac{1}{2}$ that a ballistic missile will both survive an enemy's defenses and hit its target. Further suppose that there is a $\frac{3}{4}$ probability that a missile will survive the enemy's defenses. What is the probability that a missile which survives the defenses will hit the target?

8. A study shows that in one year 2 per cent of the population indulges in a certain activity and dies. If 20 per cent of the population indulges in that activity, what is the probability that a person who indulges will die?

9. An executive had 3 successive decisions to make with an equal possibility of being right or wrong with each decision. In order to have produced the desired result, she must have been right on all 3 decisions. It turned out that the desired result was *not* obtained.
   (a) Find the probability that her first decision was wrong.
   (b) Find the probability that her second decision was wrong if one knows that her first decision was right.
   (c) Find the probability that her first decision was right if one knows that she made at least one right decision.

10. Forty per cent of all victims of a certain disease respond to a certain medicine, 70 per cent respond to physical therapy, and 20 per cent respond to both.
    (a) Find the probability that a victim who responds to the medicine will also respond to therapy.
    (b) Find the probability that a victim will *not* respond to therapy if one knows that there was no response to the medicine.

11. Suppose that a specific drug can be administered to fight a certain disease. Studies have shown that 40 per cent of the victims of the disease have received the drug and 2 per cent of the victims of this disease have both received the drug and have died. What is the probability that a victim who receives the drug will die of the disease?
12. In Exercise 11 suppose that the studies further showed that 5 per cent of the victims of the disease both did *not* receive the drug and died. What is the probability that a victim who did not receive the drug will die of the disease?
13. (a) What is the expected value for a certain game in which the player wins $2 if he flips 3 heads or 3 tails in three flips, but he loses $1 if he flips 2 heads and 1 tail in three flips?
    (b) What is the revised expected value if the first flip is shown to be a head?
14. (a) What is the expected value for a certain game in which a die is thrown twice, and the player wins $1 if the sum shows a 2, 3, 4, 9, 10, or 12, but loses $1 if the sum shows a 5, 6, 7, or 8?
    (b) What is the revised expected value if the first die shows a 6?
15. Given that $P(E) = .10$, $P(F|E) = .60$, and $P(F|E') = 30$ per cent. Find $P(E|F)$. (*Hint:* Use Example 5.)
16. In a certain company 6 per cent of the men and 10 per cent of the women have a college degree. It is known that 70 per cent of the employees are women. If an employee selected at random has a college degree, what is the probability that the employee is a man? (*Hint:* Use Example 5.)
17. Suppose that it is known that 50 per cent of all college graduates in a specified population passed a certain college entrance test and 10 per cent of all nongraduates in the population passed the same test. Also suppose that 30 per cent of the population consists of college graduates. Find the probability of graduating from college if one passes the entrance test. (*Hint:* See Example 5.)
18. Suppose that (1) 30 per cent of a population has a certain disease; (2) 1 per cent of those who have the disease registered negative on a certain medical test; and (3) 90 per cent of those who do not have the disease registered negative on the same medical test. What is the probability that a person has the disease even though he registers negative on the test? (*Hint:* See Example 5.)
19. A certain system consists of two components connected in series. If the probability that the system will perform satisfactorily is $\frac{3}{4}$, and if the first component has a $\frac{7}{8}$ probability of performing satisfactorily, what is the probability that the other component performs satisfactorily given that the first performs satisfactorily?
20. There is a $\frac{1}{2}$ probability that a certain conjunction of two simple statement forms is true. If it is known that there is a $\frac{3}{4}$ probability that the first simple statement form is true, what is the probability that the other simple statement form is true, given that the first is true?

21. Prove that if E_1 and E_2 are events of a finite sample space U of equally likely outcomes and if $n(E_1) \neq 0$, then

$$P(E_2|E_1) = \frac{n(E_2 \cap E_1)}{n(E_1)}.$$

(*Hint:* Use Definitions 1 and 5.)

22. Draw a tree diagram for the problem given in Example 5 and visualize the formula used there. (*Hint:* In the diagram, list D and D' first and A and A' second.)

4.5 Independent Events

Example 1

Suppose that we toss a pair of dice; if the dice are fair, then certainly the upward face of one die in no way affects the upward face of the second die. Thus, if we let E_1 be the event of a 6 on the first die and let E_2 be the event of a 3 on the second die, then the occurrence of E_1 will not affect the probability of E_2; hence $P(E_2|E_1) = P(E_2)$. We say that E_2 is **independent** of E_1. We now verify that $P(E_2) = P(E_2|E_1)$ for this example.

$n(U) = 36;$

$E_1 = \{\ldots\};$

$E_2 = \{\ldots\};$

$E_2 \cap E_1 = \{\ldots\};$

$P(E_2) = \dfrac{n(E_2)}{n(U)} = \dfrac{6}{36} = \dfrac{1}{6};$

$P(E_2|E_1) = \dfrac{P(E_2 \cap E_1)}{P(E_1)} = \dfrac{\frac{1}{36}}{\frac{1}{6}} = \dfrac{1}{6}.$ ∎

It can be shown (Exercise 13) that, for a given experiment, if event E_2 is independent of event E_1, then E_1 is independent of E_2.

▶ **Definition 6**
Two distinct events E_1 and E_2 [where $P(E_1) \neq 0$ and $P(E_2) \neq 0$] of a given finite sample space are said to be **independent** if

$$P(E_2|E_1) = P(E_2) \quad \text{or if} \quad P(E_1|E_2) = P(E_1).$$

Example 2

Consider 5 airlines that serve 3 cities according to the diagram shown in Figure 4.10. For the experiment of traveling from city 1 to city 3 the sample

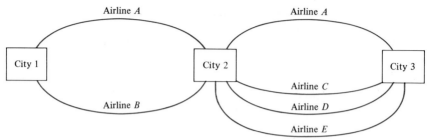

Figure 4.10. *Hypothetical airline service.*

space is
$$U = \{AA, AC, AD, AE, BA, BC, BD, BE\},$$
and we shall assume that the outcomes are equally likely. Let E_1 be the event that a person chooses airline A from city 1 to city 2; hence
$$E_1 = \{AA, AC, AD, AE\}.$$
Let E_2 be the event that a person chooses airline D from city 2 to city 3; hence
$$E_2 = \{AD, BD\}.$$
We verify that the choice of airlines from city 1 to city 2 is independent of the choice from city 2 to city 3 by showing that $P(E_2|E_1) = P(E_2)$; in other words, the probability of E_2 is the same regardless of knowledge of the condition E_1. First, since we are dealing with equally likely outcomes,
$$P(E_2) = \frac{n(E_2)}{n(U)} = \frac{2}{8} = \frac{1}{4}.$$
Second, since $E_1 \cap E_2 = \{AD\}$,
$$P(E_2|E_1) = \frac{P(E_2 \cap E_1)}{P(E_1)} = \frac{n(E_2 \cap E_1)}{n(E_1)} = \frac{1}{4}. \blacksquare$$

We are now in a position to state still another pair of formulas for finding the probability of an intersection of two events.

▶ **Theorem 6**
Let E_1 and E_2 be events of a sample space, which has a finite number of outcomes, where $P(E_1) \neq 0$ and $P(E_2) \neq 0$. Then
(i) $P(E_1 \cap E_2) = P(E_2|E_1)P(E_1) = P(E_1|E_2)P(E_2)$.
(ii) $\{P(E_1 \cap E_2) = P(E_1)P(E_2)\} \Leftrightarrow \{E_1 \text{ and } E_2 \text{ are independent events}\}$.

The proof of Theorem 6 is left as Exercises 14 and 15. ◀

Example 3
This example illustrates Theorem 6(*i*). Suppose that a certain factory produces 3 products, labeled A, B, and C. What is the probability of

selecting, at random, a defective product A if it is known that 30 per cent of the products produced at the factory are product A and 5 per cent of the A products are defective?

Solution: Let A represent the event of selecting a product A, and we are given that $P(A) = \frac{3}{10}$. Let D represent the event of selecting a defective product, and we are given that $P(D|A) = \frac{1}{20}$. We are asked to find $P(D \cap A)$. By Theorem 6(i) we have

$$P(D \cap A) = P(D|A)P(A)$$
$$= \tfrac{1}{20} \tfrac{3}{10} = \tfrac{3}{200} = 1.5 \text{ per cent.} \blacksquare$$

A visual illustration of Example 3 and Theorem 6(i) can be obtained from the tree diagram of Figure 4.11; the probability of $(D \cap A)$ is the product of the probabilities associated with the segments of the path. If, in Example 3, D had been independent of A, then $P(D|A)$ would have equaled $P(D)$.

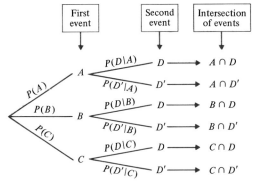

Figure 4.11. *Tree diagram of intersections of pairs of events.*

Example 4

To illustrate Theorem 6(ii), suppose that there is a 40 per cent probability that a missile will miss a given target and a 20 per cent chance that a bomber will miss the same target; assume that these two events are independent. (1) What is the probability that the target will be missed? (2) What is the probability that the target will be hit?

Solution: Let E_1 be the event that a missile misses the target and E_2 be the event that a bomber will miss the target. We are asked to find $P(E_1 \cap E_2)$. The events are independent; hence by Theorem 6(ii) we have

$$P(E_1 \cap E_2) = P(E_1)P(E_2)$$
$$= \tfrac{4}{10} \tfrac{2}{10} = \tfrac{8}{100} = 8 \text{ per cent.}$$

The probability that the target will be hit is determined by

$$P\{(E_1 \cap E_2)'\} = 1 - P(E_1 \cap E_2) = 1 - .08 = 92 \text{ per cent.} \blacksquare$$

There are some problems in which our intuition may fail us in determining whether or not a pair of events are independent; as illustrated in the next example, Theorem 6(ii) may be helpful in such cases.

Sec. 4.5] Independent Events

Example 5
Let the experiment consist of the birth of three offspring of either male or female sex. Further suppose that the following outcomes are equally likely:
$$U = \{mmm, mmf, mfm, fmm, mff, fmf, ffm, fff\}.$$
Let E_1 be the event of obtaining at least two males, and let E_2 be the event of obtaining at least one of each sex. Are the events independent?
$$E_1 = \{mmm, mmf, mfm, fmm\};$$
$$E_2 = \{mmf, mfm, fmm, mff, fmf, ffm\};$$
$$E_1 \cap E_2 = \{mmf, mfm, fmm\}.$$
Hence
$$P(E_1) = \tfrac{1}{2}, \qquad P(E_2) = \tfrac{3}{4}, \qquad P(E_1 \cap E_2) = \tfrac{3}{8}.$$
Therefore, by Theorem 6(ii), since
$$P(E_1)P(E_2) = P(E_1 \cap E_2),$$
the events E_1 and E_2 are independent. ∎

APPLICATIONS

Example 6
In Section 2.6 (Example 4) the reliability of a system was defined as the probability that the system will perform satisfactorily for a specified time interval under specified environmental conditions. In a series system, such as the one shown in Figure 4.12, the event E that the system performs

•———— p ———— q ———— s ————•

Figure 4.12. *Series system.*

satisfactorily is the intersection of the events $E_1, E_2, \ldots, E_n$ that each component functions satisfactorily. Thus the reliability R_s of the series system of Figure 4.12 is
$$R_s = P(E) = P(E_1 \cap E_2 \cap E_3).$$
We assume that the satisfactory or unsatisfactory performance of each component is independent of the performance of the other components. If $r_1, r_2,$ and r_3 represent the reliabilities of the respective components $p, q,$ and s, then it follows from Theorem 6(ii) that
$$R_s = P(E_1)P(E_2)P(E_3)$$
$$= r_1 r_2 r_3.$$

In a parallel system, such as the one shown in Figure 4.13, the reliability R_p may be calculated in two ways. First, the event E that the system performs satisfactorily is the union of the events $E_1, E_2, \ldots, E_n$ that each component functions satisfactorily. Thus the reliability R_p of the parallel system of Figure 4.13 is
$$R_p = P(E) = P(E_1 \cup E_2).$$

Figure 4.13. *Parallel system.*

By Theorem 3,
$$R_p = P(E_1) + P(E_2) - P(E_1 \cap E_2).$$
By Theorem 6(*ii*) and the assumption of independence,
$$R_p = r_1 + r_2 - r_1 r_2.$$
A second method of calculation of R_p recognizes that the event of satisfactory performance is the complement F' of the event F of failure to perform satisfactorily. Thus by Theorem 5,
$$R_p = P(F') = 1 - P(F).$$
But the event F of failure of the parallel system of Figure 4.13 is the intersection of the events F_1, F_2 of failure of each component; thus
$$P(F) = P(F_1)P(F_2)$$
and by Theorem 5,
$$P(F) = [1 - P(F_1')][1 - P(F_2')]$$
$$= (1 - r_1)(1 - r_2).$$
Therefore, substituting for $P(F)$ in the formula $R_p = 1 - P(F)$, we get
$$R_p = 1 - (1 - r_1)(1 - r_2).$$
The two formulas for R_p are algebraically equivalent. The second formula, however, is more readily generalized for parallel systems of n components, namely,
$$R_p = 1 - (1 - r_1) \cdots (1 - r_n).$$
In Exercise 16 the reader will be asked to compute the reliability of the mixed system shown in Figure 4.14, where $r_1 = \frac{1}{3}$, $r_2 = \frac{2}{3}$, and $r_3 = \frac{3}{5}$. (r_1, r_2, and r_3 are the reliabilities of p, q, and s, respectively.) ∎

Figure 4.14

EXERCISES

Suggested minimum assignment: Exercises 1, 2, 3, 6, 7, 8, and 9.

1. Let A and B be events in a finite sample space U. If $P(A) = \frac{2}{3}$, $P(B) = \frac{1}{4}$, and $P(A \cap B) = \frac{1}{8}$, determine whether or not A and B are independent by means of Definition 6 and also by Theorem 6.

2. A pair of dice are cast. Let A be the event that the first die shows an odd number. Let B be the event that the sum of both dice is 7.
   (a) Find $P(A)$.　　　(b) Find $P(B)$.
   (c) Find $P(B \cap A)$.　(d) Find $P(B|A)$.
   (e) Are A and B independent events? Give two alternative reasons.

In each of Exercises 3–8 questions are raised about the probability of intersections of events. In some cases the events are independent and in some cases they are not; in either case state the formula used to solve each exercise and specifically identify each part of the formula.

3. Suppose that 30 per cent of the population has a certain vitamin deficiency because of their diet. Ten per cent of the people with this vitamin deficiency have a certain disease. What is the probability that a person selected at random will have the disease *and* the vitamin deficiency? [*Hint:* Use Theorem 6(*i*).]
4. Suppose that there is a $\frac{5}{8}$ probability that an intercontinental missile can survive an enemy's defense system and a $\frac{3}{4}$ probability that it can hit its target if it does survive. What is the probability that a missile will survive *and* hit its target? [*Hint:* Use Theorem 6(*i*).]
5. Suppose that a certain weed killer will become ineffective on a specified variety of weed if it rains within 24 hours of application. Suppose that if it does not rain, there is an 80 per cent chance that the weed killer will be effective on the specified variety of weed. If the 24-hour weather forecast predicts that there is 10 per cent probability of rain, what is the probability of killing the specified variety of weed if the weed killer is administered now? That is, what is the probability of the weed killer's being effective and of having no rain?
6. If it rains, there is a 90 per cent probability that a certain type seed will germinate. If there is a 70 per cent chance of rain, what is the probability of both rain and germination?
7. In a deck of cards in which all the cards have been removed except the jacks, queens, and kings, what is the probability that a jack followed by a diamond will be drawn from the deck in 2 draws? Assume here that the first card is replaced before the second card is drawn. [*Hint:* Use Theorem 6(*ii*).]
8. Assume that the effect of one medicine is independent of the effect of another medicine. Suppose that experience has shown that the probability of recovery if one medicine is administered is 80 per cent, and if the other medicine is administered, the probability of recovery is 70 per cent.
   (a) What are the respective probabilities of nonrecovery for the administration of each of the medicines?
   (b) What is the probability of recovery if both medicines are administered?

9. Three fair coins are flipped: E_1 is the event that at least two tails appear, E_2 is the event that at most one tail appears, and E_3 is the event that all outcomes are heads or all are tails.
   (a) Are E_1 and E_2 independent events?
   (b) Are E_2 and E_3 independent events?
   (c) Are E_3 and E_1 independent events?
10. Two dice are cast: E_1 is the event that a 6 appears on *at least* one die, E_2 is the event that a 5 appears on *exactly* one die, and E_3 is the event that the same number appears on both dice.
    (a) Are E_1 and E_2 independent events?
    (b) Are E_2 and E_3 independent events?
    (c) Are E_3 and E_1 independent events?
11. Suppose that 2 statement forms are connected by the conjunction "and." If there is a 5 per cent probability that statement form p is false, and if there is 60 per cent probability that statement form q is false, what is the probability that the compound statement $p \wedge q$ is true? Assume independence.
12. Suppose that 2 switches are connected in series and that there is a 30 per cent chance that switch p is closed and a 20 per cent chance that switch q is closed; what is the probability that the compound switch $p \times q$ is closed? What is the probability that the compound switch is open? Assume that the events are independent.
13. Show that for a given experiment, if E_2 is independent of E_1, then E_1 is independent of E_2.
14. Prove Theorem 6(*i*).
15. Prove Theorem 6(*ii*).
16. In Example 6 determine the reliability of the system shown in Figure 4.14, where $r_1 = \frac{1}{3}$, $r_2 = \frac{2}{3}$, and $r_3 = \frac{3}{5}$.
17. (a) Find the reliability of the verbal system $p_1 \vee p_2$ if $r_1 = r_2 = \frac{1}{2}$. Assume independence and the notation of Example 6.
    (b) Find the reliability of the verbal system $p_1 \vee p_2 \vee p_3$ if $r_1 = r_2 = r_3 = \frac{1}{2}$. Assume independence and the notation of Example 6.
    (c) Find the reliability of the verbal system $p_1 \vee p_2 \vee p_3 \vee p_4$ if $r_1 = r_2 = r_3 = r_4 = \frac{1}{2}$. Assume independence and the notation of Example 6.
18. (a) Suppose that we model an oral communication from person A to person B by a series system of three switches. The first switch represents the verbalization of an idea by person A. The second switch represents the hearing of person B, and the third switch represents the interpretation by person B of what was heard. If the reliabilities of these three switches are $\frac{9}{10}$, $\frac{5}{10}$, and $\frac{6}{10}$, respectively, what is the reliability of the communication (that is, the series system)? Assume independence.
    (b) Suppose that the communication is repeated, thus creating a parallel system. What is the reliability of the repetitive communication (that is, the parallel system)? Assume independence.

4.6 Stochastic Experiments

An important class of probability problems consists of those having n repeated trials of the same activity (such as successive tosses of a coin, successive casts of a die, repetitive mass production of some product, and so on), or a sequence of n trials of different activities (such as that illustrated in Example 2). Each trial may have any finite number of outcomes; moreover, the weights of the outcomes of each trial may or may not depend on the outcomes of previous trials. Such experiments, whose trial outcomes depend on some element of chance, are called **stochastic experiments.** The next two examples illustrate a method of calculating the probability of events of stochastic experiments. Notice the distinction between a trial outcome and an experiment outcome.

Example 1

Consider three draws from an ordinary deck of cards. Each draw is made without replacement. What is the probability of the event E that a queen is drawn on exactly two, including the first, of three draws?

Solution: Let Q_1 be the trial outcome of drawing a queen on the first draw, let Q_2 be the trial outcome of drawing a queen on the second draw, let Q_3 be the trial outcome of drawing a queen on the third draw, and let N be the trial outcome of not drawing a queen on any one draw. The tree diagram of Figure 4.15 illustrates the problem. Each path of the diagram can be associated with an outcome of the total experiment. One should observe that these experiment outcomes are not equally likely outcomes. The weight of each of the experiment outcomes can be determined by taking the products of the probabilities along each path; the latter assignment of weights is consistent with a generalization of Theorem 6 and with Definition 2. The weights of the various outcomes of the stochastic experiment are calculated and shown in Figure 4.16. In this example the event space of the stochastic experiment is

$$E = \{(Q_1 Q_2 N), (Q_1 N Q_3)\};$$

thus

$$\begin{aligned} P(E) &= w(Q_1 Q_2 N) + w(Q_1 N Q_3) \\ &= \tfrac{4}{52} \cdot \tfrac{3}{51} \cdot \tfrac{48}{50} + \tfrac{4}{52} \cdot \tfrac{48}{51} \cdot \tfrac{3}{50} \\ &= \tfrac{1,152}{132,600}. \end{aligned}$$ ∎

Example 2

A certain marketing report indicates that of the male shoppers, 10 per cent will purchase brand A, 30 per cent will purchase brand B, and 60 per cent will purchase brand C. The report also shows that of the female shoppers, 40 per cent will purchase brand A, 40 per cent will purchase brand B, and 20 per cent will purchase brand C. If 40 per cent of the shoppers are male and 60 per cent are female, what is the probability that an arbitrary shopper will purchase brand C?

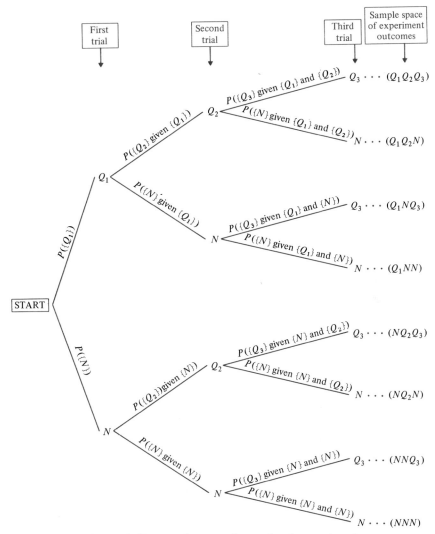

Figure 4.15. *Tree diagram of a stochastic experiment.*

Solution: The tree diagram of Figure 4.17 illustrates the problem. Consistent with Theorem 6, the weight of each nonequally likely outcome of the event space is stated at the right of each path in the tree diagram. The event space of outcomes is

$$E = \{MC, FC\}.$$

Therefore,

$$P(E) = w(MC) + w(FC)$$
$$= .24 + .12$$
$$= .36. \blacksquare$$

Sec. 4.6] Stochastic Experiments

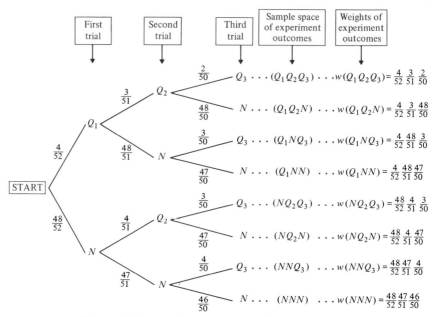

Figure 4.16. *Tree diagram of a stochastic experiment.*

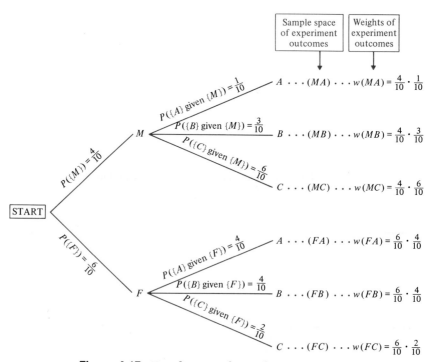

Figure 4.17. *Tree diagram of a stochastic experiment.*

Various special cases of stochastic experiments may be obtained by placing certain restrictions. One such special case is usually referred to as a *Markov chain* and will be introduced in Example 7 of Section 5.3. Another is called a *binomial experiment* and will be defined and illustrated in Section 4.10.

APPLICATIONS

Example 3

The purpose of this example is to illustrate the use of the methods of this section in analyzing a standard dice game. The player who is playing against the "house" wins if he rolls a 7 or 11 on the first throw of a standard pair of six-sided dice. He loses if he rolls a 2, 3, or 12 on the first throw. If, however, on the first throw he rolls a 4, 5, 6, 8, 9, or 10, then he continues to throw the dice until he repeats his first number or until he rolls a 7. If he repeats his number before he rolls a 7, then he wins; otherwise he loses. The tree diagram for the first two throws is shown in Figure 4.18. If E is the event of winning in two throws or less, then

$$P(E) = \frac{9 + 16 + 25 + 216 + 25 + 16 + 9 + 72}{6^4}$$

$$= \frac{388}{1296} = \frac{97}{324} \approx .3.$$

Notice how this compares with the probability of winning after one throw, which is $\frac{1}{6} + \frac{1}{18} = \frac{2}{9} \approx .22$. The actual probability of winning, regardless of the number of throws, can be calculated to be $\frac{244}{495}$ if after the first throw one ignores those results which do not produce a win or loss. In other words, if the player throws a 4 on the first throw, then there are 3 ways that another 4 eventually can be thrown and 6 ways that a 7 eventually can be thrown; thus there is a $\frac{3}{3+6}$ probability that the player will eventually win and a $\frac{6}{3+6}$ probability of losing once he obtains a 4 on the first throw. A tree diagram similar to Figure 4.18 can then be drawn to analyze the problem and obtain the solution (Exercise 9). ∎

EXERCISES

Suggested minimum assignment: Exercises 1, 3, 4, 5, and 7.

In each of the following exercises draw a tree diagram and list the event space and the sample space of the total experiment.

1. On 3 unreplaced draws from an ordinary deck of cards, what is the probability of getting a diamond only on the second and third draws?
2. On 3 unreplaced draws from an ordinary deck of cards, what is the

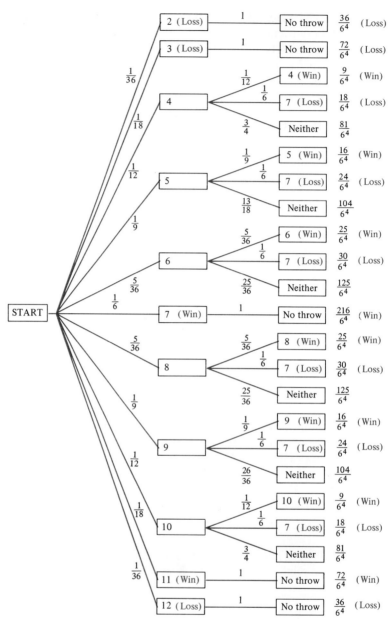

Figure 4.18. *Tree diagram of a dice game.*

probability of getting a heart on the first draw and a spade on one or both of the last 2 draws?

3. A football coach estimates that the probabilities of winning the first 3 games of the season are $\frac{1}{2}$, $\frac{3}{4}$, and $\frac{1}{3}$, respectively. What is the probability that he will have 2 wins and 1 loss after the 3 games? (Assume that no game can be tied.)

4. A political candidate for president estimates that the probabilities of winning states A, B, and C are $\frac{3}{4}$, $\frac{1}{4}$, and $\frac{1}{3}$, respectively. What is the probability that he will win exactly 2 of the 3 states? (Assume that no election can be tied.)
5. Rework Exercise 3 if the coach states that winning the first 2 games will increase his chances in the third game to $\frac{1}{2}$.
6. Rework Exercise 4 if the candidate estimates that carrying state A will increase his chances in state B to $\frac{2}{3}$.
7. If in the U.S. Senate there are 50 Republicans, 45 Democrats, and 5 independents, and if 20 per cent of the Republicans, 60 per cent of the Democrats, and 40 per cent of the independents usually favor "welfare state" proposals, what is the probability that a Senator selected at random will support a "welfare state" proposal?
8. Suppose that a certain company has 3 suppliers of an item. The company buys 20 per cent of its items from company A, 30 per cent from B, and 50 per cent from C. If 10 per cent of the items from A, 10 per cent from B, and 5 per cent from C are defective, what is the probability that an item selected at random from the imports will be defective?
9. In Example 3 verify that the probability of winning is $\frac{244}{495}$ if after one throw, one ignores those results that do not produce a win or loss.
10. In Exercise 8 suppose that a defective part has a salvage value of $1 and a nondefective part has a value of $5. What is the expected value for the choice of one item that is imported?
11. A presidential candidate estimates that if he pursues a specified campaign strategy, then there is a $\frac{1}{5}$ probability of winning state A with 20 electoral votes. If he wins state A, he thinks that there is a $\frac{1}{2}$ probability of also winning state B with 10 electoral votes; if he loses state A, however, he thinks that there is only a $\frac{1}{4}$ probability of winning state B.
    (a) What is the probability that he will win exactly one state?
    (b) What is the probability that he will win in both states?
    (c) What is the expected value of electoral votes that he will receive if he follows the specified campaign strategy?

SUPPLEMENTARY EXERCISES FOR SECTIONS 4.1–4.6

1. A pair of regular tetrahedrons (four-sided objects) are thrown on a flat surface in the same manner that one would throw a pair of dice. The sides of the tetrahedrons are numbered from 1 to 4 (like dice). Let the sum of the numbers on the downward faces of the two objects constitute an outcome of the experiment.
    (a) Find the probability of the event of throwing a sum larger than or equal to 4. (*Hint:* List the sample space and the weights of each outcome.)
    (b) If a game is designed so that one wins $3 if the event of part (a)

happens and loses $2 otherwise, determine the rate at which one might expect to win or lose over the long run.

2. A pair of ordinary dice are modified so that two faces are numbered 5 and the other four faces are numbered 1 to 4. An experiment is conducted in which the two dice are thrown on a flat surface; the outcomes of the experiment are the sums of the two numbers showing on the upward faces.
   (a) Write the sample space and the weight of each outcome, and then find the probability of the event of obtaining an even sum.
   (b) If a game is designed so that one wins $3 if the event of part (a) happens and loses $4 otherwise, determine the rate at which one might expect to win or lose over the long run.

3. An experiment is conducted in which an ordinary coin is tossed four times.
   (a) Find the probability of obtaining at least 2 heads.
   (b) A game is designed so that a person pays 2 cents to play and then wins 1 cent for each head that appears on four tosses; determine the rate at which one might expect to win or lose over the long run.

4. (a) What is the probability that a 3-judge appellate court will pass a certain motion if each possible voting pattern is considered an outcome, if the outcomes are equally likely, and if a simple majority is required for passage?
   (b) Answer the question in part (a) if the court consists of 4 judges rather than 3 judges.
   (c) Answer the question in part (a) if it is known that a certain judge will oppose the motion.

5. Four candidates are running for a single office. The pollsters agree that candidates A and B are twice as likely to win as candidate C and candidate C is twice as likely to win as candidate D. Based on the pollsters opinion, answer the following questions.
   (a) What is the probability that candidate A or C will win?
   (b) What is the answer to part (a) if candidate D withdraws from the race (that is, it is known that D will not win)? Use Definition 5 to obtain your answer.

6. If E, F, and G represent events of a sample space U having equally likely outcomes, and if

   $n(E) = 40$, $n(F) = 30$, $n(G) = 20$, $n(E \cap F) = 10$,
   $n(E \cap G) = 10$, $n(F \cap G) = 5$, $n(F \cap G \cap E) = 3$, $n(U) = 100$,

   find (a) $P((E \cup F)')$.   (b) $P((E \cup F) \cap G')$.   (c) $P(E|(G \cup F))$.

7. (a) What is the probability that when 2 dice are thrown, the numbers showing on the upward faces are the same?
   (b) Answer the question of part (a) if it is known that the sum is greater than 5.

8. A survey shows that a certain product has two parts that may be defective. On the basis of this survey a business executive estimates that

in 10 per cent of the products, part A is defective; in 5 per cent of the products, part B is defective; and in 2 per cent of the products, both parts are defective. For a person who owns one of these products, what is the probability of each of the following?
  (a) Part A is defective and part B is not defective.
  (b) Part A is defective if it is known that part B is not defective.
  (c) Neither part is defective.

9. A certain manufacturer has two machines that independently produce a certain bicycle part. Quality-control techniques have determined that 4 per cent of the parts produced by machine A are defective, and 3 per cent of the parts produced by machine B are defective. An inspector chooses one part from machine A and one part from machine B. Answer each of the following questions by means of a formula and also by means of a tree diagram.
  (a) What is the probability that both parts are defective?
  (b) What is the probability that at least one part is defective?
  (c) What is the probability that neither part is defective?

10. Suppose that the probability of contracting a certain disease for the first time is 5 per cent. Also, 20 per cent of those who contract the disease one time will have the disease a second time. Answer each of the following questions by means of a formula and also by means of a tree diagram.
  (a) What is the probability that a newborn child will have the disease twice during his or her lifetime?
  (b) The probability of having the disease the third time, given that the person had the disease the second time, is 80 per cent. What is the probability of the newborn child having the disease 3 times?
  (c) What is the probability of having the disease 3 times after the person contracts the disease once?

11. From a standard deck of cards one card is drawn; then a second card is drawn without replacing the first card. Let A be the event that the first card is a diamond and B be the event that the second card is a diamond. Are A and B independent? Justify your answer. [*Hint:* To find $P(B)$, use the fact that $B = (B \cap A) \cup (B \cap A')$, that $P(B \cap A) = P(B|A)P(A)$, and that $P(B \cap A') = P(B|A')P(A')$; a tree diagram may be useful in visualizing the problem.]

4.7 Counting Techniques: Permutations†

The problems considered so far have been limited to those for which it has been easy to find the number of elements in the sample and event spaces. Many realistic problems, however, involve sets that require more sophisticated counting techniques than we have encountered so far. We shall introduce these techniques by means of examples.

† Section 1.5 is a prerequisite for Sections 4.7–4.10.

Example 1

A certain manufacturer must decide on the design of a container that is to be used for his product. Suppose that this container consists of 3 components: the bottle, the cap, and the label. If there are 2 different bottle designs, 3 different cap designs, and 5 different labels, how many alternatives must he consider?

Solution: We exhibit two ways of approaching this problem:

1. We may use a tree diagram, shown in Figure 4.19. In the diagram one may start by choosing any one of the 2 different bottle designs; then for each

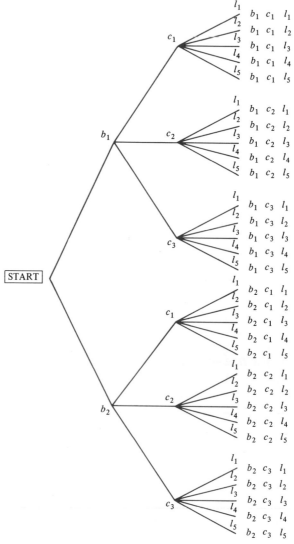

Figure 4.19. *Tree diagram of fundamental counting principle.*

bottle design there are 3 different cap designs, thus producing 2 times 3, or 6, different bottle and cap alternatives. Finally, for each of these bottle and cap alternatives there are 5 different labels. Each path of the tree represents a container that differs in some way from the others. We see that there are $2 \cdot 3 \cdot 5 = 30$ different alternatives.

2. Let B represent the set of bottle designs, let C represent the set of cap designs, and let L represent the set of label designs. The Cartesian product $B \times C \times L$ represents the set of all possible arrangements of the elements of the respective sets. Of course each element of $(B \times C \times L)$ represents a path of the tree of Figure 4.19 and is listed there. From Section 1.5 we know that

$$n(B \times C \times L) = n(B) \cdot n(C) \cdot n(L)$$
$$= 2 \cdot 3 \cdot 5$$
$$= 30. \blacksquare$$

The ideas presented in Example 1 can be generalized to produce the useful counting technique called the **fundamental counting principle:** *If one event can occur in n_1 ways, if a second event can occur in n_2 ways, if a third event can occur in n_3 ways, ..., and if a kth event can occur in n_k ways, then the number of ways the k events can occur is the product $n_1 \cdot n_2 \cdot n_3 \cdot \ldots \cdot n_k$.*

Example 2

How many license tags can be constructed if the tag numbers cannot begin with 0, if each tag number has 4 digits, and if the digits may be repeated?

Solution: There are 9 ways in which the first digit can occur, there are 10 ways in which the second digit can occur, 10 ways in which the third digit can occur, and 10 ways in which the fourth digit can occur. Hence by the fundamental counting principle, there are

$$9 \cdot 10 \cdot 10 \cdot 10 \quad \text{or} \quad 9000$$

license tags. Reconsider this solution in the following way: Let A be the set of digits $\{1, 2, 3, \ldots, 9\}$ and let B be the set of digits $\{0, 1, 2, 3, \ldots, 9\}$; then a 4-digit number with repeated digits allowed and a nonzero first digit can be considered as an element of

$$A \times B \times B \times B,$$

and, by the formulas of Section 1.5,

$$n(A \times B \times B \times B) = n(A) \cdot n(B) \cdot n(B) \cdot n(B)$$
$$= 9 \cdot 10 \cdot 10 \cdot 10$$
$$= 9000. \blacksquare$$

A very important special case of the fundamental counting principle is illustrated in the next two examples.

Example 3

In order to obtain a priority list for foreign aid, suppose that experts in the State Department are asked to rank 4 eligible nations. How many priority lists are possible?

Solution: Think of a priority list as a set of boxes.

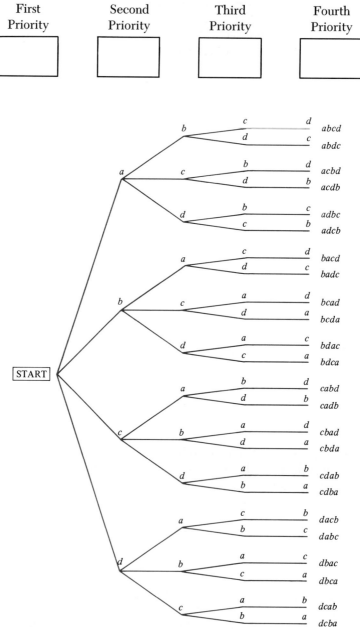

Figure 4.20. *Tree diagram of fundamental counting principle.*

There are 4 nations that can be placed in the first box. After 1 nation has been placed there, 3 nations are left for placement in the second box. After the top 2 nations have been ranked, there are 2 nations remaining for placement in the third box. Finally, after the top 3 nations are chosen, there is no choice but to place the remaining nation in the fourth box. By the fundamental counting principle we have the product

$$\boxed{4} \cdot \boxed{3} \cdot \boxed{2} \cdot \boxed{1} = 24.$$

A product of the form $n(n-1)(n-2) \cdots 3 \cdot 2 \cdot 1$ is ordinarily represented by the notation $n!$, which is read "n factorial"; hence the product $4 \cdot 3 \cdot 2 \cdot 1$ could be represented by $4!$. As in Example 1, one may draw a tree diagram as shown in Figure 4.20 for this problem, and each of the 24 paths represents a priority list. In Figure 4.20 let a, b, c, and d represent the nations. ∎

Example 4
Suppose the last example is modified so that the experts are asked to rank only 3 nations from a set of 7 nations. Again think of the priority list as a set of boxes.

First Priority   Second Priority   Third Priority

There are 7 nations that can be placed in the first box. After 1 is chosen, there are 6 left that can be placed in the second box. Finally, after the first 2 have been chosen, there are 5 left that can be placed in the third box. By the fundamental counting principle we have the product

$$\boxed{7} \cdot \boxed{6} \cdot \boxed{5} = 210.$$

Such a product can be expressed using the factorial notation, namely,

$$7 \cdot 6 \cdot 5 = \frac{7 \cdot 6 \cdot 5 \cdot 4 \cdot 3 \cdot 2 \cdot 1}{4 \cdot 3 \cdot 2 \cdot 1} = \frac{7!}{4!}. \quad ∎$$

Each priority list of Examples 3 and 4 is an example of what is commonly called a **permutation**. Each priority list is an *ordered arrangement* of a certain number of elements of a set. The ideas presented in Examples 3 and 4 are formalized in the following definition.

▶ **Definition 7**
If k of the n elements of a set are arranged in a specified order, then the *arrangement* is called a **permutation** of n elements taken k at a time.

In other words, a permutation is an *ordered arrangement* of elements; if either the elements or the order of the elements is changed, a different permutation results.

The symbols

$$P_k^n, \quad {}_nP_k, \quad \text{and} \quad P(n, k)$$

are all used to designate the *number* of permutations of n elements or objects taken k at a time, and the formula for calculating this number is

$$P_k^n = \frac{n!}{(n-k)!}.$$

Verification of this formula involves a generalization of the reasoning illustrated in Examples 3 and 4.

Example 5

$$\begin{aligned} P_3^9 &= \frac{9!}{(9-3)!} \\ &= \frac{9!}{6!} \\ &= \frac{(9 \cdot 8 \cdot 7) 6 \cdot 5 \cdot 4 \cdot 3 \cdot 2 \cdot 1}{6 \cdot 5 \cdot 4 \cdot 3 \cdot 2 \cdot 1} \\ &= 9 \cdot 8 \cdot 7 \\ &= 504. \quad \blacksquare \end{aligned}$$

When using the formula for P_k^n, one will occasionally encounter the notation $0!$, which is defined to be 1.

One of the common uses of the counting techniques presented in this section is in the calculation of probabilities, as we illustrate in the next two examples.

Example 6

In Example 4 what is the probability that nation a is first *or* that nation b is second on the priority list that is eventually chosen (assuming that all outcomes are equally likely)?

Solution: Let E_1 be the event that nation a is first, and let E_2 be the event that nation b is second. We then seek $P(E_1 \cup E_2)$. Because the outcomes are assumed to be equally likely,

$$\begin{aligned} P(E_1 \cup E_2) &= P(E_1) + P(E_2) - P(E_1 \cap E_2) \\ &= \frac{n(E_1)}{n(U)} + \frac{n(E_2)}{n(U)} - \frac{n(E_1 \cap E_2)}{n(U)}. \end{aligned}$$

We found in Example 4 that $n(U) = P_3^7$. To find $n(E_1)$, we reason that if the specified nation is first, then the problem becomes a matter of finding the number of ways that the remaining 6 nations can be ordered in the 2 positions that are left on the list. Hence $n(E_1) = P_2^6$. Similarly, $n(E_2) = P_2^6$.

If nation a is first *and* nation b is second, then there are 5 remaining nations to fill one position; thus $n(E_1 \cap E_2) = P_1^5$.

$$P(E_1 \cup E_2) = \frac{P_2^6}{P_3^7} + \frac{P_2^6}{P_3^7} - \frac{P_1^5}{P_3^7}$$

$$= \frac{6!/4! + 6!/4! - 5!/4!}{7!/4!}$$

$$= \frac{55}{210}$$

$$= \frac{11}{42}. \blacksquare$$

Example 7

If 5 people sit down in a row of 5 chairs, what is the probability that a specified pair will be seated side by side? (Assume that the outcomes are equally likely.)

Solution: The sample space U will consist of the set of permutations of the 5 people taken 5 at a time; hence $n(U) = P_5^5 = 5! = 120$. The event space E can be considered as the union of the mutually exclusive events $E_1, E_2, E_3, \ldots, E_8$, where each of the 8 events can be thought of as sets of elements illustrated in Figure 4.21. Because the events $E_1, E_2, \ldots, E_8$ are

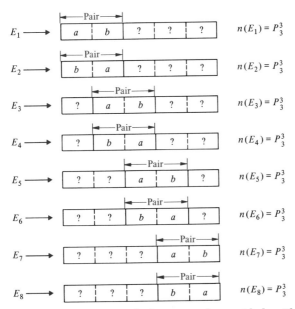

Figure 4.21. *Events in which two people are side by side.*

Sec. 4.7] Counting Techniques: Permutations

mutually exclusive and because of the assumption of equally likely outcomes,

$$P(E) = P(E_1) + P(E_2) + \cdots + P(E_8)$$
$$= 8 \frac{P_3^3}{P_5^5}$$
$$= 8 \frac{3!}{5!}$$
$$= \frac{2}{5}. \blacksquare$$

APPLICATIONS
Example 8

An interesting measure of voting power was developed in an article by Shapley and Shubik [58]. The basic idea is presented in the following illustration. Suppose there is a four-man committee that votes secretly. In case of a 2 : 2 tie the chairman has the power to break the tie. If we assume that the chairman votes the same way on both of his votes, then this method of voting amounts to 2 votes for the chairman and 1 vote for the other three members. We shall designate the members of the committee by a, b, c, d, where c is the chairman. The committee members are ordered according to their likelihood of voting yes on a certain motion. One such order might be $(c\ d\ b\ a)$; in such a permutation d is the crucial or *pivot* member because c has 2 votes, and hence d's vote could assure passage regardless of what b and a do. There are 4!, or 24, such permutations; they are listed next, with the pivot circled.

$a\ b\ ⓒ\ d$	$b\ a\ ⓒ\ d$	$c\ ⓐ\ b\ d$	$d\ a\ ⓑ\ c$
$a\ b\ ⓓ\ c$	$b\ a\ ⓓ\ c$	$c\ ⓐ\ d\ b$	$d\ a\ ⓒ\ b$
$a\ ⓒ\ b\ d$	$b\ ⓒ\ a\ d$	$c\ ⓑ\ a\ d$	$d\ b\ ⓐ\ c$
$a\ ⓒ\ d\ b$	$b\ ⓒ\ d\ a$	$c\ ⓑ\ d\ a$	$d\ b\ ⓒ\ a$
$a\ d\ ⓑ\ c$	$b\ d\ ⓐ\ c$	$c\ ⓓ\ a\ b$	$d\ ⓒ\ a\ b$
$a\ d\ ⓒ\ b$	$b\ d\ ⓒ\ a$	$c\ ⓓ\ b\ a$	$d\ ⓒ\ b\ a$

The voting power of a member is defined to be the ratio of the number of permutations in which the member is pivoted, to the total number of permutations.

Of the 24 permutations, c is pivotal in 12; therefore, the voting power of c is $\frac{1}{2}$. The voting power of each of the other three members is $\frac{1}{6}$. If this definition of voting power is applied to the Security Council of the United Nations, one finds that each of the Big Five with their veto power has a voting power of approximately .196, and each of the 10 other nations has a voting power of approximately .002. $\blacksquare$

Example 9

This example offers a simple illustration of the fundamental counting principle. In International Morse code, dots and dashes are ordered in various ways to produce a code for the letters of the English alphabet. For example, the letter a is represented by the arrangement $(\cdot -)$, the letter b is represented by the arrangement $(-\cdot\cdot\cdot)$, and so on. Repetitions are allowed. This particular code is particularly adaptable to a telegraph key. Four characters or less will produce 30 arrangements of dots and dashes as shown in Table 4.6; hence the 26 letters of our alphabet can be coded by use of these 30 arrangements. The 4 leftover arrangements not used in the Morse code are $(----)$, $(---\cdot)$, $(\cdot\cdot--)$, and $(\cdot-\cdot-)$. ∎

Table 4.6

Number of characters	Number of possible arrangements with repetitions allowed
1	2
2	4
3	8
4	16
Total	30

EXERCISES

Suggested minimum assignment: Exercises 1, 3, 5, 6, 7, 9, 10, 11, and 14.

1. (a) Suppose that a certain student has 5 different reasons for failing calculus and 4 different reasons for failing English; using 1 reason for each failure, how many stories can the student give the Dean?
   (b) Unknown to the student, the Dean is a reasonable man, and 3 of the reasons for failing calculus and 2 of the reasons for failing English would be acceptable to the Dean. Assume equally likely outcomes and calculate the probability that the student will come up with an acceptable story.
2. (a) If there are 10 different airlines serving route A, 8 different airlines serving route B, and 6 different airlines serving route C, how many different ways can one person make a trip covering all three routes?
   (b) If 6 of the airlines serving route A, 2 of those serving route B, and 4 of those serving route C use jets, find the probability that a person making a trip using all three routes will travel the entire way on jets. Assume equally likely outcomes in the choice of airlines.
3. Calculate P_2^4, P_5^7, P_3^{22}, P_8^8.
4. Calculate P_3^5, P_4^6, P_2^{19}, P_5^5.
5. In the game of Scrabble a player tries to make words from certain letters that are drawn. Suppose that a player draws the letters C, S, O, T, R, and E. In how many different ways may any 3 of the letters be arranged even if no English word is formed?

6. Suppose that a certain city decides to have automobile license tags with numbers that do not have repetitive digits. How many tags can be made if the tag number has exactly 4 digits? (Assume that the numbers beginning with 0 are permitted.)

7. In Exercise 6, if the city anticipates the need for 30,000 tags, how many digits must the numbers have in order to have enough tags to meet the demand? Assume that all tag numbers must have the same number of digits.

8. How many digits will be required in a telephone number if no digit is to be repeated, if the use of 0 is prohibited, and if 10,000 telephones are planned?

9. The ordinary telephone number has 4 digits plus a 3-digit exchange number. Digits can be repeated, but the number 0000 is prohibited. How many telephones can one exchange accommodate?

10. In relation to Exercise 9, the country is partitioned into areas and each area is assigned a 3-digit area code number. If the number 000 is prohibited from both the exchange number and the area code number, what is the maximum number of exchanges within an area, and what is the maximum number of areas?

11. (a) How many 4-letter "words" can be made from the letters $ABCDEF$ if no letters are repeated?
    (b) In part (a), what is the probability that a "word" will not have a vowel in it?
    (c) In part (a), what is the probability that a "word" will have a vowel in it?

12. (a) How many 9-man batting orders can a coach obtain from 10 men?
    (b) What is the probability that a particular man will not be in the batting order?
    (c) What is the probability that a particular man will be in the batting order?

13. What is the probability that a telephone number of 4 digits will have at least 1 repeated digit? Assume that the number 0000 is not permitted.

14. What is the probability that a license tag with 6 digits will not have any repeated digits? Assume that numbers beginning with 0 are not permitted.

15. In a 4-digit telephone number *without* repeated digits, what is the probability that a 3 will immediately follow a 2 in the number? Assume that numbers beginning with 0 are allowed.

16. In a 6-digit license number *without* repeated digits, what is the probability that all the digits 1, 2, and 3 will be in the number and adjacent to each other in some order? Assume that numbers beginning with 0 are allowed.

17. Foreign aid in the amount of $15 million for one nation, $10 million for another nation, and $5 million for still another nation is to be assigned.

The 3 nations will be selected by a panel of experts from among 6 nations.
(a) In how many ways can the nations be selected?
(b) In how many of the ways will nation A receive $15 million?
(c) In how many of the ways will nation A receive $10 million?
(d) What is the probability that nation A will receive $5 million? Assume that all selections in part (a) are equally likely.
(e) What is the expected value in foreign aid for nation A? Assume that all selections in part (a) are equally likely.

4.8 Counting Techniques: Combinations

In the last section we were concerned with counting the number of ways of selecting k elements from a set of n elements and arranging those k elements in some specified order; in this section we shall be concerned with only the selection of the k elements *without* regard to order. The next example will illustrate the method.

Example 1

In how many ways can a committee of 3 Senators be selected from a set of 4 Senators?

Solution: The order in which the Senators are selected is immaterial as far as the makeup of the committee is concerned. To explain the method of calculating the number of unordered selections, however, we shall list all permutations of 4 Senators taken 3 at a time. We shall designate the Senators with the letters a, b, c, and d.

abc	abd	acd	bcd
acb	adb	adc	bdc
bac	bad	cad	cbd
bca	bda	cda	cdb
cab	dab	dac	dbc
cba	dba	dca	dcb

By the permutation formula there are P^4_3, or 24, ways of selecting the Senators in order. Notice, however, that each column of the list of permutations consists of 3! ways of arranging the same 3 Senators. Thus if we divide the number of ways (P^4_3) that 3 Senators can be selected in order by the number of ways (3!) that each threesome can be ordered, then we have simply unordered each threesome and consequently have the number of unordered ways that 3 Senators can be selected from a set of 4 Senators. In other words, we have the number of 3-man subsets of the original set of 4 Senators; there are 4 such subsets. ∎

▶ **Definition 8**
A subset of k elements formed from a set of n elements is called a **combination** of n elements taken k at a time.

The symbols

$$C_k^n, \quad {}_nC_k, \quad C(n, k), \quad \text{and} \quad \binom{n}{k}$$

are all used to designate the *number* of combinations of n elements or objects taken k at a time, and the formula for calculating this number is

$$C_k^n = \frac{P_k^n}{k!} = \frac{n!}{k!(n-k)!}.$$

Verification of this formula can be obtained by generalizing the reasoning illustrated in Example 1.

Example 2

$$C_4^7 = \frac{P_4^7}{4!} = \frac{7!/3!}{4!} = \frac{7!}{4!3!} = \frac{7 \cdot 6 \cdot 5 \cdot 4 \cdot 3 \cdot 2 \cdot 1}{4 \cdot 3 \cdot 2 \cdot 1 \cdot 3 \cdot 2 \cdot 1} = 35. \blacksquare$$

As with permutations, one of the primary uses of combinations is their use in the calculation of probabilities.

Example 3

Ten men participate in the testing of a certain medicine. Five are given the medicine, and 5 are not given the medicine. What is the probability that a certain specified pair of men will both get the medicine?
 Solution: First, calculate the number of elements in the sample space U. The order of selection is not significant, hence

$$n(U) = C_5^{10} = \frac{P_5^{10}}{5!} = \frac{10!}{5!5!}.$$

An outcome of the event space E will consist of the 2 specified men together with 3 others selected from the remaining 8 men. Thus

$$n(E) = C_3^8 = \frac{P_3^8}{3!} = \frac{8!}{5!3!}.$$

Hence (assuming equally likely outcomes)

$$P(E) = \frac{n(E)}{n(U)} = \frac{8!/5!3!}{10!/5!5!} = \frac{8!5!5!}{10!5!3!} = \frac{2}{9}. \blacksquare$$

Example 4

Suppose in Example 3 that one seeks the probability of the 2 specified men *or* a specified third man getting the medicine. Let the event of Example 3 be designated E_1; let the event of the third man receiving the medicine be called E_2; and let the event $E_1 \cup E_2$ be designated F.

$$P(E_2) = \frac{n(E_2)}{n(U)} = \frac{C_4^9}{C_5^{10}} = \frac{9!/5!4!}{10!/5!5!} = \frac{1}{2}.$$

Now $P(F) = P(E_1 \cup E_2) = P(E_1) + P(E_2) - P(E_1 \cap E_2)$; hence we must find $P(E_1 \cap E_2)$. The event $E_1 \cap E_2$ consists of outcomes in which all 3 men get the medicine, and thus we are looking for the number of ways to fill the remaining 2 positions with the other 7 men. Hence $n(E_1 \cap E_2) = C_2^7$ and

$$P(E_2 \cap E_1) = \frac{n(E_2 \cap E_1)}{n(U)} = \frac{C_2^7}{C_5^{10}} = \frac{7!/5!2!}{10!/5!5!} = \frac{1}{12}.$$

Therefore,

$$P(F) = \frac{2}{9} + \frac{1}{2} - \frac{1}{12} = \frac{23}{36}. \blacksquare$$

Example 5

What is the probability that in a 5-card poker hand, all 5 cards will be diamonds?

Solution: The sample space U will consist of the various combinations that can be selected 5 at a time from 52 cards; that is,

$$n(U) = C_5^{52} = \frac{52!}{5!47!}.$$

The event space will consist of the various combinations that can be selected 5 at a time from 13 diamonds; that is,

$$n(E) = C_5^{13} = \frac{13!}{5!8!}.$$

Therefore,

$$P(E) = \frac{n(E)}{n(U)} = \frac{13!/5!8!}{52!/5!47!} = \frac{13!}{8!} \cdot \frac{47!}{52!} \approx .000495. \blacksquare$$

A variation of Example 5 would be to determine the probability of getting 3 aces and 2 kings in a 5-card poker hand. Of course, $n(U)$ remains the same. The event space E can be thought of as the Cartesian product of two sets A and K shown in Figure 4.22.

The number of elements of A is C_3^4 since we are selecting 3 aces from 4 in the deck, and $n(K) = C_2^4$ since we are selecting 2 kings from 4 in the deck.

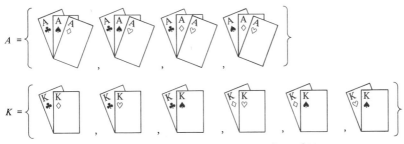

Figure 4.22. *Events of three aces and two kings.*

Sec. 4.8] Counting Techniques: Combinations

From Section 1.5 we know that
$$n(A \times K) = n(A) \cdot n(K);$$
hence
$$n(E) = n(A \times K) = 4 \cdot 6 = 24.$$
[The fact that $n(E)$ is the product of $n(A)$ and $n(K)$ can be viewed as a result of the fundamental counting principle.] Therefore,
$$P(E) = \frac{n(E)}{n(U)} = \frac{n(A \times K)}{n(U)} = \frac{24}{2,598,960} \approx .0000092.$$

Example 6
What is the probability that a committee of 3 Democrats and 2 Republicans will be selected from a group of 6 Democrats and 5 Republicans? (Assume equally likely outcomes.)

Solution: Let R be the set of all possible pairs of Republicans and let D be the set of all possible triples of Democrats. Then
$$P(E) = \frac{n(E)}{n(U)} = \frac{n(R \times D)}{n(U)} = \frac{n(R) \cdot n(D)}{n(U)} = \frac{C_2^5 C_3^6}{C_5^{11}} \approx .43. \blacksquare$$

APPLICATIONS

Example 7
In one version of the game of poker, 5 cards are dealt to each player from an ordinary deck of 52 cards. Thus there is a sample space of C_5^{52} or 2,598,960 outcomes (5-card hands). The various hands that may be dealt are classified and ranked according to Table 4.7. After all bets are made, the winner of the game is the player who remains in the game and holds the highest ranking hand. The reader can observe from Table 4.7 that the classifications of hands are ranked on the basis of the relative size of their probabilities. Some of the methods of calculation given in Table 4.7 are not easy for the beginner; hence we give a discussion of the calculation of $n(E)$, where E represents the event of receiving 3 of a kind.

One way to think of the calculation of $n(E)$ is to use the fundamental counting principle:

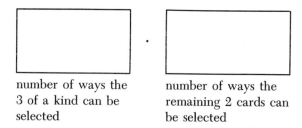

number of ways the 3 of a kind can be selected · number of ways the remaining 2 cards can be selected

To find the entry of the first box, we must select 1 kind from among 13 kinds in an ordinary deck (there are C_1^{13} such possible selections), and then we

Table 4.7. Poker hand probabilities.

Classification of hand	Definition of the event space E	Method of calculation of the number of elements $n(E)$ in E	$n(E)$	Probability of E $P(E)$†	Rank of hand
Royal flush	Ace, king, queen, jack, ten, in the same suit	C_1^4	4	.0000015	1
‡Straight flush	Any 5 cards in sequence in the same suit but not a royal flush	$(10 \cdot C_1^4) - 4$	36	.000014	2
Four of a kind	Any 4 cards all of the same kind (e.g., 4 kings)	$(C_1^{13} C_4^4)(C_1^{48})$	624	.00024	3
Full house	Any 3 cards of 1 kind and any 2 cards of another kind	$(C_1^{13} C_3^4)(C_1^{12} C_2^4)$	3,744	.0014	4
Flush	Any 5 cards of 1 suit but not a straight flush or a royal flush	$(C_5^4 C_5^{13}) - 36 - 4$	5,108	.0020	5
‡Straight	Any 5 cards in a sequence but not all in the same suit (i.e., not a flush)	$(10[C_1^4]^5) - 36 - 4$	10,200	.0039	6
Three of a kind	Any 3 cards all of the same kind (e.g., 3 kings) but not a full house	$(C_1^{13} C_3^4)(C_2^{12}[C_1^4]^2)$ or $(C_1^{13} C_3^4 C_2^{48} - 3,744)$	54,912	.0211	7
Two pairs	Any 2 cards of 1 kind and 2 more of another kind (e.g., 2 kings and 2 sixes); fifth card of a third kind	$(C_2^{13}[C_2^4]^2)/C_1^{11} C_1^4)$	123,552	.0475	8
One pair	Any 2 cards of 1 kind and each of the other cards of different kinds	$(C_1^{13} C_2^4)(C_3^{12}[C_1^4]^3)$	1,098,240	.4226	9

† $n(U) = C_5^{52} = 2,598,960$.
‡ In these calculations, an ace was considered to be high or low.

must select 3 of the 4 cards within that kind (there are C_3^4 ways of making this selection). Thus we have

$$\boxed{C_1^{13} C_3^4} \cdot \boxed{}$$

| number of ways the 3 of a kind can be selected | number of ways the remaining 2 cards can be selected |

To find the entry of the second box, we must select 2 different† kinds from the remaining 12 kinds (there are C_2^{12} such possible selections), and then we must select 1 of the 4 cards within each of the 2 kinds (there are $C_1^4 \cdot C_1^4$ ways of doing this). Thus the entry $C_2^{12} C_1^4 C_1^4$ should go in the second box.

An alternative way of calculating $n(E)$ for 3 of a kind is to fill the first box as in the first method, but in filling the second box, simply select 2 cards from the remaining 48 (after the 4 cards of 1 kind have been eliminated from consideration); however, since this result, C_2^{48}, allows pairs of the same kind, and hence a full house, we must subtract the number of possible full houses from the product $(C_1^{13} C_3^4) \cdot (C_2^{48})$.

Because classifications are mutually exclusive, the probability of being dealt a ranked hand is the sum of the probabilities in the fifth column of Table 4.7; it is interesting to observe that there is a probability of approximately $\frac{1}{2}$ of receiving a ranked hand. Of course, in many versions of poker, cards may be replaced before the betting begins, and this will affect the probabilities discussed. ∎

Example 8
This example assumes an elementary knowledge of the game of bridge. In bridge, probabilities play a very important part in developing strategies for playing hands. In one of Goren's widely read books on bridge [26] there is a table of probabilities and a discussion of various uses of the table. One surprising result is that the probability of obtaining a 3–1 or 1–3 split when 4 cards of a certain suit are out is greater than the probability of an even 2–2 split. To illustrate the use of combinations in the calculation of Goren's table, we shall calculate the probability of a 2–2 split.

At the outset of play the contract winner cannot account for 26 of the 52 cards in the deck. Of these 26 cards, 4 belong to the suit in question, say hearts. Let E be the event that exactly 2 of the 4 hearts belong to the opponent on the left. We calculate $n(E)$ by multiplying the number of ways the left opponent can hold 2 of 4 hearts (that is, C_2^4) by the number of ways the left opponent can hold 11 of 22 nonhearts (that is, C_{11}^{22}). The total number of equally likely ways that 13 of the 26 cards can be held by the left

† They must be different to keep from having a full house.

opponent is C_{13}^{26}. Thus

$$P(E) = \frac{C_2^4 C_{11}^{22}}{C_{13}^{26}} \approx .407.$$

It is left to the reader as an exercise to show that the probability of a 3–1 or 1–3 split is approximately .497. ∎

EXERCISES

Suggested minimum assignment: Exercises 1, 3, 5, 8, 9, 11, and 14.

1. Calculate C_2^6, C_3^5, and C_{18}^{20}.
2. Calculate C_4^6, C_2^5, and C_2^{20}.
3. How many ways can a team of 3 astronauts be selected from a group of 20 astronauts?
4. How many ways can a basketball team of 5 players be selected from a squad of 12 men?
5. In how many ways can a 5-man appellate court render a majority decision in favor of a motion?
6. How many different sums of money can be raised from the sale of a majority of 5 hotels that a certain man owns? Assume that each hotel has a different selling price and that each sum is different from every other sum. A majority consists of 3 or 4 or all 5 hotels.
7. What is the probability of getting 5 red cards (diamonds or hearts) in a 5-card poker hand? Express your answer in a form that uses the notation C_k^n.
8. In the game of blackjack 2 cards are dealt to each player at the outset of the game. What is the probability that a certain player will get 2 aces on the deal?
9. Suppose there are 60 Democrats and 40 Republicans in the 100-member U.S. Senate. What is the probability of selecting 3 Democrats and 2 Republicans on a 5-member committee if the outcomes of the sample space are equally likely? Express answer using the notation C_k^n.
10. A publisher decides to publish 6 college textbooks next year in the subjects of English and history. If he has 7 manuscripts in English and 8 manuscripts in history, what is the probability that he will select 3 of each? Assume that the outcomes of the sample space are equally likely.
11. An employer has 4 female applicants and 6 male applicants for 5 vacant positions.
    (a) In how many ways can he hire 3 females and 2 males?
    (b) In how many ways can he hire 4 females and 1 male?
    (c) In how many ways can he hire 5 females?
    (d) In how many ways can he hire 5 people regardless of sex?
    (e) What is the probability that the employer will hire a majority of females? (Assume equally likely hiring practices.)

12. Suppose that a certain city council has 5 Republicans and 6 Democrats.
    (a) In how many ways can one form a committee consisting of 3 Democrats and 1 Republican?
    (b) In how many ways can one form a committee consisting of 4 Democrats and 0 Republicans?
    (c) In how many ways can a 4-person committee be formed?
    (d) What is the probability that a majority of the committee members will be Democrats? (Assume equally likely assignment practices.)
13. A certain national committee of 6 people is to be composed to perform a certain task. A list of 12 names has been sent to those who must make the selection. There are 4 women among the 11 Democrats on the list, and the only Republican is a male. If the outcomes are equally likely, what is the probability that the committee will include all 4 women *or* the Republican? Leave your answer in a form using the notation C_k^n. (*Hint:* Use Theorem 3.)
14. A sample of 4 parts is to be selected from among 100 that have been manufactured in a certain period of time. If there are 2 defective, unusable parts and 2 defective but usable parts among the 100 parts, what is the probability that the 2 defective, unusable parts *or* the 2 defective, usable parts will show up in the sample? (*Hint:* Use Theorem 3.)
15. Consider the probability that a 9-person appellate court (such as the U.S. Supreme Court) will reverse a certain lower-court decision. Compute the probability of a reversal if all 9 justices vote and then compare this result with the probability of a reversal if only 8 justices vote. Assume that all outcomes of the sample space are equally likely and that a reversal requires a simple majority (*more* than half) of the votes cast.
16. Compute the probability of a reversal of a lower court decision if all 9 justices of the U.S. Supreme Court vote, and if one assumes that 1 justice is certain to favor reversal and 2 justices are certain to oppose reversal. Also, assume that all outcomes are equally likely, and that a reversal requires a simple majority of the votes cast.
17. Verify the probability of obtaining a royal flush in a 5-card poker hand as shown in Example 7; try to determine $n(E)$ before looking at the third column of Table 4.7.
18. Verify the probability of obtaining a flush in a 5-card poker hand as shown in Example 7. [*Hint:* Do not forget to subtract the number of royal flushes in calculating $n(E)$.] Try to determine $n(E)$ before looking at the third column of Table 4.7.
19. Verify the probability of obtaining 4 of a kind in a 5-card poker hand as shown in Example 7. Try to determine $n(E)$ before looking at the third column of Table 4.7.
20. Verify the probability of obtaining a full house (3 of 1 kind and 2 of another kind) in a 5-card poker hand as shown in Example 7. Try to determine $n(E)$ before looking at the third column of Table 4.7.

4.9 Counting Techniques: Partitions (Optional)

In order to unify and extend the ideas presented in the last two sections, we shall find it very useful to use the concept of splitting or partitioning a set into nonempty cells or subsets such that every element of the given set belongs to exactly one (no more, no less) of these cells or subsets. For example, an ordinary deck of cards can be partitioned into 4 cells called spades, hearts, diamonds, and clubs; notice that every card in the deck belongs to exactly one cell. Of course, the same set of cards can be partitioned in other ways such as aces, face cards, and cards numbered 2 through 10.

▶ **Definition 9**
The set of nonempty subsets $A_1, A_2, \ldots, A_n$ of a finite set U is a **partition** of U if

(i) $A_1 \cup A_2 \cup \cdots \cup A_n = U$

and

(ii) $A_i \cap A_j = \emptyset$ for $i \neq j$.

Example 1
A set of four objects $\{a, b, c, d\}$ can be partitioned into three cells, containing 1, 1, and 2 objects, in the following ways:

$$\{\{a\}, \{b\}, \{c, d\}\},$$
$$\{\{a\}, \{c\}, \{b, d\}\},$$
$$\{\{a\}, \{d\}, \{b, c\}\},$$
$$\{\{b\}, \{c\}, \{a, d\}\},$$
$$\{\{b\}, \{d\}, \{a, c\}\},$$
$$\{\{c\}, \{d\}, \{a, b\}\}.$$

The definition of a partition makes no mention of the order of the cells in the partition, and hence the sets

$$\{\{a\}, \{b\}, \{c, d\}\} \quad \text{and} \quad \{\{b\}, \{a\}, \{c, d\}\}$$

represent the same partition; to make that distinction, we define an **ordered partition**. ∎

▶ **Definition 10**
An **ordered partition** is a partition in which the order of the cells is specified.

Example 2
Consider a set of 4 children that is to be partitioned into 2 cells. The 2 members of one cell will receive an experimental vaccine, while the 2 members of the second cell will serve as a control group and do not receive the vaccine. We may picture each cell as a box, and for illustrative purposes we list the ordered partitions of this problem. Let v, w, x, and y represent the children.

	Medicine	No Medicine
Partition 1	v w	x y
Partition 2	v x	w y
Partition 3	v y	w x
Partition 4	w x	v y
Partition 5	w y	v x
Partition 6	x y	v w

Notice that while the order of the cells is specified, the elements within each cell are *not* ordered. We now develop a formula for finding that there are 6 ordered partitions. Remember that there are 4! ways of ordering the 4 children. In partition 1, there were 2! ways that v and w could have been ordered in the first cell, and there were 2! ways that x and y could have been ordered in the second cell. Thus if we divide the total number of ordered arrangements, 4!, by the number of ways the elements can be ordered within each cell, 2! and 2!, we get

$$\frac{4!}{2!2!} \quad \text{or} \quad 6$$

as the number of ordered partitions, each cell of which has unordered elements. Notice that this formula is simply the formula for finding the number of combinations of 4 elements selected 2 at a time. This should not be surprising because we can think of those elements *selected* in a combination as constituting one cell and those *not selected* as constituting a second cell. ∎

In the next example we consider an ordered partition with three cells.

Example 3
Suppose that 8 people apply for 2 jobs. If 3 people are to be accepted for job A, 2 are to be accepted for job B, and 3 are to be turned away, how many ordered partitions can be so constructed?

Solution: Think of a partition with 3 cells,

A	B	N

First, we reason that the 8 people can be arranged in order in 8! ways. Those 3 who fall in cell A can be arranged in 3! ways, those 2 in cell B can be arranged in 2! ways, and those 3 in cell N can be arranged in 3! ways. Hence, to unorder the elements within each cell, we divide the total ordered

arrangements, 8!, by the number of ordered arrangements in each cell, namely, 3!, 2!, and 3!. We get

$$\frac{8!}{3!2!3!} \quad \text{or} \quad 560$$

ordered partitions. ∎

The ideas presented in the last two examples can be generalized to produce the formula of the next theorem.

▶ **Theorem 7**
If a finite set U of n elements is partitioned into k cells such that n_1 is the number of elements in the first cell, n_2 is the number of elements in the second cell, ..., n_k is the number of elements in the kth cell, so that

$$n = n_1 + n_2 + \cdots + n_k,$$

then the number of such ordered partitions is given by the formula

$$\frac{n!}{n_1!\, n_2!\, \cdots\, n_k!}. \quad ◀$$

Example 4
Consider a set of 4 animals that have a certain disease. In testing medicines for this disease, 1 of the animals is to receive a certain chemical A, 1 more is to receive chemical B, and the remaining 2 are to receive no chemical. The set of ordered partitions of the set of animals $\{a, b, c, d\}$ can be seen in Figure 4.23. The primary problem that we are concerned with is to determine the number of ordered partitions of the set $\{a, b, c, d\}$ of 4 animals. Of course, one way is to list and count the different ordered partitions as shown

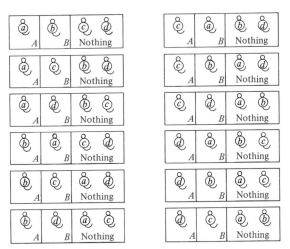

Figure 4.23. *Set of ordered partitions.*

Sec. 4.9] Counting Techniques: Partitions

in Figure 4.23. As one can see, even for this small problem, the task of listing the ordered partitions is not simple. To facilitate matters, we use the formula of Theorem 7.

First, however, in an ordered partition, we reemphasize that although the order of the cells is specified, the elements of the cells are not ordered. For example, in Figure 4.23 the interchange of the 2 animals in the last cell would not affect the ordered partition. In this problem $n = 4$, $n_1 = 1$, $n_2 = 1$, and $n_3 = 2$. Therefore, the number of ordered partitions is

$$\frac{4!}{(1!)(1!)(2!)} = \frac{(4 \cdot 3 \cdot 2 \cdot 1)}{(1) \cdot (1) \cdot (2 \cdot 1)} = 12. \blacksquare$$

Example 5

In Example 4 what is the probability that animal c, if selected at random, will receive medicine B?

Solution: The event space is pictured in Figure 4.24. From Figure 4.24

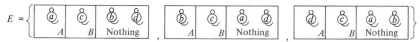

Figure 4.24

we see that $n(E) = 3$, and in Example 4 we found the number of elements in the sample space to be 12; hence

$$P(E) = \frac{n(E)}{n(U)} = \frac{3}{12} = \frac{1}{4}.$$

It is not always feasible to list the elements of the event space in such a way that we can count the elements as we did in Figure 4.24. Therefore, we use Theorem 7 as an alternative approach in finding $n(E)$; if we specify that animal c is to receive medicine B, then it is a question of finding the number of ordered partitions of the remaining set of 3 elements into the two cells "A" and "Nothing." Here we have $n = 3$, $n_1 = 1$, and $n_2 = 2$ and by the formula of Theorem 7

$$n(E) = \frac{3!}{1!2!} = 3. \blacksquare$$

Example 6

A set of 6 workers is to be partitioned in such a way that 3 workers are assigned to task A, 2 workers are assigned to task B, and 1 worker is assigned to task C. Determine the total number of ways these assignments can be made.

Solution: According to Theorem 7, $n = 6$, $n_1 = 3$, $n_2 = 2$, and $n_3 = 1$. Hence the number of ordered partitions subject to the stated requirements will be

$$\frac{6!}{3!2!1!} = \frac{6 \cdot 5 \cdot 4 \cdot 3 \cdot 2 \cdot 1}{(3 \cdot 2 \cdot 1)(2 \cdot 1)(1)} = 60. \blacksquare$$

Example 7

In Example 6 what is the probability that 2 specific workers will be assigned to task A?

Solution: The event space is too large to list; hence we rely on the formula of Theorem 7. If 2 specific workers are assigned to task A, then we must simply calculate the number of ways the remaining 4 workers can be assigned in such a way that 1 worker is assigned as the third worker to task A, 2 workers are assigned to task B, and 1 worker is assigned to task C. Hence $n = 4$, $n_1 = 1$, $n_2 = 2$, $n_3 = 1$, and, by Theorem 7,

$$n(E) = \frac{4!}{1!2!1!} = \frac{4 \cdot 3 \cdot 2 \cdot 1}{(1)(2)(1)} = 12.$$

Therefore, since from Example 6, $n(U) = 60$, we have

$$P(E) = \frac{n(E)}{n(U)} = \frac{12}{60} = \frac{1}{5}. \;\blacksquare$$

We have observed that the formula for finding the number of combinations of n elements selected k at a time can be thought of as a special case of Theorem 7. It so happens that the formula for obtaining the number of permutations of n elements selected k at a time can also be thought of as a special case of Theorem 7; think of each of the k elements selected as k separate cells while the $n - k$ unselected elements fall into a single cell. Hence from Theorem 7 we have

$$\frac{n!}{1!1! \cdots 1!(n-k)!},$$

which is equal to the formula for P_k^n.

APPLICATIONS

Example 8

In many card games, j cards are dealt to each of k players; we are now in a position to resolve the question of the number of ways this can be done. Suppose that there are n different cards in the deck and after each player receives j cards there will be $(n - kj)$ cards left undealt. (In some games all the cards are dealt and $n - kj$ is 0.) We are in essence asking for the number of ordered partitions of n cards having $k + 1$ cells (k players plus the undealt cards). Using Theorem 7, we have the number of deals equal to

$$\frac{n!}{(j!)^k(n-kj)!}.$$

For example, in the game of bridge 13 cards, from a deck of 52 cards, are dealt to each of 4 players with none left over; hence the number of bridge deals is

$$\frac{52!}{(13!)^4 0!}.$$

In the game of poker, 5 cards from a deck of 52 cards are dealt to k people. Thus we partition the 52 cards into $k + 1$ cells with 5 cards in each of k cells and $(52 - 5k)$ cards in the remaining cell. The number of ways this can be accomplished is

$$\frac{52!}{(5!)^k(52 - 5k)!}. \blacksquare$$

EXERCISES

Suggested minimum assignment: Exercises 1, 3, 4, 6, 8, and 9.

1. List all the ordered partitions of $\{a, b, c, d\}$, each partition having 2 cells with 2 elements in each cell.
2. List all the ordered partitions of $\{a, b, c, d\}$, each partition having 3 cells with 1, 2, and 1 elements in the respective cells.
3. Suppose that there are 10 cases pending in the office of a district attorney. His personnel have time to bring 5 to trial immediately, defer 3 until next month, and defer 2 until the next court session. Find the number of ways that the 10 cases can be disposed.
4. In how many ways can an 11-game football season have a record of 7 wins, 3 losses, and 1 tie?
5. How many deals of 5-card poker hands can be dealt from an ordinary deck of 52 cards in a 4-player game? (Do not try to reduce your answer to a single real number.) (*Hint:* Treat the undealt cards as a fifth hand.)
6. How many ways can 9 animals be partitioned so that 3 of them are inoculated against disease A, 3 are inoculated against disease B, and 3 are selected as the control group?
7. In Exercise 5 what is the probability that
    (a) The first hand has 4 aces?
    (b) Any of the 4 players has 4 aces?
8. In Exercise 6 what is the probability of choosing a pair of the animals that have been inoculated against
    (a) Disease A?
    (b) Any single disease?
9. A corporation has a plan by which it sends 12 of its employees to 3 specific colleges. If colleges A, B, and C will be attended by 6, 4, and 2 employees, respectively, what is the probability that 2 specific friends among the 12 will both be assigned to college A? (Assume equally likely assignments.)
10. Four 3-man space flights are planned. Twelve of 14 astronauts are to be assigned to these flights. Two will not get chosen, and none will make more than 1 flight. What is the probability that 2 specific friends will be assigned to the first flight? (Assume equally likely assignments.)
11. (a) Find the number of ways that 4 bridge hands can be dealt.

(b) Find the probability that the dealer of a bridge hand will receive all 4 aces.

(c) Find the probability that any one of the 4 players of a bridge hand will receive all 4 aces.

4.10 Binomial Experiments

In this section we again consider a method of solving probability problems having a sequence of n trials, but we add three restrictions.

We define a **binomial experiment** to be a stochastic experiment consisting of n successive trials that satisfy the following three requirements: (1) the result of any trial is independent of the result of any other trial; (2) each trial has the same 2 outcomes; and (3) the probability of either trial outcome will not change from one trial to another trial.

Example 1

Consider the experiment of flipping an ordinary coin 5 times. The trial outcomes are "heads" and "tails." The result of any trial is independent of the result of any other trial, and the probabilities of the two trial outcomes do not change for successive trials, namely,

$$P(\{\text{heads}\}) = \tfrac{1}{2} \text{ on every trial}$$

and

$$P(\{\text{tails}\}) = \tfrac{1}{2} \text{ on every trial;}$$

hence the experiment can be classified as a binomial experiment. ∎

The 2 outcomes of each trial are often viewed as a **success** or a **failure** to achieve a certain goal for each trial. The object then is to determine the probability of obtaining exactly k successes in n trials; as an illustration consider the following example.

Example 2

An assembly line production is sampled from time to time for purposes of quality control. If $\tfrac{9}{10}$ of the items are nondefective, what is the probability that a sample of 3 items will produce exactly 2 nondefective items?

Solution: In a single trial consisting of drawing 1 item from the assembly line, there is a $\tfrac{9}{10}$ probability p of success (that is, drawing a nondefective part), and there is a $\tfrac{1}{10}$ probability q of failure (that is, drawing a defective part). The 3 successive draws can be illustrated on the tree diagram shown in Figure 4.25. Each path of the diagram represents an outcome of the total experiment and may be represented by a triple such as *ssf*; we notice that there are C_2^3 or 3 paths having exactly 2 successes. Because the experiment outcomes,

sss, ssf, sfs, sff, fss, fsf, ffs, fff,

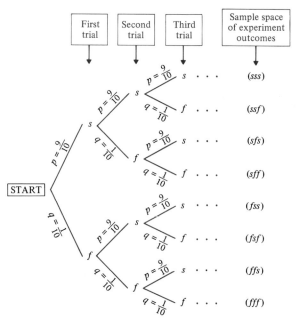

Figure 4.25. *Tree diagram of a binomial experiment.*

are not necessarily equally likely, weights must be assigned to each. Consistent with Theorem 6, each weight is defined to be the product of the probabilities of the respective *trial* outcomes that make up each *experiment* outcome. (One can prove that each of these weights is nonnegative and that the sum of the weights of all the experiment outcomes is 1 as required in Definition 2.)

Thus if we let E be the event that exactly 2 successes are achieved, then from the tree diagram we see that

$$P(E) = w(ssf) + w(sfs) + w(fss)$$
$$= (\tfrac{9}{10} \cdot \tfrac{9}{10} \cdot \tfrac{1}{10}) + (\tfrac{9}{10} \cdot \tfrac{1}{10} \cdot \tfrac{9}{10}) + (\tfrac{1}{10} \cdot \tfrac{9}{10} \cdot \tfrac{9}{10})$$
$$= C_2^3 (\tfrac{9}{10})^2 (\tfrac{1}{10}) = \tfrac{243}{1000} = .243. \blacksquare$$

The method illustrated in Example 2 is generalized by the following theorem, which is stated without proof.

▶ **Theorem 8**
If p is the probability of achieving a specified outcome on each single trial of a binomial experiment, then the probability of event E of achieving the specified outcome exactly k times out of n trials is given by the formula

$$P(E) = C_k^n p^k (1-p)^{n-k}. \blacktriangleleft$$

For the case where $n = 4$, the validity of Theorem 8 can be demonstrated by a tree diagram as shown in Figure 4.26, where $1 - p = q$. Each distinct

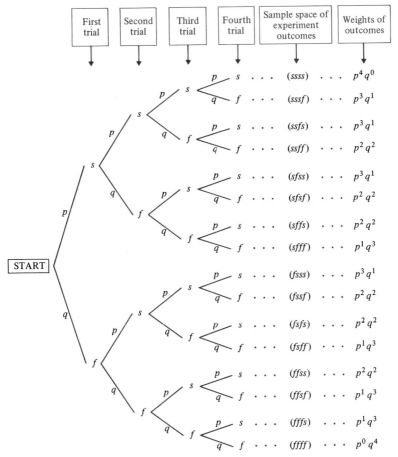

Figure 4.26. *Tree diagram of Theorem 8 for* $n = 4$.

path from "start" to a terminal on the right of the tree represents an experiment outcome (we can count 2^4 experiment outcomes in the sample space). Consistent with Theorem 6, the weight of each outcome is defined to be the product of the probabilities of the respective trial outcomes that make up each experiment outcome; hence each experiment outcome has a weight, or a probability of occurring, of $(p^k q^{n-k})$. Notice that there are C_k^4 ways in which an experiment outcome will contain exactly k specified trial outcomes; in other words, there are C_k^4 ways that one can succeed exactly k times in achieving a certain trial outcome in 4 trials.

Example 3
If 40 per cent of the population favors candidate A, what is the probability that 6 of 10 people polled will favor candidate A?
    Solution: By Theorem 8,

$$P(E) = C_6^{10} (\tfrac{4}{10})^6 (\tfrac{6}{10})^4 \approx .11. \quad \blacksquare$$

Sec. 4.10] Binomial Experiments

Example 4

In casting a fair die the probability of casting a 6 is $\frac{1}{6}$, while the probability of *not* casting a 6 is $1 - \frac{1}{6}$, or $\frac{5}{6}$. In 4 trials, the probability of event E_4 of achieving a 6 on all 4 trials is given by the formula of Theorem 8:

$$P(E_4) = C_4^4 \left(\tfrac{1}{6}\right)^4 \left(\tfrac{5}{6}\right)^0 = \tfrac{1}{1296}.$$

In 4 trials, the probability of event E_3 of achieving a 6 on exactly 3 trials is

$$P(E_3) = C_3^4 \left(\tfrac{1}{6}\right)^3 \left(\tfrac{5}{6}\right)^1 = \tfrac{20}{1296}.$$

In 4 trials, the probability of event E_2 of achieving a 6 on exactly 2 trials is

$$P(E_2) = C_2^4 \left(\tfrac{1}{6}\right)^2 \left(\tfrac{5}{6}\right)^2 = \tfrac{150}{1296}.$$

In 4 trials, the probability of event E_1 of achieving a 6 on exactly 1 trial is

$$P(E_1) = C_1^4 \left(\tfrac{1}{6}\right)^1 \left(\tfrac{5}{6}\right)^3 = \tfrac{500}{1296}.$$

In 4 trials, the probability of event E_0 of achieving a 6 on none of the trials is

$$P(E_0) = C_0^4 \left(\tfrac{1}{6}\right)^0 \left(\tfrac{5}{6}\right)^4 = \tfrac{625}{1296}. \quad\blacksquare$$

The various events $E_0, \ldots, E_4$ listed in Example 4 are mutually exclusive and exhaust all possibilities (that is, they partition the sample space); therefore, the sum of these probabilities should equal 1. That is, in Example 4,

$$1 = P(E_4) + P(E_3) + P(E_2) + P(E_1) + P(E_0),$$

or

$$1 = C_4^4\left(\tfrac{1}{6}\right)^4\left(\tfrac{5}{6}\right)^0 + C_3^4\left(\tfrac{1}{6}\right)^3\left(\tfrac{5}{6}\right)^1 + C_2^4\left(\tfrac{1}{6}\right)^2\left(\tfrac{5}{6}\right)^2 + C_1^4\left(\tfrac{1}{6}\right)^1\left(\tfrac{5}{6}\right)^3 + C_0^4\left(\tfrac{1}{6}\right)^0\left(\tfrac{5}{6}\right)^4. \quad (4.2)$$

The reason for the name "binomial experiment" can be explained as follows: From elementary algebra (or from Section 3.2) recall the binomial expansion for $(p + q)^4$,

$$(p + q)^4 = p^4 + 4p^3q + 6p^2q^2 + 4pq^3 + q^4;$$

the latter expression is the same as

$$(p + q)^4 = C_4^4 p^4 q^0 + C_3^4 p^3 q^1 + C_2^4 p^2 q^2 + C_1^4 p^1 q^3 + C_0^4 p^0 q^4. \quad (4.3)$$

If we let p be the probability of success in achieving the trial outcome described in Example 4 and if q is $1 - p$ or the probability of failure in achieving that outcome, then we can see that the expression (4.2) can be thought of as a special case of the binomial expansion (4.3), where $p = \frac{1}{6}$ and $q = \frac{5}{6}$.

In fact, we can observe that, in general, each term of the binomial expansion of $(p + q)^n$ represents the probability of success in achieving a certain trial outcome a specific number of times in n trials, where the probability of success in a given trial is p and the probability of failure is q. The terms of the expansion represent the probabilities of mutually exclusive

events, and hence the probabilities of various unions of these events can be found by simply adding the appropriate terms of the expansion.

Example 5
If a fair coin is flipped 10 times, what is the probability that a head will be flipped *at least* 8 times?

Solution: By Theorem 8 the probability of getting exactly 8 heads is $C_8^{10}(\frac{1}{2})^8(\frac{1}{2})^2$, the probability of getting exactly 9 heads is $C_9^{10}(\frac{1}{2})^9(\frac{1}{2})^1$, and the probability of getting exactly 10 heads is $C_{10}^{10}(\frac{1}{2})^{10}(\frac{1}{2})^0$. Because these are probabilities of mutually exclusive events, the probability of the union of these events is the sum of the respective probabilities, or simply the sum of the last three terms of the binomial expansion of $(p+q)^{10}$, where $p = \frac{1}{2}$ and $q = \frac{1}{2}$.

$$P(E) = C_8^{10}(\tfrac{1}{2})^8(\tfrac{1}{2})^2 + C_9^{10}(\tfrac{1}{2})^9(\tfrac{1}{2})^1 + C_{10}^{10}(\tfrac{1}{2})^{10}(\tfrac{1}{2})^0 \approx .055.\ \blacksquare$$

APPLICATIONS

Example 6
An interesting physical illustration of the binomial expansion is the fountain shown in Figure 4.27. Water runs into the top basin at a rate of, say, 1 gallon per minute; the water overflows from each basin on two sides according to a fixed ratio for all basins. The ratio chosen for Figure 4.27 was $1:2$, and 4 levels of basins are shown; this produces a physical analog of the binomial expansion for $(p+q)^n$, where $p = \frac{1}{3}$, $q = \frac{2}{3}$, and $n = 3$. $\blacksquare$

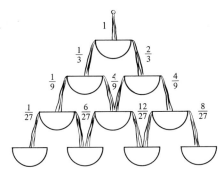

Figure 4.27. *Fountain analog of a binomial expansion.*

Example 7
Sampling is used in a variety of ways, such as political polls, opinion polls, market surveys, acceptance sampling, and so on. There are certain risks, however, in drawing conclusions based on samples; for example, suppose that a buyer makes a small sample of the items that he intends to purchase. One risk is the possibility that there *are* a very large number of defective items which will *not* appear in the sample in sufficient numbers to cause a rejection of the order. On the other hand, there may be very *few* defective

items, and yet enough of them *will* appear in the sample to cause an unfortunate rejection of the entire lot.

As an illustration, suppose that a sample of 10 items is selected with the stipulation that the order will be accepted if no more than 2 members of the sample are defective. Let p be the unknown fraction of the lot that is actually defective, and let E be the event that 0, 1, or 2 defective items will appear in the sample. By the methods of this section,

$$P(E) = C_0^{10} p^0 (1-p)^{10} + C_1^{10} p (1-p)^9 + C_2^{10} p^2 (1-p)^8$$
$$= (1-p)^{10} + 10p(1-p)^9 + 45p^2(1-p)^8.$$

A graph relating the probability of acceptance $P(E)$ to the actual fraction p of defective parts is given in Figure 4.28.

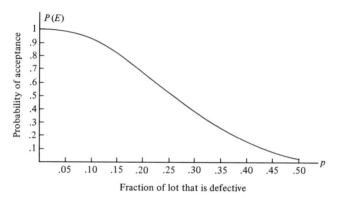

Figure 4.28. *Probability of acceptance based on the sample size of 10 items.*

Notice that there is a probability of approximately .26 that the lot will be accepted even though 35 per cent ($p = .35$) of the parts are actually defective. On the other hand, notice that when the actual percentage of defective parts is as low as 5 per cent ($p = .05$), there is a probability of approximately .01, $1 - P(E)$, that the lot will be rejected. Of course, these risks can be reduced by taking larger samples, but the buyer may not be willing to pay that price. ∎

Example 8

Suppose that unknown to the investigator, a certain drug causes a very serious reaction in 5 per cent of the people who use the drug. What is the probability that the investigator will not detect this serious reaction in a sample of 10 people who are tested with the drug? If $p = .05$ is the probability that the reaction *will* appear in 1 person (1 trial), then $q = .95$ is the probability that the reaction will *not* appear in 1 person (1 trial). In $n = 10$ trials the probability of event E of no reactions is

$$P(E) = C_0^{10} p^0 q^{10}$$
$$= 1(.05)^0 (.95)^{10}$$
$$\approx .5987.$$

If, however, a sample of 100 people were used (that is, $n = 100$ trials were used), the probability of event F of no reactions is

$$P(F) = C_0^{100} p^0 q^{100}$$
$$= 1(.05)^0(.95)^{100}$$
$$\approx .0059.$$

Thus it can be seen that an increase in the sample size from 10 to 100 can reduce the probability of nondetection of the reaction from about 60 per cent to about $\frac{3}{5}$ per cent. ∎

Example 9

Binomial experiments can be used to study the role of chance in bureaucratic decision making. For example, suppose that a certain individual in a decision-making position makes correct decisions in at least 80 per cent of his chances; is this because of the individual's wisdom, or is this a result of luck?

Under the assumption that each decision is independent of the others, that a distinction can be made between correct and incorrect decisions, and that the probability p of making a correct decision applies to all decisions and decision makers, then the probability $P(K)$ that an individual can make at least k correct decisions in n chances is

$$P(K) = C_k^n p^k (1-p)^{n-k} + \cdots + C_n^n p^n (1-p)^0.$$

If one considers m individuals as m components of a parallel system (see Example 6 of Section 4.5), then the probability $P(E)$ that at least one individual in an organization of m people will make at least k correct decisions is given by the formula

$$P(E) = 1 - [1 - P(K)]^m.$$

In other words, $P(E)$ measures the reliability of the parallel system. (Of course, one could also consider series, or mixed, systems in connection with this problem.) We consider the role of luck in achieving an 80 per cent (or better) record of correct decisions out of 10 decisions by at least 1 of 30 people; suppose that every decision was based on the flip of a fair coin, which means that $p = \frac{1}{2}$. Let $k = 8$, $n = 10$, and $m = 30$. This information produces the result

$$P(E) = 1 - (1 - .0547)^{30} \approx .815.$$

Thus in this example, under the assumptions made, there is a probability of .815 that by simply flipping a coin for each decision, at least 1 of the 30 people would have 80 per cent, or better, correct decisions.

A study similar to this example may be found in Deutsch and Madow [19]. ∎

Example 10

In certain experiments in psychology there is a time interval, called the latency period, between a stimulus and corresponding response. Binomial

experiments are used by McGill [47] to construct a model to study fluctuations in the latency period. Consider the binomial experiment shown in Figure 4.29, with outcomes response r and no response $\sim r$ having probabilities p and $1 - p$, respectively. Assume that the number of no responses occurring between 2 responses can be considered as a latency random variable, and suppose this variable is denoted x. From Figure 4.29 the probability of the event E that x is equal to k (where $k = 0, 1, 2, \cdots$) is given by the formula

$$P(E) = p(1 - p)^k. \blacksquare$$

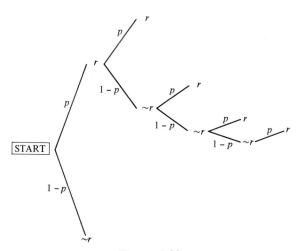

Figure 4.29

EXERCISES

Suggested minimum assignment: Exercises 1, 3, 4, 5, 7, 9, and 11.

In each of Exercises 1–4 draw a tree diagram representing the experiment, identify the sample space of experiment outcomes, state their respective weights, and calculate the requested probability from the diagram and also from the formula of Theorem 8.

1. If a fair coin is flipped 4 times, what is the probability of obtaining a head exactly 3 times? Are the experiment outcomes equally likely? Give reasons for your answer to the latter question.
2. If a die is cast 4 times, what is the probability of obtaining a number that is greater than 4 on exactly 1 throw?
3. Assume that the probability of a pair of animals having a male offspring is $\frac{4}{7}$ and the probability of having a female offspring is $\frac{3}{7}$. What is the probability of having 1 male and 3 females in 4 births?

4. Assume that the probability that a newborn member of a certain species will die before age 40 is $\frac{1}{3}$. What is the probability that exactly 2 of 4 randomly selected newborn members will live to be age 40?

5. In Exercise 3, what is the probability of obtaining at least 3 females in 4 births?

6. In Exercise 4, what is the probability that at least 2 newborn members will live to be age 40?

7. Assume that the reliability of a component of a certain system is $\frac{9}{10}$.
   (a) What is the probability that exactly 5 out of 6 of the same components are nondefective?
   (b) What is the probability that at least 5 out of 6 of the same components are nondefective?

8. (a) If 10 per cent of the parts produced by a certain factory are defective, what is the probability that a sample of 10 items will produce exactly 1 defective part?
   (b) In part (a), what is the probability that the sample will produce no more than 1 defective part?

9. Suppose that a majority of the people, say 60 per cent, favor a referendum. What is the probability that a poll of 5 people will indicate that the people oppose the referendum?

10. A multiple-choice test has 4 "answers" listed for each of 10 questions where only 1 answer is correct. What is the probability that a student who knows nothing about the material can guess at the answers and get
    (a) 10 correct answers?
    (b) 9 correct answers?
    (c) At least 9 correct answers?

11. Suppose that a certain company is trying to determine whether or not to develop a certain piece of beachfront property. Certain predictable losses will occur if a hurricane strikes the property during the fall season. The company has calculated that during the first 3 years, the company can absorb the losses incurred by 1 hurricane but that 2 or more hurricanes would cause the venture to be unprofitable. If the probability of a hurricane hitting that property in 1 year is $\frac{1}{4}$, what is the probability of a profitable venture over the 3-year period?

12. A certain person who breeds racehorses expects to produce 5 new offspring within the week. Each female will be worth $1000 and each male worth $2000. Depending upon the number of male and female births, the 5 new horses will be worth a certain total amount. Treat each possible amount (in dollars) as an outcome of an experiment.
    (a) What are the outcomes of the experiment?
    (b) If there is a $\frac{1}{3}$ probability of a female birth and a $\frac{2}{3}$ probability of a male birth for each offspring, what are the weights of the respective outcomes found in part (a)?
    (c) What is the expected value for the experiment?

13. A certain game is set up whereby a prize is given if exactly 3 "sixes" are obtained on 5 throws of a die.
    (a) What is the probability of winning?
    (b) If the sponsor of the game expects to charge 10 cents to play the game, how much can he pay as a prize if he plans an expected value of 1 cent per game?

If the restriction of 2 trial outcomes in Theorem 8 is relaxed to permit r trial outcomes with respective probabilities of $p_1, p_2, \ldots, p_r$, the formula of Theorem 8 can be generalized; the probability of event E of achieving exactly $n_1, n_2, \ldots, n_r$ of the respective trial outcomes is

$$P(E) = \frac{n!}{n_1! n_2! \cdots n_r!} p_1^{n_1} p_2^{n_2} \cdots p_r^{n_r},$$

where $n = n_1 + \cdots + n_r$. Apply this generalization to solve Exercises 14–16.

14. Assume that in a certain primitive country the probability of a live male offspring is 50 per cent, the probability of a live female offspring is 40 per cent, and the probability of a dead offspring is 10 per cent. What is the probability that in 6 births 3 will be male, 3 will be female, and 0 will be dead?

15. Suppose that in the manufacture of a certain product 10 per cent of the products will be defective and unusable, 20 per cent will be defective but usable, and 70 per cent will be nondefective. In a sample of 5 products selected at random, what is the probability that 3 products will be nondefective and 2 defective but usable?

16. Assume that the probability that a certain baseball player will get a hit is $\frac{3}{10}$, the probability that he will strike out is $\frac{2}{10}$, the probability that he will make a ground-ball out is $\frac{1}{10}$, and the probability that he will fly out is $\frac{4}{10}$. What is the probability that he will do all 4 in 4 at-bats?

N E W   V O C A B U L A R Y

sample spaces 4.1
events or event spaces 4.1
equally likely outcomes 4.1
probability of an event 4.1
weights of outcomes 4.1
a priori probabilities 4.1
a posteriori probabilities 4.1
mutually exclusive events 4.2
expected value 4.3
odds in favor of E 4.3
odds against E 4.3
conditional probability 4.4

tree diagram 4.4
path of a tree diagram 4.4
independent events 4.5
stochastic experiment 4.6
fundamental counting principle 4.7
permutation 4.7
combination 4.8
partition of a set 4.9
ordered partition of a set 4.9
binomial experiment 4.10
trial success 4.10
trial failure 4.10

5 The Algebra of Matrices

The use of matrices is becoming increasingly important in many disciplines. Current publications in agriculture, business, economics, political science, psychology, physics, chemistry, biology, and engineering reflect this fact. Actually the concept of a matrix is not a new one; some say that the matrix was first introduced by the mathematician Arthur Cayley (English, 1821–1895) in 1858. The increasing emphasis on quantitative methods in the disciplines just mentioned, together with the development of the high-speed computer, which can perform matrix operations, are two of many reasons for the recent interest in matrices.

The purpose of this chapter is to introduce the algebra of matrices sufficiently that we can apply matrix methods in a study of systems of linear equations, the characteristic value problem, and linear programming; all of these topics lead directly to a variety of applications. A further purpose is to begin building a vocabulary to enable the reader to read and understand literature that uses the language of matrix algebra.

With the exception of the "Applications," the material of this chapter is not dependent upon the material of the previous chapters, although Section 3.10 would be helpful.

Prerequisites: high school algebra. Chapter 4 is needed for many of the applications.
Suggested Sections for Moderate Emphasis: 5.1–5.8.
Suggested Sections for Minimum Emphasis: 5.1–5.3, 5.5.

5.1 Matrices

In recent years there has been a dramatic increase in the use of rectangular arrays of numbers in applied problems. These arrays, such as

$$\begin{bmatrix} 2 & 3 & 4 \\ 6 & 9 & 2 \end{bmatrix},$$

are treated as single entities and are subject to various operations, some of which we shall define in this chapter. Evidence of the usefulness of arrays may be found by observing the very prominent position they occupy in some computer languages such as APL. In this section we shall define a matrix and a few operations involving matrices.

▶ **Definition 1**
Let m and n be positive integers; a rectangular array of *real* numbers arranged in m rows and n columns,

$$A = \begin{bmatrix} a_{11} & a_{12} & \cdots & a_{1n} \\ a_{21} & a_{22} & \cdots & a_{2n} \\ \vdots & \vdots & & \vdots \\ a_{m1} & a_{m2} & \cdots & a_{mn} \end{bmatrix}$$

is called an **m by n matrix**.

Example 1
The rectangular arrays

$$\begin{bmatrix} 2 & 4 \\ 3 & -2 \end{bmatrix}, \quad \begin{bmatrix} \sqrt{3} \\ 0 \\ 1 \end{bmatrix}, \quad \begin{bmatrix} 3 & \tfrac{2}{3} \\ 4 & 1 \\ 6 & 0 \end{bmatrix}, \quad [-2 \ \ 0 \ \ 1], \quad \begin{bmatrix} \sqrt{3} & 4 & 2 \\ 1 & \tfrac{7}{2} & 9 \\ 0 & 6 & 4 \end{bmatrix}$$

are all examples of matrices. ∎

The phrase "m by n" in Definition 1 is called the **order** of the matrix; for instance, in Example 1 the order of the first matrix is 2 by 2, while the order of the second matrix is 3 by 1. Notce that the number of rows is always listed first. The individual numbers in the array are called the **entries** of the matrix.

If the number of rows is the same as the number of columns of a matrix, then the matrix is said to be a **square matrix**. The **main diagonal** of a square matrix consists of those entries with the same row and column subscript; that is, $a_{11}, a_{22}, \ldots, a_{ii}$.

Example 2
The main diagonal entries of the following 3 by 3 matrix are enclosed in parentheses for emphasis.

$$A = \begin{bmatrix} (2) & 3 & 4 \\ 2 & (3) & 0 \\ 1 & 4 & (5) \end{bmatrix}. \ \blacksquare$$

In this book we shall consider matrices whose entries are real numbers, but the reader should be aware that studies of matrices can be made where the entries are other than real numbers. *In this book when the word "matrix" is used, we shall mean a matrix with real number entries.*

Rather than write out the array shown in Definition 1, it is customary to abbreviate the representation of a general m by n matrix by

$$A = [a_{ij}]_{(m,n)},$$

where it is understood that the row subscript i assumes all integral values from 1 to m and the column subscript j assumes all integral values from 1 to n.

Example 3
Using abbreviated notation, we can represent the matrix

$$\begin{bmatrix} a_{11} & a_{12} & a_{13} & a_{14} \\ a_{21} & a_{22} & a_{23} & a_{24} \\ a_{31} & a_{32} & a_{33} & a_{34} \end{bmatrix} \text{ by } [a_{ij}]_{(3,4)}.\ \blacksquare$$

We now use the abbreviated notation to define matrix equality, matrix addition, multiplication of a matrix by a real number, and matrix subtraction.

▶ **Definition 2**
Two matrices of the same order $A = [a_{ij}]_{(m,n)}$ and $B = [b_{ij}]_{(m,n)}$ are **equal** if $a_{ij} = b_{ij}$ for all i and j.

In other words, two matrices of the same order are equal when all their corresponding entries are equal. By the notation $A \neq B$ we mean that A is not equal to B.

Example 4

$$\begin{bmatrix} 2 & 3 & \frac{4}{2} \\ \sqrt{4} & 0 & -1 \end{bmatrix} = \begin{bmatrix} 2 & 3 & 2 \\ 2 & 0 & -1 \end{bmatrix},$$

$$\begin{bmatrix} 2 & 3 & 4 \\ 7 & 1 & 9 \end{bmatrix} \neq \begin{bmatrix} 2 & 3 & 4 \\ 7 & 1 & 4 \end{bmatrix}.\ \blacksquare$$

▶ **Definition 3**
The **sum** of two matrices of the same order $A = [a_{ij}]_{(m,n)}$ and $B = [b_{ij}]_{(m,n)}$ is defined to be

$$A + B = [a_{ij} + b_{ij}]_{(m,n)}.$$

In other words, two matrices of the same order can be added by adding corresponding entries. When two matrices are of the same order, they are said to be **conformable for addition**.

Example 5

$$\begin{bmatrix} 1 & 3 & -2 \\ \frac{2}{3} & -4 & 0 \end{bmatrix} + \begin{bmatrix} 2 & 0 & 4 \\ \frac{4}{3} & 1 & 6 \end{bmatrix} = \begin{bmatrix} (1+2) & (3+0) & (-2+4) \\ (\frac{2}{3}+\frac{4}{3}) & (-4+1) & (0+6) \end{bmatrix}$$

$$= \begin{bmatrix} 3 & 3 & 2 \\ 2 & -3 & 6 \end{bmatrix}.\ \blacksquare$$

In Exercises 15 and 16 we will prove that matrix addition has the following properties.

▶ **Theorem 1**

(Commutative Property) If A and B are matrices of the same order, then
$$A + B = B + A. \blacktriangleleft$$

▶ **Theorem 2**

(Associative Property) If A, B, and C are matrices of the same order, then
$$(A + B) + C = A + (B + C). \blacktriangleleft$$

The reader will be asked to illustrate these properties in Exercises 13 and 14.

Now consider the sum $A + A + A + A$; the following definition is made in such a way that we can say that this sum is equal to $4A$.

▶ **Definition 4**

If $A = [a_{ij}]_{(m,n)}$ is a matrix and c is a real number, then
$$cA = [ca_{ij}]_{(m,n)} \quad \text{and} \quad Ac = [a_{ij}c]_{(m,n)}.$$

In other words, any matrix can be multiplied by a real number by multiplying each of the entries of the matrix by the real number.

Example 6

If $A = \begin{bmatrix} 2 & -3 & 4 \\ 6 & 9 & 0 \end{bmatrix}$ and $c = 2$, then

$$cA = (2)\begin{bmatrix} 2 & -3 & 4 \\ 6 & 9 & 0 \end{bmatrix} = \begin{bmatrix} 4 & -6 & 8 \\ 12 & 18 & 0 \end{bmatrix}. \blacksquare$$

Various properties of the operation defined in Definition 4 are given in Exercise 18. One such property is $cA = Ac$.

Now that the meaning of $(-1)A$ has been established by Definition 4, we can define **subtraction** of two matrices of the same order by

$$B - A = B + (-1)A.$$

APPLICATIONS

Example 7

Diagrams are often used to exhibit certain defined relations among persons, nations, animals, plants, genetic characteristics, or various other components of a system. As an illustration of one such relation, consider that of influence among 4 people. In Figure 5.1 let the circles represent the persons (components) involved and let an arrow from person i to person j indicate that person i has influence over person j; suppose that within the group of 4 people, influence can be exerted as shown in the figure. Figure 5.1 can be characterized by the matrix

$$A = \begin{array}{c} \\ \\ \\ \\ \end{array} \overbrace{\begin{bmatrix} 0 & 1 & 0 & 0 \\ 1 & 0 & 0 & 1 \\ 0 & 0 & 0 & 1 \\ 1 & 1 & 0 & 0 \end{bmatrix}}^{\text{Influenced}} \left.\begin{array}{c} 1 \\ 2 \\ 3 \\ 4 \end{array}\right\} \text{Influencer}$$

where $a_{ij} = 1$ if i influences j and $a_{ij} = 0$ if i does *not* influence j. We assume that no person influences himself, and hence the entries on the main diagonal are 0.

A very interesting use of this model was made in a study by Ulmer [64] of influence among the Justices of the Michigan Supreme Court. Similar models may apply to dominance of biological characteristics or sociological behavior, or to communication systems between nations or business associates. Of course, for large communication systems, diagrams like Figure 5.1 are very unwieldy, whereas the matrix representation of the system is a single entity that is subject to the various useful matrix operations that will be developed later. The concept of a mathematical model is discussed in Sections 12.4 and 12.5. ∎

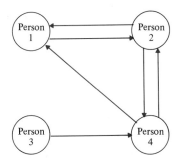

Figure 5.1. *Influence diagram.*

Example 8

The illustration that is begun in this example will be continued in later sections of this chapter as the reader attains more information about matrices. The basic mathematical model that is developed has widespread use in a variety of applications. Suppose that at specified time intervals, a certain system can be in one of two states: It either operates satisfactorily or it does not operate satisfactorily (that is, has values of 1 or 0). Further suppose that this system has a self-correcting device that activates if needed at the specified time intervals. Experience has shown that when the system is operating satisfactorily at the beginning of one time interval, then 60 per cent of the time it will continue to operate satisfactorily and 40 per cent of the time it will not operate satisfactorily at the beginning of the next time interval. If, however, the system is not operating satisfactorily, then 20 per cent of the time it will remain in this state and 80 per cent of the time it

will operate satisfactorily at the beginning of the next time interval. This information is given in Figure 5.2. As we shall see in future examples, it is advantageous to display these probabilities in matrix form.

$$A = \begin{bmatrix} .6 & .4 \\ .8 & .2 \end{bmatrix} \begin{matrix} \text{satisfactory} \\ \text{unsatisfactory} \end{matrix} \bigg\} \text{From}$$

with column headings "To: satisfactory, unsatisfactory".

An n by n matrix whose ijth entry represents the probability that a system in state i will, after a specified time interval, be in state j is called a **transition matrix** of the system. Other uses of the same basic model include (1) the transition matrix of consumer purchasing probabilities of brands on successive purchases, (2) the transition matrix of probabilities of a disease progressing from one state to another (for an illustration see Alling [3]), and (3) the transition matrix of probabilities of voting behavior. ∎

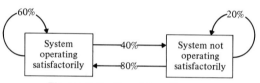

Figure 5.2. *Transition diagram.*

Example 9

This example is an illustration of the use of transition matrices in the field of genetics. For a particular gene locus with possible alleles A and a, assume that within a population, allele A is present with probability p and allele a is present with probability $1 - p$. Suppose that we consider a parent and his offspring; for each possible genotype of the parent, the probabilities of the genotype of the child are given in the tree diagrams of Figure 5.3 and also in the transition diagram of Figure 5.4. The same information can be displayed in the matrix

$$A = \begin{bmatrix} p & 1-p & 0 \\ \tfrac{1}{2}p & \tfrac{1}{2} & \tfrac{1}{2}(1-p) \\ 0 & p & 1-p \end{bmatrix} \begin{matrix} AA \\ Aa \\ aa \end{matrix} \bigg\} \begin{matrix} \text{Genotype} \\ \text{of parent} \end{matrix}$$

with column headings "Genotype of child: AA, Aa, aa".

One advantage of the matrix display over the displays of Figures 5.3 and 5.4 will be illustrated in Section 5.3 when the probabilities of the genotypes of later generations will be calculated by means of matrix multiplication. ∎

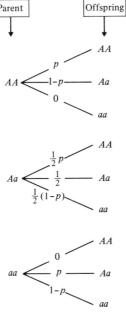

Figure 5.3. Tree diagrams of transitions.

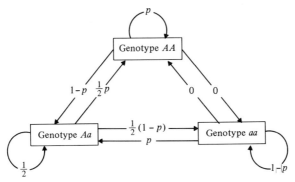

Figure 5.4. Transition diagram.

EXERCISES

Suggested minimum assignment: Exercises 1, 3, 5, 7, 9, 12, 14, and 18(a).

In Exercises 1–6 find the sum of each pair of matrices if possible.

1. $\begin{bmatrix} 1 & 2 \\ 4 & 0 \end{bmatrix}, \begin{bmatrix} 0 & 3 \\ 2 & 4 \end{bmatrix}.$

2. $\begin{bmatrix} 1 \\ 3 \end{bmatrix}, \begin{bmatrix} 2 \\ -4 \end{bmatrix}.$

3. $[2 \ 3 \ 4], [7 \ 9 \ 1].$

4. $\begin{bmatrix} 4 & 3 \\ 2 & 1 \end{bmatrix}, \begin{bmatrix} 0 & 0 \\ 0 & 0 \end{bmatrix}.$

5. $\begin{bmatrix} 2 & 1 & 2 \\ 4 & 3 & 1 \end{bmatrix}$, $\begin{bmatrix} 2 & 4 \\ 2 & 1 \end{bmatrix}$.

6. $\begin{bmatrix} 2 & 1 \\ 3 & 4 \\ 6 & -2 \end{bmatrix}$, $\begin{bmatrix} 1 \\ 3 \\ 2 \end{bmatrix}$.

7. Let $A = \begin{bmatrix} 2 & 3 & -4 \\ 6 & 9 & 1 \end{bmatrix}$, $B = \begin{bmatrix} 0 & -2 & 4 \\ -1 & 3 & 1 \end{bmatrix}$, and $k = 4$.

   (a) Find $A + kB$.
   (b) What is the order of the result of part (a)?
   (c) If $C = A - B$, find c_{12}.

8. Let $A = \begin{bmatrix} 2 & 8 & 1 \\ 1 & 0 & -2 \\ 2 & 1 & 0 \end{bmatrix}$, $B = \begin{bmatrix} 0 & -4 & 1 \\ 2 & 3 & 5 \\ 0 & 0 & 1 \end{bmatrix}$, and $k = 3$.

   (a) Find $A + kB$.
   (b) What is the order of the result of part (a)?
   (c) If $C = A - B$, find c_{23}.

9. Express $A = \begin{bmatrix} a_{11} & a_{12} & a_{13} \\ a_{21} & a_{22} & a_{23} \end{bmatrix}$ in abbreviated notation.

10. Write the rectangular array for which
$$A = [a_{ij}]_{(4,2)}.$$

11. Two matrices can be added only when they have the same order; is there any restriction on when a matrix can be multiplied by a real number?

12. Are $A = [a_{ij}]_{(3,4)}$ and $B = [b_{ij}]_{(3,5)}$ conformable for addition? Give the reason for your answer.

13. Illustrate Theorem 1 with the matrices
$$A = \begin{bmatrix} 1 & 2 \\ 3 & 4 \end{bmatrix} \quad \text{and} \quad B = \begin{bmatrix} 0 & -1 \\ 4 & 2 \end{bmatrix}.$$

14. Illustrate Theorem 2 with the matrices
$$A = \begin{bmatrix} 2 & -4 \\ 6 & 9 \end{bmatrix}, \quad B = \begin{bmatrix} 2 & 4 \\ 0 & 3 \end{bmatrix}, \quad \text{and} \quad C = \begin{bmatrix} 2 & 0 \\ 7 & 1 \end{bmatrix}.$$

15. Prove Theorem 1. (*Hint:* The statements are given, and the reader should complete the proof by supplying the reasons.)
$$\begin{aligned} A + B &= [a_{ij}]_{(m,n)} + [b_{ij}]_{(m,n)} \\ &= [a_{ij} + b_{ij}]_{(m,n)} \\ &= [b_{ij} + a_{ij}]_{(m,n)} \\ &= [b_{ij}]_{(m,n)} + [a_{ij}]_{(m,n)} \\ &= B + A. \end{aligned}$$

16. Prove Theorem 2. (*Hint:* See Exercise 15.)

17. Prove: If A, B, and C have the same order, then
$$A + (B + C) = (B + A) + C.$$

18. For the operation defined in Definition 4, the following properties are valid. If A and B are matrices of the same order and if c and d are real numbers, then
$$cA = Ac,$$
$$(cd)A = c(dA),$$
$$c(A + B) = cA + cB,$$
$$(c + d)A = cA + dA.$$

(a) If $A = \begin{bmatrix} 2 & 1 \\ 4 & 0 \end{bmatrix}$, $B = \begin{bmatrix} 3 & 4 \\ 2 & 1 \end{bmatrix}$, $c = 2$, and $d = 3$, illustrate each property.

(b) Prove the first property. (*Hint:* Let the first statement be $cA = c[a_{ij}]_{(m,n)}$.)

(c) Prove the second property.

(d) Prove the third property.

(e) Prove the fourth property.

5.2 Vectors

There is often the need for expressing certain information by means of a list of numbers that are arranged in a specific order; for example, consider the alphabetical price list ($10, $20, $50, $25, $60) of 5 items in a store or the coordinates (3, 1, 4) of a point in 3-space (see Section 3.10). Such ordered lists of numbers are known as **vectors**. We shall limit our discussion of vectors to ordered lists of *real* numbers.

▶ **Definition 5**

An ordered list $(a_1, a_2, \ldots, a_n)$ of n real numbers is a **vector** of **dimension** n. A vector also may be displayed in the following two ways:

1. As a 1 by n matrix $[a_1 \quad a_2 \quad \cdots \quad a_n]$, which is called a **row vector**, and

2. As a n by 1 matrix $\begin{bmatrix} a_1 \\ a_2 \\ \vdots \\ a_n \end{bmatrix}$, which is called a **column vector**.

The numbers $a_1, a_2, \ldots, a_n$ are called the **components** of the vector $(a_1, a_2, \ldots, a_n)$. If all the components are 0, then the vector is called a **zero vector**. If at least one component is nonzero, then the vector is a **nonzero vector**. Ordinarily, we shall designate vectors by the Greek letters α, β, γ, and so on.

Example 1
The list of closing prices for the alphabetized stocks listed on the New York Stock Exchange is a vector. The dimension of the vector is the number of stocks listed. ∎

Example 2
The ordered list of coordinates (3, 6, 7) of a point in 3-space (see Figure 5.5) is a vector of dimension 3. The vector also can be displayed in either of the two forms

$$[3 \quad 6 \quad 7] \quad \text{or} \quad \begin{bmatrix} 3 \\ 6 \\ 7 \end{bmatrix}.$$

An arrow originating at the origin of coordinates and terminating at the point (3, 6, 7) (see Figure 5.5) may be used as a geometric representation of the row vector [3 6 7] or of the corresponding column vector. Such an interpretation is very useful because an arrow indicates both length and direction; as shown later in Example 12 of this section, these ideas lead to several applications. ∎

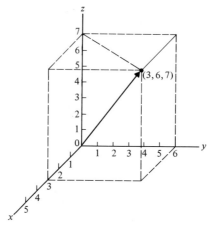

Figure 5.5

Example 3
Each column of a given m by n matrix is a column vector of dimension m, and each row of the given matrix is a row vector of dimension n. ∎

We now define equality of vectors and several operations.

▶ **Definition 6**
Two vectors of the same dimension are **equal** if their corresponding components are equal.

▶ **Definition 7**

The **sum** of two vectors of the same dimension, $\alpha = (a_1, a_2, \ldots, a_n)$ and $\beta = (b_1, b_2, \ldots, b_n)$, is defined by

$$\alpha + \beta = (a_1 + b_1, a_2 + b_2, \ldots, a_n + b_n).$$

In other words, two vectors can be added by adding corresponding components.

▶ **Definition 8**

If $\alpha = (a_1, a_2, \ldots, a_n)$ is a vector and c is a real number, then

$$c\alpha = c(a_1, a_2, \ldots, a_n) = (ca_1, ca_2, \ldots, ca_n).$$

Subtraction of vectors can be defined according to

$$\alpha - \beta = \alpha + (-1)\beta.$$

Example 4

Under what condition is $c\alpha - \beta = \gamma$ if $\alpha = (1, 2)$, $\beta = (4, 3)$, $c = 2$, and $\gamma = (x, y)$?

$$c\alpha - \beta = 2(1, 2) + (-1)(4, 3) = (2, 4) + (-4, -3)$$
$$= (-2, 1).$$

Therefore, $\gamma = c\alpha - \beta$ if $x = -2$ and $y = 1$. ∎

At this point it is natural to wonder about the possibility of multiplying two vectors. Actually there are several types of products of vectors. The one that we shall define here is known as the **dot product.**

▶ **Definition 9**

If α and β are two vectors of the same dimension, where $\alpha = (a_1, a_2, \ldots, a_n)$ and $\beta = (b_1, b_2, \ldots, b_n)$, then the **dot product** $\alpha \cdot \beta$ is the real number determined by

$$\alpha \cdot \beta = a_1 b_1 + a_2 b_2 + \cdots + a_n b_n.$$

Example 5

Let $\alpha = (3, 2, 0)$ and $\beta = (2, 4, 6)$. Then according to Definition 9,

$$\alpha \cdot \beta = 3 \cdot 2 + 2 \cdot 4 + 0 \cdot 6 = 14. \blacksquare$$

Example 6

Let $\alpha = \begin{bmatrix} 2 \\ 6 \end{bmatrix}$ and $\beta = \begin{bmatrix} 3 \\ 4 \end{bmatrix}$. Then

$$\alpha \cdot \beta = 2 \cdot 3 + 6 \cdot 4 = 30. \blacksquare$$

Sec. 5.2] Vectors

Example 7
Suppose that a stockbroker ordered shares of four different stocks in quantities revealed by the respective components of
$$\alpha = [100 \quad 200 \quad 200 \quad 300].$$
The respective prices of these stocks are given by the components of
$$\beta = [\$60 \quad \$50 \quad \$20 \quad \$40].$$
The total value of the purchase is
$$\alpha \cdot \beta = 100 \cdot 60 + 200 \cdot 50 + 200 \cdot 20 + 300 \cdot 40$$
$$= \$32,000. \blacksquare$$

Example 8
A primary application of the dot product of vectors is its use in writing a linear expression. The dot product of $[a_1 \quad a_2 \quad \cdots \quad a_n]$ and $[x_1 \quad x_2 \quad \cdots \quad x_n]$ is the linear expression $a_1 x_1 + a_2 x_2 + \cdots + a_n x_n$. $\blacksquare$

▶ **Definition 10**
The **magnitude** (or **length**) of a vector α is denoted by $\|\alpha\|$ and is equal to the real number $\sqrt{\alpha \cdot \alpha}$. That is, if $\alpha = (a_1, a_2, \ldots, a_n)$, then
$$\|\alpha\| = \sqrt{a_1^2 + a_2^2 + \cdots + a_n^2}.$$

Example 9
Let $\alpha = (3, 6, 7)$. Then by Definition 10,
$$\|\alpha\| = \sqrt{(3, 6, 7) \cdot (3, 6, 7)}$$
$$= \sqrt{3^2 + 6^2 + 7^2}$$
$$= \sqrt{94}.$$
Geometrically, $\|\alpha\|$ can be interpreted as the length of the arrow from the origin to the point $(3, 6, 7)$ as shown in Figure 5.6. $\blacksquare$

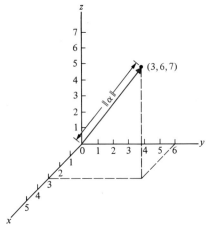

Figure 5.6. *Geometric interpretation of* $\|\alpha\|$.

APPLICATIONS

Example 10

Construct an oversimplified model economy in which there are 3 industries: the farming industry, the clothing industry, and the construction industry. There are 6 consumers: the general public, the government, those outside the economy who consume the exports, and the 3 industries themselves. These consumers exert certain demands on each of the industries. For example, suppose that in one unit of time the farming industry uses 3 units of its own production, 1 unit of production from the clothing industry, and 2 units of production from the construction industry. The demand vector then for the farming industry is expressed as $\alpha_f = (3, 1, 2)$. In the same manner the demand vectors for the industries are specified in the form

$$\alpha = (\text{farm products, clothing products, construction products}),$$

and suppose they are

farming industry	$\alpha_f = (3, 1, 2)$,
clothing industry	$\alpha_c = (1, 1, 2)$,
construction industry	$\alpha_b = (2, 2, 1)$,
general public	$\alpha_p = (2, 2, 2)$,
government	$\alpha_g = (1, 1, 1)$,
exports	$\alpha_e = (4, 1, 1)$.

The total demand on all industries then is

$$\alpha_{\text{total}} = \alpha_f + \alpha_c + \alpha_b + \alpha_p + \alpha_g + \alpha_e = (13, 8, 9).$$

If the price of one unit of farm production is $3, one unit of the clothing industry is $4, and one unit of the construction industry is $5, then a price vector β can be written

$$\beta = (3, 4, 5).$$

If the industries produce exactly what is demanded of them, the income of the farming industry is 13 units times a price of $3 per unit, or $39. Its operating costs, however, are

$$\alpha_f \cdot \beta = (3, 1, 2) \cdot (3, 4, 5) = \$23.$$

Therefore, the profit of the farming industry is

$$\$39 - \$23 = \$16.$$

A calculation of the profits or losses of the other industries is left for Exercise 19. ∎

Example 11

Suppose that a sum of money is to be distributed by four arbitrators A, B, C, and D into 3 different projects. Suppose that the arbitrators' initial positions

on the allocation of the items are represented by the vectors

$$\alpha = (a_1, a_2, a_3),$$
$$\beta = (b_1, b_2, b_3),$$
$$\gamma = (c_1, c_2, c_3),$$
$$\delta = (d_1, d_2, d_3).$$

Let the actual compromise allocation be represented by the vector

$$\pi = (p_1, p_2, p_3).$$

The four vectors $\pi - \alpha$, $\pi - \beta$, $\pi - \gamma$, and $\pi - \delta$ represent both the magnitude and direction of change in the positions of each arbitrator. The magnitudes $||\pi - \alpha||$, $||\pi - \beta||$, $||\pi - \gamma||$, and $||\pi - \delta||$ may be used to rank the degree of position changes of the arbitrators, and, over a period of time, such information may be used to study such things as leadership, influence, and extreme positions. Of course, the basic mathematical model can be interpreted in many ways, such as 3 allocations of property by a committee of four realtors, or appropriations to 3 budget items by a governmental body of four people or four agencies, or the allocation of 3 chemicals for control of a disease by a group of four scientists. ∎

Example 12

The purpose of this example is to point out a few of the many ways that vectors can be applied in the life sciences. All these applications are summarized in greater detail in Batschelet [6] pp. 426–432, where further references may be found. Most of these applications result from the fact that a vector exhibits both magnitude and direction; hence concepts such as force and velocity can be quantified through the use of vectors. Certain information from elementary trigonometry and elementary vector analysis is assumed in the derivation of the equations of this example; an inability to follow these derivations should not preclude the reader from achieving the purpose of the example as stated in the first sentence.

1. *Animal navigation.* Birds and fish are particularly vulnerable to air and water currents, respectively. Consider a bird that attempts to fly in a certain direction with a wind blowing. Let β be the velocity of the bird with respect to the air, and let ω be the velocity of the wind with respect to the ground. If the direction and magnitudes are as indicated in Figure 5.7, then the actual velocity α of the bird with respect to the ground is given by $\alpha = \beta + \omega$. Because of the wind, the bird's efforts must be oriented in the direction of β to achieve the direction of α. The ground speed of the bird is $||\alpha||$.

2. *Limb traction.* If pulleys are arranged as shown in Figure 5.8, the stretching force (vector) σ is the sum of the forces (vectors) σ_1 and σ_2. If μ is a vector of length 1 parallel to σ, then the magnitude of the stretching force σ can be calculated to be $2(\sigma_1 \cdot \mu) = 2||\sigma_1|| \cos \theta$, where $||\sigma_1||$ equals w. The stretching force can be increased by making the angle θ smaller.

3. *Limb action.* In Figure 5.9 one can see that the force (vector) α

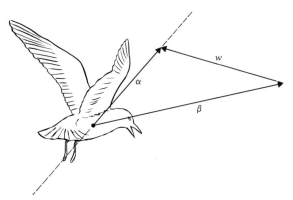

Figure 5.7. *Wind effect on bird flight.*

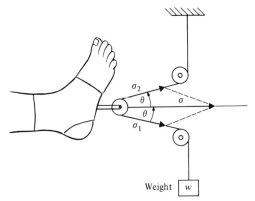

Figure 5.8. *Force of traction.*

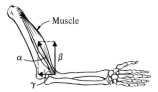

Figure 5.9. *Forces exerted upon flex of the arm muscle.*

exerted by the muscle that helps lift the horizontal forearm must be as large or larger than the lifting force β because $\alpha = \beta + \gamma$; how much larger depends on the angle of the arm at the elbow. The force γ is exerted on the elbow with corresponding stress on the bones and ligaments therein. Similar studies can be made to investigate forces causing broken bones and torn ligaments.

4. *Animal locomotion.* Consider a grasshopper whose center of gravity moves from point A to point B during the time his feet are on the ground. See Figure 5.10. Represent the directed line segment from A to B by the vector β. Let the average force exerted by the grasshopper to get from A to

Sec. 5.2] Vectors

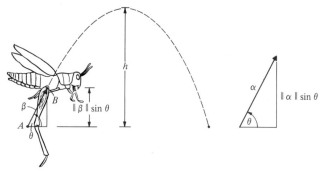

Figure 5.10. *Forces and motion involved in a grasshopper jump.*

B be designated by the vector α. The product of the vertical components of these two vectors is the amount of kinetic energy that is transformed into potential energy. Neglecting air resistance, we have

$$(\|\alpha\| \sin \theta)(\|\beta\| \sin \theta) = mgh,$$

where m is the mass of the grasshopper, g is the acceleration due to gravity, and h is the maximum height of the jump above level ground. The height of the jump then is given by

$$h = \frac{1}{mg} \|\alpha\| \, \|\beta\| \sin^2 \theta. \ \blacksquare$$

EXERCISES

Suggested minimum assignment: Exercises 1, 3, 5, 7, 9, 12, 14, and 18.

In each of Exercises 1–8 let $\alpha = (1, 4, -2)$ and $\beta = (6, 0, 2)$.

1. Find $\alpha + \beta$.
2. Find 3α.
3. Find $3\alpha - \beta$.
4. Find $4\alpha - 2\beta$.
5. Find $\alpha \cdot \beta$.
6. Find $\beta \cdot 2\alpha$.
7. Find $\|\alpha\|$.
8. Find $\|\beta\|$.

9. If $\alpha = (2, 3, 5)$ represents the coordinates of a point in 3-space, graph α and interpret $\|\alpha\|$ geometrically.
10. If $\beta = (4, 3, 7)$ represents the coordinates of a point in 3-space, graph β and interpret $\|\beta\|$ geometrically.
11. Find the sum and the dot product of $\alpha = (6, 3, 4)$ and the zero vector of the same dimension.
12. If we define vectors $\alpha = (2, 1, 4)$ and $\gamma = (3, 2)$, find the sum and the dot product of α and γ if possible. If not possible, give reason.
13. If an arbitrary nonzero vector α is multiplied by the real number $1/\|\alpha\|$, then the resulting vector has a magnitude of 1 unit. Verify that this is the case for $\alpha = [1 \ \ 3 \ \ 5]$.

14. If
$$A = [2\ 3\ 4] \quad \text{and} \quad B = \begin{bmatrix} 1 \\ 4 \\ 2 \end{bmatrix},$$
find $A \cdot B$.
15. Prove that the addition of vectors of the same dimension is commutative.
16. Prove that the dot product of vectors of the same dimension is commutative.
17. Is subtraction of vectors of the same dimension commutative? Why?
18. In Example 7 of this section suppose that one week later the stockbroker sold all the same shares that he had purchased at prices given by the vector
$$[\$70\ \$40\ \$10\ \$50].$$
Use the dot product to calculate his profit or loss.
19. In Example 10 find the profits or losses of the other two industries.
20. In a situation similar to Example 11, suppose that 4 councilmen must allocate $2000 among two budgetary items. Suppose that the initial positions of the councilmen were
$$\alpha_1 = (\$1000, \$1000), \quad \alpha_2 = (\$1500, \$500),$$
$$\alpha_3 = (\$1400, \$600), \quad \alpha_4 = (\$900, \$1100).$$
The final allocation was ($1200, $800). Represent the change in position of each councilman by a vector. If the magnitudes of the vectors just calculated are used as the criteria, which councilman made the greatest change in position and which made the least change in position?

5.3 Matrix Multiplication

We now turn our attention to another matrix operation that has proved to be very useful in a variety of applications. The **product** of two matrices is simply the result of a succession of dot products of vectors.

▶ **Definition 11**
Let A be an m by p matrix, and let B be a p by n matrix. The **product** $C = AB$ is an m by n matrix, each entry of which is determined by the following dot product:
$$c_{ij} = (i\text{th row of } A) \cdot (j\text{th column of } B).$$

Example 1
Let
$$A = \begin{bmatrix} 2 & 1 \\ 3 & 4 \end{bmatrix} \quad \text{and} \quad B = \begin{bmatrix} 0 & 1 \\ 5 & 7 \end{bmatrix}.$$

Then the product AB can be found by taking four dot products as shown below:

1. $\begin{bmatrix} \boxed{2\ 1} \\ 3\ 4 \end{bmatrix} \begin{bmatrix} \boxed{0} & 1 \\ \boxed{5} & 7 \end{bmatrix} = \begin{bmatrix} (2\cdot 0 + 1\cdot 5) & (\quad) \\ (\quad) & (\quad) \end{bmatrix}.$

2. $\begin{bmatrix} \boxed{2\ 1} \\ 3\ 4 \end{bmatrix} \begin{bmatrix} 0 & \boxed{1} \\ 5 & \boxed{7} \end{bmatrix} = \begin{bmatrix} (\quad) & (2\cdot 1 + 1\cdot 7) \\ (\quad) & (\quad) \end{bmatrix}.$

3. $\begin{bmatrix} 2\ 1 \\ \boxed{3\ 4} \end{bmatrix} \begin{bmatrix} \boxed{0} & 1 \\ \boxed{5} & 7 \end{bmatrix} = \begin{bmatrix} (\quad) & (\quad) \\ (3\cdot 0 + 4\cdot 5) & (\quad) \end{bmatrix}.$

4. $\begin{bmatrix} 2\ 1 \\ \boxed{3\ 4} \end{bmatrix} \begin{bmatrix} 0 & \boxed{1} \\ 5 & \boxed{7} \end{bmatrix} = \begin{bmatrix} (\quad) & (\quad) \\ (\quad) & (3\cdot 1 + 4\cdot 7) \end{bmatrix}.$

Thus

$$C = AB = \begin{bmatrix} 5 & 9 \\ 20 & 31 \end{bmatrix}. \blacksquare$$

Example 2
Consider the matrix equation

$$\begin{bmatrix} 2 & 3 & 1 \\ 1 & 0 & -1 \\ 1 & 1 & 1 \end{bmatrix} \begin{bmatrix} x \\ y \\ z \end{bmatrix} = \begin{bmatrix} 2 \\ 3 \\ 4 \end{bmatrix}.$$

Multiplying the matrices on the left side, we get

$$\begin{bmatrix} (2x + 3y + z) \\ (x\quad\ -z) \\ (x + y + z) \end{bmatrix} = \begin{bmatrix} 2 \\ 3 \\ 4 \end{bmatrix}.$$

From the definition of equality of matrices,

$$\begin{aligned} 2x + 3y + z &= 2, \\ x\quad\quad - z &= 3, \\ x + y + z &= 4. \end{aligned}$$

Thus we see that matrix multiplication allows the use of a single matrix equation to represent a system of linear equations. ∎

Because each entry of the product AB is a dot product of a row of A with a column of B, the product exists only when the number of components of a row of A is the same as the number of components of a column of B. In other words, the number of columns of A must equal the number of rows of B in order for AB to exist.

Example 3
Let
$$A = \begin{bmatrix} 3 & 2 & 1 \\ 4 & 6 & 1 \end{bmatrix} \quad \text{and} \quad B = \begin{bmatrix} 2 & 1 \\ 3 & 2 \end{bmatrix}.$$

The product AB does not exist because we cannot find the dot product of a row of A with a column of B. For instance,

$$[3 \quad 2 \quad 1] \cdot \begin{bmatrix} 2 \\ 3 \end{bmatrix}$$

does not exist. The product BA does exist, however, and is

$$BA = \begin{bmatrix} 2 & 1 \\ 3 & 2 \end{bmatrix} \begin{bmatrix} 3 & 2 & 1 \\ 4 & 6 & 1 \end{bmatrix} = \begin{bmatrix} 10 & 10 & 3 \\ 17 & 18 & 5 \end{bmatrix}.$$

We say that **B is conformable to A for multiplication** when the product BA exists. ∎

Example 3 illustrates the very important fact that, in general, $AB \neq BA$. In other words, multiplication is *not* commutative. Exercises 1–8 further illustrate this fact. Multiplication is associative, however, as the following theorem states.

▶ **Theorem 3**
(Associative Property) If A, B, and C are matrices, and if A is conformable to B and B is conformable to C for multiplication, then

$$A(BC) = (AB)C.$$

A proof of Theorem 3, for the case where A, B, and C are 2 by 2 matrices, is left for Exercise 14. A proof for the general case (using abbreviated notation) is similar. ◀

Another important property that can be stated now is the distributive property, which involves both addition and multiplication.

▶ **Theorem 4**
(Distributive Property) If A, B, and C are matrices, and if the necessary conformability for multiplication and addition exists, then
 (i) $A(B + C) = AB + AC$.
 (ii) $(A + B)C = AC + BC$.

A proof of Theorem 4(i), for the case where A, B, and C are 2 by 2 matrices, is left for Exercise 15. A proof of each part for the general case (using abbreviated notation) is similar. ◀

Example 4
The purpose of this example is to illustrate Theorems 3 and 4. Let

$$A = \begin{bmatrix} 2 & 1 \\ 0 & 7 \end{bmatrix}, \quad B = \begin{bmatrix} 4 & 3 \\ 2 & 1 \end{bmatrix}, \quad \text{and} \quad C = \begin{bmatrix} 0 & 2 \\ 1 & 4 \end{bmatrix}.$$

$$(AB)C = \left(\begin{bmatrix} 2 & 1 \\ 0 & 7 \end{bmatrix}\begin{bmatrix} 4 & 3 \\ 2 & 1 \end{bmatrix}\right)\begin{bmatrix} 0 & 2 \\ 1 & 4 \end{bmatrix} = \begin{bmatrix} 10 & 7 \\ 14 & 7 \end{bmatrix}\begin{bmatrix} 0 & 2 \\ 1 & 4 \end{bmatrix} = \begin{bmatrix} 7 & 48 \\ 7 & 56 \end{bmatrix};$$

$$A(BC) = \begin{bmatrix} 2 & 1 \\ 0 & 7 \end{bmatrix}\left(\begin{bmatrix} 4 & 3 \\ 2 & 1 \end{bmatrix}\begin{bmatrix} 0 & 2 \\ 1 & 4 \end{bmatrix}\right) = \begin{bmatrix} 2 & 1 \\ 0 & 7 \end{bmatrix}\begin{bmatrix} 3 & 20 \\ 1 & 8 \end{bmatrix} = \begin{bmatrix} 7 & 48 \\ 7 & 56 \end{bmatrix}.$$

Notice that $(AB)C = A(BC)$. Also,

$$A(B+C) = \begin{bmatrix} 2 & 1 \\ 0 & 7 \end{bmatrix}\left(\begin{bmatrix} 4 & 3 \\ 2 & 1 \end{bmatrix} + \begin{bmatrix} 0 & 2 \\ 1 & 4 \end{bmatrix}\right) = \begin{bmatrix} 2 & 1 \\ 0 & 7 \end{bmatrix}\begin{bmatrix} 4 & 5 \\ 3 & 5 \end{bmatrix} = \begin{bmatrix} 11 & 15 \\ 21 & 35 \end{bmatrix};$$

$$AB + AC = \begin{bmatrix} 2 & 1 \\ 0 & 7 \end{bmatrix}\begin{bmatrix} 4 & 3 \\ 2 & 1 \end{bmatrix} + \begin{bmatrix} 2 & 1 \\ 0 & 7 \end{bmatrix}\begin{bmatrix} 0 & 2 \\ 1 & 4 \end{bmatrix}$$

$$= \begin{bmatrix} 10 & 7 \\ 14 & 7 \end{bmatrix} + \begin{bmatrix} 1 & 8 \\ 7 & 28 \end{bmatrix} = \begin{bmatrix} 11 & 15 \\ 21 & 35 \end{bmatrix}.$$

Notice that $A(B+C) = AB + AC$. ∎

Now that a type of multiplication of matrices has been defined, we can introduce positive integral powers of square matrices. If A is a square matrix, then

$$A^1 = A, \quad A^2 = AA, \quad A^3 = AA^2, \quad \ldots, \quad A^{k+1} = AA^k,$$

where k is a positive integer.

APPLICATIONS

Example 5
Consider a particular ecological system having vegetation providing food for herbivorous animals. Construct a matrix

$$A = \begin{matrix} & \overbrace{\begin{matrix} 1 & 2 & & p \end{matrix}}^{\text{Species of herbivorous animals}} & \\ \begin{bmatrix} a_{11} & a_{12} & \cdots & a_{1p} \\ a_{21} & a_{22} & \cdots & a_{2p} \\ \vdots & \vdots & & \vdots \\ a_{m1} & a_{m2} & \cdots & a_{mp} \end{bmatrix} & \begin{matrix} 1 \\ 2 \\ \\ m \end{matrix} \right\} \text{Species of vegetation} \end{matrix}$$

where a_{ij} represents the average amount of species i of vegetation eaten by each individual of species j of herbivorous animals during a season. Also

construct a matrix

$$B = \begin{bmatrix} b_{11} & b_{12} & \cdots & b_{1n} \\ b_{21} & b_{22} & \cdots & b_{2n} \\ \vdots & \vdots & & \vdots \\ b_{p1} & b_{p2} & \cdots & b_{pn} \end{bmatrix} \begin{matrix} 1 \\ 2 \\ \\ p \end{matrix}$$

with column labels "Species of carnivorous animals" ($1, 2, \ldots, n$) and row labels "Species of herbivorous animals" ($1, 2, \ldots, p$),

where b_{ij} represents the average amount of species i of herbivorous animals eaten by each individual of species j of carnivorous animals during a season.

Carnivorous species 1, by feeding on herbivorous species 1, indirectly consumes $a_{11}b_{11}$ amount of plant species 1. Also, the same carnivorous species 1, by feeding on herbivorous species 2, indirectly consumes $a_{12}b_{21}$ amount of plant species 1. In general, an individual of carnivorous species 1, by feeding on all p of the herbivorous species, will indirectly consume

$$a_{11}b_{11} + a_{12}b_{21} + \cdots + a_{1p}b_{p1}$$

units of plant species 1. The latter result is simply the first entry of the product AB. Other entries of AB will similarly produce the amounts of each of various plant species indirectly consumed by an individual of each of the various carnivorous species.

Similar uses of matrix multiplication may be made in representing the interaction of components in an environmental system. ∎

Example 6

A **probability vector** can be defined as a row matrix with nonnegative entries whose sum is 1. A matrix consisting of rows of probability vectors is called a **stochastic matrix**. As the name suggests, these matrices can be used to display the probabilities of a stochastic experiment illustrated by the tree diagram of Figure 5.11. The probabilities of the first trial can be displayed in the stochastic row matrix $P = [p_{11} \quad p_{12} \quad p_{13}]$; the probabilities of the second trial can be displayed in the stochastic matrix

$$Q = \begin{bmatrix} q_{11} & q_{12} \\ q_{21} & q_{22} \\ q_{31} & q_{32} \end{bmatrix},$$

where q_{ij} is the probability that trial outcome Q_j will follow trial outcome P_i. Consider the matrix product

$$PQ = [p_{11} \quad p_{12} \quad p_{13}] \begin{bmatrix} q_{11} & q_{12} \\ q_{21} & q_{22} \\ q_{31} & q_{32} \end{bmatrix}$$

$$= [(p_{11}q_{11} + p_{12}q_{21} + p_{13}q_{31}) \quad (p_{11}q_{12} + p_{12}q_{22} + p_{13}q_{32})].$$

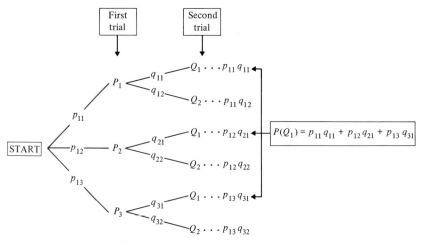

Figure 5.11. *Stochastic experiment.*

Notice that PQ is a 1 by 2 matrix, where the first entry is the probability that after 2 trials, outcome Q_1 will be obtained (see Figure 5.11), and the second entry is the probability that after 2 trials, outcome Q_2 will be obtained (see Figure 5.11). The use of these ideas in the study of anthropology may be found in Kay [36], pp. 185–190. ∎

Example 7

In this example we consider a special type of stochastic experiment, where a finite set of trial outcomes (called **states**) is the same for every trial, and where the probability p_{ij} that trial outcome a_j will follow trial outcome a_i is the same for every trial. Such stochastic experiments are called **Markov chains**. As an illustration, reconsider Example 8 of Section 5.1, where the set of trial outcomes or states was

{system performing satisfactorily, system not performing satisfactorily}.

The probabilities of changing or not changing states is displayed in the transition matrix

$$A = \begin{matrix} & \overbrace{\text{satis-} \quad \text{unsatis-}}^{\text{To}} \\ & \text{factory} \quad \text{factory} \\ & \begin{bmatrix} .6 & .4 \\ .8 & .2 \end{bmatrix} \begin{matrix} \text{satisfactory} \\ \text{unsatisfactory} \end{matrix} \Big\} \text{From} \end{matrix}$$

A row matrix showing the probabilities of the respective elements of the set of trial outcomes is called a **state vector**. If there is a probability of .9 that the system begins in the state of performing satisfactorily, then the initial state vector B_0 is [.9 .1] and the state vector B_1 after one period is given by the product

$$B_1 = B_0 A = [.9 \quad .1] \begin{bmatrix} .6 & .4 \\ .8 & .2 \end{bmatrix}$$
$$= [(.54 + .08)(.36 + .02)] = [.62 \quad .38].$$

This result is illustrated in Figure 5.12. Next, to find the state vector B_2 after 2 periods, the process is repeated.

$$B_2 = B_1 A = [.62 \quad .38] \begin{bmatrix} .6 & .4 \\ .8 & .2 \end{bmatrix}$$
$$= [(.372 + .304)(.248 + .076)]$$
$$= [.676 \quad .324].$$

An alternative way of calculating B_2 is to recognize that

$$B_2 = B_1 A = (B_0 A)A = B_0(AA) = B_0 A^2.$$

In other words, A^2 represents a transition matrix of a 2-stage change. Using this same approach, one can develop a formula for B_k:

$$B_k = B_0 A^k. \quad \blacksquare$$

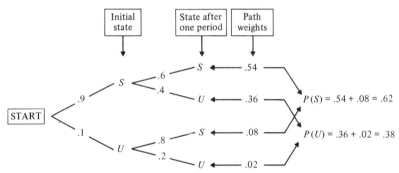

Figure 5.12. *Tree diagram computation of components of the state vector.*

Example 8
In Example 9 of Section 5.1 we found that the matrix

$$A = \begin{array}{c} \\ \\ \end{array} \overbrace{\begin{bmatrix} p & 1-p & 0 \\ \tfrac{1}{2}p & \tfrac{1}{2} & \tfrac{1}{2}(1-p) \\ 0 & p & 1-p \end{bmatrix}}^{\text{Genotype of child}} \begin{array}{l} AA \\ Aa \\ aa \end{array} \Big\} \begin{array}{l} \text{Genotype} \\ \text{of parent} \end{array}$$

can be used to display the various probabilities of offspring genotypes for each parental genotype. Suppose that an investigator wants to determine the probabilities of the various genotypes of the grandchildren of a member of the same population used in Example 9 of Section 5.1; using the information of that example, we can use a tree diagram and the methods of

Chapter 4 to calculate the desired probabilities as shown in Figure 5.13. (In Figure 5.13 let $q = 1 - p$.) An alternative way of obtaining these desired probabilities is to recognize that A is a transition matrix of a Markov chain, and hence the probabilities after 2 transitions are displayed as the entries of A^2. Let $q = 1 - p$; then

$$A^2 = \begin{bmatrix} p^2 + \tfrac{1}{2}pq & pq + \tfrac{1}{2}q & \tfrac{1}{2}q^2 \\ \tfrac{1}{2}p^2 + \tfrac{1}{4}p & pq + \tfrac{1}{4} & \tfrac{1}{4}q + \tfrac{1}{2}q^2 \\ \tfrac{1}{2}p^2 & \tfrac{1}{2}p + pq & \tfrac{1}{2}pq + q^2 \end{bmatrix} \begin{matrix} AA \\ Aa \\ aa \end{matrix} \Bigg\} \text{Genotype of parent}$$

$$\overbrace{}^{\text{Genotype of grandchild}}$$

Notice that the last row of A^2 provides the same information that can be calculated from the tree diagram of Figure 5.13. The first two rows of A^2 provide information that can be illustrated by tree diagrams similar to Figure 5.13. Probabilities of genotypes of later generations of offspring can be calculated by considering higher powers of A. ∎

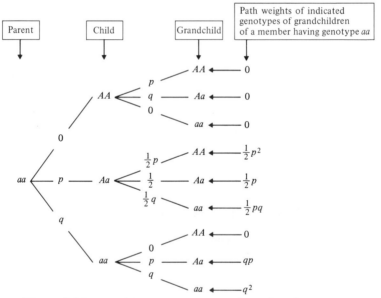

Figure 5.13. *Tree diagram computation of weights of genotypes.*

EXERCISES

Suggested minimum assignment: Exercises 1, 3, 5, 7, 9, 12, 13, and 16.

In Exercises 1–8 find AB and BA if possible. If AB or BA does not exist, give reasons.

1. $A = \begin{bmatrix} 2 & 1 \\ 3 & 0 \end{bmatrix}$, $B = \begin{bmatrix} 0 & 1 \\ 3 & -4 \end{bmatrix}$.

2. $A = \begin{bmatrix} 3 & 0 & 0 \\ 0 & 2 & 1 \\ 6 & 9 & 0 \end{bmatrix}$, $B = \begin{bmatrix} 0 & 1 & 4 \\ 0 & 2 & 3 \\ -1 & 0 & 5 \end{bmatrix}$.

3. $A = \begin{bmatrix} 2 & 1 & 3 \\ 0 & 4 & 2 \end{bmatrix}$, $B = \begin{bmatrix} 1 & 2 \\ 1 & 5 \end{bmatrix}$.

4. $A = \begin{bmatrix} 3 & 1 & 4 \\ 0 & 2 & 1 \\ 6 & 5 & 0 \end{bmatrix}$, $B = \begin{bmatrix} 3 & 2 \\ 0 & 1 \\ -3 & 5 \end{bmatrix}$.

5. $A = \begin{bmatrix} 1 & 3 \end{bmatrix}$, $B = \begin{bmatrix} 8 \\ 3 \end{bmatrix}$.

6. $A = \begin{bmatrix} 1 \\ 4 \\ 3 \end{bmatrix}$, $B = \begin{bmatrix} 2 & 1 & 5 \end{bmatrix}$.

7. $A = \begin{bmatrix} 2 & 3 \\ 4 & 5 \end{bmatrix}$, $B = \begin{bmatrix} x \\ y \end{bmatrix}$.

8. $A = \begin{bmatrix} 3 & 4 & 5 \\ 6 & 9 & 2 \\ 3 & 5 & 2 \end{bmatrix}$, $B = \begin{bmatrix} x \\ y \\ z \end{bmatrix}$.

9. Find A^2 if $A = \begin{bmatrix} 2 & 3 \\ 4 & 6 \end{bmatrix}$.

10. Find A^3 if $A = \begin{bmatrix} 3 & 2 \\ 1 & 0 \end{bmatrix}$.

11. Illustrate the associative property of multiplication with the matrices
$$A = \begin{bmatrix} 2 & 3 \\ 4 & 5 \end{bmatrix}, \quad B = \begin{bmatrix} 0 & 1 \\ 2 & 3 \end{bmatrix}, \quad C = \begin{bmatrix} 4 & 2 \\ 0 & 1 \end{bmatrix}.$$

12. Illustrate Theorem 4(i) with the matrices of Exercise 11.

13. Given: $A = [a_{ij}]_{(m,n)}$ and $B = [b_{ij}]_{(p,q)}$.
    (a) Under what conditions does AB exist?
    (b) If AB exists, what is its order?
    (c) Under what conditions does BA exist?
    (d) If BA exists, what is its order?
    (e) Under what conditions does A^2 exist?
    (f) If A^2 exists, what is its order?

14. Prove Theorem 3 for the case where A, B, and C are 2 by 2 matrices.

15. Prove Theorem 4(i) for the case where A, B, and C are 2 by 2 matrices.

16. The production per day of 2 products by 3 factories is given by the following matrix:

$$M = \begin{bmatrix} 1 & 3 & 4 \\ 6 & 9 & 2 \end{bmatrix} \begin{matrix} P \\ Q \end{matrix} \text{ Products}$$

overbrace{A\ B\ C}^{Factories}

If the entries of

$$X = \begin{bmatrix} 3 \\ 2 \\ 1 \end{bmatrix}$$

represent the number of days that the respective factories operate, find MX and state what the entries represent.

17. If 5 judges on an appellate court influence each other according to Figure 5.14, write a matrix that shows the number of ways 1 judge can influence another judge through 1 intermediary. (*Hint:* Reconsider Example 7 of Section 5.1, and then investigate the significance of the product AA, where A is an influence matrix.)

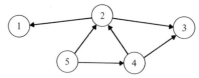

Figure 5.14

18. Use the data of Exercise 17 to write a matrix that shows the number of ways that 1 judge can influence another through *at most* 1 intermediary. Rank the judges according to their influence through *at most* 1 intermediary.

19. A certain factory manufactures two products which we label A and B. Each product is known to contain three different materials which we label X, Y, and Z. The manufacture of one unit of each material requires the expenditure of three forms of energy: electricity, oil, and gas. The entries of Table 5.1 indicate the number of units of each material that is required for one unit of each product. The entries of Table 5.2 indicate

Table 5.1

	Products	
	A	B
Material X	2	3
Material Y	1	1
Material Z	0	1

Table 5.2

	\multicolumn{3}{c}{Materials}		
	X	Y	Z
Electricity in kwh	1	0	3
Oil in gallons	0	4	6
Gas in btu	1	0	4

the number of units of each form of energy that is required to manufacture one unit of each type of material. Use matrix multiplication to determine a matrix that indicates the number of units of each form of energy that is indirectly used to manufacture one unit of each product. (*Hint:* See Example 5.)

5.4 Special Matrices

The purpose of this section is to introduce some special matrices that are important in the theory and applications of matrices.

▶ **Definition 12**
An m by n matrix with entries that are all 0's is called a **zero matrix** and is designated by **0**.

Example 1
Each of the following is a zero matrix:

$$\begin{bmatrix} 0 & 0 & 0 \\ 0 & 0 & 0 \end{bmatrix}, \quad \begin{bmatrix} 0 \\ 0 \end{bmatrix}, \quad [0 \ 0 \ 0], \quad \begin{bmatrix} 0 & 0 \\ 0 & 0 \end{bmatrix}. \blacksquare$$

A zero matrix has the property that when it is added to any other matrix A of the same order, the sum is the same matrix A; that is,

$$A + 0 = 0 + A = A.$$

Because of this property the zero matrix is called the **identity element for matrix addition**. The special matrix I_n that is defined next has the property that when it is *multiplied* on either side by a matrix A of the same order, the product is A; that is,

$$AI_n = I_n A = A.$$

Because of this property the **identity matrix** I_n is called the **identity element for matrix multiplication**.

▶ **Definition 13**
An n by n matrix with 1's on the main diagonal and 0's elsewhere is called an **identity matrix** and is designated by I_n. The columns of I_n are designated $I_{(1)}, I_{(2)}, \ldots, I_{(n)}$.

Example 2

$$I_2 = \begin{bmatrix} 1 & 0 \\ 0 & 1 \end{bmatrix}, \quad I_3 = \begin{bmatrix} 1 & 0 & 0 \\ 0 & 1 & 0 \\ 0 & 0 & 1 \end{bmatrix}, \quad I_4 = \begin{bmatrix} 1 & 0 & 0 & 0 \\ 0 & 1 & 0 & 0 \\ 0 & 0 & 1 & 0 \\ 0 & 0 & 0 & 1 \end{bmatrix}.$$

Moreover,

$$AI_2 = \begin{bmatrix} 2 & 1 \\ 3 & 4 \end{bmatrix} \begin{bmatrix} 1 & 0 \\ 0 & 1 \end{bmatrix} = \begin{bmatrix} 2 & 1 \\ 3 & 4 \end{bmatrix} = A$$

and

$$I_2 A = \begin{bmatrix} 1 & 0 \\ 0 & 1 \end{bmatrix} \begin{bmatrix} 2 & 1 \\ 3 & 4 \end{bmatrix} = \begin{bmatrix} 2 & 1 \\ 3 & 4 \end{bmatrix} = A.$$

Also, for I_3 we have

$$I_{(1)} = \begin{bmatrix} 1 \\ 0 \\ 0 \end{bmatrix}, \quad I_{(2)} = \begin{bmatrix} 0 \\ 1 \\ 0 \end{bmatrix}, \quad I_{(3)} = \begin{bmatrix} 0 \\ 0 \\ 1 \end{bmatrix}. \blacksquare$$

Notice that the identity matrix I_n with respect to multiplication corresponds to the real number 1 with respect to real number multiplication, and the zero matrix **0** with respect to matrix addition corresponds to the real number 0 with respect to real number addition.

▶ **Definition 14**
If A is a square matrix with $a_{ij} = 0$ for all $i \neq j$, then A is a **diagonal matrix**.

Example 3
The following matrices are diagonal.

$$\begin{bmatrix} 2 & 0 & 0 \\ 0 & 1 & 0 \\ 0 & 0 & 3 \end{bmatrix}, \quad \begin{bmatrix} 2 & 0 \\ 0 & 1 \end{bmatrix}, \quad \begin{bmatrix} 0 & 0 & 0 \\ 0 & 3 & 0 \\ 0 & 0 & 2 \end{bmatrix}, \quad I_n, \quad \begin{bmatrix} 0 & 0 & 0 \\ 0 & 0 & 0 \\ 0 & 0 & 0 \end{bmatrix}. \blacksquare$$

Those diagonal matrices that have the same entries on the main diagonal are called **scalar matrices**; an identity matrix multiplied by any real number is an example of a scalar matrix.

▶ **Definition 15**
Let A be a square matrix. If $a_{ij} = 0$ for all $i > j$ or for all $i < j$, then A is a **triangular matrix**.

Example 4
The following matrices are triangular. For emphasis, the 0's required to make each matrix triangular are placed in parentheses.

$$\begin{bmatrix} 8 & 2 & 3 \\ (0) & 4 & 2 \\ (0) & (0) & 6 \end{bmatrix}, \quad \begin{bmatrix} 2 & (0) \\ 3 & 4 \end{bmatrix}, \quad \begin{bmatrix} 0 & 2 & 1 \\ (0) & 4 & 3 \\ (0) & (0) & 5 \end{bmatrix}, \quad \begin{bmatrix} 0 & (0) & (0) \\ 0 & 0 & (0) \\ 0 & 2 & 0 \end{bmatrix}. \blacksquare$$

▶ **Definition 16**
If some rows or columns of a matrix A are deleted, the remaining array is called a **submatrix** of A. Also, it is customary to treat A as a submatrix of itself.

Example 5
For the matrix
$$A = \begin{bmatrix} 2 & 1 & 3 \\ 5 & 0 & 4 \end{bmatrix},$$
the following matrices are submatrices of A:

$$\begin{bmatrix} 2 & 1 & 3 \\ 5 & 0 & 4 \end{bmatrix}, \begin{bmatrix} 2 & 1 \\ 5 & 0 \end{bmatrix}, \begin{bmatrix} 2 & 3 \\ 5 & 4 \end{bmatrix}, \begin{bmatrix} 1 & 3 \\ 0 & 4 \end{bmatrix}, \begin{bmatrix} 2 \\ 5 \end{bmatrix}, \begin{bmatrix} 1 \\ 0 \end{bmatrix}, \begin{bmatrix} 3 \\ 4 \end{bmatrix},$$

$[2\ 1\ 3]$, $[5\ 0\ 4]$, $[2\ 1]$, $[2\ 3]$, $[1\ 3]$,
$[5\ 0]$, $[5\ 4]$, $[0\ 4]$, $[2]$, $[1]$, $[3]$, $[5]$, $[0]$, $[4]$. $\blacksquare$

In some problems, it is helpful to separate a matrix into submatrices by means of dashed lines. For example, we can write

$$A = \begin{bmatrix} A_{11} & | & A_{12} \\ \hline A_{21} & | & A_{22} \end{bmatrix} = \begin{bmatrix} a_{11} & | & a_{12} & a_{13} \\ \hline a_{21} & | & a_{22} & a_{23} \\ a_{31} & | & a_{32} & a_{33} \end{bmatrix}.$$

When this has been done, we say that A has been **partitioned**, and the partitioned submatrices, designated by A_{ij}, are called **blocks of A**; A is then called a **block matrix**. In Exercises 12 and 13 diagonal and triangular block matrices are defined.

A very simple but useful operation that can be performed on a single matrix is that of interchanging the rows and columns of the matrix. This operation will lead to more special matrices.

▶ **Definition 17**
The **transpose** of an m by n matrix A is the n by m matrix A^T that is formed when the ith row of A becomes the ith column of A^T ($i = 1, 2, 3, \ldots, m$).

Example 6

If $A = \begin{bmatrix} 3 & 2 & 1 \\ 6 & 4 & 5 \end{bmatrix}$, then $A^T = \begin{bmatrix} 3 & 6 \\ 2 & 4 \\ 1 & 5 \end{bmatrix}$. $\blacksquare$

Once a new operation has been defined, it is natural to wonder about its properties. A few of the properties of this operation are stated in Exercise 7.

▶ **Definition 18**
A square matrix A is said to be **symmetric** if $A = A^T$, and **skew-symmetric** if $A = -A^T$.

The names "symmetric" and "skew-symmetric" arise because of the physical symmetry that exists with respect to the main diagonal, as shown in Figures 5.15 and 5.16.

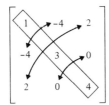

Figure 5.15. *Symmetric matrix.*

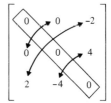

Figure 5.16. *Skew-symmetric matrix.*

Example 7
The matrix

$$A = \begin{bmatrix} 4 & 2 & 1 \\ 2 & 7 & 3 \\ 1 & 3 & 8 \end{bmatrix}$$

is symmetric because $A = A^T$. The matrix

$$B = \begin{bmatrix} 0 & 1 & 6 \\ -1 & 0 & -2 \\ -6 & 2 & 0 \end{bmatrix}$$

is not symmetric because $B \neq B^T$, but B is skew-symmetric because $B = -B^T$. The matrix

$$\begin{bmatrix} 2 & 1 \\ 3 & 4 \end{bmatrix}$$

is neither symmetric nor skew-symmetric. ∎

A few of the many properties of symmetric and skew-symmetric matrices will be given in Exercises 14 and 16.
At this point it may be helpful to review the special matrices presented in this section:

1. The *zero* and *identity matrices* serve as identity elements for addition and multiplication of n by n matrices and in this sense correspond to the numbers 0 and 1 in the real number system.
2. A *diagonal matrix* has the property that off-diagonal entries must all be 0, and the special diagonal matrix where all the diagonal entries are the same is called a *scalar matrix*.
3. A *triangular matrix* has the property that all entries either above or below the main diagonal are 0's.
4. A *submatrix* of a matrix is obtained by deleting some rows or columns of the given matrix. Also a matrix is a submatrix of itself.
5. *Symmetric* and *skew-symmetric matrices* have the respective properties that $A = A^T$ and $A = -A^T$.

APPLICATIONS

Example 8
Many applications of matrix algebra in accounting may be found in Ijiri [31]. Background for this example may be found in Chapter 5 of [31]. Consider the problem of computing the closing balance for a firm that has an initial balance sheet given in Table 5.3. The firm, which is in the

Table 5.3. INITIAL BALANCE SHEET.

Debit balances		Credit balances	
Cash	$30	Accounts payable	$15
Inventories	20	Equity	35
	$50		$50

merchandising business, buys on credit and makes only cash sales; we assume that the only accounts in use are those listed in the table. Arrange the initial balances as a column matrix,

$$U = \begin{bmatrix} u_1 \\ u_2 \\ u_3 \\ u_4 \end{bmatrix} = \begin{bmatrix} 30 \\ 20 \\ -15 \\ -35 \end{bmatrix}.$$

Over a certain time period suppose that the following transactions took place:

$40	purchases of goods on credit
$35	cash payments to suppliers
$15	fixed costs paid in cash
$25	contribution margin, cash sales
$30	cost of goods sold
$ 5	goods returned to suppliers

The dollar representations of the transactions over a period of time are arranged into a "spread sheet matrix" W with a row for each account and a column for each account. The rows represent the accounts to be debited, and the columns represent accounts to be credited with the transactions of the time period.

The first transaction listed above is $w_{23} = 40$ because, by accounting convention, item 2 (inventory) is debited and item 3 (accounts payable) is credited. The second transaction is $w_{31} = 35$ because item 3 (accounts payable) is debited and item 1 (cash) is credited. In the same manner the remaining transactions are represented by $w_{41} = 15$, $w_{14} = 25$, $w_{12} = 30$, and $w_{32} = 5$. The spread matrix W is therefore

$$W = \begin{bmatrix} 0 & 30 & 0 & 25 \\ 0 & 0 & 40 & 0 \\ 35 & 5 & 0 & 0 \\ 15 & 0 & 0 & 0 \end{bmatrix}.$$

The main diagonal of W consists of all 0's because it is not customary to recognize a debit and a credit to the same account as a genuine transaction. By transposing matrix W, we form matrix W^T. Then the change in U, called ΔU, is equal to the expression $(W - W^T)E$, where E is a column vector that has all its elements equal to 1 and has as many elements as there are accounts in the system. (Notice that $W - W^T$ is a skew-symmetric matrix.)

$$\Delta U = (W - W^T)E = \left(\begin{bmatrix} 0 & 30 & 0 & 25 \\ 0 & 0 & 40 & 0 \\ 35 & 5 & 0 & 0 \\ 15 & 0 & 0 & 0 \end{bmatrix} - \begin{bmatrix} 0 & 0 & 35 & 15 \\ 30 & 0 & 5 & 0 \\ 0 & 40 & 0 & 0 \\ 25 & 0 & 0 & 0 \end{bmatrix} \right) \begin{bmatrix} 1 \\ 1 \\ 1 \\ 1 \end{bmatrix}$$

$$= \begin{bmatrix} 0 & 30 & -35 & 10 \\ -30 & 0 & 35 & 0 \\ 35 & -35 & 0 & 0 \\ -10 & 0 & 0 & 0 \end{bmatrix} \begin{bmatrix} 1 \\ 1 \\ 1 \\ 1 \end{bmatrix} = \begin{bmatrix} 5 \\ 5 \\ 0 \\ -10 \end{bmatrix}.$$

The closing balance U' can now be found by

$$U' = U + \Delta U = \begin{bmatrix} 30 \\ 20 \\ -15 \\ -35 \end{bmatrix} + \begin{bmatrix} 5 \\ 5 \\ 0 \\ -10 \end{bmatrix} = \begin{bmatrix} 35 \\ 25 \\ -15 \\ -45 \end{bmatrix},$$

from which the closing balance sheet can be written; see Table 5.4. ∎

Table 5.4. CLOSING BALANCE SHEET.

Debit balances		Credit balances	
Cash	\$35	Accounts payable	\$15
Inventories	25	Equity	45
	\$60		\$60

Example 9

Suppose that each of 8 patients suffering from a certain psychological disorder is given 3 different treatments. Furthermore, suppose that the response of these patients to each of these treatments is classified as improvement, no change, or deterioration. The response of each patient to each treatment can be characterized by three matrices. Suppose that the "improvement" response matrix is

$$\begin{array}{c} \text{Patients} \\ \overbrace{\begin{array}{cccccccc} 1 & 2 & 3 & 4 & 5 & 6 & 7 & 8 \end{array}} \end{array}$$

$$R_1 = \begin{bmatrix} 1 & 0 & 1 & 0 & 1 & 0 & 0 & 1 \\ 1 & 1 & 1 & 0 & 0 & 1 & 0 & 0 \\ 0 & 0 & 1 & 1 & 0 & 1 & 0 & 0 \end{bmatrix} \left.\begin{matrix} 1 \\ 2 \\ 3 \end{matrix}\right\} \text{Treatment}$$

where a 1 in the ij position means that improvement was shown when treatment i was given to patient j. Otherwise, a 0 is given. Next suppose that the "no change" response matrix is

$$\begin{array}{c} \text{Patients} \\ \overbrace{\begin{array}{cccccccc} 1 & 2 & 3 & 4 & 5 & 6 & 7 & 8 \end{array}} \end{array}$$

$$R_2 = \begin{bmatrix} 0 & 0 & 0 & 1 & 0 & 0 & 1 & 0 \\ 0 & 0 & 0 & 1 & 1 & 0 & 1 & 1 \\ 1 & 0 & 0 & 0 & 0 & 0 & 0 & 0 \end{bmatrix} \left.\begin{matrix} 1 \\ 2 \\ 3 \end{matrix}\right\} \text{Treatment}$$

where a 1 in the ij position means that no change was shown when treatment i was given to patient j. Otherwise, a 0 is given. Finally, suppose that the "deterioration" response matrix is

$$\begin{array}{c} \text{Patients} \\ \overbrace{\begin{array}{cccccccc} 1 & 2 & 3 & 4 & 5 & 6 & 7 & 8 \end{array}} \end{array}$$

$$R_3 = \begin{bmatrix} 0 & 1 & 0 & 0 & 0 & 1 & 0 & 0 \\ 0 & 0 & 0 & 0 & 0 & 0 & 0 & 0 \\ 0 & 1 & 0 & 0 & 1 & 0 & 1 & 1 \end{bmatrix} \left.\begin{matrix} 1 \\ 2 \\ 3 \end{matrix}\right\} \text{Treatment}$$

where the ij entry is 1 if deterioration occurs when treatment i is given to patient j, and is 0 otherwise. Now form the partitioned matrix

$$R = \begin{bmatrix} R_1 \\ \hline R_2 \\ \hline R_3 \end{bmatrix},$$

and then the product

$$A = RR^T = \begin{bmatrix} R_1 \\ \hline R_2 \\ \hline R_3 \end{bmatrix} [R_1^T \mid R_2^T \mid R_3^T] = \begin{bmatrix} R_1R_1^T & R_1R_2^T & R_1R_3^T \\ \hline R_2R_1^T & R_2R_2^T & R_2R_3^T \\ \hline R_3R_1^T & R_3R_2^T & R_3R_3^T \end{bmatrix}$$

$$= \begin{bmatrix} 4 & 2 & 1 & 0 & 2 & 1 & 0 & 0 & 2 \\ 2 & 4 & 2 & 0 & 0 & 1 & 2 & 0 & 1 \\ 1 & 2 & 3 & 1 & 1 & 0 & 1 & 0 & 0 \\ \hline 0 & 0 & 1 & 2 & 2 & 0 & 0 & 0 & 1 \\ 2 & 0 & 1 & 2 & 4 & 0 & 0 & 0 & 3 \\ 1 & 1 & 0 & 0 & 0 & 1 & 0 & 0 & 0 \\ \hline 0 & 2 & 1 & 0 & 0 & 0 & 2 & 0 & 1 \\ 0 & 0 & 0 & 0 & 0 & 0 & 0 & 0 & 0 \\ 2 & 1 & 0 & 1 & \underline{3} & 0 & 1 & 0 & 4 \end{bmatrix}$$

(Column groups under Responses: improve / no change / deteriorate, each over Treatment 1 2 3. Row groups on the right: Treatment 1,2,3 under improve / no change / deteriorate, collectively labeled Responses.)

It can be shown that the ijth entry of the submatrix $R_h R_k^T$ of A represents the number of patients in which treatment i produced response h and treatment j produced response k. For instance, from matrix A we can determine at a glance that in exactly 3 patients, treatment 2 produced no change and treatment 3 produced deterioration. (The entries that convey this fact are underlined; there are two such entries because A is a symmetric matrix.) Also the ijth entry of the sum $R_1 R_1^T + R_2 R_2^T + R_3 R_3^T$, which in this problem is

$$\begin{bmatrix} 8 & 4 & 2 \\ 4 & 8 & 2 \\ 2 & 2 & 8 \end{bmatrix},$$

represents the number of times that treatments i and j produced the *same* response; for instance, treatments 1 and 2 produced the same response in 4

Chap. 5] The Algebra of Matrices

of the 8 patients. Although the matrix method illustrated in this example derives results that can be obtained by other methods, this matrix method has the advantage of displaying and manipulating vast amounts of data in a systematic way. Of course, an added advantage is that computer programs needed to perform the necessary matrix operations are readily available. Another application of this method to a study of responses to various appeals by the judges of an appellate court may be found in Campbell [13], pp. 83–84, 207–209. ∎

Example 10
The matrix method of the last example applies to any problem in which m objects yield p responses when subjected to n stimuli; moreover, the responses need not be mutually exclusive as they were in the last example. Also, it can be observed that objects, stimuli, and responses need only to be listed, numbered, and counted. No other quantification is necessary. This matrix method might apply to such diverse problems as (1) studies of responses of certain business or economic indicators when subjected to various national or international events; (2) studies of responses of legislatures or legislators, or nations of the UN, when subjected to votes on certain issues; (3) studies of responses of animals or plants when subjected to certain diseases, medication, weather, or food, and so on; and (4) psychological studies of responses of individuals or groups of individuals when subjected to certain emotional stresses or physical stresses or mental testing. ∎

Example 11
In this example we further extend Example 7 of Section 5.3 to illustrate the use of the transpose of a matrix to achieve a desired result. Recall that the transition matrix from one stage to another stage was

$$A = \begin{bmatrix} .6 & .4 \\ .8 & .2 \end{bmatrix} \begin{matrix} \text{satisfactory} \\ \text{unsatisfactory} \end{matrix} \text{From}$$

with columns labeled "To: satisfactory, unsatisfactory".

Suppose that there is some value, say profit, associated with each transition; these profits are given by the matrix

$$V = \begin{bmatrix} \$100 & \$20 \\ \$30 & -\$50 \end{bmatrix},$$

where v_{ij} is the value or profit to the owner of the system for a transition from the ith state to the jth state. The two main diagonal entries of the product

$$AV^T = \begin{bmatrix} .6 & .4 \\ .8 & .2 \end{bmatrix} \begin{bmatrix} 100 & 30 \\ 20 & -50 \end{bmatrix} = \begin{bmatrix} (60+8) & (\quad) \\ (\quad) & (24-10) \end{bmatrix} = \begin{bmatrix} 68 & (\quad) \\ (\quad) & 14 \end{bmatrix}$$

Sec. 5.4] Special Matrices

represent the expected profit to the owner for each of the two respective trial outcomes after the first transition.

In other words, if the system is working at the outset, then the expected profit in one transition is $68, but if the system is not working at the outset, the expected profit is $14. A more complete discussion, including that of expected values after repeated transitions and optimal decisions, may be found in Searle and Hausman [57], pp. 201–204. ∎

EXERCISES

Suggested minimum assignment: Exercises 1, 2, 3, 5, 6, and 7.

1. Given:

$$A = \begin{bmatrix} 3 & 0 & 0 \\ 0 & 2 & 0 \\ 0 & 0 & 0 \end{bmatrix}, \quad B = \begin{bmatrix} 1 & 0 & 0 \\ 0 & 1 & 0 \\ 0 & 0 & 1 \end{bmatrix}, \quad C = \begin{bmatrix} 3 & 2 & 0 \\ 2 & 1 & 0 \\ 0 & 0 & 0 \end{bmatrix},$$

$$D = \begin{bmatrix} 1 & 1 & 0 \\ 0 & 0 & 0 \\ 0 & 0 & 1 \end{bmatrix}, \quad E = \begin{bmatrix} 0 & 0 & 0 \end{bmatrix}, \quad F = \begin{bmatrix} 1 & 0 & 0 \\ 0 & 1 & 0 \end{bmatrix}.$$

(a) Which matrices, if any, are zero matrices?
(b) Which matrices, if any, are identity matrices for multiplication?
(c) Which matrices, if any, are diagonal matrices?
(d) Which matrices, if any, are scalar matrices?
(e) Which matrices, if any, are triangular matrices?
(f) I_2 is a submatrix of which matrices?
(g) Which matrices, if any, are symmetric?
(h) Which matrices, if any, are skew-symmetric?

2. Answer the same questions raised in Exercise 1 for the matrices

$$A = \begin{bmatrix} 0 & 0 & 6 \\ 0 & 3 & 0 \\ 6 & 0 & 0 \end{bmatrix}, \quad B = \begin{bmatrix} 1 & 0 \\ 0 & 0 \end{bmatrix}, \quad C = \begin{bmatrix} 0 & 0 \\ 0 & 0 \\ 0 & 0 \end{bmatrix},$$

$$D = \begin{bmatrix} 1 \\ 0 \\ 0 \end{bmatrix}, \quad E = \begin{bmatrix} 2 & 0 \\ 0 & 2 \end{bmatrix}, \quad F = \begin{bmatrix} 3 & 0 & 0 \\ 1 & 4 & 0 \\ 3 & 2 & 0 \end{bmatrix}.$$

3. Verify that $I_2 A = A I_2 = A$ if

$$A = \begin{bmatrix} a_{11} & a_{12} \\ a_{21} & a_{22} \end{bmatrix}.$$

4. Verify that $A + 0 = 0 + A = A$ if 0 is a 2 by 2 zero matrix and

$$A = \begin{bmatrix} a_{11} & a_{12} \\ a_{21} & a_{22} \end{bmatrix}.$$

5. Partition the matrix
$$\begin{bmatrix} 3 & 5 & 1 \\ 4 & 0 & 6 \\ 9 & 2 & 1 \end{bmatrix}$$
so that $[5 \ \ 1]$ is one block and $\begin{bmatrix} 4 \\ 9 \end{bmatrix}$ is another block.

6. Tell whether each of the following is true or false. If false, give an example illustrating your conclusion.
   (a) The transpose of a triangular matrix is a triangular matrix.
   (b) Every diagonal matrix is a triangular matrix.
   (c) Every scalar matrix is square.
   (d) Every scalar matrix is also a diagonal matrix.
   (e) The product of two diagonal matrices of the same order is a diagonal matrix.

7. A few of the properties of the transpose of a matrix follow: Let A and B be matrices, let k be a real number, and let the orders of A and B be such that conformability requirements are met. Then
   (i) $(A^T)^T = A$.
   (ii) $(A + B)^T = A^T + B^T$.
   (iii) $(AB)^T = B^T A^T$.
   (iv) $(kA)^T = kA^T$.

   Illustrate these properties with
   $$A = \begin{bmatrix} 1 & 4 \\ 3 & 5 \end{bmatrix}, \quad B = \begin{bmatrix} 6 & 0 \\ 2 & 3 \end{bmatrix}, \quad \text{and} \quad k = 3.$$

8. Prove property (i) of Exercise 7.
9. Prove property (ii) of Exercise 7. (*Hint:* Let $C = A + B$. Show that the ijth entry of C^T is the same as the ijth entry of $A^T + B^T$.)
10. Prove property (iii) of Exercise 7. [*Hint:* Let $C = AB$. Show that the ijth entry of C^T is the same as the ijth entry of $B^T A^T$. The first two steps are

    (ijth entry of C^T) = (jith entry of C)
    = (jth row of A) · (ith column of B).]

11. Prove property (iv) of Exercise 7.
12. Let A_{ij} represent blocks of A. If a square matrix A can be partitioned in such a way
    $$A = \begin{bmatrix} A_{11} & A_{12} & \cdots & A_{1m} \\ \vdots & \vdots & & \vdots \\ A_{m1} & A_{m2} & \cdots & A_{mm} \end{bmatrix}$$
    that all A_{ij} are square when $i = j$, and all A_{ij} are zero matrices when $i \neq j$, then A is a **diagonal block matrix**. Which, if any, of the matrices in Exercise 1 are diagonal block matrices?

Sec. 5.4] Special Matrices

13. Let A_{ij} represent blocks of A. If a square matrix A can be partitioned in such a way

$$A = \begin{bmatrix} A_{11} & A_{12} & \cdots & A_{1m} \\ \vdots & \vdots & & \vdots \\ A_{m1} & A_{m2} & \cdots & A_{mm} \end{bmatrix}$$

that all A_{ii} are square, and if, for $i > j$ (or $i < j$), all A_{ij} are zero matrices, then A is a **triangular block matrix**. Which, if any, of the matrices in Exercise 1 are triangular block matrices?

14. Construct a pair of 2 by 2 matrices and illustrate the following properties of symmetric matrices. If A and B are symmetric matrices of the same order and if k is a real number, then
    (a) $A + B$ is symmetric.
    (b) kA is symmetric.
    (c) A^2 is symmetric.
    (d) $AA^T = A^TA$ and is symmetric.

15. Prove that each of the properties listed in Exercise 14 is valid. Use properties given in Exercise 7.

16. Construct a pair of 2 by 2 matrices and illustrate the following properties of skew-symmetric matrices. If A and B are skew-symmetric matrices of the same order and if k is a real number, then
    (a) $A + B$ is skew-symmetric.
    (b) kA is skew-symmetric.
    (c) A^2 is symmetric.
    (d) $AA^T = A^TA$ and is symmetric.

17. Prove that each of the properties listed in Exercise 16 is valid. Use properties given in Exercise 7.

18. Prove that if A is an m by n matrix, then AA^T is symmetric. Use properties given in Exercise 7.

5.5 Systems of Linear Equations Having a Unique Solution

A system of linear equations over the real numbers is a set of m equations in n unknowns of the form

$$\begin{aligned} a_{11}x_1 + a_{12}x_2 + \cdots + a_{1n}x_n &= b_1, \\ a_{21}x_1 + a_{22}x_2 + \cdots + a_{2n}x_n &= b_2, \\ \vdots \quad \vdots \quad \quad \vdots \quad \quad \vdots& \\ a_{m1}x_1 + a_{m2}x_2 + \cdots + a_{mn}x_n &= b_m, \end{aligned} \quad (5.1)$$

where the coefficients $a_{11}, a_{12}, \cdots, a_{mn}$ and the right-hand numbers $b_1, b_2, \cdots, b_m$ are real numbers. A **solution** of the linear system (5.1) is a vector $(c_1, c_2, \cdots, c_n)$ such that m identities result when its n components are substituted for the respective unknowns.

For the linear system (5.1) the matrix of coefficients

$$A = \begin{bmatrix} a_{11} & a_{12} & \cdots & a_{1n} \\ a_{21} & a_{22} & \cdots & a_{2n} \\ \vdots & \vdots & & \vdots \\ a_{m1} & a_{m2} & \cdots & a_{mn} \end{bmatrix}$$

is known as the **coefficient matrix** of the system, and this matrix augmented by the members of the right side, namely,

$$[A \mid B] = \left[\begin{array}{cccc|c} a_{11} & a_{12} & \cdots & a_{1n} & b_1 \\ a_{21} & a_{22} & \cdots & a_{2n} & b_2 \\ \vdots & \vdots & & \vdots & \vdots \\ a_{m1} & a_{m2} & \cdots & a_{mn} & b_m \end{array} \right],$$

is known as the **augmented matrix** of the system.

Example 1

The vector $(3, 4)$ is a solution of the linear system

$$x + y = 7,$$
$$2x - y = 2,$$

because if $x = 3$ and $y = 4$ are substituted in the system, we get the identities

$$(3) + (4) = 7,$$
$$2(3) - (4) = 2.$$

A graph of the system with the solution is shown in Figure 5.17; the graph of each equation is a line, and the intersection of the two lines represents the solution. The coefficient matrix of the given system is

$$\begin{bmatrix} 1 & 1 \\ 2 & -1 \end{bmatrix},$$

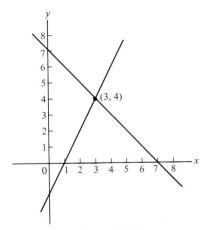

Figure 5.17

Sec. 5.5] Systems of Linear Equations Having a Unique Solution

and the augmented matrix of the given system is

$$\begin{bmatrix} 1 & 1 & | & 7 \\ 2 & -1 & | & 2 \end{bmatrix}. \blacksquare$$

The primary purpose of this section is to introduce a method for finding a solution of a linear system if a solution exists.

From the old addition–subtraction method of elementary algebra the reader probably remembers that if, on a linear system, we perform the following operations, then the resulting linear system will have the same solution as the original linear system.

1. Interchange any two equations.
2. Multiply the terms of any equation by a nonzero real number. (5.2)
3. Add to the terms of any equation k times the corresponding terms of any other equation (k is a real number).

In other words, the solution of a system remains unchanged when the system is modified by these three operations. (A proof of this assertion can be found in Campbell [13], pp. 18–20.)

Example 2
Find a solution, if one exists, of the system

$$\begin{aligned} x_1 + 2x_2 - 6x_3 &= 4, \\ -3x_1 - 6x_2 + 9x_3 &= -3, \\ x_2 - 3x_3 &= 1. \end{aligned}$$

Solution: We shall solve this system by applying the operations (5.2) to the given system; we shall also apply corresponding operations to the augmented matrix of the system in order to introduce the method that we are developing in this section. We start with

$$\begin{aligned} x_1 + 2x_2 - 6x_3 &= 4, \\ -3x_1 - 6x_2 + 9x_3 &= -3, \\ x_2 - 3x_3 &= 1, \end{aligned} \qquad \begin{bmatrix} 1 & 2 & -6 & | & 4 \\ -3 & -6 & 9 & | & -3 \\ 0 & 1 & -3 & | & 1 \end{bmatrix}.$$

Add 3 times the terms of the first equation to the corresponding terms of the second equation (denoted by $3R_1 + R_2$, where R_i represents the ith equation or the ith row of the augmented matrix).

$$\begin{aligned} x_1 + 2x_2 - 6x_3 &= 4, \\ -9x_3 &= 9, \\ x_2 - 3x_3 &= 1, \end{aligned} \qquad \begin{bmatrix} (1) & 2 & -6 & | & 4 \\ (0) & 0 & -9 & | & 9 \\ (0) & 1 & -3 & | & 1 \end{bmatrix}.$$

Parentheses are used in the matrix to indicate when an entry is in the desired form. Next interchange the second and third equations (denoted by $R_2 \leftrightarrow R_3$).

Chap. 5] The Algebra of Matrices

$$x_1 + 2x_2 - 6x_3 = 4, \quad \begin{bmatrix} (1) & 2 & -6 & | & 4 \\ (0) & (1) & -3 & | & 1 \\ (0) & (0) & -9 & | & 9 \end{bmatrix}.$$
$$x_2 - 3x_3 = 1,$$
$$-9x_3 = 9,$$

Add -2 times the terms of the second equation to the corresponding terms of the first equation (denoted $-2R_2 + R_1$).

$$x_1 \qquad\qquad = 2, \quad \begin{bmatrix} (1) & (0) & 0 & | & 2 \\ (0) & (1) & -3 & | & 1 \\ (0) & (0) & -9 & | & 9 \end{bmatrix}.$$
$$x_2 - 3x_3 = 1,$$
$$-9x_3 = 9,$$

Multiply the terms of the third equation by $-\frac{1}{9}$ (denoted $-\frac{1}{9}R_3$).

$$x_1 \qquad\qquad = 2, \quad \begin{bmatrix} (1) & (0) & 0 & | & 2 \\ (0) & (1) & -3 & | & 1 \\ (0) & (0) & (1) & | & -1 \end{bmatrix}.$$
$$x_2 - 3x_3 = 1,$$
$$x_3 = -1,$$

Add 3 times the terms of the third equation to the corresponding terms of the second equation (denoted $3R_3 + R_2$).

$$x_1 \qquad = 2, \quad \begin{bmatrix} (1) & (0) & (0) & | & 2 \\ (0) & (1) & (0) & | & -2 \\ (0) & (0) & (1) & | & -1 \end{bmatrix}.$$
$$x_2 = -2,$$
$$x_3 = -1.$$

Because only the three allowed operations were used, we can say that the solution of the original system is the same as the obvious solution of the last system. Notice that this solution is prominently displayed in the last column of the augmented matrix. Geometrically, this solution represents the intersection of three planes in 3-space at a single point—namely $(2, -2, -1)$. ∎

By performing the operations on the augmented matrix in each step of Example 2, we are simply illustrating the fact that the operations of (5.2) are in reality performed on the coefficients of the system.

The operations performed on the augmented matrices of Example 2 are called **elementary row operations**. They obviously correspond to the operations (5.2) that are used to solve linear systems of equations.

▶ **Definition 19**
The following operations performed on matrices are called **elementary row operations**:
1. The interchange of any two rows.
2. The multiplication of the entries of any row by a nonzero real number.
3. The addition to the entries of any row of k times the corresponding entries of any other row, where k is a real number.

▶ **Definition 20**
If a matrix A can be transformed into a matrix B by means of one or more elementary row operations, then we say that A is **row equivalent** to B; this

relation will be denoted by

$$A \overset{\text{row}}{\sim} B.$$

Example 3
If the first and second rows of the matrix

$$A = \begin{bmatrix} 2 & 9 & | & 6 \\ 1 & 4 & | & 2 \end{bmatrix}$$

are interchanged, denoted $(R_1 \leftrightarrow R_2)$, we get

$$B = \begin{bmatrix} 1 & 4 & | & 2 \\ 2 & 9 & | & 6 \end{bmatrix}.$$

By Definition 20 we can say that $A \overset{\text{row}}{\sim} B$. Furthermore, if we add -2 times the entries of the first row to the corresponding entries of the second row, denoted $(-2R_1 + R_2)$, we get

$$C = \begin{bmatrix} 1 & 4 & | & 2 \\ 0 & 1 & | & 2 \end{bmatrix}.$$

Thus we say that $B \overset{\text{row}}{\sim} C$. Moreover, $A \overset{\text{row}}{\sim} C$. Finally, if we add -4 times the entries of the second row to the corresponding entries of the first row (denoted $-4R_2 + R_1$), we get

$$D = \begin{bmatrix} 1 & 0 & | & -6 \\ 0 & 1 & | & 2 \end{bmatrix}.$$

We can say that $C \overset{\text{row}}{\sim} D$ and also that $A \overset{\text{row}}{\sim} D$. Without the explanations the work above can be expressed very simply as

$$\begin{bmatrix} 2 & 9 & | & 6 \\ 1 & 4 & | & 2 \end{bmatrix} \overset{\text{row}}{\underset{R_1 \leftrightarrow R_2}{\sim}} \begin{bmatrix} (1) & 4 & | & 2 \\ 2 & 9 & | & 6 \end{bmatrix} \overset{\text{row}}{\underset{-2R_1 + R_2}{\sim}} \begin{bmatrix} (1) & 4 & | & 2 \\ (0) & 1 & | & 2 \end{bmatrix} \overset{\text{row}}{\underset{-4R_2 + R_1}{\sim}} \begin{bmatrix} (1) & (0) & | & -6 \\ (0) & (1) & | & 2 \end{bmatrix}.$$

Therefore, if we had been asked to solve the system

$$2x_1 + 9x_2 = 6,$$
$$x_1 + 4x_2 = 2,$$

with the corresponding augmented matrix

$$\begin{bmatrix} 2 & 9 & | & 6 \\ 1 & 4 & | & 2 \end{bmatrix},$$

we could read the answer in the last column of the row equivalent augmented matrix

$$\begin{bmatrix} (1) & (0) & | & -6 \\ (0) & (1) & | & 2 \end{bmatrix}. \blacksquare$$

In summary, we have illustrated a method of solution of systems of linear equations that have a single solution. The reader should be aware, however,

that some systems of linear equations have more than one solution and some have no solution; a method for dealing with such systems is similar to that used in this section and will be discussed later.

APPLICATIONS

Example 4

In this example we introduce a model that has been used extensively since its invention in 1936 by Leontief. The model, which is discussed further in Section 12.5, is frequently referred to as the Leontief input–output model. Consider an economy (system) with n industries (components) that are dependent upon each other. Each industry enjoys a demand for its products—an *internal* demand from the other industries and an *external* demand from outside the economy. Once an agreement has been made concerning a time interval and the units of measure (such as dollar value) of each of the products, then the following notation can be introduced. Let

x_j = output of industry j,
c_{ij} = number of units of output of industry i used in the production of one unit of output of industry j,
d_i = external demand on industry i.

If c_{51} represents the number of units of output of industry 5 used in the production of *one* unit of output of industry 1, and if x_1 represents the number of units of output of industry 1, then the product $c_{51}x_1$ represents the demand on industry 5 by industry 1, the product $c_{52}x_2$ represents the demand on industry 5 by industry 2, and so on. The sum of these products,

$$c_{51}x_1 + c_{52}x_2 + \cdots + c_{5n}x_n,$$

represents the total internal demand on industry 5. This internal demand plus the external demand d_5 should equal the production, x_5. Thus we have the equation

$$c_{51}x_1 + c_{52}x_2 + \cdots + c_{5n}x_n + d_5 = x_5$$

for the fifth industry. Analogous equations for the other industries produce the system

$$\begin{aligned}
c_{11}x_1 + c_{12}x_2 + \cdots + c_{1n}x_n + d_1 &= x_1, \\
c_{21}x_1 + c_{22}x_2 + \cdots + c_{2n}x_n + d_2 &= x_2, \\
\vdots \quad \vdots \quad \vdots \quad \vdots \quad & \\
c_{n1}x_1 + c_{n2}x_2 + \cdots + c_{nn}x_n + d_n &= x_n.
\end{aligned}$$

If rearranged in standard form, the system is

$$\begin{aligned}
(1 - c_{11})x_1 - c_{12}x_2 - \cdots - c_{1n}x_n &= d_1, \\
-c_{21}x_1 + (1 - c_{22})x_2 - \cdots - c_{2n}x_n &= d_2, \\
\vdots \quad \vdots \quad \vdots \quad & \\
-c_{n1}x_1 - c_{n2}x_2 - \cdots + (1 - c_{nn})x_n &= d_n. \blacksquare
\end{aligned}$$

Example 5

In a wide variety of applications it is desirable to generate polynomial equations that will approximate a possible relationship between two variables when pairs of corresponding values of these variables are known. One method of obtaining such an equation is the **method of least squares.** We shall illustrate the method by means of an example in which we generate a first-degree polynomial equation $y = mx + b$ subject to the observed data of Table 5.5. The least-squares method requires that we find numbers m and b in the equations

$$\begin{aligned} y_1 &= mx_1 + b + r_1, \\ y_2 &= mx_2 + b + r_2, \\ y_3 &= mx_3 + b + r_3, \\ y_4 &= mx_4 + b + r_4, \end{aligned} \tag{5.3}$$

in such a way that $S = r_1^2 + r_2^2 + r_3^2 + r_4^2$ is a minimum (whence the name "least squares"). The quantities $r_1, r_2, r_3,$ and r_4 are called **residuals** and are

Table 5.5

x	1	3	4	6
y	3	7	8	8

pictured in Figure 5.18. The four equations of (5.3) can be expressed in matrix notation as

$$Y = C\alpha + \rho,$$

where

$$Y = \begin{bmatrix} y_1 \\ y_2 \\ y_3 \\ y_4 \end{bmatrix}, \quad C = \begin{bmatrix} x_1 & 1 \\ x_2 & 1 \\ x_3 & 1 \\ x_4 & 1 \end{bmatrix}, \quad \alpha = \begin{bmatrix} m \\ b \end{bmatrix}, \quad \rho = \begin{bmatrix} r_1 \\ r_2 \\ r_3 \\ r_4 \end{bmatrix}.$$

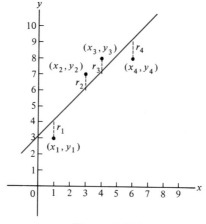

Figure 5.18

Using calculus, one can prove (Campbell [13], p. 145) that S assumes its minimum value only if

$$(C^T C)\alpha = C^T Y.$$

Using the data of Table 5.5, we get

$$\begin{bmatrix} 1 & 3 & 4 & 6 \\ 1 & 1 & 1 & 1 \end{bmatrix} \begin{bmatrix} 1 & 1 \\ 3 & 1 \\ 4 & 1 \\ 6 & 1 \end{bmatrix} \begin{bmatrix} m \\ b \end{bmatrix} = \begin{bmatrix} 1 & 3 & 4 & 6 \\ 1 & 1 & 1 & 1 \end{bmatrix} \begin{bmatrix} 3 \\ 7 \\ 8 \\ 8 \end{bmatrix},$$

or

$$\begin{bmatrix} 62 & 14 \\ 14 & 4 \end{bmatrix} \begin{bmatrix} m \\ b \end{bmatrix} = \begin{bmatrix} 104 \\ 26 \end{bmatrix}.$$

By the methods of this section the solution of this system is

$$\begin{bmatrix} m \\ b \end{bmatrix} = \begin{bmatrix} 1 \\ 3 \end{bmatrix}.$$

Hence if S is a minimum, then the desired equation of the line of Figure 5.18 is

$$y = x + 3.$$

In summary, data such as Table 5.5 generally produce a system $C\alpha = Y$ that has no solutions, but by multiplying both sides of this matrix equation by C^T and by solving for α, we obtain the least-squares solution of the first-degree polynomial equation of best fit. Generalizations of the method, including weighted residuals and multiple variables, are outlined in Campbell [13], pp. 144–145. ∎

EXERCISES

Suggested minimum assignment: Exercises 1, 5, 7, 9, 13, and 15.

In Exercises 1–4 write the coefficient and augmented matrices of the given system.

1. $x_1 - x_2 = 2,$
    $2x_1 + 4x_2 = 1.$

2. $x_1 + 3x_2 = 6,$
    $2x_1 - x_2 = 1.$

3. $x_1 + 3x_2 + x_3 - 2 = 0,$
    $-x_2 - 2x_3 + 3 = 0.$

4. $x_1 - 3x_2 - x_3 + 5 = 0,$
    $x_1 \qquad + x_3 - 7 = 0.$

In Exercises 5–8 find the solution of the given system by following the format of Example 2. That is, solve the given system by using the operations of (5.2), and by performing the corresponding elementary row operations on the associated augmented matrices. Check each answer by substituting it in the original system to see if m identities are obtained.

5. $x_1 + 3x_2 = 3,$
   $2x_1 + x_2 = 6.$

6. $x_1 - x_2 = 4,$
   $-3x_1 + x_2 = 6.$

7. $x_1 - x_3 = 2,$
   $x_1 - 2x_2 = 7,$
   $ x_2 + 3x_3 = 1.$

8. $x_1 + 2x_2 - x_3 = 6,$
   $ -5x_2 + 5x_3 = -25,$
   $3x_1 - 2x_2 + 3x_3 = -16.$

In Exercises 9–12 find the solution of the given system by using elementary row operations on augmented matrices. Check your answers.

9. $x_1 + 2x_2 = 5,$
   $-2x_1 + 4x_2 = 6.$

10. $2x_1 + 3x_2 = 8,$
    $x_1 + x_2 = 2.$

11. $ x_2 + x_3 = 1,$
    $x_1 + x_2 + x_3 = 2,$
    $ x_2 - x_3 = 5.$

12. $x_1 + 4x_3 = 12,$
    $2x_1 + x_2 + 4x_3 = 13,$
    $-2x_1 + 2x_3 = 6.$

13. Graph the equations of the system in Exercise 9 and give a geometric interpretation of the solution.
14. Graph the equations of the system in Exercise 10 and give a geometric interpretation of the solution.
15. In finding the solution of Exercise 9, one must write a sequence of row equivalent matrices. Each of these matrices represents an augmented matrix of a corresponding system; graph the equations of each of these systems. Give a geometric interpretation of the algebraic procedure used to find the solution.
16. In finding the solution of Exercise 10, one must write a sequence of row equivalent matrices. Each of these matrices represents an augmented matrix of a corresponding system; graph the equations of each of these systems. Give a geometric interpretation of the algebraic procedure used to find the solution.

5.6 The Inverse Matrix

In this section we shall present an alternative method to that given in the last section for solving a system of m linear equations in n unknowns when that system has a single solution, and when $m = n$.

As we illustrated in Example 2 of Section 5.3, multiplication can be used to express a system of linear equations

$$a_{11}x_1 + a_{12}x_2 + \cdots + a_{1n}x_n = b_1,$$
$$a_{21}x_1 + a_{22}x_2 + \cdots + a_{2n}x_n = b_2,$$
$$\vdots \qquad \vdots \qquad \vdots \qquad \vdots$$
$$a_{m1}x_1 + a_{m2}x_2 + \cdots + a_{mn}x_n = b_m,$$

by means of a single matrix equation

$$AX = B,$$

where

$$A = \begin{bmatrix} a_{11} & a_{12} & \cdots & a_{1n} \\ a_{21} & a_{22} & \cdots & a_{2n} \\ \vdots & \vdots & & \vdots \\ a_{m1} & a_{m2} & \cdots & a_{mn} \end{bmatrix}, \quad X = \begin{bmatrix} x_1 \\ x_2 \\ \vdots \\ x_n \end{bmatrix}, \quad \text{and} \quad B = \begin{bmatrix} b_1 \\ b_2 \\ \vdots \\ b_m \end{bmatrix}.$$

In the algebra of real numbers, when one was faced with the question of solving a single equation

$$ax = b, \quad \text{where } a \neq 0,$$

the standard approach was to premultiply both sides by $1/a$ or a^{-1} to obtain

$$a^{-1}(ax) = a^{-1}b,$$

which by the associative law for multiplication of real numbers is

$$(a^{-1}a)x = a^{-1}b.$$

Since $(a^{-1}a) = 1$, we have

$$1 \cdot x = a^{-1}b,$$

and because 1 is the identity element for multiplication we have

$$x = a^{-1}b.$$

It is therefore natural to wonder if we can use an analogous procedure to solve for X in the matrix equation $AX = B$. In order to provide an answer to this speculation, we must be sure that the term $1/a$ or a^{-1} is clearly understood and then extend this concept to matrix algebra. If for a given real number a there is a real number $1/a$ or a^{-1} such that

$$a \cdot a^{-1} = a^{-1} \cdot a = 1,$$

we say that a^{-1} is the **multiplicative inverse** of a. The reader should remember that all real numbers except 0 have a multiplicative inverse. For example, the multiplicative inverse of 3 is $\frac{1}{3}$ because $3 \cdot \frac{1}{3} = \frac{1}{3} \cdot 3 = 1$. We shall define a multiplicative inverse of a matrix in a similar way.

▶ **Definition 21**
If for a given n by n matrix A, there is a matrix designated A^{-1} such that

$$AA^{-1} = A^{-1}A = I_n,$$

then A^{-1} is an **inverse** of A with respect to multiplication.

Example 1

For the matrix $A = \begin{bmatrix} 3 & 1 \\ 5 & 2 \end{bmatrix}$, the matrix $\begin{bmatrix} 2 & -1 \\ -5 & 3 \end{bmatrix}$ can be designated as A^{-1} because

$$\begin{bmatrix} 3 & 1 \\ 5 & 2 \end{bmatrix} \begin{bmatrix} 2 & -1 \\ -5 & 3 \end{bmatrix} = \begin{bmatrix} 2 & -1 \\ -5 & 3 \end{bmatrix} \begin{bmatrix} 3 & 1 \\ 5 & 2 \end{bmatrix} = \begin{bmatrix} 1 & 0 \\ 0 & 1 \end{bmatrix}. \quad \blacksquare$$

Example 2

Consider the system

$$3x_1 + x_2 = 3,$$
$$5x_1 + 2x_2 = 2,$$

which can be expressed as the matrix equation

$$AX = B,$$

where

$$A = \begin{bmatrix} 3 & 1 \\ 5 & 2 \end{bmatrix}, \quad X = \begin{bmatrix} x_1 \\ x_2 \end{bmatrix}, \quad \text{and} \quad B = \begin{bmatrix} 3 \\ 2 \end{bmatrix}.$$

From Example 1 we know that

$$A^{-1} = \begin{bmatrix} 2 & -1 \\ -5 & 3 \end{bmatrix};$$

hence we can premultiply both sides of $AX = B$ by A^{-1} to get

$$A^{-1}(AX) = A^{-1}B,$$
$$(A^{-1}A)X = A^{-1}B,$$
$$I_2 X = A^{-1}B,$$
$$X = A^{-1}B.$$

Thus a solution of the given system is

$$X = A^{-1}B = \begin{bmatrix} 2 & -1 \\ -5 & 3 \end{bmatrix} \begin{bmatrix} 3 \\ 2 \end{bmatrix} = \begin{bmatrix} 4 \\ -9 \end{bmatrix}. \blacksquare$$

From the last two examples, three rather obvious questions emerge. (1) Do all matrices have multiplicative inverses? (2) How does one calculate a multiplicative inverse if one exists? (3) Is there more than one multiplicative inverse of a matrix? Answers to these questions are provided by the next two theorems.

▶ **Theorem 5**

An n by n matrix A has a multiplicative inverse if and only if $A \overset{\text{row}}{\sim} I_n$. Moreover, if $[A \mid I_n] \overset{\text{row}}{\sim} [I_n \mid P]$, then P is a multiplicative inverse of A.

A proof of Theorem 5 is beyond the scope of this text and may be found in more advanced texts. (One source is Section 3.4 of Campbell [13].) ◀

Example 3

Calculate an inverse of $A = \begin{bmatrix} 1 & 4 \\ 2 & 9 \end{bmatrix}$, if possible.

Solution: According to Theorem 5, we construct the matrix $[A \mid I_2]$ and then determine if we can find a matrix $[I_2 \mid P]$ to which $[A \mid I_2]$ is row equivalent. If we can find such a matrix, then $A^{-1} = P$.

$$[A \mid I_2] = \begin{bmatrix} 1 & 4 & | & 1 & 0 \\ 2 & 9 & | & 0 & 1 \end{bmatrix} \xrightarrow[-2R_1 + R_2]{\text{row}} \begin{bmatrix} 1 & 4 & | & 1 & 0 \\ 0 & 1 & | & -2 & 1 \end{bmatrix}$$

$$\xrightarrow[-4R_2 + R_1]{\text{row}} \begin{bmatrix} 1 & 0 & | & 9 & -4 \\ 0 & 1 & | & -2 & 1 \end{bmatrix} = [I_2 \mid P].$$

Therefore,
$$A^{-1} = P = \begin{bmatrix} 9 & -4 \\ -2 & 1 \end{bmatrix}. \blacksquare$$

Example 4

Calculate an inverse of $A = \begin{bmatrix} 1 & 2 \\ 3 & 6 \end{bmatrix}$, if possible.

Solution

$$[A \mid I_2] = \begin{bmatrix} 1 & 2 & | & 1 & 0 \\ 3 & 6 & | & 0 & 1 \end{bmatrix} \xrightarrow[-3R_1 + R_2]{\text{row}} \begin{bmatrix} 1 & 2 & | & 1 & 0 \\ 0 & 0 & | & -3 & 1 \end{bmatrix}.$$

We quickly see that A is *not* row equivalent to I_2; hence, by Theorem 5, A does *not* have a multiplicative inverse. ∎

A square matrix that has a multiplicative inverse is called an **invertible** matrix, and a square matrix that does not have a multiplicative inverse is called a **noninvertible** matrix.

▶ **Theorem 6**

If a square matrix is invertible, then the inverse is unique.

Proof: Assume that there are two multiplicative inverses of an n by n matrix A and show that they must be equal. Assume that B and C are inverses of A; hence

$$AB = BA = I_n \quad \text{and} \quad AC = CA = I_n.$$

Statements: $C = CI_n = C(AB) = (CA)B = I_n B = B$. The reasons for the statements of the proof are left as Exercise 20. ◀

Other properties of the inverse matrix are stated in Exercise 15.

Because of Theorem 6, we can be sure that in Example 2 the given multiplicative inverse is the only multiplicative inverse, and hence the solution of the system given in Example 2 is unique. We now conclude the material of this section with the following important theorem.

▶ **Theorem 7**

If A is invertible, then the linear system $AX = B$ has the unique solution

$$X = A^{-1}B.$$

The proof of Theorem 7 is left as Exercise 21. ◀

Sec. 5.6] The Inverse Matrix

APPLICATIONS

Example 5

Theorem 7 is particularly useful in those applications where the coefficient matrix remains fixed and the constants on the right-hand side vary. For example, suppose that a certain company has two machines, A and B, that manufacture two products, C and D. The time needed by each machine to produce one unit of each product is given in the following chart:

	Products	
	C	D
Machine A	5 hours used	3 hours used
Machine B	1 hour used	7 hours used

The management wishes to decide how many units (or fractions of units) of each product to make if each machine is to be used exactly 16 hours per day. (The remaining hours are reserved for the maintenance of the machines.) They also want to know the effect on production if additional machines are purchased.

Solution: Let x_1 be the number of units of product C produced each day, and let x_2 be the number of units of product D produced each day. The time spent by machine A is

(number of hours per unit on product C) (number of units of C)
+ (number of hours per unit on product D) (number of units of D)

or

$$5x_1 + 3x_2,$$

and the amount must be equal to 16 hours. Hence we have the equation

$$5x_1 + 3x_2 = 16.$$

For machine B we have the equation

$$1x_1 + 7x_2 = 16.$$

These two equations can be expressed as the single matrix equation

$$AX = B,$$

where

$$A = \begin{bmatrix} 5 & 3 \\ 1 & 7 \end{bmatrix}, \quad X = \begin{bmatrix} x_1 \\ x_2 \end{bmatrix}, \quad \text{and} \quad B = \begin{bmatrix} 16 \\ 16 \end{bmatrix}.$$

By Theorem 7, if A^{-1} exists,

$$X = A^{-1}B,$$

and by use of Theorem 5 we find that

$$A^{-1} = \begin{bmatrix} \frac{7}{32} & \frac{-3}{32} \\ \frac{-1}{32} & \frac{5}{32} \end{bmatrix}.$$

Hence

$$X = \begin{bmatrix} \frac{7}{32} & \frac{-3}{32} \\ \frac{-1}{32} & \frac{5}{32} \end{bmatrix} \begin{bmatrix} 16 \\ 16 \end{bmatrix} = \begin{bmatrix} 2 \\ 2 \end{bmatrix}.$$

Therefore, two units of product C and two units of product D should be produced each day to satisfy the conditions stated initially. Now suppose that the management wants to know the effect on production if more machines are purchased. For example, if one additional machine of type A is purchased, and three additional machines of type B are purchased, then 32 hours are available on a type A machine and 64 hours are available on a type B machine.

We now recalculate $A^{-1}B$, noticing that A^{-1} does not change.

$$X = \begin{bmatrix} \frac{7}{32} & \frac{-3}{32} \\ \frac{-1}{32} & \frac{5}{32} \end{bmatrix} \begin{bmatrix} 32 \\ 64 \end{bmatrix} = \begin{bmatrix} 1 \\ 9 \end{bmatrix}. \blacksquare$$

Example 6

In this example we illustrate an application of the Leontief input–output model presented in Example 4 of Section 5.5. Suppose that a certain utility corporation has three components of operation: it mines coal, produces gasoline, and generates electricity. Each component makes use of the products of the other components as shown in Table 5.6. The entries of Table 5.6 form what is called a consumption matrix, C. If for a given time interval, we let x_1 be the output of coal, x_2 be the output of gasoline, and x_3 be the output of electricity, then the product

$$CX = \begin{bmatrix} 0 & 0 & \frac{1}{5} \\ 1 & \frac{1}{5} & \frac{2}{5} \\ 1 & \frac{2}{5} & \frac{1}{5} \end{bmatrix} \begin{bmatrix} x_1 \\ x_2 \\ x_3 \end{bmatrix}$$

Table 5.6. CONSUMPTION TABLE.

	To produce one unit of		
	coal requires	gasoline requires	electricity requires
	0 units of coal	0 units of coal	$\frac{1}{5}$ unit of coal
	1 unit of gasoline	$\frac{1}{5}$ unit of gasoline	$\frac{2}{5}$ unit of gasoline
	1 unit of electricity	$\frac{2}{5}$ unit of electricity	$\frac{1}{5}$ unit of electricity

represents the internal consumption of each of the products of the 3 components. Finally, we let $d_1 = 100$, $d_2 = 100$, and $d_3 = 100$ be the external consumption of the products of each component; then we have

(production) = (internal consumption) + (external consumption)

or

$$X = CX + D \quad \text{or} \quad (I_3 - C)X = D.$$

The latter matrix equation can be solved for X provided that $(I_3 - C)^{-1}$ exists. We get

$$X = (I_3 - C)^{-1}D,$$

which for our data becomes

$$X = \left(\begin{bmatrix} 1 & 0 & 0 \\ 0 & 1 & 0 \\ 0 & 0 & 1 \end{bmatrix} - \begin{bmatrix} 0 & 0 & \frac{1}{5} \\ 1 & \frac{1}{5} & \frac{2}{5} \\ 1 & \frac{2}{5} & \frac{1}{5} \end{bmatrix} \right)^{-1} \begin{bmatrix} 100 \\ 100 \\ 100 \end{bmatrix} = \begin{bmatrix} 300 \\ 1000 \\ 1000 \end{bmatrix}.$$

Once the internal consumption is established and $(I_3 - C)^{-1}$ is calculated, then variations in X will be a function of D only. ∎

Example 7

In Chapter 6 of Johnston et al. [34] the reader may find several variations of the model presented in Example 6; these variations are applied to problems in business and economics. Particular attention is called to the model for allocation of service charges in accounting that is begun on p. 262 of [34]. Students of business and economics are encouraged to at least glance at Chapter 6 of [34] to assure them of the relevance of the material they are studying in this chapter. The bibliography and bibliographical notes on pp. 272–273 of [34] may also be helpful. ∎

EXERCISES

Suggested minimum assignment: Exercises 1, 4, 6, 8, 10, 12, and 14.

1. Write the multiplicative inverses of the real numbers $7, \frac{2}{5}$, and $\sqrt{2}$. Why does the real number 0 not have an inverse?
2. Write the multiplicative inverses of the 1 by 1 matrices $[7]$, $[\frac{2}{5}]$, and $[\sqrt{2}]$. Why does the matrix $[0]$ not have an inverse?

In Exercises 3–10 calculate the inverses of the given matrices, if possible. If not possible, give the reason. Check your answer by using Definition 21.

3. $\begin{bmatrix} 2 & 1 \\ 6 & 2 \end{bmatrix}$.

4. $\begin{bmatrix} 4 & 5 \\ 6 & 7 \end{bmatrix}$.

5. $\begin{bmatrix} 1 & -2 \\ -3 & 6 \end{bmatrix}$.

6. $\begin{bmatrix} -1 & 3 \\ -3 & 9 \end{bmatrix}$.

7. $\begin{bmatrix} 2 & 1 \\ 3 & 2 \\ 1 & 4 \end{bmatrix}$.

8. $\begin{bmatrix} 1 & 2 & 6 \\ 0 & 1 & 5 \end{bmatrix}$.

9. $\begin{bmatrix} 1 & 1 & 7 \\ 0 & 1 & 2 \\ 2 & 4 & 17 \end{bmatrix}$.

10. $\begin{bmatrix} 1 & 2 & 4 \\ 2 & 4 & 7 \\ 0 & 1 & 3 \end{bmatrix}$.

In Exercises 11 and 12 find the solution of the system by finding and using the multiplicative inverse of the coefficient matrix. Check your solution by substituting it into the original system.

11. $x_1 + x_2 = 5,$
    $2x_1 - x_2 = 1.$

12. $x_1 - 2x_2 = -5,$
    $3x_1 + 5x_2 = 18.$

In Exercises 13 and 14 find the solution of the systems $AX = B$, where A^{-1} is given. Check your solution by substituting it into the original system.

13. $2x_1 + 4x_2 = 60,$
    $ 2x_2 + x_3 = 20,$
    $3x_1 + 2x_3 = 40.$

$A^{-1} = \begin{bmatrix} \frac{4}{20} & \frac{-8}{20} & \frac{4}{20} \\ \frac{3}{20} & \frac{4}{20} & \frac{-2}{20} \\ \frac{-6}{20} & \frac{12}{20} & \frac{4}{20} \end{bmatrix}.$

14. $x_1 - x_2 - x_3 = 0,$
    $5x_1 + 20x_3 = 50,$
    $ 10x_2 - 20x_3 = 30.$

$A^{-1} = \begin{bmatrix} \frac{4}{7} & \frac{3}{35} & \frac{4}{70} \\ \frac{-2}{7} & \frac{2}{35} & \frac{5}{70} \\ \frac{-1}{7} & \frac{1}{35} & \frac{-1}{70} \end{bmatrix}.$

15. A few of the properties involving invertible and noninvertible matrices follow. (In all four cases A and B are square matrices.)
    (i) If A is an invertible matrix, then $(A^{-1})^{-1} = A$.
    (ii) If A and B are invertible matrices, then $(AB)^{-1} = B^{-1}A^{-1}$.
    (iii) If $AB = 0$, then $A = 0$, or $B = 0$, or both A and B are noninvertible.
    (iv) If $AB = I_n$, then $B = A^{-1}$, $A = B^{-1}$, and $AB = BA$.

    (a) Illustrate (i) and (ii) with $A = \begin{bmatrix} 1 & 3 \\ 2 & 7 \end{bmatrix}$, $B = \begin{bmatrix} 1 & 3 \\ 0 & 1 \end{bmatrix}$.

    (b) Does $(AB)^{-1} = A^{-1}B^{-1}$ and why?

    (c) Illustrate (iii) with $A = \begin{bmatrix} 1 & 1 \\ 2 & 2 \end{bmatrix}$, $B = \begin{bmatrix} 1 & -3 \\ -1 & 3 \end{bmatrix}$.

Sec. 5.6] The Inverse Matrix

(d) Illustrate (iv) with $A = \begin{bmatrix} 1 & 3 \\ 0 & 1 \end{bmatrix}$, $B = \begin{bmatrix} 1 & -3 \\ 0 & 1 \end{bmatrix}$.

(e) If A and B are nonzero matrices and $AB = 0$, is $B \stackrel{\text{row}}{\sim} I_n$ and why?

16. Prove property (i) of Exercise 15. (Hint: Let $B = A^{-1}$ and hence $AB = BA = I_n$; then $A = B^{-1}$.)
17. Prove property (ii) of Exercise 15. [Hint: Show that $(AB)(B^{-1}A^{-1}) = I_n$ and that $(B^{-1}A^{-1})(AB) = I_n$; then use Definition 21 and Theorem 6.]
18. Prove property (iii) of Exercise 15. (Hint: Show that the existence of A^{-1} implies $B = 0$; show that the existence of B^{-1} implies $A = 0$, and then consider the only other possibility.)
19. Prove property (iv) of Exercise 15 under the assumption that A is invertible. [Hint: Start with $B = I_n B = (A^{-1}A)B$ and continue on to show that $B = A^{-1}$, and then use Definition 21 to show that $AB = BA$.]
20. Give reasons for the statements of the proof of Theorem 6.
21. Prove Theorem 7. (Hint: See Example 2.)
22. Rework Example 5 if there are 4 type A machines and 2 type B machines.

5.7 Systems of Linear Equations Not Having a Unique Solution

In general, a system of linear equations over the real numbers may have (1) a unique solution, (2) an infinite number of solutions, or (3) no solution.

In the last two sections we have concentrated on the first of these three possibilities; in Section 5.5 we used elementary row operations on the augmented matrix to obtain the unique solution, and in Section 5.6 we used the inverse matrix and multiplication to determine the unique solution. In this and the next section we shall extend and formalize the use of elementary operations on the augmented matrix of a linear system in such a way that the same procedure may be used on any linear system, regardless of the nature of the solution. We shall begin by illustrating the method in the following two examples.

Example 1
Consider the following linear system over the real numbers:

(1) $x_1 + x_2 - 2x_3 = -1$,
(2) $x_1 + x_2 + 2x_3 = 3$,
(3) $2x_1 + 2x_2 + x_3 = 3$.

We shall find that this system has an infinite number of solutions; this case is illustrated in Figure 5.19, where the graphs of the equations of the system are planes which are seen to intersect along a line.

The points of the line of intersection correspond to the solutions of the system; algebraically, we can find those solutions in the following way. The augmented matrix of the system is

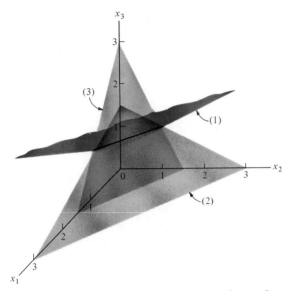

Figure 5.19. *Intersection of three planes along a line.*

$$\begin{bmatrix} 1 & 1 & -2 & | & -1 \\ 1 & 1 & 2 & | & 3 \\ 2 & 2 & 1 & | & 3 \end{bmatrix}.$$

First we perform the elementary row operations necessary to make the first entry in the first row a 1 and to make the other entries in the first column a 0. The operations $(-R_1 + R_2)$ and $(-2R_1 + R_3)$ will accomplish this task, and we get the row equivalent matrix

$$\begin{bmatrix} (1) & 1 & -2 & | & -1 \\ (0) & 0 & 4 & | & 4 \\ (0) & 0 & 5 & | & 5 \end{bmatrix}.$$

We now move to the second row and perform the elementary row operations necessary to make the first nonzero entry of the second row a 1. The operation $(\frac{1}{4}R_2)$ will do the job, and we obtain the row equivalent matrix

$$\begin{bmatrix} (1) & 1 & -2 & | & -1 \\ (0) & 0 & (1) & | & 1 \\ (0) & 0 & 5 & | & 5 \end{bmatrix}.$$

We then make the other entries in the third column be 0. The operations $(2R_2 + R_1)$ and $(-5R_2 + R_3)$ will produce the row equivalent matrix

$$\begin{bmatrix} (1) & 1 & (0) & | & 1 \\ (0) & 0 & (1) & | & 1 \\ (0) & 0 & (0) & | & 0 \end{bmatrix}.$$

Sec. 5.7] Linear Equations Not Having a Unique Solution

Because there are no other nonzero rows, the process is complete, and we write the system for which the latter matrix is the augmented matrix;

$$\begin{aligned} x_1 + x_2 &= 1, \\ x_3 &= 1, \\ 0x_1 + 0x_2 + 0x_3 &= 0. \end{aligned} \qquad (5.4)$$

The last equation of (5.4) will be an identity no matter what values are assigned to the unknowns, and hence (5.4) has the same solution as the system

$$\begin{aligned} x_1 + x_2 &= 1, \\ x_3 &= 1. \end{aligned} \qquad (5.5)$$

Obviously, there are more unknowns than there are equations in (5.5). If there are r equations in n unknowns at this stage, then we can *express r unknowns* (called **basic variables**) *in terms of the remaining $n-r$ unknowns*, which are considered to be arbitrary parameters. System (5.5) can then be expressed as

$$\begin{aligned} x_1 &= 1 - x_2, \\ x_3 &= 1 - 0x_2, \end{aligned}$$

which is known as a **complete solution** of the original system. If particular values are assigned to the parameters (x_2 in this example), then a **particular solution** is obtained. For example, if we let $x_2 = 1$, then we obtain the particular solution (0, 1, 1), which can be interpreted as the list of coordinates of a single point on the line of intersection shown in Figure 5.19. ∎

The next example will illustrate how we can use elementary row operations to reveal the case in which no solution exists for a linear system.

Example 2
Consider the system

$$\begin{aligned} (1) \quad x_1 \quad\quad + 2x_3 &= 4, \\ (2) \quad\quad + x_2 - 2x_3 &= 2, \\ (3) \quad x_1 + x_2 \quad\quad &= 7, \end{aligned}$$

whose graph (see Figure 5.20) consists of 3 planes oriented in such a way that no point is common to all 3 planes. Construct the augmented matrix of the system and perform the indicated elementary row operations,

$$\begin{bmatrix} 1 & 0 & 2 & | & 4 \\ 0 & 1 & -2 & | & 2 \\ 1 & 1 & 0 & | & 7 \end{bmatrix} \xrightarrow[-R_1 + R_3]{\text{row}} \begin{bmatrix} (1) & 0 & 2 & | & 4 \\ (0) & 1 & -2 & | & 2 \\ (0) & 1 & -2 & | & 3 \end{bmatrix}$$

$$\xrightarrow[-R_2 + R_3]{\text{row}} \begin{bmatrix} (1) & (0) & 2 & | & 4 \\ (0) & (1) & -2 & | & 2 \\ (0) & (0) & 0 & | & 1 \end{bmatrix} \xrightarrow[-4R_3 + R_1]{\text{row} \atop -2R_3 + R_2} \begin{bmatrix} (1) & (0) & 2 & | & (0) \\ (0) & (1) & -2 & | & (0) \\ (0) & (0) & 0 & | & (1) \end{bmatrix}.$$

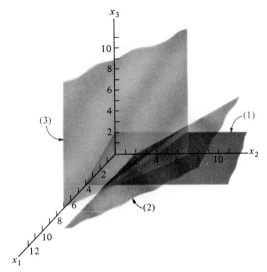

Figure 5.20. *Three planes having no point in common.*

The last matrix is the augmented matrix of the system

$$x_1 + 2x_3 = 0,$$
$$x_2 - 2x_3 = 0,$$
$$0x_1 + 0x_2 + 0x_3 = 1,$$

which we know must have the same solution as the original system. But the last system has no solution because there are no real numbers for x_1, x_2, and x_3 such that $0x_1 + 0x_2 + 0x_3 = 1$. Therefore, the original system has no solution. ∎

APPLICATIONS

Example 3

Consider the system of linear equations

$$p_{11}x_1 + p_{12}x_2 + p_{13}x_3 = \lambda x_1,$$
$$p_{21}x_1 + p_{22}x_2 + p_{23}x_3 = \lambda x_2,$$
$$p_{31}x_1 + p_{32}x_2 + p_{33}x_3 = \lambda x_3,$$

where λ is a constant, p_{ij} is the probability that a purchaser who bought brand j of a certain product last time will buy brand i next time, and where x_j represents the numbers of buyers that bought brand j the first time. If the probabilities p_{ij} are known, then we have a linear system

$$PX = \lambda X,$$

whose solution would be the number of initial buyers of each brand that would result in the same proportion of buyers of each brand at the next purchase time. A stability in the proportions of buyers of each brand would

be in effect as long as the probabilities did not change. To illustrate, suppose that the probabilities are as given in Figure 5.21, where an arrow extending from brand j to brand i represents p_{ij}. For some values of λ the system will have meaningful solutions; the means for determining such values of λ will be presented in Section 5.10. For the given probabilities, $\lambda = 1$ will produce meaningful solutions. The resulting system is

$$.3x_1 + .6x_2 + .3x_3 = x_1,$$
$$.6x_1 + .2x_2 + .2x_3 = x_2,$$
$$.1x_1 + .2x_2 + .5x_3 = x_3,$$

or

$$-.7x_1 + .6x_2 + .3x_3 = 0,$$
$$.6x_1 - .8x_2 + .2x_3 = 0,$$
$$.1x_1 + .2x_2 - .5x_3 = 0.$$

By the methods of this section we find that the system has an infinite number of solutions and that a complete solution is

$$x_1 = \tfrac{9}{5}x_3,$$
$$x_2 = \tfrac{8}{5}x_3.$$

Therefore, $X = [\tfrac{9}{5} \;\; \tfrac{8}{5} \;\; 1]^T x_3$ is a complete solution of the original system; for this solution there is a stability in the proportion of buyers of each brand. Observe that P^T is a stochastic matrix. ∎

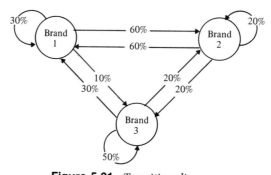

Figure 5.21. *Transition diagram.*

Example 4
The matrix model developed in Example 3 can also be used to find stable proportions of n genetic characteristics of a population with respect to certain breeding probabilities, or to find stable proportions of n political affiliations of a population of voters with respect to certain probabilities of voting transition. ∎

EXERCISES

Suggested minimum assignment: Exercises 1, 2, 5, 9, and 11.

1. Find a complete solution and a particular solution of the system

$$x_1 + 2x_2 + 2x_3 = 4,$$
$$2x_1 + x_3 = 4.$$

Interpret your complete and particular solutions geometrically.

2. Show algebraically that the system

$$x_1 + 2x_2 = 6,$$
$$2x_1 + 4x_2 = 8,$$

has no solution. Interpret your algebraic results geometrically.

In each of Exercises 3–10, by the methods of this section, solve the given system or determine that there is no solution. If the system has an infinite number of solutions, find a complete solution and one particular solution.

3. $x_1 + 2x_2 + x_3 = 4,$
   $2x_1 + x_2 + 6x_3 = 5.$

4. $x_1 + 3x_2 + x_3 = 4,$
   $2x_1 + 6x_2 + 2x_3 = 6.$

5. $x_1 - x_2 + x_3 = 4,$
   $x_1 + x_2 + 2x_3 = 2,$
   $ -2x_2 - x_3 = 2.$

6. $x_1 + x_2 + x_3 = 4,$
   $ x_2 - 2x_3 = 1,$
   $2x_1 + 3x_2 = 9.$

7. $x_1 + x_2 + x_3 = 4,$
   $x_1 - x_2 = 3,$
   $2x_1 + x_3 = 8.$

8. $x_1 + 3x_2 + x_3 = 4,$
   $x_1 + 4x_2 + 5x_3 = 2.$

9. $x_1 + 2x_2 + x_3 + x_4 = 0,$
   $x_1 - x_2 + x_3 = 7.$

10. $x_1 + x_2 - x_3 + x_4 = 0,$
    $x_1 + 2x_2 + x_3 + x_4 = 0.$

11. In order to control a certain crop disease, someone determines that it is necessary to use 12 units of chemical A and 16 units of chemical B. One barrel of commercial spray P contains 1 unit of A and 2 units of B. One barrel of spray Q contains 1 unit of A and 1 unit of B, and one barrel of spray R contains 3 units of A and 1 unit of B. How many barrels of each type of spray could be used to spread the exact amount of chemicals needed to control the disease?

12. Three types of trucks are equipped to haul 2 different types of machines per load according to the following chart:

Machine	Truck 1	Truck 2	Truck 3
A	3	1	2
B	2	1	1

How many trucks of each type should be sent to haul exactly 16 of the type A machines and 12 of the type B machines? (Assume that each truck is fully loaded.)

Sec. 5.7] Linear Equations Not Having a Unique Solution

5.8 The Gauss–Jordan Method

As we mentioned before, in general, a linear system over the real numbers has (1) a unique solution, (2) an infinite number of solutions, or (3) no solution. In previous sections we have illustrated by means of examples how elementary row operations can be applied to the augmented matrix of any linear system; we now formalize the method of solution used in those examples. In the first two examples of Section 5.7, which had three linear equations in three unknowns, elementary row operations were used to obtain the columns

$$I_{(1)} = \begin{bmatrix} 1 \\ 0 \\ 0 \end{bmatrix}, \quad I_{(2)} = \begin{bmatrix} 0 \\ 1 \\ 0 \end{bmatrix}, \quad \text{and, if possible,} \quad I_{(3)} = \begin{bmatrix} 0 \\ 0 \\ 1 \end{bmatrix}$$

in the augmented matrix; these columns $I_{(1)}$, $I_{(2)}$, and $I_{(3)}$ need not be adjacent to each other, nor need they appear in that order, although in practice they frequently do. We also observe in Example 1 of Section 5.7 that when $I_{(3)}$ did not appear, the last row consisted entirely of 0's. In general, we formalize these observations with the following definition.

▶ **Definition 22**
Let A be an m by n matrix, and let r be an integer such that $1 \le r \le m$. If A has $I_{(1)}, I_{(2)}, \ldots, I_{(r)}$ among its columns and if A has $(m - r)$ rows consisting entirely of 0 entries, then A is said to be a **reduced matrix**.

Example 1
The following matrices *are* reduced matrices:

$$\begin{bmatrix} 1 & 0 & 0 & 2 \\ 0 & 1 & 0 & 3 \\ 0 & 0 & 1 & 4 \end{bmatrix}, \quad \begin{bmatrix} 1 & 0 & 4 & 6 \\ 0 & 1 & 3 & 2 \\ 0 & 0 & 0 & 0 \end{bmatrix}, \quad \begin{bmatrix} 1 & 0 & 3 & 0 \\ 0 & 1 & 2 & 0 \\ 0 & 0 & 0 & 1 \end{bmatrix},$$

$$\begin{bmatrix} 1 & 0 & 2 & 0 & 2 \\ 0 & 1 & 3 & 0 & 3 \\ 0 & 0 & 4 & 1 & 2 \end{bmatrix}, \quad \begin{bmatrix} 1 & 0 & 3 \\ 0 & 1 & 2 \\ 0 & 0 & 0 \end{bmatrix}, \quad \begin{bmatrix} 2 & 0 & 5 & 1 & 2 \\ 3 & 1 & 6 & 0 & 3 \\ 0 & 0 & 0 & 0 & 0 \end{bmatrix}.$$

The matrix

$$\begin{bmatrix} 1 & 0 & 2 & 2 \\ 0 & 1 & 3 & 1 \\ 0 & 0 & 1 & 5 \end{bmatrix}$$

is *not* a reduced matrix because $I_{(3)}$ is not a column and the third row does not have all 0 entries; one or the other must be true for a matrix having 3 rows. ∎

Because it can be proved that every nonzero matrix is row equivalent to a reduced matrix and that the resulting reduced matrix is unique for specific positions of $I_{(1)}, I_{(2)}, \ldots, I_{(r)}$, we can define the general method of solution as a sequence of elementary row operations necessary to transform the augmented matrix of a linear system to a row equivalent reduced matrix; we shall call this method of solving a linear system the **Gauss–Jordan method**.

Linear systems $AX = B$ in which $B = 0$ are called **homogeneous systems**; such systems always have a solution because the zero vector always satisfies the system. The question when solving a homogeneous system is whether the system has only a unique solution, namely, the zero vector, or an infinite number of solutions.

Example 2

Solve the homogeneous system

$$x_1 + x_2 - x_3 = 0,$$
$$2x_2 - x_3 = 0,$$
$$4x_1 + 6x_2 - 5x_3 = 0,$$

by the Gauss–Jordan method.

Solution: Elementary row operations necessary to change the augmented matrix of the given system to a reduced matrix are as follows:

$$\begin{bmatrix} 1 & 1 & -1 & | & 0 \\ 0 & 2 & -1 & | & 0 \\ 4 & 6 & -5 & | & 0 \end{bmatrix} \underset{-4R_1 + R_3}{\overset{\text{row}}{\sim}} \begin{bmatrix} (1) & 1 & -1 & | & 0 \\ (0) & 2 & -1 & | & 0 \\ (0) & 2 & -1 & | & 0 \end{bmatrix} \underset{\frac{1}{2}R_2}{\overset{\text{row}}{\sim}} \begin{bmatrix} (1) & 1 & -1 & | & 0 \\ (0) & (1) & -\frac{1}{2} & | & 0 \\ (0) & 2 & -1 & | & 0 \end{bmatrix}$$

$$\underset{\substack{-2R_2 + R_3 \\ -R_2 + R_1}}{\overset{\text{row}}{\sim}} \begin{bmatrix} (1) & (0) & -\frac{1}{2} & | & 0 \\ (0) & (1) & -\frac{1}{2} & | & 0 \\ (0) & (0) & 0 & | & 0 \end{bmatrix}.$$

The reduced matrix represents the augmented matrix of a system that has the same solution as that of the original system. The system is

$$x_1 \qquad - \tfrac{1}{2}x_3 = 0,$$
$$x_2 - \tfrac{1}{2}x_3 = 0,$$
$$0x_1 + 0x_2 + 0x_3 = 0,$$

and a complete solution of this homogeneous system is

$$x_1 = \tfrac{1}{2}x_3,$$
$$x_2 = \tfrac{1}{2}x_3.$$

Thus we have a system with an infinite number of solutions, depending upon the value that is assigned to the parameter x_3. If we let the parameter x_3 equal 2, then a particular solution is (1, 1, 2). ∎

APPLICATIONS

Example 3

A primary and significant use of this and the last section occurs in solving linear programming problems. This use, and applications of linear programming problems, will be demonstrated in Chapters 6 and 7. ∎

Example 4

In Example 7 of Section 5.3 we learned that the state vector B_k after k transitions of a Markov chain is found by

$$B_n = B_{n-1}A \quad \text{or by} \quad B_n = B_0 A^n,$$

where A is the transition matrix. In some applications, for large values of n, B_n will differ very little from B_{n-1}. In such applications it is natural to wonder, to what vector the sequence of vectors $B_1, B_2, B_3, \ldots, B_n, \ldots$ will approach for increasing values of n; it can be proved that if there is such a row vector X, it is the solution of the equation

$$X = XA$$

and is called a **steady-state vector** of the matrix A.

In this example we shall illustrate the calculation of a steady-state vector by again using the example in which a system performs satisfactorily or unsatisfactorily at given time intervals (Example 7 of Section 5.3). Thus

$$X = XA$$

becomes

$$[x_1 \quad x_2] = [x_1 \quad x_2] \begin{bmatrix} .6 & .4 \\ .8 & .2 \end{bmatrix},$$

which is equivalent to the system

$$x_1 = .6x_1 + .8x_2,$$
$$x_2 = .4x_1 + .2x_2,$$

or

$$.4x_1 - .8x_2 = 0,$$
$$-.4x_1 + .8x_2 = 0.$$

The latter system has a complete solution $x_1 = 2x_2$; since x_1 and x_2 must add up to 1, we choose $x_2 = \frac{1}{3}$, and hence $x_1 = \frac{2}{3}$. Thus in the long run after a sufficient number of transitions, and independent of the initial state, the state vector will eventually approach

$$[\tfrac{2}{3} \quad \tfrac{1}{3}].$$

This means that in the long run the system will operate satisfactorily $\frac{2}{3}$ of the time and unsatisfactorily $\frac{1}{3}$ of the time. It can be proved that a steady-state vector exists for any Markov chain having a transition matrix with all positive entries.

Because the sum of the components of X must be 1, some efficiency can be

gained by applying the Gauss–Jordan method initially to the nonhomogeneous system

$$(I_n - A)X^T = 0,$$
$$[1\ 1\ \cdots\ 1]X^T = [1].\ \blacksquare$$

Example 5
The use of matrices in the study of Markov chains has been illustrated by several examples in this chapter. For a more comprehensive study of Markov chains and their applications, the reader is referred to Kemeny et al. [38], pp. 215–253. Markov decision processes with applications are discussed in Chapter 8 of [38]. $\blacksquare$

EXERCISES

Suggested minimum assignment: Exercises 1, 4, 5, and 7.

1. Which of the following matrices are reduced matrices? If they are not reduced matrices, state the reason.

   (a) $\begin{bmatrix} 1 & 0 & 0 & 3 \\ 2 & 1 & 0 & 2 \\ 0 & 0 & 1 & 4 \end{bmatrix}$.
   (b) $\begin{bmatrix} 1 & 4 & 0 & 4 & 1 \\ 0 & 0 & 1 & 3 & 6 \\ 0 & 0 & 0 & 0 & 0 \end{bmatrix}$.
   (c) $\begin{bmatrix} 1 & 3 & 0 & 4 \\ 0 & 0 & 1 & 5 \\ 0 & 0 & 0 & 1 \end{bmatrix}$.

2. Which of the following are reduced matrices? If they are not reduced matrices, state the reason.

   (a) $\begin{bmatrix} 2 & 0 & 0 & 3 \\ 0 & 1 & 0 & 2 \\ 0 & 0 & 1 & 4 \end{bmatrix}$.
   (b) $\begin{bmatrix} 0 & 1 & 0 & 3 \\ 1 & 0 & 0 & 2 \\ 0 & 0 & 1 & 1 \end{bmatrix}$.
   (c) $\begin{bmatrix} 1 & 3 & 2 & 6 \\ 0 & 0 & 0 & 4 \\ 0 & 0 & 0 & 0 \end{bmatrix}$.

In each of Exercises 3–8 use the Gauss–Jordan method to solve the given system or to determine that there is no solution. If the system has an infinite number of solutions, find a complete solution and one particular solution.

3. $x_1 + x_2 \qquad\qquad = 4,$
   $x_1 - x_2 + 2x_3 = 2,$
   $\qquad x_2 + x_3 = 5.$

4. $x_1 \qquad\ + x_3 = 5,$
   $\quad 2x_2 + 4x_3 = 22,$
   $x_1 - 2x_2 \qquad = -2.$

5. $x_1 + 2x_2 + x_3 = 0,$
   $\qquad x_2 + x_3 = 0,$
   $3x_1 + 5x_2 + 2x_3 = 0.$

6. $x_1 + x_2 - x_3 - 1 = 0,$
   $\quad 2x_2 + 4x_3 - 3 = 0,$
   $x_1 + x_2 + 5x_3 - 2 = 0.$

7. $x_1 - x_2 \qquad\ + x_4 = 2,$
   $\quad x_2 + x_3 + x_4 = 5,$
   $-x_1 + 2x_2 + x_3 \qquad = 3,$
   $-x_1 + 3x_2 + 2x_3 + x_4 = 8.$

8. $\qquad\quad 2x_2 + 2x_3 \qquad = 2,$
   $x_1 \qquad\qquad\quad + x_4 = 2,$
   $2x_1 \qquad\qquad\ + 2x_4 = 6,$
   $\qquad\qquad\ x_3 + x_4 = 7.$

9. Which of Exercises 3–8 is a homogeneous system?
10. If A is a square matrix, prove that the homogeneous system $AX = 0$ has *only* the solution $X = 0$ if and only if A is invertible.

Sec. 5.8] The Gauss–Jordan Method

5.9 The Determinant of a Matrix (Optional†)

The reader should have noticed that the definition of a matrix does *not* assign a numerical value to a matrix. The purpose of this section is to define an operation on a matrix in such a way that to every square matrix A there corresponds a single real number. The result of this operation is known as the **determinant of A** and is designated by $|A|$ or det A.

We shall begin by defining the **determinant of a 1 by 1 matrix** to be equal to the single entry of that matrix.

Example 1
$$\det [6] = 6, \quad \det [-3] = -3, \quad \det [0] = 0. \blacksquare$$

The **determinant of a 2 by 2 matrix** is defined by
$$\det \begin{bmatrix} a_{11} & a_{12} \\ a_{21} & a_{22} \end{bmatrix} = a_{11}a_{22} - a_{12}a_{21}.$$

Example 2
$$\det \begin{bmatrix} 3 & 2 \\ 4 & 6 \end{bmatrix} = 3 \cdot 6 - 2 \cdot 4 = 10;$$

$$\det \begin{bmatrix} 0 & -2 \\ 6 & 3 \end{bmatrix} = 0 \cdot 3 - (-2) \cdot 6 = 12. \blacksquare$$

Before defining the determinant of a 3 by 3 matrix, we first define a **minor** and a **cofactor** of an entry.

▶ **Definition 23**
If the ith row and the jth column of a 3 by 3 matrix A are deleted, then the determinant of the resulting 2 by 2 submatrix is called the **minor** of the entry a_{ij} and is denoted by M_{ij}. The quantity $A_{ij} = (-1)^{i+j} M_{ij}$ is known as the **cofactor** of a_{ij}.

Example 3
Let
$$A = \begin{bmatrix} 2 & 3 & 0 \\ 4 & 2 & 1 \\ 6 & 5 & 2 \end{bmatrix}.$$

---

† This section is required for the optional Section 5.10.

$$M_{11} = \det \begin{bmatrix} 2 & 3 & 0 \\ 4 & 2 & 1 \\ 6 & 5 & 2 \end{bmatrix} = \det \begin{bmatrix} 2 & 1 \\ 5 & 2 \end{bmatrix} = 2 \cdot 2 - 1 \cdot 5 = -1;$$

$$M_{12} = \det \begin{bmatrix} 2 & 3 & 0 \\ 4 & 2 & 1 \\ 6 & 5 & 2 \end{bmatrix} = \det \begin{bmatrix} 4 & 1 \\ 6 & 2 \end{bmatrix} = 4 \cdot 2 - 1 \cdot 6 = 2;$$

$$M_{13} = \det \begin{bmatrix} 2 & 3 & 0 \\ 4 & 2 & 1 \\ 6 & 5 & 2 \end{bmatrix} = \det \begin{bmatrix} 4 & 2 \\ 6 & 5 \end{bmatrix} = 4 \cdot 5 - 2 \cdot 6 = 8;$$

$$A_{11} = (-1)^{1+1} M_{11} = (+1)(-1) = -1;$$
$$A_{12} = (-1)^{1+2} M_{12} = (-1)(2) = -2;$$
$$A_{13} = (-1)^{1+3} M_{13} = (+1)(8) = 8.$$

The **vector of cofactors** corresponding to the first row of A is

$$(-1, -2, 8). \blacksquare$$

▶ **Definition 24**
The **determinant of a 3 by 3 matrix** A is the dot product of the ith row (or jth column) of A with the vector of cofactors corresponding to the ith row (or jth column) of A.†

The validity of Definition 24 depends upon the fact that det A is the same no matter which row i (or column j) is chosen. The latter assertion is illustrated in Exercise 3.

Example 4
If we use the matrix

$$A = \begin{bmatrix} 2 & 3 & 0 \\ 4 & 2 & 1 \\ 6 & 5 & 2 \end{bmatrix}$$

of Example 3, and if we let $i = 1$ in Definition 24, then, using the cofactors found in Example 3, we get

$$\begin{aligned} \det A &= (a_{11}, a_{12}, a_{13}) \cdot (A_{11}, A_{12}, A_{13}) \\ &= a_{11} A_{11} + a_{12} A_{12} + a_{13} A_{13} \\ &= 2(-1) + 3(-2) + 0(8) \\ &= -8. \end{aligned}$$

The dot product of any other row or column with its vector of cofactors will also produce the number -8. $\blacksquare$

† Notice that the definition of the determinant of a 2 by 2 matrix can be expressed in a manner that is analogous to Definition 24.

We have defined the determinant of matrix A, where A was 1 by 1, 2 by 2, or 3 by 3. In general, if n is any positive integer such that the determinant of an $(n-1)$ by $(n-1)$ matrix is defined, then the **determinant of an n by n matrix A** is defined as the dot product of the ith row (or jth column) of A with the vector of cofactors corresponding to the ith row (or jth column); the cofactor of an entry of an n by n matrix is defined in the manner of Definition 23. Thus if A is an n by n matrix,

$$\det A = (a_{i1}, a_{i2}, \ldots, a_{in}) \cdot (A_{i1}, A_{i2}, \ldots, A_{in})$$
$$= a_{i1}A_{i1} + a_{i2}A_{i2} + \cdots + a_{in}A_{in}, \tag{5.6}$$

or

$$\det A = (a_{1j}, a_{2j}, \ldots, a_{nj}) \cdot (A_{1j}, A_{2j}, \ldots, A_{nj})$$
$$= a_{1j}A_{1j} + a_{2j}A_{2j} + \cdots + a_{nj}A_{nj}. \tag{5.7}$$

Expression (5.6) is said to be the **expansion** of $\det A$ about the ith row, whereas expression (5.7) is said to be the **expansion** of $\det A$ about the jth column. In Example 4, $\det A$ was calculated by means of the expansion about the first row; any other row or column could be chosen to expand about, and the value of $\det A$ would be the same. The first row was chosen because the presence of a zero in the upper right-hand corner shortened the work.

Elementary row operations have been used frequently in this chapter; it seems reasonable, therefore, to ascertain what effect, if any, these operations have on the value of the determinant of a matrix.

▶ **Theorem 8**

If a matrix B is formed from a square matrix A by the interchange of two rows, then $\det B = -\det A$. ◀

▶ **Theorem 9**

If a matrix B is formed from a square matrix A by multiplying each entry of any row of A by a real number k, then $\det B = k \det A$. ◀

Notice the difference in $\det kA$ and $k \det A$.

▶ **Theorem 10**

If a matrix B is formed from a square matrix A by adding to the entries of any row of A a constant real number multiple of the corresponding entries of another row, then $\det B = \det A$.

In Exercises 20–22 the reader will be asked to prove Theorems 8–10 for the case where $n = 3$. ◀

In practice, the calculation of the determinant of a matrix often can be handled more efficiently if the following theorem is used.

▶ **Theorem 11**
The determinant of a triangular matrix is the product of the entries on the main diagonal.

The proof of Theorem 11 is left as Exercise 23. ◀

We now illustrate the use of Theorem 11 in calculating the determinant of a matrix.

Example 5
If a matrix is triangular, then by Theorem 11,

$$\det \begin{bmatrix} 4 & 2 & 3 & 4 \\ 0 & 6 & 2 & 4 \\ 0 & 0 & 1 & 7 \\ 0 & 0 & 0 & 3 \end{bmatrix} = 4 \cdot 6 \cdot 1 \cdot 3 = 72. \blacksquare$$

Example 6
If a matrix is not triangular, we can first repeatedly use Theorem 10 (and Theorem 8 if needed†) to triangularize it. Using the elementary row operations $(-2R_1 + R_3)$, $(2R_2 + R_3)$, and $(-R_3 + R_4)$ in that order, we evaluate the following determinant;

$$\det \begin{bmatrix} 1 & 3 & 0 & 2 \\ 0 & 2 & 1 & 3 \\ 2 & 2 & 1 & 3 \\ 0 & 0 & 3 & -2 \end{bmatrix} \underset{-2R_1 + R_3}{=} \det \begin{bmatrix} 1 & 3 & 0 & 2 \\ 0 & 2 & 1 & 3 \\ 0 & -4 & 1 & -1 \\ 0 & 0 & 3 & -2 \end{bmatrix}$$

$$\underset{2R_2 + R_3}{=} \det \begin{bmatrix} 1 & 3 & 0 & 2 \\ 0 & 2 & 1 & 3 \\ 0 & 0 & 3 & 5 \\ 0 & 0 & 3 & -2 \end{bmatrix} \underset{-R_3 + R_4}{=} \det \begin{bmatrix} 1 & 3 & 0 & 2 \\ 0 & 2 & 1 & 3 \\ 0 & 0 & 3 & 5 \\ 0 & 0 & 0 & -7 \end{bmatrix}$$

$$= 1 \cdot 2 \cdot 3 \cdot -7$$
$$= -42. \blacksquare$$

APPLICATIONS

Example 7
The determinant of a matrix can be used in an alternative method for calculating the multiplicative inverse of a square matrix. If

$$A = \begin{bmatrix} a_{11} & a_{12} & \cdots & a_{1n} \\ a_{21} & a_{22} & \cdots & a_{2n} \\ \vdots & \vdots & & \vdots \\ a_{n1} & a_{n2} & \cdots & a_{nn} \end{bmatrix},$$

† Theorem 8 is needed if a zero appears on the main diagonal and nonzeros appear in the same column below the main diagonal.

then the **cofactor matrix** of A, designated by $\operatorname{cof} A$, is defined by

$$\operatorname{cof} A = \begin{bmatrix} A_{11} & A_{12} & \cdots & A_{1n} \\ A_{21} & A_{22} & \cdots & A_{2n} \\ \vdots & \vdots & & \vdots \\ A_{n1} & A_{n2} & \cdots & A_{nn} \end{bmatrix}.$$

It can be proved (Campbell [12], Section 6.3) that a square matrix A is invertible if and only if $\det A \neq 0$, and moreover,

$$A^{-1} = \left(\frac{1}{\det A}\right)(\operatorname{cof} A)^T.$$

Let

$$A = \begin{bmatrix} 1 & 0 & 4 \\ 0 & 0 & 1 \\ 0 & 2 & 3 \end{bmatrix}.$$

By Theorems 8 and 11,

$$\det A = -\det \begin{bmatrix} 1 & 0 & 4 \\ 0 & 2 & 3 \\ 0 & 0 & 1 \end{bmatrix} = -(1 \cdot 2 \cdot 1) = -2.$$
$$\scriptstyle R_2 \leftrightarrow R_3$$

Then

$$\operatorname{cof} A = \begin{bmatrix} \det\begin{bmatrix}0&1\\2&3\end{bmatrix} & -\det\begin{bmatrix}0&1\\0&3\end{bmatrix} & \det\begin{bmatrix}0&0\\0&2\end{bmatrix} \\ -\det\begin{bmatrix}0&4\\2&3\end{bmatrix} & \det\begin{bmatrix}1&4\\0&3\end{bmatrix} & -\det\begin{bmatrix}1&0\\0&2\end{bmatrix} \\ \det\begin{bmatrix}0&4\\0&1\end{bmatrix} & -\det\begin{bmatrix}1&4\\0&1\end{bmatrix} & \det\begin{bmatrix}1&0\\0&0\end{bmatrix} \end{bmatrix} = \begin{bmatrix} -2 & 0 & 0 \\ 8 & 3 & -2 \\ 0 & -1 & 0 \end{bmatrix}.$$

Therefore, $A^{-1} = \left(\dfrac{1}{-2}\right)\begin{bmatrix} -2 & 0 & 0 \\ 8 & 3 & -2 \\ 0 & -1 & 0 \end{bmatrix}^T = \begin{bmatrix} 1 & -4 & 0 \\ 0 & -\frac{3}{2} & \frac{1}{2} \\ 0 & 1 & 0 \end{bmatrix}.$ ∎

EXERCISES

Suggested minimum assignment: Exercises 1, 3, 5, and 9.

1. Find: (a) $\det\begin{bmatrix}2&1\\3&6\end{bmatrix}$.   (b) $\det\begin{bmatrix}-3&2\\-9&0\end{bmatrix}$.   (c) $\det\begin{bmatrix}-1&-2\\6&12\end{bmatrix}$.

2. Find: (a) $\det\begin{bmatrix}0&2\\1&4\end{bmatrix}$.   (b) $\det\begin{bmatrix}2&-1\\3&4\end{bmatrix}$.   (c) $\det\begin{bmatrix}2&1\\8&4\end{bmatrix}$.

3. Given:
$$A = \begin{bmatrix} 1 & 3 & 4 \\ 0 & 1 & 1 \\ 2 & 4 & 3 \end{bmatrix}.$$

Without the aid of the theorems of this section:
(a) Write the expansion of det A about the first row and evaluate det A.
(b) Write the expansion of det A about the second column and evaluate det A.
(c) Write the expansion of det A about the second row and evaluate det A.
(d) Write the expansion of det A about the first column and evaluate det A.

4. Given:
$$A = \begin{bmatrix} 3 & 2 & 1 \\ 4 & 2 & 1 \\ 6 & 7 & 2 \end{bmatrix}.$$

(a) Write the minor of the entry in the first row and second column.
(b) Write the cofactor of the entry in the first row and second column.
(c) Write the minor of the entry in the second row and third column.
(d) Write the cofactor of the entry in the second row and third column.

In Exercises 5–10 use the theorems of this section to evaluate the determinants of the following matrices.

5. $\begin{bmatrix} 2 & 6 & 5 \\ 0 & 3 & 2 \\ 4 & 12 & 1 \end{bmatrix}.$

6. $\begin{bmatrix} 2 & 1 & 2 \\ 4 & 3 & 1 \\ 6 & 7 & 4 \end{bmatrix}.$

7. $\begin{bmatrix} 1 & 3 & 2 \\ 2 & 6 & 7 \\ 1 & 2 & 4 \end{bmatrix}.$

8. $\begin{bmatrix} 2 & 4 & 3 \\ 4 & 8 & 4 \\ 4 & 2 & 1 \end{bmatrix}.$

9. $\begin{bmatrix} 2 & 1 & 0 & 2 \\ 0 & 1 & 3 & 4 \\ 2 & 4 & 13 & 3 \\ 0 & 2 & 2 & 4 \end{bmatrix}.$

10. $\begin{bmatrix} 3 & 3 & 2 & 1 \\ 6 & 7 & 3 & 2 \\ 0 & 0 & 1 & 4 \\ 9 & 2 & 1 & 0 \end{bmatrix}.$

11. An important property is that the determinant of the product of two square matrices of the same order is the product of the determinants of the two matrices; that is,

$$\det(AB) = (\det A)(\det B).$$

Illustrate this property for

$$A = \begin{bmatrix} 2 & 1 \\ 3 & 2 \end{bmatrix} \quad \text{and} \quad B = \begin{bmatrix} 3 & 1 \\ 6 & 0 \end{bmatrix}.$$

Sec. 5.9] The Determinant of a Matrix

12. Prove, using the definition of a determinant, that if all the entries of a given row or column of a square matrix A are 0's, then $\det A = 0$.
13. (a) Use Theorem 8 to prove that if any two rows of a square matrix A are identical, then $\det A = 0$.
    (b) Prove that if any two rows of a square matrix A are proportional, then $\det A = 0$. [*Hint:* Use Theorem 9 and part (a).]
14. A property of determinants is that if A is square, then $\det A^T = \det A$. Prove the validity of this property for $n = 1, 2, 3$.
15. Prove that in Theorems 8, 9, and 10, the word "row" may be replaced by the word "column." (*Hint:* Use Exercise 14.)

In Exercises 16–19 calculate A^{-1} by the formula given in Example 7.

16. $A = \begin{bmatrix} 2 & 1 \\ 6 & 2 \end{bmatrix}$.

17. $A = \begin{bmatrix} 1 & 4 \\ 2 & 10 \end{bmatrix}$.

18. $A = \begin{bmatrix} 1 & 2 & 0 \\ 0 & 1 & 0 \\ 1 & 0 & 2 \end{bmatrix}$.

19. $A = \begin{bmatrix} 0 & 0 & 1 \\ 1 & 3 & 0 \\ 0 & 2 & 1 \end{bmatrix}$.

20. Prove Theorem 8 for the case $n = 3$. (*Hint:* Expand about the row not involved in the switch.)
21. Prove Theorem 9 for the case $n = 3$. (*Hint:* Expand about the row that is multiplied by the number k.)
22. Prove Theorem 10 for the case $n = 3$. [*Hint:* Expand about the row that is changed by the row operation, and use Exercise 13(a).]
23. Prove Theorem 11.

5.10 The Characteristic Value Problem (Optional)

A problem that is common in many applications of matrix algebra is that of finding a real number λ and a nonzero column vector X such that

$$AX = \lambda X,$$

where A is a square matrix that is known.

Example 1

If $A = \begin{bmatrix} 2 & 1 \\ 3 & 4 \end{bmatrix}$, find a real number λ and a nonzero column vector X such that

$$AX = \lambda X.$$

In other words, because

$$\begin{bmatrix} 2 & 1 \\ 3 & 4 \end{bmatrix} \begin{bmatrix} x_1 \\ x_2 \end{bmatrix} = \lambda \begin{bmatrix} x_1 \\ x_2 \end{bmatrix}$$

is the matrix representation of the linear system

$$2x_1 + x_2 = \lambda x_1,$$
$$3x_1 + 4x_2 = \lambda x_2,$$

we must first find λ such that nonzero solutions of the system then can be found.

Solution: After transposing the terms of the latter system from the right side to the left side and combining terms, we obtain

$$(2 - \lambda)x_1 + x_2 = 0,$$
$$3x_1 + (4 - \lambda)x_2 = 0. \qquad (5.8)$$

The last system is a linear homogeneous system

$$CX = 0,$$

where

$$C = \begin{bmatrix} 2 - \lambda & 1 \\ 3 & 4 - \lambda \end{bmatrix} \quad \text{and} \quad X = \begin{bmatrix} x_1 \\ x_2 \end{bmatrix}.$$

In Section 5.8 we learned that a homogeneous system such as

$$CX = 0$$

has either a unique solution, which is the zero vector, or an infinite number of solutions. Because we are looking for nonzero solutions, we must find those values of λ in matrix C that will make the system $CX = 0$ have an infinite number of solutions. It can be proved that the homogeneous system $CX = 0$ has solutions other than the zero vector if and only if $\det C = 0$. Thus we write

$$\det \begin{bmatrix} 2 - \lambda & 1 \\ 3 & 4 - \lambda \end{bmatrix} = 0,$$

and solve for λ. Expansion of the determinant about the first row gives us

$$(2 - \lambda)(4 - \lambda) - 1 \cdot 3 = 0,$$
$$8 - 6\lambda + \lambda^2 - 3 = 0,$$
$$\lambda^2 - 6\lambda + 5 = 0,$$
$$(\lambda - 1)(\lambda - 5) = 0,$$
$$\lambda = 1, \quad \lambda = 5.$$

These values of λ that we have just found are called **characteristic values** (or **eigenvalues**) of matrix A and are substituted into the system (5.8) so that we can solve that system for x_1 and x_2. With $\lambda = 1$ the system (5.8) becomes

$$(2 - 1)x_1 + x_2 = 0,$$
$$3x_1 + (4 - 1)x_2 = 0,$$

or

$$x_1 + x_2 = 0,$$
$$3x_1 + 3x_2 = 0.$$

By the Gauss–Jordan method a complete solution of the latter system is

$$x_1 = -x_2.$$

Hence
$$X = \begin{bmatrix} x_1 \\ x_2 \end{bmatrix} = \begin{bmatrix} -x_2 \\ x_2 \end{bmatrix} = \begin{bmatrix} -1 \\ 1 \end{bmatrix} x_2,$$
where x_2 is a parameter that can be assigned an arbitrary nonzero real number value. If we let $x_2 = 1$, then
$$X = \begin{bmatrix} -1 \\ 1 \end{bmatrix}.$$
The column vector X is called a **characteristic vector** of matrix A.

Next, we substitute the other characteristic value $\lambda = 5$ into the system (5.8) to get
$$(2-5)x_1 + x_2 = 0,$$
$$3x_1 + (4-5)x_2 = 0,$$
or
$$-3x_1 + x_2 = 0,$$
$$3x_1 - x_2 = 0.$$
By the Gauss–Jordan method a complete solution of the latter system is
$$x_1 = \tfrac{1}{3} x_2.$$
Hence
$$X = \begin{bmatrix} x_1 \\ x_2 \end{bmatrix} = \begin{bmatrix} \tfrac{1}{3} x_2 \\ x_2 \end{bmatrix} = \begin{bmatrix} \tfrac{1}{3} \\ 1 \end{bmatrix} x_2,$$
where x_2 is a parameter that can be assigned an arbitrary nonzero real number value. If we let $x_2 = 3$, then
$$X = \begin{bmatrix} 1 \\ 3 \end{bmatrix},$$
and we have a second characteristic vector of matrix A. To distinguish between the two characteristic vectors that we have found for A, we label them X_1 and X_2, respectively, where the subscript identifies the corresponding characteristic value. Thus
$$X_1 = \begin{bmatrix} -1 \\ 1 \end{bmatrix} \quad \text{and} \quad X_2 = \begin{bmatrix} 1 \\ 3 \end{bmatrix},$$
together with the appropriate λ, satisfy the system
$$AX = \lambda X.$$
That is,
$$\begin{bmatrix} 2 & 1 \\ 3 & 4 \end{bmatrix} \begin{bmatrix} -1 \\ 1 \end{bmatrix} = (1) \begin{bmatrix} -1 \\ 1 \end{bmatrix},$$
and also
$$\begin{bmatrix} 2 & 1 \\ 3 & 4 \end{bmatrix} \begin{bmatrix} 1 \\ 3 \end{bmatrix} = (5) \begin{bmatrix} 1 \\ 3 \end{bmatrix}. \blacksquare$$

In general, we may summarize the characteristic value problem as follows. First, for a given n by n matrix A and for an unknown column vector

$$X = \begin{bmatrix} x_1 \\ x_2 \\ \vdots \\ x_n \end{bmatrix},$$

find values of the real number λ for which the matrix equation

$$AX = \lambda X$$

has nonzero solutions; that is, $X \neq \mathbf{0}$. The matrix equation $AX = \lambda X$, which can be rewritten as

$$(A - \lambda I_n)X = \mathbf{0},$$

is a homogeneous linear system having nonzero solutions if and only if

$$\det(A - \lambda I_n) = 0.$$

The equation $\det(A - \lambda I_n) = 0$ is called the **characteristic equation,** and its roots λ_i are called **characteristic values** of the matrix A. Once the characteristic values have been found, then each homogeneous system

$$(A - \lambda_i I_n)X_i = \mathbf{0}, \qquad i = 1, 2, \ldots, n,$$

can be solved for nonzero X_i. The resulting nonzero column vectors X_i are called **characteristic vectors.**

At this point the reader should be reminded that although the characteristic equation will have n *complex* roots, it may have fewer *real* roots (possibly none). Thus when *real* characteristic values and vectors are sought, there can be from 0 to n of them. (See Exercise 9 for an illustration of a characteristic value problem with less than n real characteristic values.) When a matrix is symmetric, however, the roots are always real. (See Exercise 11.)

APPLICATIONS

Example 2
Consider a population of objects living in a certain environment in which

$x_{i,t}$ = number of living objects of age i at time t,
p_i = probability that an object of age i at time t will survive to time $t + 1$,
f_i = number of new objects created per old object of age i,
n = maximum age.

A column matrix showing the population of each age group at time $t + 1$ can be determined by the following matrix equation:

$$\begin{bmatrix} x_{0,t+1} \\ x_{1,t+1} \\ x_{2,t+1} \\ \vdots \\ x_{n,t+1} \end{bmatrix} = \begin{bmatrix} f_0 & f_1 & f_2 & \cdots & f_{n-1} & f_n \\ p_0 & 0 & 0 & \cdots & 0 & 0 \\ 0 & p_1 & 0 & \cdots & 0 & 0 \\ \vdots & \vdots & \vdots & & \vdots & \vdots \\ 0 & 0 & 0 & \cdots & p_{n-1} & 0 \end{bmatrix} \begin{bmatrix} x_{0,t} \\ x_{1,t} \\ x_{2,t} \\ \vdots \\ x_{n,t} \end{bmatrix}.$$

Let the respective matrices in the latter equation be designated X_{t+1}, A, and X_t so that the equation can be expressed as

$$X_{t+1} = AX_t.$$

Suppose that we want to find a population matrix X_t that will produce a proportional population matrix X_{t+1} one unit of time later in the given environment. In other words, we want to find an X_t such that

$$X_{t+1} = \lambda X_t.$$

In this case we can say that there is a stability of the relative size of age groups. The equation $X_{t+1} = AX_t$ becomes

$$AX_t = \lambda X_t,$$

which is the characteristic value problem. In [8], Bernadelli describes a hypothetical beetle living in a certain environment in which the A matrix is

$$\begin{bmatrix} 0 & 0 & 6 \\ \frac{1}{2} & 0 & 0 \\ 0 & \frac{1}{3} & 0 \end{bmatrix}.$$

The only real characteristic value is 1 and a corresponding characteristic vector of A is

$$X_t = \begin{bmatrix} 6 \\ 3 \\ 1 \end{bmatrix};$$

therefore, any population of these beetles with an age distribution proportional to $6:3:1$ should keep a stable age distribution in the given environment. ∎

EXERCISES

Suggested minimum assignment: Exercises 1, 5, and 7.

In each of Exercises 1–4 find the characteristic equation, the real characteristic values, and corresponding characteristic vectors for the given matrix.

1. $\begin{bmatrix} 1 & 2 \\ -1 & 4 \end{bmatrix}.$
2. $\begin{bmatrix} 5 & -2 \\ -2 & 5 \end{bmatrix}.$
3. $\begin{bmatrix} 0 & 3 \\ 3 & 0 \end{bmatrix}.$
4. $\begin{bmatrix} 2 & 3 \\ 1 & 4 \end{bmatrix}.$

In each of Exercises 5-8 find a nonzero real solution, if possible, of the equation $AX = \lambda X$, where A is given and X is a corresponding column vector of unknowns.

5. $\begin{bmatrix} 1 & -1 \\ 3 & 5 \end{bmatrix}.$
6. $\begin{bmatrix} 0 & 1 \\ 3 & 2 \end{bmatrix}.$
7. $\begin{bmatrix} 0 & 0 & 2 \\ 0 & 1 & 0 \\ 2 & 0 & 0 \end{bmatrix}.$
8. $\begin{bmatrix} 1 & 1 & 1 \\ 0 & 3 & 1 \\ 0 & 0 & 4 \end{bmatrix}.$

9. In Example 2 verify that the real characteristic value and characteristic vector are as stated for the hypothetical beetle matrix

$$\begin{bmatrix} 0 & 0 & 6 \\ \tfrac{1}{2} & 0 & 0 \\ 0 & \tfrac{1}{3} & 0 \end{bmatrix}.$$

10. Prove: If A is square and if a system of linear homogeneous equations $AX = \mathbf{0}$ over the real numbers has a nonzero solution, then $\det A = 0$. (*Hint:* Either $\det A = 0$ or $\det A \neq 0$; assume that $\det A \neq 0$, and show that this leads to a contradiction of the hypothesis. Also use Example 7 of Section 5.9.)

11. Many applications involving the characteristic value problem require the characteristic values of a symmetric matrix. It is important in such problems to realize that the characteristic values of a symmetric matrix are always real numbers. Prove this assertion for the 2 by 2 symmetric matrix

$$A = \begin{bmatrix} a & b \\ b & c \end{bmatrix}.$$

NEW VOCABULARY

matrix 5.1
order of a matrix 5.1
entry of a matrix 5.1
square matrix 5.1
main diagonal 5.1
equal matrices 5.1
sum of two matrices 5.1
conformable for addition 5.1

multiplication of a matrix by a real number 5.1
subtraction of two matrices 5.1
vector 5.2
dimension of a vector 5.2
components 5.2
row vector 5.2
column vector 5.2

Sec. 5.10] The Characteristic Value Problem

zero vector 5.2
nonzero vector 5.2
equal vectors 5.2
sum of two vectors 5.2
multiplication of a vector by a
  real number 5.2
subtraction of two vectors 5.2
dot product of two vectors 5.2
magnitude or length of a
  vector 5.2
matrix product 5.3
conformable for
  multiplication 5.3
zero or null matrix 5.4
identity element for matrix
  addition 5.4
identity element for matrix
  multiplication 5.4
identity matrix 5.4
diagonal matrix 5.4
scalar matrix 5.4
triangular matrix 5.4
submatrix 5.4
partitioned matrix 5.4
blocks of a matrix 5.4
block matrix 5.4
transpose of a matrix 5.4
symmetric matrix 5.4

skew-symmetric matrix 5.4
system of linear equations over
  the real numbers 5.5
solution of a linear system 5.5
coefficient matrix 5.5
augmented matrix 5.5
elementary row operations 5.5
row equivalent matrices 5.5
multiplicative inverse of a
  real number 5.6
inverse of a matrix 5.6
invertible matrix 5.6
noninvertible matrix 5.6
basic variables 5.7
complete solution 5.7
particular solution 5.7
reduced matrix 5.8
Gauss–Jordan method 5.8
homogeneous system 5.8
determinant of a matrix 5.9
minor of an entry 5.9
cofactor of an entry 5.9
vector of cofactors 5.9
expansion of a determinant 5.9
characteristic values (or
  eigenvalues) 5.10
characteristic vector 5.10
characteristic equation 5.10

6 Linear Programming

The purpose of this chapter is to study a special type of optimization problem known as a linear programming problem. Such problems, which are frequently concerned with decision making, have attracted considerable attention in recent years. As an illustration of a linear programming problem, consider the example: Minimize $(3x_1 + 5x_2)$ subject to the linear conditions or constraints

$$3x_1 + 4x_2 \geq 340,$$
$$x_1 + x_2 \geq 100, \quad \text{and} \quad x_1 \geq 0,$$
$$x_1 + 5x_2 \geq 150, \qquad\qquad x_2 \geq 0.$$

The word "programming" does not refer to the programming of a computer; it refers to the programming or allocation of items or entities. The unknowns x_1 and x_2 represent a measure of those allocations. The word "linear" is used to describe the fact that we wish to optimize a linear function subject to linear constraints.

Prerequisites: 1.1, 1.2, 3.5–3.8, 5.1–5.3, 5.5 if only 6.1 and 6.2 are to be covered; 1.1, 1.2, 3.5–3.8, 5.1–5.8 if 6.1–6.7 are to be covered.
Suggested Sections for Moderate Emphasis: 6.1–6.4.
Suggested Sections for Minimum Emphasis: 6.1, 6.2.

6.1 The Feasible Set of a Linear Programming Problem

The basic ideas of this section will be introduced by means of the following example.

Example 1

Consider a hypothetical situation in which a dietician wishes to serve food that provides the necessary calories, vitamins, and minerals. The following chart indicates the amounts of calories, vitamins, and minerals to be provided, and the amounts found in foods F_1 and F_2.

	Food F_1	Food F_2	Units needed
Calories	3 units/pound	4 units/pound	340
Vitamins	1 unit/pound	1 unit/pound	100
Minerals	1 unit/pound	5 units/pound	150

Suppose that foods F_1 and F_2 cost \$3 and \$5, respectively. Further suppose that the dietician wants to minimize the total cost of the foods purchased, subject to the dietary requirements shown in the chart. In other words, how many pounds of each food should be purchased in order to meet the dietary needs at least cost? The solution of this problem requires a determination of the possible combinations of foods that can be used to meet the dietary requirements, and from among these, a choice of that combination of foods that costs the least.

The remainder of this example will treat only the problem of meeting the dietary requirements; the minimization of cost will be the subject of Section 6.2. We now construct a mathematical model of the dietary needs.

Let x_1 be the number of pounds of food F_1 that is ordered, and let x_2 be the number of pounds of food F_2 that is ordered. The number of units of *calories* obtained will be

$$\underbrace{(3 \text{ units per pound}) \cdot (x_1 \text{ pounds})}_{\text{from } F_1} + \underbrace{(4 \text{ units per pound}) \cdot (x_2 \text{ pounds})}_{\text{from } F_2}.$$

Thus from both foods we obtain $(3x_1 + 4x_2)$ units of calories; the fact that we should have at least 340 units of calories produces the first constraint,

$$(1) \quad 3x_1 + 4x_2 \geq 340.$$

Similarly, we obtain a constraint corresponding to the requirement for *vitamins*,

$$(2) \quad x_1 + x_2 \geq 100,$$

and a constraint for the requirement for *minerals*,

$$(3) \quad x_1 + 5x_2 \geq 150.$$

The fact that we must purchase a nonnegative number of pounds of each food produces the constraints

$$(4) \quad x_1 \geq 0 \quad \text{and} \quad (5) \quad x_2 \geq 0.$$

Hence the mathematical model of the dietary requirements is the linear system of constraints

$$\begin{array}{ll} (1) & 3x_1 + 4x_2 \geq 340, \\ (2) & x_1 + x_2 \geq 100, \\ (3) & x_1 + 5x_2 \geq 150, \end{array} \quad \text{and} \quad \begin{array}{ll} (4) & x_1 \geq 0, \\ (5) & x_2 \geq 0. \end{array}$$

We shall determine values of x_1 and x_2 that satisfy this system. Geometrically, the constraint $3x_1 + 4x_2 \geq 340$ represents the set of all points on or above the line

$$3x_1 + 4x_2 = 340,$$

as shown by the colored region in Figure 6.1. Similarly, the other constraints can be graphed, and the set of points whose coordinates satisfy all the constraints simultaneously is shown as the colored region in Figure 6.2. ∎

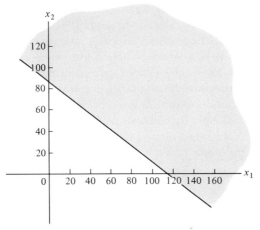

Figure 6.1. *Geometric representation of the inequality* $3x_1 + 4x_2 \geq 340$.

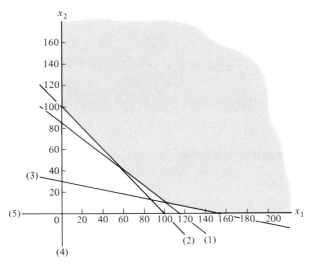

Figure 6.2. *Geometric representation of a feasible set.*

Any one point in the colored region of Figure 6.2 represents a solution of the system of constraints given in Example 1. The set of solutions is known as the **feasible set of the system.** In Figure 6.2 the colored region is the geometric representation of the feasible set of the given system.

▶ **Definition 1**
The **feasible set of a system of linear constraints** (which may include equations as well as inequalities) is the set of all vectors that satisfy all the constraints simultaneously. Elements of the feasible set are called **feasible solutions.**

Sec. 6.1] The Feasible Set of a Linear Programming Problem

It is customary in linear programming problems to require that all the unknowns used in the constraints be nonnegative. If problems are encountered where there may be negative values for an unknown x_i, then x_i can be expressed as $x_i = u_i - v_i$, where u_i and v_i are nonnegative. The requirements that the unknowns be nonnegative are appropriately called the **nonnegativity constraints**. All the other constraints are called **structural constraints**. Structural constraints may be in the form of equations or inequalities. In Example 1, constraints (1), (2), and (3) were structural constraints, and constraints (4) and (5) were nonnegativity constraints.

EXERCISES

Suggested minimum assignment: Exercises 1, 3, 5, and 7.

In Exercises 1–4 graph the feasible set for each of the given systems of linear constraints.

1. $x_1 + x_2 \leq 6,$ and $x_1 \geq 0,$
   $3x_1 - x_2 \leq 10,$ $x_2 \geq 0.$

2. $x_1 - x_2 \leq 4,$ and $x_1 \geq 0,$
   $x_2 \leq 3,$ $x_2 \geq 0.$

3. $2x_1 + x_2 \geq 8,$
   $x_1 + x_2 \geq 6,$ and $x_1 \geq 0,$
   $x_1 + 2x_2 \geq 9,$ $x_2 \geq 0.$

4. $x_1 + x_2 \geq 1,$
   $x_1 - x_2 \geq -1,$ and $x_1 \geq 0,$
   $x_1 + 2x_2 \geq 4,$ $x_2 \geq 0.$

5. Identify the structural and nonnegativity constraints of Exercises 1 and 3.
6. Identify the structural and nonnegativity constraints of Exercises 2 and 4.

In Exercises 7 and 8 graph the feasible set for each of the given systems of linear constraints.

7. $4x_1 + 3x_2 \geq 18,$ and $x_1 \geq 0,$
   $2x_1 + 5x_2 \geq 16,$ $x_2 \geq 0.$

8. $-x_1 + x_2 \leq 1,$ and $x_1 \geq 0,$
   $x_1 + x_2 \leq 3,$ $x_2 \geq 0.$

6.2 A Geometric Method of Solution

So far we have discussed only the *constraints* of a linear programming problem. In various applied problems one often wishes to optimize some linear function subject to such a system of linear constraints.

Example 1

In the dietary problem (Example 1) of Section 6.1, remember that foods F_1 and F_2 cost \$3 and \$5 per pound, respectively. The dietician wants to minimize the total cost of the foods purchased, subject to the constraints imposed by the dietary requirements discussed in the last section. We let x_1 and x_2 represent the number of pounds of the respective foods that are to be purchased. Hence the total cost will be

$$\underbrace{(\$3 \text{ per pound})(x_1 \text{ pounds})}_{\text{cost of } F_1} + \underbrace{(\$5 \text{ per pound})(x_2 \text{ pounds})}_{\text{cost of } F_2}.$$

Therefore, the dietician seeks to minimize $(3x_1 + 5x_2)$ subject to

(1) $3x_1 + 4x_2 \geq 340,$
(2) $x_1 + x_2 \geq 100,$ and (4) $x_1 \geq 0,$
(3) $x_1 + 5x_2 \geq 150,$ (5) $x_2 \geq 0.$

Values of the cost $(3x_1 + 5x_2)$ will be represented by the letter f. We now illustrate a geometric method of solution of this linear programming problem.

If arbitrary values are assigned to f in the equation $f = 3x_1 + 5x_2$, we obtain a family of straight lines imposed on the feasible set as shown by the dashed lines in Figure 6.3. Notice that for increasing values of f, the corresponding lines appear farther upward and away from the origin. From Figure 6.3 it appears that the line corresponding to $f = 350$ is the "lowest" line that has a point in common with the shaded region; any lower line does not intersect the shaded region, and higher lines represent larger values of f than the minimum value. Hence the point $(100, 10)$, which is the intersection of

(1) $3x_1 + 4x_2 = 340$ and (3) $x_1 + 5x_2 = 150,$

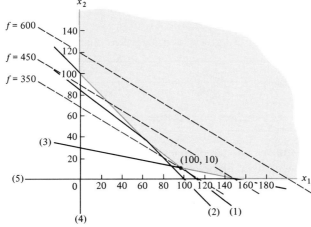

Figure 6.3. *Geometric representation of the objective equation showing minimum f.*

Sec. 6.2] A Geometric Method of Solution

gives the minimum f, namely,
$$f = 3(100) + 5(10) = 350.$$

Thus if 100 pounds of F_1 and 10 pounds of F_2 are purchased, the dietary requirements will be met at less cost than with any other combination.

Notice that if $x_1 = 100$ and $x_2 = 10$, then $x_1 + x_2 = 110$, or 110 units of vitamins are acquired. This is more than the minimum requirement of vitamins, and at first it seems strange that a surplus of some item can be obtained at minimum cost. ∎

In general the linear function, whose values are determined by
$$c_1 x_1 + c_2 x_2 + \cdots + c_n x_n$$
and which is to be optimized, is called the **objective function**; the corresponding equation
$$f = c_1 x_1 + c_2 x_2 + \cdots + c_n x_n$$
is called the **objective equation**. [Because of the number of independent variables that may be involved, we shall use f rather than $f(x_1, x_2, \ldots, x_n)$ to represent the value of the objective function at $(x_1, x_2, \ldots, x_n)$.]

A further illustration of a linear programming problem and the geometric method of solution follows.

Example 2

A local television network has found that program A, with 20 minutes of music and 1 minute of advertisement, draws 30,000 viewers; and program B, with 10 minutes of music and 1 minute of advertisement, draws 10,000 viewers. Within one week the advertiser insists that at least 6 minutes be devoted to his advertisement, and the network can afford no more than 80 minutes of music. How many times per week should each program be given to obtain the maximum number of viewers while satisfying the requirements of the advertiser and the network?

Solution: Arrange the given information in chart form.

	A	B	Requirements
Music	20	10	at most 80
Advertisement	1	1	at least 6
Viewers	30,000	10,000	

Let x_1 represent the number of times program A is put on the air, and let x_2 represent the number of times program B is put on the air. From the information in the chart we form the constraints

$$\begin{matrix} 20x_1 + 10x_2 \leq 80, \\ x_1 + x_2 \geq 6, \end{matrix} \quad \text{and} \quad \begin{matrix} x_1 \geq 0, \\ x_2 \geq 0. \end{matrix}$$

The objective function to be maximized, subject to the constraints, has

values given by $(30{,}000x_1 + 10{,}000x_2)$. The feasible set and the family of lines representing the objective function are graphed in Figure 6.4. From the graph we see that f increases as the lines representing the objective function move away from the origin. The member of the family of lines representing the largest f that intersects the feasible set is the line that passes through the point $(2, 4)$. Therefore, we have

$$\text{maximum } f = 100{,}000 \text{ at } (2, 4),$$

which means that program A should be offered twice, and program B should be offered four times to obtain a maximum audience of 100,000 viewers. ∎

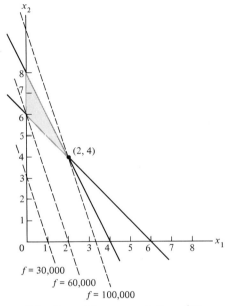

Figure 6.4. *Geometric representation of Example 2.*

At this point we remark that the optimum of the objective function will always occur at a vertex of the feasible set. For the case where the graph of the objective equation is parallel to a certain edge of the feasible set, then the whole edge, including both vertices, produces an optimum. Also, an optimum always exists if the feasible set is bounded, as it was in Example 2; but an optimum may or may not exist if the feasible set is not bounded (the feasible set was not bounded in Example 1).

APPLICATIONS

In order that the reader may obtain some idea of how linear programming can be applied to various fields, the basic mathematical model of Example 1 will be given five different interpretations.

Sec. 6.2] A Geometric Method of Solution

Example 3
Two factories can manufacture three different grades of paper at known rates. The company that owns the factories has contracts to supply certain quantities of each grade of paper. How many days should each factory be operated in order to satisfy the demand and yet minimize the cost of production?

	Factory 1	Factory 2	Tons needed
High grade	3 tons/day	4 tons/day	340
Medium grade	1 ton/day	1 ton/day	100
Low grade	1 ton/day	5 tons/day	150
Cost per day (thousands of dollars)	3	5	

Let x_1 be the number of days that factory 1 is to be operated. Let x_2 be the number of days that factory 2 is to be operated. ∎

Example 4
The minimum requirements of the amount of certain chemicals necessary to grow a certain crop successfully have been found. Knowing the content and cost of two special types of commercial fertilizers, packaged in 100-pound bags, a grower wants to know how many bags of each type should be applied to his crop to assure proper growth at minimum cost.

	Fertilizer 1	Fertilizer 2	Units needed
Chemical A	3 units/bag	4 units/bag	340
Chemical B	1 unit/bag	1 unit/bag	100
Chemical C	1 unit/bag	5 units/bag	150
Cost per bag	$3	$5	

Let x_1 be the number of bags of fertilizer 1 that are applied. Let x_2 be the number of bags of fertilizer 2 that are applied. ∎

Example 5
A fruit buyer needs a certain number of units of various varieties of fruit. Two suppliers can supply his needs but will sell only in full truckloads, consisting of a specified number of units of each variety (let 1 unit equal 10 boxes). How many loads should the buyer order from each supplier in order to save time and money by holding the total shipping distance to a minimum? (See the chart at the top of the next page.)

Let x_1 be the number of truckloads ordered from supplier 1. Let x_2 be the number of truckloads ordered from supplier 2. ∎

	Supplier 1	Supplier 2	Units needed
Variety A	3 units/load	4 units/load	340
Variety B	1 unit/load	1 unit/load	100
Variety C	1 unit/load	5 units/load	150
Distance (hundreds of miles)	3	5	

Example 6

An office manager must assign his 2 groups of employees in such a way that each of three tasks is performed a certain number of times and the time to do them is held to a minimum.

	Group 1	Group 2	Times needed
Task A	3 times/assignment	4 times/assignment	340
Task B	1 time/assignment	1 time/assignment	100
Task C	1 time/assignment	5 times/assignment	150
Hours taken per assignment	3	5	

Let x_1 be the number of assignments given to group 1. Let x_2 be the number of assignments given to group 2. ∎

Example 7

A nutritionist working for a spacecraft designer is faced with the problem of minimizing weight subject to certain nutritional requirements. He is considering 2 foods which are packaged in tubes.

	Food 1	Food 2	Units needed
Carbohydrates	3 units/tube	4 units/tube	340
Fats	1 unit/tube	1 unit/tube	100
Proteins	1 unit/tube	5 units/tube	150
Weight per tube (ounces)	3	5	

Let x_1 be the number of tubes of food 1 sent on the spacecraft. Let x_2 be the number of tubes of food 2 sent on the spacecraft. ∎

The reader should observe that in Examples 5 and 7 the solution vector had to have integral components; when applied problems of this type are encountered, and if such a solution vector is not obtained by our methods, then techniques beyond the scope of this text must be employed. The study of such problems is a very interesting branch of linear programming known as *integer programming*.

Sec. 6.2] A Geometric Method of Solution

EXERCISES

Suggested minimum assignment: Exercises 1, 3, 5, 7, and 9.

In Exercises 1–4 solve the given linear programming problem (the feasible set of each problem was graphed in the corresponding exercise of the last section).

1. Maximize $(x_1 + 2x_2)$ subject to $\begin{array}{l} x_1 + x_2 \leq 6, \\ 3x_1 - x_2 \leq 10, \end{array}$ and $\begin{array}{l} x_1 \geq 0, \\ x_2 \geq 0. \end{array}$

2. Maximize $(x_1 + x_2)$ subject to $\begin{array}{l} x_1 - x_2 \leq 4, \\ x_2 \leq 3, \end{array}$ and $\begin{array}{l} x_1 \geq 0, \\ x_2 \geq 0. \end{array}$

3. Minimize $(2x_1 + 3x_2)$ subject to $\begin{array}{l} 2x_1 + x_2 \geq 8, \\ x_1 + x_2 \geq 6, \\ x_1 + 2x_2 \geq 9, \end{array}$ and $\begin{array}{l} x_1 \geq 0, \\ x_2 \geq 0. \end{array}$

4. Minimize $(2x_1 + x_2)$ subject to $\begin{array}{l} x_1 + x_2 \geq 1, \\ x_1 - x_2 \geq -1, \\ x_1 + 2x_2 \geq 4, \end{array}$ and $\begin{array}{l} x_1 \geq 0, \\ x_2 \geq 0. \end{array}$

In Exercises 5 and 6 solve the given linear programming problem by maximizing the negative of the objective function, and thereby illustrate that minimization problems in linear programming can be transformed easily to maximization problems.

5. Minimize $(x_1 + x_2)$ subject to $\begin{array}{l} x_1 + 2x_2 \leq 7, \\ x_1 - 3x_2 \geq 2, \end{array}$ and $\begin{array}{l} x_1 \geq 0, \\ x_2 \geq 0. \end{array}$

6. Minimize $(3x_1 + x_2)$ subject to $\begin{array}{l} x_1 + 2x_2 \geq 5, \\ 2x_1 + x_2 \geq 4, \end{array}$ and $\begin{array}{l} x_1 \geq 0, \\ x_2 \geq 0. \end{array}$

In Exercises 7 and 8 show that more than one feasible solution will optimize the objective function (the feasible set of each problem was graphed in the corresponding exercise of the last section).

7. Minimize $(2x_1 + 5x_2)$ subject to $\begin{array}{l} 4x_1 + 3x_2 \geq 18, \\ 2x_1 + 5x_2 \geq 16, \end{array}$ and $\begin{array}{l} x_1 \geq 0, \\ x_2 \geq 0. \end{array}$

8. Maximize $(x_1 + x_2)$ subject to $\begin{array}{l} -x_1 + x_2 \leq 1, \\ x_1 + x_2 \leq 3, \end{array}$ and $\begin{array}{l} x_1 \geq 0, \\ x_2 \geq 0. \end{array}$

In Exercises 9 and 10 show that the given linear programming problem has no solution.

9. Maximize $(4x_1 - 5x_2)$ subject to the constraints of Exercise 7.
10. Minimize $(-x_1 + x_2)$ subject to the constraints of Exercise 3.

11. There are 2 factories that manufacture three different grades of paper. There is some demand for each grade. The company that controls the factories has contracts to supply 16 tons of low-grade, 5 tons of medium-grade, and 20 tons of high-grade paper. It costs $1000 per day to operate the first factory and $2000 per day to operate the second

factory. Factory 1 produces 8 tons of low-grade, 1 ton of medium-grade, and 2 tons of high-grade paper in 1 day's operation. Factory 2 produces 2 tons of low-grade, 1 ton of medium-grade, and 7 tons of high-grade paper per day. How many days should each factory be operated in order to fill the orders most economically?

12. Two oil refineries produce three grades of gasoline: A, B, and C. At each refinery the various grades of gasoline are produced in a single operation so that they are in fixed proportions. Assume that 1 operation at refinery 1 produces 1 unit of A, 3 units of B, and 1 unit of C. One operation at refinery 2 produces 1 unit of A, 4 units of B, and 5 units of C. Refinery 1 charges \$300 for what is produced in 1 operation, and refinery 2 charges \$500 for the production of 1 operation. A consumer needs 100 units of A, 340 units of B, and 150 units of C. How should the orders be placed if the consumer is to meet his needs most economically?

13. Suppose that a certain company has 2 methods, M_1 and M_2, of manufacturing 3 automobile gadgets G_1, G_2, and G_3. The first method will produce 1 of each gadget in 3 hours. The second method will produce 3 G_1's and 1 G_3 in 4 hours. The company has an order for 6 G_1's, 2 G_2's, and 4 G_3's. How many times should we employ each method to fill the order and minimize the time we spend in production?

14. A fruit dealer ships 800 boxes of fruit north on a certain truck. If he must ship at least 200 boxes of oranges at \$0.20 per box profit, at least 100 boxes of grapefruit at \$0.10 per box profit, and at *most* 200 boxes of tangerines at \$0.30 per box profit, how should he load his truck for maximum profit? (*Hint:* Use a three-dimensional graph, or use the constraint that a total of 800 boxes are shipped, to reduce the number of unknowns in the linear programming model from three to two.)

6.3 An Algebraic Method of Solution

In Section 6.2 a geometric method of solution was illustrated. Obviously, if the number of unknowns exceeds 3, such a method is inadequate. In the next two sections we shall introduce an algebraic method of solution of a linear programming problem with any number of unknowns.

Example 1
Consider the linear programming problem:

Maximize $(6x_1 + x_3)$ subject to

$$2x_1 + x_2 + x_3 = 10,$$
$$x_1 + 4x_2 \leq 12,$$

and

$$x_1 \geq 0,$$
$$x_2 \geq 0,$$
$$x_3 \geq 0.$$

First, write the structural constraints as a system of linear equations. If there is a structural constraint that is an inequality, then it can be changed

to an equation by adding or subtracting a new nonnegative variable that will be called a **slack variable**. For example, in our illustration the left side of the constraint,

$$x_1 + 4x_2 \leq 12,$$

is smaller than or equal to the right side; hence there is some unknown nonnegative quantity x_4 that can be added to the left side so that

$$(x_1 + 4x_2) + x_4 = 12.$$

If the sense of the inequality had been reversed, we simply would have subtracted, rather than added, x_4. The problem can be expressed now as maximize $(6x_1 + 0x_2 + x_3 + 0x_4)$ subject to

$$\begin{aligned} 2x_1 + x_2 + x_3 &= 10, \\ x_1 + 4x_2 \qquad\quad + x_4 &= 12, \end{aligned} \quad \text{and} \quad \begin{aligned} x_1 &\geq 0, \\ x_2 &\geq 0, \\ x_3 &\geq 0, \\ x_4 &\geq 0. \end{aligned}$$

Notice that the introduction of a slack variable has no actual effect on values of the objective function because the coefficient of the slack variable is zero. ∎

▶ **Definition 2**

A **slack variable** is a nonnegative variable x_q that changes an inequality

$$a_{i1}x_1 + \cdots + a_{ip}x_p \leq p_{i0},$$

or

$$a_{i1}x_1 + \cdots + a_{ip}x_p \geq p_{i0},$$

to an equation

$$a_{i1}x_1 + \cdots + a_{ip}x_p + x_q = p_{i0},$$

or

$$a_{i1}x_1 + \cdots + a_{ip}x_p - x_q = p_{i0},$$

respectively.

Example 1 (continued)

If we designate values of the objective function as f, the linear programming problem can be restated as: Find nonnegative values of x_1, x_2, x_3, and x_4 of the linear system

$$\begin{aligned} 2x_1 + x_2 + x_3 &= 10, \\ x_1 + 4x_2 \qquad\quad + x_4 &= 12, \end{aligned} \tag{6.1}$$

so that $f = 6x_1 + x_3$ is a maximum.

Because there may be an infinite number of solutions of system (6.1), the task of finding one satisfying the specified conditions seems at first to be an impossible task. Fortunately, however, it can be proved that the solution of this problem, if a solution exists, is a *nonnegative* particular solution of

system (6.1), where the parameters of a complete solution are assigned the value 0; solutions of (6.1) in which the parameters of a complete solution are assigned a value of 0 are known as **basic solutions**. Because there is always a finite number of basic solutions, the task of solving a problem with a solution may be long but not impossible.

The calculation of a couple of basic solutions of (6.1) will now be illustrated. An obvious complete solution of (6.1) is

$$x_3 = 10 - 2x_1 - x_2,$$
$$x_4 = 12 - x_1 - 4x_2.$$

If we let the parameters x_1 and x_2 be 0, we have a basic solution

$$(0, 0, 10, 12),$$

with a corresponding f equal to 10. Suppose we now seek another complete solution by replacing x_2 as a parameter. We perform the elementary row operations of $(\tfrac{1}{4}R_2)$ and $(-R_2 + R_1)$ in that order on the augmented matrix of (6.1); the following equivalent system will be obtained:

$$\tfrac{7}{4}x_1 \qquad + x_3 - \tfrac{1}{4}x_4 = 7,$$
$$\tfrac{1}{4}x_1 + x_2 \qquad + \tfrac{1}{4}x_4 = 3.$$
(6.2)

An obvious complete solution corresponding to (6.2) is

$$x_3 = 7 - \tfrac{7}{4}x_1 + \tfrac{1}{4}x_4,$$
$$x_2 = 3 - \tfrac{1}{4}x_1 - \tfrac{1}{4}x_4.$$

If we let the new parameters x_1 and x_4 be zero, we have another basic solution,

$$(0, 3, 7, 0),$$

with a corresponding f equal to 7. We can continue in this manner to find the other nonnegative basic solutions corresponding to complete solutions of the structural constraints. They are $(5, 0, 0, 7)$ with $f = 30$ and $(4, 2, 0, 0)$ with $f = 24$. The largest value of f corresponding to the four nonnegative basic solutions is 30, and hence if there is a maximum value of f that satisfies (6.1), then that value must be 30 and occur at

$$(5, 0, 0, 7).$$

(There can be as many as $C_2^4 = 6$ ways of identifying 2 parameters from 4 unknowns. In this example there are two other basic solutions, but it turns out that they are *not* nonnegative; hence they are not considered because of the nonnegativity constraints.)

The basic solutions of the original system of constraints (6.1) can be found by simply letting every possible pair of unknowns be zero and solving for the remaining pair. However, the procedure of using elementary row operations to find the basic solutions from successive complete solutions (as outlined in this example) is important in understanding the simplex method, which is to be introduced in Section 6.6. ∎

Sec. 6.3] An Algebraic Method of Solution

We now give formal definitions to some of the concepts that were introduced in the previous illustration.

▶ **Definition 3**

Consider a system of m linear equations in n unknowns ($n \geq m$) having a complete solution with $n - m$ parameters. If the parameters are assigned the value 0, the resulting particular solution is called a **basic solution**. The m unknowns or variables *not* serving as parameters in the complete solution associated with a basic solution are called **basic variables**. The $n - m$ parameters that *are* assigned the value 0 are called **nonbasic variables**. Those basic solutions of a linear system of structural constraints (equations) that are feasible, and hence have no negative coordinates, are called **basic feasible solutions**.

Throughout this book we shall assume that every system of m constraints (equations) will have a basic solution; that is, there will be a complete solution with exactly $n - m$ parameters. The case for which this assumption is not valid is discussed briefly in item (4), p. 387, of Campbell [13].

▶ **Definition 4**

An **optimal solution** is a feasible solution at which the linear objective function is optimum.

EXERCISES

Suggested minimum assignment: Exercises 1, 3, 5, and 8.

In Exercises 1 and 2 express the given structural constraints of a linear programming problem as a system of linear equations. What are the new nonnegativity constraints?

1. $x_1 + 2x_2 \leq 7,$
   $x_1 - 3x_2 + x_3 \geq 2,$
   $2x_1 + x_2 + x_3 = 4.$

2. $x_1 - x_2 - x_3 = 2,$
   $2x_1 - x_2 + x_4 \leq 1,$
   $x_1 + 2x_2 + x_4 \geq 3.$

In Exercises 3–6 use the method of Example 1 to find two complete solutions and corresponding basic solutions of the given structural constraints. Are your basic solutions feasible?

3. $x_1 + 2x_2 + x_3 = 14,$
   $3x_1 + x_2 + x_4 = 12.$

4. $2x_1 + 4x_2 + x_3 = 10,$
   $4x_1 + x_2 + x_4 = 6.$

5. $x_1 + 2x_2 \leq 7,$
   $x_1 - 3x_2 \geq 2.$

6. $x_1 + 2x_2 \geq 5,$
   $2x_1 + x_2 \geq 4.$

7. By the method begun in Example 1, find the remaining complete solutions of Example 1 and verify that the associated nonnegative basic solutions are as stated.

8. In this section we have stressed the calculation of basic solutions of structural constraints from complete solutions; this has been done in anticipation of the simplex method, which is to come later. Basic solutions, however, can be calculated by simply selecting m appropriate variables to be basic variables, then assigning the remaining variables (nonbasic variables) the value of zero, and solving the resulting system.
    (a) Use this method to find all the basic solutions of Exercise 3.
    (b) Which of the basic solutions are feasible?
    (c) If we wish to maximize $(x_1 + 3x_2)$, subject to the given constraints and if we are told that an optimal solution will occur at a basic feasible solution, then find an optimal solution.
9. Repeat Exercise 8 by using Exercise 4 rather than Exercise 3.

6.4 An Algebraic Method of Solution (Continued)

We now state the theorem that is the cornerstone of the algebraic method of solving linear programming problems because it assures us that if there is an optimal solution of a linear programming problem, then that solution can be found among the basic feasible solutions. Because there are a finite number of basic feasible solutions, the task of finding an existing, optimal solution is possible.

▶ **Theorem 1**
If there is an optimal solution of a linear programming problem, then an optimal solution can be found among the basic feasible solutions. Moreover, there may be more than one optimal solution.

A proof of Theorem 1 is beyond the scope of this text (but an outline may be found in Campbell [13] pp. 372–373). ◀

The following example will illustrate the use of Theorem 1, under the assumption that an optimal solution does exist. We shall also give a geometric interpretation of the basic feasible solutions.

Example 1
Maximize $(2x_1 + x_2)$ subject to

$$x_1 + x_2 \le 5, \quad \text{and} \quad x_1 \ge 0,$$
$$x_1 + 3x_2 \le 9, \qquad\qquad x_2 \ge 0.$$

Solution: Introduce slack variables x_3 and x_4 to change the structural constraints from inequalities to equations, and let $f = 2x_1 + x_2$. Form the system

$$\begin{aligned} x_1 + x_2 + x_3 &= 5, \\ x_1 + 3x_2 \phantom{{}+{}} + x_4 &= 9, \end{aligned} \tag{6.3}$$

By Theorem 1, maximum f, if it exists, will occur at a basic feasible solution

of (6.3). In Exercise 3 the reader will be asked to verify that

$$\begin{array}{c}(0, 0, 5, 9) \text{ with } f = 0,\\ (5, 0, 0, 4) \text{ with } f = 10,\\ (0, 3, 2, 0) \text{ with } f = 3,\\ (3, 2, 0, 0) \text{ with } f = 8,\end{array} \quad (6.4)$$

are the basic feasible solutions of (6.3). If we know that an optimal solution exists, then by Theorem 1 that basic feasible solution must be

$$(5, 0, 0, 4)$$

because 10 is the largest value of f from among the 4 candidates. Of course, in the original statement of the problem we were not concerned with the slack variables; hence the solution of the original problem is: Maximum $(2x_1 + x_2) = 10$ at $(5, 0)$. In Figure 6.5, notice that each basic feasible solution of (6.3) corresponds to a vertex of the feasible set (colored region) of the original constraints. Each vertex is an intersection of the bounding lines of the solution sets of two of the constraints of the original problem. Notice, also, that the point $(9, 0)$ that lies on the intersection of the bounding lines of the solution sets of the constraints

$$x_1 + 3x_2 \leq 9 \quad \text{and} \quad x_2 \geq 0,$$

corresponds to the basic solution $(9, 0, -4, 0)$ of (6.3). The point $(9, 0)$, as well as its corresponding basic solution, is *not* feasible, however, since it fails to satisfy *all* the constraints. ∎

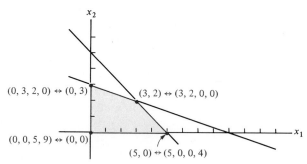

Figure 6.5. *Correspondence of basic feasible solutions of (6.3) with vertices of the feasible set of the original constraints.*

Unless one is certain that an optimal solution exists, the method presented in this section has rather obvious deficiencies. Moreover, the calculation of all basic feasible solutions can be very time consuming in a large problem—even for a computer. The simplex method, which is a refinement of the method presented in this section, overcomes these difficulties and is the subject of Sections 6.6 and 6.7.

APPLICATIONS

Example 2
A survey of applications of linear programming may be found in Gass [25], pp. 11–18. This survey includes the following areas: agriculture, contract awards, economics, military practice, personnel assignment, production scheduling, inventory control, structural design, traffic analysis, transportation, and network theory. Industrial applications from the following kinds of industries are also discussed: chemical, coal, commercial aviation, communication, iron and steel, paper, petroleum, and railroad. On pp. 325–337 of [25] a bibliography is given for applications in each of the categories listed above. The reader is strongly encouraged to at least glance through the material in [25] in order to appreciate properly the role of linear programming in applications of mathematics in our society; such an appreciation should help to motivate the student to study the simplex method, which is to be presented in Sections 6.6 and 6.7 and which is the primary method of solution of linear programming problems. ∎

Example 3
The usefulness of linear programming in management accounting is made quite evident by Ijiri in [31]. Any serious student of business should at least glance through [31] to observe first-hand the degree of application of linear programming in accounting; other references that illustrate the use of linear programming in accounting are [16], [21], and [32]. ∎

EXERCISES

Suggested minimum assignment: Exercises 1 and 4.

1. (a) On a two-dimensional graph illustrate the feasible set determined by the constraints

$$x_1 + 3x_2 \leq 9, \quad \text{and} \quad x_1 \geq 0,$$
$$2x_1 + x_2 \leq 8, \qquad\qquad x_2 \geq 0.$$

   (b) The feasible set in part (a) has 4 bounding lines; find the 6 points, each of which is common to some 2 of these bounding lines.

   (c) Introduce slack variables to make the structural constraints a system of linear equations. Find the 6 basic solutions and show that they correspond to the 6 points of intersection found in part (b). (You may wish to find these basic solutions by the shorter method outlined in Exercise 8 of the last section.)

   (d) Find the 4 basic *feasible* solutions from among the 6 basic solutions of part (c), and find the 4 vertices of the feasible set of part (a). Observe the correspondence.

   (e) Which basic feasible solution is the optimal solution if we seek maximum $(x_1 + x_2)$? Give a reason for your answer.

Sec. 6.4] An Algebraic Method of Solution

2. Repeat Exercise 1 if the constraints are

$$x_1 + 4x_2 \leq 13, \quad \text{and} \quad x_1 \geq 0,$$
$$3x_1 + x_2 \leq 6, \qquad\qquad x_2 \geq 0.$$

3. In Example 1 of this section verify that (6.4) are the basic feasible solutions of (6.3).

In Exercises 4 and 5 solve the given linear programming problem by applying Theorem 1 after finding all the basic feasible solutions. Be careful about the sense of each inequality and the consequent addition or subtraction of the slack variable. (The reader was asked to solve each of the exercises by the geometric method in Exercises 5 and 6 of Section 6.2.)

4. Minimize $(x_1 + x_2)$ subject to

$$x_1 + 2x_2 \leq 7, \quad \text{and} \quad x_1 \geq 0,$$
$$x_1 - 3x_2 \geq 2, \qquad\qquad x_2 \geq 0.$$

5. Minimize $(3x_1 + x_2)$ subject to

$$x_1 + 2x_2 \geq 5, \quad \text{and} \quad x_1 \geq 0,$$
$$2x_1 + x_2 \geq 4, \qquad\qquad x_2 \geq 0.$$

6. A trucking company owns two types of trucks. Type A has 20 cubic yards of refrigerated space and 30 cubic yards of nonrefrigerated space. Type B has 20 cubic yards of refrigerated space and 10 cubic yards of nonrefrigerated space. A customer wants to transport some produce a certain distance; he will require 160 cubic yards of refrigerated space and 120 cubic yards of nonrefrigerated space. The trucking company figures it will take 300 gallons of gas for the type A truck to make the trip and 200 gallons of gas for the type B truck. Find the number of trucks of each type that the company should allow for the job in order to minimize gas consumption.

7. An insurance consultant must provide the necessary life and disability insurance for a client. Suppose that policies P_1 and P_2 provide the following monthly income to the beneficiary per thousand dollars of insurance.

	P_1	P_2
Life	$20	$40
Disability	$50	$20

At least $800 of life insurance income and $600 of disability insurance income must be provided. If the costs of P_1 and P_2 are $1 and $.80 per thousand, respectively, how many thousands of each type of policy should be bought to meet the desired coverage while minimizing the total cost of the insurance purchased?

6.5 Matrix Notation

In the literature that deals with linear programming, it is customary to express linear programming problems using matrix notation; to do this, the following definition is needed.

▶ **Definition 5**
A matrix A is "**greater than or equal to**" ($\geq$) matrix B of the same order when each entry of A is greater than, or equal to, the corresponding entry of B ($<$, $>$, or $\leq$ can be substituted for $\geq$ with corresponding changes in meaning).

Example 1
By Definition 5,
$$\begin{bmatrix} 2 & 3 \\ 4 & 5 \end{bmatrix} \geq \begin{bmatrix} 1 & 3 \\ 0 & 2 \end{bmatrix}$$
because each entry of the matrix on the left is larger than or equal to the corresponding entry on the right. ∎

Example 2
Consider the diet problem of Section 6.2: Minimize $(3x_1 + 5x_2)$ subject to
$$\begin{array}{c} 3x_1 + 4x_2 \geq 340, \\ x_1 + x_2 \geq 100, \\ x_1 + 5x_2 \geq 150, \end{array} \quad \text{and} \quad \begin{array}{c} x_1 \geq 0, \\ x_2 \geq 0. \end{array}$$

Using matrix multiplication and Definition 5, we can rewrite the problem as

$$\text{Minimize } \begin{bmatrix} 3 & 5 \end{bmatrix} \begin{bmatrix} x_1 \\ x_2 \end{bmatrix} \text{ subject to}$$

$$\begin{bmatrix} 3 & 4 \\ 1 & 1 \\ 1 & 5 \end{bmatrix} \begin{bmatrix} x_1 \\ x_2 \end{bmatrix} \geq \begin{bmatrix} 340 \\ 100 \\ 150 \end{bmatrix} \quad \text{and} \quad \begin{bmatrix} x_1 \\ x_2 \end{bmatrix} \geq \begin{bmatrix} 0 \\ 0 \end{bmatrix},$$

or as

Minimize CX subject to $AX \geq P_0$ and $X \geq \mathbf{0}$, where

$$C = \begin{bmatrix} 3 & 5 \end{bmatrix}, \quad X = \begin{bmatrix} x_1 \\ x_2 \end{bmatrix}, \quad A = \begin{bmatrix} 3 & 4 \\ 1 & 1 \\ 1 & 5 \end{bmatrix}, \quad \text{and} \quad P_0 = \begin{bmatrix} 340 \\ 100 \\ 150 \end{bmatrix}.$$

Although technically CX is a 1 by 1 matrix, it is permissible here to use the matrix and its single entry interchangeably. ∎

Example 3
If nonnegative slack variables are subtracted from each of the constraints of Example 2 in order to make equations, then the resulting problem can be written:

Minimize CX subject to $AX = P_0$ and $X \geq 0$, where
$$C = [3 \quad 5 \quad 0 \quad 0 \quad 0],$$

$$X = \begin{bmatrix} x_1 \\ x_2 \\ x_3 \\ x_4 \\ x_5 \end{bmatrix}, \quad A = \begin{bmatrix} 3 & 4 & -1 & 0 & 0 \\ 1 & 1 & 0 & -1 & 0 \\ 1 & 5 & 0 & 0 & -1 \end{bmatrix}, \quad P_0 = \begin{bmatrix} 340 \\ 100 \\ 150 \end{bmatrix}. \blacksquare$$

In expressing constraints in matrix notation, one must be careful that the constraints are all of the same type; that is, they must all be ($\leq$) inequalities or ($\geq$) inequalities or equations. The constraints

$$x_1 + x_2 \geq 1,$$
$$x_1 - x_3 \leq 4,$$

are not all of the same type but can be made so by multiplying the last constraint by (-1). The result will then be

$$x_1 + x_2 \geq 1,$$
$$-x_1 + x_3 \geq -4,$$

which is suitable for matrix expression.

EXERCISES

Suggested minimum assignment: Exercises 1, 3, 5, and 7.

In each of Exercises 1 and 2 express the given problem using matrices and *without* the introduction of slack variables. Do *not* solve the problem.

1. Maximize $(x_1 + x_2)$ subject to $\begin{aligned} 2x_1 + x_2 &\leq 4, \\ x_1 - x_2 &\geq -5, \end{aligned}$ and $\begin{aligned} x_1 &\geq 0, \\ x_2 &\geq 0. \end{aligned}$

2. Maximize $(3x_1 - x_2)$ subject to $\begin{aligned} x_1 - x_2 &\geq 5, \\ x_1 + 4x_2 &\leq -3, \end{aligned}$ and $\begin{aligned} x_1 &\geq 0, \\ x_2 &\geq 0. \end{aligned}$

In each of Exercises 3 and 4 express the given problem using matrices and *with* the introduction of slack variables. Do *not* solve the problem.

3. Minimize $(5x_1 + x_2)$ subject to $\begin{aligned} 2x_1 + x_2 &= 3, \\ x_1 + x_2 &\leq 4, \\ x_1 - x_2 &\geq 2, \end{aligned}$ and $\begin{aligned} x_1 &\geq 0, \\ x_2 &\geq 0. \end{aligned}$

4. Minimize $(x_1 - x_2)$ subject to $\begin{array}{l} x_1 + x_2 \geq 2, \\ 2x_1 - x_2 \leq 4, \\ x_1 - 3x_2 = 2, \end{array}$ and $\begin{array}{l} x_1 \geq 0, \\ x_2 \geq 0. \end{array}$

In each of Exercises 5 and 6 find the optimal solution, if one exists, by the geometric method.

5. Maximize CX subject to $AX \leq P_0$ and $X \geq 0$, where $C = [2 \quad 1]$,

$$X = \begin{bmatrix} x_1 \\ x_2 \end{bmatrix}, A = \begin{bmatrix} 1 & 1 \\ 3 & 2 \end{bmatrix}, \text{ and } P_0 = \begin{bmatrix} 5 \\ 12 \end{bmatrix}.$$

6. Minimize CX subject to $AX \geq P_0$ and $X \geq 0$, where $C = [2 \quad -1]$,

$$X = \begin{bmatrix} x_1 \\ x_2 \end{bmatrix}, A = \begin{bmatrix} 1 & 2 \\ -1 & 1 \end{bmatrix}, \text{ and } P_0 = \begin{bmatrix} 2 \\ -1 \end{bmatrix}.$$

In each of Exercises 7 and 8 find the optimal solution, given that one exists, by use of the algebraic method.

7. Minimize CX subject to $AX = P_0$ and $X \geq 0$, where

$$C = [1 \quad 2 \quad 0 \quad 0], X = \begin{bmatrix} x_1 \\ x_2 \\ x_3 \\ x_4 \end{bmatrix}, A = \begin{bmatrix} 1 & 1 & 1 & 0 \\ 3 & 1 & 0 & 1 \end{bmatrix}, \text{ and } P_0 = \begin{bmatrix} 6 \\ 12 \end{bmatrix}.$$

8. Maximize CX subject to $AX = P_0$ and $X \geq 0$, where

$$C = [-3 \quad 1 \quad 0 \quad 0], X = \begin{bmatrix} x_1 \\ x_2 \\ x_3 \\ x_4 \end{bmatrix}, A = \begin{bmatrix} 2 & 1 & 1 & 0 \\ -1 & 1 & 0 & 1 \end{bmatrix}, \text{ and } P_0 = \begin{bmatrix} 8 \\ 5 \end{bmatrix}.$$

6.6 An Introduction to the Simplex Method

In Section 6.4 we learned that an optimal solution, if one exists, will be among the basic feasible solutions of the structural constraints. Some difficulties remain, however. First, there must be some way to determine whether or not an optimal solution exists, and second, an efficient procedure must be developed for finding the optimal solution without having to find all the basic feasible solutions of the structural constraints; in a large problem, the job of calculating all the basic feasible solutions can be very time consuming—even for a computer. The **simplex method**, developed by George Dantzig in 1947, will overcome these difficulties. A brief history of the development of the simplex method may be found in Dantzig [18], pp. 12–31.

Although the execution of the simplex method is rather involved, the basic idea is reasonably easy. That basic idea is to use the objective function

to identify, one at a time, only those basic feasible solutions that can enlarge (or maintain) the value of the objective function of a maximum problem. This process avoids the need to find all the basic feasible solutions and results in considerable efficiency. Moreover, the objective function can be used to determine when an optimal solution is reached or to determine that no solution exists. We shall illustrate these basic ideas of the simplex method in the next example.

Example 1

In Example 1 of Section 6.3 we found maximum $(6x_1 + x_3)$ by evaluating $(6x_1 + x_3)$ at *all* the basic feasible solutions of the structural constraints

$$2x_1 + x_2 + x_3 = 10,$$
$$x_1 + 4x_2 + x_4 = 12.$$

The simplex method begins with only one basic feasible solution as determined by the obvious complete solution

$$x_3 = 10 - 2x_1 - x_2,$$
$$x_4 = 12 - x_1 - 4x_2.$$

In this initial complete solution, the basic variables are x_3 and x_4 and the nonbasic variables (parameters) are x_1 and x_2. At the outset we express the value of the objective function in terms of only the nonbasic variables; the objective equation

$$f = 6x_1 + x_3$$

becomes

$$f = 6x_1 + (10 - 2x_1 - x_2),$$

or when simplified

$$f = 10 + 4x_1 - x_2.$$

When the nonbasic variables (parameters) x_1 and x_2 are set equal to 0, then $f = 10$ at the initial basic feasible solution $(0, 0, 10, 12)$. The question then arises: Would f be any larger than 10 if one of the nonbasic variables x_1 or x_2 became a basic variable? If, for instance, x_1 replaced x_3 as a basic variable and if x_1 assumed a positive value, then certainly

$$f = 10 + 4x_1 - x_2$$

would be larger because the coefficient 4 of x_1 is positive. We then proceed to make x_1 a basic variable by finding a new complete solution. In order to do so, as well as to find the corresponding new expression for f, we apply certain elementary row operations to the augmented matrix of the system

$$\begin{array}{rl} 2x_1 + x_2 + x_3 & = 10, \\ x_1 + 4x_2 \phantom{{}+{}} + x_4 & = 12, \\ \hline 4x_1 - x_2 \phantom{{}+{}} & = f - 10, \end{array}$$

in such a way as to maintain the nonnegative numbers on the right side of

the structural constraints (this step ensures that the new basic solution will be feasible). The augmented matrix is

$$M_1 = \begin{bmatrix} 2 & 1 & 1 & 0 & | & 10 \\ 1 & 4 & 0 & 1 & | & 12 \\ \hline 4 & -1 & 0 & 0 & | & f-10 \end{bmatrix}.$$

If x_1 is to become a basic variable and if the numbers on the right side of the first equations are to remain nonnegative, then the successive elementary row operations ($\frac{1}{2}R_1$), ($-R_1 + R_2$), and ($-4R_1 + R_3$) should be performed on M_1 (methods for determining these operations will be explained in the next section). The resulting matrix is called M_2.

$$M_2 = \begin{bmatrix} 1 & \frac{1}{2} & \frac{1}{2} & 0 & | & 5 \\ 0 & \frac{7}{2} & -\frac{1}{2} & 1 & | & 7 \\ \hline 0 & -3 & -2 & 0 & | & f-30 \end{bmatrix}.$$

The latter matrix is the augmented matrix of the system

$$\begin{aligned} x_1 + \tfrac{1}{2}x_2 + \tfrac{1}{2}x_3 &= 5, \\ \tfrac{7}{2}x_2 - \tfrac{1}{2}x_3 + x_4 &= 7, \\ \hline -3x_2 - 2x_3 &= f - 30. \end{aligned}$$

From this system we see that the new complete solution of the structural constraints is

$$x_1 = 5 - \tfrac{1}{2}x_2 - \tfrac{1}{2}x_3,$$
$$x_4 = 7 - \tfrac{7}{2}x_2 + \tfrac{1}{2}x_3,$$

and the new expression for f in terms of only the nonbasic variables x_2 and x_3 is

$$f = 30 - 3x_2 - 2x_3.$$

When the nonbasic variables are set equal to 0, then $f = 30$ at the second basic feasible solution (5, 0, 0, 7). Again the question is raised: Would f be any larger than 30 if one of the nonbasic variables x_2 or x_3 became basic? Notice that if either x_2 or x_3 could assume positive values, then f would not be larger than 30 because the coefficients of x_2 and x_3 are negative. It can be proved that no further improvement in f is possible and hence maximum $f = 30$ at (5, 0, 0, 7). Notice that we arrived at the optimal solution by calculating only two of the basic feasible solutions; moreover, we had a method of determining that the final result was indeed the maximum. ∎

In summary, the simplex method is an iterative procedure that begins with one basic feasible solution of the structural constraints (equations) of the maximum problem and then, by means of elementary row operations, replaces the old basic feasible solution by a new basic feasible solution that will increase (or in some cases only maintain) the value of the objective

function. The procedure is repeated until an optimal solution is found or until it is determined that no optimal solution exists.

In this section we have been concerned primarily with the basic rationale behind the simplex method. Some of the finer points in the execution of the simplex method will be discussed in Section 6.7.

EXERCISES

Suggested minimum assignment: Exercise 1.

1. Consider the problem: Maximize $(2x_1 + x_3)$ subject to

$$\begin{aligned} -x_1 + x_2 + x_3 &= 1, \\ x_1 + 2x_2 + x_4 &= 5, \end{aligned} \quad \text{and} \quad X \geq 0.$$

  (a) Find an initial complete solution with associated basic solution that is feasible.
  (b) Express the value f of the objective function in terms of only the nonbasic variables.
  (c) Can the initial value of f be enlarged by making one of the nonbasic variables basic? In other words, can the value of the objective function be enlarged if one of the nonbasic variables could take on positive values? If your answer is yes, name such a variable.
  (d) Write the augmented matrix of the system consisting of the structural constraints and your answer to part (b). Call the result M_1.
  (e) Use elementary row operations to find a new complete solution (of the structural constraints) that introduces your answer of part (c) as a basic variable. Remember that the first two entries of the last column of M_1 must remain nonnegative.
  (f) Express the value f of the objective function in terms of the new nonbasic variables.
  (g) Can the value of the objective function be enlarged by making one of the current nonbasic variables basic?
  (h) If your answer to part (g) was yes, what nonbasic variable should be made basic? If your answer to part (g) was no, what is maximum f and at what basic feasible solution does maximum f occur?
2. Repeat Exercise 1 for the problem: Maximize $(-2x_1 - x_3)$ subject to

$$\begin{aligned} 3x_1 + x_2 + x_3 &= 9, \\ x_1 + 2x_2 + x_4 &= 8, \end{aligned} \quad \text{and} \quad X \geq 0.$$

6.7 The Simplex Method

In this section, formal statements of the steps of the simplex method will be given and illustrated. The plausibility of some of these steps will be discussed, but the reader should be aware of the fact that most of the assertions are not proved. Such justifications may be found in more advanced books on linear programming.

The steps formally listed below apply to the linear programming problem: Maximize CX subject to $AX = P_0$ and $X \geq 0$. Moreover, two restrictions must be imposed. First, $P_0 \geq 0$. This restriction is easily satisfied by multiplying both sides of the appropriate constraint equations by (-1) to make all the entries of P_0 nonnegative. Second, the m by n matrix A must have all m columns of the identity matrix I_m among its columns (not necessarily in any order). This restriction is not always easy to satisfy, but there are several ways of overcoming the difficulty when such a problem is encountered; one way is mentioned in this section and other ways are discussed in [12] (Campbell). The purpose of these two restrictions is simply to ensure an obvious initial complete solution (of the structural constraints) with an associated basic solution that is feasible. This situation is necessary in order to start the simplex method.

If a minimum problem is encountered, it can be changed to a maximum problem by simply maximizing $(-CX)$ instead of minimizing CX; the constraints remain the same.

Each of the following steps will be illustrated by the same example used in Section 6.6.

Step A: *Write the $m + 1$ by $n + 1$ augmented matrix $M_0 = \begin{bmatrix} A & | & P_0 \\ \hline C & | & f \end{bmatrix}$ of the system*

$$AX = P_0,$$
$$CX = [f].$$

For the linear programming problem: Maximize $(6x_1 + x_3)$ subject to

$$\begin{aligned} 2x_1 + x_2 + x_3 &= 10, \\ x_1 + 4x_2 + x_4 &= 12, \end{aligned} \quad \text{and} \quad X \geq 0,$$

we have

$$M_0 = \begin{bmatrix} 2 & 1 & 1 & 0 & | & 10 \\ 1 & 4 & 0 & 1 & | & 12 \\ \hline 6 & 0 & 1 & 0 & | & f \end{bmatrix}.$$

Step B: *Use the appropriate elementary row operations on M_0 to produce a matrix that will have the first m columns of I_{m+1} among its columns. Call the result M_1.*

Since we have required that the submatrix A of matrix M_0 has among its columns all the columns of I_m, step B simply requires that those entries in the last row of M_0 just below the identity columns of A must be 0. This action has the effect of changing the corresponding objective equation so that f is expressed in terms of only nonbasic variables.

For the matrix M_0, stated above, the elementary row operation $(-R_1 + R_3)$ will make the third column become $\begin{bmatrix} 1 \\ 0 \\ 0 \end{bmatrix}$; the fourth column of

M_0 already is the required identity column $\begin{bmatrix} 0 \\ 1 \\ 0 \end{bmatrix}$. Thus after applying $(-R_1 + R_3)$ to M_0, the first two columns of I_3 will be among the columns of the new matrix

$$M_1 = \left[\begin{array}{cccc|c} 2 & 1 & 1 & 0 & 10 \\ 1 & 4 & 0 & 1 & 12 \\ \hline 4 & -1 & 0 & 0 & f-10 \end{array}\right].$$

From the last row of M_1 we have a new objective equation

$$4x_1 - x_2 = f - 10 \quad \text{or} \quad f = 10 + 4x_1 - x_2.$$

The purpose of step B is to rewrite the initial objective equation in such a way that only nonbasic variables are included; then in the next step, we can determine which, if any, nonbasic variables will cause an increase in the value of f if that nonbasic variable were to become a basic variable.

Step C: *Excluding the final entry, locate any positive number among the entries of the last row of M_1 and designate the column in which that number appears as the **pivot column**. If there are no such positive numbers, then the simplex method terminates, and the current basic feasible solution is an optimal solution.*

For our example, the only such positive number in the last row of M_1 is 4 and it appears in the first column. Therefore, the first column is the pivot column.

The purpose of step C is to locate any nonbasic variable that might cause the value of the objective function to be enlarged if that nonbasic variable were to become a basic variable. Notice that the first four entries of the last row of M_1 are simply the coefficients of the variables in the current objective equation.

Step D: *Divide each positive entry of the pivot column (except the final entry) into the corresponding entry of the last column of M_1 and select the smallest of these ratios. The row of M_1 that produces this smallest ratio is designated the **pivot row**.† The entry that belongs to both the pivot row and the pivot column is called the **pivot**. If there are no positive entries in the pivot column above the last row, then the simplex method terminates and there is no optimal solution of the problem.*

For our example, the ratios are formed in Figure 6.6, and since $\frac{10}{2}$ is the smaller of the ratios, and since this ratio was produced from the first row of M_1, then the first row is the pivot row, and the entry in the first row and first column of M_1 is the pivot.

---

† It is possible to have a "tie" for the smallest ratio; this situation is symptomatic of what is known as degeneracy. Although degeneracy causes considerable theoretical difficulties, the practical difficulties are few. Arbitrarily pick any one of the candidates as the pivot row and continue.

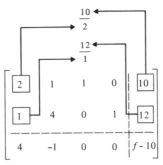

Figure 6.6. *Formation of ratios of step D.*

The purpose of step D is to ensure that the nonnegative entries of the last column remain nonnegative when the elementary row operations of the next step are performed. Therefore, the new basic solution will be feasible.

Step E: *Perform the following elementary row operations on M_1: Multiply the pivot row by a real number that will reduce the pivot to a 1. Add multiples of the pivot row to the other rows to reduce all other entries of the pivot column to a 0. Call the result M_2.*

In our example the operation $\frac{1}{2}R_1$ will make the pivot a 1. Then the operation $(-R_1 + R_2)$ will produce a 0 in the second entry of the pivot column, and the operation $(-4R_1 + R_3)$ will produce a 0 in the third entry of the pivot column. Notice that the interchange of two rows is *not* permitted in step E. The pivot is identified by an asterisk in matrix M_1.

$$
\underbrace{\begin{bmatrix} 2^* & 1 & 1 & 0 & | & 10 \\ 1 & 4 & 0 & 1 & | & 12 \\ \hline 4 & -1 & 0 & 0 & | & f-10 \end{bmatrix}}_{M_1} \underset{\sim}{\text{row}} \underbrace{\begin{bmatrix} 1 & \frac{1}{2} & \frac{1}{2} & 0 & | & 5 \\ 0 & \frac{7}{2} & -\frac{1}{2} & 1 & | & 7 \\ \hline 0 & -3 & -2 & 0 & | & f-30 \end{bmatrix}}_{M_2}.
$$

The purpose of step E is to generate a new complete solution of the structural constraints, together with a new expression for f in terms of the new nonbasic variables.

Step F: *Return to step C for M_2 and continue until the method terminates.*

When step C is reconsidered for M_2, we see that all the entries of the last row (excluding the final entry) are nonpositive, and hence, according to step C, the simplex method terminates. Thus the maximum $f = 30$ is obtained from $f = 30 - 3x_2 - 2x_3$ at the basic feasible solution $(5, 0, 0, 7)$, as found from the last complete solution

$$x_1 = 5 - \tfrac{1}{2}x_2 - \tfrac{1}{2}x_3,$$
$$x_4 = 7 - \tfrac{7}{2}x_2 + \tfrac{1}{2}x_3.$$

A version of the simplex method has been stated and illustrated. It

Sec. 6.7] The Simplex Method

requires an initial complete solution (of the structural constraints in equation form) with an associated basic solution that is feasible. This situation is ensured by requiring that $P_0 \geq \mathbf{0}$ and that all the columns of the identity matrix I_m must be among the columns of the m by n matrix A. The following example illustrates the solution of a problem that does not initially satisfy these two restrictions. The reader is cautioned, however, that the procedure for satisfying the second restriction that is about to be illustrated may be either impossible (see Exercise 13) or very difficult to implement. Alternative procedures are presented in Sections 10.7 and 10.9 of [12] (Campbell).

Example 1

Minimize $(2x_1 + x_2)$ subject to

$$-x_1 - x_2 \leq -1,$$
$$2x_1 + 3x_2 \leq 6, \quad \text{and} \quad X \geq \mathbf{0}.$$

First, we observe that to minimize $(2x_1 + x_2)$ is to maximize $(-2x_1 - x_2)$. Then we must write the structural constraints in the form $AX = P_0$, where $P_0 \geq \mathbf{0}$. To do so, add slack variables to the left side and then multiply the first equation by -1. We obtain

$$x_1 + x_2 - x_3 = 1,$$
$$2x_1 + 3x_2 + x_4 = 6.$$

Now notice that the matrix

$$A = \begin{bmatrix} 1 & 1 & -1 & 0 \\ 2 & 3 & 0 & 1 \end{bmatrix}$$

does not have the identity columns $\begin{bmatrix} 1 \\ 0 \end{bmatrix}$ and $\begin{bmatrix} 0 \\ 1 \end{bmatrix}$ among its columns, and therefore, we do *not* have an immediate complete solution that will provide an initial basic *feasible* solution. Observe that the obvious complete solution

$$x_3 = -1 + x_1 + x_2,$$
$$x_4 = 6 - 2x_1 - 3x_2,$$

is not suitable because the corresponding basic solution is not feasible.

Therefore, we shall use elementary row operations on $[A \mid P_0]$ to try to develop a row equivalent matrix that *will* have $\begin{bmatrix} 1 \\ 0 \end{bmatrix}$ and $\begin{bmatrix} 0 \\ 1 \end{bmatrix}$ among its columns while maintaining nonnegative entries in the last column. This *attempt* can be accomplished by repeatedly designating some column of A as a pivot column and proceeding in the manner of steps D and E until $\begin{bmatrix} 1 \\ 0 \end{bmatrix}$ and $\begin{bmatrix} 0 \\ 1 \end{bmatrix}$ become columns of a matrix that is row equivalent to A. In general, for the procedure just described, there is no unique way to proceed and there is no guarantee of success. If, however, we choose the first column of

A as the pivot column and perform the operation $(-2R_1 + R_2)$ as indicated by steps D and E, we obtain the suitable matrix

$$[A' \mid P'_0] = \begin{bmatrix} 1 & 1 & -1 & 0 & | & 1 \\ 0 & 1 & 2 & 1 & | & 4 \end{bmatrix}.$$

The original problem then can be restated as: Maximize $(-2x_1 - x_2)$ subject to

$$x_1 + x_2 - x_3 = 1,$$
$$x_2 + 2x_3 + x_4 = 4, \quad \text{and} \quad X \geq 0,$$

and now that the restrictions are satisfied we can proceed with the simplex method.

Step A. Form

$$M_0 = \begin{bmatrix} A' & | & P'_0 \\ \hline C & | & f \end{bmatrix} = \begin{bmatrix} 1 & 1 & -1 & 0 & | & 1 \\ 0 & 1 & 2 & 1 & | & 4 \\ \hline -2 & -1 & 0 & 0 & | & f \end{bmatrix}.$$

Step B. To obtain 0 for the last entry in the first column, we apply the elementary row operation $(2R_1 + R_3)$. Therefore,

$$M_1 = \begin{bmatrix} 1 & 1 & -1 & 0 & | & 1 \\ 0 & 1 & 2 & 1 & | & 4 \\ \hline 0 & 1 & -2 & 0 & | & f+2 \end{bmatrix}.$$

Observe that the associated objective equation

$$x_2 - 2x_3 = f + 2 \quad \text{or} \quad f = -2 + x_2 - 2x_3$$

does *not* include either of the basic variables x_1 or x_4.

Step C. The second column of M_1 must be the pivot column because the second column contains the only positive entry among the first four entries of the last row.

Step D. Form the ratios shown in Figure 6.7; since the smallest ratio is produced by the first row, that row becomes the pivot row.

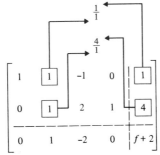

Figure 6.7. *Formation of ratios of step D.*

Step E. Perform the operations $(-R_1 + R_2)$ and $(-R_1 + R_3)$ on M_1 to make the nonpivot entries of the second column 0.

$$M_1 \qquad\qquad M_2$$

$$\begin{bmatrix} 1 & 1° & -1 & 0 & | & 1 \\ 0 & 1 & 2 & 1 & | & 4 \\ 0 & 1 & -2 & 0 & | & f+2 \end{bmatrix} \xrightarrow{\text{row}} \begin{bmatrix} 1 & 1 & -1 & 0 & | & 1 \\ -1 & 0 & 3 & 1 & | & 3 \\ -1 & 0 & -1 & 0 & | & f+1 \end{bmatrix}.$$

Step F. (Return to step C.) None of the first four entries of the last row of M_2 are positive, hence maximum $f = -1$ at $(0, 1, 0, 3)$. Since maximum $(-2x_1 - x_2) = -1$, then minimum $(2x_1 + x_2) = +1$. ∎

APPLICATIONS

Example 2

An application of linear programming to a probability problem is described in this example. Suppose that two types of missiles (say A and B) are available to fire at two types of targets, say, target 1 and target 2. Let

x_1 = number of A missiles fired at target 1,
x_2 = number of B missiles fired at target 1,
x_3 = number of A missiles fired at target 2,
x_4 = number of B missiles fired at target 2,
$Q_1 = .10$ = probability that a single A missile will fail to destroy target 1,
$Q_2 = .20$ = probability that a single B missile will fail to destroy target 1,
$Q_3 = .15$ = probability that a single A missile will fail to destroy target 2,
$Q_4 = .25$ = probability that a single B missile will fail to destroy target 2,
$C_1 = \$20{,}000$ = cost of a single A missile,
$C_2 = \$15{,}000$ = cost of a single B missile,
$E = \$1{,}000{,}000$ = maximum amount that can be spent on a given engagement.

If no more than 20 missiles of each type, and no more than 30 missiles altogether can be expended on a given engagement, determine the number of missiles (quantities $x_1, x_2, x_3,$ and x_4) that will minimize the overall probability of failure, which is

$$Q = Q_1^{x_1} Q_2^{x_2} Q_3^{x_3} Q_4^{x_4}.$$

Because of certain properties of logarithms,† the last equation can be transformed to

$$\ln Q = x_1(\ln Q_1) + x_2(\ln Q_2) + x_3(\ln Q_3) + x_4(\ln Q_4),$$

which for constant values of Q produces a family of linear equations. Since Q decreases if and only if $\ln Q$ decreases, minimum Q occurs when $\ln Q$ is a minimum. Thus the original mathematical model has been converted into

† Logarithms and their properties are discussed in Chapter 14.

a linear programming model, namely,

$$\text{Minimize}^\dagger \ (-2.3x_1 - 1.6x_2 - 1.9x_3 - 1.4x_4), \text{ subject to}$$

$$20{,}000x_1 + 15{,}000x_2 + 20{,}000x_3 + 15{,}000x_4 \le 1{,}000{,}000,$$
$$x_1 + x_2 + x_3 + x_4 \le 30,$$
$$x_1 + x_3 \le 20,$$
$$x_2 + x_4 \le 20,$$

and $X \ge 0$.

After dividing the first inequality by 1000, adding four slack variables, and changing to a maximum problem (minimum f = maximum $-f$), we obtain the initial augmented matrix M_0:

$$\begin{bmatrix} 20 & 15 & 20 & 15 & 1 & 0 & 0 & 0 & | & 1000 \\ 1 & 1 & 1 & 1 & 0 & 1 & 0 & 0 & | & 30 \\ 1 & 0 & 1 & 0 & 0 & 0 & 1 & 0 & | & 20 \\ 0 & 1 & 0 & 1 & 0 & 0 & 0 & 1 & | & 20 \\ \hline +2.3 & +1.6 & +1.9 & +1.4 & 0 & 0 & 0 & 0 & | & f-0 \end{bmatrix}.$$

After three iterations it is possible to determine that the number of missiles that will minimize the overall probability of failure is

$$x_1 = 20, \quad x_2 = 10, \quad x_3 = 0, \quad \text{and} \quad x_4 = 0.$$

The conversion of models illustrated in this example illustrates one of the strengths of mathematical modeling. ∎

Example 3

The last example need not be restricted to missiles fired at targets. Actually, the model will serve any similar situation in which m "programs" are available for use to solve n "problems," where the probabilities of failure are known and where the constraints are linear. For example, one may have m chemicals available to attack n diseases or m plans available to apply to n social problems. ∎

EXERCISES

Suggested minimum assignment: Exercises 1, 5, 7, 9, and 13.

In each of Exercises 1–8 apply the simplex method to solve the problem.

1. Maximize $(x_1 + 2x_2)$ subject to

$$x_1 + x_2 + x_3 = 4,$$
$$x_1 + 4x_2 + x_4 = 7,$$ and $X \ge 0$.

---

† The coefficients of the objective function are obtained from a table of natural logarithms: $\ln(.10) \approx -2.3$; $\ln(.20) \approx -1.6$; $\ln(.15) \approx -1.9$; $\ln(.25) \approx -1.4$.

2. Maximize $(x_1 - 3x_2)$ subject to

$$\begin{aligned} -x_1 + x_2 + x_3 &= 1, \\ x_1 + x_2 \phantom{{}+x_3} + x_4 &= 1, \end{aligned} \quad \text{and} \quad X \geq 0.$$

3. Maximize $(2x_1 + x_2 + 6x_3 + x_4)$ subject to

$$\begin{aligned} x_1 + 3x_2 + x_3 + x_4 &\leq 4, \\ x_1 \phantom{{}+3x_2} + x_3 + 2x_4 &\leq 5, \\ x_2 + x_3 \phantom{{}+2x_4} &\leq 2, \end{aligned} \quad \text{and} \quad X \geq 0.$$

4. Maximize $(4x_1 - 6x_2 + 5x_3)$ subject to

$$\begin{aligned} -x_1 + x_2 \phantom{{}+2x_3} &\leq 1, \\ x_2 + 2x_3 &\leq 4, \\ 2x_1 \phantom{{}+x_2} + x_3 &\leq 6, \\ 2x_2 + x_3 &\leq 1, \end{aligned} \quad \text{and} \quad X \geq 0.$$

5. Maximize $(2x_1 + x_4)$ subject to

$$\begin{aligned} x_1 + x_2 + 2x_3 + x_4 &= 2, \\ x_1 + 2x_2 + x_3 &\leq 5, \end{aligned} \quad \text{and} \quad X \geq 0.$$

6. Minimize $(3x_1 + x_3)$ subject to

$$\begin{aligned} x_1 + x_2 + x_3 &= 3, \\ x_1 + 2x_2 \phantom{{}+x_3} + x_4 &= 7, \end{aligned} \quad \text{and} \quad X \geq 0.$$

7. Minimize $(2x_1 + x_2)$ subject to

$$\begin{aligned} x_1 + x_2 &\geq 1, \\ -x_1 + x_2 &\leq 1, \end{aligned} \quad \text{and} \quad X \geq 0.$$

8. Minimize $(x_1 + x_2 + 4x_3 + x_4)$ subject to

$$\begin{aligned} -x_1 - x_2 + x_3 + x_4 &= 2, \\ -x_1 + x_2 + 2x_3 &\geq 1, \end{aligned} \quad \text{and} \quad X \geq 0.$$

(*Hint:* Subtract a slack variable from the left side of the second constraint.)

In each of Exercises 9–12 show that the simplex method determines that there is no optimal solution. (*Hint:* Read the last sentence of step *D*.) In each case demonstrate that this determination is correct by means of a graph.

9. Maximize $(2x_1 + x_2)$ subject to

$$\begin{aligned} -x_1 + x_2 &\leq 1, \\ x_1 - 2x_2 &\leq 2, \end{aligned} \quad \text{and} \quad X \geq 0.$$

10. Maximize $(x_1 + x_2)$ subject to

$$\begin{aligned} -2x_1 + x_2 &\leq 4, \\ x_1 - 3x_2 &\leq 6, \end{aligned} \quad \text{and} \quad X \geq 0.$$

11. Minimize $(x_1 - 2x_2)$ subject to

$$x_1 - x_2 \leq 2,$$
$$3x_1 - x_2 \geq -3,$$ and $X \geq 0.$

12. Minimize $(x_1 - x_2)$ subject to

$$2x_1 - 5x_2 \leq 0,$$
$$-3x_1 + x_2 \leq 6,$$ and $X \geq 0.$

13. (a) Show that the restrictions cannot be met for the problem:
Maximize $(-x_1 + x_2)$ subject to

$$x_1 + x_2 \leq 3,$$
$$x_1 - x_2 \geq 5,$$ and $X \geq 0.$

(b) Illustrate graphically that the feasible set is empty.

NEW VOCABULARY

feasible set of a system of linear constraints 6.1
feasible solution 6.1
nonnegativity constraints 6.1
structural constraints 6.1
objective function 6.2
objective equation 6.2
slack variable 6.3
basic solution 6.3

basic variables 6.3
nonbasic variables 6.3
basic feasible solution 6.3
optimal solution 6.3
inequality of matrices 6.5
simplex method 6.6
pivot column 6.7
pivot row 6.7
pivot 6.7

7 Game Theory

Game theory is a relatively new area of mathematics that is of considerable interest in some disciplines, particularly the social and managerial sciences. One reason for this interest is that game theory furnishes methods for quantification and analysis of certain conflict situations in which two or more adversaries compete. Examples of such conflict situations include political, economic, and personal competition, plus warfare and competition against nature. The analysis of the behavior of the competitors themselves can be of some interest in psychology and sociology. Many of the problems to which game theory may apply, however, are very complex, and much work remains to be done, both in developing new theory and in applying present theory to applied problems.

Prerequisites: 1.1, 1.2, 4.1–4.3, 5.1–5.3, 6.1, 6.2; some Exercises and Applications in 7.3 and 7.4 require 6.3–6.6, but those Exercises and Applications can be omitted.
Suggested Sections for Moderate Emphasis: 7.1–7.3.
Suggested Sections for Minimum Emphasis: 7.1, 7.2.

7.1 Games and Strategies

In this introduction to the subject of game theory, we shall limit our consideration to a specific class of games known as two-person, zero-sum games; Example 1 is an example of such a game.

Example 1

One of the simplest of all games is that of matching pennies. The game consists of two players, who we will call player R (for row) and player C (for column), and a set of rules: (1) Each player has a penny which he must lay on a flat surface so that one side of the coin is exposed; (2) if both players produce a head, or if both produce a tail, on their respective coins, then R wins a penny from C; (3) if either player turns up a head and the other player turns up a tail, then R will lose a penny to C. In order to analyze the game just described, we shall construct a mathematical model of the game and analyze it (the concept of a mathematical model is discussed further in Sections 12.4 and 12.5). The model is called a **2 by 2 matrix game** which includes two players and a **payoff matrix** (or **game matrix**). The payoffs of the penny matching game are exhibited as entries in the payoff matrix

$$A = \begin{bmatrix} +1 & -1 \\ -1 & +1 \end{bmatrix} \begin{matrix} \text{Heads} \\ \text{Tails} \end{matrix} \Bigg\} \text{Player } R.$$

$$\overbrace{\begin{matrix} \text{Heads} & \text{Tails} \end{matrix}}^{\text{Player } C}$$

A **play** of the matrix game is a choice of a row by R (the row player) and the choice of a column by C (the column player). Notice that the entries of the payoff matrix represent **payoffs** from the *row player's point of view*; therefore, the entry -1 represents a loss to R and a gain for C. It is customary in game theory to always express the payoffs from the row player's point of view. After each play, R receives a payoff from C of an amount equal to the value of the entry that is common to the chosen row and column. If that entry is negative, then one can think of the payoff as a bill to R from C of money owed to C. As an illustration suppose that R chooses row 1 and C chooses column 2 of the payoff matrix; then the payoff to R is -1, which means that R owes C 1 penny. In many games, such as the penny-matching game, it is agreed that R's loss is C's gain, and vice versa; such a game is called a **zero-sum game**. ∎

From here on we shall concentrate on matrix games that serve as mathematical models for other zero-sum, two-person games. If in a matrix game the column choice of player C is predictable, then it is obvious that R, if intelligent, will adjust his choice so that he increases his payoff if possible. But if C is also intelligent, we can assume that he too will make a choice to his advantage. The natural question to ask is whether there is some way that a player could determine his play so as to optimize the payoff against an opponent whose responses are unpredictable. The answer to this question is the subject of the remainder of this chapter and is based on the concepts of **expected value** (Section 4.3) and **strategy** (explained and defined next).

Example 2
In the game of matching pennies one common way of determining the play of "heads" or "tails" is to flip the penny and let the choice be determined by chance. A determination of one's play in this manner is an illustration of a **strategy:** For such a strategy there is a probability of $\frac{1}{2}$ of obtaining a "head" and a probability of $\frac{1}{2}$ of obtaining a "tail"; hence this strategy can be represented by the vector $(\frac{1}{2}, \frac{1}{2})$. ∎

▶ **Definition 1**
Let p_i represent the probability that the ith row of an m by n payoff matrix will be played by the row player. The probability vector $(p_1, p_2, \ldots, p_m)$, where $p_1 + p_2 + \cdots + p_m = 1$, is called the **strategy** of the row player. Let q_j represent the probability that the jth column of an m by n payoff matrix will be played by the column player. The probability vector $(q_1, q_2, \ldots, q_n)$, where $q_1 + q_2 + \cdots + q_n = 1$, is called the **strategy** of the column player.

Example 3
In the penny-matching game of Example 1, suppose that the row player adopts a strategy of $(\frac{1}{4}, \frac{3}{4})$. Such a strategy simply means that R decides to determine his play in such a way that there is $\frac{1}{4}$ probability that he will choose a head and a $\frac{3}{4}$ probability that he will choose a tail. Such a strategy can be executed by the construction of an experiment consisting of the selection of one of two outcomes having respective probabilities of $\frac{1}{4}$ and $\frac{3}{4}$. Experiments having such outcomes are easily designed; for example, a simple spinner similar to that found in children's games and shown in Figure 7.1 can serve as the instrument of such an experiment. An unrehearsed glance at the location of the second hand on a wristwatch will serve as a make-shift spinner (see Figure 7.2). For example, if the spinner (second hand) is on 1–3, R plays heads; if it is on 4–12 he plays tails. ∎

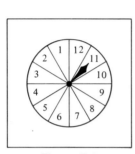

Figure 7.1   **Figure 7.2**

At first it may seem irresponsible to entrust the selection of a play of a game to a chance device. This is not the case at all, however. As we shall illustrate later, a strategy based on a chance device affords a means of optimizing a player's expectation, regardless of the strategy of the opponent. Furthermore, in most games it is vital that the opponent not be able to anticipate your play; a chance device assures the element of surprise.

By **an expected value of the game to a player,** we mean an expected value for an experiment having outcomes that are the potential payoffs resulting from a single play of the opponent.

Example 4
With a strategy of $(\frac{1}{4}, \frac{3}{4})$ in the penny-matching game, what is an expected value of the game to the row player? The answer to this question depends upon the column choice of the opponent. If the opponent's choice is the first column,

$$\begin{bmatrix} +1 \\ -1 \end{bmatrix},$$

then the possible outcomes are $(+1)$ and (-1) and the expected value to R with strategy $(\frac{1}{4}, \frac{3}{4})$ is

$$\tfrac{1}{4}(+1) + \tfrac{3}{4}(-1) = -\tfrac{1}{2}.$$

If, however, the opponent chooses the second column

$$\begin{bmatrix} -1 \\ +1 \end{bmatrix},$$

then the possible outcomes are (-1) and $(+1)$ and the expected value to R with strategy $(\frac{1}{4}, \frac{3}{4})$ is

$$\tfrac{1}{4}(-1) + \tfrac{3}{4}(+1) = +\tfrac{1}{2}.$$

Although there are two expected values for R's given strategy in this game, it is the smaller of the two expected values that R is concerned about. In Section 7.3 we shall develop a method for determining a strategy that will maximize the smallest expected value to R. ∎

In our discussion of game theory it is assumed that the primary objective of a player is to obtain a strategy that will make his worst expected value as favorable as possible.

▶ **Definition 2**
A strategy of the row player that maximizes his worst (smallest) expected value is called an **optimal strategy** of the row player. A strategy of the column player that minimizes his worst (largest) expected value is called an **optimal strategy** of the column player.

Formulas for calculating optimal strategies for *certain* 2 by 2 matrix games are developed in Exercise 8. General methods for calculating optimal strategies of larger zero-sum matrix games are discussed in the remaining sections of this chapter.

APPLICATIONS

Example 5
The electoral votes of two large states A and B are at stake in a presidential election. Candidate R is leading over candidate C in both states having respective electoral votes of 40 and 30 votes each. As the campaign draws to a close, each candidate has the time and resources to visit exactly one of these states. Candidate R can be expected to successfully defend either state in a face-to-face confrontation, but candidate R can be expected to lose either state that C alone visits. The problem facing candidate R is what strategy will maximize his least expected value against an intelligent opponent. The following matrix game is a mathematical model of the real game situation. The payoff matrix† is

---

† To assure a zero-sum game, the assumption is made that C originally had all the electoral votes; thus when a state is carried by R, the payoff represents R's gain and C's loss, but if a state is not carried by R, then there is no movement of electoral votes.

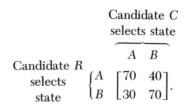

The reader will be asked to verify in Exercise 8 that the best strategy for R is $(\frac{4}{7}, \frac{3}{7})$ and that the best strategy for C is $(\frac{3}{7}, \frac{4}{7})$. The reader should recognize that similar models can be constructed to simulate more complicated situations in which any number n of states are visited by the candidates. ∎

Example 6

A special type of game situation is one in which an individual is competing against a set of possible events rather than against another thinking individual with options. For example, suppose that a person wishes to make a long distance telephone call; the night charge for a station-to-station call is $3 and the night charge for a successful person-to-person call is $5. If the caller is unsuccessful in getting any answer that night, he knows that he can reach the individual the next day using a station-to-station call at a rate of $4. If, however, someone other than the desired person answers that night, then the caller can, through the operator, determine a time when a subsequent station-to-station night call will be successful.

The problem facing the caller is what strategy will minimize his expected expenditure. The following matrix game is a mathematical model of the real game situation.

		Individual answers	No one answers	Another person answers
Caller chooses	Person-to-person call	−$5	−$4	−$3
	Station-to-station call	−$3	−$4	−$6

In this particular game, the row player (the caller) is competing against a set of three uncoordinated events; in such games it is customary to refer to such a nonhuman opponent as "Nature." Assume that the game is a zero-sum game: This assumption requires that the opponent includes the recipient of the payoff (the phone company).

The reader will be asked to verify in Exercise 10 of Section 7.3 that the best strategy for the caller is $(\frac{3}{5}, \frac{2}{5})$. The latter strategy assumes that the caller has no knowledge of the probabilities of occurrence of each of the three columns. ∎

Example 7

On pp. 434–445 of Kemeny et al. [39] the author introduces the model of an expanding economy which was originally proposed by John von Neumann.

The presentation attempts to illustrate how finite mathematics in general, and game theory in particular, can be used in an economic problem. A further reference to this model is Kemeny et al. [37]. ∎

EXERCISES

Suggested minimum assignment: Exercises 1, 2, 3, and 4.

1. Consider a penny-matching game similar to Example 1 except that player R wins 4 cents if both coins show heads and loses 2 cents if both coins show tails; if the coins differ, R loses 1 cent as before.
    (a) Construct the payoff matrix of a matrix game that serves as a mathematical model of the matching game.
    (b) If on one play of the game R chooses the first row and C chooses the second column, what is the payoff? To whom is the payoff made?
    (c) If player R adopts a strategy of $(\frac{1}{2}, \frac{1}{2})$, what are the expected values to R for each column?
    (d) If player R adopts a strategy of $(1, 0)$, what are the expected values to R for each column?
    (e) If player R adopts a strategy of $(0, 1)$, what are the expected values to R for each column?
    (f) Which of the three strategies in parts (c)–(e) is the "safest" for R to use? (That is, which has the most favorable, least expected value?)

2. Two adversaries are engaged in a competitive situation in which they compete for two positions A and B, with payoffs from loser to winner of 3 and 4 units, respectively. Both adversaries attack exactly one position. If C attacks the position that R does not choose to attack, then C will prevail; R will prevail at every position that R attacks. Positions that are not attacked remain neutral.
    (a) Construct the payoff matrix of a matrix game that serves as a mathematical model of the game just described.
    (b) If on one play of the matrix game R chooses the second row and C chooses the second column, what is the payoff? To whom is the payoff made?
    (c) If R adopts a strategy of $(\frac{1}{3}, \frac{2}{3})$, what are the expected values to R for each column?
    (d) If R adopts a strategy of $(1, 0)$, what are the expected values to R for each column?
    (e) If R adopts a strategy of $(0, 1)$, what are the expected values to R for each column?
    (f) Which of the three strategies in parts (c)–(e) is the "safest" for R to use? (That is, which has the most favorable, least expected value?)

3. Use Exercise 1 to determine the expected values to C for each row if player C adopts a strategy of
    (a) $(\frac{1}{2}, \frac{1}{2})$.   (b) $(1, 0)$.   (c) $(0, 1)$.
    (d) Which of the strategies of parts (a)–(c) is the "safest" for C to use? (That is, which has the most favorable, worst expected value?)

4. Use Exercise 2 to determine the expected value to C for each row if C adopts a strategy of
    (a) $(\frac{1}{2}, \frac{1}{2})$.   (b) $(1, 0)$.   (c) $(0, 1)$.
    (d) Which of the strategies of parts (a)–(c) is the "safest" for C to use? (That is, which has the most favorable, worst expected value?)
5. In Example 5 of the text determine the expected value to R for each column if the strategy designated below is adopted by R.
    (a) $(\frac{1}{2}, \frac{1}{2})$.   (b) $(1, 0)$.   (c) $(0, 1)$.   (d) $(\frac{4}{7}, \frac{3}{7})$.
6. In Example 6 of the text determine the expected value to R for each column if the strategy designated below is adopted by R.
    (a) $(\frac{1}{3}, \frac{2}{3})$.   (b) $(1, 0)$.   (c) $(0, 1)$.   (d) $(\frac{3}{5}, \frac{2}{5})$.
7. In Example 5 of the text determine the expected value to C for each row if the strategy designated below is adopted by C.
    (a) $(\frac{1}{4}, \frac{3}{4})$.   (b) $(1, 0)$.   (c) $(0, 1)$.   (d) $(\frac{3}{7}, \frac{4}{7})$.
8. Consider a 2 by 2 zero-sum matrix game with payoff matrix

$$\begin{bmatrix} a_{11} & a_{12} \\ a_{21} & a_{22} \end{bmatrix}.$$

If an unspecified strategy of the row player is $(p, 1 - p)$, then R's two expected values (depending upon C's play) are

$$a_{11}p + a_{21}(1 - p) \quad \text{and} \quad a_{12}p + a_{22}(1 - p).$$

Provided that the optimal strategy is not $(1, 0)$ or $(0, 1)$, one can show (after studying Section 7.3) that an optimal strategy occurs when these two expected values are equal. Thus we have

$$a_{11}p + a_{21}(1 - p) = a_{12}p + a_{22}(1 - p).$$

Solve this equation for p and for $1 - p$, and call the results p° and $1 - p^\circ$. Thus we have

(optimal strategy for the row player) = $(p^\circ, 1 - p^\circ)$

$$= \left(\frac{a_{22} - a_{21}}{a_{11} - a_{12} - a_{21} + a_{22}}, \frac{a_{11} - a_{12}}{a_{11} - a_{12} - a_{21} + a_{22}} \right).$$

We can find the optimal strategy for the column player in the same manner [provided that it is not $(1, 0)$ or $(0, 1)$].

(optimal strategy for the column player) = $(q^\circ, 1 - q^\circ)$

$$= \left(\frac{a_{22} - a_{12}}{a_{11} - a_{12} - a_{21} + a_{22}}, \frac{a_{11} - a_{21}}{a_{11} - a_{12} - a_{21} + a_{22}} \right).$$

Use the formulas just developed to determine the optimal strategy for the row player and the optimal strategy for the column player for the matrix game of Example 5.

9. Use the formulas of Exercise 8 to verify that the optimal strategy of both players in Example 1 is $(\frac{1}{2}, \frac{1}{2})$. In other words, verify that the best

way to play the game of matching pennies with payoff matrix $\begin{bmatrix} 1 & -1 \\ -1 & 1 \end{bmatrix}$ is to determine one's play by flipping the coin.

10. Two adversaries are engaged in a competitive situation in which they compete for two positions A and B worth 3 and 4 units, respectively.† Assume that C holds both positions initially, but can defend only one of them. Assume that R can attack only one position. If R attacks the position that C does not choose to defend, then R will prevail; otherwise C will maintain control. (a) Construct the payoff matrix of a matrix game that serves as a mathematical model of the game just described. (b) Use Exercise 8 to determine the optimal strategy of the row player. (c) Use Exercise 8 to determine the optimal strategy of the column player.

7.2 Pure Strategies

A strategy having 1 as one component and 0 for all the other components is called a **pure strategy**. Strategies that are not pure are called **mixed strategies**. Obviously, a pure strategy dictates the choice of a specific row (or column) on a single play. There are some matrix games in which a pure strategy is an optimal strategy. By that we mean that the pure strategy will maximize the smallest expected value to the row player (or minimize the largest expected value to the column player); the primary purpose of this section is to investigate games in which a pure strategy is optimal.

Example 1
From the point of view of the row player R, consider the matrix game

$$A = \begin{bmatrix} 3 & 2 & 4 \\ 5 & 1 & 0 \\ 7 & 1 & -4 \end{bmatrix}.$$

If R selects the pure strategy $(1, 0, 0)$, then the expected values, depending upon his opponent's choice of columns, are

$$1(3) + 0(5) + 0(7) = 3$$

and

$$1(2) + 0(1) + 0(1) = 2$$

and

$$1(4) + 0(0) + 0(-4) = 4.$$

Notice that the expected values are equal to the respective entries of the first row. Thus if R adopts a pure strategy of $(1, 0, 0)$, the smallest payoff

† There is a distinction between this exercise and Exercise 2, in that here the position itself is the payoff and hence must be assumed to belong to one of the players in order that one player's gain is the other's loss (zero-sum requirement).

would be 2, and could be larger if his opponent chooses a column other than the second column. If, however, R selects a pure strategy of $(0, 1, 0)$, then the expected values to R similarly can be shown to be equal to the respective entries of the second row, and the smallest of these values is 0. Finally, if R selects the remaining pure strategy $(0, 0, 1)$, then the expected values are equal to the respective entries of the third row and the smallest of these values is -4. These smallest expected values corresponding to each of the three pure strategies are illustrated in Figure 7.3.

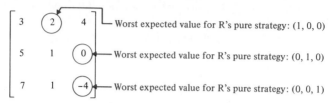

Figure 7.3. *Identification of worst expected value for each pure strategy of R.*

Notice in Figure 7.3 that the smallest or worst expected value of each pure strategy is simply the smallest entry in the respective row, and each represents the worst that can happen to R for the respective pure strategy. Obviously from R's standpoint, the best of the pure strategies is the one that produces the largest of these worst expected values, namely, 2. In other words, R can assure himself a minimum expected value of 2 if he will choose that pure strategy that produces the largest of the worst expected values shown in Figure 7.3. Such a strategy will be called **R's best pure strategy.**

Next consider the same game from C's standpoint, remembering that positive numbers represent losses to C. The worst expected value that C would suffer if he chose the pure strategy $(1, 0, 0)$ is 7 (actually a loss of 7 from C's point of view), which is the largest entry in the first column. The worst expected values to C corresponding to all of the pure strategies of C are shown in Figure 7.4.

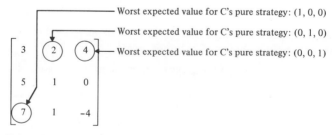

Figure 7.4. *Identification of worst expected value for each pure strategy of C.*

In Figure 7.4 notice that for each pure strategy, the largest or worst expected value to C is simply the largest entry in the respective column and represents the worst that can happen to C for the respective pure strategy chosen. Obviously, from C's standpoint the best of the pure strategies is the one that produces the smallest of these worst expected values, namely, 2. In

other words, C can assure himself of an expected value of no worse than 2 (minimum loss to C) if he will choose that pure strategy that produces the least of the worst expected values shown in Figure 7.4. Such a strategy will be called **C's best pure strategy**. ∎

Notice that for the best pure strategies of Example 1, the corresponding expected values for both R and C were the same, namely, 2; as we shall see in the next example, it is not always the case that the expected values corresponding to the best pure strategies of R and C are the same, but when it *is* the case we say that the game is **strictly determined**.

▶ **Definition 3**
If a matrix has an entry that is the common expected value for the best pure strategies of both players, then the game is said to be **strictly determined**.

Strictly determined games are important because it can be proved that in such games the best pure strategies of R and C are optimal strategies. In other words, in a strictly determined game, the common expected value corresponding to the best pure strategy of each player is as good as the worst expected value produced by any pure or mixed strategy of each player. In Example 1 the game was strictly determined because the best expected value for the best pure strategies of both players was the same, namely, $a_{12} = 2$. Because the game in Example 1 is strictly determined, then we know that the optimal strategies are the pure strategies $(1, 0, 0)$ for R and $(0, 1, 0)$ for C.

In the next example we illustrate a matrix game that is not strictly determined.

Example 2
Consider the matrix game having a payoff matrix

$$\begin{bmatrix} 3 & 2 & 1 \\ 0 & -2 & 4 \\ 7 & -3 & 5 \end{bmatrix}.$$

The worst (smallest) expected value to the row player R for each pure strategy is circled in Figure 7.5; the best (largest) of these circled expected values is underlined. For the column player C, the worst (largest) expected

$$\begin{bmatrix} 3 & 2 & ① \\ 0 & ㊀2 & 4 \\ 7 & ㊂3 & 5 \end{bmatrix} \begin{matrix} \text{Row} \\ \text{minima} \\ 1 \\ -2 \\ -3 \end{matrix}$$

Figure 7.5. *Worst expected value for each pure strategy of R.*

value for each pure strategy is boxed in Figure 7.6; the best (smallest) of these boxed values is underlined. For the pure strategies of both players, the best of the worst of the expected values are *not* the same; hence the game is not strictly determined, and it can be shown that the optimal strategy is a mixed strategy rather than a pure strategy. ∎

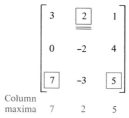

$$\begin{bmatrix} 3 & \boxed{2} & 1 \\ 0 & -2 & 4 \\ \boxed{7} & -3 & \boxed{5} \end{bmatrix}$$

Column
maxima   7   2   5

Figure 7.6. *Worst expected value for each pure strategy of C.*

APPLICATIONS

Example 3

Consider an arrangement of 4 towns and 3 shopping centers as shown in Figure 7.7. Each of two firms is planning to open competitive stores in some one of the 3 shopping centers. Both firms, called R and C, agree that the store built by R will get 90 per cent of the customers who are much closer to R's store and 20 per cent of the customers who are much closer to C's store. Of the customers who are about the same distance from both stores, R will get 60 per cent. The question facing each firm is where to locate each store. A matrix game can be constructed to model the decision facing manage-

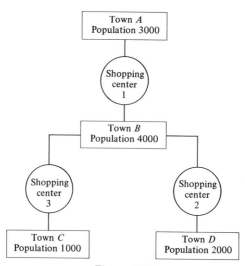

Figure 7.7

ment. The payoff matrix of such a game is

$$A = \begin{bmatrix} 6000 & 6100 & 6500 \\ 5400 & 6000 & 6200 \\ 5100 & 5500 & 6000 \end{bmatrix}.$$

To illustrate the calculations of the entries of matrix A, consider the entry a_{12}, which assumes the location of R's store in shopping center 1 and C's store in shopping center 2.

$$a_{12} = (90\% \text{ of } 3000) + (60\% \text{ of } 5000) + (20\% \text{ of } 2000)$$
$$= 6100.$$

In Figure 7.8 the worst expected value corresponding to each pure strategy of R is circled. Remember that for each pure strategy the expected values are simply the entries of the chosen row and the worst expected value in each row is the smallest entry in each row. In Figure 7.9 the worst expected value corresponding to each pure strategy of C is circled.

$$\begin{bmatrix} \text{\textcircled{6000}} & 6100 & 6500 \\ \text{\textcircled{5400}} & 6000 & 6200 \\ \text{\textcircled{5100}} & 5500 & 6000 \end{bmatrix}$$

Figure 7.8. *Worst expected value for each pure strategy of R.*

$$\begin{bmatrix} \text{\textcircled{6000}} & \text{\textcircled{6100}} & \text{\textcircled{6500}} \\ 5400 & 6000 & 6200 \\ 5100 & 5500 & 6000 \end{bmatrix}$$

Figure 7.9. *Worst expected value for each pure strategy of C.*

For the pure strategies of firm R the best of the worst expected values is 6000, which came from the entry a_{11}. For the pure strategies of firm C the best of the worst expected values is also 6000, which again came from the entry a_{11}. Because these two values are the same, the game is strictly determined and the pure strategy $(1, 0, 0)$ is optimal for both firms. Hence both firms should place their stores in shopping center 1. ∎

Example 4
A patient has an undiagnosed disease which is one of 3 types. Two types of medication are available. The probabilities of a cure under the administration of the medications are given in the payoff matrix of the corresponding matrix game. Assume that the game is a zero-sum game; thus without medication there is 0 probability of a cure.

$$\text{Medication} \begin{cases} & \begin{array}{c} \text{Disease} \\ 1 \quad 2 \quad 3 \end{array} \\ & \begin{array}{c} 1 \\ 2 \\ \text{Both} \end{array} \begin{bmatrix} 1 & \frac{1}{3} & \frac{1}{2} \\ \frac{3}{4} & \frac{3}{4} & \frac{2}{3} \\ 1 & \frac{1}{2} & \frac{1}{2} \end{bmatrix} \end{cases}$$

The reader will be asked to verify in Exercise 9 that when there is no knowledge of the probability of the diseases, then the best strategy for the doctor is the pure strategy $(0, 1, 0)$. Thus the probability for cure is at least $\frac{2}{3}$, which will be the case if medication 2 is administered. The reader is reminded that similar models can be constructed for similar real problems with known probabilities of success when m various methods are used to attack n types of problems. ∎

Example 5

Two partners in business decide to dissolve their partnership by having one sell out to the other. They agree that the business is worth $60,000. Each partner will produce a sealed bid at a prescribed time stating the amount that he will pay the other for his share of the business. The bids will be made in increments of $10,000 with the understanding that the larger bid will be accepted. The maximum bid allowed will be $40,000. In case the bids are the same, they agree to flip a coin with the winner paying the loser $30,000 for full ownership. In calculating the payoff matrix from R's point of view, a successful bid by R must be subtracted from the $30,000 value of the acquired share of the business; a successful bid by C must be added to the $-$$30,000 value of the disposed share of the business. In units of $10,000, the payoff matrix is

$$R\text{'s bid} \begin{cases} & \begin{array}{c} C\text{'s bid} \\ 0 \quad 1 \quad 2 \quad 3 \quad 4 \end{array} \\ & \begin{array}{c} 0 \\ 1 \\ 2 \\ 3 \\ 4 \end{array} \begin{bmatrix} 0 & -2 & -1 & 0 & 1 \\ 2 & 0 & -1 & 0 & 1 \\ 1 & 1 & 0 & 0 & 1 \\ 0 & 0 & 0 & 0 & 1 \\ -1 & -1 & -1 & -1 & 0 \end{bmatrix} \end{cases}$$

This example illustrates a case in which more than one pure strategy may be optimal. In Exercise 10 the reader is asked to verify that a pure strategy of $(0, 0, 1, 0, 0)$ or $(0, 0, 0, 1, 0)$ is optimal for both bidders. The former, $(0, 0, 1, 0, 0)$, is actually preferable for each bidder if the opponent does not choose an optimal strategy. ∎

EXERCISES

Suggested minimum assignment: Exercises 1, 3, 5, and 7.

In each of the following payoff matrices, determine whether the corresponding game is strictly determined or not. If the game is strictly determined, state the optimal strategy of each player and the corresponding payoff.

1. $\begin{bmatrix} 6 & 3 & 4 \\ 1 & 0 & -2 \\ 2 & -1 & 7 \end{bmatrix}$.

2. $\begin{bmatrix} -1 & 7 & -3 \\ -2 & 3 & 5 \\ 0 & 4 & 1 \end{bmatrix}$.

3. $\begin{bmatrix} 4 & 2 & 3 \\ 0 & -1 & 2 \\ 3 & 1 & 6 \end{bmatrix}$.

4. $\begin{bmatrix} 3 & 4 & 5 \\ 2 & 9 & 6 \\ 1 & 8 & 7 \end{bmatrix}$.

5. $\begin{bmatrix} 3 & 5 & 2 & 7 \\ 6 & 3 & 2 & 2 \\ 1 & 4 & 3 & 5 \\ 6 & 4 & 4 & 1 \end{bmatrix}$.

6. $\begin{bmatrix} 3 & 9 & 1 & 6 \\ 0 & 2 & 1 & 5 \\ 7 & 3 & 2 & 1 \\ 1 & 0 & 8 & 4 \end{bmatrix}$.

7. A manufacturer must use a certain component in a device that he is producing. He can purchase a cheap component for $6, and if it goes bad, it will cost him an *additional* $10 to correct the problem. He can, however, purchase a guaranteed component for $8, and if it goes bad, it will cost him only an *additional* $2 labor to fix the defect.
   (a) Write the payoff matrix for a corresponding matrix game. (Assume that the game is a zero-sum game in which the manufacturer is competing against a set of two uncoordinated events; it is customary to refer to such a nonhuman opponent as "nature." Of course, nature includes the recipients of any payoffs from the manufacturer.)
   (b) Is the game strictly determined?
   (c) If the game is strictly determined, what is the optimal strategy of the manufacturer?
8. Suppose that the manufacturer of Exercise 7 considers a third option in which the component can be purchased for $9, and if found defective, it will be corrected at no cost to the manufacturer; moreover, the $9 cost will be refunded.
   (a) Write the payoff matrix for a corresponding matrix game.
   (b) Is the game strictly determined?
9. Verify that the pure strategy (0, 1, 0) is the doctor's optimal strategy for Example 4.
10. In Example 5 verify that the pure strategies (0, 0, 1, 0, 0) or (0, 0, 0, 1, 0) are optimal for each player. Why is the former preferable for each bidder in case the opponent does not use an optimal strategy?

Sec. 7.2] Pure Strategies

7.3 An Optimal Strategy for the Row Player

The purpose of this section is to establish a general procedure for determining an optimal strategy (mixed or pure) for the row player in an m by n zero-sum matrix game. We shall introduce the procedure by means of an example.

Example 1
Consider a matrix game with payoff matrix

$$\begin{bmatrix} 5 & 1 \\ 2 & 4 \end{bmatrix}.$$

Suppose that the row player R adopts a mixed strategy of $(\frac{3}{4}, \frac{1}{4})$; that is, there is a $\frac{3}{4}$ probability that R will choose the first row, and a $\frac{1}{4}$ probability that R will choose the second row. The choice for a single play will be determined by a suitable experiment having two outcomes with probabilities of $\frac{3}{4}$ and $\frac{1}{4}$. Thus *on a single play*, if the opponent chooses the first column, then the expected value of the game to R is

$$\tfrac{3}{4}(5) + \tfrac{1}{4}(2) = \tfrac{17}{4}.$$

If, however, the opponent chooses the second column, then the expected value of the game to R is

$$\tfrac{3}{4}(1) + \tfrac{1}{4}(4) = \tfrac{7}{4}.$$

Hence with the strategy $(\frac{3}{4}, \frac{1}{4})$, R anticipates an expected value of at least $\frac{7}{4}$ on a single play no matter what column the column player selects. It is only natural for R to wonder if there is another strategy that will produce a least expected value larger than $\frac{7}{4}$. Suppose that R lets his strategy be an unspecified (p_1, p_2). For a single play, if the opponent chooses the first column, then the expected value of the game to R is

$$p_1(5) + p_2(2).$$

If, however, the opponent chooses the second column, then the expected value of the game to R is

$$p_1(1) + p_2(4).$$

If we let v be the smaller of these two expected values, then with strategy (p_1, p_2), R anticipates an expected value of at least v no matter which column the opponent selects. Obviously, R would like to maximize v. Of course, this maximization is subject to the constraints that each expected value must be no smaller than v; that is,

$$p_1(5) + p_2(2) \geq v,$$
$$p_1(1) + p_2(4) \geq v.$$

We know that p_1 and p_2 must be nonnegative and the sum of the components of the strategy vector must be 1; that is, $p_1 \geq 0$, $p_2 \geq 0$, and

$p_1 + p_2 = 1$. Moreover, because of these constraints on p_1 and p_2 and because the entries of the payoff matrix are positive, we can say that $v \geq 0$ (actually $v > 0$). Thus we have the linear programming problem:

Maximize v subject to

$$\begin{aligned} 5p_1 + 2p_2 - v &\geq 0, & & & p_1 &\geq 0, \\ p_1 + 4p_2 - v &\geq 0, & & \text{and} & p_2 &\geq 0, \\ p_1 + p_2 &= 1, & & & v &\geq 0. \end{aligned} \quad (7.1)$$

The problem can be solved by the methods of Chapter 6. There is a way, however, to reduce both the number of constraints and the number of independent variables. Because we know that $v > 0$, we can divide each constraint of (7.1) by v and rearrange the terms to obtain

$$5\left(\frac{p_1}{v}\right) + 2\left(\frac{p_2}{v}\right) \geq 1,$$

$$\left(\frac{p_1}{v}\right) + 4\left(\frac{p_2}{v}\right) \geq 1,$$

$$\left(\frac{p_1}{v}\right) + \left(\frac{p_2}{v}\right) = \frac{1}{v}.$$

Since maximizing positive v can be accomplished by minimizing $1/v$, and because we can let $x_1 = p_1/v$ and $x_2 = p_2/v$, the original problem can be restated as:

Minimize $(x_1 + x_2)$ subject to

$$\begin{aligned} 5x_1 + 2x_2 &\geq 1, & & & x_1 &\geq 0, \\ x_1 + 4x_2 &\geq 1, & & \text{and} & x_2 &\geq 0. \end{aligned}$$

Notice that the coefficient matrix of the constraints is simply the transpose of the payoff matrix. Graphically, the solution can be found by Figure 7.10 to be minimum $1/v = \frac{1}{3}$ at $(x_1, x_2) = (\frac{1}{9}, \frac{2}{9})$.

In terms of the original variables p_1, p_2, and v, we find maximum $v = 3$ at $(\frac{1}{3}, \frac{2}{3})$. We designate an optimum v with the symbol v^* and call v^* the **row value** of the game. The strategy at which v^* is obtained is the optimal strategy of the row player and is designated by $(p_1^*, \ldots, p_m^*)$. In summary, if R will use the strategy $(\frac{1}{3}, \frac{2}{3})$, he will always come out with an expected value of at least three per play no matter what column the opponent chooses. ∎

In the procedure introduced in Example 1 it is required that $v > 0$; we can require† $v > 0$ if the payoff matrix has nonnegative entries and does not have a column of zeros. If, however, one encounters a game matrix that does not satisfy these conditions, then the difficulty can be resolved by simply

---

† If the payoff matrix has only *positive* entries, then for *every* strategy the corresponding value of v must be positive. Actually, we only need to be able to constrain v to be positive, and this can be done if $v^* > 0$. (For an illustration, see Example 3, p. 499 of Kemeny [38].) The sufficient condition assumed here is particularly appropriate for the simplex method.

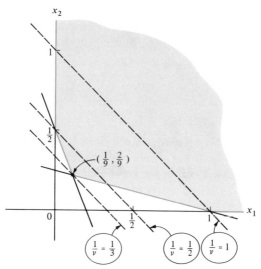

Figure 7.10. *Geometric solution of corresponding linear programming problem.*

adding to every entry a sufficiently large real number; this resolution is possible because it can be proved (Exercise 13) that if the real number k is added to every entry of the original payoff matrix, then the optimal row strategy will remain unchanged, and the row value of the new game will be the row value of the original game plus k.

Thus in general, for a payoff matrix (of the type described in the last paragraph)

$$A = \begin{bmatrix} a_{11} & \cdots & a_{1n} \\ \vdots & & \vdots \\ a_{m1} & \cdots & a_{mn} \end{bmatrix},$$

the optimal row strategy $(p_1^*, \ldots, p_m^*)$ is the solution of the linear programming problem:

Maximize v subject to
$$a_{11}p_1 + \cdots + a_{m1}p_m - v \geq 0,$$
$$\vdots \qquad \vdots \qquad \vdots$$
$$a_{1n}p_1 + \cdots + a_{mn}p_m - v \geq 0, \qquad \text{and}$$
$$p_1 + \cdots + p_m = 1,$$
$$p_1 \geq 0,$$
$$\vdots$$
$$p_m \geq 0,$$
$$v \geq 0,$$

or, if $v > 0$, the corresponding problem where $x_i = p_i/v$:

Minimize $(x_1 + \cdots + x_m)$ subject to
$$a_{11}x_1 + \cdots + a_{m1}x_m \geq 1,$$
$$\vdots \qquad \vdots \qquad \vdots$$
$$a_{1n}x_1 + \cdots + a_{mn}x_m \geq 1,$$
$$x_1 \geq 0,$$
$$\vdots$$
$$x_m \geq 0.$$

The latter formulation of the associated linear programming problem can be expressed in matrix notation as follows. Let

$$C = [1 \quad 1 \quad \cdots \quad 1], \quad X = \begin{bmatrix} x_1 \\ \vdots \\ x_m \end{bmatrix}, \quad \text{and} \quad P_0 = \begin{bmatrix} 1 \\ \vdots \\ 1 \end{bmatrix}.$$

Minimize CX subject to $A^T X \geq P_0$ and $X \geq 0$. If minimum CX occurs at $X = X^*$, then the row value v^* of the associated matrix game with payoff matrix $A \geq 0$ (without a column of zeros) is given by

$$v^* = \frac{1}{\min CX} \quad \text{at} \quad (p_1^*, \ldots, p_m^*) = v^*(x_1^*, \ldots, x_m^*).$$

APPLICATIONS

Example 2

Consider a conflict situation where each adversary has m objects which can be deployed in n positions. The adversary having the greater number of objects in a position will prevail in that position. In case of a tie suppose that player R prevails. As an illustration consider a presidential election campaign in which each candidate has 3 units of money to distribute to two states A and B having 40 and 60 electoral votes, respectively. Let the notation (a, b) denote that a units of money were distributed to state A and b units were distributed to state B. Under the assumption that C originally holds the electoral votes of the states, the payoff matrix in terms of electoral votes gained by R is

		Distributed by C			
		(3, 0)	(2, 1)	(1, 2)	(0, 3)
Distributed by R	(3, 0)	100	40	40	40
	(2, 1)	60	100	40	40
	(1, 2)	60	60	100	40
	(0, 3)	60	60	60	100

In Exercise 11 the reader can verify that for candidate R the row value is $64\frac{12}{13}$ at the optimal strategy $(\frac{8}{65}, \frac{12}{65}, \frac{18}{65}, \frac{27}{65})$. Notice that the optimal strategy for R is preferable to the pure strategy of $(0, 0, 0, 1)$ because an intelligent opponent might select one of the first three columns and thereby reduce the least expected value to R to 60. The optimal strategy of C is left as Exercise 10 of Section 7.4.

Example 3

On p. 101 of Williams [67] an application is given in which a man has 3 options for investing a certain amount of money. The man has estimated the payoffs for each investment for 3 different future economic and political situations. An optimal strategy of the row player is given with the suggestion that the row player could play the odds or invest a mixture of $\frac{5}{17}$ of his principal in one investment and $\frac{12}{17}$ of his principal in another. The author

then gives a brief discussion of the ramifications of the latter departure from the established practice of making a clear-cut choice of a single row with a suitable chance device. ∎

Example 4

A hypothetical military situation is developed on p. 47 of Williams [67]. In that illustration there are two planes on a bombing mission; one carries the bomb and the other carries various electronic equipment. The planes fly in a formation such that one plane is more protected than the other. The alternatives for the offense are

1. To put the bomb in the most favored position.
2. To put the bomb in the least favored position.

The alternatives for the defense, which can get off only one attack, are

1. To attack the most favored position.
2. To attack the least favored position.

The supposition is made that there is a 60 per cent chance the bomb carrier will survive if attacked in the least favored position, a 80 per cent chance of survival if attacked in the most favored position, and a 100 per cent chance of survival if not attacked. It turns out that the optimal strategy for the offense is $(\frac{2}{3}, \frac{1}{3})$ and the row value is $86\frac{2}{3}$ per cent (Exercise 12). Observe that the chance of survival is $6\frac{2}{3}$ per cent in excess of the intuitive action of placing the bomb in the most favored position. ∎

EXERCISES

Suggested minimum assignment: Exercises 1, 3, 5, and 7.

In Exercises 1–4, for the given payoff matrices, formulate a linear programming problem that will maximize the least expected value to R; that is, formulate the linear programming problem that will determine the row value. Do *not* solve the linear programming problem.

1. $\begin{bmatrix} 2 & 1 \\ 3 & 2 \end{bmatrix}$.

2. $\begin{bmatrix} 8 & 1 \\ 4 & 7 \end{bmatrix}$.

3. $\begin{bmatrix} 3 & 0 & 1 \\ 0 & 4 & 1 \\ 1 & 2 & 6 \end{bmatrix}$.

4. $\begin{bmatrix} 1 & 2 & 3 \\ 0 & 4 & 6 \\ 9 & 2 & 1 \end{bmatrix}$.

For the given payoff matrices in Exercises 5–8 determine the row value of the game and the optimal strategy of the row player. Use the graphical method to solve the associated linear programming problem. The theorem stated in Exercise 13 will be needed to obtain the correct result in Exercises 7 and 8.

5. $\begin{bmatrix} 2 & 1 \\ 1 & 2 \end{bmatrix}$.

6. $\begin{bmatrix} 3 & 0 \\ 1 & 4 \end{bmatrix}$.

7. $\begin{bmatrix} 1 & -2 \\ -1 & 2 \end{bmatrix}.$
8. $\begin{bmatrix} 0 & 3 \\ 3 & -1 \end{bmatrix}.$

9. In Example 5 of Section 7.1 use linear programming to verify that $(\frac{4}{7}, \frac{3}{7})$ is the best strategy for R.
10. In Example 6 of Section 7.1 use linear programming to verify that $(\frac{3}{5}, \frac{2}{5})$ is the best strategy for R.
11. Verify that $(\frac{8}{65}, \frac{12}{65}, \frac{18}{65}, \frac{27}{65})$ is the optimal strategy for candidate R in Example 2. [*Hint:* Subtract 40 from every entry of the payoff matrix (see Exercise 13) to shorten the work in this lengthy problem.]
12. In Example 4 of this section write the payoff matrix; also verify the optimal strategy of the offense and the row value.
13. Prove that if the real number k is added to every entry of the original payoff matrix, then the optimal strategy of R will remain unchanged and the row value of the new game will be k plus the row value of the old game. (Answer provided.)

7.4 An Optimal Strategy for the Column Player

In this section we shall establish a general procedure for finding an optimal strategy for the column player in an m by n zero-sum matrix game. As an illustration we shall use the same 2 by 2 matrix game that was used in Example 1 of the last section.

Example 1
Consider the matrix game with payoff matrix

$$\begin{bmatrix} 5 & 1 \\ 2 & 4 \end{bmatrix}.$$

Suppose that the column player C adopts a mixed strategy of $(\frac{1}{3}, \frac{2}{3})$; that is, there is a $\frac{1}{3}$ probability that C will choose the first column and a $\frac{2}{3}$ probability that C will choose the second column. The choice for a single play will be determined by a suitable experiment with a pair of outcomes having probabilities $\frac{1}{3}$ and $\frac{2}{3}$. Thus for a *single play*, if the opponent chooses the first row, then the expected value of the game to C is

$$\tfrac{1}{3}(5) + \tfrac{2}{3}(1) = \tfrac{7}{3},$$

or if the opponent selects the second row, then the expected value of the game to C is

$$\tfrac{1}{3}(2) + \tfrac{2}{3}(4) = \tfrac{10}{3}.$$

Hence with strategy $(\frac{1}{3}, \frac{2}{3})$ C has an expected value of at worst $\frac{10}{3}$ on a single play, no matter which row the row player selects. Notice that we said "at worst $\frac{10}{3}$"; we must remember that to C a payoff of $+\frac{10}{3}$ means a $\frac{10}{3}$ loss because C does the paying.

Of course, C is going to wonder if there is another strategy that will produce a smaller "worst expected value" than $\frac{10}{3}$. Suppose that C lets his strategy be an unspecified (q_1, q_2); for a single play, if the opponent chooses the first row, then the expected value of the game to C is

$$q_1(5) + q_2(1),$$

or if the opponent chooses the second row, then the expected value of the game to C is

$$q_1(2) + q_2(4).$$

If we let w be the *larger* of these two expected values, then with strategy (q_1, q_2), C has an expected value of at worst w no matter which row the opponent selects. Because a positive w represents a loss to C, he obviously would like to minimize w. This minimization is subject to the constraints similar to those of the last section, namely, that each expected value be no larger than w; that q_1, q_2, and w be nonnegative; and that $q_1 + q_2 = 1$. Thus we have another linear programming problem:

Minimize w subject to

$$\begin{array}{lll} 5q_1 + q_2 \leq w, & & q_1 \geq 0, \\ 2q_1 + 4q_2 \leq w, & \text{and} & q_2 \geq 0, \\ q_1 + q_2 = 1, & & w \geq 0. \end{array} \qquad (7.2)$$

As before, we can reduce both the number of constraints and the number of independent variables by dividing each constraint of (7.2) by positive w, by letting $y_1 = q_1/w$, $y_2 = q_2/w$, and by recognizing that minimizing positive w can be accomplished by maximizing $1/w$.

A corresponding linear programming problem is

Maximize $(y_1 + y_2)$ subject to

$$\begin{array}{ll} 5y_1 + y_2 \leq 1, & y_1 \geq 0, \\ 2y_1 + 4y_2 \leq 1, & \text{and} \quad y_2 \geq 0. \end{array}$$

Notice that the coefficient matrix of the constraints is simply the payoff matrix. Graphically, the solution can be found by Figure 7.11 to be

$$\text{maximum } \frac{1}{w} = \frac{1}{3} \text{ at } (y_1, y_2) = \left(\frac{1}{6}, \frac{1}{6}\right).$$

In terms of the original variables q_1, q_2, and w we find minimum $w = 3$ at $(\frac{1}{2}, \frac{1}{2})$. We designate an optimal w with the symbol w^* and call w^* the **column value** of the game. The strategy at which w^* is obtained is the optimal strategy of the column player and is designated $(q_1^*, \ldots, q_n^*)$. In summary, if C will use the strategy $(\frac{1}{2}, \frac{1}{2})$, he will always come out with an expected value (loss to the column player) of no worse than 3 per play no matter what row the opponent chooses. ∎

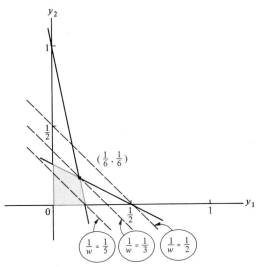

Figure 7.11. *Geometric solution of corresponding linear programming problem.*

In general, the optimal column strategy for the matrix game with payoff matrix $A \geq 0$ (without a row of zeros) can be found as follows: Let

$$C = [1 \ 1 \ \cdots \ 1], \quad Y = \begin{bmatrix} y_1 \\ y_2 \\ \vdots \\ y_n \end{bmatrix}, \quad \text{and} \quad P_0 = \begin{bmatrix} 1 \\ 1 \\ \vdots \\ 1 \end{bmatrix}.$$

Then maximize CY subject to $AY \leq P_0$ and $Y \geq 0$. If maximum CY occurs at $Y = Y^*$, then the column value w^* of the associated matrix game with payoff matrix $A \geq 0$ (without a row of zeros) is given by

$$w^* = \frac{1}{\max CY} \text{ at } (q_1^*, \ldots, q_n^*) = w^*(y_1^*, \ldots, y_n^*).$$

At this point it is natural to wonder whether or not there is any guarantee that every matrix game has an optimal strategy for each player. A famous theorem of game theory states that for every matrix game, there exist optimal strategies for both players. (The proof of this theorem is beyond the scope of this book and may be found in McKinsey [48], pp. 32–37.) Moreover, the theorem provides the remarkable result that the corresponding row and column values are the same; this common value is known as the **value V of the game**. In fact, the value V of the matrix game with payoff matrix A can be calculated by the equation†

$$V = P^*AQ^*,$$

where P^* is the row player's optimal strategy in row matrix form and Q^* is

---

† Although technically P^*AQ^* is a 1 by 1 matrix, it is permissible here to use the matrix and its single entry interchangeably.

Sec. 7.4] An Optimal Strategy for the Column Player

the column player's optimal strategy in column matrix form. (A may or may not have all nonnegative entries.) We reemphasize that when both players use their optimal strategies, then the corresponding expected values of the game coincide. When V is 0, the game is called a **fair game.** When V is positive, the row player has the advantage; when V is negative, the column player has the advantage.

The presentation of game theory presented in this chapter has been necessarily elementary and many useful concepts have been left out. For an elementary discussion of some of these concepts the reader is referred to Chapter 5 of Williams [67]. In that chapter the topics, approximations of optimal strategies (particularly important), dominance, simple solutions, multiple solutions, measurements of payoffs, qualitative payoffs, symmetric games, and non-zero-sum games are discussed in a readable style. In fact, Williams's book [67] serves as a very enjoyable primer in game theory.

APPLICATIONS

Example 2

Consider a network of streets shown in Figure 7.12. Suppose that a crime has been committed at point B and the criminals leave the scene of the crime in a car to avoid capture. The police, on the other hand, who are initially at point A, would like to cross paths with the criminals in order to effect their capture. Assume that each car moves from intersection to intersection in the same time and can make 90° turns but not 180° turns. Further assume that if the criminals can avoid the police for three blocks, they will escape. A meeting on any street or intersection will result in capture. The various paths that each car can travel are identified for reference in Figure 7.13. Using the figure, we calculate the payoff matrix of the associated matrix game by letting a 1 represent a capture and a 0 represent an escape. The resulting payoff matrix follows:

$$P_1 = \begin{bmatrix} 1 & 0 & 0 & 0 & 0 & 0 & 0 & 0 & 0 & 1 \\ 1 & 0 & 0 & 0 & 0 & 1 & 0 & 0 & 1 & 1 \\ 0 & 1 & 1 & 1 & 0 & 0 & 1 & 1 & 1 & 0 \\ 0 & 1 & 1 & 1 & 0 & 0 & 1 & 1 & 1 & 0 \\ 0 & 1 & 1 & 1 & 0 & 1 & 1 & 1 & 1 & 0 \\ 0 & 0 & 0 & 0 & 1 & 1 & 0 & 0 & 0 & 0 \\ 0 & 1 & 1 & 1 & 0 & 0 & 1 & 1 & 1 & 0 \\ 0 & 1 & 1 & 1 & 0 & 0 & 1 & 1 & 1 & 0 \\ 0 & 1 & 1 & 1 & 0 & 0 & 1 & 1 & 1 & 1 \\ 0 & 0 & 0 & 1 & 1 & 1 & 0 & 0 & 0 & 1 \end{bmatrix} \begin{matrix} 1 \\ 2 \\ 3 \\ 4 \\ 5 \\ 6 \\ 7 \\ 8 \\ 9 \\ 10 \end{matrix}$$

with columns 1–10 labeled "Path of criminals" and rows 1–10 labeled "Path of police".

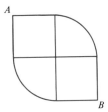

Figure 7.12. *Street arrangement.*

Because each entry of the second row is greater than or equal to the corresponding entry of the first row, the police would never choose row 1 in any optimal strategy. In cases like this we say that row 2 *dominates* row 1 or that row 1 is a *recessive* row. In any optimal strategy the row 1 component therefore has to be 0. Hence row 1 can be deleted from consideration. Row 10 dominates row 6 and row 5 dominates rows 3, 4, 7, and 8. Thus if the corresponding strategy vector components of the recessive rows are assigned the value 0 and the rows are deleted from consideration, the size of the payoff matrix can be reduced.

Next consider the criminal's point of view. He observes that each entry of column 5 is less than or equal to each entry of column 6; hence he would always choose column 5 in preference to column 6. In other words, from the criminal's point of view column 5 dominates column 6 and column 1 dominates column 10. Columns 4 and 9 are dominated by columns 2, 3, 7, and 8, all of which are the same. Thus if he chooses column 2 as the representative of these equal columns, the components of the strategy vector that correspond to columns 3, 4, 6, 7, 8, 9, and 10 are 0, and these columns can be deleted. The result of these considerations of dominance of both rows and columns is that the original payoff matrix can be drastically simplified. It is now

$$P_2 = \begin{matrix} & \begin{matrix} 1 & 2 & 5 \end{matrix} & \\ & \begin{bmatrix} 1 & 0 & 0 \\ 0 & 1 & 0 \\ 0 & 1 & 0 \\ 0 & 0 & 1 \end{bmatrix} & \begin{matrix} 2 \\ 5 \\ 9 \\ 10 \end{matrix} \end{matrix}$$

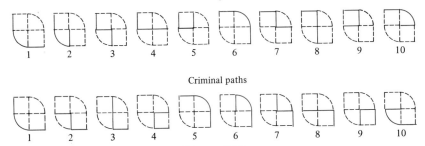

Figure 7.13

Sec. 7.4] An Optimal Strategy for the Column Player

From the reduced game we see that the middle two rows are the same, and hence we can say that either one dominates the other. Consequently, the game can be reduced further to the matrix game for which the payoff matrix is

$$P_3 = \begin{matrix} & 1 & 2 & 5 & \\ & \begin{bmatrix} 1 & 0 & 0 \\ 0 & 1 & 0 \\ 0 & 0 & 1 \end{bmatrix} & \begin{matrix} 2 \\ 5 \\ 10 \end{matrix} \end{matrix}.$$

The methods of this section and the last section can be used to verify that the optimal strategy in this reduced game is $(\frac{1}{3}, \frac{1}{3}, \frac{1}{3})$ for both players (Exercise 11). From this result the original game has an optimal strategy of

$$(0, \tfrac{1}{3}, 0, 0, 0, 0, 0, 0, \tfrac{1}{3}, \tfrac{1}{3}) \quad \text{or} \quad (0, \tfrac{1}{3}, 0, 0, \tfrac{1}{3}, 0, 0, 0, 0, \tfrac{1}{3})$$

for the police. Two optimal strategies for the police arise from the fact that two of the rows of P_2 were the same and undominated by the other rows of P_2. One of the optimal strategies for the criminals is

$$(\tfrac{1}{3}, \tfrac{1}{3}, 0, 0, \tfrac{1}{3}, 0, 0, 0, 0, 0).$$

Four optimal strategies for the criminals arise from the fact that four of the columns of P_1 were the same and were undominated by other columns. The value of the game is $\frac{1}{3}$.

The ideas presented in this example can be extended to apply to larger networks of paths of various designs. ∎

Although a deeper study of game theory ordinarily is required to develop applications, the titles of the following books will indicate to the reader some areas of applications and furnish references as well.

BELLMAN, R., and BLACKWELL, D., "Red Dog, Blackjack Poker," *Scientific American*, 184, 1951, pp. 44–47.

GALE, DAVID, *The Theory of Linear Economic Models*, McGraw-Hill Book Company, New York, 1960.

LUCE, R. D., and RAIFFA, HOWARD, *Games and Decisions: Introduction and Critical Survey*, John Wiley & Sons, Inc., New York, 1957.

MACDONALD, J., *Strategy in Poker, Business, and War*, W. W. Norton & Company, New York, 1950.

RAPOPORT, ANATOL, *Fights, Games, and Debates*, University of Michigan Press, Ann Arbor, Mich., 1960.

RAPOPORT, ANATOL, *Two Person-Game Theory: The Essential Ideas*, University of Michigan Press, Ann Arbor, Mich., 1966.

SHUBIK, MARTIN, *Strategy and Market Structure*, John Wiley & Sons, Inc., New York, 1959.

VON NEUMANN, J., and MORGENSTERN, OSKAR, *Theory of Games and Economic Behavior*, 3rd ed., Princeton University Press, Princeton, N.J., 1953.

EXERCISES

Suggested minimum assignment: Exercises 1, 3, 5, and 7.

In Exercises 1–4, for the given payoff matrices, formulate a linear programming problem that will minimize the greatest expected value to C; that is, formulate a linear programming problem that will determine the column value. Do *not* solve the linear programming problem.

1. $\begin{bmatrix} 2 & 1 \\ 3 & 2 \end{bmatrix}$.

2. $\begin{bmatrix} 8 & 1 \\ 4 & 7 \end{bmatrix}$.

3. $\begin{bmatrix} 3 & 0 & 1 \\ 0 & 4 & 1 \\ 1 & 2 & 6 \end{bmatrix}$.

4. $\begin{bmatrix} 1 & 2 & 3 \\ 0 & 4 & 6 \\ 9 & 2 & 1 \end{bmatrix}$.

In Exercises 5–8, for the given payoff matrices, determine the column value of the game and the optimal strategy of the column player. Using the answers to the same matrix games of Exercises 5–8 of Section 7.3, verify that the column values and row values of each game are the same. A theorem comparable to that given in Exercise 13 of Section 7.3 is needed in Exercises 7 and 8.

5. $\begin{bmatrix} 2 & 1 \\ 1 & 2 \end{bmatrix}$.

6. $\begin{bmatrix} 3 & 0 \\ 1 & 4 \end{bmatrix}$.

7. $\begin{bmatrix} 1 & -2 \\ -1 & 2 \end{bmatrix}$.

8. $\begin{bmatrix} 0 & 3 \\ 3 & -1 \end{bmatrix}$.

9. In Example 5 of Section 7.1 use the method of this section to verify that $(\frac{3}{7}, \frac{4}{7})$ is the optimal strategy for C.
10. In Example 2 of Section 7.3 determine the optimal strategy of C and verify that the row value is the same as the column value of the game. (*Hint:* Subtract 40 from every entry of the payoff matrix to shorten the work in this lengthy problem.)
11. Verify the optimal strategies given in Example 2 of this section.

NEW VOCABULARY

2 by 2 matrix game 7.1
payoff matrix or game matrix 7.1
play of a game 7.1
zero-sum game 7.1
strategy or strategy vector 7.1
an expected value of the game to a player 7.1
optimal strategy 7.1
pure strategy 7.2

mixed strategy 7.2
R's best pure strategy 7.2
C's best pure strategy 7.2
strictly determined games 7.2
row value of the game 7.3
column value of the game 7.4
value of the game 7.4
fair game 7.4

8 Statistics[†]

According to some writers the English haberdasher John Graunt (1620–1674) was the first to realize the usefulness of statistics as a method of academic inquiry. In 1662 Graunt published a book that surveyed vast amounts of data concerning the mortality of various segments of the population. His observations and methods are said to have initiated both statistics and the use of the scientific method in the social sciences. Graunt's work was supported by the writings of his friend Sir William Petty (1623–1685) in a book entitled Political Arithmetic. *The development of insurance companies at the end of the seventeenth century added more stimulus to this new subject. Much later, in 1835, the Belgian L. A. J. Quetelet (1796–1874) revived interest when he published his* Essay on Social Physics; *this work was inspired by the success of the deductive approach in the physical sciences and the lack of success of the same approach in the social sciences. From the time of Quetelet the development of statistics has been dramatic, and today the statistical method is a "standard tool of the trade" in many occupations. In this chapter we offer only a brief introduction to statistics; the interested reader will find some elaboration and a good list of references on pp. 613–635 of [41] Kline.*

Prerequisites: 1.1, 1.2, 3.5–3.9.
Suggested Sections for Moderate Emphasis: 8.1–8.4.
Suggested Sections for Minimum Emphasis: 8.1–8.3.

8.1 Introduction

Statistics may be described as the science of collecting, analyzing, and interpreting data; it is also commonly described as the science of decision making in the face of uncertainty. A statistical novice will often make incorrect interpretations from a set of data, whereas a good statistician knows what methods to use in various situations in order to reach sound conclusions. In this chapter we shall present some of the basic methods that can be used to analyze data in a manner that will lead to the determination of valid conclusions.

[†] Although this chapter is not a prerequisite to the text of the later chapters, those readers who plan to study the applications in the later chapters of calculus to statistics and probability are advised to at least read this chapter.

The data that we obtain must come from a predetermined sample space (universal set) U. For example, if a teacher is analyzing the grades made on a test by a class of 25 students, then U may be considered to be the set of 25 students. We may think of a function X that associates with each student $u \in U$ the test grade $X(u)$ of that student. Such a function that associates a *real number* to each element $u \in U$ is called a **random variable** or **measurement** (for a more detailed discussion, see Example 6 of Section 3.7). In this chapter *we are interested in data that are in the form of real numbers* (the statistician is interested in analyzing the 25 test grades rather than in knowing the names of the 25 students). In statistics the original totality of elements U from which data can be obtained is referred to as a **population**. The set of data themselves (in the form of real numbers) is also often called the population. (We shall find that these two uses of the word "population" will not cause confusion and will enable us to avoid reference to a random variable X in each problem.)

Often it is impossible or impractical to obtain all the data from a population. In this case we work with a carefully chosen finite nonempty subset, called a **sample,** of the population. We analyze the sample in order to draw inferences about the whole population. If a biologist were studying characteristics of a common type of animal, he would use a sample consisting of just a few of the animals. A quality-control worker would select a sample of a few items coming off a production line and make the required observations, rather than try to test every item of the population. A political pollster obtains opinions from a small sample (percentage-wise) of a voting population. On the other hand, a teacher analyzing test grades for a class would probably use the whole population rather than a few sample students from the class. In census-type work an attempt is made to analyze the whole population.

The method of obtaining a sample that is representative of a whole population is a science in itself, and we do not consider this question in depth. Some method of **random sampling** is advisable—that is, some method of selecting a sample in such a way that each element of the population is equally likely to be chosen in the sample.† As with populations, corresponding to the elements of a sample are data in the form of real numbers; a collection of such sample data is also often referred to as a sample. From now on, we shall start with given sample data in the form of real numbers (which we assume have been chosen in a proper manner) and learn something about analyzing the data. A sample may or may not be the whole population. A population may be either finite or infinite, but a given sample will be assumed to consist of a *finite* number n (where n is a positive integer) of real numbers. Each of the n numbers in a sample is also called an **observation**.

In the remainder of this section we shall discuss some methods of displaying sample data in tabular and graphical manners.

† This condition describes what is commonly referred to as simple random sampling.

Example 1

Suppose that a salary survey is made of 32 workers who perform a wide variety of jobs for a given corporation. Each worker is asked to state his or her annual salary in thousands of dollars; salaries are to be rounded to the nearest thousand with any salary such as $10,500 rounded up rather than down. Assume that the following sample of $n = 32$ real numbers is obtained:

$$\begin{array}{cccccccc}
17 & 11 & 7 & 26 & 15 & 6 & 9 & 12 \\
5 & 24 & 9 & 16 & 29 & 9 & 18 & 12 \\
17 & 6 & 9 & 22 & 9 & 33 & 12 & 8 \\
11 & 13 & 15 & 7 & 10 & 30 & 20 & 17
\end{array}$$

The data are displayed in Table 8.1. Each salary from the lowest to the highest is listed and a tally is made of the **frequency** of occurrence of each

Table 8.1. FREQUENCY AND CUMULATIVE FREQUENCY.

Salary (to nearest thousand)	Tally marks	Frequency	Cumulative frequency
5	|	1	1
6	||	2	3
7	||	2	5
8	|	1	6
9	ⅢⅡ	5	11
10	|	1	12
11	||	2	14
12	|||	3	17
13	|	1	18
14		0	18
15	||	2	20
16	|	1	21
17	|||	3	24
18	|	1	25
19		0	25
20	|	1	26
21		0	26
22	|	1	27
23		0	27
24	|	1	28
25		0	28
26	|	1	29
27		0	29
28		0	29
29	|	1	30
30	|	1	31
31		0	31
32		0	31
33	|	1	32

salary. In the last column the frequencies less than or equal to the current salary under consideration are totaled, and these results are called **cumulative frequencies**.

The frequencies shown in Table 8.1 can be displayed graphically. That is done in Figure 8.1 in what may be called a **line chart**. In Figure 8.1 the distribution of the frequencies of the various salaries can be visualized—that is, the **frequency distribution** can be seen.

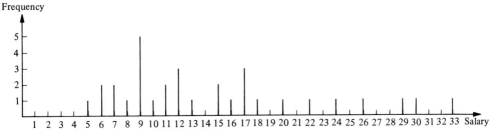

Figure 8.1. *Line chart.*

The cumulative frequencies in Table 8.1 can also be displayed graphically; one method of doing this is to sketch the **cumulative frequency distribution function** shown in Figure 8.2. Note in Table 8.1 that when the salary is 15, the corresponding cumulative frequency is 20; therefore, we plot the point (15, 20) in Figure 8.2, and this point indicates that there are 20 people whose salary in thousands is less than or equal to 15. The ordinate remains at 20 until the salary of 16 is reached—then the ordinate jumps to 21 because there are 21 people whose salary in thousands is less than or equal to 16. An alternative method of presenting the information in Figure 8.2 is to use probabilities ($\frac{1}{32}$, $\frac{2}{32}$, $\frac{3}{32}$, etc.) on the vertical axis instead of cumulative frequencies—this is an important method and will be introduced in Example 8 of Section 9.3. (The percentiles and median shown in Figure 8.2 will be discussed in Section 8.3.) ∎

EXERCISES

Suggested minimum assignment: Exercise 1.

For the sample data in Exercises 1–4, make a table like Table 8.1. Also draw a line chart and a cumulative frequency distribution function.

1. Thirty-six male college freshmen are measured and their height to the nearest inch is recorded.

69	71	68	66	68	64	70	67	69
73	70	67	68	69	61	70	71	69
67	74	68	75	73	68	72	80	71
63	65	65	76	72	64	66	67	68

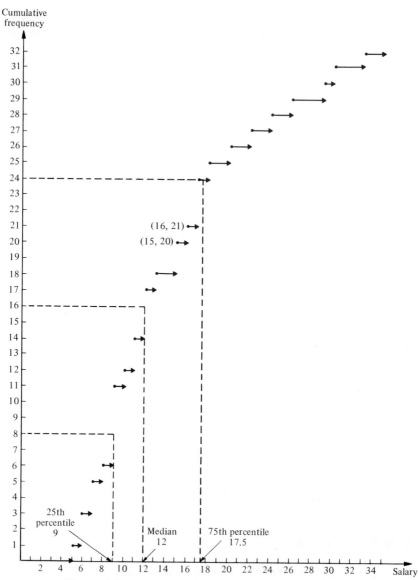

Figure 8.2. *Cumulative frequency distribution function.*

2. A nurse weighs 40 newborn children and records their weight to the nearest tenth of a pound.

6.8	7.3	5.9	8.0	7.4	6.7	6.9	6.1
8.5	7.7	6.1	4.8	5.2	7.3	7.5	6.5
6.3	6.4	7.6	7.0	6.9	5.8	6.0	6.6
8.2	7.7	7.7	6.5	8.9	6.3	6.2	7.0
5.7	6.2	6.3	6.8	7.1	7.8	7.6	8.2

Chap. 8] Statistics

3. The diameter in inches of 20 buttons made by a certain machine is recorded to the nearest hundredth of an inch.

.50 .52 .53 .49 .50 .49 .51 .50 .49 .48
.50 .51 .51 .50 .49 .50 .51 .52 .48 .49

4. Thirty students take a test and their grades are recorded.

85 89 57 75 80 90 100 78 81 89
73 76 100 95 93 84 86 67 78 67
89 77 98 93 91 65 89 72 76 82

5. Discuss methods that a political pollster might use in order to select a random sample from the voting population.

8.2 Grouped Data and Summation Notation

The methods of Example 1 of Section 8.1 are good when the number n of elements in the sample does not exceed about 25. When n is larger, these methods are somewhat unsatisfactory as well as tedious, so in this case it is better to group the data into about 10 or 15 intervals (or cells) before constructing any tables or graphs. We illustrate this procedure by using the data of Example 1 of Section 8.1.

Example 1

Again assume that the sample consists of the following $n = 32$ real numbers:

17 11 7 26 15 6 9 12
5 24 9 16 29 9 18 12
17 6 9 22 9 33 12 8
11 13 15 7 10 30 20 17

Refer to Table 8.2. The salaries range from a low of 5 to a high of 33, and we have grouped the salaries into 10 intervals of length 3. It is customary to make the intervals of equal length, and of course the intervals are chosen so

Table 8.2. GROUPED FREQUENCY AND CUMULATIVE FREQUENCY.

Interval	Midpoint x_i of interval	Tally marks	Frequency f_i	Cumulative frequency
4.5–7.5	6	ЖЖ	5	5
7.5–10.5	9	ЖЖ ‖	7	12
10.5–13.5	12	ЖЖ ‖	6	18
13.5–16.5	15	‖‖‖	3	21
16.5–19.5	18	‖‖‖‖	4	25
19.5–22.5	21	‖‖	2	27
22.5–25.5	24	‖	1	28
25.5–28.5	27	‖	1	29
28.5–31.5	30	‖‖	2	31
31.5–34.5	33	‖	1	32

that each of the 32 observations in the sample falls in exactly one interval. The first interval has endpoints 4.5 and 7.5, and the **midpoint** x_1 of this interval is the point halfway from 4.5 to 7.5, namely,

$$x_1 = \frac{4.5 + 7.5}{2} = 6.$$

We let the endpoints be measured to an extra decimal place so that none of the 32 given numbers equals an endpoint. Also, an effort should be made to select the endpoints in a manner that makes the lengths of the intervals and the midpoints easy numbers to handle (rather than complicated fractions). In Table 8.2 the distribution of the frequencies among the 10 intervals can be seen in the frequency column; since the data were first grouped into intervals, we say that Table 8.2 shows the **grouped frequency distribution**. In Example 1 of Section 8.1 we worked with **ungrouped data,** whereas in this example we are dealing with **grouped data** since the data are grouped into intervals. In analyzing grouped data we are really assuming that each element of the sample in a given interval equals the midpoint of that interval. We could represent the frequencies in Table 8.2 graphically by another line chart with the vertical lines drawn at the 10 midpoints, but with grouped data it is more customary to use what is called a **histogram** or **bar graph**, as shown in Figure 8.3.

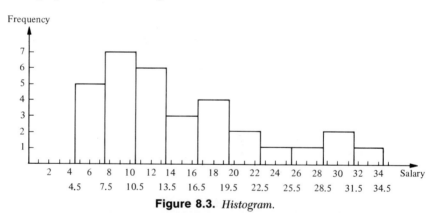

Figure 8.3. *Histogram.*

In Figure 8.4 the cumulative frequencies of Table 8.2 are represented graphically by what is called a **cumulative frequency polygon.** Observe how the points are plotted; for example, the right end of the third interval is 13.5 and the cumulative frequency through three intervals is 18, and hence we plot the point (13.5, 18) in Figure 8.4. The cumulative frequency polygon is obtained by joining these points together with line segments. (The percentiles and median shown in Figure 8.4 will be discussed in Section 8.3.) ∎

For the remainder of this section we shall discuss **summation notation,** which we shall find useful in the next two sections. The sum

$$3 + 6 + 9 + 12 + 15 + 18$$

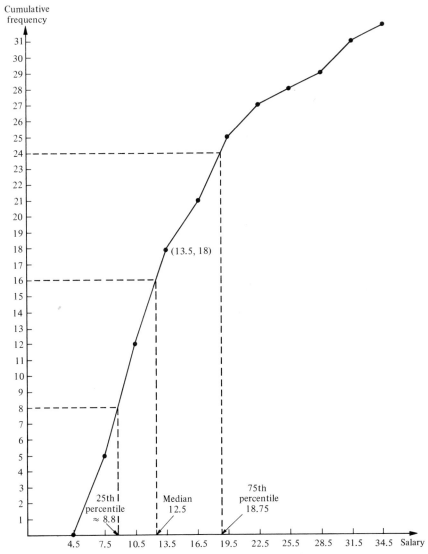

Figure 8.4. *Cumulative frequency polygon.*

can be written briefly as

$$\sum_{i=1}^{6} 3i,$$

which is an example of the summation notation. The Greek letter Σ (sigma) suggests the word sum. The notation $\Sigma_{i=1}^{6} 3i$ indicates that the values 1, 2, 3, 4, 5, and 6 are to be substituted successively for i in the expression $3i$ and that the results are to be added. The letter i is called the **index of summation,** but any other letter may be used; $\Sigma_{j=1}^{6} 3j$ represents the same sum.

Sec. 8.2] Grouped Data and Summation Notation

Example 2

(a) $\sum_{i=2}^{4} (i^2 + 1) = (2^2 + 1) + (3^2 + 1) + (4^2 + 1) = 32.$

(b) $\sum_{i=1}^{5} x_i = x_1 + x_2 + x_3 + x_4 + x_5.$

(c) The expected value for an experiment (defined in Chapter 4)

$$p_1 a_1 + p_2 a_2 + \cdots + p_n a_n$$

can be written

$$\sum_{i=1}^{n} p_i a_i. \blacksquare$$

Three important properties of the summation notation, which are not difficult to prove, are stated in the next theorem and then illustrated with examples.

▶ **Theorem 1**
Assume that n is a positive integer, c is a real number, and f and g are functions of the index of summation i. Then the following properties are valid:

(i) $\sum_{i=1}^{n} cf(i) = c \sum_{i=1}^{n} f(i).$

(ii) $\sum_{i=1}^{n} [f(i) \pm g(i)] = \sum_{i=1}^{n} f(i) \pm \sum_{i=1}^{n} g(i).$

(iii) $\sum_{i=1}^{n} c = nc.$

The proof of Theorem 1 is left as Exercise 12. ◀

Example 3

(a) $\sum_{i=1}^{3} 4i^2 = 4(1)^2 + 4(2)^2 + 4(3)^2$

$= 4(1^2 + 2^2 + 3^2)$

$= 4 \sum_{i=1}^{3} i^2.$

This illustrates Theorem 1(*i*) for $n = 3$, $c = 4$, and $f(i) = i^2$.

(b) $\sum_{i=1}^{3} \left(i^3 - \frac{1}{i}\right) = \left(1^3 - \frac{1}{1}\right) + \left(2^3 - \frac{1}{2}\right) + \left(3^3 - \frac{1}{3}\right)$

$= (1^3 + 2^3 + 3^3) - \left(\frac{1}{1} + \frac{1}{2} + \frac{1}{3}\right)$

$= \sum_{i=1}^{3} i^3 - \sum_{i=1}^{3} \frac{1}{i}.$

This illustrates Theorem 1(*ii*) for $n = 3$, $f(i) = i^3$, $g(i) = 1/i$, and a minus sign between $f(i)$ and $g(i)$.

(c) $\sum_{i=1}^{5} 8 = 8 + 8 + 8 + 8 + 8 = 40 = (5)(8).$

This illustrates Theorem 1(*iii*) for $n = 5$ and $c = 8$. ∎

Theorem 1(*ii*) can be generalized to sums or differences of any finite number of functions. For example,

$$\sum_{i=1}^{n} [f(i) - g(i) + h(i)] = \sum_{i=1}^{n} f(i) - \sum_{i=1}^{n} g(i) + \sum_{i=1}^{n} h(i)$$

and a reference to Theorem 1(*ii*) will be considered to be sufficient justification of this equality.

Example 4

Show that $\sum_{i=1}^{4} (2 - x_i)(3 - x_i) = 24 - 5 \sum_{i=1}^{4} x_i + \sum_{i=1}^{4} x_i^2.$

$\sum_{i=1}^{4} (2 - x_i)(3 - x_i) = \sum_{i=1}^{4} (6 - 5x_i + x_i^2)$ by multiplication

$= \sum_{i=1}^{4} 6 - \sum_{i=1}^{4} 5x_i + \sum_{i=1}^{4} x_i^2$ by Theorem 1(*ii*)

$= 24 - \sum_{i=1}^{4} 5x_i + \sum_{i=1}^{4} x_i^2$ by Theorem 1(*iii*)

$= 24 - 5 \sum_{i=1}^{4} x_i + \sum_{i=1}^{4} x_i^2$ by Theorem 1(*i*). ∎

EXERCISES

Suggested minimum assignment: Exercises 1, 5, 7, 8, and 13.

In Exercises 1–4 group the data into intervals and make a table like Table 8.2. Also, sketch a histogram and a cumulative frequency polygon.

1. Use the data of Exercise 1 of Section 8.1. Let the intervals be 60.5–62.5, 62.5–64.5, ... , 78.5–80.5.
2. Use the data of Exercise 2 of Section 8.1.
3. Use the data of Exercise 3 of Section 8.1.
4. Use the data of Exercise 4 of Section 8.1.
5. (a) In Section 8.1 (ungrouped data) the sample data can be labeled $x_1, x_2, \ldots, x_n$. Is it possible for some two of the numbers $x_1, x_2, \ldots, x_n$ to be the same?
   (b) In Section 8.2 (grouped data) the midpoints of the intervals can be labeled $x_1, x_2, \ldots, x_k$. Is it possible for some two of the numbers $x_1, x_2, \ldots, x_k$ to be the same?
6. Write the sum represented by each of the following.
   (a) $\sum_{i=1}^{6} (2i + 3)$.   (b) $\sum_{j=2}^{5} \frac{1}{j^2}$.
7. Write the sum represented by each of the following.
   (a) $\sum_{i=1}^{4} (i^2 + 3i)$.   (b) $\sum_{i=1}^{5} f(c_i) \Delta x_i$.
8. Write in summation notation.
   (a) $2 + 4 + 6 + 8 + 10 + 12$.   (b) $x_1^2 + x_2^2 + x_3^2 + x_4^2$.
9. Write in summation notation.
   (a) $3 + 5 + 7 + 9 + 11$.   (b) $x_1 f_1 + x_2 f_2 + \cdots + x_k f_k$.
10. Show that $\sum_{i=1}^{3} (x_i + y_i) = \sum_{i=1}^{3} x_i + \sum_{i=1}^{3} y_i$.
11. Show that $\sum_{i=1}^{4} 2i^3 = 2 \sum_{i=1}^{4} i^3$.
12. (a) Prove Theorem 1(i).
    (b) Prove Theorem 1(ii).
    (c) Prove Theorem 1(iii).
13. Show that $\sum_{i=1}^{3} (4 - x_i)(5 - x_i) = 60 - 9 \sum_{i=1}^{3} x_i + \sum_{i=1}^{3} x_i^2$.

    Use Theorem 1 and give a reason for each step.

14. If $x_1 = 7$, $x_2 = 9$, $x_3 = 5$, and $x_4 = 10$, calculate $\sum_{i=1}^{4} x_i^2$.

15. Show that $\sum_{i=-2}^{2} i^2 = 2 \sum_{i=1}^{2} i^2$.

8.3 Measures of Central Tendency

When analyzing a sample, one often wishes to find a number that in some sense represents the "center" of the set of data. In this section we shall discuss three measures of central tendency, called the mode, mean, and median. Assume until stated otherwise that we are dealing with *ungrouped data*.

▶ **Definition 1**
The **mode** of a sample is the real number that occurs with greatest frequency. There may be more than one mode.

Example 1
From the frequency column of Table 8.1 on p. 356 it is easy to see that the mode in our salary example is 9. ∎

Example 2
Suppose that the test scores made by 10 students are

$$95, \ 95, \ 92, \ 87, \ 85, \ 83, \ 78, \ 78, \ 78, \text{ and } 7.$$

The mode is 78 since the score of 78 was earned by 3 students and no other one score was made by as many as 3 students. ∎

If one more grade of 95 had been recorded in Example 2, then both 78 and 95 would have been modes and we would say that the new sample is **bimodal**. One disadvantage of using the mode as a measure of central tendency is that there may be more than one mode. Another disadvantage is that the very largest or smallest element of the sample may be a mode, and such a number usually is not a typical representative of the sample. On the other hand, a mode is easy to find and may have significant meaning for sets of data containing a very large number of elements. However, the mode is probably the least used of our three measures of central tendency. (Further discussion of the mode is given in Example 6 of Section 11.4.)

Probably the most commonly used measure of central tendency of a sample is the **mean**. There is more than one kind of mean, but when we use the word mean we are referring to what is also called the *arithmetic mean* or the *average*. The mean of a sample will be denoted by $\bar{x}$ (read x bar).

▶ **Definition 2**
Suppose that a sample consists of the real numbers $x_1, x_2, \ldots, x_n$. The **mean** of the sample is given by

$$\bar{x} = \frac{\sum_{i=1}^{n} x_i}{n}.$$

Example 3
For the 10 scores

$$95, \ 95, \ 92, \ 87, \ 85, \ 83, \ 78, \ 78, \ 78, \text{ and } 7$$

of Example 2 the mean is

$$\bar{x} = \frac{\sum_{i=1}^{10} x_i}{10} = \frac{95 + 95 + 92 + \cdots + 7}{10} = \frac{778}{10} = 77.8. \ \blacksquare$$

The mean has the advantage of taking into consideration all the observations in a sample. One disadvantage is that the mean can be strongly influenced by a few unusually high or low observations. For instance, in Example 3 the one low test score of 7 caused the mean to be smaller than 9 of the 10 test scores.

While the symbol $\bar{x}$ is commonly used to denote the mean of a sample that is taken from a population, the symbol μ is often used to represent the mean of the whole population. Suppose that a sample consists of n real numbers. If the sample has been selected wisely, $\bar{x}$ should be a good approximation of μ. A better approximation of μ might be obtained by finding the means of several samples, each of size n, taken from the population, and then computing the mean of these means. (More discussion of the mean is given in Example 5 of Section 13.2.)

Example 4
In Example 1 of Section 8.1 we have $x_1 = 17$, $x_2 = 11$, $x_3 = 7, \ldots,$ $x_{32} = 17$. The mean is given by

$$\bar{x} = \frac{\sum_{i=1}^{32} x_i}{32} = \frac{17 + 11 + 7 + \cdots + 17}{32} = \frac{464}{32} = 14.5. \ \blacksquare$$

Example 5
In Example 1 of Section 8.2 we computed the midpoint of an interval by adding the two endpoints (4.5 and 7.5) and dividing by 2. We can now say that the midpoint of an interval is the mean of the two endpoints. ∎

The third measure of central tendency is called the **median**.

▶ **Definition 3**
Suppose that a sample of n real numbers is given and that these numbers are listed in order of size. The **median** of the sample is
  (i) The middle number in the list if n is odd.
  (ii) The mean of the two middle numbers in the list if n is even.

Example 6
(a) Suppose that a sample consists of the five numbers

$$21, \quad 27, \quad 34, \quad 38, \text{ and } 38.$$

Here, n is odd ($n = 5$) and the median is the middle number 34. (Be sure that the numbers are listed in either increasing or decreasing order before reading the middle number.)

(b) For our sample of 10 test scores

$$95, \quad 95, \quad 92, \quad 87, \quad 85, \quad 83, \quad 78, \quad 78, \quad 78, \text{ and } 7,$$

the number of elements in the sample is even ($n = 10$). The two middle numbers are 85 and 83, and by Definition 3(ii) the median is

$$\frac{85 + 83}{2} = 84.$$

(c) In the salary example of Section 8.1, the sample consisted of an even number ($n = 32$) of elements. The two middle numbers in order of size are both 12, and therefore the median is 12. ∎

The median of 84 in Example 6(b) is probably a better measure of central tendency than the mean $\bar{x} = 77.8$. The median is not influenced as greatly by the extremely low score of 7 as is the mean. On the other hand, if the test scores are

$$95, \quad 95, \quad 78, \quad 78, \text{ and } 78,$$

then the mean $\bar{x} = 84.8$ seems to be a better measure of central tendency than the median of 78. Clearly, judgment must often be exercised in choosing the best measure of central tendency for a sample.

In standardized educational tests, grades are often reported in terms of **percentiles,** which is a way of dividing the scores as nearly as possible into 100 equal parts. If a student's grade is above the 99th percentile, it is an indication that the student scored in the top 1 per cent on the test. In terms of percentiles, the median is the 50th percentile. The 25th percentile is called the **first quartile** or the **lower quartile,** and the 75th percentile is called the **third quartile** or **upper quartile.**

Example 7
For ungrouped data a valid method for finding any desired percentile is indicated in Figure 8.2 on p. 358. First, sketch the cumulative frequency distribution function. Suppose that we want the 25th percentile in Figure

8.2. Since $n = 32$, we convert $\frac{25}{100}$ to $\frac{8}{32}$. Start at the ordinate 8 and move to the right until arriving at a point directly below one of the red dots [in this case we arrive at $(9, 8)$ which is directly below the dot at $(9, 11)$]. The abscissa 9 is the 25th percentile. An exception occurs if, when moving to the right, one arrives exactly at one of the dots instead of below one of the dots. This exception is encountered in finding the 75th percentile. For the 75th percentile, convert $\frac{75}{100}$ to $\frac{24}{32}$ and move to the right from the cumulative frequency 24 on the vertical axis. One arrives at the dot at $(17, 24)$. When this happens, find the mean of the abscissas of $(17, 24)$ and the next dot to the right, which is at $(18, 25)$. The 75th percentile is the mean of the abscissas 17 and 18; that is,

$$\frac{17 + 18}{2} = 17.5. \blacksquare$$

In Example 7 a graphical method for finding percentiles is explained. A more technical description of how to find percentiles will be given in Example 4 of Section 13.6.

For the remainder of this section we shall illustrate the methods for finding the mode, mean, median, and percentiles for the case of *grouped data*.

Example 8

For the grouped data in Example 1 of Section 8.2, find the mode, mean, median, and 25th percentile.

Refer to Table 8.2 on p. 359. The mode is the midpoint of the interval that has the greatest frequency; the greatest frequency is $f_2 = 7$, and hence the mode is the midpoint $x_2 = 9$ of the second interval. It is possible for more than one mode to exist (depending upon the data and how it is grouped). The modes for ungrouped and grouped data do not have to be the same, although they are in this example. In fact, we shall find that all of our measures of central tendency may be changed when the data are grouped.

Let k be the number of intervals into which the data are grouped—in our example $k = 10$. The mean is given by the formula

$$\bar{x} = \frac{\sum_{i=1}^{k} x_i f_i}{n}. \tag{8.1}$$

From Table 8.2 we can read the midpoints x_i and the frequencies f_i, and hence we obtain

$$\bar{x} = \frac{\sum_{i=1}^{10} x_i f_i}{32} = \frac{(6)(5) + (9)(7) + \cdots + (33)(1)}{32}$$

$$= \frac{468}{32} = 14\frac{5}{8}.$$

Note that all five salaries in the first interval are counted as 6 in (8.1), whereas in reality the salaries are 5, 6, 6, 7, and 7. This is typical of what happens for grouped data and explains why $\bar{x} = 14\frac{5}{8}$ instead of $\bar{x} = 14.5$ (see Example 4). It is interesting to observe that

$$\sum_{i=1}^{k} f_i = n, \tag{8.2}$$

and we shall use this formula in the next section.

In Figure 8.4 on p. 361 we have indicated that the median is 12.5. Since the median is the 50th percentile, we convert $\frac{50}{100}$ to $\frac{16}{32}$. From a cumulative frequency of 16 on the vertical axis in Figure 8.4 we move to the right until the cumulative frequency polygon is intersected. The abscissa of this point of intersection is 12.5, which is the median. It is possible to calculate the median from Table 8.2 instead of Figure 8.4. Through two intervals the cumulative frequency is 12, and after three intervals it is 18. The cumulative frequency of 16 (50 per cent of 32) is $\frac{2}{3}$ of the way from 12 to 18. Therefore, we calculate the number $\frac{2}{3}$ of the way from the left endpoint 10.5 of the third interval to the right endpoint 13.5. This number is

$$10.5 + \frac{2}{3}(13.5 - 10.5) = 12.5.$$

For purposes of calculating the median it is customary to assume that the 6 observations in the interval 10.5–13.5 are spread evenly through the interval, and hence the calculation above is an instance of what is called *interpolation*.

Following the procedure just discussed, we compute the 25th percentile (see Table 8.2) to be

$$7.5 + \frac{8-5}{12-5}(10.5 - 7.5) = 7.5 + \frac{3}{7}(3) \approx 8.8.$$

Notice that this agrees with Figure 8.4. ∎

EXERCISES

Suggested minimum assignment: Exercises 1, 5(a), and 9.

In Exercises 1–4 find the mode, mean, and median of the given sample data. Also, tell which of these measures of central tendency seems to be best.

1. 204   201   201   123   107
2. 52   52   48   46   43   39   2
3. 80   80   60   50   49   49
4. 7.5   3.8   3.8   3.8   3.4   3.3

5. Find $\bar{x}$ for each of the following ungrouped data.
   (a) The data of Exercise 1 of Section 8.1.
   (b) The data of Exercise 2 of Section 8.1.

Sec. 8.3] **Measures of Central Tendency**

(c) The data of Exercise 3 of Section 8.1.
(d) The data of Exercise 4 of Section 8.1.

6. A biologist weighs 10 male cats and records their weights in pounds as follows:

    14.3   16.2   15.5   15.3   15.4   16.5   15.9   15.8   15.0   14.7.

    Find the median weight of these 10 cats.

7. Let $x_1, x_2, \ldots, x_n$ be the n observations in a sample, and let $\bar{x}$ be the mean of these n numbers.

    (a) Prove that $\sum_{i=1}^{n} (x_i - \bar{x}) = 0$.

    (b) Let c be a real number and let $y_i = x_i + c$, for $i = 1, 2, \ldots, n$.

    If $\bar{y} = \dfrac{\sum_{i=1}^{n} y_i}{n}$, prove that $\bar{y} = \bar{x} + c$.

    (c) How can the result of part (b) be useful sometimes as a timesaver in computing the mean of a sample?

8. Given the ungrouped test scores

    60   65   65   68   70   74   74   78   81   87.

    (a) Sketch the cumulative frequency distribution function.
    (b) From your sketch determine each of the following: 25th percentile, 40th percentile, median, and 75th percentile.

9. (a) Refer to the cumulative frequency distribution function that was sketched in Exercise 1 of Section 8.1 and determine each of the following from your sketch: 25th percentile, median, 75th percentile.
    (b) Refer to the cumulative frequency polygon that was sketched in Exercise 1 of Section 8.2 (now we are dealing with grouped data). From your sketch determine the 25th percentile, the median, and the 75th percentile.
    (c) For the grouped data of Exercise 1 of Section 8.2 find the mode and the mean.

10. From Figure 8.2 on p. 358 find the 13th percentile and the 37.5th percentile.

11. Figure 8.4 on p. 361 shows that the 75th percentile in that example is 18.75. Verify this result by a method that does not depend on Figure 8.4. (Hint: Use the interpolation procedure discussed at the end of Example 8.)

12. Refer to your solution of Exercise 2 of Section 8.2. Determine the mode, mean, median, and 25th percentile. (The answers may depend upon how you grouped the data into intervals.)

8.4 Measures of Variation

Two sets of sample data may have the same mean or the same median and yet be very different in other ways. For example, suppose that one sample consists of the numbers

$$48, \quad 49, \quad 50, \quad 51, \quad \text{and} \quad 52$$

and a second sample consists of the numbers

$$0, \quad 25, \quad 50, \quad 75, \quad \text{and} \quad 100.$$

Each sample has 50 for both its mean and its median. However, the data in the second sample are scattered out much greater from 50 than are the data in the first sample. In this section we shall discuss methods of measuring the amount of variation, or scatter, or spread, or dispersion that a sample has.

The **range** of a sample is the largest number in the sample minus the smallest number. For the sample whose elements are

$$48, \quad 49, \quad 50, \quad 51, \quad \text{and} \quad 52,$$

the range is

$$52 - 48 = 4.$$

However, if the sample consists of

$$48, \quad 49, \quad 50, \quad 51, \quad \text{and} \quad 108,$$

the range is

$$108 - 48 = 60,$$

which is a drastic change. Generally, the range is not considered to be a very good measure of variation because only the largest and smallest observations are used in the calculation.

Another measure of variation is the **interquartile range,** which is defined to be the 75th percentile minus the 25th percentile. The advantage of the interquartile range is that it is not influenced greatly by any extremely large or small observations, and thus it usually gives a more stable measure of variation than the range. The interquartile range is an indication of the amount of scatter of the middle 50 per cent of the observations, and it can be computed for either ungrouped data or grouped data. It is desirable to have at least $n = 25$ observations in the sample so that the 75th and 25th percentiles are quite surely meaningful numbers. For the ungrouped data in our salary example of Section 8.1, the interquartile range (see Figure 8.2 on p. 358) is

$$17.5 - 9 = 8.5,$$

whereas after grouping the data (see Figure 8.4 on p. 361) the interquartile range is approximately

$$18.75 - 8.8 = 9.95.$$

Probably the most important measures of variation of a sample are the **variance,** which will be denoted by s^2, and its nonnegative square root s, which is called the **standard deviation.**

▶ **Definition 4**

For a sample consisting of n real numbers with mean $\bar{x}$, the **variance** s^2 is given by

$$s^2 = \frac{1}{n} \sum_{i=1}^{n} (x_i - \bar{x})^2. \tag{8.3}$$

If the data are grouped into k intervals,† then

$$s^2 = \frac{1}{n} \sum_{i=1}^{k} (x_i - \bar{x})^2 f_i. \tag{8.4}$$

▶ **Definition 5**

If s^2 is the variance of a sample, then the **standard deviation** s is given by

$$s = \sqrt{s^2}.$$

Example 1

Calculate the variance and standard deviation for the sample consisting of the $n = 5$ observations

$$44, \quad 47, \quad 49, \quad 53, \quad \text{and} \quad 57.$$

First, calculate the mean $\bar{x}$ as follows:

$$\bar{x} = \frac{\sum_{i=1}^{5} x_i}{5} = \frac{44 + 47 + 49 + 53 + 57}{5} = \frac{250}{5} = 50.$$

From (8.3) the variance is

$$s^2 = \frac{1}{5} \sum_{i=1}^{5} (x_i - 50)^2$$

$$= \frac{1}{5}[(44 - 50)^2 + (47 - 50)^2 + (49 - 50)^2 + (53 - 50)^2 + (57 - 50)^2]$$

$$= \frac{1}{5}(104) = 20.8$$

By Definition 5 the standard deviation is

$$s = \sqrt{20.8} \approx 4.6. \quad \blacksquare$$

† Theoretically, statisticians claim that $1/(n-1)$ should appear in (8.3) and (8.4) instead of $1/n$ because there are only $n - 1$ degrees of freedom associated with the sum of squares $\sum_{i=1}^{n}(x_i - \bar{x})^2$. We shall use $1/n$ for simplicity as most elementary texts do; it makes little difference in most applied problems where n is large.

Example 2
For the sample data
$$50, \ 50, \ 50, \ 50, \ \text{and} \ 50$$
the mean is $\bar{x} = 50$. Therefore, each $x_i - \bar{x}$ is 0, and hence the variance and standard deviation are equal to 0. ∎

One big advantage of the variance and standard deviation as measures of variation is that every observation in the sample is used in the computations. The only time that s^2 and s will equal 0 is when every element of the sample is the same (see Example 2). In general, larger values of s^2 and s (for a fixed value of n) correspond to a greater spread of the data from the mean. When we calculate s^2 and s for a sample taken from a population, we are estimating the variance and standard deviation for the whole population, which are designated by σ^2 and σ, respectively (We shall discuss σ^2 and σ further in Example 6 of Section 13.2 and Example 5 of Section 14.7.)

The calculations in Examples 1 and 2 were simple because the samples were small and because the means were integers. There are alternative formulas for s^2 that reduce the amount of calculation in practical problems, where $\bar{x}$ is likely to be a "messy" fraction. Before we find s^2 and s for a problem with grouped data, we shall present a formula in Theorem 2 which will reduce the amount of calculation.

▶ **Theorem 2**
Suppose that the n observations in a sample have been grouped into k intervals. Then
$$s^2 = \frac{1}{n}\left(\sum_{i=1}^{k} x_i^2 f_i\right) - \bar{x}^2.$$

Proof

	Statement	Reason
1.	$s^2 = \dfrac{1}{n}\sum_{i=1}^{k}(x_i - \bar{x})^2 f_i$	1. Definition 4.
2.	$= \dfrac{1}{n}\sum_{i=1}^{k}(x_i^2 f_i - 2x_i \bar{x} f_i + \bar{x}^2 f_i)$	2. By multiplication.
3.	$= \dfrac{1}{n}\left[\sum_{i=1}^{k} x_i^2 f_i - \sum_{i=1}^{k} 2x_i \bar{x} f_i + \sum_{i=1}^{k} \bar{x}^2 f_i\right]$	3. Theorem 1(ii).
4.	$= \dfrac{1}{n}\left[\sum_{i=1}^{k} x_i^2 f_i - 2\bar{x}\sum_{i=1}^{k} x_i f_i + \bar{x}^2 \sum_{i=1}^{k} f_i\right]$	4. Theorem 1(i).

5. $= \dfrac{1}{n}\left[\sum_{i=1}^{k} x_i^2 f_i - 2\bar{x}(n\bar{x}) + \bar{x}^2 \sum_{i=1}^{k} f_i\right]$ 5. Equation (8.1).

6. $= \dfrac{1}{n}\left[\sum_{i=1}^{k} x_i^2 f_i - 2\bar{x}(n\bar{x}) + \bar{x}^2 n\right]$ 6. Equation (8.2).

7. $= \dfrac{1}{n}\left[\sum_{i=1}^{k} x_i^2 f_i - n\bar{x}^2\right]$ 7. Combine like terms.

8. $= \dfrac{1}{n}\left(\sum_{i=1}^{k} x_i^2 f_i\right) - \bar{x}^2.$ 8. By multiplication. ◀

Example 3
Calculate the variance s^2 and the standard deviation s for the grouped data of the salary example in Section 8.2.

We found in Section 8.3 that $\bar{x} = 14\tfrac{5}{8}$. To use Theorem 2, the work is arranged as shown in Table 8.3. (The basic data can be found in Table 8.2.)

By Theorem 2 the variance s^2 is given by

$$s^2 = \tfrac{1}{32}(8658) - (14\tfrac{5}{8})^2$$
$$\approx 270.56 - 213.89$$
$$= 56.67.$$

Therefore, the standard deviation s is given by (see Definition 5)

$$s = \sqrt{s^2} \approx 7.53. \blacksquare$$

The calculation of s^2 in Example 3 can be accomplished with (8.4) instead of Theorem 2; the work is more tedious and is left as Exercise 11.

Table 8.3

Interval number	x_i	x_i^2	f_i	$x_i^2 f_i$
1	6	36	5	180
2	9	81	7	567
3	12	144	6	864
4	15	225	3	675
5	18	324	4	1296
6	21	441	2	882
7	24	576	1	576
8	27	729	1	729
9	30	900	2	1800
$k = 10$	33	1089	1	1089
			$\sum_{i=1}^{10} f_i = 32 = n$	$8658 = \sum_{i=1}^{10} x_i^2 f_i$

APPLICATIONS

*Example 4

In order to give more meaning to standard deviation, we wish to discuss briefly what is called the **normal distribution.** A more thorough discussion can be given after we study calculus (see Example 5 of Section 14.7). Suppose that one weighs a large sample of washers that are produced by a certain machine and records the weight to the nearest hundredth of a gram. If a line chart is drawn that shows frequencies of various weights, one might expect something similar to Figure 8.5. Probably, the washers will not all have the same weight, but instead the weights will show a certain amount of "normal" variation from the mean. Hopefully, the mean, median, and mode will all equal the weight that the washers are supposed to have. By weighing a sample of the washers, a quality-control expert may be able to detect a malfunction in the machine. Since the weights are rounded to the nearest hundredth of a gram, we are thinking of the weight w as a **discrete variable**—that is, there are gaps between the allowable values of the variable. If the weights are not rounded, then the weight may have any real value within some interval; in this case we say that w is a **continuous variable,** and in this illustration the **normal curve** in Figure 8.6 replaces the line chart in Figure 8.5. We hasten to point out that not every sample or population is distributed normally. Actually, the graph of a frequency distribution can

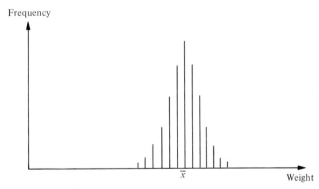

Figure 8.5. *Line chart.*

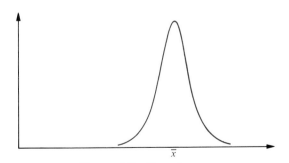

Figure 8.6. *Normal curve.*

Sec. 8.4] Measures of Variation

have one of many shapes, but in practice the normal curve is the most common.

A normal curve has a bell-shaped appearance with the highest point at the mean. There are many normal curves for the same mean, because the shape also depends upon the amount of variation as measured by the standard deviation. However, when the population has a normal distribution, it is known that about 68.3 per cent of the elements of the population are within one standard deviation of the mean, 95.4 per cent are within two standard deviations of the mean, and 99.7 per cent are within three standard deviations of the mean. See Figure 8.7.

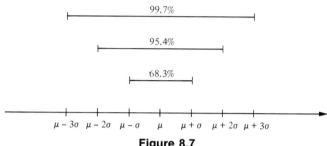

Figure 8.7

As an example, suppose that a company produces a certain type of light bulb. Experience has shown that the life in hours of this type of light bulb is normally distributed with mean 1000 hours and standard deviation 40 hours. If one of the light bulbs is selected at random, the probability is .683 that its life is between 960 hours and 1040 hours. ∎

★Example 5†

The assumption of normality is frequently made in many practical situations. Suppose that in naval gunnery tests a 20-pound charge weight of a certain type of propellant is expected to produce a projectile velocity of 2650 feet per second in a new 5-inch 38-caliber gun. Projectile velocity is measured continuously over a .6-second interval by means of a gun-mounted velocimeter. This device operates on a radar Doppler technique. The measurement time interval begins approximately .1 second after the projectile exits the gun barrel and extends to about .7 second after exit. Velocimeter readings are taken every .05 second during this span and used to estimate the "initial" projectile velocity (PV) for this firing. Experimental conditions prohibit the direct measurement of initial velocity.

Clearly, PV is a random variable since each time a round is fired in the above setup a different value of PV may result. If a test engineer desired to make probability statements concerning PV, he would require knowledge of the distribution that this random variable follows. Under the above firing conditions it is usually assumed, based on past experience, that PV follows a

† This example was furnished by John R. Crigler, a statistician at the Naval Surface Weapons Center, Dahlgren, Virginia.

normal distribution with mean $\mu = 2650$ feet per second and standard deviation $\sigma = 13$ feet per second. Note that this distributional assumption reflects the test engineer's feeling that the PV associated with any randomly chosen test firing is equally likely to be above or below 2650 feet per second. Utilizing this information, we can expect 99.7 per cent of all firings under the above conditions to produce an initial projectile velocity that lies between 2611 feet per second and 2689 feet per second, 95.4 per cent to result in a PV that lies between 2624 feet per second and 2676 feet per second, and 68.3 per cent to yield a PV that lies between 2637 feet per second and 2663 feet per second. ∎

EXERCISES

Suggested minimum assignment: Exercises 1, 4, 5, 7, and 9.

1. For the ungrouped data of Exercise 1 of Section 8.1, find the range and the interquartile range. [See Exercise 9(a) of Section 8.3.]
2. For the ungrouped data of Exercise 2 of Section 8.1, find the range and the interquartile range.
3. Find the range for the sample data given in Exercise 6 of Section 8.3.
4. Can the variance or standard deviation ever be negative?
5. If a sample consists of the five numbers 38, 36, 31, 27, and 22, calculate the variance and standard deviation.
6. Calculate the variance and standard deviation for the data of Exercise 3 of Section 8.3.
7. Calculate the variance and standard deviation for the data of Exercise 1 of Section 8.1. [See Exercise 5(a) of Section 8.3.] How many of the 36 heights are within one standard deviation of the mean?
8. (a) Start with (8.3) and derive the formula

$$s^2 = \frac{1}{n}\left(\sum_{i=1}^{n} x_i^2\right) - \bar{x}^2.$$

(b) Use the formula derived in part (a) to calculate the variance for the data 38, 36, 31, 27, and 22. Compare your answer with that obtained in Exercise 5.
9. Use Theorem 2 to calculate the variance for the grouped data of Exercise 1 of Section 8.2. [*Hint:* The mean was found in Exercise 9(c) of Section 8.3.] Then find the standard deviation.
10. Use Theorem 2 to calculate the variance for the grouped data of Exercise 2 of Section 8.2. Then find the standard deviation.
11. In Example 3 of this section use (8.4) instead of Theorem 2 in order to calculate the variance.
12. Given a sample of ungrouped data consisting of the numbers $x_1, x_2, \ldots, x_n$, suppose that we construct a new sample consisting of the numbers $x_1 + c, x_2 + c, \ldots, x_n + c$ by adding a real number c to each obser-

vation. How does the variance of the new sample compare with the variance of the original sample?
13. A population of female college students is thought to have mean height of 65 inches with a standard deviation of 2 inches. (Assume that the heights are normally distributed.)
    (a) About 68.3 per cent of the population is between what two heights?
    (b) What is the probability that a member of the population selected at random will be between 61 and 69 inches tall?
14. A machine produces gadgets with mean thickness of 1.42 centimeters and standard deviation .005 centimeter. (Assume that the thicknesses are normally distributed.) A customer purchases these gadgets and throws away any of them that are not between 1.41 and 1.43 centimeters in thickness. About what percentage of the gadgets must the customer throw away?

NEW VOCABULARY

statistics 8.1
random variable or measurement 8.1
population 8.1
sample 8.1
random sampling 8.1
observation 8.1
frequency 8.1
cumulative frequency 8.1
line chart 8.1
frequency distribution 8.1
cumulative frequency distribution function 8.1
midpoint of an interval 8.2
grouped frequency distribution 8.2
ungrouped data 8.1, 8.2

grouped data 8.2
histogram or bar graph 8.2
cumulative frequency polygon 8.2
summation notation 8.2
index of summation 8.2
mode 8.3
bimodal 8.3
mean 8.3
median 8.3
percentiles 8.3
first quartile or lower quartile 8.3
third quartile or upper quartile 8.3
range of a sample 8.4
interquartile range 8.4
variance 8.4
standard deviation 8.4

9 Limits and Continuity

This chapter is devoted to a brief introduction to the limit concept. This is necessary because two of the basic concepts of the calculus—the derivative and the definite integral—are defined as limits. Historically, the attainment of precision and clarity in the presentation of the basic ideas concerning limits proved to be very elusive. More than one hundred years elasped after the discovery of calculus before the underlying notion of a limit was refined to the extent that the calculus was placed on a rigorous foundation; many of these refinements were made by Augustin Louis Cauchy (1789-1857) in the early part of the 19th century.

Some immediate applications of these topics will be given; moreover, the reader should recognize that several applications which occur in later chapters also depend to a large extent upon the topics in this chapter.

Prerequisites: 1.1, 1.2, 3.1–3.9.
Suggested Sections for Moderate Emphasis: 9.1–9.4 should be covered; 9.1 and 9.2 can be combined into one lesson.
Suggested Sections for Minimum Emphasis: 9.1–9.4 should be covered. For a short course with emphasis on techniques, 9.1–9.4 can be skimmed in two lessons.

9.1 Introduction to Calculus

In this section we will present a brief introduction to the mathematical subject called **calculus.** Then in Sections 9.2–9.4 we shall discuss some of the fundamental concepts that are needed for a study of calculus in Chapters 10–15.

As participants in a physical, biological, economic, political, and social environment characterized by continuing change, we find it helpful to be able to study these changes and associated rates of change through the construction of mathematical models (Sections 12.4 and 12.5) and other quantifying techniques. Calculus is invaluable in this endeavor. There are two major classifications of problems that are particularly susceptible to the techniques of calculus.

1. A relationship between two variables is known; find a second relationship that involves the original two variables and the *rate of change* of one variable with respect to the other.
2. The converse of classification 1, where the second relationship is known and the first relationship is sought.

As examples of these two types of problems consider a population of objects, such as people, plants, voters, dollars, or blood cells. Let y represent the number of objects at a given time t. As time t passes, y may vary. In fact, it may be possible to establish an equation that will model the relationship between y and t. Suppose that this is the case and that this relationship is expressed by an equation

$$y = f(t).$$

Now if t_1 and t_2 are two different measures of time, then $f(t_1)$ and $f(t_2)$ are measures of the corresponding populations. It is customary to represent the change $t_2 - t_1$ in time by the symbol Δt, and likewise, the change in population by

$$\Delta y = f(t_2) - f(t_1).$$

If the change in population is divided by the change in time, then the result is referred to as the *average rate of change* of population with respect to time over the time interval from t_1 to t_2. This average rate of change of y with respect to t can be expressed as

$$\frac{\Delta y}{\Delta t} = \frac{f(t_2) - f(t_1)}{t_2 - t_1}.$$

Under certain conditions,† if t_1 remains unchanged, and smaller and smaller changes $t_2 - t_1$ in time are considered, then the average rate of change

$$\frac{f(t_2) - f(t_1)}{t_2 - t_1}$$

will approximate what might be called the *immediate* or *instantaneous rate of change* of population with respect to time at time t_1. In other words, we are raising the following question: What value does $\Delta y/\Delta t$ approach as Δt approaches zero? The answer to this question is so important that we give it a special name; we say that it is the value of the **derivative** of y with respect to t at time t_1. The question of finding the immediate or instantaneous rate of change of one variable with respect to another variable, when the relationship between them is known, occurs so frequently in applied problems that it is difficult to overemphasize its importance.

Other very important applications arise from the inverse question of finding the relationship between two variables when the instantaneous rate of change is known. For example, if we reason that the rate of change of a population y with respect to time t is proportional to the current population, then $\Delta y/\Delta t$ approaches ky as Δt approaches zero, where k is the constant of proportionality. From this assumption concerning the rate of change, the techniques of calculus can be used to find an equation representing the relationship between the population y and the time t; this relationship can be used as a model to make projections into the future.

† One condition is that the range of f is assumed to be an interval of real numbers.

In a serious approach to the solution of the many applied problems similar to those just mentioned, we must gain some precision in our understanding of the terms and concepts that have been mentioned so far. Hence in this chapter we shall deal with those topics that are essential to the main concept of the derivative that is to be presented in Chapter 10.

Primary credit for the discovery of calculus is usually given to Isaac Newton (1642–1727) and Gottfried Leibniz (1646–1716). The student who wishes to read an interesting historical article about calculus before pursuing the study of the subject in detail is referred to Newman's book [50], pp. 53–62.

EXERCISES

Suggested minimum assignment: Exercises 1, 2, and 4.

1. Suppose that the number n of bacteria in a culture after t hours is given by Table 9.1. By guessing a formula relating n and t, determine a mathematical model that expresses the number of bacteria at any given time. Be sure that your formula checks for the data in Table 9.1.

Table 9.1

n	100	150	200	250	300
t	0	1	2	3	4

2. Suppose that a child attempts to measure the area of a circle. Let A represent the area and r represent the radius of the circle. The child uses the formula $A = 3r^2$ as the mathematical model of the relationship between the area and radius of a circle. What are the advantages and disadvantages of this model over the formula $A = \pi r^2$?

3. Let R_0 be the sensation evoked by a stimulus of intensity S_0. Suppose that ΔR is the increase in sensation caused by an increase of ΔS in the stimulus. It has been shown that

$$\frac{\Delta S}{S_0} = \Delta R.$$

Therefore, for a given increase of ΔS in the stimulus, the increase in sensation is greater for a relatively small original stimulus S_0 than for a relatively large original stimulus. Verify this fact for the case that $\Delta S = 2$ by computing ΔR under the assumption that $S_0 = 50$ and also under the assumption that $S_0 = 100$. For further discussion see pp. 23–29 of *Quantification in Psychology* by Hays [28].

4. An automobile manufacturer wants to test a new car to find certain information about its performance. Let y be the distance in feet that the car travels from the starting point in t seconds. Suppose that $y = 10$ when $t = 1$ and $y = 100$ when $t = 3$. What is the change Δy in distance

corresponding to the change in time of $\Delta t = 3 - 1 = 2$ seconds? Also find $\Delta y/\Delta t$ and interpret its meaning.

5. Find in your school library an article related to the history of calculus, and write a short summary of what you find.

9.2 Limits

The concept of a limit is one of the most important concepts in mathematics and is fundamental in the development of calculus. We introduce the concept by means of two examples.

Example 1
Recall from Section 3.9 that if $10 million is deposited at 6 per cent compounded n times a year, then the amount at the end of 1 year is

$$f(n) = (10{,}000{,}000)\left(1 + \frac{.06}{n}\right)^n.$$

In Table 9.2 the amount $f(n)$ is shown for several values of n (dollar signs omitted). It is not surprising that $f(n)$ gets larger as n gets larger, because if interest is compounded more frequently, one obtains more benefit from "interest on interest." What may be surprising is that the numbers near the bottom of the $f(n)$ column in Table 9.2 appear to show very little change; these numbers appear to be getting close to some **limit** that $f(n)$ cannot exceed, no matter how large n becomes. In Chapter 14 we shall be able to show that, as n gets very large, $f(n)$ approaches but never exceeds the limit $L = 10{,}618{,}365.47$. This is often expressed symbolically as

$$f(n) \to L \quad \text{as } n \to \infty,$$

where $f(n) \to L$ is read "$f(n)$ approaches L" and $n \to \infty$ is understood to mean that n increases without bound and is read "n becomes infinite." The same idea is also expressed by the notation

$$\lim_{n \to \infty} f(n) = L,$$

Table 9.2

n	$f(n)$
1 (annually)	10,600,000.00
2 (semiannually)	10,609,000.00
4 (quarterly)	10,613,635.51
12 (monthly)	10,616,778.12
52 (weekly)	10,617,998.20
365 (daily)	10,618,313.11
1000	10,618,346.35
1,000,000	10,618,365.44
1,500,000	10,618,365.45

which says that "the limit of $f(n)$ as n becomes infinite is L." Note that "lim" is an abbreviation of the word "limit." (The value $L = 10{,}618{,}365.47$ corresponds to compounding interest "continuously," as more and more banks are doing—further discussion will appear in Section 14.1. Notice that compounding continuously gives only slightly more interest than compounding daily.) ∎

Example 2
Consider again the sequence

$$\frac{2}{3}, \frac{3}{5}, \frac{4}{7}, \frac{5}{9}, \ldots, \frac{n+1}{2n+1}, \ldots$$

of Example 1 of Section 3.9. The 100th term of the sequence is $\frac{101}{201}$, and we shall show later in this section that the terms of the sequence get very close to the limit $L = \frac{1}{2}$ for large values of n. We write

$$f(n) \to \frac{1}{2} \quad \text{as } n \to \infty$$

or

$$\lim_{n \to \infty} f(n) = \lim_{n \to \infty} \frac{n+1}{2n+1} = \frac{1}{2}.$$

In Figure 9.1 the situation may be viewed geometrically. For the type of function known as a sequence, the independent variable n is called a **discrete variable**; on the graph there are gaps between the allowable values (positive integers) of the discrete independent variable. We also want to study limits of values of functions of a **continuous variable**, in which case the independent variable takes on all real values in an interval. Consider

$$g(x) = \frac{x+1}{2x+1},$$

where the domain of g is the set of all real numbers except $-\frac{1}{2}$. A portion of the graph of g is shown in Figure 9.2. In this case also, $g(x)$ can be made as close to $\frac{1}{2}$ as desired by choosing x large enough, and we write

$$\lim_{x \to \infty} g(x) = \lim_{x \to \infty} \frac{x+1}{2x+1} = \frac{1}{2}. \quad ∎$$

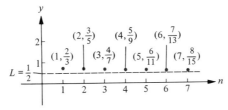

Figure 9.1. $y = f(n) = \dfrac{n+1}{2n+1}$.

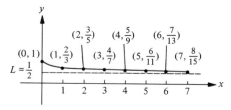

Figure 9.2. $y = g(x) = \dfrac{x+1}{2x+1}$.

In Definition 1 the meaning of $\lim_{x \to \infty} f(x) = L$ is given. It will suffice for the purposes of this text, but one should recognize that some precision can be gained by the introduction of more symbols, as shown in Exercise 6.

▶ **Definition 1**
Suppose that the domain of f includes arbitrarily large numbers. The statement "**the limit of $f(x)$, as x becomes positively infinite, is the real number L** [written $\lim_{x \to \infty} f(x) = L$]" means that $|f(x) - L|$ can be made smaller than any given positive number by requiring that x (in the domain of f) be greater than a sufficiently large positive number.

Definition 1 applies whether x is a discrete variable or a continuous variable. In trying to understand Definition 1, the reader might review Figures 9.1 and 9.2 again, keeping in mind that the number L in Definition 1 is equal to $\frac{1}{2}$ in each case. Notice, for example, in Figure 9.2 that $|g(x) - \frac{1}{2}| < \frac{1}{10}$ whenever $x > 2$ [$\frac{1}{10}$ is the difference between $g(2) = \frac{3}{5}$ and $L = \frac{1}{2}$]. Moreover, $|g(x) - \frac{1}{2}|$ can be made even smaller by requiring that x be larger (see Exercise 11).

Not more than one number L can be the limit in Definition 1. Whenever the limit equals one real number (such as $L = \frac{1}{2}$), we say that the **limit exists**. Sometimes $\lim_{x \to \infty} f(x)$ does not exist—for example, $\lim_{x \to \infty} x^2$ does not exist because x^2 does not get close to any one number L as x gets large. Another example is the sequence

$$1, 2, 1, 2, 1, 2, \ldots, a_n, \ldots,$$

whose terms alternate between 1 and 2; the terms do not *all* get close to any one number L as one lets the term number n get large, and we say that the sequence does not have a limit (in other words, $\lim_{n \to \infty} a_n$ does not exist).

Incidentally, a formula for a_n above is

$$a_n = \frac{3 + (-1)^n}{2}.$$

Theorem 1 is very useful in the calculation of limits. Its proof, however, is beyond the scope of this text.

▶ **Theorem 1**

If $\lim_{x\to\infty} f(x)$ and $\lim_{x\to\infty} g(x)$ exist, then

(i) $\lim_{x\to\infty} [f(x) + g(x)] = \lim_{x\to\infty} f(x) + \lim_{x\to\infty} g(x)$.

(ii) $\lim_{x\to\infty} [f(x)g(x)] = \left[\lim_{x\to\infty} f(x)\right]\left[\lim_{x\to\infty} g(x)\right]$.

(iii) $\lim_{x\to\infty} \dfrac{f(x)}{g(x)} = \dfrac{\lim_{x\to\infty} f(x)}{\lim_{x\to\infty} g(x)}$, provided that $\lim_{x\to\infty} g(x) \neq 0$.

(iv) $\lim_{x\to\infty} k = k$, where k is a real number.

(v) $\lim_{x\to\infty} [kf(x)] = k \lim_{x\to\infty} f(x)$, where k is a real number.

(vi) $\lim_{x\to\infty} \dfrac{1}{x^m} = 0$, provided that $m > 0$.

(vii) $\lim_{n\to\infty} q^n = 0$, provided that $-1 < q < 1$ and n takes only positive integral values. ◀

Example 3

Again consider $\lim_{x\to\infty} [(x + 1)/(2x + 1)]$ from Example 2. In order to evaluate this limit of values of a rational function, first divide numerator and denominator by the highest power of x in the fraction, and then apply various parts of Theorem 1 as follows:

$$\lim_{x\to\infty} \frac{x+1}{2x+1} = \lim_{x\to\infty} \frac{1 + (1/x)}{2 + (1/x)}$$

$$= \frac{\lim_{x\to\infty}[1 + (1/x)]}{\lim_{x\to\infty}[2 + (1/x)]} \qquad \text{by Theorem 1}(iii)$$

$$= \frac{\lim_{x\to\infty} 1 + \lim_{x\to\infty}(1/x)}{\lim_{x\to\infty} 2 + \lim_{x\to\infty}(1/x)} \qquad \text{by Theorem 1}(i)$$

$$= \frac{1 + \lim_{x\to\infty}(1/x)}{2 + \lim_{x\to\infty}(1/x)} \qquad \text{by Theorem 1}(iv)$$

$$= \frac{1 + 0}{2 + 0} \qquad \text{by Theorem 1}(vi)$$

$$= \frac{1}{2}.$$

The above calculations are also valid for showing that

$$\lim_{n \to \infty} \frac{n+1}{2n+1} = \frac{1}{2},$$

where n takes only positive integral values (see the discussion in Example 2). ∎

Part (*i*) of Theorem 1 also holds for subtraction as well as for addition, and parts (*i*) and (*ii*) can be extended to more than two functions.

Example 4

$$\lim_{x \to \infty} \left[5 - 3\left(\frac{x+1}{2x+1}\right) + \frac{4}{x^2} \right] = 5 - 3\left(\frac{1}{2}\right) + 0 = \frac{7}{2}$$

by Theorem 1(*i*), (*iv*), (*v*), and (*vi*) and the result of Example 3. ∎

Example 5
If n takes only positive integral values,

$$\lim_{n \to \infty} (\tfrac{1}{2})^n = 0 \quad \text{and} \quad \lim_{n \to \infty} (-\tfrac{1}{2})^n = 0$$

by Theorem 1(*vii*). ∎

One can also consider limits of values of functions in which the independent variable x becomes negatively infinite instead of positively infinite. This concept can be defined by making appropriate changes in Definition 1 (see Exercise 12). For example,

$$\lim_{x \to -\infty} \frac{1}{x} = 0 \tag{9.1}$$

is correct because as x takes on values such as -10, -100, and -1000, the expression $1/x$ becomes closer and closer to 0. Parts (*i*) through (*v*) of Theorem 1 hold if $x \to -\infty$ instead of $x \to \infty$; part (*vi*) holds for values of m for which $1/x^m$ has meaning [see (9.1)]; but part (*vii*) is *not* valid for $n \to -\infty$. By following steps exactly like those in Example 3, it can be shown that

$$\lim_{x \to -\infty} \frac{x+1}{2x+1} = \frac{1}{2}.$$

APPLICATIONS

Example 6
Members of legislative bodies are often called on to respond with respect to a certain issue on repeated occasions. In time these legislators may alter

their positions for a variety of reasons. On pp. 538–540 of [9] a formula is developed for the probability that a legislator will vote yes on the nth yes–no roll call pertaining to the same issue. The formula is

$$p_n = (1-2p)^n p_0 + \frac{1-(1-2p)^n}{2}$$

or

$$p_n = \frac{1}{2} + \left(p_0 - \frac{1}{2}\right)(1-2p)^n, \qquad (9.2)$$

where $n =$ number of roll calls taken;
$p_n =$ probability that the legislator will vote yes on the nth roll call;
$p =$ probability that the legislator will reverse his position on successive roll calls;
$p_0 =$ probability that the legislator favors the issue upon initial consideration.

If $0 < p < 1$, then $-1 < (1-2p) < 1$; hence $(1-2p)^n$ gets closer to zero as n gets larger, by Theorem 1(vii). In other words,

$$\lim_{n \to \infty} (1-2p)^n = 0.$$

By the use of various parts of Theorem 1, the limit of p_n as $n \to \infty$ is found from (9.2) to be

$$\lim_{n \to \infty} p_n = \lim_{n \to \infty} \left[\tfrac{1}{2} + (p_0 - \tfrac{1}{2})(1-2p)^n\right]$$
$$= \left(\lim_{n \to \infty} \tfrac{1}{2}\right) + \left[\lim_{n \to \infty} (p_0 - \tfrac{1}{2})\right]\left[\lim_{n \to \infty} (1-2p)^n\right]$$
$$= (\tfrac{1}{2}) + [(p_0 - \tfrac{1}{2})](0)$$
$$= \tfrac{1}{2}.$$

The result indicates that, subject to the assumptions made in the derivation of the formula, the probability of a legislator's voting yes on a given issue approaches $\tfrac{1}{2}$ as the number of roll calls increases. Notice that the limit is $\tfrac{1}{2}$ regardless of the values of p and p_0. This result helps explain the rationality of perseverance by a minority in favor of a certain issue. ∎

Example 7
An expression

$$a_1 + a_2 + a_3 + \cdots + a_n + \cdots, \qquad (9.3)$$

in which addition of an infinite number of real numbers is indicated, is called an **infinite series**. For example,

$$\frac{1}{2} + \frac{1}{4} + \frac{1}{8} + \frac{1}{16} + \cdots + \frac{1}{2^n} + \cdots \qquad (9.4)$$

is an infinite series. This particular infinite series was encountered in a previous application concerning probability (Example 6 of Section 3.9). In

this example we shall develop a formula for finding what we shall call the **sum** of (9.4); in the process the reader will be introduced to the subject of infinite series, which has many applications.

In (9.3) let
$$s_1 = a_1,$$
$$s_2 = a_1 + a_2,$$
$$s_3 = a_1 + a_2 + a_3,$$

and, in general, for any positive integer n,
$$s_n = a_1 + a_2 + a_3 + \cdots + a_n.$$

The sequence $\{s_n\}$ is called the **sequence of partial sums** of the infinite series (9.3). If $\lim_{n \to \infty} s_n$ exists and equals S, then S is called the **sum** of the infinite series, and we write
$$S = a_1 + a_2 + a_3 + \cdots + a_n + \cdots.$$

An infinite series of the form
$$a + ar + ar^2 + ar^3 + \cdots + ar^{n-1} + \cdots \tag{9.5}$$

is called a **geometric series.** The number a is the first term of the infinite series, and the number r is called the **common ratio.** For the case that $-1 < r < 1$, let us try to find, if possible, the sum S of (9.5). Observe that
$$s_n = a + ar + ar^2 + ar^3 + \cdots + ar^{n-1}$$
and hence
$$rs_n = ar + ar^2 + ar^3 + \cdots + ar^{n-1} + ar^n.$$

Thus, by subtraction,
$$s_n - rs_n = a - ar^n.$$
Therefore,
$$s_n(1 - r) = a(1 - r^n)$$
and
$$s_n = \frac{a}{1-r}(1 - r^n).$$

Using $S = \lim_{n \to \infty} s_n$ and Theorem 1, we find that
$$S = \lim_{n \to \infty} \frac{a}{1-r}(1 - r^n)$$
$$= \frac{a}{1-r}\left(\lim_{n \to \infty} 1 - \lim_{n \to \infty} r^n\right)$$
$$= \frac{a}{1-r}(1 - 0)$$
$$= \frac{a}{1-r}.$$

Chap. 9] Limits and Continuity

Notice especially that $\lim_{n \to \infty} r^n = 0$ by Theorem 1(vii), since we assumed that $-1 < r < 1$.

Let us demonstrate how the formula $S = a/(1 - r)$ can be used to find the sum of (9.4). The infinite series (9.4) is a geometric series with first term $a = \frac{1}{2}$ and common ratio $r = \frac{1}{2}$. Therefore,

$$S = \frac{a}{1-r} = \frac{\frac{1}{2}}{1-\frac{1}{2}} = 1. \blacksquare$$

EXERCISES

Suggested minimum assignment: Exercises 1, 2, 5, and 9.

1. Plot the first six terms of the sequence $\left\{\dfrac{3n}{n+1}\right\}$ in a manner similar to that used in Figure 9.1. Estimate the value of $\lim_{n\to\infty} \dfrac{3n}{n+1}$.

2. If $g(x) = \dfrac{3x}{x+1}$, provided that $x \neq -1$, graph the portion of g for which $x > 0$ (see Exercise 1). Estimate the value of $\lim_{x\to\infty} \dfrac{3x}{x+1}$.

3. Repeat Exercise 1 for $\left\{\dfrac{n+3}{n+2}\right\}$.

4. If $g(x) = \dfrac{x+3}{x+2}$, provided that $x \neq -2$, graph the portion of g for which $x > 0$ (see Exercise 3). As $x \to \infty$, $g(x) \to L$. Estimate L.

5. Plot enough points to estimate the limit of each of the following sequences, provided that a limit exists.
   (a) $2, -2, 2, -2, 2, -2, \ldots, (-1)^{n+1}(2), \ldots$
   (b) $\left\{\dfrac{n^2 + 3}{n^2 + 2}\right\}$.
   (c) $\left\{\dfrac{n}{n^2 + 2}\right\}$.
   (d) $\dfrac{1}{2}, -\dfrac{1}{3}, \dfrac{1}{4}, -\dfrac{1}{5}, \dfrac{1}{6}, -\dfrac{1}{7}, \ldots, (-1)^{n+1}\dfrac{1}{n+1}, \ldots$
   (e) $5, 10, 15, 20, 25, \ldots, 5n, \ldots$

6. The meaning of $\lim_{x\to\infty} f(x) = L$ in Definition 1 can be stated more precisely as follows: For any $\varepsilon > 0$ there exists a number $N > 0$ such that $|f(x) - L| < \varepsilon$ whenever x is in the domain of f and $x > N$. Draw a sketch to illustrate this definition and label ε, L, and N on your sketch.

7. Explain why $\lim_{x\to\infty} [f(x) - g(x)] = \lim_{x\to\infty} f(x) - \lim_{x\to\infty} g(x)$ follows from Theorem 1(i) and (v).

Sec. 9.2] Limits

8. Use Theorem 1 to evaluate each of the following if possible.

(a) $\lim\limits_{x \to \infty} \dfrac{2x + 7}{x + 3}$.   (b) $\lim\limits_{x \to \infty} \dfrac{x^2 + 1}{x + 2x^2}$.

(c) $\lim\limits_{x \to \infty} (2x - 3)$.   (d) $\lim\limits_{x \to \infty} \dfrac{x^2}{x^3 - 3}$.

9. Use Theorem 1 (or its counterpart for $x \to -\infty$) to evaluate each of the following, if possible.

(a) $\lim\limits_{x \to \infty} \dfrac{-3}{x^2}$.   (b) $\lim\limits_{x \to \infty} \dfrac{x^2}{x^2 + 2}$.

(c) $\lim\limits_{x \to -\infty} \dfrac{2x + 7}{x}$.   (d) $\lim\limits_{x \to -\infty} \dfrac{2x + 7}{x} \cdot \dfrac{x^3}{1975 + 2x^3}$.

(e) $\lim\limits_{n \to \infty} (-\tfrac{1}{3})^n$, where n takes only positive integral values.

10. If digital computer facilities are available, verify in Table 9.2 that $f(1000) = 10{,}618{,}346.35$ is correct.

11. In Figure 9.2 $|g(x) - \tfrac{1}{2}| < \tfrac{1}{30}$ whenever $x > N$. What is the smallest number that N can equal?

12. State the meaning of $\lim\limits_{x \to -\infty} f(x) = L$ by making the appropriate changes in Definition 1.

13. Find the sum of each of the following geometric series.

(a) $3 + \tfrac{9}{4} + \tfrac{27}{16} + \cdots$.
(b) $\tfrac{1}{2} - \tfrac{1}{4} + \tfrac{1}{8} - \tfrac{1}{16} + \cdots$.

14. Change the repeating decimal $.272727\ldots$ to a rational number by finding the sum of the geometric series

$$.27 + .0027 + .000027 + \cdots.$$

9.3 Limits (Continued)

In Section 9.2 we considered limits of values of functions as the independent variable became either positively or negatively infinite. Many applied problems involve the limit of function values as the independent variable approaches a finite number.

Example 1
In the population problem of Section 9.1 the average rate of change of population $y = f(t)$ with respect to time over a time interval $[t_1, t_2]$ was found to be

$$\frac{\Delta y}{\Delta t} = \frac{f(t_2) - f(t_1)}{t_2 - t_1}.$$

The immediate, or instantaneous, rate of change of population with respect to time at $t = t_1$ can be found by finding the value to which

$$\frac{f(t) - f(t_1)}{t - t_1}$$

gets close as t gets close to t_1. Suppose that $f(t) = t^2$, and we want to find the instantaneous rate of change of population with respect to time at $t = 3$.

$$\frac{f(t) - f(t_1)}{t - t_1} = \frac{t^2 - 3^2}{t - 3} = \frac{(t + 3)(t - 3)}{t - 3} = t + 3 \quad \text{when } t \neq 3.$$

Intuitively, one expects $(t + 3)$ to get close to 6 as t gets close to 3. This can be stated

$$t + 3 \to 6 \quad \text{as } t \to 3$$

or

$$\lim_{t \to 3}(t + 3) = 6. \quad \blacksquare$$

Before giving a formal definition of the meaning of $\lim_{x \to c} f(x) = L$, where c and L are real numbers, we shall try to impart an intuitive understanding of its meaning by means of more examples.

Example 2
Suppose that a function f whose domain is the set of all real numbers is defined by

$$f(x) = \begin{cases} 2x + 1 & \text{if } x \neq 2, \\ \frac{5}{2} & \text{if } x = 2. \end{cases}$$

The graph of f is shown in Figure 9.3 (recall that $y = 2x + 1$ is a line with slope 2 and y-intercept 1). Despite the fact that $f(2) = \frac{5}{2}$, it is correct to write

$$\lim_{x \to 2} f(x) = 5$$

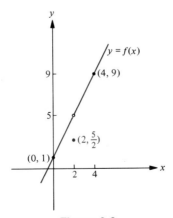

Figure 9.3

because the closer x is to 2 (but *not* equal to 2), the closer $f(x)$ is to 5. The value of f at 2 makes no difference at all in determining the limit of $f(x)$ as x approaches 2; even if $f(2)$ were undefined, $\lim_{x \to 2} f(x)$ would still equal 5. Also from Figure 9.3 it can be seen that if x is very close to 4, then $f(x)$ is very close to 9, and we write

$$f(x) \to 9 \quad \text{as } x \to 4$$

or

$$\lim_{x \to 4} f(x) = 9. \blacksquare$$

Example 3
Referring back to Example 2 of Section 3.9 and to Figure 3.23, we can estimate from the figure that

$$\lim_{x \to 40} f(x) = 8.$$

However,

$$\lim_{x \to 10} f(x)$$

does not exist; this is because $f(x)$ is 6 when x is slightly less than 10, but $f(x)$ is 8 when x is slightly greater than 10. In order for the limit as x approaches 10 to exist, $f(x)$ must be near to some *one* value whenever x is *both* slightly less than 10 *and* slightly greater than 10. $\blacksquare$

In Definition 2 the meaning of $\lim_{x \to c} f(x) = L$ is given. It will suffice for the purposes of this text, but again one should recognize that precision can be gained by the introduction of more symbols, as shown in Exercise 4.

▶ **Definition 2**
Suppose that a function f is defined at every number in some open interval containing the number c, except possibly at the number c. The statement "**the limit of $f(x)$, as x approaches c, is the real number L** [written $\lim_{x \to c} f(x) = L$]" means that $|f(x) - L|$ can be made smaller than any given positive number by requiring that x be sufficiently close to c (but not equal to c).

As we did in Section 9.2, we present a theorem without proof that will be used in the calculation of limits.

▶ **Theorem 2**
Let c be a real number and suppose that $\lim_{x \to c} f(x)$ and $\lim_{x \to c} g(x)$ exist. Then

(i) $\lim_{x \to c} [f(x) + g(x)] = \lim_{x \to c} f(x) + \lim_{x \to c} g(x)$.

(ii) $\lim_{x \to c} [f(x) g(x)] = \left[\lim_{x \to c} f(x) \right] \left[\lim_{x \to c} g(x) \right]$.

(iii) $\lim_{x \to c} \dfrac{f(x)}{g(x)} = \dfrac{\lim_{x \to c} f(x)}{\lim_{x \to c} g(x)}$, provided that $\lim_{x \to c} g(x) \neq 0$.

(iv) $\lim_{x \to c} k = k$, where k is a real number.

(v) $\lim_{x \to c} [k f(x)] = k \lim_{x \to c} f(x)$, where k is a real number.

(vi) $\lim_{x \to c} x = c$.

(vii) $\lim_{x \to c} \sqrt[n]{f(x)} = \sqrt[n]{\lim_{x \to c} f(x)}$, provided that either n is a positive odd integer, or $\lim_{x \to c} f(x) > 0$ and n is a positive even integer. ◀

Theorem 2(i) also holds for subtraction as well as for addition; parts (i) and (ii) can be extended to more than two functions.

Example 4

Evaluate $\lim_{x \to 2} \dfrac{3x^2 + 4}{x}$.

$$\lim_{x \to 2} \dfrac{3x^2 + 4}{x} = \dfrac{\lim_{x \to 2}(3x^2 + 4)}{\lim_{x \to 2} x} \qquad \text{by Theorem 2}(iii)$$

$$= \dfrac{\lim_{x \to 2}(3x^2 + 4)}{2} \qquad \text{by Theorem 2}(vi)$$

$$= \dfrac{\lim_{x \to 2} 3x^2 + \lim_{x \to 2} 4}{2} \qquad \text{by Theorem 2}(i)$$

$$= \dfrac{3 \lim_{x \to 2} x^2 + \lim_{x \to 2} 4}{2} \qquad \text{by Theorem 2}(v)$$

$$= \dfrac{3 \left(\lim_{x \to 2} x\right)\left(\lim_{x \to 2} x\right) + \lim_{x \to 2} 4}{2} \qquad \text{by Theorem 2}(ii)$$

$$= \dfrac{(3)(2)(2) + 4}{2} \qquad \text{by Theorem 2}(iv) \text{ and } (vi)$$

$$= 8. \quad \blacksquare$$

Example 5

$\lim_{x \to 9} \sqrt{x} = \sqrt{\lim_{x \to 9} x}$ by Theorem 2(vii)

$\qquad\; = \sqrt{9}$ by Theorem 2(vi)

$\qquad\; = 3.$ ∎

Sec. 9.3] Limits

Example 6
The limit $\lim_{x \to 0} 1/x^2$ does not exist. Rather than approach any one real number, the expression $1/x^2$ becomes larger and larger as x gets closer and closer to zero. ∎

In Example 1 we were concerned with $\lim_{t \to 3} [(t^2 - 9)/(t - 3)]$. If t is close to 3, both numerator and denominator of $(t^2 - 9)/(t - 3)$ are close to zero, and Theorem 2(*iii*) does not apply. When this happens, try to obtain a cancellation—it is permissible to cancel $(t - 3)$ from numerator and denominator because in evaluating the limit, t approaches 3 but is not equal to 3, and hence we are not dividing by zero when the cancellation is made. Example 7 is similar to ones that we shall encounter in Chapter 10. Notice that the numerator and denominator each approach zero again, so a cancellation is made to overcome this difficulty.

Example 7

$$\lim_{\Delta x \to 0} \frac{(2 + \Delta x)^2 - 4}{\Delta x}$$

$$= \lim_{\Delta x \to 0} \frac{4\Delta x + (\Delta x)^2}{\Delta x} \qquad \text{by expansion of } (2 + \Delta x)^2 \text{ and simplification}$$

$$= \lim_{\Delta x \to 0} (4 + \Delta x) \qquad \text{by cancellation of } \Delta x, \text{ since } \Delta x \neq 0$$

$$= \lim_{\Delta x \to 0} 4 + \lim_{\Delta x \to 0} \Delta x \qquad \text{by Theorem 2}(i)$$

$$= 4 + 0 \qquad \text{by Theorem 2}(iv) \text{ and } (vi)$$

$$= 4.$$

The reader should think of (Δx) as a single quantity in this example and realize that we evaluated a limit as $\Delta x \to 0$ of a function of Δx, whereas previous examples considered a limit as $x \to c$ of a function of x. ∎

APPLICATIONS

*Example 8 Cumulative Probability Distribution Function
In Example 6 of Section 3.9 an experiment was discussed in which a coin was flipped until a head appeared. A random variable or measurement was defined over the sample space U. The set of values (the range) of this discrete measurement constituted a new sample space $S = \{1, 2, 3, \ldots\}$. The probability of an event E, where E is a subset of S, was expressed in terms of the weighting function w; for example, if $E = \{1, 2, 3\}$, $P(E) = w(1) + w(2) + w(3)$. Recall that

$$w(s) = \begin{cases} \dfrac{1}{2} & \text{if } s = 1, \\ \dfrac{1}{4} & \text{if } s = 2, \\ \vdots & \\ \dfrac{1}{2^n} & \text{if } s = n, \\ \vdots & \end{cases}$$

In order to consider a second means of expressing the probability of an event of the type $E = \{1, 2, \ldots, k\}$, we shall introduce a function F whose domain is the set of *all* real numbers. Let F be related to w in such a way that $F(x)$ equals the sum of the weights of those measurement values that are less than or equal to x. In other words, the value of F for any real number x is the cumulative weight of those elements of S that are less than or equal to x; the function F is called the **cumulative probability distribution function**. In our illustration

$$F(x) = \begin{cases} 0 & \text{if } x < 1, \\ \dfrac{1}{2} & \text{if } 1 \le x < 2, \\ \dfrac{3}{4} & \text{if } 2 \le x < 3, \\ \vdots & \\ \dfrac{2^n - 1}{2^n} & \text{if } n \le x < n + 1, \\ \vdots & \end{cases}$$

Portions of the graphs of w and F are given in Figures 9.4 and 9.5. Notice that the domain of F is the set of *all* real numbers, whereas the domain of w is the sample space S. The probability of the event $E = \{1, 2\}$, which is to say the probability that $s \le 2$, can be expressed as

$$P(E) = F(2) = \tfrac{3}{4}.$$

In general, if $E = \{s : s \le k\}$, then $P(E) = F(k)$.

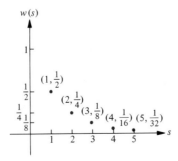

Figure 9.4. *Weighting function.*

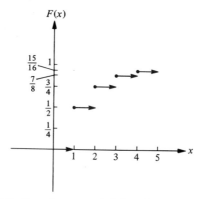

Figure 9.5. *Cumulative probability distribution function.*

A cumulative probability distribution function F may be defined independently of its relation to w. A **cumulative probability distribution function** is any function F, whose domain is the set of all real numbers, possessing the following four properties:

1. $\lim_{x \to -\infty} F(x) = 0$.
2. $\lim_{x \to \infty} F(x) = 1$.
3. If $x_1 < x_2$, then $F(x_1) \leq F(x_2)$.
4. $\lim_{\Delta x \to 0} F(x + \Delta x) = F(x)$ for any real number x, where $\Delta x > 0$.

This abstract definition is consistent with our previous discussion of F, and the key point in understanding this consistency is to associate $F(k)$ with $P(E)$, where $E = \{s : s \leq k\}$.

Next we explain why the function F of the illustration used in this example possesses all four properties required of a cumulative probability distribution function.

1. $\lim_{x \to -\infty} F(x) = 0$ because $F(x) = 0$ for $x < 1$.
2. $\lim_{x \to \infty} F(x) = \lim_{n \to \infty} [(2^n - 1)/2^n] = \lim_{n \to \infty} 1 - \lim_{n \to \infty} (\tfrac{1}{2})^n = 1 - 0 = 1$.
3. The graph of F never "falls" as one moves from left to right along the curve. In other words, if $y = F(x)$, then y never decreases as x increases. This is clearly true from the definition of F as illustrated in Figure 9.5.
4. Notice that the graph of F consists of portions of *horizontal* lines, each of which is defined on an interval that does not include the right endpoint. For example, $F(x) = \tfrac{1}{2}$ on the interval $[1, 2)$, and that interval does not include the right endpoint 2. Therefore, for any real number x, a sufficiently small positive change Δx in x will cause no change in the value of F. Hence we can say that

$$\lim_{\Delta x \to 0} F(x + \Delta x) = \lim_{\Delta x \to 0} F(x) = F(x), \quad \text{where } \Delta x > 0.$$

If the intervals above had included the right endpoint, this property would not be satisfied when x is equal to a right endpoint.

In this example we have illustrated the concept of a cumulative probability distribution function for the case of a discrete measurement. An example for the case of a continuous measurement will be given in Example 7 of Section 10.2. ∎

EXERCISES

Suggested minimum assignment: Exercises 3, 5, 7, 8, and 9(e).

1. By the observation of Figure 3.24 on p. 000, estimate the values, if possible, of each of the following.
   (a) $\lim_{x \to 0} f(x)$.   (b) $\lim_{x \to 2} f(x)$.

2. (a) Sketch the function f defined by
$$f(x) = \begin{cases} -x & \text{if } x \leq -2, \\ 2x & \text{if } x > -2. \end{cases}$$
   (b) Estimate the value of $\lim_{x \to -2} f(x)$, if it exists.
   (c) Estimate the value of $\lim_{x \to 1} f(x)$, if it exists.

3. From Figure 9.6 estimate the value of each of the following or state that it does not exist.
   (a) $f(-5)$.   (b) $\lim_{x \to -5} f(x)$.   (c) $f(-3)$.   (d) $\lim_{x \to -3} f(x)$.
   (e) $\lim_{x \to -2} f(x)$.   (f) $\lim_{x \to 1} f(x)$.   (g) $\lim_{x \to 3} f(x)$.

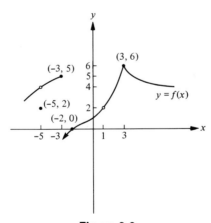

Figure 9.6

4. A more precise definition of the meaning of $\lim_{x \to c} f(x) = L$ than that given in Definition 2 is the following: For any $\varepsilon > 0$ there exists a $\delta > 0$ such

that $|f(x) - L| < \varepsilon$ whenever $0 < |x - c| < \delta$. Draw a sketch illustrating the statement above, showing ε, δ, c, and L on your figure.

5. Use Theorem 2 to evaluate each of the following.

   (a) $\lim_{x \to 4} \dfrac{x^2 + 3}{5x + 2}$.   (b) $\lim_{x \to 4} (2x - \sqrt[3]{16x})$.

6. Use Theorem 2 to evaluate each of the following.

   (a) $\lim_{x \to -2} \dfrac{x}{3x + 1}$.   (b) $\lim_{x \to 3} \dfrac{x^2 + 3x}{x^3 + 1}$.

7. In Example 1 find the instantaneous rate of change of population with respect to time at $t = 4$ by evaluating $\lim_{t \to 4} \dfrac{t^2 - 16}{t - 4}$.

8. In Example 1 an alternative way of writing $\dfrac{f(t) - f(t_1)}{t - t_1}$ is to let $t = t_1 + \Delta t$; then the expression becomes $\dfrac{f(t_1 + \Delta t) - f(t_1)}{\Delta t}$. Use the latter expression to rework Exercise 7, after noting that $\Delta t \to 0$ as $t \to t_1$; that is, evaluate $\lim_{\Delta t \to 0} \dfrac{(4 + \Delta t)^2 - 4^2}{\Delta t}$.

9. Evaluate each of the following, if possible.

   (a) $\lim_{x \to 5} \dfrac{x^2 - 25}{x^2 - 5x}$.   (b) $\lim_{x \to 2} \dfrac{3x^2 - 7x + 2}{x - 2}$.

   (c) $\lim_{x \to 5} \dfrac{x - 5}{x + 5}$.   (d) $\lim_{x \to 5} \dfrac{x + 5}{x - 5}$.

   (e) $\lim_{\Delta x \to 0} \dfrac{(1 + \Delta x)^2 - 1}{\Delta x}$.

10. Evaluate $\lim_{t \to -2} \dfrac{t^3 + 8}{t + 2}$.

11. Describe the cumulative probability distribution function F that corresponds to the weighting function given in Table 3.3.

9.4 Continuity

Many results in calculus are valid if a particular function in question possesses a property known as **continuity**. In this section we shall distinguish between continuous and discontinuous functions.

Example 1

Company A sells an item by the pound (or fraction thereof). In order to encourage large purchases, the company charges $1.10 per pound if less than 8 pounds are ordered but charges only $1.00 per pound on all orders of

8 or more pounds. Hence, if x is the number of pounds purchased, a cost function C is described by

$$C(x) = \begin{cases} 1.1x & \text{if } 0 < x < 8, \\ x & \text{if } x \geq 8. \end{cases}$$

The graph of C is shown in Figure 9.7. We say that the function C has a discontinuity at $x = 8$ because of the sudden change in cost. A customer would be foolish to buy 7.5 pounds of this item from company A, since he can purchase 8 pounds at a lower cost. Observe from Figure 9.7 that $\lim_{x \to 8} C(x)$ does not exist, whereas $C(8) = 8$. Because $\lim_{x \to 8} C(x)$ does *not* equal $C(8)$, we say that C is discontinuous (not continuous) at $x = 8$. ∎

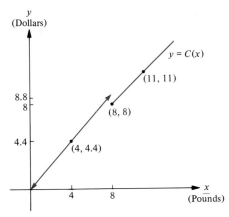

Figure 9.7

Example 2
Company B sells a competitive product by the pound (see Example 1). Its cost function f is given by

$$f(x) = \begin{cases} 1.2x & \text{if } 0 < x \leq 5, \\ 6 + .9(x - 5) & \text{if } x > 5. \end{cases}$$

In other words, company B charges $1.20 per pound for the first 5 pounds, and for purchases over 5 pounds it charges $6.00 plus $.90 for each pound over 5 pounds. See Figure 9.8. Even though company B changes its method of computing the cost at 5 pounds, the function f is continuous at $x = 5$. This is because there is no sudden jump in the cost at $x = 5$. It can be shown that $\lim_{x \to 5} f(x) = 6$ and also that $f(5) = 6$; since

$$\lim_{x \to 5} f(x) = f(5),$$

we say that f is continuous at $x = 5$. ∎

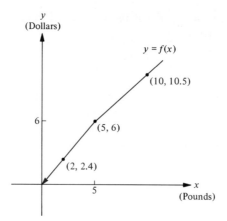

Figure 9.8

▶ **Definition 3**

A function f is said to be **continuous** at a number c if

$$\lim_{x \to c} f(x) = f(c).$$

Otherwise, f is said to be **discontinuous** at $x = c$.

In order for a function f to have a chance to be continuous at $x = c$, the domain of f must include c because of the presence of the symbol $f(c)$ in Definition 3. Similarly, $\lim_{x \to c} f(x)$ must exist in order for f to be continuous at $x = c$. Definition 3 says that $f(c)$ and $\lim_{x \to c} f(x)$ not only must each exist in order for f to be continuous at $x = c$, but that they must be *equal*. Recall from Section 9.3 that $\lim_{x \to c} f(x)$ can exist whether or not $f(c)$ is defined. See Figures 9.9 and 9.10 for some typical situations.

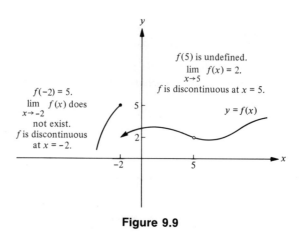

Figure 9.9

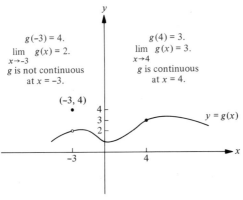

Figure 9.10

It is important that one be able to recognize whether certain familiar functions are continuous and also be able to detect any points of discontinuity without a great deal of effort. For example, is the quadratic function defined by $f(x) = x^2 + 1$ continuous at $x = 3$? We can answer the question affirmatively by showing that $\lim_{x \to 3} f(x) = f(3)$ as follows:

$$f(3) = (3)^2 + 1 = 10,$$
$$\lim_{x \to 3} (x^2 + 1) = \left(\lim_{x \to 3} x\right)\left(\lim_{x \to 3} x\right) + \lim_{x \to 3} 1$$
$$= 3(3) + 1 = 10.$$

This example indicates how Theorem 2 and Definition 3 can be used to prove Theorem 3.

▶ **Theorem 3**
    (*i*) A polynomial function is continuous for all real numbers.
    (*ii*) A rational function is continuous except at a number for which the denominator is zero.

As indicated above, the proof follows readily from Theorem 2 and Definition 3. ◀

Example 3
The polynomial function f defined by
$$f(x) = 7x^4 - 5x^3 + 2x^2 - 8x + 3$$
is continuous for all real numbers by Theorem 3(*i*). ■

Example 4
The rational function g defined by
$$g(x) = \frac{x^2 + 5}{x^2 - 9}$$

is continuous for any real number except $x = 3$ or $x = -3$ by Theorem 3(*ii*). Observe that $g(3)$ and $g(-3)$ do not exist. ∎

So far we have discussed the continuity of a function at a point; later in the text, the reader will need an understanding of the continuity of a function on some type of interval. Therefore, we shall discuss continuity on various types of intervals for the remainder of this section.

▶ **Definition 4**
A function f is **continuous on an open interval** if it is continuous at each number in the interval.

Example 5
In Example 1 the function C is continuous at any positive number except $x = 8$ (see Figure 9.7). Therefore, by Definition 4, C is continuous on any open interval in its domain that does not include the number 8. Thus C is continuous on the intervals $(2, 5)$, $(5, 8)$, and $(8, 10)$. However, C is not continuous on the intervals $(5, 10)$ or $(7.9, 8.1)$. ∎

▶ **Definition 5**
A function f is **continuous on a closed interval** $[a, b]$, provided that the following three conditions are satisfied:
  (*i*) f is continuous on the open interval (a, b).
  (*ii*) If as $x \to a$ from within (a, b), then $f(x) \to f(a)$.
  (*iii*) If as $x \to b$ from within (a, b), then $f(x) \to f(b)$.
Continuity on $[a, b)$ is defined in the same way, except that only (*i*) and (*ii*) are required, and continuity on $(a, b]$ requires only (*i*) and (*iii*).

Example 6
A function f with domain $[2, 7]$ is continuous on the open interval $(2, 7)$, and has a graph as shown in Figure 9.11. Definition 5(*i*) is satisfied. As $x \to 2$ from within $(2, 7)$, $f(x) \to 5$. However, $f(2) = 3$, and hence $f(x)$ does not

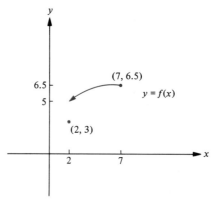

Figure 9.11

approach $f(2)$ as x approaches 2 through values larger than 2. Thus Definition 5(*ii*) is not satisfied. As $x \to 7$ from within $(2, 7)$, $f(x) \to f(7)$, where $f(7) = 6.5$. Hence Definition 5(*iii*) is satisfied. We can conclude that f is continuous on $(2, 7]$, but that f is not continuous on $[2, 7]$ or $[2, 7)$. As one might expect intuitively, the jump in the function value at the endpoint $x = 2$ prevents f from being continuous on $[2, 7]$ or $[2, 7)$. ■

Example 7
Suppose that an automobile manufacturer tests a new model car to determine the number y of miles per gallon the car can attain when driven at x miles per hour. Exhaustive tests are made for values of x in the closed interval $[30, 80]$; the results are shown in Figure 9.12. Certainly, y is a function of x, say $y = f(x) = -\frac{1}{550}x^2 + \frac{293}{11}$, and suppose that the nature of the practical situation has restricted the domain of f to the closed interval $[30, 80]$. The function f is continuous on the closed interval $[30, 80]$ because it can be verified that all three parts of Definition 5 are satisfied. ■

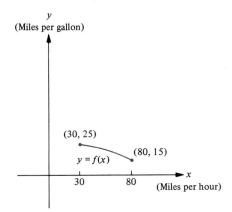

Figure 9.12

Because of practical restrictions, the domain of a function encountered in applications is often a closed interval $[a, b]$ as in Example 7. For future reference, important properties of a function that is continuous on a closed interval are stated in Exercises 12 and 13.

APPLICATIONS

Example 8
Suppose that a biologist is studying the effect of a new drug. He injects a known amount of the drug into a person and then measures the concentration y of the drug in the blood after x hours. Measurements are made for several people and for several values of x. The biologist seeks to find a mathematical model of the situation in the form of a function f such that $y = f(x)$. Since the concentration of the drug in the blood should be 0 at the beginning of the experiment, $f(0)$ should equal 0. The function f should be

continuous for $x \geq 0$ because the change in concentration is expected to be gradual rather than something that suddenly jumps from one value to another. After a period of time the effect of the drug should be negligible, and we should require that

$$\lim_{x \to \infty} f(x) = 0$$

be true. For the people tested, the concentration after 1 hour is about $\frac{2}{3}$ per cent, after 2 hours is between $\frac{3}{4}$ and $\frac{4}{5}$ per cent, and after 4 hours is down to about $\frac{2}{3}$ per cent again. Show that the function f given by

$$y = f(x) = \frac{x}{20x^2 + 50x + 80}, \qquad x \geq 0,$$

possesses the properties described above.

$$f(0) = \frac{0}{0 + 0 + 80} = 0.$$

The denominator $20x^2 + 50x + 80$ is always positive, as can be seen by completing the square and writing it in the form $20(x + \frac{5}{4})^2 + \frac{195}{4}$. Hence by Theorem 3(ii) of this section there are no values of x for which f is discontinuous. By use of the methods of Section 9.2 we find that

$$\lim_{x \to \infty} \frac{x}{20x^2 + 50x + 80} = \lim_{x \to \infty} \frac{1/x}{20 + (50/x) + (80/x^2)}$$

$$= \frac{0}{20 + 0 + 0} = 0.$$

By substitution, $f(1)$ and $f(4)$ can be shown to equal exactly $\frac{2}{3}$ per cent and $f(2) \approx .0077$, which is between $\frac{3}{4}$ and $\frac{4}{5}$ per cent.

There are still some other properties that f must possess in order for it to be satisfactory for the biologist to use as a means of predicting concentration of the drug in the blood. These properties will be discussed in Example 4 of Section 11.2. ∎

EXERCISES

Suggested minimum assignment: Exercises 2, 3, 6, 7, and 9(b).

1. Suppose that a cost function C is given by

$$C(x) = \begin{cases} 5x & \text{if } 0 < x < 5, \\ 6x & \text{if } x \geq 5. \end{cases}$$

   Is C continuous at $x = 5$? Why?

2. Suppose that a revenue function R is given by

$$R(x) = \begin{cases} 3x & \text{if } 0 \leq x < 10, \\ 30 + 2x & \text{if } x \geq 10. \end{cases}$$

   Is R continuous at $x = 10$? Why? Draw a graph.

3. Suppose that a profit function P is given by
$$P(x) = \tfrac{1}{5}x - 3, \quad \text{provided that } 0 \le x \le 1000.$$
Is P continuous at $x = 25$? Why?

4. Why is it not appropriate to discuss continuity for the type of function known as a sequence?

5. Refer to the function f in Example 2 on p. 128.
   (a) Is f continuous at $x = 10$?
   (b) Is f continuous at $x = 40$?

6. Refer to the function f in Example 2 on p. 391.
   (a) Is f continuous at $x = 2$? Why?
   (b) Is f continuous on $(0, 2)$?
   (c) Is f continuous on $[0, 2]$? Why?

7. Let f be defined by
$$\begin{cases} f(x) = \dfrac{x-1}{x^2-1} & \text{if } x \ne 1, \\ f(1) = 3. \end{cases}$$

   (a) Evaluate $\lim\limits_{x \to 1} f(x)$.
   (b) Explain why f is discontinuous at $x = 1$.
   (c) Does f have any other points of discontinuity?

8. Sketch the graph of the function f defined by
$$f(x) = \begin{cases} x + 5 & \text{if } x < 2, \\ 6 & \text{if } x \ge 2. \end{cases}$$

   (a) Is f continuous on $[0, 2)$?
   (b) Is f continuous on $(1, 3)$? Why?

9. State all values of x for which each of the following rational functions is discontinuous.

   (a) $f(x) = \dfrac{x}{x-1}$.
   (b) $g(x) = \dfrac{x^2 - 16}{x^2 - 5x + 6}$.
   (c) $h(x) = \dfrac{x^2 + 3x + 1}{x^2 + 3x}$.

10. Show that f is continuous at $x = 2$ by using the procedure illustrated just prior to Theorem 3, if $f(x) = x^2 + 7x + 2$.

11. If $y = f(x)$ and $y = g(x)$ are each continuous at $x = c$, use Theorem 2(i) to show that $y = f(x) + g(x)$ is continuous at $x = c$. In other words, show that the sum of two continuous functions is continuous. (Similar statements can be made about products, differences, and quotients, provided that in a quotient the denominator is not zero at $x = c$.)

12. (a) Draw a sketch to illustrate the following property of a continuous function on a closed interval: If f is continuous on $[a, b]$ and $f(a) \ne f(b)$, then for any number K between $f(a)$ and $f(b)$ there is at least one value of x, say $x = c$, such that $f(c) = K$.

(b) If $f(a)$ and $f(b)$ have opposite signs in part (a), what can be said about a root of the equation $f(x) = 0$?

13. (a) Suppose that f is not a constant function and that f is continuous on a closed interval $[a, b]$. Draw a sketch which illustrates the property that $f(x)$ must take on a least value m and a greatest value M on $[a, b]$.

    (b) Draw a sketch to illustrate that the statement of part (a) might not be true if $[a, b]$ were replaced by (a, b).

14. Suppose that f is continuous at $x = c$ and $f(c) > 0$. Show, by use of Definitions 2 and 3, that there is an open interval that contains c throughout which $f(x) > 0$. [It is also true that if f is continuous at $x = c$ and $f(c) < 0$, then there is an open interval that contains c, throughout which $f(x) < 0$. The facts stated in this exercise will be helpful in proving some theorems in Chapter 11.]

NEW VOCABULARY

discrete variable 9.2
continuous variable 9.2
limit of values of a function (as independent variable becomes infinite) 9.2
limit of values of a function (as independent variable approaches finite number) 9.3
continuous function 9.4
discontinuous function 9.4
continuity on an interval 9.4

10 Differentiation

Scholars generally regard the development of calculus in the latter part of the 17th century, by Isaac Newton (1642–1727) and Gottfried Leibniz (1646–1716), as one of the most significant advances in the history of human thought. Two quotations attributed to Newton reveal the perspective in which he viewed his monumental achievement. In one quotation Newton stated that if he had seen farther than other men, it was only because he had stood on the shoulders of giants. He also said, "I do not know what I may appear to the world; but to myself I seem to have been only like a boy playing on the seashore, and diverting myself in now and then finding a smoother pebble or a prettier shell than ordinary, while the great ocean of truth lay all undiscovered before me." Time has shown that one of the more potent tools for the discovery of unknown truths has been Newton's own calculus.

Ordinarily, the study of calculus is organized into two parts—the differential calculus and the integral calculus. Fundamental to the study of the differential calculus is "the derivative of a function." The purpose of this chapter is to define, illustrate, and apply the idea of the derivative of a function.

Prerequisites: 1.1, 1.2, 3.1–3.9, Chapter 9; 10.6 is not a prerequisite to optional Section 10.7.
Suggested Sections for Moderate Emphasis: 10.1–10.6.
Suggested Sections for Minimum Emphasis: 10.1–10.5.

10.1 Instantaneous Rates of Change and Tangent Lines

Let $y = f(x)$ express the relationship of the daily profit y of a theater to the price x of each ticket, where the units in both cases are in dollars. The theater manager wants to analyze the effect of price changes on the profit.

Assume that the domain of the function f consists of all real values of x in the open interval $(0, 8)$. The domain represents the set of ticket prices that the manager can charge.† Let x_1 be some value of x in the domain. Starting at x_1, we change the ticket price x by an amount $\Delta x \neq 0$ to get the number $x_1 + \Delta x$, which must also be in $(0, 8)$. The increment Δx can be either a positive or negative number. When x changes from x_1 to $x_1 + \Delta x$, y changes

† In reality, of course, the ticket price x will be some positive integral multiple of $0.01.

from $f(x_1)$ to $f(x_1 + \Delta x)$, as shown in Figure 10.1. The change in y that corresponds to a change of Δx in x is denoted by Δy and represents the change in the profit. Therefore,

$$\Delta y = f(x_1 + \Delta x) - f(x_1).$$

Of course, Δy can be positive, negative, or zero. In Figure 10.1 Δx is positive and Δy is negative.

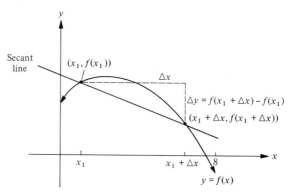

Figure 10.1

If Δy is divided by Δx, the result represents the **average rate of change** in profit with respect to a change in the ticket price between x_1 and $x_1 + \Delta x$. We can write

$$\frac{\Delta y}{\Delta x} = \frac{f(x_1 + \Delta x) - f(x_1)}{\Delta x}.$$

Geometrically, this average rate of change can be represented by the slope of a line that passes through the two points $(x_1, f(x_1))$ and $(x_1 + \Delta x, f(x_1 + \Delta x))$, as shown in Figure 10.1; this line is called a **secant line**. Obviously, the location of this line, and hence its slope, will depend on the value x_1 and the change in x, namely, Δx.

Example 1

Suppose that the profit function is given by $y = f(x) = -x^2 + 8x - 5$; the initial ticket price is \$1; and the price change is an increase of \$4. What is the corresponding change in profit, and what is the average rate of change of profit on the interval $[1, 5]$?

We are given $x_1 = 1$, $\Delta x = 4$, and asked to find the corresponding values of Δy and $\Delta y/\Delta x$.

$$\begin{aligned}
\Delta y &= f(x_1 + \Delta x) - f(x_1) \\
&= f(1 + 4) - f(1) \\
&= f(5) - f(1) \\
&= [-(5)^2 + 8(5) - 5] - [-(1)^2 + 8(1) - 5] \\
&= 10 - 2 \\
&= 8.
\end{aligned}$$

Therefore,
$$\frac{\Delta y}{\Delta x} = \frac{8}{4} = 2.$$

Thus the change in profit is $8, and the average rate of change in profit per unit change in ticket price on $[1, 5]$ is $2. The graph of the profit function, along with the secant line passing through $(1, 2)$ and $(5, 10)$ with slope of 2, is shown in Figure 10.2. The graph is a portion of a parabola. ∎

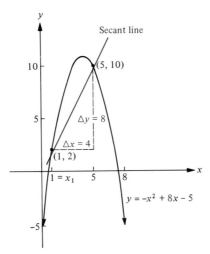

Figure 10.2

Example 2
As in Example 1, let $y = f(x) = -x^2 + 8x - 5$. This time, however, we shall let the initial ticket price be $x_1 = 2$ and let $\Delta x \neq 0$ be unspecified. Again we seek Δy and $\Delta y/\Delta x$.

$$\begin{aligned}
\Delta y &= f(x_1 + \Delta x) - f(x_1) \\
&= f(2 + \Delta x) - f(2) \\
&= [-(2 + \Delta x)^2 + 8(2 + \Delta x) - 5] - [-(2)^2 + 8(2) - 5] \\
&= \{-[4 + 4\Delta x + (\Delta x)^2] + 8(2 + \Delta x) - 5\} - 7 \\
&= -4 - 4\Delta x - (\Delta x)^2 + 16 + 8\Delta x - 5 - 7 \\
&= 4\Delta x - (\Delta x)^2,
\end{aligned}$$

and
$$\frac{\Delta y}{\Delta x} = 4 - \Delta x,$$

which completes the required calculation. ∎

For various values of Δx in Example 2 the corresponding values of $\Delta y/\Delta x$ are shown in Table 10.1. The interesting thing about the table is that $\Delta y/\Delta x$ is rather close to 4 when Δx is near 0 (notice particularly the values $\Delta x = -.01$ and $\Delta x = .01$). This suggests the idea of a limit and leads to the

Sec. 10.1] Instantaneous Rates of Change and Tangent Lines

Table 10.1

Δx	-2	-1	$-.1$	$-.01$	$.01$	$.1$	1	2
$\dfrac{\Delta y}{\Delta x}$	6	5	4.1	4.01	3.99	3.9	3	2

conjecture that, for $x_1 = 2$,

$$\frac{\Delta y}{\Delta x} \to 4 \quad \text{as} \quad \Delta x \to 0$$

or

$$\lim_{\Delta x \to 0} \frac{\Delta y}{\Delta x} = 4.$$

Using a previous limit theorem, we can verify this conjecture:

$$\lim_{\Delta x \to 0} \frac{\Delta y}{\Delta x} = \lim_{\Delta x \to 0} (4 - \Delta x) = \lim_{\Delta x \to 0} 4 - \lim_{\Delta x \to 0} \Delta x = 4 - 0 = 4.$$

Thus, for $x_1 = 2$, the average rate of change of profit with respect to the ticket price approaches 4 as Δx approaches 0. The limiting value $\left[\lim_{\Delta x \to 0} (\Delta y/\Delta x) = 4\right]$ represents the **instantaneous rate of change** of profit at $x_1 = 2$, as opposed to the average rate of change of profit between 2 and $2 + \Delta x$, which is represented by the ratio $\Delta y/\Delta x$.

The ideas just presented may also be interpreted geometrically. In Figure 10.3 two secant lines are drawn through (2, 7), corresponding to two of the values of Δx in Table 10.1, namely, -2 and 2. The slopes of these lines as

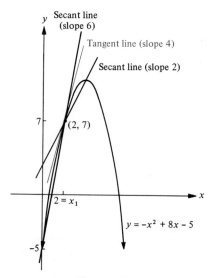

Figure 10.3

also given in Table 10.1 are 6 and 2, respectively. A third line, in color in Figure 10.3, is also drawn through (2, 7); it has a slope of 4, which is the limiting value of $\Delta y/\Delta x$ as $\Delta x \to 0$. The latter line is known as the **tangent line**.

We have just presented an intuitive introduction to the idea of the instantaneous rate of change of one variable with respect to another variable at a given point and to the geometric counterpart—the slope of the tangent line at the point. Let us now define precisely these important concepts.

▶ **Definition 1**
Suppose that a function f is defined by $y = f(x)$ throughout some open interval that contains the number x_1. The **instantaneous rate of change** of y with respect to x at $x = x_1$ is

$$\lim_{\Delta x \to 0} \frac{\Delta y}{\Delta x} = \lim_{\Delta x \to 0} \frac{f(x_1 + \Delta x) - f(x_1)}{\Delta x}$$

if this limit exists. The **tangent line** to the curve at $(x_1, f(x_1))$ is the line through $(x_1, f(x_1))$ with slope

$$\lim_{\Delta x \to 0} \frac{\Delta y}{\Delta x} = \lim_{\Delta x \to 0} \frac{f(x_1 + \Delta x) - f(x_1)}{\Delta x}$$

if this limit exists.

Figure 10.4 is presented to demonstrate that the slope of the tangent line is the limit of the slopes of the secant lines as Δx approaches zero, provided that the limit exists.

A notation is needed to show that the concepts just defined—the instantaneous rate of change and the slope of the tangent line—depend upon both the function f and the value x_1; hence, if the limit exists, we choose to write

$$f'(x_1) = \lim_{\Delta x \to 0} \frac{\Delta y}{\Delta x} = \lim_{\Delta x \to 0} \frac{f(x_1 + \Delta x) - f(x_1)}{\Delta x}.$$

The way to read $f'(x_1)$ is "f prime of x_1."

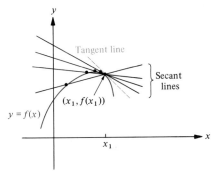

Figure 10.4. *Slopes of secant lines approach the slope of the tangent line as $\Delta x \to 0$.*

Example 3 illustrates how the equation of the tangent line to a curve at a point can be determined.

Example 3
In the discussion following Example 2, we found that the tangent line through (2, 7) had a slope of $f'(2) = 4$, so its equation can be written using the point-slope form of the equation of a line,

$$y - 7 = 4(x - 2),$$

or, in general form,

$$4x - y - 1 = 0. \blacksquare$$

In the majority of examples that we shall consider, the instantaneous rate of change of y with respect to x at $x = x_1$ will exist, and hence it will be possible to draw a tangent line to $y = f(x)$ at $x = x_1$. There are exceptions, however. We shall learn in Section 10.4 that if f is *discontinuous* at $x = x_1$, then $f'(x_1)$ will *not* exist; although f is discontinuous at $x = x_1$, it is still possible to introduce increments $\Delta x \neq 0$ and Δy and to discuss the idea of secant line, but there will be no tangent line because the limit in Definition 1 will not exist. On the other hand, if f is *continuous* at $x = x_1$, then $f'(x_1)$ may or may not exist, as we illustrate next. First consider statement (10.1) which is of theoretical importance (an outline of its proof is given in Exercise 16).

If f is continuous at x_1 and Δx approaches zero, then Δy also approaches zero. (10.1)

A consequence of (10.1) is that the limit of Definition 1, $\lim_{\Delta x \to 0} (\Delta y/\Delta x)$, is a type of limit in which the numerator and denominator each approach zero as the limiting process takes place. Example 7 of Section 9.3 and the discussion prior to that example illustrate that it is quite possible for $\lim_{\Delta x \to 0} (\Delta y/\Delta x)$ to exist even though the numerator and denominator each approach zero; Example 4, however, illustrates that it is also possible for $\lim_{\Delta x \to 0} (\Delta y/\Delta x)$ *not* to exist when the numerator and denominator each approach zero.

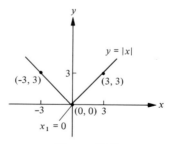

Figure 10.5

Example 4
If $y = f(x) = |x|$, show that the instantaneous rate of change of y with respect to x at $x = 0$ does *not* exist. (See Figure 10.5.)

$$\frac{\Delta y}{\Delta x} = \frac{f(0 + \Delta x) - f(0)}{\Delta x} = \frac{|\Delta x| - 0}{\Delta x} = \frac{|\Delta x|}{\Delta x}.$$

If $\Delta x > 0$, $\Delta y/\Delta x$ is always equal to 1 but, if $\Delta x < 0$, $\Delta y/\Delta x$ is always equal to -1. In order for $f'(0) = \lim_{\Delta x \to 0} (\Delta y/\Delta x)$ to exist, $\Delta y/\Delta x$ would have to approach the same value whether Δx approaches zero through positive or negative values. However, $\Delta y/\Delta x \to 1$ if $\Delta x > 0$ and $\Delta x \to 0$, whereas $\Delta y/\Delta x \to -1$ if $\Delta x < 0$ and $\Delta x \to 0$. Therefore, the instantaneous rate of change of y with respect to x at $x = 0$ does not exist and, as one might guess from Figure 10.5, there is no tangent line to the curve at the origin. Note that f is continuous at $x = 0$ because $\lim_{x \to 0} f(x) = f(0) = 0$. ∎

APPLICATIONS

Example 5
One reason that a theater manager might be interested in a mathematical analysis such as that begun in Examples 1 and 2 would be to determine the ticket price x_1 that would maximize (make largest) the profit y. In Chapter 11 we shall find that if a maximum exists for a polynomial function over an open interval, then the slope $f'(x_1)$ of the tangent line must be zero. For the profit function given by

$$y = f(x) = -x^2 + 8x - 5,$$

we can calculate

$$\begin{aligned} \Delta y &= f(x_1 + \Delta x) - f(x_1) \\ &= [-(x_1 + \Delta x)^2 + 8(x_1 + \Delta x) - 5] - (-x_1^2 + 8x_1 - 5) \\ &= [-x_1^2 - 2x_1 \Delta x - (\Delta x)^2 + 8x_1 + 8 \Delta x - 5] - (-x_1^2 + 8x_1 - 5) \\ &= -2x_1 \Delta x - (\Delta x)^2 + 8 \Delta x \end{aligned}$$

and

$$\frac{f(x_1 + \Delta x) - f(x_1)}{\Delta x} = -2x_1 - \Delta x + 8.$$

The slope $f'(x_1)$ of the tangent line at $x = x_1$ is computed as follows:

$$\begin{aligned} f'(x_1) &= \lim_{\Delta x \to 0} \frac{f(x_1 + \Delta x) - f(x_1)}{\Delta x} \\ &= \lim_{\Delta x \to 0} (-2x_1 - \Delta x + 8) \\ &= \lim_{\Delta x \to 0} (-2x_1) - \lim_{\Delta x \to 0} \Delta x + \lim_{\Delta x \to 0} 8 \\ &= -2x_1 + 8. \end{aligned}$$

We wish to find that value x_1 for which the slope of the tangent line is zero. Thus

$$0 = -2x_1 + 8,$$
$$2x_1 = 8,$$
$$x_1 = 4.$$

This means that if there is a maximum profit, then that maximum profit will occur at a ticket price of $4. (In Exercise 6 of Section 11.4 it can be shown that maximum profit *will* occur at a ticket price of $4.) ∎

Example 6

This example illustrates a use of an average rate of change for sequences. In a paper entitled "Innovation in the States: A Diffusion Study," Virginia Gray [27] examines the diffusion (spread) of ideas from state to state through the legal adoption of those ideas by the respective states. The policy areas selected for study are education, welfare, and civil rights.

Let A_t represent the cumulative proportion of adopters of a particular type of legislation (such as Fair Housing) t years after innovation of the law by some state. Since t is assumed to be a positive integer, $\{A_t\}$ is a sequence of numbers. The terms of $\{A_t\}$ are never greater than 1 because not over 100 per cent of the states can adopt legislation of a given type. Let L represent the maximum possible proportion of adopters of a certain type of legislation. If $L = 1$, all states have the potential for adoption; however, for some types of legislation there are reasons why $L < 1$.

Starting at a particular time t_1 and changing t by an amount Δt, we can denote the corresponding change in A_t by ΔA_t (Δt and ΔA_t play the roles of Δx and Δy, respectively, that were discussed earlier in this section). $\Delta A_t/\Delta t$ is the average rate of change of the cumulative proportion of adopters with respect to time from t_1 to $t_1 + \Delta t$ ($\Delta A_t/\Delta t$ is comparable to $\Delta y/\Delta x$).

One object of Gray's paper is to predict the rate of diffusion among the states—in other words, if A_1 is known for a particular type of legislation, we want to predict A_2, A_3, A_4, and so on. In order to predict A_2, A_3, A_4, and so on, which correspond to time changes of 1 year each, let us choose $\Delta t = 1$. The basic assumption made is that $\Delta A_t/\Delta t$ is directly proportional to A_t and $L - A_t$. Thus the mathematical model is

$$\frac{\Delta A_t}{\Delta t} = bA_t(L - A_t),$$

where b is the constant of proportionality. Because $\Delta t = 1$, the equation becomes

$$\Delta A_t = bA_t(L - A_t).$$

Such an equation involving a sequence and increments is known as a difference equation. Since $\Delta t = 1$, $\Delta A_t = A_{t+1} - A_t$, and the difference equation becomes

$$A_{t+1} = bA_t(L - A_t) + A_t. \tag{10.2}$$

Since A_{t+1} is expressed in terms of A_t, (10.2) makes it possible to find any term of $\{A_t\}$ if the previous term is known. For a particular type of legislation, b can be estimated using experience from other related types of legislation, L can be estimated, and after 1 year A_1 is known. Then (10.2) is used to find A_2. Once A_2 is known, (10.2) is used again to determine A_3, and this process can be continued. In fact, a computer program can be written to generate as many terms of the sequence $\{A_t\}$ as desired. ∎

EXERCISES

Suggested minimum assignment: Exercises 1(a), 3, 7, 9, 11, and 13.

1. Suppose that $y = x^2 - 4x + 7$ and $x_1 = 3$.
   (a) If $\Delta x = .5$, find Δy.   (b) If $\Delta x = -.1$, find Δy.
2. Suppose that $y = \dfrac{1}{x+5}$ and $x_1 = -2$.
   (a) If $\Delta x = \frac{1}{3}$, find Δy.   (b) If $\Delta x = -1$, find Δy.
3. Let f be defined by $f(x) = 2 - x^2$. Suppose that $x_1 = 1$ and $\Delta x = .5$. Plot the secant line joining $(x_1, f(x_1))$ and $(x_1 + \Delta x, f(x_1 + \Delta x))$, and find the slope of this secant line.
4. Explain why the secant line between two points on the graph of a function is never a vertical line.
5. Suppose that $y = x^3$ and $x_1 = 1$.
   (a) If $\Delta x = .2$, find Δy. Draw the secant line and find its slope.
   (b) If $\Delta x = -1$, find Δy. Draw the secant line and find its slope.
   (c) Find the average rate of change of y with respect to x between $x = 1$ and $x = 1.1$.
6. Suppose that $y = x^2$ and $x_1 = 1$. Let Δx equal $-1, -.1, -.01, .01, .1,$ and 1. Construct a table like Table 10.1.
7. Suppose that $y = f(x) = x^3$ and $x_1 = 2$.
   (a) If $\Delta x = .01$, find Δy.   (b) If $\Delta x = -.01$, find Δy.
   (c) Make a conjecture as to the value of $\lim\limits_{\Delta x \to 0} \dfrac{\Delta y}{\Delta x} = f'(2)$.
8. Suppose that the number y of bacteria in a culture after t hours is given by $y = f(t) = t^2 + 5t + 100$, provided that $0 \le t \le 24$. Let the initial time be $t_1 = 3$, and let $\Delta t \ne 0$ be unspecified. Find Δy. Find $\Delta y/\Delta t$. Find the instantaneous rate of change of y with respect to t at $t = 3$.
9. If $y = x^2 - 5x$, find the average rate of change of y with respect to x between $x = -1$ and $x = 1$. Also find the instantaneous rate of change of y with respect to x at $x = -1$.
10. Let a profit function f be given by $y = f(x) = -x^2 + 10x - 3$, provided that $0 < x < 9$. Find the instantaneous rate of change of y with respect to x at $x = 3$.
11. If $y = f(x) = x^3$, show that $f'(2) = 12$. (See Exercise 7.)
12. Show that the slope of the tangent line to $y = x^2$ at $(1, 1)$ is 2. (See Exercise 6.)

Sec. 10.1] Instantaneous Rates of Change and Tangent Lines

13. Find the equation of the tangent line to $y = x + x^2$ at $x = -2$.
14. Find the equation of the tangent line to $y = 6 - x^2$ at $x = 3$.
15. Sketch the graph of $y = f(x) = |x - 1|$. Explain why there is no tangent line to this curve at $(1, 0)$.
16. Prove that if f is continuous at x_1 and Δx approaches zero, then Δy approaches zero. [*Hint:* $\lim_{x \to x_1} [f(x) - f(x_1)] = 0$ follows from $\lim_{x \to x_1} f(x) = f(x_1)$. Let $x = x_1 + \Delta x$, and let Δx approach zero. Then demonstrate that $\lim_{\Delta x \to 0} [f(x_1 + \Delta x) - f(x_1)] = 0$.]
17. Use the method of Example 5 on the profit function given in Exercise 10 to find the value x_1 of x at which maximum profit may occur.

10.2 The Derivative and Its Interpretations

Much of the material presented in Chapters 3 and 9 has been given as a foundation so that the reader will be able to understand the concept that will be defined in this section. This concept is important in such diverse ways as the study of marginal analysis in economics, the study of rates of tissue change in biology, and the study of population problems in sociology.

In the last section we learned that the average rate of change of $y = f(x)$ with respect to x between $x = x_1$ and $x = x_1 + \Delta x$ is given by the ratio $\Delta y/\Delta x$ or $[f(x_1 + \Delta x) - f(x_1)]/\Delta x$. If Δx is allowed to approach zero in this ratio, we get the average rate of change over a smaller and smaller interval. In the limit we get what is called the instantaneous rate of change of y with respect to x at $x = x_1$. In other words, if $y = f(x)$, then the instantaneous rate of change of y with respect to x at $x = x_1$ is given by

$$f'(x_1) = \lim_{\Delta x \to 0} \frac{\Delta y}{\Delta x} = \lim_{\Delta x \to 0} \frac{f(x_1 + \Delta x) - f(x_1)}{\Delta x}, \qquad (10.3)$$

provided that this limit exists. The phrase **rate of change** will often be used instead of instantaneous rate of change and will be understood to have the same meaning.

We also learned in Section 10.1 that a geometric interpretation of (10.3), provided that the limit exists, is the slope of the tangent line to $y = f(x)$ at the point $(x_1, f(x_1))$.

Example 1

Find the rate of change of $y = f(x) = x^2 - x$ with respect to x at $x = 3$; also find the slope of the tangent line to $y = f(x) = x^2 - x$ at $(3, 6)$.

By (10.3) with $x_1 = 3$ we can find both; we have

$$f'(3) = \lim_{\Delta x \to 0} \frac{\Delta y}{\Delta x} = \lim_{\Delta x \to 0} \frac{f(3 + \Delta x) - f(3)}{\Delta x}$$

$$= \lim_{\Delta x \to 0} \frac{[(3 + \Delta x)^2 - (3 + \Delta x)] - [(3)^2 - 3]}{\Delta x}$$

$$= \lim_{\Delta x \to 0} \frac{9 + 6\Delta x + (\Delta x)^2 - 3 - \Delta x - 6}{\Delta x}$$

$$= \lim_{\Delta x \to 0} \frac{5\Delta x + (\Delta x)^2}{\Delta x}$$

$$= \lim_{\Delta x \to 0} (5 + \Delta x)$$

$$= 5.$$

See Figure 10.6 for a graph. ∎

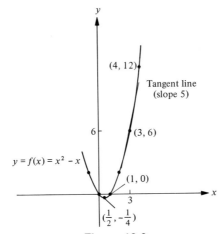

Figure 10.6

The limit given in (10.3) and illustrated in Example 1 leads to the definition of the **derivative** of a function, one of the most important concepts in calculus.

▶ **Definition 2**
The **derivative** (or **derived function**) of the function f is the function f' defined by

$$f'(x) = \lim_{\Delta x \to 0} \frac{f(x + \Delta x) - f(x)}{\Delta x},$$

where x is any real number for which this limit exists.

The limit given in Definition 2 and the limit given in (10.3) differ in that Definition 2 defines a function f', whereas (10.3) yields a value of f' at the

Sec. 10.2] The Derivative and Its Interpretations

specific value $x = x_1$. It is customary to yield to the convention of also using the term derivative to represent the value $f'(x)$.

Notice in Definition 2 that $f(x)$ appears in the definition of $f'(x)$, and therefore $f'(x)$ cannot exist unless $f(x)$ exists—in other words, the domain of f' is a subset of the domain of f. The function f is said to be **differentiable** at $x = x_1$ if $f'(x_1)$ exists.

Example 2

Let the function f be defined by $y = f(x) = x^2 - x$. Find $f'(x)$. Find $f'(3)$.

$$f'(x) = \lim_{\Delta x \to 0} \frac{f(x + \Delta x) - f(x)}{\Delta x}$$

$$= \lim_{\Delta x \to 0} \frac{[(x + \Delta x)^2 - (x + \Delta x)] - (x^2 - x)}{\Delta x}$$

$$= \lim_{\Delta x \to 0} \frac{x^2 + 2x\,\Delta x + (\Delta x)^2 - x - \Delta x - x^2 + x}{\Delta x}$$

$$= \lim_{\Delta x \to 0} \frac{2x\,\Delta x + (\Delta x)^2 - \Delta x}{\Delta x}$$

$$= \lim_{\Delta x \to 0} (2x + \Delta x - 1)$$

$$= 2x - 1.$$

The calculation above is valid for any real number x—that is, the domain of f' is the set of all real numbers. Hence $f'(3)$ can be found by substituting $x = 3$ as follows:

$$f'(3) = 2(3) - 1 = 5. \blacksquare$$

One could calculate $f'(3)$ in Example 2 without first finding $f'(x)$; this was done in Example 1 of this section. However, by making the general calculation of $f'(x)$ first, it is possible to find $f'(x_1)$ quickly for *any* number x_1 in the domain of f'. For example, $f'(1) = 2(1) - 1 = 1$. The graph of f is given in Figure 10.7, and the graph of f' is given in Figure 10.8; notice the relationship between the two graphs as illustrated by $f'(3) = 5$ and $f'(1) = 1$.

Let $y = f(x) = x^2 - x$ be used again to demonstrate some other notation which is often used in working with the derivative. Each of the following has the same meaning:

$$f'(x) = 2x - 1,$$
$$D_x y = 2x - 1,$$
$$D_x f(x) = 2x - 1,$$
$$y' = 2x - 1,$$
$$\frac{dy}{dx} = 2x - 1.$$

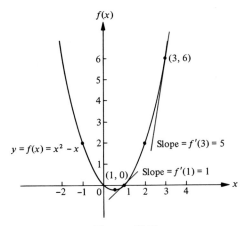

Figure 10.7

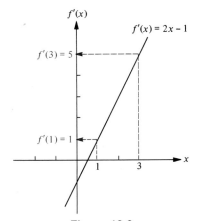

Figure 10.8

The symbols $D_x y$ and dy/dx are read "the derivative of y with respect to x"; the dy/dx notation will be explained more fully in Section 10.6. If the derivative is to be evaluated at $x = 3$, any of the following notations may be used:

$$f'(3) = 5,$$
$$D_x y|_{x=3} = 5,$$
$$D_x f(x)|_{x=3} = 5,$$
$$y'|_{x=3} = 5,$$
$$\frac{dy}{dx}\bigg|_{x=3} = 5.$$

The symbol $D_x f(x)|_{x=3}$ is read "the derivative of $f(x)$ with respect to x evaluated at $x = 3$."

Sec. 10.2] The Derivative and Its Interpretations

Example 3
Find y' if $y = 1/2x$.

$$y' = \lim_{\Delta x \to 0} \frac{1/[2(x + \Delta x)] - (1/2x)}{\Delta x}$$

$$= \lim_{\Delta x \to 0} \frac{x - (x + \Delta x)}{2x(x + \Delta x)} \frac{1}{\Delta x}$$

$$= \lim_{\Delta x \to 0} \frac{-\Delta x}{2x(x + \Delta x)(\Delta x)}$$

$$= \lim_{\Delta x \to 0} \frac{-1}{2x(x + \Delta x)}$$

$$= -\frac{1}{2x^2}. \ \blacksquare$$

Geometrically, we have seen that the derivative of a function at a point may be interpreted as a slope. In general, the derivative gives the rate of change of the dependent variable with respect to the independent variable. In the next two examples we indicate some specialized interpretations.

Example 4
In this example we point out an economic interpretation of the derivative of a function. Suppose that $R(x)$ and $C(x)$ are given, where R is a revenue function and C is a cost function. **Marginal revenue** is defined as the derivative of R with respect to x. **Marginal cost** is defined as the derivative of C with respect to x. **Marginal profit** [recall that $P(x) = R(x) - C(x)$] is defined as the derivative of P with respect to x.

Suppose, for example, that

$$R(x) = x^2 - x.$$

Then, from Example 2 of this section, the marginal revenue $R'(x)$ is given by

$$R'(x) = 2x - 1.$$

The marginal revenue function R' can be evaluated at a number, say, $x = 3$, in its domain to obtain

$$R'(3) = 5.$$

In Example 2 of Section 10.6 we shall attempt to explain the economic meaning of marginal revenue, marginal cost, and marginal profit. In this example we are just trying to indicate one important use of the derivative. $\blacksquare$

Example 5
Another interpretation of the derivative occurs in the study of the motion of a particle along a line. The particle might be a spacecraft going to the

moon, an object on a production line, or an automobile in a traffic pattern. The line along which the particle moves is called the *s*-axis and can be visualized as a horizontal line with positive direction to the right (see Figure 10.9). The location of the particle at various times is given by an **equation of motion**,

$$s = f(t).$$

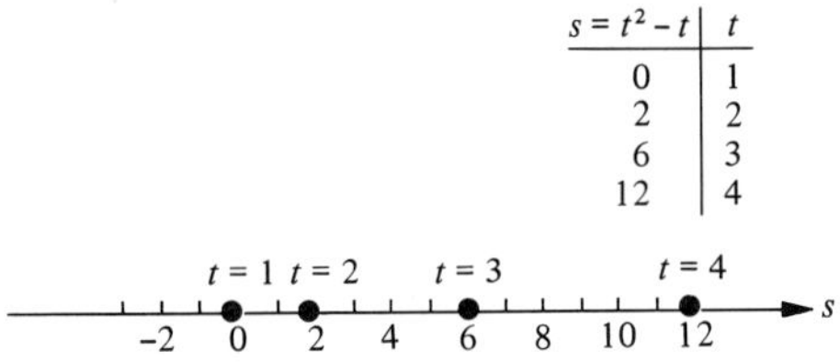

$s = t^2 - t$	t
0	1
2	2
6	3
12	4

Figure 10.9

Assume in this example that the independent variable t is measured in seconds and that the dependent variable s is measured in feet. Starting at a particular time $t = t_1$, we suppose that t is changed by the amount $\Delta t \neq 0$ and that the corresponding change in s is denoted by Δs. It is natural to call

$$\frac{\Delta s}{\Delta t} = \frac{f(t_1 + \Delta t) - f(t_1)}{\Delta t}$$

the **average velocity** of the particle between $t = t_1$ and $t = t_1 + \Delta t$ and to call

$$f'(t_1) = \lim_{\Delta t \to 0} \frac{f(t_1 + \Delta t) - f(t_1)}{\Delta t},$$

if it exists, the **instantaneous velocity** of the particle at $t = t_1$. If $s = t^2 - t$, then $D_t s = 2t - 1$ (see Example 2). The instantaneous velocity at $t = 3$ is given by

$$D_t s|_{t=3} = 2(3) - 1 = 5 \text{ feet per second.}$$

Instantaneous velocity will often be referred to as just **velocity**. A positive velocity indicates that the particle is moving to the right along the *s*-axis, whereas a negative velocity signifies movement to the left. ∎

APPLICATIONS

Example 6
In this section we have learned that the derivative of a function may be interpreted in a general sense as a rate of change of one variable with respect to another variable. We have also discussed well-known geometric, economic, and physical interpretations of the derivative. Other specialized interpretations arise in various disciplines; in this example we give an illustration adapted from pp. 18–19 of Thrall et al. [63].

Suppose that a substance is kept at constant pressure but that its temperature is allowed to vary. The specific heat of the substance is a function of the temperature (the specific heat of a substance is defined as the amount of heat needed to raise the temperature of a unit mass of that substance 1 degree). The derivative of the specific heat of a substance with respect to the temperature gives the rate of change of specific heat with respect to the temperature, and might be useful in problems concerning heats of reaction.

Let T be the temperature of ethyl alcohol in degrees Celsius, and let y be the specific heat of ethyl alcohol. For $0 \leq T \leq 60$, it has been found experimentally that y can be approximated by

$$y = (5.068)(10)^{-1} + (2.68)(10)^{-3}T + (5.4)(10)^{-6}T^2.$$

Definition 2 can be used to show (after some calculation) that

$$D_T y = (2.68)(10)^{-3} + (1.08)(10)^{-5}T.$$

This result can be evaluated for various values of T if desired. For example,

$$D_T y|_{T=20} = (2.68)(10)^{-3} + (1.08)(10)^{-5}(20)$$
$$= (2.896)(10)^{-3}.$$

This means that if ethyl alcohol were being heated, at the *instant* the temperature reached 20 degrees Celsius, the specific heat would be increasing by $(2.896)(10)^{-3}$ unit per degree of increase in the temperature. ∎

*Example 7   Probability Density Function

Consider a certain type of machine that has a life span of no more than 20 years. Thus a sample space of the life span of this type of machine is $U = \{t : 0 \leq t \leq 20\}$, where t is a real number. A measurement is defined on U in such a way that to each value of t there is associated a real number s representing the proportion of maximum age at the time of breakdown. The range of the measurement is $S = \{s : 0 \leq s \leq 1\}$ and provides a new sample space. Here we have a *continuous* measurement because the range S is an interval of real numbers. Suppose that a fairly accurate cumulative probability distribution function F has been found to be given by

$$F(x) = \begin{cases} 0 & \text{if } x < 0, \\ x^2 & \text{if } 0 \leq x \leq 1, \\ 1 & \text{if } x > 1. \end{cases}$$

The graph of F is given in Figure 10.10. Notice from Figure 10.10 that F has the necessary properties to be a cumulative probability distribution function. That is, $\lim_{x \to -\infty} F(x) = 0$, $\lim_{x \to \infty} F(x) = 1$, the graph of F never decreases as x increases, and the fourth requirement that $\lim_{\Delta x \to 0} F(x + \Delta x) = F(x)$ for any real number x, where $\Delta x > 0$, is satisfied. Suppose that we seek the probability of event $E = \{s : .3 < s \leq .5\}$. Let $E_1 = \{s : s \leq .5\}$ and $E_2 = \{s : s \leq .3\}$ so that $P(E) = P(E_1) - P(E_2)$. Recalling Example 8 of

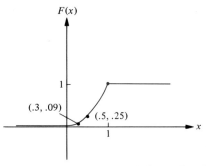

Figure 10.10. *Graph of the cumulative probability distribution function F.*

Section 9.3, we can say $P(E_1) = F(.5)$ and $P(E_2) = F(.3)$. Therefore,

$$\begin{aligned} P(E) = P(E_1) - P(E_2) &= F(.5) - F(.3) \\ &= (.5)^2 - (.3)^2 \\ &= .25 - .09 \\ &= .16. \end{aligned}$$

Thus we have found that the probability is .16 that a machine will break down between 30 and 50 per cent of its maximum life span.

Now suppose for our machine illustration that we think of probability as one unit of some sort of quantity that is distributed in some manner over the sample space $S = [0, 1]$. The probability $P(E)$, where $E = \{s : c < s \leq d\}$, can then be thought of as the fraction of that one unit of quantity that is distributed over the subinterval $(c, d]$ of $[0, 1]$. If we divide this subquantity $P(E) = F(d) - F(c)$ by $d - c$ (the length of the interval), we have a measure of the concentration or density of the quantity that is distributed to the interval $(c, d]$. In other words,

$$\begin{pmatrix} \text{the probability} \\ \text{density in } (c, d] \end{pmatrix} = \frac{F(d) - F(c)}{d - c}.$$

Suppose that we let the interval $(c, d]$ get smaller and smaller in length by considering intervals of the form $(c, x]$, where $c < x < d$ and $x \to c$. We say that

$$\lim_{x \to c} \frac{F(x) - F(c)}{x - c},$$

if it exists, represents the concentration or density of probability at $x = c$. Let $c + \Delta x = x$; then $\Delta x \to 0$ as $x \to c$ and

$$\lim_{x \to c} \frac{F(x) - F(c)}{x - c} = \lim_{\Delta x \to 0} \frac{F(c + \Delta x) - F(c)}{\Delta x}.$$

From the material of this section we recognize the latter limit as the derivative of F evaluated at $x = c$.

There is no assurance that F' exists for all possible values $x = c$, but if it

does exist, we call $F'(c) = p(c)$ the value of the **probability density function** p at $x = c$. For any values of x for which F' does not exist, the function p is assigned values so that the graph of p does not have any isolated points. The domain of p is the set of all real numbers.

In our illustration it can be shown that

$$\lim_{x \to c} \frac{F(x) - F(c)}{x - c} \text{ is } \begin{cases} 2c & \text{if } 0 \leq c < 1, \\ \text{nonexistent} & \text{if } c = 1, \\ 0 & \text{otherwise.} \end{cases}$$

We choose to define function p according to

$$p(x) = \begin{cases} 2x & \text{if } 0 \leq x \leq 1, \\ 0 & \text{otherwise.} \end{cases}$$

The graph of p is given in Figure 10.11. We could have assigned $p(1)$ the value 0 instead of the value 2, but our choice makes p continuous on $[0, 1]$. The area of the trapezoid in Figure 10.11 is .16, representing the fact that .16 unit of probability is distributed to $(.3, .5]$.

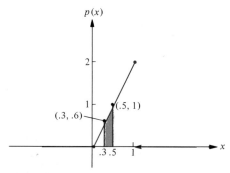

Figure 10.11. *Graph of the probability density function p.*

The probability density function p plays a role for continuous measurements that is analogous to the role of the weighting function w for discrete measurements; this is true in the sense that p and w each display how the total probability of 1 is distributed over the sample space S. ∎

EXERCISES

Suggested minimum assignment: Exercises 1, 3, 5, 7, 13, and 15(b).

1. Suppose that $y = f(x) = 2x^2$.
    (a) Find the rate of change of y with respect to x at $x = -1$.
    (b) What is the slope of the tangent line at $(-1, 2)$?
2. Suppose that $y = f(x) = x^3 - x$.
    (a) Find the average rate of change of y with respect to x between $x = 2$ and $x = 2.5$.

Chap. 10] Differentiation

(b) Find the instantaneous rate of change of y with respect to x at $x = 2$.

3. If $f(x) = 3x^2 - 5x$, find $f'(x)$ using Definition 2. What is the domain of the function f'?

4. If $f(x) = \dfrac{1}{x-1}$, find $f'(x)$. What is the domain of the function f'?

5. Let f be defined by $f(x) = \dfrac{x}{x-1}$. Is f differentiable at $x = 4$? If so, find $f'(4)$. Is f differentiable at $x = 1$? If so, find $f'(1)$.

6. If $y = \dfrac{1}{x} - 3x$, find y'.

7. If $y = 1 - x - x^2$, find $D_x y$. Also find $D_x y\big|_{x=0}$.

8. Suppose that $y = f(x) = 3 - x^2$. Sketch the graph of f and also sketch the graph of f'. Explain the relationship between the graphs, focusing your attention at the point where $x = 2$.

9. If $f(x) = \dfrac{x-3}{2x+1}$, find $D_x f(x)$.

10. Explain why $f'(x_1)$, if it exists, could be calculated using the formula
$$f'(x_1) = \lim_{x \to x_1} \frac{f(x) - f(x_1)}{x - x_1}.$$

11. If $f(x) = 3x^2 - 8$, calculate $f'(-2)$ using the formula in Exercise 10.

12. A revenue function is given by $R(x) = 5x - \frac{1}{6}x^2$. Find the marginal revenue $R'(x)$.

13. A cost function is given by $C(x) = 2x + 1$. Find the marginal cost $C'(x)$.

14. A profit function is given by $P(x) = 10x - x^2$. Find the marginal profit $P'(x)$. Also find $P'(3)$.

15. A particle moves along a line according to the equation of motion $s = t^3$, where s is measured in feet and t in seconds.
    (a) Find the velocity of the particle at $t = 1$.
    (b) Find the velocity of the particle at $t = 2$.
    (c) Find the average velocity of the particle between $t = 1$ and $t = 2$.

16. A particle moves along a line according to the equation of motion $s = 8 - t^2$, where s is measured in meters and t in seconds. Find ds/dt and $ds/dt\big|_{t=2}$. Is the particle moving to the right or to the left along the s-axis when $t = 2$?

17. A ball is thrown directly upward at time $t = 0$ seconds. Its height s in feet above the point where it is released is given by
$$s = 64t - 16t^2,$$
provided that $0 \leq t \leq 4$. How long does it take for the ball to reach its highest point? (*Hint:* The velocity is 0 at its highest point.)

18. A cumulative probability distribution function F is given by

$$F(x) = \begin{cases} 0 & \text{if } x < 0, \\ \frac{1}{4}x^2 & \text{if } 0 \leq x \leq 2, \\ 1 & \text{if } x > 2. \end{cases}$$

Find the corresponding probability density function p.

19. In Example 7 what is the probability that a machine will break down between 60 and 70 per cent of its maximum life span?

10.3 Basic Differentiation Formulas

When $f(x)$ is given, the process of finding $f'(x)$, if it exists, is called **differentiation**. In Section 10.2 the differentiation of some carefully chosen functions was carried out by directly calculating the limit—a rather tedious method. Rather than apply the definition of the derivative of a function each time a function is to be differentiated, we can use formulas that make it possible to write the answer quickly. In this section we shall state some of these formulas, give some examples to illustrate their use, and then discuss briefly the method of obtaining the formulas. The process will be repeated in the next section for some additional formulas. In Theorem 1 five basic differentiation formulas are stated—the formulas are valid if it is assumed that all indicated derivatives exist.

▶ **Theorem 1**
  (i) If $f(x) = k$, where k is a real number, then $D_x f(x) = 0$.
  (ii) If n is a real number, $D_x x^n = nx^{n-1}$.
  (iii) If k is a real number, then $D_x(kf(x)) = kD_x f(x)$.
  (iv) $D_x(f(x) + g(x)) = D_x f(x) + D_x g(x)$.
  (v) Let $u = f(x)$. If $y = u^n$, where n is a real number, then $D_x y = nu^{n-1} D_x u$ (**power rule for differentiation**).

A discussion of the proof of this theorem is given at the end of this section. ◀

Example 1

Consider the constant function defined by $f(x) = 5$. By Theorem 1(i), $D_x f(x) = 0$. Alternatively, we can write $f'(x) = 0$, or, if $y = 5$ is given, we can write $y' = 0$ or $dy/dx = 0$. Part (i) says that the derivative of a constant function is zero—this agrees with the fact that the graph is a horizontal line that has slope zero at every point. ∎

Example 2

Use Theorem 1(ii) to differentiate (a) $y = x^8$, (b) $f(x) = x^{-\pi}$, (c) $f(x) = \sqrt{x^3}$, and (d) $y = x^0$.

Solutions

(a) The exponent 8 is multiplied by x to an exponent that is 1 less than the original exponent of x. Hence $y' = 8x^7$.
(b) $D_x f(x) = -\pi x^{-\pi-1}$.

(c) Rewrite the stated problem as $f(x) = x^{3/2}$. Then $f'(x) = \frac{3}{2}x^{1/2} = \frac{3}{2}\sqrt{x}$.
(d) $y' = 0x^{-1} = 0$. Since $x^0 = 1$, part (i) instead of (ii) could also be used to get the same answer. ∎

Example 3
Let a revenue function be given by $R(x) = 8x^3$. Find the marginal revenue $R'(x)$.

By Theorem 1(iii), $R'(x) = 8D_x x^3$. Then, by part (ii), $R'(x) = 8(3x^2) = 24x^2$. Part (iii) says that the derivative of a real number times a function is the real number times the derivative of the function. ∎

Part (iv) of Theorem 1 will be illustrated in Example 4. Part (iv) says that the derivative of the sum of two functions is the sum of their derivatives, but part (iv) can easily be extended to the case where there are any finite number of functions rather than just two functions.

Example 4
Find (a) $D_x(x^{5/2} + x)$, (b) $D_x(x^3 - 8x^2)$, and (c) $D_x(3x^4 + x^6 - x^2)$.
Solutions
(a) $D_x(x^{5/2} + x^1) = \frac{5}{2}x^{3/2} + 1x^0 = \frac{5}{2}x^{3/2} + 1$.
(b) $D_x(x^3 - 8x^2) = 3x^2 - 16x$.
(c) $D_x(3x^4 + x^6 - x^2) = 12x^3 + 6x^5 - 2x$.
Notice in (b) we used the fact that
$$x^3 - 8x^2 = x^3 + (-8x^2),$$
and hence that Theorem 1(iv) applies to differences of functions as well as sums of functions. ∎

Example 5
If $y = (3x + 1)^8$, find $D_x y$.
If we let $u = 3x + 1$, then $y = u^8$ and Theorem 1(v) applies with $n = 8$.
$$D_x y = 8u^7 D_x u = 8u^7(3) = 24u^7 = 24(3x + 1)^7.$$
With a little practice the problem can be done without even writing $u = 3x + 1$, but the reader must note that $8(3x + 1)^7$ is *not* the answer— remember to multiply by $D_x u$. ∎

If $u = x$ in Theorem 1(v), then $D_x u = 1$ and part (v) reduces to part (ii). Hence part (ii) could be omitted from Theorem 1, and the power rule for differentiation, part (v), could be used instead in Examples 2, 3, and 4.

Example 6
Find the slope of the tangent line to $y = (x^3 - x^2 + x)^3$ at $x = 1$.
By Theorem 1(v),
$$y' = 3(x^3 - x^2 + x)^2 D_x(x^3 - x^2 + x).$$

Then by parts (ii), (iii), and (iv),
$$y' = 3(x^3 - x^2 + x)^2(3x^2 - 2x + 1).$$
The slope of the tangent line is
$$y'|_{x=1} = 3(1 - 1 + 1)^2(3 - 2 + 1) = 6. \blacksquare$$

Example 7
Find the rate of change of y with respect to x at $x = 2$ if $y = (x^2 + 2)^2/x^2$.
First write
$$y = \frac{x^4 + 4x^2 + 4}{x^2} = x^2 + 4 + 4x^{-2}.$$
Then
$$D_x y = 2x + 0 - 8x^{-3} = 2x - \frac{8}{x^3},$$
and the rate of change of y with respect to x at $x = 2$ is
$$D_x y|_{x=2} = 2(2) - \frac{8}{2^3} = 3. \blacksquare$$

The remainder of this section is devoted to discussion of the proof of Theorem 1. Although some parts are harder to prove than others, the main idea is to use the definition of derivative (Definition 2 of Section 10.2). Part (i) is proved as follows:

$$D_x f(x) = \lim_{\Delta x \to 0} \frac{f(x + \Delta x) - f(x)}{\Delta x} = \lim_{\Delta x \to 0} \frac{k - k}{\Delta x}$$
$$= \lim_{\Delta x \to 0} \frac{0}{\Delta x} = \lim_{\Delta x \to 0} (0) = 0.$$

Proofs of parts (iii) and (iv) are similar and also require the use of Definition 2; they are left as exercises. If n is a positive integer, part (ii) can be proved in a similar manner by using the binomial theorem to expand $(x + \Delta x)^n$—see Exercise 22; the case in which n is a negative integer can be handled after Theorem 2 of the next section has been studied (Exercise 19 of Section 10.4). If $n = 0$, part (ii) is a special case of part (i). Thus we have indicated how to prove part (ii) if n is any integer. The proof of part (ii) for any real number n will be given in Section 14.3, provided that $x > 0$. Part (v) will be proved in Example 5 of the next section.

APPLICATIONS

Example 8
In many applied problems it is necessary to solve various types of equations. In this example we discuss a general method, called **Newton's method,** for approximating a real root of an equation of the form

$$f(x) = 0.$$

We shall see that one of the prerequisites for using Newton's method is the ability to find $f'(x)$ by using rules studied in this section.

Suppose that a portion of the graph of $y = f(x)$ is as shown in Figure 10.12. A root r of $f(x) = 0$ is a number r for which $f(r) = 0$; such a number is the x-coordinate of a point of intersection of $y = f(x)$ with the x-axis. We first make an estimate x_1 of the desired root—this number should be fairly close to the root r. As a second, and usually better, estimate of r we choose the x-coordinate x_2 of the point where the tangent line to $y = f(x)$ at $(x_1, f(x_1))$ crosses the x-axis. The slope of this tangent line is $f'(x_1)$, and the tangent line passes through $(x_1, f(x_1))$; thus its equation in point-slope form is

$$y - f(x_1) = f'(x_1)(x - x_1).$$

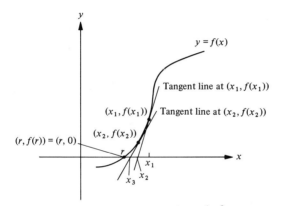

Figure 10.12. *Newton's method.*

Since $(x_2, 0)$ satisfies this equation, the second estimate x_2 can be found as follows:

$$0 - f(x_1) = f'(x_1)(x_2 - x_1),$$

$$x_2 - x_1 = -\frac{f(x_1)}{f'(x_1)},$$

$$x_2 = x_1 - \frac{f(x_1)}{f'(x_1)}.$$

The process is repeated by letting the third estimate x_3 of the root r be the x-coordinate of the point at which the tangent line at $(x_2, f(x_2))$ intersects the x-axis. This yields

$$x_3 = x_2 - \frac{f(x_2)}{f'(x_2)}.$$

Sec. 10.3] Basic Differentiation Formulas

In general, successive estimates of r are given by

$$x_n = x_{n-1} - \frac{f(x_{n-1})}{f'(x_{n-1})},$$

for $n = 2, 3, 4, \ldots$.

If we assume that the successive estimates of r get very near to r, it follows that r can be approximated to any number of decimal places of accuracy by repeating the estimation process enough times. The repetitive nature of the procedure makes it possible to program a digital computer to do Newton's method. We have not stated the precise conditions under which Newton's method is valid, but clearly f must be differentiable at points near r and $f'(x_1)$, $f'(x_2)$, etc., cannot be zero.

Let the procedure be illustrated by estimating a root of the equation

$$f(x) = x^3 - x - 4 = 0.$$

Since

$$f(1) = 1 - 1 - 4 = -4 < 0$$

and

$$f(2) = 8 - 2 - 4 = 2 > 0,$$

and the polynomial function f is continuous, there is a root of the equation between 1 and 2 (see Exercise 12 of Section 9.4). Let the first approximation of this root be

$$x_1 = 2.$$

Noting that

$$f'(x) = 3x^2 - 1,$$

we substitute to find x_2 as follows:

$$x_2 = 2 - \frac{f(2)}{f'(2)} = 2 - \frac{2}{3(2)^2 - 1} = \frac{20}{11}.$$

Suppose that one more estimate x_3 is computed:

$$x_3 = \frac{20}{11} - \frac{f(\frac{20}{11})}{f'(\frac{20}{11})}$$

$$= \frac{20}{11} - \frac{(\frac{20}{11})^3 - \frac{20}{11} - 4}{3(\frac{20}{11})^2 - 1}.$$

After some arithmetic we find that

$$x_3 = \frac{21{,}324}{11{,}869},$$

which is very near the exact root r. In fact, $x_3 \approx 1.7966$, whereas the correct value of r to four decimal places can be shown to be 1.7963 by repeating the method more times.

Newton's method can be used to estimate roots of equations other than polynomial equations. For example, after we learn how to differentiate 2^x,

we can approximate the root between 1 and 2 of the equation
$$2^x + x - 4 = 0.$$
(See Exercise 9 of Section 14.3.)

Newton's method can be used as a technique for calculating square roots of positive real numbers to any desired accuracy (see Exercise 26). ∎

EXERCISES

Suggested minimum assignment: Exercises 1, 3, 7, 9, 13, 15, 16, and 19.

In Exercises 1–6 find $D_x f(x)$.

1. $f(x) = 8$.
2. $f(x) = \pi$.
3. $f(x) = x^4$.
4. $f(x) = x^{-3} + x^{\sqrt{3}}$.
5. $f(x) = \sqrt[3]{x} + x$.
6. $f(x) = x^2(2x^3 + 3)$.

In Exercises 7–10 find y'.

7. $y = 5x^3 - 8x^2 + 3x - 4$.
8. $y = \dfrac{1}{x^2} + \dfrac{2x}{3}$.
9. $y = (x^2 + x)^5$.
10. $y = (x^4 + 3)^4$.

In Exercises 11–13 find the slope of the tangent line to the given curve at the given value of x.

11. $y = 3x^3 - 8x + 2$, $x = 2$.
12. $y = (1 - 3x)^4$, $x = -1$.
13. $y = \dfrac{1}{x}$, $x = 3$.

In Exercises 14 and 15 find the rate of change of y with respect to x at the given value of x.

14. $y = \dfrac{x^2 + x^4}{x}$, $x = 1$.
15. $y = \sqrt{3x + 1}$, $x = 5$.

16. For the cost function given by $C(x) = (x^2 + 10)/5$, find the marginal cost $C'(x)$.
17. For the revenue function given by $R(x) = 20x - (x^2/10)$, find the marginal revenue $R'(x)$.
18. For the profit function given by $P(x) = x + 2x^2 - x^3$, find the marginal profit $P'(x)$.

In Exercises 19–21 a particle is moving along a line according to the given equation of motion which is assumed to be valid if $t \geq 1$. Find the velocity of the particle when $t = 4$.

19. $s = \dfrac{(t^3 + 1)^2}{t^3}$.
20. $s = \sqrt{t^2 + 9}$.
21. $s = (t^2 - t)^2$.

Sec. 10.3] Basic Differentiation Formulas

22. Prove Theorem 1(*ii*) for the case that n is a positive integer.
23. Prove Theorem 1(*iii*).
24. Prove Theorem 1(*iv*).
25. Use Newton's method (Example 8) to approximate the root between 1 and 2 of the equation
$$x^3 + x^2 - x - 2 = 0.$$
Use $x_1 = 1$ and find x_2 and x_3.
26. Use Newton's method (Example 8) to approximate $\sqrt{5}$ by approximating the root between 2 and 3 of the equation
$$x^2 - 5 = 0.$$
Use $x_1 = 2$ and find x_2 and x_3. (As a special project, if digital computer facilities are available, one could attempt to write a program for finding successive approximations to $\sqrt{5}$ by Newton's method.)

10.4 Basic Differentiation Formulas (Continued)

In this chapter we have learned that the ability to find the derivative of a function is useful in problems concerning slope of a tangent line to a curve, rate of change, marginal cost, marginal revenue, marginal profit, or velocity. The derivative can be found more quickly using the formulas of Theorem 1 than by applying Definition 2. In Theorem 2 we state three more very important differentiation formulas, which are valid under the assumption that all indicated derivatives exist and all denominators are not zero.

▶ **Theorem 2**

(i) $D_x[f(x)g(x)] = f(x) D_x g(x) + g(x) D_x f(x).$ (product rule)

(ii) $D_x\left[\dfrac{f(x)}{g(x)}\right] = \dfrac{g(x) D_x f(x) - f(x) D_x g(x)}{[g(x)]^2}.$ (quotient rule)

(iii) If $y = g(u)$ and $u = f(x)$, then $\dfrac{dy}{dx} = \dfrac{dy}{du}\dfrac{du}{dx}.$ (chain rule)

A discussion of the proof of this theorem is given at the end of this section. ◀

Example 1
Apply Theorem 2(*i*) to differentiate the product $(x^3 + x^2)(x^2 + 8)$.
$$\begin{aligned} D_x[(x^3 + x^2)(x^2 + 8)] &= (x^3 + x^2) D_x(x^2 + 8) + (x^2 + 8) D_x(x^3 + x^2) \\ &= (x^3 + x^2)(2x) + (x^2 + 8)(3x^2 + 2x) \\ &= 5x^4 + 4x^3 + 24x^2 + 16x. \end{aligned}$$

Theorem 2(*i*) says that *the derivative of the product of two functions is the first times the derivative of the second plus the second times the derivative of*

the first. It is possible to avoid using the product rule in Example 1 by multiplying the two functions together at the beginning.

Theorem 2(ii) can be used whenever the function to be differentiated can be expressed as a quotient of two functions. In words, it says that *the derivative of the quotient of two functions is the denominator times the derivative of the numerator minus the numerator times the derivative of the denominator, all divided by the square of the denominator.* Observe the use of this quotient rule in the next example.

Example 2
Find the slope of the tangent line to $y = x^2/(3x + 1)$ at $x = 1$.

By Theorem 2(ii),

$$y' = \frac{(3x + 1) D_x(x^2) - x^2 D_x(3x + 1)}{(3x + 1)^2}$$

$$= \frac{(3x + 1)(2x) - x^2(3)}{(3x + 1)^2}$$

$$= \frac{3x^2 + 2x}{(3x + 1)^2}.$$

Another procedure that will lead to the same result for y' is to first write $y = x^2(3x + 1)^{-1}$ and then to use the product rule.

The slope of the tangent line at $x = 1$ is

$$y'|_{x=1} = \frac{3(1)^2 + 2(1)}{[3(1) + 1]^2}$$

$$= \frac{5}{16}. \blacksquare$$

Example 3
If $y = g(u) = 3u + 1$ and $u = f(x) = x^2 + 4$, use Theorem 2(iii) to find dy/dx.

$$\frac{dy}{dx} = \frac{dy}{du}\frac{du}{dx}$$

$$= [D_u(3u + 1)][D_x(x^2 + 4)]$$

$$= 3(2x)$$

$$= 6x. \blacksquare$$

Some additional discussion and explanation is needed in connection with Example 3. If y is a function of u, say $y = g(u)$, and u is a function of x, say $u = f(x)$, then a combination of these functional relationships makes y a function of x. In general we have

$$y = g(u) = g(f(x)).$$

A function given by $y = g(f(x))$ is called a **composite function**. The domain

of the composite function given by $y = g(f(x))$ consists of the elements x in the domain of f such that $f(x)$ is in the domain of g. In Example 3 $g(u) = 3u + 1$ and $f(x) = x^2 + 4$; therefore,

$$y = g(f(x)) = g(x^2 + 4) = 3(x^2 + 4) + 1 = 3x^2 + 13,$$

and the domain of this composite function is the set of all real numbers. The answer to Example 3 can be checked as follows:

$$\frac{dy}{dx} = D_x\, g(f(x)) = D_x\, (3x^2 + 13) = 6x.$$

However, we want to emphasize the method used in Example 3, since the chain rule will be useful in several sections later in this text.

Example 4
If $z = (w^2 - 1)^3$ and $w = 3/(t - 1)$, find the rate of change of z with respect to t when $t = 2$.

By the chain rule,

$$\frac{dz}{dt} = \frac{dz}{dw}\frac{dw}{dt}.$$

dz/dw can be found by the power rule for differentiation:

$$\frac{dz}{dw} = 3(w^2 - 1)^2(2w).$$

dw/dt can be found by the quotient rule:

$$\frac{dw}{dt} = \frac{(t-1)(0) - 3(1)}{(t-1)^2} = \frac{-3}{(t-1)^2}.$$

Hence

$$\frac{dz}{dt} = [6w(w^2 - 1)^2]\frac{-3}{(t-1)^2}. \qquad (10.4)$$

When $t = 2$, $w = 3/(t-1) = 3/(2-1) = 3$. Substituting $t = 2$ and $w = 3$ in (10.4), we find that

$$\left.\frac{dz}{dt}\right|_{t=2} = 6(3)(9-1)^2\frac{-3}{(2-1)^2} = -3456. \blacksquare$$

Example 5
Suppose that $y = u^n$, where n is a real number, and $u = f(x)$. Find $D_x y$.

By the chain rule,

$$\frac{dy}{dx} = \frac{dy}{du}\frac{du}{dx}$$

or

$$D_x y = D_u y\, D_x u.$$

By Theorem 1(ii),
$$D_u y = nu^{n-1}.$$
Therefore,
$$D_x y = nu^{n-1} D_x u.$$
This is the power rule for differentiation, which is just a special case of the chain rule for the case when $y = u^n$. ∎

Before we can discuss the proof of Theorem 2, we need to make the statement of a very well known theorem.

▶ **Theorem 3**
If a function f is differentiable at $x = x_1$, then f is continuous at $x = x_1$.

A hint for the proof of this important relationship between differentiability and continuity is given in Exercise 24. ◀

One consequence of Theorem 3 is that if a function f is discontinuous at $x = x_1$, then f is not differentiable at $x = x_1$. The converse of Theorem 3 is not always true; in Example 4 of Section 10.1 an example is given of a function that is continuous at a point but not differentiable at that point.

The remainder of this section is devoted to some discussion of the proof of Theorem 2. Theorem 2(i) can be proved as follows:

$$D_x[f(x)g(x)] = \lim_{\Delta x \to 0} \frac{f(x + \Delta x)g(x + \Delta x) - f(x)g(x)}{\Delta x}$$

$$= \lim_{\Delta x \to 0} \frac{f(x + \Delta x)g(x + \Delta x) - f(x + \Delta x)g(x) + f(x + \Delta x)g(x) - f(x)g(x)}{\Delta x}$$

$$= \lim_{\Delta x \to 0} f(x + \Delta x) \left[\frac{g(x + \Delta x) - g(x)}{\Delta x} \right] + \lim_{\Delta x \to 0} g(x) \left[\frac{f(x + \Delta x) - f(x)}{\Delta x} \right]$$

$$= f(x) D_x g(x) + g(x) D_x f(x).$$

An explanation of why $\lim_{\Delta x \to 0} f(x + \Delta x) = f(x)$ needs to be made. We are assuming that f is differentiable at x; therefore, by Theorem 3, f is continuous at x. By (10.1) of Section 10.1, $\lim_{\Delta x \to 0} [f(x + \Delta x) - f(x)] = 0$, and therefore $\lim_{\Delta x \to 0} f(x + \Delta x) = f(x)$.

The proof of Theorem 2(ii) is similar to the proof of part (i) and is left as Exercise 18.

For Theorem 2(iii) suppose that x is given an increment $\Delta x \neq 0$. This causes a change Δu in u, which in turn causes a change Δy in y. If $\Delta u \neq 0$,

$$\frac{\Delta y}{\Delta x} = \frac{\Delta y}{\Delta u} \frac{\Delta u}{\Delta x}$$

and
$$\frac{dy}{dx} = \lim_{\Delta x \to 0} \frac{\Delta y}{\Delta x} = \lim_{\Delta x \to 0} \frac{\Delta y}{\Delta u} \frac{\Delta u}{\Delta x}$$
$$= \left(\lim_{\Delta x \to 0} \frac{\Delta y}{\Delta u}\right)\left(\lim_{\Delta x \to 0} \frac{\Delta u}{\Delta x}\right).$$

Since we are assuming that f is differentiable, f is also continuous and hence $\Delta u \to 0$ as $\Delta x \to 0$. Therefore,
$$\frac{dy}{dx} = \left(\lim_{\Delta u \to 0} \frac{\Delta y}{\Delta u}\right)\left(\lim_{\Delta x \to 0} \frac{\Delta u}{\Delta x}\right) = \frac{dy}{du}\frac{du}{dx}.$$

Unfortunately, this is not a complete proof of the chain rule because we assumed that $\Delta u \neq 0$. If $\Delta u = 0$, the proof is more complicated and the reader is referred to a more advanced text.

APPLICATIONS

Example 6

Let $C(x)$ give the cost of producing x items. The average cost per item is $C(x)/x$. The function f defined by
$$f(x) = \frac{C(x)}{x}$$
is called the **average cost function.** It is of some importance in economics because a manufacturer wants the average cost to be as small as possible. We shall learn in the next chapter that in order to find when the average cost function is smallest, it is necessary to find its derivative (when the derivative exists). By the quotient rule of this section,
$$f'(x) = \frac{xC'(x) - C(x)}{x^2}.$$

Suppose that x_1 items have been produced at a cost of $C(x_1)$. The fraction
$$\frac{[C(x_1 + \Delta x) - C(x_1)]/C(x_1)}{\Delta x/x_1}$$
is the ratio of the percentage change in total cost to a given percentage change in quantity produced. The ratio is called the elasticity of cost on the interval between x_1 and $x_1 + \Delta x$ and may also be written
$$\frac{\Delta C/\Delta x}{C(x_1)/x_1}.$$
If $\Delta x \to 0$, the ratio becomes
$$\frac{C'(x_1)}{C(x_1)/x_1},$$

which is the marginal cost at x_1 divided by the average cost of producing x_1 items and is called the **elasticity of cost at** $x = x_1$. In general, the function E which represents the elasticity of cost is determined by

$$E(x) = \frac{C'(x)}{C(x)/x}.$$

Further interpretation of the meaning of elasticity of cost will be given in Example 5 of Section 10.6. ∎

EXERCISES

Suggested minimum assignment: Exercises 1, 5, 7, 11, 12, 15, and 17.

Find y' in Exercises 1–6 using Theorem 2(i) or (ii).

1. $y = x^2(2x^3 + 3)$.
2. $y = x(2x + 3)^4$.
3. $y = x\sqrt{2x + 1}$.
4. $y = \dfrac{x}{2x + 7}$.
5. $y = \dfrac{x^2 + 1}{x^2 + 2}$.
6. $y = \dfrac{1}{x^2 - 1}$.

Find dy/dx in Exercises 7 and 8 using Theorem 2(iii).

7. $y = 1 - 2u$, $u = x^3 - 2x^2 + 6$.
8. $y = \dfrac{1}{u}$, $u = x^2 - 8x + 2$.
9. Find the slope of the tangent line to $y = (x^2 + 2x - 3)(x^3 - x^2 + 6)$ at $x = -1$.
10. By repeated use of the product rule show that if $y = f(x)g(x)h(x)$, then $y' = f'(x)g(x)h(x) + f(x)g'(x)h(x) + f(x)g(x)h'(x)$.
11. If $y = 2/(x - 3)$, find the rate of change of y with respect to x at $x = 5$.
12. Suppose that a cost function is given by $C(x) = \dfrac{x^2 + 5}{2x + 1}$. Find the marginal cost $C'(x)$.
13. Suppose that a revenue function is given by $R(x) = \dfrac{1}{x + 1}(5x^2 + 2)$. Find the marginal revenue $R'(x)$.
14. Suppose that a profit function is given by $P(x) = \sqrt{x}(x - 1)$. Find the marginal profit $P'(x)$.
15. A particle moves along a line according to the given equation of motion. Find the velocity when $t = 4$.

    (a) $s = t(t^2 - 15)^4$. (b) $s = \dfrac{t}{\sqrt{2t + 1}}$.

16. Find the rate of change of y with respect to x when $x = 4$ if $y = \sqrt{2u - 1}$ and $u = \sqrt{2x + 1}$.

Sec. 10.4] Basic Differentiation Formulas

17. If $y = g(u) = u^2 + 1$ and $u = f(x) = 3x + 7$, find $g(f(x))$ and then find $D_x g(f(x))$. Check by using the chain rule.
18. Prove Theorem 2(ii).
19. Prove Theorem 1(ii) of Section 10.3 for the case that n is a negative integer. (*Hint:* Write x^n as $1/x^{-n}$, where $-n$ is a positive integer, and use the quotient rule.)
20. Find $D_x z$ if $z = \dfrac{y}{y+1}$ and $y = x + \dfrac{1}{x}$.
21. Find $D_t s$ if $s = \dfrac{(t^3 + 1)^2}{t}$.
22. Find $D_v w$ if $w = \dfrac{v}{(v^2 + 2)^3}$.
23. Sketch $y = f(x) = |x| + 1$. Is f continuous at $x = 0$? Is f differentiable at $x = 0$?
24. Prove Theorem 3. [*Hint:*

$$\lim_{x \to x_1} [f(x) - f(x_1)] = \lim_{x \to x_1} (x - x_1) \frac{f(x) - f(x_1)}{x - x_1} = (0)[f'(x_1)] = 0.$$

See Exercise 10 of Section 10.2.]
25. Suppose that a cost function is given by $C(x) = x^2 + 1$. Find the derivative of the average cost function. Also find the elasticity of cost $E(x)$. (See Example 6.)

10.5 Higher Derivatives

In Definition 2 of Section 10.2 the derivative f' of a function f was defined. The function f' is also called the **derived function** or the **first derivative** of f. One can begin again with the function f' and define its derivative, which is denoted by f'' and which is called the **second derivative** of f. In other words, the second derivative is the derivative of the first derivative—the limit defining $f''(x)$ is

$$f''(x) = \lim_{\Delta x \to 0} \frac{f'(x + \Delta x) - f'(x)}{\Delta x},$$

provided that this limit exists. The process can be continued—the **third derivative** is the derivative of the second derivative, the **fourth derivative** is the derivative of the third derivative, and in general for any positive integer n the **nth derivative** is the derivative of the $(n - 1)$st derivative. Derivatives beyond the first derivative are called **higher derivatives**. Example 1 illustrates the correct notation to use when writing higher derivatives (except that the higher derivative notation corresponding to the dy/dx notation will not be presented until the next section).

Example 1
Suppose that $y = f(x) = x^6$. Then

$$y' = f'(x) = D_x y = D_x f(x) = 6x^5,$$
$$y'' = f''(x) = D_x^2 y = D_x^2 f(x) = 30x^4,$$
$$y''' = f'''(x) = D_x^3 y = D_x^3 f(x) = 120x^3,$$
$$y^{(4)} = f^{(4)}(x) = D_x^4 y = D_x^4 f(x) = 360x^2,$$
$$y^{(5)} = f^{(5)}(x) = D_x^5 y = D_x^5 f(x) = 720x,$$
$$y^{(6)} = f^{(6)}(x) = D_x^6 y = D_x^6 f(x) = 720,$$
$$y^{(7)} = f^{(7)}(x) = D_x^7 y = D_x^7 f(x) = 0.$$

All derivatives of x^6 higher than the sixth derivative of x^6 are equal to zero. ∎

As an illustration of how to read the notation in Example 1, $D_x^4 y$ is read "the fourth derivative of y with respect to x." One should think of $D_x^4 y$ as a single symbol rather than trying to assign any meaning to the component parts—notice where the dependent variable, y, is written in this notation and observe also the location of the independent variable, x. Since the fourth derivative is the derivative of the third derivative, the following notations have the same meaning as $D_x^4 y$:

$$D_x y''', \quad D_x[f'''(x)], \quad D_x(D_x^3 y), \quad D_x[D_x^3 f(x)].$$

If the fourth derivative is evaluated at $x = 1$ in Example 1, the correct notations are

$$y^{(4)}|_{x=1} = f^{(4)}(1) = D_x^4 y|_{x=1} = D_x^4 f(x)|_{x=1} = 360.$$

Next we mention two interpretations that can be associated with the second derivative; the reader can be assured that several other uses of higher derivatives will be encountered later in this text. The second derivative is the rate of change of the first derivative; since the first derivative gives the slope of a tangent line to a curve, we may say that the second derivative gives the rate of change of slope. Also, for a particle moving along a line, the second derivative gives the rate of change of the velocity of the particle, which is called the **acceleration** of the particle.

Example 2
Find the rate of change of slope at the point $(3, 4)$ on the semicircle $y = \sqrt{25 - x^2}$.

By the power rule for differentiation,

$$y' = \left(\frac{1}{2}\right)(25 - x^2)^{-1/2}(-2x) = \frac{-x}{(25 - x^2)^{1/2}}.$$

By the quotient rule,

$$y'' = \frac{(25-x^2)^{1/2}(-1) - (-x)(\tfrac{1}{2})(25-x^2)^{-1/2}(-2x)}{25-x^2}$$

$$= \frac{[-(25-x^2) - x^2]/(25-x^2)^{1/2}}{25-x^2}$$

$$= \frac{-25}{(25-x^2)^{3/2}}.$$

Therefore,

$$y''|_{x=3} = \frac{-25}{(16)^{3/2}} = -\frac{25}{64}.$$

This means that if one could move from left to right along the semicircle, then at the instant the point (3, 4) is reached, the slope is becoming smaller by $\tfrac{25}{64}$ unit per each unit of change in x. See Figure 10.13. ∎

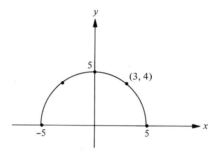

Figure 10.13. $y = \sqrt{25 - x^2}$.

Example 3
A particle moves along a line according to the equation of motion $s = t^3 - 15t^2$. Find the velocity and acceleration at $t = 6$ seconds. (Assume that s is measured in feet and t in seconds.)

$$D_t s = 3t^2 - 30t,$$
$$D_t^2 s = 6t - 30.$$

Since $D_t s|_{t=6} = 3(36) - 180 = -72$, the velocity is -72 feet per second. The units for acceleration are feet per second per second, and therefore the acceleration is $D_t^2 s|_{t=6} = 6$ feet per second per second. If the particle is visualized as moving back and forth along a horizontal line with positive direction to the right, the positive acceleration means that a force is being exerted on the particle in the positive direction when $t = 6$; however, because the velocity is negative, the particle is actually moving to the left, but it is slowing down, since the acceleration is positive. In a situation in which velocity and acceleration have opposite signs, the particle is slowing down, whereas the particle is speeding up when these signs are the same. ∎

Example 4
If $D_x^4 y = (x^2 + 3)^5$, find $D_x^6 y$.
By the power rule for differentiation
$$D_x^5 y = D_x(x^2 + 3)^5 = 5(x^2 + 3)^4(2x) = (10x)(x^2 + 3)^4.$$
$D_x^6 y$ is the derivative of $D_x^5 y$ and is found by applying the product rule, the power rule for differentiation, and simplifying the result.
$$\begin{aligned} D_x^6 y &= D_x[(10x)(x^2 + 3)^4] \\ &= (10x) D_x(x^2 + 3)^4 + (x^2 + 3)^4 D_x(10x) \\ &= (10x)(4)(x^2 + 3)^3(2x) + 10(x^2 + 3)^4 \\ &= 10(x^2 + 3)^3[8x^2 + (x^2 + 3)] \\ &= 10(x^2 + 3)^3(9x^2 + 3) \\ &= 30(x^2 + 3)^3(3x^2 + 1). \ \blacksquare \end{aligned}$$

APPLICATIONS

Example 5
Suppose that a manufacturer sells an item by the ton (or partial ton). If he sells x tons per week, he has found that the price $p(x)$ that he must charge per ton depends upon the number x of tons sold. His weekly revenue is given by
$$R(x) = xp(x).$$
Using the product rule of Section 10.4, the marginal revenue $R'(x)$ is given by
$$\begin{aligned} R'(x) &= xp'(x) + [p(x)](1) \\ &= xp'(x) + p(x). \end{aligned}$$
The rate of change of the marginal revenue $R'(x)$ with respect to x is the derivative of $R'(x)$, which is designated $R''(x)$. If we apply the product rule again, we find that
$$\begin{aligned} R''(x) &= xp''(x) + [p'(x)](1) + p'(x) \\ &= xp''(x) + 2p'(x). \end{aligned}$$
We could call $R'(x)$ the velocity in revenue and $R''(x)$ the acceleration in revenue. Under normal conditions the velocity in revenue is positive, because the more tons that are sold, the greater the gross revenue from sales. Then if the acceleration in revenue is also positive, it means that the rate of increase of revenue is speeding up as more tons are sold; a negative acceleration in revenue indicates a slowing down in the rate of increase of revenue. ∎

Sec. 10.5] **Higher Derivatives**

EXERCISES

Suggested minimum assignment: Exercises 1, 2, 8, 11, 13, and 15(b).

1. If $y = 7x^4$, find y', y'', and y'''.
2. If $f(x) = 7x^3 - x^5$, find $f'(2)$, $f''(2)$, and $f'''(2)$.
3. If $y = \sqrt{x}$, find $D_x y$, $D_x^2 y$, $D_x^3 y$, and $D_x^4 y$.
4. If $y = u^3$ and $u = x + \sqrt{x}$, find $D_x y$ and $D_x^2 y$.
5. If $f(x) = 3/(x - 4)$, find $f'(x)$ and $f''(x)$.
6. If $f(x) = x(2x + 3)^6$, find $f'(x)$ and $f''(x)$.
7. If $y = \frac{1}{8}x^8 + 4000x^3 + 89x^2$, find $y^{(4)}$.
8. If $f(x) = (2x + 1)^{10}$, find $D_x^4 f(x)$.
9. If $y = \sqrt{7x + 8}$, find $y''|_{x=4}$.
10. If $f'(x) = 3x^2$, find $f''(x)$ by using $f''(x) = \lim_{\Delta x \to 0} \dfrac{f'(x + \Delta x) - f'(x)}{\Delta x}$.
11. If $D_x^7 y = \dfrac{1}{2x}$, find $D_x^8 y$.
12. If $y = x^5 - x^4$, find $D_x(D_x^3 y)$.
13. If $y = 2x^4$, find the rate of change of slope at $x = 1$.
14. If $y = \sqrt{x^2 + 3}$, find the rate of change of slope at $(1, 2)$.
15. In each of the following, a particle moves along a line according to the given equation of motion. Find the velocity and acceleration at $t = 3$ seconds. Assume that s is measured in meters.
    (a) $s = t^3 - 3t^2 - 8t + 31$.   (b) $s = (t^2 - 7)^4$.
16. In each of the following, a particle moves along a line according to the given equation of motion. At $t = 2$ seconds, tell whether the particle is moving in the positive or negative direction along the s-axis and tell whether it is speeding up or slowing down.
    (a) $s = t^2$.   (b) $s = t^3 - 10t^2$.
    (c) $s = 30t - t^3$.
17. A revenue function is given by $R(x) = x^3 + x^2$. Find the rate of change of marginal revenue at $x = 2$. At $x = 2$, is the rate of increase of revenue speeding up or slowing down?

10.6 Differentials

In Example 2 of this section we shall try to gain a better understanding of marginal revenue, marginal cost, and marginal profit. Before this can be done, we need to introduce a concept of considerable importance.

▶ **Definition 3**

Suppose that $y = f(x)$ and that f is differentiable at x. If x is changed by an increment $\Delta x \neq 0$, then the **differential** dx of the independent variable x is defined by

$$dx = \Delta x,$$

and the **differential** dy of the dependent variable y is defined by

$$dy = f'(x) \, \Delta x \quad \text{or} \quad dy = f'(x) \, dx.$$

Definition 3 involves two separate definitions; in order to show that these two definitions are consistent with each other, we examine the function defined by $y = f(x) = x$. Then $dy = f'(x) \, \Delta x = (1) \, \Delta x = \Delta x$. However, $dy = dx$, since $y = x$. Hence $dx = \Delta x$.

The geometric interpretation of dy is important (see Figure 10.14). Since

$$dy = f'(x) \, dx,$$

we obtain, by dividing by $dx \neq 0$, the equation

$$f'(x) = \frac{dy}{dx}.$$

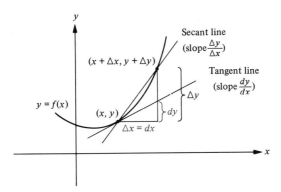

Figure 10.14

Since dy/dx is the slope of the tangent line to $y = f(x)$ at (x, y) and since $dx = \Delta x$, it follows that in Figure 10.14 dy must be the distance from the horizontal up to the tangent line.

As we learned in Section 10.1, Δy is the exact change in y corresponding to the change of Δx in x. In general, both Δy and dy can be either positive, negative, or zero, depending upon the figure. Notice that, in general, Δy does *not* equal dy, but by definition Δx *does* equal dx. If $|\Delta x|$ is relatively small, then dy and Δy usually are almost the same and dy can be calculated as a close approximation to Δy, as we illustrate in Example 1.

Example 1

If $y = f(x) = x^2$, calculate Δy and dy if $x = 2$ and $\Delta x = dx = .01$.

$$\begin{aligned}
\Delta y &= f(x + \Delta x) - f(x) \\
&= f(2.01) - f(2) \\
&= (2.01)^2 - (2)^2 \\
&= 4.0401 - 4 \\
&= .0401.
\end{aligned}$$

$$dy = f'(x) \, dx = 2x \, dx = 2(2)(.01) = .04.$$

Notice that dy is easier to calculate than Δy and that dy is a close approximation to Δy when $|\Delta x|$ is a relatively small number; several useful applications of dy result from the latter fact. ∎

Example 2
Suppose that a motorcycle manufacturer produces x motorcycles on a single work day. The number x depends upon several factors, such as the number of workers available and the frequency of breakdown of machinery, but x is always an integer from 0 to 100 inclusive. The cost in dollars of producing x motorcycles in one day is known to be given by the cost function C, where

$$C(x) = 2000 + 300x - \frac{x^2}{2}.$$

If 50 motorcycles come off the assembly line in 1 day, the cost (dollar signs omitted) is

$$C(50) = 2000 + (300)(50) - \frac{(50)^2}{2} = 15{,}750.00.$$

The manager wonders what the cost would have been if $x = 51$, instead of $x = 50$, and so he computes

$$C(51) = 2000 + (300)(51) - \frac{(51)^2}{2} = 15{,}999.50.$$

The difference is

$$C(51) - C(50) = 249.50,$$

which would have been the exact cost of one additional motorcycle.

Let us also calculate the marginal cost at $x = 50$ as follows (we assume now that x is any real number so that C can be differentiated):

$$C(x) = 2000 + 300x - \frac{x^2}{2},$$
$$C'(x) = 300 - x,$$
$$C'(50) = 250.$$

The answer for $C'(50)$ is almost the same as for $C(51) - C(50)$ and is much easier to calculate—let us see why this is no coincidence. If

$$y = C(x),$$

with $x = 50$ and $\Delta x = 1$, then

$$\Delta y = C(x + \Delta x) - C(x) = C(51) - C(50).$$

Also, using Definition 3 of this section,

$$dy = C'(x)\,\Delta x = [C'(50)](1) = C'(50).$$

Since dy is an approximation to Δy, it follows that $C'(50)$ is an approxima-

tion to $C(51) - C(50)$. Hence the marginal cost $C'(50)$ gives approximately the cost of the next, or 51st, motorcycle.

In general, the marginal cost $C'(k)$ gives the approximate cost of the $(k + 1)$st item. In a similar way it can be shown that the marginal revenue $R'(k)$ is approximately the revenue that will be received for the $(k + 1)$st item, assuming that the first k items have already been sold. The marginal profit $P'(k)$ gives the approximate profit derived from the sale of the $(k + 1)$st item and

$$P'(k) = R'(k) - C'(k).$$

Suppose in our motorcycle problem that the revenue function R is given by

$$R(x) = 400x + \frac{x^2}{10}.$$

Then the marginal revenue is

$$R'(x) = 400 + \frac{x}{5},$$

and at $x = 50$ the marginal revenue is

$$R'(50) = 400 + \frac{50}{5} = 410.$$

The approximate profit for the sale of the 51st motorcycle is the marginal profit $P'(50)$, where

$$P'(50) = R'(50) - C'(50) = 410 - 250 = 160. \blacksquare$$

If $y = f(x)$, we learned at the beginning of this section that

$$f'(x) = \frac{dy}{dx},$$

which justifies our previous use of dy/dx as a notation for "the derivative of y with respect to x." Since meanings have been assigned to dy and dx, the symbol dy/dx can also be thought of as a quotient. If dy/dx is used for the first derivative of y with respect to x, then the notation for higher derivatives is

$$\frac{d^2y}{dx^2}, \frac{d^3y}{dx^3}, \ldots.$$

No meaning is assigned to d^2y, dx^2, d^3y, dx^3, ..., so that the symbols for higher derivatives *cannot* be thought of as quotients but must be regarded as entire symbols; for example, the symbol d^2y/dx^2 is "the second derivative of y with respect to x," even though d^2y and dx^2 by themselves have no meaning. Observe carefully the notation used in the next example.

Sec. 10.6] Differentials

Example 3
If $y = f(x) = x^{-1}$, find dy, dy/dx, d^2y/dx^2, and d^3y/dx^3.

$$dy = f'(x)\, dx = -x^{-2}\, dx = -\frac{1}{x^2}\, dx,$$

$$\frac{dy}{dx} = f'(x) = -x^{-2} = -\frac{1}{x^2},$$

$$\frac{d^2y}{dx^2} = \frac{d}{dx}\left(\frac{dy}{dx}\right) = \frac{d}{dx}(y') = \frac{dy'}{dx} = \frac{d(-1/x^2)}{dx} = -(-2)x^{-3} = \frac{2}{x^3},$$

$$\frac{d^3y}{dx^3} = \frac{d}{dx}\left(\frac{d^2y}{dx^2}\right) = \frac{d}{dx}(y'') = \frac{dy''}{dx} = \frac{d(2/x^3)}{dx} = 2(-3)x^{-4} = -\frac{6}{x^4}. \blacksquare$$

Example 4
Show that the differential of the product of two differentiable functions is the first times the differential of the second plus the second times the differential of the first.

Let
$$h(x) = f(x)g(x).$$

By the product rule,
$$h'(x) = f(x)g'(x) + g(x)f'(x).$$

Therefore,
$$d(h(x)) = h'(x)\, dx = [f(x)][g'(x)\, dx] + [g(x)][f'(x)\, dx]$$
$$= [f(x)][d(g(x))] + [g(x)][d(f(x))].$$

Other rules for derivatives carry over in a similar way to corresponding rules for differentials. $\blacksquare$

APPLICATIONS

Example 5
In Example 6 of Section 10.4 the elasticity of cost at a point x_1 was defined as the marginal cost at x_1 divided by the average cost of producing x_1 items. Now that we have a better understanding of the meaning of marginal cost, let us use the cost function of Example 2 to try to gain a better understanding of elasticity of cost. The average cost of producing a motorcycle after 50 have been produced is

$$\frac{C(50)}{50} = \frac{15{,}750}{50} = 315.$$

The approximate cost of making one additional motorcycle is the marginal cost

$$C'(50) = 250,$$

which is somewhat less than the average cost of the first 50 motorcycles. The ratio

$$E(50) = \frac{C'(50)}{C(50)/50} = \frac{250}{315} \approx .8$$

is the elasticity of cost at $x = 50$ and indicates that the cost of the next motorcycle is only about $\frac{8}{10}$ of the average cost of the first 50 motorcycles. In general, an elasticity of cost that is less than 1 indicates that the next item will cost less than the average of the cost of the items already produced, and therefore that the average cost will become smaller if another item is produced. An elasticity of cost greater than 1 indicates that production of another item will raise the average cost per item. A more comprehensive discussion of the role of elasticity in economics may be found in Chapter 6 of Kim [40]. ∎

EXERCISES

Suggested minimum assignment: Exercises 1(f), 3, 9, 10, and 11.

Find dy in each of the following.

(a) $y = x^2$. (b) $y = \dfrac{x+3}{x}$. (c) $y = x$.

(d) $y = (1 + x^2)^3$. (e) $y = 7$. (f) $y = x^3 - 4x^2$.

2. Draw several figures similar to Figure 10.14 and label Δx, Δy, and dy on each. Be sure to include cases in which Δy is negative and cases in which Δx is negative.
3. If $y = f(x) = x^3$, calculate Δy and dy if $x = 2$ and $\Delta x = dx = .1$.
4. If $y = f(x) = 3 - 4x + 2x^2$, calculate Δy and dy if $x = -2$ and $\Delta x = dx = -.1$.
5. If $V = \frac{4}{3}\pi r^3$, calculate ΔV and dV if $r = 1$ and $\Delta r = .2$.
6. The volume of a cube with side x is given by $V = x^3$. If the side, originally thought to equal 10 inches, is found to be 10.3 inches, use differentials to find approximately how much is added to the volume by the discovery of the change in x. Also find the exact change ΔV.
7. If $y = 4x^5 - 8x^3 + 7x^2 - 8x + 3$, find $\dfrac{dy}{dx}$, $\dfrac{d^2y}{dx^2}$, and $\dfrac{d^3y}{dx^3}$.
8. A cost function is given by $C(x) = \dfrac{5x^3}{x^2 + 2}$. Find $C'(8)$ and give an interpretation of your answer.
9. A manufacturer of storm doors produces x storm doors on a single work day, where x is an integer from 0 to 50 inclusive. Suppose that the cost function for this job is given by

$$C(x) = 200 + 40x - \frac{x^2}{4}.$$

Sec. 10.6] Differentials

Find $C'(10)$. What is the exact cost of the 11th storm door? Suppose also that the revenue function is given by

$$R(x) = 50x + \frac{x^2}{20}.$$

Find $P'(10)$, the approximate profit on the sale of the 11th storm door.

10. The revenue from the sale of x items is given by $R(x) = 5x + \dfrac{10}{x+1}$.
    Use differentials to find the approximate revenue from the sale of the seventh item. Also find the exact revenue from the sale of the seventh item.

11. If $y = (3x + 4)^9$, find $\dfrac{dy}{dx}$ and $\dfrac{d^2y}{dx^2}$.

12. Show that the differential of the sum of two differentiable functions is the sum of their differentials.

13. Approximate $\sqrt{82}$ without tables by calculating $y + dy$ if $y = \sqrt{x}$, $x = 81$, and $\Delta x = 1$.

14. Suppose that $y = g(u)$ and $u = f(x)$, where g and f are differentiable. Show that $dy = g'(u)\, du$ is valid even though u is dependent upon x. (*Hint:* Use the chain rule.)

15. If $y = k$, where k is a real number, does dy always equal zero? Why?

10.7 Implicit Differentiation (Optional)

In the last few sections we have begun with an equation of the form

$$y = f(x),$$

in which y has been expressed as a function of x, and we have learned rules for calculating dy/dx and higher derivatives of y with respect to x. When $y = f(x)$ is given, we say that y is expressed **directly** or **explicitly** as a function of x. Many equations, such as

$$x^2 + 4y^2 = 64,$$
$$y^3 + x^2y = 5,$$

and

$$y + \sqrt{x^2 + y^2} = x + 6,$$

are not solved for y in terms of x, but certainly a relationship between x and y is implied. Even though we may not attempt to solve for y, we shall assume that equations of this type do define one or more differentiable functions f such that $y = f(x)$; we say that y is defined **implicitly** as a function of x. If

$$x^2 + 4y^2 = 64,$$

it is easy to solve for y in terms of x in order to see that there are two possible ways in which the equation defines y as a differentiable function of x; we obtain

and
$$y = +\tfrac{1}{2}\sqrt{64 - x^2}$$
$$y = -\tfrac{1}{2}\sqrt{64 - x^2}.$$

Geometrically, the two equations above represent the upper and lower halves of an oval-shaped curve called an **ellipse** (see Figure 10.15). We are interested in finding dy/dx when y is defined implicitly as a function of x, but we cannot depend upon being able to solve a given equation for y, as this may be difficult or impossible. In an equation in which there are a mixture of x's and y's, we may treat x as the independent variable and y as the dependent variable and calculate dy/dx by the following technique, known as **implicit differentiation:**

> Differentiate each term in the equation with respect to x, treating y as a function of x. Then solve for $\dfrac{dy}{dx}$.

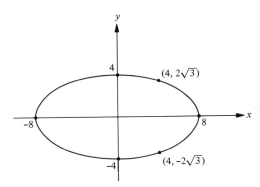

Figure 10.15. $x^2 + 4y^2 = 64$.

The reader will find that the hardest thing to understand in Examples 1 through 3, which follow, is the technique to be used when the dependent variable y occurs in a term to be differentiated. For example, suppose that the term y^2 is encountered; let $z = y^2$, where y in turn is a function of x, and apply the chain rule to obtain

$$\frac{d}{dx}(y^2) = \frac{dz}{dx} = \frac{dz}{dy}\frac{dy}{dx} = 2y\frac{dy}{dx}.$$

As another example, suppose that the term $2y/x$ is encountered. To differentiate this term with respect to x, the quotient rule and chain rule are

applied as follows:

$$D_x\left(\frac{2y}{x}\right) = \frac{x\,D_x(2y) - 2y\,D_x x}{x^2}$$

$$= \frac{x[D_y(2y)]\,(dy/dx) - 2yD_x x}{x^2}$$

$$= \frac{x(2)\,(dy/dx) - 2y(1)}{x^2}$$

$$= \frac{2x\,(dy/dx) - 2y}{x^2}.$$

Example 1

Find the slope of the tangent line to the ellipse $x^2 + 4y^2 = 64$ at the point $(4, 2\sqrt{3})$.

First find dy/dx by implicit differentiation:

$$2x + 8y\frac{dy}{dx} = 0,$$

$$\frac{dy}{dx} = -\frac{x}{4y}.$$

Then substitute $(4, 2\sqrt{3})$ to find that the desired slope is

$$\left.\frac{dy}{dx}\right|_{(4,2\sqrt{3})} = -\frac{4}{4(2\sqrt{3})} = -\frac{1}{2\sqrt{3}}\frac{\sqrt{3}}{\sqrt{3}} = -\frac{\sqrt{3}}{6}.\ \blacksquare$$

Notice that in Example 1 the expression for dy/dx involves *both* x and y—this happens frequently with implicit differentiation. If only the value of x is given in Example 1, the slope cannot be determined because the slope depends on whether the point is on the upper or lower part of the ellipse. However,

$$\frac{dy}{dx} = -\frac{x}{4y}$$

can also be used to find the correct slope at $(4, -2\sqrt{3})$, and implicit differentiation in general will give the correct derivative of *each* differentiable function defined by a given equation. The next example illustrates how higher derivatives can be determined after dy/dx has been found by implicit differentiation (some uses for higher derivatives were given in Section 10.5 and others will arise in Chapter 11).

Example 2

Find d^2y/dx^2 and d^3y/dx^3 if $x^2 + 4y^2 = 64$.

From Example 1,

$$\frac{dy}{dx} = -\frac{x}{4y}.$$

Differentiate again, substitute for dy/dx, and simplify:

$$\frac{d^2y}{dx^2} = -\frac{(4y)(1) - x(4)(dy/dx)}{(4y)^2}$$

$$= -\frac{4y - 4x(-x/4y)}{16y^2}$$

$$= -\frac{16y^2 + 4x^2}{64y^3}$$

$$= -\frac{4(4y^2 + x^2)}{64y^3}$$

$$= -\frac{4(64)}{64y^3} \quad \text{(from the given equation)}$$

$$= -\frac{4}{y^3}.$$

In a similar manner d^3y/dx^3 is computed:

$$\frac{d^3y}{dx^3} = \frac{d}{dx}\left(\frac{d^2y}{dx^2}\right) = \frac{d}{dx}(-4y^{-3})$$

$$= \left[\frac{d}{dy}(-4y^{-3})\right]\frac{dy}{dx}$$

$$= 12y^{-4}\frac{dy}{dx}$$

$$= 12y^{-4}\left(-\frac{x}{4y}\right)$$

$$= -\frac{3x}{y^5}. \blacksquare$$

Example 3

If $y^3 + x^2y = 5$, calculate the rate of change of y with respect to x at $(2, 1)$.

Notice that $(2, 1)$ does satisfy the given equation. First calculate dy/dx by implicit differentiation and then evaluate at $(2, 1)$; we choose to use the notation y' instead of dy/dx.

$$3y^2y' + x^2y' + y(2x) = 0,$$
$$y'(3y^2 + x^2) = -2xy,$$
$$y' = -\frac{2xy}{3y^2 + x^2}.$$

Therefore,

$$y'|_{(2,1)} = -\frac{2(2)(1)}{3 + 4} = -\frac{4}{7}. \blacksquare$$

When an equation contains a mixture of x's and y's, there is no reason why we cannot choose x instead of y as the dependent variable. In Example 4

each term is differentiated with respect to y, and x is treated as a function of y.

Example 4
Find $dx/dy|_{(2,1)}$ if $y^3 + x^2y = 5$.

$$3y^2 + x^2(1) + y\left(2x\frac{dx}{dy}\right) = 0,$$

$$2xy\frac{dx}{dy} = -(3y^2 + x^2),$$

$$\frac{dx}{dy} = -\frac{3y^2 + x^2}{2xy},$$

$$\left.\frac{dx}{dy}\right|_{(2,1)} = -\frac{3+4}{4} = -\frac{7}{4}. \blacksquare$$

The reader probably observed that the answers to Examples 3 and 4 are reciprocals of each other. From geometrical considerations it should not be surprising that

$$\frac{dx}{dy} = \frac{1}{dy/dx}, \quad \text{provided that } \frac{dy}{dx} \neq 0,$$

since it would seem that the rate of change of x with respect to y should be the reciprocal of the rate of change of y with respect to x.

APPLICATIONS

Example 5
The marketing department of a large corporation has determined an equation that expresses the annual demand z for its product as a function of its price x per item and the price y per item of a rival synthetic product. Suppose that the equation is

$$z = 3000 - 30x^2 + 20y^2,$$

provided that $1 \leq x \leq 10$ and $1 \leq y \leq 8$. Both x and y are measured in dollars per item, and z is measured in units of production; x, y, and z are each functions of time t. Notice from the equation above that the demand for the product will increase if its price is decreased and will also increase if the price of the competing synthetic product is increased. Suppose that the rates of change of the prices x and y are the maximum allowed by federal guidelines, say 4 and 3 per cent, respectively, per year. Find the expected rate of change (per year) of the demand when $x = \$6$ and $y = \$4$. In other words, find dz/dt when $x = 6$, $y = 4$, $dx/dt = 6(.04) = .24$ and $dy/dt = 4(.03) = .12$.

If the given equation is differentiated implicitly with respect to t, we get

$$\frac{dz}{dt} = 0 - 60x\frac{dx}{dt} + 40y\frac{dy}{dt},$$

which can be evaluated as

$$\frac{dz}{dt} = -60(6)(.24) + 40(4)(.12)$$
$$= -67.2 \text{ units per year.}$$

The significance of the negative answer is that the demand is decreasing under the given pricing practices. ∎

EXERCISES

Suggested minimum assignment: Exercises 3, 7, 11, 17, and 23.

In Exercises 1–10 find $\frac{dy}{dx}$ by implicit differentiation.

1. $x^2 + y^2 = 9$.
2. $x^2 - y^2 = 7$.
3. $xy = 5$.
4. $xy^2 + yx^2 = 2$.
5. $y^2 - 8x = xy$.
6. $y^2 - 8x = \frac{x}{y}$.
7. $y^5 + xy^3 + x^4 - 3 = 0$.
8. $2x^2 + 3y^2 = 12$.
9. $\sqrt{x} + \sqrt{y} = 4$.
10. $(x^2 + y^2)^5 - 32 = 0$.

In Exercises 11–14 find $\frac{dy}{dx}$ and $\frac{d^2y}{dx^2}$. Use implicit differentiation.

11. $x^2 + y^2 = 4$.
12. $x^2 + xy + y^2 - 3 = 0$.
13. $xy + y^2 = 12$.
14. $x^2 + 4y^2 = 5$.

15. If $y^2 - x^2 = 16$, find $D_x y$, $D_x^2 y$, and $D_x^3 y$.

In Exercises 16–18 find the slope of the tangent line at the given point.

16. $x^2 y^2 = 12 - x$; $(3, 1)$.
17. $y + \sqrt{x^2 + y^2} = x + 6$; $(3, 4)$.
18. $x^{2/3} + y^{2/3} = 2$; $(1, 1)$.

19. Find $D_t s$ if $s^2 + t^2 + st = 10$.
20. Explain why $x^2 + y^2 = 16$ defines two differentiable functions f_1 and f_2 such that $y = f_1(x)$ and $y = f_2(x)$.
21. The location of a particle moving along a line is determined by the equation $st^2 - st^3 + 8s - 1 = 0$. Find s when $t = 1$. Then use implicit differentiation to find the velocity of the particle when $t = 1$.
22. If $x^3 + y^3 - 3xy = 13$, verify that the rate of change of y with respect to x at $(-1, 2)$ is the reciprocal of the rate of change of x with respect to y.

23. If $2\pi r^2 + 2\pi rh = 15$, find $\dfrac{dh}{dr}$ by implicit differentiation. Also find $\dfrac{dr}{dh}$ by implicit differentiation. Verify that $\dfrac{dr}{dh}$ is the reciprocal of $\dfrac{dh}{dr}$.

NEW VOCABULARY

average rate of change 10.1
secant line 10.1
instantaneous rate of change 10.1
tangent line 10.1
rate of change 10.2
derivative 10.2
derived function 10.2
differentiable 10.2
marginal revenue 10.2
marginal cost 10.2
marginal profit 10.2
equation of motion 10.2
average velocity 10.2
instantaneous velocity 10.2
velocity 10.2
differentiation 10.3
power rule for differentiation 10.3

product rule 10.4
quotient rule 10.4
chain rule 10.4
composite function 10.4
first derivative 10.5
second derivative 10.5
higher derivatives 10.5
acceleration 10.5
differential dx 10.6
differential dy 10.6
explicit functional relationship 10.7
implicit functional relationship 10.7
ellipse 10.7
implicit differentiation 10.7

11 Maximum and Minimum Problems

In many applied problems there are functions that must be maximized or minimized. Such problems are called optimization problems and are approached in a variety of ways, depending upon the nature of the problem. Some optimization problems, such as linear programming problems (which were discussed in Chapter 6), are usually solved by an algebraic algorithm. Other optimization problems, such as those illustrated in this chapter, are particularly vulnerable to the techniques of differentiation. Still others, which will be studied in Chapter 15, will require further knowledge of the calculus.

Historically, optimization problems have provided considerable incentive for the discovery of new mathematics. There is evidence that questions pertaining to (1) the maximum height of a projectile in free flight near the earth, and (2) the maximum distance from one planet to another or to the sun provided some motivation for the discovery of calculus. Today optimization problems are pertinent to many disciplines other than the physical sciences, and the need for solving these problems contributes significantly to the interest of these disciplines in the techniques of calculus.

Prerequisites: 1.1, 1.2, 3.1–3.9, Chapter 9, 10.1–10.5. Section 11.4 is not a prerequisite to 11.5 and 11.6. Sections 11.4–11.6 are not needed in optional section 11.7. The only prerequisites to the first part of 11.6 (symmetry and asymptotes) are Chapters 3 and 9.

Suggested Sections for Moderate Emphasis: 11.1–11.4 and the first part of 11.6.

Suggested Sections for Minimum Emphasis: 11.1–11.3 and the first part of 11.6.

11.1 Introduction

Problems in which one tries to maximize profits or to minimize expenses are illustrations of optimization problems. According to the dictionary, an optimum is a most favorable point, degree, or amount. In this chapter we shall discuss various uses of the derivative in solving some optimization problems.

Example 1

As an illustration we introduce a problem that will be solved in Section 11.4. A store manager sells an item, say a certain style of small street lamp, for x dollars, where $5 \leq x \leq 15$. Each lamp costs him $5, and hence his profit per lamp is $(x - 5)$ dollars. The number of lamps sold depends upon the selling price x. By keeping records, the manager has observed that $(150 - 10x)$ is a good estimate of the number of lamps per month which are sold, and hence the monthly profit is given by

$$P(x) = (\text{profit per lamp})(\text{number sold per month})$$

or

$$P(x) = (x - 5)(150 - 10x).$$

If the selling price is $8, then the profit is $3 per lamp on sales of $150 - 10(8) = 70$ lamps, for a profit of $210. If the selling price is $11.50, the profit per lamp is $6.50, but the number sold is only $150 - (10)(11.5) = 35$ lamps. However, the profit is

$$P(11.50) = (6.50)(35) = \$227.50,$$

which is better than when $x = 8$. The problem is to find, if possible, the value of x in the closed interval $[5, 15]$ that will maximize the profit. ∎

In this section we shall present some background material that is needed in order to solve maximum and minimum problems similar to the one just discussed. In Section 3.5 we studied closed intervals such as $[4, 7]$, open intervals such as $(4, 7)$, and half-open intervals such as $(4, 7]$ and $[4, 7)$; these intervals are called **finite intervals** because their length $(7 - 4)$ is a finite number. On the other hand, the set of numbers

$$\{x : x \geq 5\}$$

is an example of an **infinite interval** and will be denoted by

$$[5, \infty).$$

The plus infinity symbol (∞) does not represent a specific real number, but the $[5, \infty)$ notation is used to indicate that the set includes *all* real numbers that are 5 or greater. See Figure 11.1. Examples of infinite open intervals are

$$(-\infty, 2),$$

which designates the set

$$\{x : x < 2\},$$

and

$$(-3, \infty),$$

Figure 11.1. *Infinite interval* $[5, \infty)$.

which designates the set
$$\{x : x > -3\}.$$
The set of all real numbers can be denoted by $(-\infty, \infty)$.

In Example 2 we introduce the idea of calculating the sign of an expression on various intervals. This process will be repeated frequently in this chapter.

Example 2
Suppose that $y = f(x) = x^2 - 2x - 8$. For what values of x is $f(x)$ positive, and for what values of x is $f(x)$ negative?

First write $f(x)$ in the factored form
$$f(x) = (x + 2)(x - 4).$$

Since $f(x)$ is the product of the factors $(x + 2)$ and $(x - 4)$, we can determine the sign of $f(x)$ for a given value of x by finding the sign (plus or minus) of each factor. The factor $(x + 2)$ is negative $(-)$ if $x < -2$ and positive $(+)$ if $x > -2$, and we say the factor $(x + 2)$ **changes sign** at $x = -2$. The factor $(x - 4)$ changes sign at $x = 4$ because it is negative if $x < 4$, but it has the opposite sign (positive) if $x > 4$. We use the points -2 and 4 to divide the real line into the intervals $(-\infty, -2), (-2, 4)$, and $(4, \infty)$. (We shall always list the intervals from left to right along the real line.) Instead of considering a large number of particular values of x, we need to deal with only the three cases represented by the three intervals above—a factor can change sign at an endpoint (-2 or 4) but not in the interior of any of these intervals. Our method of solving the problem is to construct Table 11.1, from which it can be seen how $f(x)$ changes sign as the individual factors change sign. ∎

Table 11.1

Interval	Calculation of sign of $f(x) = (x + 2)(x - 4)$	Result	Conclusion
$(-\infty, -2)$	$(-)(-)$	$+$	$f(x)$ positive
$(-2, 4)$	$(+)(-)$	$-$	$f(x)$ negative
$(4, \infty)$	$(+)(+)$	$+$	$f(x)$ positive

In Example 2 the polynomial function f is continuous for all real numbers, and the points at which a factor changes sign are the roots of the equation $f(x) = 0$. In general, a continuous function cannot change sign (either from $+$ to $-$ or from $-$ to $+$) without taking on the value zero; however, a continuous function does *not* always change sign at a point where its value is zero. Figures 11.2 and 11.3 are presented to clarify these facts (see also Exercise 12 of Section 9.4). For a function f that is continuous for all real numbers, we can conclude that the only possible points at which a sign change *may* occur are obtained by solving the equation $f(x) = 0$.

Next we learn how to find the sign of $f(x)$ on various intervals, where f is a rational function (remember that a polynomial function is a special type

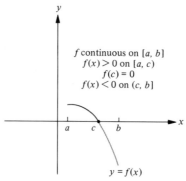

Figure 11.2. $f(x)$ changes sign from $+$ to $-$ at $x = c$.

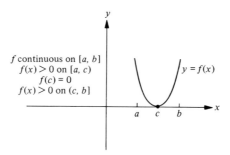

Figure 11.3. $f(x)$ does not change sign at $x = c$.

of rational function). *Unless otherwise stated, we shall consider only the rational functions in this chapter.* Consider the function f defined by $f(x) = (x-1)/(x-2)$. The only solution of $f(x) = 0$ is $x = 1$, and it can be shown that $f(x)$ changes sign at $x = 1$ [note that $f(.9) > 0$ and that $f(1.1) < 0$]. However, $f(x)$ also changes sign at $x = 2$ [note that $f(1.9) < 0$ and that $f(2.1) > 0$], despite the fact that 2 is not in the domain of f. In general, a rational function f *may* change sign at any point for which the numerator is zero [which is a solution of $f(x) = 0$], and also it *may* change sign at any point for which the denominator is zero (which is a point of discontinuity). Points at which either the numerator or denominator is zero are called **dividing points** of f because in making a table like Table 11.1 the intervals are determined by "cutting" the real line at the dividing points.

Given a rational function f in this chapter, we often shall need to make tables like Table 11.1 in order to determine the sign of either $f'(x)$ or $f''(x)$ on various intervals. The functions f' and f'' will also be rational functions [$f'(x)$ and $f''(x)$ can be obtained by the quotient rule], and the dividing points for f' and f'' are found again by finding points at which either the numerator or denominator is zero. When we are given a rational function f, the points of discontinuity will be the same for f, f', and f''; therefore, we have a function that is continuous and whose derivatives f' and f'' exist and are continuous except at points for which the denominator of $f(x)$ is zero.

Many functions met in applications are of the type discussed above, but often the domain is restricted for practical reasons to some finite or infinite interval (such as $[0, 5]$ or $(0, \infty)$). This must be remembered in making a table like Table 11.1. For example, suppose that $f(x) = (x - 1)/(x - 2)$ for $0 \leq x \leq 5$, provided that $x \neq 2$, and we want to compute the sign of $f(x)$ on various intervals. The dividing points of f are 1 and 2. The intervals that are considered are the open intervals $(0, 1)$, $(1, 2)$, and $(2, 5)$—we always elect not to include the endpoints or dividing points in the intervals.

Example 3
Suppose that
$$f(x) = \frac{(x + 1)^2(3 - x)}{(2x - 9)(x^2 + 4)}.$$

Find the open intervals on which $f(x)$ is positive and the open intervals on which $f(x)$ is negative.

Any number that makes any factor equal to zero (in either the numerator or denominator) is a dividing point of f. The dividing points are -1, 3, and $\frac{9}{2}$; there is no dividing point for $(x^2 + 4)$, since this quantity is never zero. The problem is solved in Table 11.2. One way to get the signs in Table 11.2 is to work with the individual factors. For example, $2x - 9 < 0$ when $x < \frac{9}{2}$, and hence $(-)$ appears for this factor in all intervals for which $x < \frac{9}{2}$; if, however, $x > \frac{9}{2}$, then $2x - 9 > 0$, and therefore $(+)$ appears for $(2x - 9)$ in the interval $(\frac{9}{2}, \infty)$. Another method of getting the signs is to substitute for x any chosen number in a particular interval. If $x = 0$, $2x - 9 = -9 < 0$, and therefore $(-)$ appears for $(2x - 9)$ in the interval $(-1, 3)$ that includes $x = 0$. ∎

Table 11.2

Interval	Calculation of sign of $f(x)$	Result	Conclusion
$(-\infty, -1)$	$\dfrac{(-)^2(+)}{(-)(+)}$	$-$	$f(x)$ negative
$(-1, 3)$	$\dfrac{(+)^2(+)}{(-)(+)}$	$-$	$f(x)$ negative
$(3, \frac{9}{2})$	$\dfrac{(+)^2(-)}{(-)(+)}$	$+$	$f(x)$ positive
$(\frac{9}{2}, \infty)$	$\dfrac{(+)^2(-)}{(+)(+)}$	$-$	$f(x)$ negative

APPLICATIONS

Example 4
A company is taken over by a new president, who immediately embarks on a cost-reduction program. He finds that several items are produced at one plant in a complicated manufacturing operation. The plant has been making

very little profit, so he orders that a study be made to determine which items are profitable and which ones are not. Item A is analyzed by a cost accountant and an engineer. Item A has been sold during the past 2 years for prices that have fluctuated frequently between $5.95 and $8.95. The cost accountant determines that the daily sales S depend primarily upon the price p, and that S can be estimated quite well by the formula

$$S = 46 + 9p - p^2, \quad 5.95 \le p \le 8.95.$$

The cost accountant also determines that the company has been losing money on item A whenever it is sold for less than $7.40. The engineer studies the efficiency of the production operation, and he decides that more than 60 of item A should be manufactured daily—otherwise item A should be discontinued so that the labor and machinery can be used for another item. Should the new president order that item A be discontinued?

In order that sales keep pace with production, it is necessary that

$$46 + 9p - p^2 > 60,$$
$$-14 + 9p - p^2 > 0,$$
$$p^2 - 9p + 14 < 0,$$
$$(p - 2)(p - 7) < 0.$$

In other words, the sign of $(p - 2)(p - 7)$ must be negative. By making a table similar to Tables 11.1 and 11.2 (see Exercise 9), it can be shown that the restriction $(p - 2)(p - 7) < 0$ requires the price p to be in the open interval $(2, 7)$. In order to produce and sell more than 60 of item A daily (to satisfy the engineer), the price p must be less than $7; in order to make a profit (to satisfy the cost accountant), p must be at least $7.40. Hence the new president discontinues production of item A. ∎

EXERCISES

Suggested minimum assignment: Exercises 1, 2(a) (c), 3(b), 4, 5(a) (c), and 7.

1. In Example 1 find the monthly profit if the selling price of each lamp is $9.30.
2. Change each of the following intervals to set notation. Show the set on a drawing of the real line.
   (a) $(-\infty, 3)$.   (b) $(0, \infty)$.
   (c) $[1, \infty)$.   (d) $(-\infty, -8]$.
3. Change from set notation to interval notation.
   (a) $\{x : x \ge 6\}$.   (b) $\{x : x > -1\}$.
   (c) $\{x : x < 20\}$.   (d) $\{x : x \le 15\}$.
4. Sketch the graph of $y = f(x) = x^2 - 2x - 8$ (see Example 2). Explain how your graph agrees with the conclusion column of Table 11.1.

In Exercises 5 and 6 find the open intervals on which $f(x)$ is positive and the open intervals on which $f(x)$ is negative. Make a table similar to Tables 11.1 and 11.2.

5. (a) $f(x) = (x-3)(x-5)$, provided that $x \geq 0$.
   (b) $f(x) = (x-1)(x-2)(x-3)$, provided that $0 \leq x \leq 5$.
   (c) $f(x) = \dfrac{(x-4)(1-x)}{x+8}$.

6. (a) $f(x) = x^2 - 7x - 8$. (b) $f(x) = \dfrac{(x-5)(6-x)}{x-7}$.
   (c) $f(x) = \dfrac{(x-1)^2(x-2)}{2x+1}$.

7. Suppose that $f(x) = \dfrac{2x}{x-1}$.

   (a) Find the dividing points of f.
   (b) Find $f'(x)$ and then find the dividing points of f'.
   (c) Find $f''(x)$ and then find the dividing points of f''.
   (d) The functions f, f', and f'' all exist and are continuous, except for one value of x. What is that value?
   (e) Does $f'(x)$ change sign at $x = 1$?

8. If $f(x) = \dfrac{(x-1)(x-2)^2}{(x-3)^3}$, find the dividing points of f. Then find the points at which $f(x)$ changes sign.

9. Find the values of p for which $f(p) = (p-2)(p-7)$ is negative.

11.2 Increasing and Decreasing Functions

Let us assume that consumer demand for a certain item depends only upon the price of the item. If a store reduces the price of an item, the quantity of the item demanded by customers over a certain time period is ordinarily expected to increase. On the other hand, if the price is raised, the demand is ordinarily expected to decrease. In general, if two variables are related by $y = f(x)$ and if y decreases when x increases, we say that f is a **decreasing function**. If, however, y increases when x increases, we say that f is an **increasing function**. A study of increasing and decreasing functions in this section will lead directly to the solution of maximum and minimum problems.

Example 1

If the relationship between a store's selling price of an item and the number of the item demanded by consumers over a certain period of time is plotted, the result is known as a **demand curve**. It is customary in economics to plot price on the y-axis and demand on the x-axis. A typical demand curve $y = f(x)$ is shown in Figure 11.4, and f is a *decreasing* function throughout its domain. The point $(1000, 35)$ represents the fact that 1000 of the item can be sold at \$35 each. The point $(0, 50)$ represents the fact that no one will

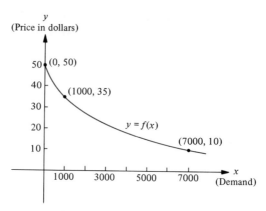

Figure 11.4. *Demand curve.*

pay $50 for the item. Since price and demand are nonnegative quantities, a demand curve is restricted to the first quadrant.

The quantity of an item that various producers will supply to a store over a certain period of time also depends upon the selling price of the item. A higher selling price is expected to spur greater production of the item because the producers can make more profit. In this situation it is also customary to plot price on the *y*-axis, with supply on the *x*-axis. The resulting graph is called a **supply curve** and is restricted to the first quadrant; ordinarily, the graph reveals an *increasing* function, such as the one shown in Figure 11.5. (Figures 11.4 and 11.5 are for *different* items.) For the item studied in Figure 11.5, a quantity of 400 will be supplied if the selling price is $300 per item, but none will be supplied by any manufacturer if the selling price is $100 or less. ∎

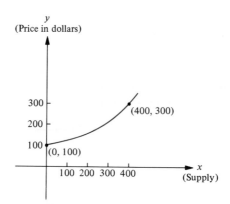

Figure 11.5. *Supply curve.*

We now define precisely what is meant by an increasing, or decreasing, function on an interval.

▶ **Definition 1**

Suppose that f is defined throughout an interval. Then:

(i) f is said to be **increasing on the interval** if $f(x_1) < f(x_2)$ for each pair of numbers x_1 and x_2 in the interval such that $x_1 < x_2$.

(ii) f is said to be **decreasing on the interval** if $f(x_1) > f(x_2)$ for each pair of numbers x_1 and x_2 in the interval such that $x_1 < x_2$.

Although the interval referred to in Definition 1 may be open, closed, or half-open, and may or may not be an infinite interval, we shall be especially interested in the case when it is an open interval. Geometrically, Definition 1 says that a function is increasing on an interval if its graph is rising as x increases, whereas a function is decreasing on an interval if its graph is falling as x increases. In Figure 11.6 a function f is assumed to be defined on $[-2, \infty)$; f is increasing on the intervals $(-2, -1)$ and $(2, \infty)$, whereas f is decreasing on the interval $(-1, 2)$.

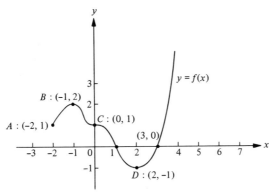

Figure 11.6

Theorem 1 is of practical importance in determining on what open intervals a function is increasing and on what open intervals it is decreasing.

▶ **Theorem 1**

(i) If $f'(x) > 0$ at each point of (a, b), then f is increasing on (a, b).

(ii) If $f'(x) < 0$ at each point of (a, b), then f is decreasing on (a, b).

Theorem 1 seems reasonable from geometric considerations. If $f'(x) > 0$ at each point of (a, b), then the curve has a positive slope at each point of (a, b). See Figure 11.7. Hence it seems that the values of f should increase as x increases in (a, b). The argument above, especially when accompanied by a picture, may seem convincing; however, it does not constitute a proof. A proof will be given in optional Section 11.7. ◀

In Examples 2 and 3 we determine the sign of $f'(x)$ on various open intervals, using methods learned in Section 11.1; then we apply Theorem 1 to determine intervals on which f is increasing and intervals on which f is decreasing. (It is valid to apply Theorem 1 to infinite intervals as well as finite intervals.)

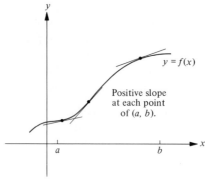

Figure 11.7

Example 2
Determine on which open intervals f is increasing and on which open intervals f is decreasing if $y = f(x) = \frac{2}{3}x^3 - \frac{1}{2}x^2 - 6x + 3$.

First find $f'(x)$ in factored form as follows:

$$f'(x) = 2x^2 - x - 6 = (2x + 3)(x - 2).$$

Hence $-\frac{3}{2}$ and 2 are the dividing points of f' at which a change in sign of $f'(x)$ may occur. The problem is solved in Table 11.3; notice that to use Theorem 1 we calculate the *sign of $f'(x)$* rather than the sign of $f(x)$. ∎

Table 11.3

Interval	Calculation of sign of $f'(x)$	Result	Conclusion
$(-\infty, -\frac{3}{2})$	$(-)(-)$	$+$	f increasing
$(-\frac{3}{2}, 2)$	$(+)(-)$	$-$	f decreasing
$(2, \infty)$	$(+)(+)$	$+$	f increasing

Example 3
Repeat Example 2 for $y = f(x) = x^3$.

$$f'(x) = 3x^2 = (3x)(x).$$

The only dividing point of f' at which a change in sign of $f'(x)$ may occur is zero. The problem is solved in Table 11.4, and a graph of $y = x^3$ is shown in Figure 11.8. ∎

Table 11.4

Interval	Calculation of sign of $f'(x)$	Result	Conclusion
$(-\infty, 0)$	$(-)(-)$	$+$	f increasing
$(0, \infty)$	$(+)(+)$	$+$	f increasing

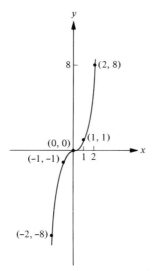

Figure 11.8. $y = f(x) = x^3$.

In Example 3 observe that the slope of the curve at the origin is not positive because

$$f'(0) = 3(0)^2 = 0.$$

Notice, however, that Definition 1(i) is satisfied for any open interval in the domain of f, including those that contain $x = 0$. These observations illustrate that for each part of Theorem 1 the converse is not always true.

APPLICATIONS

Example 4

In Example 8 of Section 9.4 we showed that the function f, defined by

$$y = f(x) = \frac{x}{20x^2 + 50x + 80}, \qquad x \geq 0,$$

possessed several desirable properties of a function that would model the concentration y of a drug in the blood x hours after an injection of a known amount of the drug. In this example we shall first show, by use of the methods described in Section 11.1, that $f(x)$ is positive when x is positive (the function would be a poor model if it ever gave a negative concentration). Because the denominator is always positive (explained in Example 8 of Section 9.4), it is not difficult to construct Table 11.5.

Table 11.5

Interval	Calculation of sign of $f(x)$	Result	Conclusion
$(0, \infty)$	$\dfrac{+}{+}$	$+$	$f(x)$ positive

Sec. 11.2] Increasing and Decreasing Functions

It is quite important that the function f be an increasing function from the time of the injection ($x = 0$) until a later time when the effect of the drug begins to wear off, after which f should be a decreasing function. Let us apply the methods of this section to determine on which open intervals f is increasing and on which open intervals f is decreasing. By the quotient rule,

$$f'(x) = \frac{(20x^2 + 50x + 80)(1) - x(40x + 50)}{(20x^2 + 50x + 80)^2}$$

$$= \frac{-20x^2 + 80}{(20x^2 + 50x + 80)^2}$$

$$= \frac{(-20)(x + 2)(x - 2)}{(20x^2 + 50x + 80)^2}.$$

As we stated before, $20x^2 + 50x + 80$ is always positive, so the only values of x at which a change in sign of $f'(x)$ may occur are $x = -2$ and $x = 2$. Since the domain of f is $\{x : x \geq 0\}$, we choose 2 as the only dividing point of f' (rather than -2 and 2). See Table 11.6. According to the table the given model indicates that the concentration of the drug in the blood increases for the first 2 hours and decreases after that. This conclusion agrees, of course, with our previous calculations of the concentration after 0, 1, 2, and 4 hours. ∎

Table 11.6

Interval	Calculation of sign of $f'(x)$	Result	Conclusion
$(0, 2)$	$\dfrac{(-)(+)(-)}{(+)^2}$	$+$	f increasing
$(2, \infty)$	$\dfrac{(-)(+)(+)}{(+)^2}$	$-$	f decreasing

EXERCISES

Suggested minimum assignment: Exercises 1, 3, 13, 16, and 20.

In each of Exercises 1–17 determine on which open intervals f is increasing and on which open intervals f is decreasing; in each case make a table similar to Tables 11.3 and 11.4.

1. $y = f(x) = x^2 - 8x, x > 0$.
2. $y = f(x) = \frac{1}{2}x^2 + 6x - 10$.
3. $y = f(x) = \frac{1}{3}x^3 - \frac{9}{2}x^2 + 20x - 4$.
4. $y = f(x) = 2x^3 + 1$.
5. $y = f(x) = 1 - 2x^3$.
6. $y = f(x) = x^4 - 2x^2 - 3$.
7. $y = f(x) = \dfrac{x^4}{2} - \dfrac{2x^3}{3} - 6x^2$.

8. $y = f(x) = x^4, x \geq -2$.
9. $y = f(x) = x^5 + 2$.
10. $y = f(x) = 7x + 3, 0 \leq x \leq 10$.
11. $y = f(x) = \frac{1}{3}x^3 + 2x^2 + x - 6$. [Hint: $f'(x) = (x - r_1)(x - r_2)$, where r_1 and r_2 can be found by quadratic formula.]
12. $y = f(x) = x^3 - 9x + 5$.
13. $y = f(x) = \frac{1}{4}x^4 - \frac{8}{3}x^3 + 8x^2 - 10$.
14. $y = f(x) = \dfrac{3}{x}$.
15. $y = f(x) = \dfrac{1}{x^2}$.
16. $y = f(x) = \dfrac{x}{x^2 + 1}$.
17. $y = f(x) = (x - 1)^3(x + 1)^2$.

18. Find an example of a function f such that f is decreasing on any open interval and such that $f'(0) = 0$.
19. The equation of a demand curve is $y = 100/(x + 1)$, $0 \leq x \leq 20$. Sketch this curve and show that y' is negative on $(0, 20)$.
20. The equation of a supply curve is

$$y = 50 + \frac{x^2}{100}, \qquad 0 \leq x \leq 100.$$

Sketch this curve, and show that y' is positive on $(0, 100)$.

11.3 First Derivative Test for Maximum and Minimum Points

We now have the background to study maximum and minimum problems. Ordinarily, the graph of a function defined by $y = f(x)$ rises or falls as x increases. *We seek to find those points whose y-coordinates are larger (or smaller) than the y-coordinates of any other nearby points.* In this section we shall learn techniques of finding these **maximum points** and **minimum points**. In other words, we shall find all **maxima** and **minima** (these terms are plural). Otherwise stated, we shall find all **extreme points** (either a maximum point or a minimum point is called an extreme point). Then, in Section 11.4, we shall use the techniques of this section to solve some applied problems.

As a preview of what is to come, consider the function f with domain $[a, b]$ that is shown in Figure 11.9. The points $(x_1, f(x_1))$ and $(b, f(b))$ are maximum points, since their y-coordinates are greater than the y-coordinates of all other points on the curve in the immediate vicinity. The maximum point $(b, f(b))$ is called an **endpoint maximum** because it occurs at an endpoint of the domain of f. Similarly, $(x_2, f(x_2))$ is a minimum point and $(a, f(a))$ is an **endpoint minimum**.

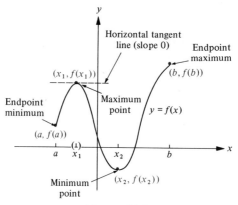

Figure 11.9

Because of the methods used to determine maximum and minimum points, these points will be studied in two groups; such points will occur at either

1. Interior points of the domain (such as x_1 and x_2 above), or
2. Endpoints of the domain (such as a and b above).

The domain of the rational functions that we consider will be a specified interval (with the omission of points at which the denominator is zero). If the interval does not include any endpoints, then obviously the function cannot have an endpoint maximum or an endpoint minimum. We shall focus our attention first at finding all maxima and minima that occur at interior points of the domain.

▶ **Definition 2**

Suppose that x_1 is an interior point of the domain of a function f. The point $(x_1, f(x_1))$ is a **relative maximum point** if there exists an open interval containing x_1 such that $f(x_1) > f(x)$ for all $x \neq x_1$ in this open interval. The point $(x_1, f(x_1))$ is a **relative minimum point** if there exists an open interval containing x_1 such that $f(x_1) < f(x)$ for all $x \neq x_1$ in this open interval.

The word *relative* is included above to emphasize that the y-coordinate of a relative maximum point does *not* necessarily equal the greatest value of the function throughout its domain, but only that it equals the greatest function value in the immediate vicinity of the point. With this understanding *we shall omit the word relative in the future;* for instance, when we say maximum point, we shall mean relative maximum point.

As stated previously, the point $(x_1, f(x_1))$ in Figure 11.9 is a maximum point. On the x-axis in Figure 11.9 we have shown one possible open interval (containing x_1) of the type referred to in Definition 2. Any such open interval in the domain of f will suffice as long as $f(x_1)$ is the largest function value taken on throughout the interval. At the maximum point $(x_1, f(x_1))$ in Figure 11.9 the slope of the curve is zero because the tangent line at this

point is horizontal—this fact is more than a mere coincidence, according to Theorem 2.

▶ **Theorem 2**
Suppose that x_1 is an interior point of the domain of a rational function f, and suppose that $(x_1, f(x_1))$ is either a maximum or minimum point. Then $f'(x_1) = 0$.

Outline of Proof: Since x_1 is an interior point of the domain of the rational function f, $f'(x_1)$ exists. Suppose that $(x_1, f(x_1))$ is a maximum point. Because

$$f'(x_1) = \lim_{\Delta x \to 0} \frac{f(x_1 + \Delta x) - f(x_1)}{\Delta x},$$

it can be shown that $f'(x_1) \leq 0$ and $f'(x_1) \geq 0$ by consideration of the cases that $\Delta x > 0$ and $\Delta x < 0$, respectively. Therefore, $f'(x_1) = 0$. If $(x_1, f(x_1))$ is a minimum point, the proof is similar. ◀

From Theorem 2 we know that points at which the slope is zero are important; therefore, they are given a special name.

▶ **Definition 3**
If $f'(x_1) = 0$, where x_1 is an interior point of the domain of f, then x_1 is called a **critical number** of the function f. The corresponding point $(x_1, f(x_1))$ is called a **critical point** of the function f.

Example 1
Refer to Figure 11.6. The critical numbers are -1, 0, and 2, because the slopes $f'(-1)$, $f'(0)$, and $f'(2)$ are zero. The corresponding points B, C, and D are critical points. Notice that B is a maximum point, D is a minimum point, and C is neither a maximum nor a minimum point. (Point A is an endpoint minimum.) ∎

Theorem 2 tells us that the critical numbers are the only numbers in the interior of the domain of a rational function that are *candidates* for consideration in looking for maximum or minimum points. At a critical number there may be a maximum point, there may be a minimum point, or there may be neither a maximum nor a minimum point (as we saw in Example 1). If the function increases just to the left of a critical point and decreases just to the right, then the critical point is a maximum point; if the function decreases just to the left of a critical point and increases just to the right, then the critical point is a minimum point. If the function increases (or if it decreases) on both sides of a critical point, then the critical point is neither a maximum nor a minimum point. These facts are stated formally in Theorem 3, which is called the **first derivative test** for maximum and minimum points.

▶ **Theorem 3**

Let x_1 be a critical number of a rational function f. Let $x \neq x_1$ belong to an open interval (in the domain of f) that contains x_1 as its only critical number.

(i) $(x_1, f(x_1))$ is a maximum point if $f'(x) > 0$ for $x < x_1$ and $f'(x) < 0$ for $x > x_1$.

(ii) $(x_1, f(x_1))$ is a minimum point if $f'(x) < 0$ for $x < x_1$ and $f'(x) > 0$ for $x > x_1$.

(iii) $(x_1, f(x_1))$ is neither a maximum nor minimum point if $f'(x) > 0$ for both $x < x_1$ and $x > x_1$, or if $f'(x) < 0$ for both $x < x_1$ and $x > x_1$.

The proof of Theorem 3 is left as Exercise 18. ◀

Figures 11.10, 11.11, and 11.12 illustrate Theorem 3 and should be studied to give the reader an intuitive understanding of the theorem.

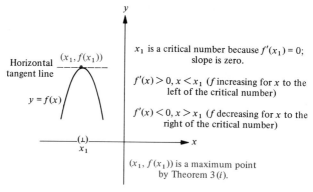

Figure 11.10. *Illustration of Theorem 3(i).*

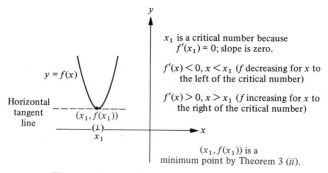

Figure 11.11. *Illustration of Theorem 3(ii).*

Example 2
Find all maximum and minimum points of f if

$$y = f(x) = \frac{x^4}{4} - \frac{4x^3}{3} + 2x^2 + 3.$$

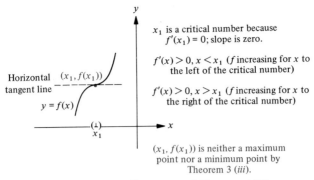

Figure 11.12. Illustration of Theorem 3(iii).

The domain of f is the set of all real numbers. We calculate $f'(x)$ as follows:

$$f'(x) = x^3 - 4x^2 + 4x = x(x^2 - 4x + 4) = x(x-2)(x-2).$$

The equation $f'(x) = x(x-2)(x-2) = 0$ yields the critical numbers $x = 0$ and $x = 2$, which are also the only dividing points of f'. The corresponding values of y are

$$f(0) = 3 \quad \text{and} \quad f(2) = \frac{16}{4} - \frac{4(8)}{3} + 2(4) + 3 = \frac{13}{3},$$

and therefore $(0, 3)$ and $(2, \frac{13}{3})$ are the only points that are candidates for maximum or minimum points. See Table 11.7. ∎

Table 11.7

Interval	Calculation of sign of $f'(x)$	Result	Conclusion
$(-\infty, 0)$	$(-)(-)(-)$	$-$; f decreasing	
			$(0, 3)$ is minimum point
$(0, 2)$	$(+)(-)(-)$	$+$; f increasing	
			$(2, \frac{13}{3})$ is neither maximum nor minimum point
$(2, \infty)$	$(+)(+)(+)$	$+$; f increasing	

Since we have the advantage of knowing that *all* maximum and minimum points have been located in Example 2, a reasonably accurate graph of $y = f(x)$ can be obtained by plotting only a very few points (see Figure 11.13).

We now turn our attention to finding endpoint extreme points. Suppose that the domain of a rational function f is an interval that includes the left endpoint—$[a, b]$, $[a, b)$, or $[a, \infty)$. Then $f(x) \to f(a)$ as $x \to a$ from within the interval, and f must be continuous on an interval of which a is the left

Sec. 11.3] **First Derivative Test**

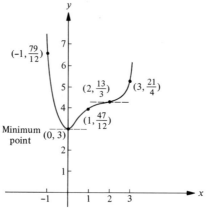

Figure 11.13. $y = \dfrac{x^4}{4} - \dfrac{4x^3}{3} + 2x^2 + 3$.

endpoint (see Definition 5 of Section 9.4). If f is decreasing just to the right of a, then $(a, f(a))$ is an endpoint (relative) maximum because it is "higher" than any nearby point [(1) of Figure 11.14]. If f is increasing just to the right of a, then $(a, f(a))$ is an endpoint (relative) minimum [(2) of Figure 11.14]. Endpoint maximum or minimum points at the right endpoint of the domain of a rational function are detected in an analogous manner [(3) and (4) of

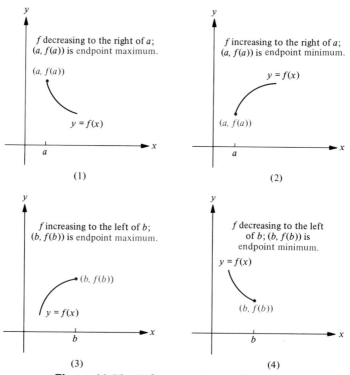

Figure 11.14. *Endpoint maxima and minima.*

Chap. 11] Maximum and Minimum Problems

Figure 11.14]. Any of these situations can be spotted in the process of making our standard table (like Table 11.7), and all endpoint maxima and minima can be included, as we demonstrate in Example 3.

Example 3
Find all maximum and minimum points of f if

$$y = f(x) = \frac{x^2 - 5}{x - 3}, \qquad 0 \leq x \leq 4.$$

The domain of the rational function f has been restricted to $[0, 4]$, and $x = 3$ must be excluded from the domain so that the denominator is not zero. Notice that the left endpoint 0 and the right endpoint 4 are included in the domain. By the quotient rule,

$$\begin{aligned} f'(x) &= \frac{(x - 3)(2x) - (x^2 - 5)(1)}{(x - 3)^2} \\ &= \frac{x^2 - 6x + 5}{(x - 3)^2} \\ &= \frac{(x - 1)(x - 5)}{(x - 3)^2}. \end{aligned}$$

The number 1 is the only critical number because the only root of $f'(x) = 0$ in the domain is $x = 1$ (recall that 5 is not in the domain). The only interior point that is a candidate for being a maximum or minimum point is $(1, f(1)) = (1, 2)$. Also it will have to be determined whether $(0, \frac{5}{3})$ is an endpoint maximum or endpoint minimum, and the same thing will have to be determined for $(4, 11)$. The only dividing points of f' in $[0, 4]$ are $x = 1$ [where the numerator of $f'(x)$ is zero] and $x = 3$ [where the denominator of $f'(x)$ is zero]. The problem is solved in Table 11.8 (sketch postponed until Exercise 5 of Section 11.6). ∎

Table 11.8

Interval	Calculation of sign of $f'(x)$	Result	Conclusion
			$(0, \frac{5}{3})$ is endpoint minimum
$(0, 1)$	$\frac{(-)(-)}{(-)^2}$	$+$; f increasing	
			$(1, 2)$ is maximum point
$(1, 3)$	$\frac{(+)(-)}{(-)^2}$	$-$; f decreasing	
			No maximum or minimum point at $x = 3$, since 3 not in domain
$(3, 4)$	$\frac{(+)(-)}{(+)^2}$	$-$; f decreasing	
			$(4, 11)$ is endpoint minimum

Sec. 11.3] First Derivative Test

The following steps summarize the method that can always be used to find maximum and minimum points of a rational function f:

1. After noting the domain of f, calculate $f'(x)$.
2. Solve the equation $f'(x) = 0$ to find all critical numbers in the domain of f. Calculate $f(x_1)$ for each critical number x_1, and list all points $(x_1, f(x_1))$ that are candidates for being maximum or minimum points in the interior of the domain of f. Also list points, if any, that can be endpoint maxima or minima.
3. Note all dividing points of f' [critical numbers together with numbers for which the denominator of $f'(x)$ is zero] together with any finite endpoints, in order to determine the intervals to use in the first column of tables such as Tables 11.7 and 11.8. Complete such a table; use Theorem 3 and the explanation in Figure 11.14 to detect all maximum and minimum points.

APPLICATIONS

Example 4

In this example we present one application of the material in this section; several more examples will appear in the next section.

The density of water at its freezing point is 1 gram per cubic centimeter, but the density varies slightly as the temperature increases. Assume that the temperature is measured on the Celsius scale. According to Thrall et al. [63], pp. 20–21, the specific weight (density) s of water at temperature t is given by

$$s = f(t) = 1 + (5.3)(10)^{-5}t - (6.53)(10)^{-6}t^2 + (1.4)(10)^{-8}t^3,$$

provided that $0 \leq t < 100$.

Let us find all maximum and minimum points of f on $[0, 100)$.

$$f'(t) = (5.3)(10)^{-5} - (2)(6.53)(10)^{-6}t + (3)(1.4)(10)^{-8}t^2$$
$$= (4.2)(10)^{-8}\left(t^2 - \frac{1306}{4.2}t + \frac{5300}{4.2}\right).$$

By the quadratic formula the roots of $f'(t) = 0$ are approximately $t = 4.1$ and $t = 306.8$, and therefore $f'(t)$ may be written

$$f'(t) \approx (.000000042)(t - 4.1)(t - 306.8).$$

The number 4.1 is the only critical number, because 306.8 is outside the domain. If $t = 4.1$, s can be calculated to be approximately 1.0001085, and hence (4.1, 1.0001085) is a critical point. If $t = 0$, then $s = 1$ and the endpoint (0, 1) must be considered. The only dividing point of f' in $[0, 100)$ is 4.1. See Table 11.9 for the answers to the problem.

The maximum point at the temperature of about 4.1°C is of interest. As water is warmed starting at $t = 0$, the breaking down of some hydrogen bonds enables water molecules to crowd more closely together, which

Table 11.9

Interval	Calculation of sign of $f'(t)$	Result	Conclusion
			$(0, 1)$ is endpoint minimum
$(0, 4.1)$	$(+)(-)(-)$	$+$; f increasing	
			$(4.1, 1.0001085)$ is maximum point
$(4.1, 100)$	$(+)(+)(-)$	$-$; f decreasing	

makes the density tend to increase. However, the increased energy of the water molecules causes them to spread apart, which makes the density tend to decrease. These two phenomena have opposite effects. Up to $t = 4.1$ the first phenomenon is dominant and the density increases to a maximum; above $t = 4.1$ the second phenomenon is dominant and the density of water decreases. ∎

EXERCISES

Suggested minimum assignment: Exercises 1, 3, 5, 7, 11, and 13.

In Exercises 1-4 find all critical numbers of the given function.

1. $y = f(x) = x^3 - 12x$.
2. $y = f(x) = x^4 - 4x^3 + 1$.
3. $y = f(x) = \dfrac{x^2}{x - 1}$.
4. $y = f(x) = \dfrac{x^2}{x^2 - 9}$, $-2 < x < 5$.
5. Find all maximum and minimum points in Exercise 1. Sketch.
6. Find all maximum and minimum points in Exercise 2. Sketch.
7. Find all maximum and minimum points in Exercise 3.
8. Find all maximum and minimum points in Exercise 4.

In Exercises 9-14 find all maximum and minimum points of the given function. Make a table similar to Tables 11.7 and 11.8.

9. $y = f(x) = x^4$, $-1 \le x \le 1$.
10. $y = f(x) = 3 - x^4$, $x \ge -1$.
11. $y = f(x) = (x - 2)^5$, $0 \le x \le 3$.
12. $y = f(x) = (2 - x)^6$.
13. $y = P(x) = \dfrac{40x}{x^2 + 4}$, $0 \le x \le 4$, where P is a profit function.
14. $y = f(x) = (x + 2)^2(x - 3)^3$.

In Exercises 15–17 find all maximum and minimum points of the given function. Sketch. Make a table similar to Tables 11.7 and 11.8.

15. $y = f(x) = x^3 - \frac{3}{2}x^2 - 18x + 8$.
16. $y = f(x) = \frac{x^4}{4} - 3x^3 + 9x^2 - 1$.
17. $y = f(x) = x^4 - \frac{20}{3}x^3$.
18. Prove Theorem 3.
19. The graph of the constant function defined by $y = f(x) = 3$ is a horizontal line. Some authors consider every point on the line to be both a maximum point and a minimum point, whereas others say that the function has no maximum or minimum points. According to Definition 2, which viewpoint do we take in this text? What minor changes could be made in Definition 2 to force us to adopt the other viewpoint?

11.4 Applied Maximum and Minimum Problems

Many of the important applications of mathematics involve maximizing or minimizing some quantity. If a formula for such a quantity can be written as a rational function of just one variable, then the procedures of Section 11.3 can be used.

Example 1

Return to Example 1 of Section 11.1. A store manager wants to maximize his monthly profit that is given by

$$P(x) = (x - 5)(150 - 10x),$$

where the domain of P is the closed interval $[5, 15]$.

$$P(x) = -10x^2 + 200x - 750,$$
$$P'(x) = -20x + 200 = (-20)(x - 10).$$

Equating $P'(x)$ to zero yields the critical number $x = 10$. Then

$$P(10) = (10 - 5)[150 - (10)(10)] = 250,$$

and hence $(10, 250)$ is a candidate for being a maximum or minimum point. The points $(5, 0)$ and $(15, 0)$ may be endpoint maximum or minimum points. We see from Table 11.10 that maximum profit will be attained if the selling price of each lamp is $10. This makes the monthly profit $250, which is a larger profit than can be attained with any other selling price. Notice that there is zero profit when the selling price is either $5 or $15. ∎

The behavior of the profit function in Example 1 is typical, in the sense that a profit function usually has a maximum point. Up to a certain point, the greater the quantity produced of an item (during a certain period of time), the greater the profit. Beyond that point the extra quantity that is produced may not be sold or may be sold at a loss. As we have learned

Table 11.10

Interval	Calculation of sign of $P'(x)$	Result	Conclusion
			$(5, 0)$ is endpoint minimum
$(5, 10)$	$(-)(-)$	$+$; P increasing	
			$(10, 250)$ is maximum point
$(10, 15)$	$(-)(+)$	$-$; P decreasing	
			$(15, 0)$ is endpoint minimum

previously, a profit function P for an item is often given by

$$P(x) = R(x) - C(x),$$

where R is the revenue function and C is the cost function. The maximum profit, if it exists, occurs at a critical number obtained by solving the equation $P'(x) = 0$. If $P'(x) = 0$, then

$$R'(x) - C'(x) = 0$$

and

$$R'(x) = C'(x).$$

Therefore, *maximum profit, if it exists, occurs only if marginal revenue equals marginal cost.* This result is an important theorem in economics.

Example 2
The equation of the demand curve for an item is

$$y = f(x) = 14 - \tfrac{1}{3}x^2.$$

Assume throughout this example that $0 \leq x \leq 5$. The revenue is the number of the item sold times the price received per item; that is,

$$\begin{aligned} R(x) &= xf(x) \\ &= x(14 - \tfrac{1}{3}x^2) \\ &= 14x - \tfrac{1}{3}x^3. \end{aligned}$$

Suppose that the cost function is given by

$$C(x) = 5x + 2.$$

We seek the value of x (critical number) at which maximum profit may occur, and therefore we solve the equation $R'(x) = C'(x)$ as follows:

$$\begin{aligned} 14 - x^2 &= 5, \\ x^2 &= 9, \\ x &= 3. \end{aligned}$$

(The solution $x = -3$ of $x^2 = 9$ is omitted because -3 is not a permissible value of x.) In Exercise 4 the reader is asked to verify that the profit function P really has a *maximum* point at $x = 3$. ∎

Sec. 11.4] Applied Maximum and Minimum Problems

In Examples 3 and 4 the quantity to be maximized or minimized will be formulated first in terms of two variables, but an auxiliary equation will make possible the elimination of one variable.

Example 3

A land corporation is willing to spend $1200 to fence in a rectangular portion of its property (see Figure 11.15). Its property already has a fence along one side; this existing fence will be used for one side of the fenced rectangle. On the sides of length x (as shown in Figure 11.15) the corporation wants wire fence that costs $2 per foot to buy and install, and on the side of length y, a wood fence that costs $3 per foot to buy and install is desired. Find x and y so that the enclosed area will be as large as possible.

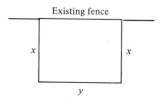

Figure 11.15

The quantity to be maximized is the area

$$A = xy,$$

where $x > 0$ and $y > 0$. An auxiliary equation can be written expressing the total cost of the fence, namely,

$$2x + 2x + 3y = 1200,$$

from which we obtain

$$y = \frac{1200 - 4x}{3}.$$

Therefore,

$$A = x\left(\frac{1200 - 4x}{3}\right) = 400x - \frac{4}{3}x^2.$$

Now A is expressed as a function of a single variable x, and the domain of the function is (0, 300). (If $x \geq 300$, there is not enough money to purchase both types of fencing material.)

$$\frac{dA}{dx} = 400 - \frac{8}{3}x = \left(-\frac{8}{3}\right)(x - 150).$$

Setting dA/dx equal to zero to obtain the critical number, we find that

$$x = 150 \text{ feet.}$$

If $x = 150$, then

$$y = \frac{1200 - (4)(150)}{3} = \frac{1200 - 600}{3} = 200 \text{ feet}$$

and
$$A = 150(200) = 30{,}000 \text{ square feet.}$$

The solution (see Table 11.11) is $x = 150$ feet and $y = 200$ feet, which yields a maximum possible enclosed area of 30,000 square feet under the restriction that only \$1200 can be spent. ∎

Table 11.11

Interval	Calculation of sign of $\dfrac{dA}{dx}$	Result	Conclusion
(0, 150)	$(-)(-)$	$+$; A increasing	(150, 30,000) is maximum point
(150, 300)	$(-)(+)$	$-$; A decreasing	

Example 4

A certain company decides to manufacture aluminum cans (in the shape of a right circular cylinder). Each can is to have a volume of 16π cubic inches. Find the radius r and height h of each can so as to make the surface area (and thus the cost) as small as possible. See Figure 11.16.

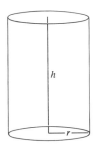

Figure 11.16

The total surface area S that is to be minimized is the sum of the areas of the top, the bottom, and the side. Therefore,

$$S = \pi r^2 + \pi r^2 + 2\pi rh,$$

where $r > 0$ and $h > 0$. The quantity S to be minimized is expressed in terms of both r and h, but the formula $V = \pi r^2 h$ for the volume of a right circular cylinder will make possible the elimination of one of these variables.

$$V = 16\pi = \pi r^2 h$$

or
$$h = \frac{16}{r^2}.$$

Sec. 11.4] Applied Maximum and Minimum Problems

Therefore,
$$S = 2\pi r^2 + 2\pi r \frac{16}{r^2}$$
$$= 2\pi r^2 + \frac{32\pi}{r}$$
$$= \frac{2\pi r^3 + 32\pi}{r}.$$

Now S is expressed as a function of a single variable r; the domain of this function is $(0, \infty)$, since the radius of the can must be a positive number.

$$\frac{dS}{dr} = \frac{r(6\pi r^2) - (2\pi r^3 + 32\pi)(1)}{r^2}$$
$$= \frac{4\pi r^3 - 32\pi}{r^2}$$
$$= 4\pi \frac{r^3 - 8}{r^2}$$
$$= 4\pi \frac{(r-2)(r^2 + 2r + 4)}{r^2}$$
$$= \frac{(4\pi)(r-2)[(r+1)^2 + 3]}{r^2}.$$

The only critical number is
$$r = 2 \text{ inches},$$
obtained by setting dS/dr equal to zero. If $r = 2$, then
$$h = \frac{16}{(2)^2} = 4 \text{ inches}$$
and
$$S = \frac{2\pi(2)^3 + 32\pi}{2} = 24\pi \text{ square inches.}$$

Next we construct Table 11.12. The solution to the problem is to make the radius of each can 2 inches and the height 4 inches, which gives each can the least possible surface area under the restriction that the volume be 16π cubic inches. ∎

Table 11.12

Interval	Calculation of sign of $\frac{dS}{dr}$	Result	Conclusion
$(0, 2)$	$\frac{(+)(-)(+)}{(+)}$	$-$; S decreasing	
$(2, \infty)$	$\frac{(+)(+)(+)}{(+)}$	$+$; S increasing	$(2, 24\pi)$ is minimum point

Chap. 11] Maximum and Minimum Problems

APPLICATIONS

Example 5

Suppose that a company sells a quantity q of a particular item each year. The company buys the item and stores it before selling it, even though money is spent to store the item. Assume that adequate storage space is available. It also costs money to reorder the item, but the item can be reordered frequently if necessary. Assume that sales are distributed *evenly* over the year and that all costs remain unchanged. Suppose that the company buys x items each time it reorders, and that this can be done in such a manner that a shipment arrives just as the previous inventory is depleted—thus since sales are evenly distributed, the average inventory is $x/2$. The problem is to find x so as to minimize the total annual cost. If x is too large, the storage cost (also known as carrying cost) is too great; but if x is too small, the reorder costs are too great.

Let p be the unit price of each item, r be the reorder cost associated with each shipment, and s be the storage cost of one item for a year. For 1 year the basic cost of the items is pq. The number of shipments each year is q/x, and hence the total annual reorder cost is rq/x. The annual storage cost is $(s)(x/2)$, since $x/2$ is the average inventory. The total cost is

$$C(x) = pq + \frac{rq}{x} + \frac{sx}{2},$$

where x must be positive and p, q, r, and s are positive constants.

$$C'(x) = -\frac{rq}{x^2} + \frac{s}{2}$$

$$= \frac{-2rq + sx^2}{2x^2}$$

$$= \frac{s(x^2 - 2rq/s)}{2x^2}$$

$$= \frac{s(x + \sqrt{2rq/s})(x - \sqrt{2rq/s})}{(2x)(x)}.$$

If $C'(x) = 0$, then $x = \pm\sqrt{2rq/s}$. The number $x = \sqrt{2rq/s}$ is the only critical number (the negative of this number is not in the domain of C). If $x = 0$, the denominator of $C'(x)$ is zero, but $x = 0$ is not a dividing point of C' because it is not within the domain $(0, \infty)$ of C. The only candidate for a maximum or minimum point is $(\sqrt{2rq/s}, C(\sqrt{2rq/s}))$. See Table 11.13. From the table the total annual cost is minimized if the size of each shipment is $x = \sqrt{2rq/s}$; this formula is often called the **optimal lot-size formula**, and it does not depend upon the price p of each item. ∎

*Example 6 The Mode of a Measurement

When a measurement (whose domain is a sample space) is defined, it is useful to locate numbers that in some manner are representative of the

Table 11.13

Interval	Calculation of sign of $C'(x)$	Result	Conclusion
$\left(0, \sqrt{\dfrac{2rq}{s}}\right)$	$\dfrac{(+)(+)(-)}{(+)(+)}$	$-$; C decreasing	
			$\left(\sqrt{\dfrac{2rq}{s}}, C\left(\sqrt{\dfrac{2rq}{s}}\right)\right)$ is minimum point
$\left(\sqrt{\dfrac{2rq}{s}}, \infty\right)$	$\dfrac{(+)(+)(+)}{(+)(+)}$	$+$; C increasing	

values of the measurement. We shall define three such representatives, which are normally referred to as **measures of central tendency** (also see Section 8.3). These measures, the mode, the mean (average), and the median, are often used as references to which the other values are compared. In this example we shall be concerned only with the mode of a measurement; the mean and the median will be defined later. For a *discrete* measurement the **mode** is defined to be the measurement value (or values) that has the largest associated weight. For example, consider a sample space U of 10 students who take a test. A measurement, with domain U, is defined by assigning to each student his test grade. Suppose that the grade of 75 is assigned to 5 students, 80 to 3 students, and 65 to 2 students. The weighting function w that has domain $\{65, 75, 80\}$ can be defined as follows: $w(65) = \frac{2}{10}$, $w(75) = \frac{5}{10}$, $w(80) = \frac{3}{10}$. Because $\frac{5}{10}$ is the largest weight, 75 is the mode.

For a *continuous* measurement the **mode** is defined to be the measurement value (or values) for which the associated probability density function is a maximum. As an illustration consider a measurement whose values measure the life span (in decades) of a certain type of machine. Suppose that the probability density function of this measurement is given by

$$p(x) = \begin{cases} \frac{3}{2}x - \frac{3}{4}x^2 & \text{if } 0 \leq x \leq 2, \\ 0 & \text{otherwise.} \end{cases}$$

By the methods of this chapter one can find that p has a maximum point at $x = 1$, and hence one decade is the mode life span; that is, a life span of one

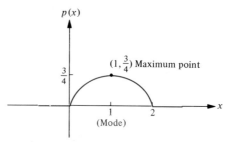

Figure 11.17. *Graph of probability density function p.*

decade has a greater density of probability than that of any other unit of time in the immediate vicinity of $x = 1$. See Figure 11.17.

In both the discrete and continuous cases, there can be more than one mode. ∎

EXERCISES

Suggested minimum assignment: Exercises 1, 5, 9, 13, and 15.

1. The owner of an apple orchard estimates that if he begins picking apples today, he will have a crop of 1000 bushels and can make a profit of $1 per bushel. He could wait as long as 25 days in order to begin picking; for each day he waits he estimates that he will have 50 extra bushels, but his profit per bushel will decrease by 2 cents for each day he waits. How many days should he wait to begin picking his crop so that he can earn a maximum profit? [*Hint:* Let x be the number of days that the owner waits; $P(x) =$ (profit per bushel)(number of bushels).]
2. Suppose that each apple tree in an orchard has an annual yield of 20 bushels, provided that there are no more than 30 trees per acre. For each additional tree per acre above 30, the annual yield decreases by $\frac{2}{5}$ bushel per tree. How many trees per acre will give the maximum annual yield?
3. An alumni group wants to charter a bus for a football game. The bus company offers a bus that seats 100 persons and insists that at least 40 persons make the trip. The company will charge $15 per person if exactly 40 members of the alumni group make the trip, but the company will decrease the price by 10 cents per person for each person above 40 (for example, if 50 persons make the trip, the charge is $14 per person). How many persons would have to make the trip in order to maximize the revenue received by the bus company?
4. In Example 2, $P(x) = R(x) - C(x) = (14x - \frac{1}{3}x^3) - (5x + 2)$, provided that $0 \le x \le 5$. Make a table (like Table 11.10) to find all maximum and minimum points of P. (Be sure that your maximum point agrees with Example 2.)
5. For a certain item the equation of the demand curve is $y = f(x) = 70 - \frac{3x}{2} - \frac{x^2}{15}$, and the cost function is given by $C(x) = 50x + 5$. Assume that $0 \le x \le 15$. The profit function is given by $P(x) = xf(x) - C(x)$. For what value of x is the profit largest?
6. Suppose that a profit function P is given by

$$y = P(x) = -x^2 + 8x - 5, \quad 0 < x < 8.$$

Show that maximum profit is attained at $x = 4$ and find the maximum profit. (See Example 5 of Section 10.1.)

7. The revenue and cost functions for an item are given by

$$R(x) = 12x - \frac{x^2}{12} - \frac{x^3}{6} \quad \text{and} \quad C(x) = 2 + 4x - \frac{x^2}{12},$$

where $0 \leq x < 6$.

For what value of x is the profit greatest?

8. Three sides of a rectangular plot of ground are to be fenced so that the total area of the plot is 5000 square feet. Find the length and width of the plot so that the number of feet of fencing is minimized.

9. Rework Example 3 if the corporation is willing to spend $1800.

10. The space within a 440-yard racetrack for track meets is to consist of a rectangle with a semicircle at each end. Find the length and width of the rectangle so that the area of the rectangle is maximized.

11. Rework Example 4 if the volume of each can is 54π cubic inches.

12. Rework Example 4 if the volume of each can is 16 cubic inches.

13. Find the radius and height of an open (no top) right circular cylinder with minimum surface area if the volume of the cylinder is to be 8π cubic feet. (Hint: Surface area $S = \pi r^2 + 2\pi rh$.)

14. (a) Find two positive numbers x and y, if possible, whose product is 100, such that the sum of the two numbers is minimized.

    (b) Find two positive numbers x and y, if possible, whose product is 100, such that the sum of the two numbers is maximized.

15. A rectangular-shaped advertisement sign with total area of 50 square feet is planned. The sign is to have margins of 6 inches at the top and at the bottom with margins of 1 foot at the left and right. Find the dimensions of the sign so that the area in the center for printed matter is maximum.

16. Suppose that the concentration y of a drug in the blood x hours after an injection is given by $y = x/(5x^2 + 80)$. How many hours after the injection does the concentration reach its maximum value?

17. Suppose that the reaction r of the body to a dose d of a drug can be represented by

$$r = f(d) = d^2 \left(\frac{c}{2} - \frac{d}{3} \right),$$

where $c > 0$ is the greatest amount of the drug that may be given. The rate of change of r with respect to d is given by

$$f'(d) = cd - d^2.$$

Find the value of d for which $f'(d)$ is greatest. The answer will tell when the rate of change in r is greatest, and it is sometimes useful in determining the exact dose of a drug to be given. For more discussion, see pp. 216–218 of Thrall et al. [63].

18. An open box is manufactured from a piece of cardboard 24 inches × 24 inches by cutting equal squares out of the corners and turning up the

sides. How large a square should be cut out of each corner to obtain a box with maximum volume?

19. An average cost function (see Example 6 of Section 10.4) is given by

$$f(x) = \frac{C(x)}{x} = \frac{\frac{1}{10}x^2 + 7x + 10}{x}, \qquad 0 < x \le 30.$$

For what value of x is the average cost minimized?

20. A company sells 3000 of an item each year, and the conditions of Example 5 are met. The reorder cost of each shipment is \$125, and the storage cost of one item for a year is \$3. How many of the item should the company order at one time? How often should it reorder this particular item?

21. (a) In Example 6 of Section 3.9 a discrete measurement and the weighting function were given for the experiment of flipping a coin until a head appears. What is the mode of the measurement?

(b) In Example 7 of Section 10.2 a continuous measurement and the probability density function were given for a machine problem. What is the mode of the measurement?

11.5 Concavity and Second Derivative Test

The population of a particular type of object over a certain time interval is graphed in Figure 11.18. From time 0 to time 2 the population is increasing, and the slope, which represents the rate of change of population, is also increasing. For $2 < t < 4$, however, the slope starts decreasing, even though the population continues to rise; in other words, the population continues to rise but at a decreasing rate. Eventually, at $t = 4$ the continuing decreasing slope (rate) causes the population itself to start decreasing. Then at $t = 5$ something else happens to cause the slope (rate) to start increasing again; this causes the population to decrease at a slower rate and eventually (at $t = 6$) even begin to increase. It is important to distinguish between the case of decreasing slope (rate) and the case of increasing slope (rate). When the

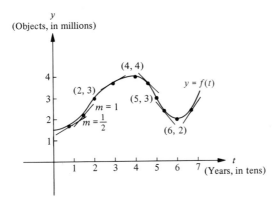

Figure 11.18

slope is increasing on the intervals (0, 2) and (5, 7), notice that a tangent line will be below the curve; it is customary to say that the curve is **concave upward** in this case. When the slope is decreasing on the interval (2, 5), notice that a tangent line will be above the curve; it is customary to say that the curve is **concave downward** in this case.

▶ **Definition 4**
Suppose that a function f is differentiable at each point of an interval. Then
(i) f is said to be **concave upward on the interval** if the tangent line at each point of the interval is below the graph of f.
(ii) f is said to be **concave downward on the interval** if the tangent line at each point of the interval is above the graph of f.

The interval referred to in Definition 4 may be open, closed, or half-open and may or may not be an infinite interval; we shall be especially interested in the case in which it is an open interval.

Concave upward and concave downward are the two types of **concavity**. The following theorem, whose proof is omitted, gives a practical method of determining on which open intervals a function possesses a particular type of concavity.

▶ **Theorem 4**
(i) If $f''(x) > 0$ at each point of an open interval, then f is concave upward on the interval.
(ii) If $f''(x) < 0$ at each point of an open interval, then f is concave downward on the interval. ◀

Example 1
Suppose that a function f is defined by
$$y = f(x) = \tfrac{1}{16}(-x^3 + 12x^2 - 32).$$
Use Theorem 4 to find open intervals on which f is concave upward and open intervals on which f is concave downward.
$$f'(x) = \tfrac{1}{16}(-3x^2 + 24x),$$
$$f''(x) = \tfrac{1}{16}(-6x + 24)$$
$$= (-\tfrac{3}{8})(x - 4).$$
According to Theorem 4 the problem is solved by calculating the sign of $f''(x)$ on various open intervals. The only dividing point of f'' is 4, at which a change in sign of $f''(x)$ may occur. The work is shown in Table 11.14—let it be emphasized that in determining the concavity, it is necessary to calculate the *sign of* $f''(x)$. The graph of f is shown in Figure 11.19—observe that the curve is drawn so that the concavity agrees with Table 11.14. The maximum and minimum points indicated in Figure 11.19 are found by the methods of Section 11.3. ∎

Table 11.14

Interval	Calculation of sign of $f''(x)$	Result	Conclusion
$(-\infty, 4)$	$(-)(-)$	$+$	f concave upward
$(4, \infty)$	$(-)(+)$	$-$	f concave downward

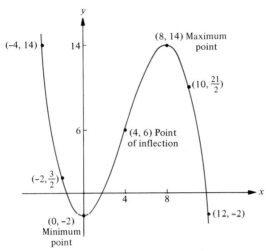

Figure 11.19. $y = \frac{1}{16}(-x^3 + 12x^2 - 32)$.

The function f in Example 1 is differentiable at $x = 4$, and the point $(4, 6)$ separates a portion of the curve that is concave upward from a portion that is concave downward—such a point is called a **point of inflection** of the function. If the tangent line to the curve at the point of inflection $(4, 6)$ were drawn, it would cross the curve rather than be either entirely above or entirely below the curve in the immediate vicinity of $(4, 6)$.

A point of inflection of a function must occur at an interior point of the domain of the function. It cannot occur at an endpoint because a function is defined on only one side (left or right) of an endpoint; therefore, an endpoint cannot separate a portion of the curve that is concave upward from a portion that is concave downward. In Example 1 there is a point of inflection at $x = 4$ and $f''(4) = 0$. Theorem 5 says that for a rational function the value of the second derivative at a point of inflection *must* be zero. The significance of this statement is that the only *candidates* for points of inflection of a rational function are those in the interior of the domain at which the value of the second derivative is zero.

▶ **Theorem 5**

If $(x_1, f(x_1))$ is a point of inflection of a rational function f, then $f''(x_1) = 0$.

A hint for the proof of Theorem 5 will be given in Exercise 18. ◀

Example 2
Find all points of inflection of the function defined by
$$y = f(x) = x^5 + \tfrac{5}{2}x^4 - 2.$$
We begin by finding $f''(x)$ in factored form.
$$f'(x) = 5x^4 + 10x^3,$$
$$f''(x) = 20x^3 + 30x^2$$
$$= (20x)(x)(x + \tfrac{3}{2}).$$

The second derivative is zero if $x = 0$ or if $x = -\tfrac{3}{2}$. Therefore, $(0, f(0))$ and $(-\tfrac{3}{2}, f(-\tfrac{3}{2}))$ are the only candidates for points of inflection. We calculate
$$f(0) = 0 + 0 - 2 = -2$$
and
$$f(-\tfrac{3}{2}) = -\tfrac{243}{32} + \tfrac{5(81)}{2(16)} - 2 = \tfrac{49}{16}.$$

According to Table 11.15, $(-\tfrac{3}{2}, \tfrac{49}{16})$ is a point of inflection because it separates portions of the curve with different concavity. The candidate $(0, -2)$ is not a point of inflection because the concavity does not change at this point. See Figure 11.20. ∎

Table 11.15

Interval	Calculation of sign of $f''(x)$	Result	Conclusion
$(-\infty, -\tfrac{3}{2})$	$(-)(-)(-)$	$-$; concave downward	
			$(-\tfrac{3}{2}, \tfrac{49}{16})$ is point of inflection
$(-\tfrac{3}{2}, 0)$	$(-)(-)(+)$	$+$; concave upward	
			$(0, -2)$ is *not* point of inflection
$(0, \infty)$	$(+)(+)(+)$	$+$; concave upward	

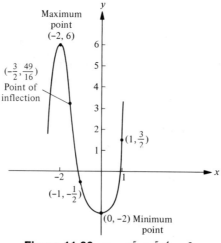

Figure 11.20. $y = x^5 + \tfrac{5}{2}x^4 - 2.$

Let us return to the problem of determining maximum and minimum points of a rational function f. Suppose that x_1 is an interior point of the domain of f, and suppose that $f'(x_1) = 0$. The first derivative test, Theorem 3 of Section 11.3, is effective in deciding whether the critical point $(x_1, f(x_1))$ is a maximum point, a minimum point, or neither; Theorem 3 *always* yields a definite conclusion. There is another test that is often easier to apply that *usually* gives a definite conclusion about a critical point. This test is called the **second derivative test** and is stated in Theorem 6.

▶ **Theorem 6**
Suppose that $(x_1, f(x_1))$ is a critical point of a rational function f.
   (i) $(x_1, f(x_1))$ is a minimum point if $f''(x_1) > 0$.
   (ii) $(x_1, f(x_1))$ is a maximum point if $f''(x_1) < 0$.
[If $f''(x_1) = 0$, no conclusion can be made; use Theorem 3 in this case.]

Proof: We shall prove (i) and leave the similar proof of (ii) as Exercise 19. Since x_1 is an interior point of the domain of the *rational* function f, the function f'' is continuous at x_1. Because f'' is continuous at x_1 and $f''(x_1) > 0$, there is an open interval containing x_1 throughout which $f''(x) > 0$ by Exercise 14 of Section 9.4. This interval can be chosen so that x_1 is the only critical number in the interval. By Theorem 1, f' is increasing on this open interval. Since $f'(x_1) = 0$, $f'(x)$ changes from negative to zero to positive on this interval. By Theorem 3(ii), $(x_1, f(x_1))$ is a minimum point. ◀

Example 3
Find all maximum and minimum points of f if

$$y = f(x) = x^5 + \tfrac{5}{2}x^4 - 2.$$

This is the same function as given in Example 2.

$$f'(x) = 5x^4 + 10x^3 = 5x^3(x + 2).$$

It follows that $f'(-2) = 0$ and $f'(0) = 0$. The only candidates for maximum or minimum points are $(-2, 6)$ and $(0, -2)$. Since $f''(x) = 20x^3 + 30x^2$,

$$f''(-2) = 20(-8) + 30(4) = -40 < 0,$$

and hence $(-2, 6)$ is a maximum point by Theorem 6. Also

$$f''(0) = 20(0) + 30(0) = 0,$$

and in this case Theorem 6 does not apply. In Theorem 6, if $f''(x_1) = 0$, then no conclusion can be made—in other words, if $f''(x_1) = 0$, then $(x_1, f(x_1))$ may be either a maximum point, a minimum point, or neither. In this problem $(0, -2)$ can be shown to be a minimum point by applying Theorem 3. Our conclusions agree with Figure 11.20. ∎

APPLICATIONS

Example 4
In Example 1 we showed (see Table 11.14) that $f''(x) > 0$ if $x < 4$. From Theorem 1 of Section 11.2 it follows that f' is increasing if $x < 4$. However, if $x > 4$, then $f''(x) < 0$ and f' is decreasing. Hence, at the inflection point $(4, 6)$, the *derivative* must attain a maximum value—in Figure 11.19 the slope is greater at $(4, 6)$ than at any nearby point. At a point of inflection the slope is a (relative) maximum or minimum.

Suppose that a profit function P is given by

$$P(x) = \tfrac{1}{16}(-x^3 + 12x^2 - 32), \qquad 0 \le x \le 10.$$

(Except for the domain, P is the same function as in Example 1.) $P'(x)$ is the marginal profit. From our discussion above, the marginal profit is maximum at $x = 4$, since P changes from concave upward to concave downward at the point of inflection $(4, 6)$. Therefore, the additional profit obtained by production of one more item is greatest at $x = 4$. ∎

Example 5
A book that contains a wealth of applications of calculus in economics is *Quantitative Economic Theory* by Brems [11]. For example, on pp. 37–41, Brems discusses the problem of finding the optimal useful life of consumer durables. If an item such as a television set, washing machine, or car is replaced frequently, repair costs are not very great but the cost of the new item is incurred too often; if one waits too long to replace the item, the repair costs become excessive. Brems makes certain reasonable assumptions, and then finds the useful life that will minimize cost; he uses the second derivative test (Theorem 6) to prove that the cost has been *minimized*. The optimal useful life can also be used to predict demand. A glance at Brems's book should convince the student of the importance of calculus in economics, although many of his examples depend upon material that will be presented in Chapters 14 and 15 of this text. ∎

EXERCISES

Suggested minimum assignment: Exercises 1, 5, 9, 11, 14(b), and 15.

In Exercises 1–4 use Theorem 4 to find open intervals on which f is concave upward and open intervals on which f is concave downward. Make a table like Table 11.14.

1. $y = f(x) = \dfrac{x^5}{20} - \dfrac{x^3}{6}$.

2. $y = f(x) = x^4 + 2x^3 - 12x^2 + x + 3$.

3. $y = f(x) = \dfrac{x}{x^2 + 3}$.

4. $y = f(x) = x^4 + 1$.

In Exercises 5–8 find all points of inflection. Make a table like Table 11.15.

5. $y = f(x) = x^3 + 6x^2 + 4x - 6$.
6. $y = f(x) = x^3$.
7. $y = f(x) = x^4 - 24x^2$.
8. $y = f(x) = \dfrac{x}{x^2 + 3}$. (See Exercise 3.)

9. Given $y = f(x) = 1/x$. On what open intervals is f concave upward and on what open intervals is f concave downward? Why is there no point of inflection at $x = 0$?

In Exercises 10–12 find all maximum and minimum points. Use Theorem 6 if possible—otherwise use Theorem 3. Also sketch the graph of f.

10. $y = f(x) = \frac{2}{3}x^3 + \frac{3}{2}x^2 + x - 1$.
11. $y = f(x) = x^2 - x^3$.
12. $y = f(x) = 1 - (x - 1)^6$.

13. For each of the following profit functions, find the value of x for which the profit is maximum.
    (a) $P(x) = -x^3 + 3x^2 + 9x - 4$, $0 \le x \le 4$.
    (b) $P(x) = -\frac{1}{3}x^3 + 2x^2 + 96x - 10$, $0 \le x \le 16$.

14. Sketch a graph of each of the given functions after doing the following: find the open intervals where f is increasing and the open intervals where f is decreasing; find all maximum and minimum points; find the open intervals where f is concave upward and the open intervals where f is concave downward; and find all points of inflection.
    (a) $y = f(x) = \frac{1}{3}x^3 - x^2 - 8x + 2$.
    (b) $y = f(x) = x^4 - 4x^3$.

15. Let revenue and cost functions be given by
    $$R(x) = -\tfrac{1}{6}x^3 + 4x^2 + 110x, \qquad 0 \le x \le 14,$$
    and
    $$C(x) = \tfrac{1}{6}x^3 + 2x^2 + 14x + 10, \qquad 0 \le x \le 14.$$
    Find the value x_1 at which the profit is maximum. Show that $P''(x_1) < 0$. Also verify that marginal revenue equals marginal cost at the point of maximum profit.

16. Repeat Exercise 15 for
    $$R(x) = 2x^2 - \tfrac{1}{32}x^3, \qquad 0 \le x \le 10,$$
    and
    $$C(x) = \tfrac{1}{32}x^3 + \tfrac{5}{4}x^2 + 2, \qquad 0 \le x \le 10.$$
    Also explain the relationship of this exercise to Figure 11.19.

17. The rate y of a certain autocatalytic reaction is given by
    $$y = kx(a - x), \qquad 0 \le x \le a,$$
    where $a > 0$ and $k > 0$. Use Theorem 6 to find the value of x at which

the rate y is greatest. (In an autocatalytic reaction an amount a of one substance is converted into a new substance, called the product, in such a way that the product catalyzes its own formation. The amount of the product is denoted by x; $x = 0$ when the reaction begins. The equation $y = kx(a - x)$ says that the rate of the reaction is proportional to the quantity $x(a - x)$, where $k > 0$ is the constant of proportionality. For more discussion, see pp. 33–34 of Thrall et al. [63].)

18. Prove Theorem 5. [*Hint:* Show that $f''(x_1) > 0$ and $f''(x_1) < 0$ are impossible—use Exercise 14 of Section 9.4, Theorem 4, and the definition of a point of inflection.]
19. Prove Theorem 6(*ii*).
20. At what point does the function f in Exercise 5 have minimum slope? What is the slope at this point?
21. For each of the profit functions in Exercise 13, find the value of x for which the marginal profit is maximum.

11.6 Curve Sketching

Pictures are often worth many words in mathematics as in other areas of life. In this section we shall recall several ideas from previous sections that are helpful in sketching the graph of an equation. We shall be aided by the new concepts of **symmetry** and **asymptotes.**

First we discuss symmetry—*the discussion of symmetry applies to equations in general and is not restricted to an equation of a rational function.* In Figure 11.21 the graph of the equation $x = 9 - y^2$ is shown. For any point (x, y) on the graph of the equation, there exists a point $(x, -y)$ directly across the x-axis which also satisfies the equation, and therefore we say that the graph is **symmetric to the x-axis.** To show that this is true, one can start with the original equation $x = 9 - y^2$ [we are assuming that (x, y) satisfies this equation when we write the equality] and determine whether $(x, -y)$ also satisfies the equation:

$x = 9 - y^2$ (original equation).
$x = 9 - (-y)^2$ [replace (x, y) with $(x, -y)$].
$x = 9 - y^2$ (simplify).

Since the original equation is obtained again, we have shown that $(x, -y)$ satisfies the equation whenever (x, y) does, and therefore the symmetry to the x-axis has been proved. Whenever symmetry to the x-axis can be detected at the outset, it is an aid in sketching the curve. Incidentally, the graph of $x = ay^2 + by + c$, $a \neq 0$, is called a parabola and opens to the left if $a < 0$, and to the right if $a > 0$ (parabolas that open upward or downward were mentioned in Section 3.9).

In a similar way the graph of an equation can be shown to be **symmetric to the y-axis** if replacement of (x, y) by the point $(-x, y)$ yields the original equation again. For example, the graph of $y = x^4 - 2x^2 - 1$ is symmetric to the y-axis because $y = (-x)^4 - (2)(-x)^2 - 1$ reduces to $y = x^4 -$

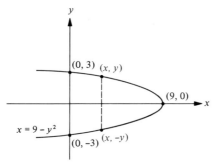

Figure 11.21. *Curve symmetric to x-axis.*

$2x^2 - 1$. Observe in Figure 11.22 that the portion of the curve on one side of the y-axis is a "reflection" across the y-axis of the portion on the other side. The symmetry of a curve to any line can be defined by this type of reflection property.

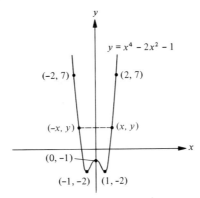

Figure 11.22. *Curve symmetric to y-axis.*

In addition to symmetry about a line such as the x-axis or y-axis, there is another type of symmetry that we shall discuss briefly. We are referring to symmetry to a point, and, in particular, symmetry to the origin. The graph of an equation is said to be **symmetric to the origin** if for any point (x, y) that satisfies the equation, the point $(-x, -y)$ also satisfies the equation. For example, the graph of $x = \frac{1}{2}y^3$ is symmetric to the origin because $-x = \frac{1}{2}(-y)^3$ can be simplified to yield the original equation. See Figure 11.23.

Example 1

The circle $x^2 + y^2 = 9$ is symmetric to the x-axis, the y-axis, and the origin. The parabola $y = x^2 + 4$ is symmetric to the y-axis but is not symmetric to the x-axis or to the origin. The graph of the rational function defined by $y = x/(x^2 + 1)$ is symmetric to the origin but is not symmetric to the x-axis or y-axis. The line $y = 2x + 1$ is not symmetric to the x-axis, the y-axis, or the origin. ■

Sec. 11.6] Curve Sketching

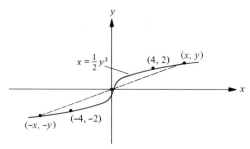

Figure 11.23. *Curve symmetric to the origin.*

For the remainder of this section we shall consider only equations that define *rational functions*—in other words, we shall discuss equations of the form

$$y = \frac{g(x)}{h(x)},$$

where g and h are polynomial functions. We shall introduce the ideas of **vertical asymptotes** and **horizontal asymptotes** in Example 2.

Example 2
Consider the function f defined by

$$y = f(x) = \frac{g(x)}{h(x)} = \frac{2x}{x-3}.$$

Notice that $f(3)$ is undefined. However, for values of x that are close to 3, the corresponding values of y are large in absolute value as seen in Table 11.16. For values of x even closer to 3 than those shown in Table 11.16, the corresponding y values are still larger in absolute value, and the points on the curve are very close to the vertical line $x = 3$. In this case, the line $x = 3$ is called a **vertical asymptote** (see Figure 11.24). Vertical asymptotes can be identified according to the following rule: If $h(x_1) = 0$ and $g(x_1) \neq 0$, then the line $x = x_1$ is a vertical asymptote of $y = g(x)/h(x)$. The rule is a result of the fact that $|y| \to \infty$ as $x \to x_1$.

Next, calculate

$$\lim_{x \to \infty} \frac{2x}{x-3} = \lim_{x \to \infty} \frac{2}{1-(3/x)} = \frac{2}{1-0} = 2.$$

Table 11.16

x	2.9	2.99	2.999	3.1	3.01	3.001
y	-58	-598	-5998	62	602	6002

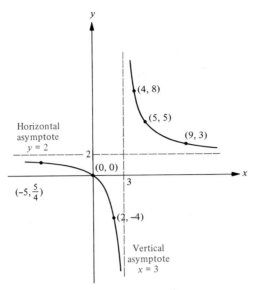

Figure 11.24. $y = \dfrac{2x}{x-3}$.

Hence $y \to 2$ as $x \to \infty$, and the horizontal line $y = 2$ is a **horizontal asymptote** in this example. For large values of x, the curve approaches the line $y = 2$ (see Figure 11.24). Similarly, it can be shown that

$$\lim_{x \to -\infty} \frac{2x}{x-3} = 2,$$

and therefore the curve also approaches the line $y = 2$ as $x \to -\infty$. If either $\lim_{x \to \infty} [g(x)/h(x)] = k$ or $\lim_{x \to -\infty} [g(x)/h(x)] = k$, where k is a real number, then the line $y = k$ is a horizontal asymptote of $y = g(x)/h(x)$. ∎

To draw a sufficiently accurate sketch of the graph of a rational function f, there are several things we have studied that one may attempt to find; in a given problem it may not be practical or necessary to find all of them.
1. Domain and range (Section 3.7).
2. Intercepts: x-intercepts and y-intercepts (Section 3.6).
3. Symmetry (Section 11.6).
4. Vertical and horizontal asymptotes (Section 11.6).
5. Intervals where f is increasing or decreasing (Section 11.2).
6. Maximum and minimum points (Sections 11.3 and 11.5).
7. Intervals where f is concave upward or concave downward (Section 11.5).
8. Points of inflection (Section 11.5).
9. Plot points other than those found in items 2, 6, and 8 (Section 3.6).
10. Slope at key points (Chapter 10).

Sec. 11.6] Curve Sketching

Example 3

Sketch the graph of the equation

$$y = f(x) = \frac{1}{x^2 - 4}.$$

The information used to aid in drawing the sketch is shown in Table 11.17, and a discussion of how to obtain some of the information is given below. The curve is drawn in Figure 11.25.

The remainder of this example is devoted to a discussion of how some of the information in Table 11.17 is obtained. The denominator is zero at

Table 11.17. INFORMATION FOR $y = 1/(x^2 - 4)$.

Domain: set of all real numbers except -2 and 2. Range: set of all real numbers except those in the interval $(-\frac{1}{4}, 0]$.
Intercepts: x-intercepts—none; y-intercepts—$-\frac{1}{4}$.
Symmetric to y-axis. Not symmetric to x-axis or origin.
Vertical asymptotes: $x = -2$, $x = 2$. Horizontal asymptotes: $y = 0$.
Increasing: $x < -2$, $-2 < x < 0$. Decreasing: $0 < x < 2$, $x > 2$.
Maximum points: $(0, -\frac{1}{4})$. Minimum points: none.
Concave upward: $x < -2$, $x > 2$. Concave downward: $-2 < x < 2$.
Points of inflection: none.
Other points: $(-3, \frac{1}{5})$, $(-1, -\frac{1}{3})$, $(\frac{7}{4}, -\frac{16}{15})$, $(\frac{21}{10}, \frac{100}{41})$.
Slope at $(-3, \frac{1}{5})$: $\frac{6}{25}$.

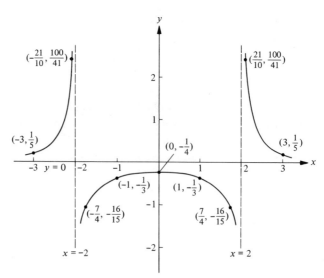

Figure 11.25. $y = \dfrac{1}{x^2 - 4}.$

$x = -2$ and $x = 2$. Hence the domain consists of all real numbers except ± 2, and the only points of discontinuity are at $x = \pm 2$. Since the numerator is not zero at these points, the lines $x = \pm 2$ are vertical asymptotes. If the equation is solved for x, the result is

$$x = \pm \sqrt{\frac{1 + 4y}{y}}.$$

If $-\frac{1}{4} < y < 0$, we are led to the square root of a negative number, and therefore numbers in $(-\frac{1}{4}, 0)$ must be excluded from the range. Also, because division by zero is undefined, y cannot equal zero. Because

$$\lim_{x \to \infty} \frac{1}{x^2 - 4} = 0,$$

the line $y = 0$ is a horizontal asymptote. Note that $\lim_{x \to -\infty} [1/(x^2 - 4)] = 0$ also. After differentiating and simplifying, one can obtain

$$f'(x) = \frac{(-2)(x)}{(x^2 - 4)^2} = \frac{(-2)(x)}{(x + 2)^2(x - 2)^2}$$

and

$$f''(x) = \frac{6x^2 + 8}{(x^2 - 4)^3} = \frac{6x^2 + 8}{(x + 2)^3(x - 2)^3},$$

which lead to Tables 11.18 and 11.19. ∎

Table 11.18

Interval	Calculation of sign of $f'(x)$	Result	Conclusion
$(-\infty, -2)$	$\dfrac{(-)(-)}{(-)^2(-)^2}$	$+$; f increasing	
			No maximum or minimum point at $x = -2$, since -2 not in domain
$(-2, 0)$	$\dfrac{(-)(-)}{(+)^2(-)^2}$	$+$; f increasing	
			$(0, -\frac{1}{4})$ is maximum point
$(0, 2)$	$\dfrac{(-)(+)}{(+)^2(-)^2}$	$-$; f decreasing	
			No maximum or minimum point at $x = 2$, since 2 not in domain
$(2, \infty)$	$\dfrac{(-)(+)}{(+)^2(+)^2}$	$-$; f decreasing	

Sec. 11.6] Curve Sketching

Table 11.19

Interval	Calculation of sign of $f''(x)$	Result	Conclusion
$(-\infty, -2)$	$\dfrac{+}{(-)^3(-)^3}$	$+$; concave upward	
			No point of inflection at $x = -2$, since -2 not in domain
$(-2, 2)$	$\dfrac{+}{(+)^3(-)^3}$	$-$; concave downward	
			No point of inflection at $x = 2$, since 2 not in domain
$(2, \infty)$	$\dfrac{+}{(+)^3(+)^3}$	$+$; concave upward	

APPLICATIONS

Example 4

In Example 6 of Section 10.4 we defined the average cost function f for an item by

$$f(x) = \frac{C(x)}{x},$$

where C is the cost function for the item. If C is a polynomial function, then f is a rational function. If the fixed cost $C(0) \neq 0$, the line $x = 0$ is a vertical asymptote of f.

Suppose, as an example, that the cost function for an item is given by

$$C(x) = x^2 + 10x + 36, \qquad 0 \leq x \leq 12.$$

The equation of the average cost function is

$$y = f(x) = \frac{C(x)}{x} = \frac{x^2 + 10x + 36}{x}, \qquad 0 < x \leq 12.$$

Let us sketch the graph of f after some preliminary analysis. The function f has $x = 0$ as a vertical asymptote, which is typical for an average cost function, as we suggested above. There are no horizontal asymptotes, since the domain is $(0, 12]$; it must be possible to let $x \to \infty$ or $x \to -\infty$ before there is a chance of a horizontal asymptote. It can be shown by the quotient rule that

$$f'(x) = \frac{x^2 - 36}{x^2} = \frac{(x+6)(x-6)}{x^2}.$$

The only critical number in the domain of f is $x = 6$. It is easily seen that f is decreasing for $0 < x < 6$ and increasing for $6 < x < 12$, which means

that by Theorem 3 there is a minimum point at $x = 6$. The fact that the average cost is minimized at $x = 6$ is probably the most significant feature in this example. The quotient rule can be applied again to yield

$$f''(x) = \frac{72}{x^3}.$$

Since $f''(x) > 0$ for $0 < x < 12$, f is concave upward on the interval $(0, 12)$. Also note that $f''(6) > 0$, which shows by Theorem 6 that f has a minimum point at $x = 6$. Complete information is given in Table 11.20, and the graph of the average cost function is shown in Figure 11.26. ∎

Table 11.20. INFORMATION FOR

$$y = \frac{x^2 + 10x + 36}{x}, \quad 0 < x \le 12.$$

Domain: $\{x : 0 < x \le 12\}$. Range: $\{y : y \ge 22\}$.
Intercepts: x-intercepts—none; y-intercepts—none.
Not symmetric to x-axis, y-axis, or origin.
Vertical asymptotes: $x = 0$. Horizontal asymptotes: none.
Increasing: $6 < x < 12$. Decreasing: $0 < x < 6$.
Minimum point: $(6, 22)$. Endpoint maximum: $(12, 25)$.
Concave upward: $0 < x < 12$.
Points of inflection: none.
Other points: $(1, 47)$, $(2, 30)$, $(3, 25)$, $(9, 23)$.
Slope at $(2, 30)$: -8.

EXERCISES

Suggested minimum assignment: Exercises 1(b) (e) (g) (i), 2(c), 3(a), 4(c), and 4(g).

1. Determine whether the graph of each of the following equations is symmetric to (1) the x-axis, (2) the y-axis, and (3) the origin.
   (a) $y = \dfrac{x^2 - 4}{x^2 - 9}$.
   (b) $y^2 = \dfrac{x^2 - 4}{x^2 - 9}$.
   (c) $x^2 - y^2 = 16$.
   (d) $y = x^6 + x^4 + x^2$.
   (e) $y = x^3$.
   (f) $x^2 + 4y^2 = 5$.
   (g) $y = |x|$.
   (h) $xy = 8$.
   (i) $x^3 = y^2$.
   (j) $y = x^2 + 3x$.

2. Find the equations of all vertical asymptotes.
   (a) $y = \dfrac{1}{x^2 - 4}$.
   (b) $y = \dfrac{x - 2}{x^2 - 4}$.
   (c) $y = \dfrac{x + 1}{x^2 - 3x}$.
   (d) $y = \dfrac{x^2 - 1}{x^2 + 1}$.

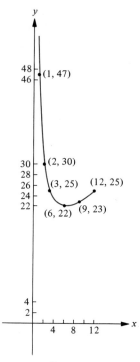

Figure 11.26. $y = \dfrac{x^2 + 10x + 36}{x}$, $0 < x \leq 12$.

3. Find the equations of all horizontal asymptotes.

   (a) $y = \dfrac{2x^2}{3x^2 - 12}$.   (b) $y = \dfrac{x}{5x - 10}$.

   (c) $y = \dfrac{2x}{x^2 - 1}$.   (d) $y = \dfrac{x^2 + 4}{x - 1}$.

4. Construct a table similar to Table 11.17 for the graph of each of the following equations. Then make a sketch.

   (a) $y = \dfrac{3}{x - 1}$.   (b) $y = \dfrac{x - 2}{x + 2}$.

   (c) $y = \dfrac{5}{x^2 + 3}$.   (d) $y = \dfrac{5}{x^2 - 1}$.

   (e) $y = \dfrac{4x}{x^2 + 9}$.   (f) $y = \dfrac{4x}{x^2 - 9}$.

   (g) $y = \dfrac{2x^2}{x^2 - 4}$.   (h) $y = \dfrac{x^2 - 9}{x^2 - 4}$.

5. Sketch the function in Example 3 of Section 11.3.
6. Suppose that the concentration y of a drug in the blood after x hours is given by

Chap. 11] Maximum and Minimum Problems

$$y = f(x) = \frac{x}{20x^2 + 50x + 80}, \qquad x \geq 0.$$

Sketch the graph of f. (See Example 8 of Section 9.4 and Example 4 of Section 11.2.)

7. Suppose that the equation of a demand curve is

$$y = f(x) = \frac{100}{x + 10}, \qquad 0 \leq x \leq 40.$$

Sketch the graph of f.

8. Suppose that the equation of a supply curve is

$$y = f(x) = \frac{2x + 400}{x + 800}, \qquad 0 \leq x \leq 1200.$$

Sketch the graph of f.

9. Sketch the graph of the average cost function given by

$$y = f(x) = \frac{C(x)}{x} = \frac{\frac{1}{10}x^2 + 7x + 10}{x}, \qquad 0 < x \leq 30.$$

(See Exercise 19 of Section 11.4 and Example 4 of this section.)

11.7 Mean Value Theorem for Derivatives (Optional)

Earlier in this chapter (Theorem 1) we stated the fact that if $f'(x) > 0$ at each point of an open interval (a, b), then f is increasing on (a, b). This result will be proved in this section. In order to accomplish the proof of Theorem 1, however, we need the **mean value theorem for derivatives** that will be stated in Theorem 8. The mean value theorem for derivatives is also needed to prove Theorem 1 of Chapter 12. This section will be more theoretical in nature than other sections in this text; if one is willing to accept without proof the results referred to above, then this section can be omitted. *The validity of the results of this section is not restricted to rational functions.*

In order to prove the mean value theorem for derivatives, we shall use Theorem 7, which is known as **Rolle's theorem**. The geometric interpretation in Figure 11.27 will aid in understanding Theorem 7 and its proof.

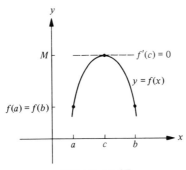

Figure 11.27

▶ **Theorem 7**

If a function f is continuous on $[a, b]$, differentiable on (a, b), and if $f(a) = f(b)$, then there exists at least one number c, where $a < c < b$, such that $f'(c) = 0$.

Outline of Proof: If f is a constant function on $[a, b]$, then $f'(c) = 0$ for any number c in (a, b), and the theorem is proved. If, however, f is not a constant function on $[a, b]$, then, by the statement of Exercise 13(a) of Section 9.4, $f(x)$ must take on a least value m and a greatest value M on $[a, b]$. Either $M > f(a)$ or $m < f(a)$. Suppose that f attains the maximum value M at $x = c$, where $a < c < b$. Since f is differentiable on (a, b), $f'(c)$ exists. If f is a rational function, then $f'(c) = 0$ by Theorem 2 of Section 11.3. The proof of Theorem 2 is also valid for functions other than rational functions if $f'(c)$ is known to exist. Hence $f'(c) = 0$ is valid whether or not f is a rational function. ◀

If f is not a constant function on $[a, b]$, then under the hypotheses of Rolle's theorem there must be a maximum or minimum point between a and b at which the slope is zero (see Figure 11.27).

A geometric interpretation of the next theorem, known as the **mean value theorem for derivatives,** can be obtained from Figure 11.28. In that figure we notice that the secant line (see Section 10.1) between $(a, f(a))$ and $(b, f(b))$ is parallel to the tangent line at $x = c$. The slope of the secant line is $[f(b) - f(a)]/(b - a)$, and the slope of the tangent line is $f'(c)$. It seems evident geometrically that if the function f satisfies certain conditions, then there is at least one point c between a and b at which the slope of the tangent line is equal to the slope $[f(b) - f(a)]/(b - a)$ of the secant line.

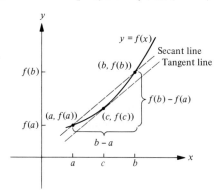

Figure 11.28

▶ **Theorem 8**

If a function f is continuous on $[a, b]$ and differentiable on (a, b), then there exists at least one number c, where $a < c < b$, such that

$$\frac{f(b) - f(a)}{b - a} = f'(c).$$

Outline of Proof: Create a function F with domain $[a, b]$ such that

$$F(x) = f(x) - f(a) - \frac{f(b) - f(a)}{b - a}(x - a).$$

It can be shown that F satisfies the hypotheses of Rolle's theorem (Theorem 7). Therefore, there exists a number c, where $a < c < b$, such that $F'(c) = 0$. Since we can calculate

$$F'(x) = f'(x) - \frac{f(b) - f(a)}{b - a},$$

the equation $F'(c) = 0$ becomes

$$f'(c) - \frac{f(b) - f(a)}{b - a} = 0$$

or

$$\frac{f(b) - f(a)}{b - a} = f'(c). \blacktriangleleft$$

Example 1
If $f(x) = x^3 - x + 2$, $a = 0$, and $b = 3$, find one number that can serve as c in Theorem 8.

Because f is a polynomial function, the hypotheses of Theorem 8 are satisfied. Since

$$f(b) = f(3) = 27 - 3 + 2 = 26$$

and

$$f(a) = f(0) = 0 - 0 + 2 = 2,$$

then

$$\frac{f(b) - f(a)}{b - a} = \frac{26 - 2}{3 - 0} = 8.$$

Also $f'(x) = 3x^2 - 1$, and therefore

$$f'(c) = 3c^2 - 1.$$

Theorem 8 guarantees that the equation $f'(c) = [f(b) - f(a)]/(b - a)$, which in this problem is

$$3c^2 - 1 = 8,$$

is satisfied by a number c in $(0, 3)$. When we solve the last equation for c, we find that this number is

$$c = \sqrt{3}. \blacksquare$$

Theorem 1 of Section 11.2 is restated below. Its proof can be given now that the mean value theorem for derivatives is available.

Sec. 11.7] Mean Value Theorem for Derivatives

▶ **Theorem 1**
  (i) If $f'(x) > 0$ at each point of (a, b), then f is increasing on (a, b).
  (ii) If $f'(x) < 0$ at each point of (a, b), then f is decreasing on (a, b).

Proof: For part (i), assume that $f'(x) > 0$ at each point of (a, b). Let x_1 and x_2 be arbitrary numbers such that $a < x_1 < x_2 < b$. According to Definition 1 of Section 11.2, we need to show that $f(x_1) < f(x_2)$. Since f is differentiable on (a, b), f is continuous on (a, b). Therefore, the hypotheses of the mean value theorem for derivatives are satisfied on $[x_1, x_2]$, and hence by Theorem 8 there exists a number c, where $x_1 < c < x_2$, such that

$$\frac{f(x_2) - f(x_1)}{x_2 - x_1} = f'(c).$$

Since $f'(c)$ is positive by hypothesis and $x_2 - x_1 > 0$, it follows that

$$f(x_2) - f(x_1) > 0,$$

and therefore

$$f(x_1) < f(x_2).$$

The proof of part (ii) requires only minor modifications of the proof above and is left as Exercise 5. ◀

Theorem 9 is another consequence of Theorem 8. Theorem 9 will be used to prove Theorem 10, which in turn leads immediately to the first theorem of Chapter 12.

▶ **Theorem 9**
If $f'(x) = 0$ for all values of x in (a, b), then f is a constant function on (a, b).

Proof: Let x_1 and x_2 be arbitrary numbers such that $a < x_1 < x_2 < b$. To show that f is a constant function on (a, b), we must prove that $f(x_1) = f(x_2)$. Theorem 8 applies on $[x_1, x_2]$; therefore, there exists a number c, where $x_1 < c < x_2$, such that

$$\frac{f(x_2) - f(x_1)}{x_2 - x_1} = f'(c).$$

Since $f'(c) = 0$ by hypothesis and $x_1 \neq x_2$, it follows that $f(x_1) = f(x_2)$. ◀

▶ **Theorem 10**
If g and h are differentiable functions that have equal derivatives on (a, b), then g and h differ by a constant on (a, b).

Proof: We are given that

$$\frac{d}{dx}[g(x)] = \frac{d}{dx}[h(x)] \text{ on } (a, b).$$

Therefore,
$$\frac{d}{dx}[g(x) - h(x)] = 0 \text{ on } (a, b).$$
By Theorem 9,
$$g(x) - h(x) = C \text{ on } (a, b),$$
where C is a constant. ◀

If continuity on $[a, b]$ is assumed for the functions in Theorems 1, 9, and 10, then the conclusions of those theorems are valid on $[a, b]$ instead of (a, b).

APPLICATIONS

The primary applications of the material in this section are indirect and numerous. The theorems stated and proved here occupy a prominent position in the theory of calculus; hence many applications that follow directly from other sections can be considered as rooted in the results of this section. One direct application, however, is given in the following example.

Example 2

Suppose that a cost function C satisfies the hypotheses of the mean value theorem for derivatives (Theorem 8) on the interval $[0, 100]$. Also assume that the fixed cost is zero—in other words, $C(0) = 0$. From Theorem 8 it can be concluded that there is at least one number c, where $0 < c < 100$, such that

$$C'(c) = \frac{C(100) - C(0)}{100 - 0} = \frac{C(100)}{100}. \tag{11.1}$$

If 100 items are produced, then $C(100)/100$ is called the average cost per item (see Example 6 of Section 10.4). Therefore, (11.1) may be interpreted as meaning that there is at least one level of production between 0 and 100 at which the marginal cost equals the average cost of producing 100 items. The result above might not hold if $C(0) > 0$ [instead of $C(0) = 0$]; in that case the marginal cost might always remain less than $C(100)/100$. ∎

EXERCISES

Suggested minimum assignment: Exercises 1(c), 3(b), 4(a), and 7.

1. By finding the x-coordinates of the maximum and minimum points of f, find all numbers that can serve as c in Theorem 7. [Observe that $f(a) = f(b)$.]
   (a) $f(x) = x^2 - 4$, $a = -2$, $b = 2$.
   (b) $f(x) = x^3 + x^2$, $a = -1$, $b = 0$.
   (c) $f(x) = \frac{x^4}{4} - 2x^2$, $a = -3$, $b = 3$.

2. In each of the following there is no number c, where $a < c < b$, such that $f'(c) = 0$. Explain in each case why Theorem 7 is not contradicted.
   (a) $f(x) = x^2 + 3$, $a = 1$, $b = 2$.
   (b) $f(x) = |x|$, $a = -1$, $b = 1$.
3. Find all numbers that can serve as c in Theorem 8.
   (a) $f(x) = x^2 + 5x + 2$, $a = 0$, $b = 2$.
   (b) $f(x) = x^3 - x$, $a = -1$, $b = 2$.
4. Find the x-coordinates of all points between a and b at which the tangent line is parallel to the secant line joining $(a, f(a))$ and $(b, f(b))$. Make a sketch.
   (a) $y = f(x) = (x - 1)^2$, $a = 0$, $b = 3$.
   (b) $y = f(x) = 2x^3 - 6x$, $a = -2$, $b = 2$.
5. Prove Theorem 1(ii).
6. If $f'(x) = 0$ on the interval $(2, 7)$, what can be said about f?
7. Suppose that F is a function with the property that $F'(x) = f(x) = 2x$ on an interval (a, b). Let G be any function defined on (a, b) for which $G'(x) = 2x$. Show, using Theorem 10, that $G(x)$ is of the form $F(x) + C$ on (a, b), where C is constant. (This result illustrates Theorem 1 of Chapter 12.) What is C if $F(x) = x^2 + 3$ and $G(x) = x^2 + 7$?
8. Suppose that a cost function C is given by $C(x) = 200x - x^2$, provided that $0 \le x \le 100$. Find a value of x between 0 and 100 at which the marginal cost equals the average cost of producing 100 items. (See Example 2.)

NEW VOCABULARY

finite interval 11.1
infinite interval 11.1
sign change of a factor 11.1
dividing point 11.1
demand curve 11.2
supply curve 11.2
function increasing (decreasing) on an interval 11.2
maxima, minima 11.3
extreme point 11.3
(relative) maximum point 11.3
(relative) minimum point 11.3
endpoint (relative) maximum 11.3
endpoint (relative) minimum 11.3
critical number 11.3

critical point 11.3
first derivative test 11.3
concave upward (downward) on an interval 11.5
concavity 11.5
point of inflection 11.5
second derivative test 11.5
symmetric to the x-axis 11.6
symmetric to the y-axis 11.6
symmetric to the origin 11.6
vertical asymptote 11.6
horizontal asymptote 11.6
Rolle's theorem 11.7
mean value theorem for derivatives 11.7

12 Antiderivatives and Mathematical Models

As stated earlier, the study of calculus is ordinarily divided into two parts—the differential calculus and the integral calculus. The last two chapters have been concerned with the differential calculus. The purpose of Chapters 12 and 13 is to introduce the other part—the integral calculus. In the differential calculus, problems were encountered where functions were known and instantaneous rates of change of the functions were sought. Illustrations in this chapter, however, will show that in many applied situations the inverse problem is encountered; the instantaneous rates of change are known and functions relating the variables are sought. The operation that determines a function f whose derivative f' is known is called integration. Many historians credit Isaac Barrow (1630–1677), Newton's teacher, with the exploitation of the inverse relationship between the operations of differentiation and integration.

The last two sections of the chapter are devoted to a discussion of the technique of constructing mathematical models to represent real-world phenomena. In particular, antiderivatives are used to determine theoretic results from a model consisting of equations involving derivatives.

Prerequisites: 1.1, 1.2, 3.1–3.9, Chapter 9, 10.1–10.6, 11.1, 11.2.
Suggested Sections for Moderate Emphasis: 12.1, 12.2.
Suggested Sections for Minimum Emphasis: 12.1, 12.2.

12.1 Introduction

Suppose that we are given the function F defined by

$$y = F(x) = \tfrac{1}{3}x^3,$$

where the domain of F is the set of all real numbers. In Chapter 10 we learned that the derivative of F is given by

$$y' = F'(x) = x^2,$$

and that the differential dy is given by

$$dy = F'(x)\,dx = x^2\,dx.$$

Observe that functions that differ from F by only a constant have the same

derivative and differential that F has; for example, if $y = \frac{1}{3}x^3 + C$, where C is any real number, then $y' = x^2$ as before and $dy = x^2\, dx$ as before.

In this section we shall be concerned with the inverse problem in which either $F'(x)$ or $F'(x)\, dx$ is known, and the object is to find $F(x)$. If we use our experience with differentiation and work backward, it may not be too difficult to find by trial and error that if $F'(x) = x^2$, then $F(x) = \frac{1}{3}x^3 + C$, where C is any real number. If $F'(x) = f(x) = x^2$ and $F(x) = \frac{1}{3}x^3 + 5$, then F is called an **antiderivative** of f according to Definition 1.

▶ **Definition 1**
If F is a function whose derivative is the function f, then F is called an **antiderivative** of f.

It is understood in the definition above that F is differentiable and that $F'(x) = f(x)$ holds for all x in a suitable interval of the real numbers, although this interval will not be specifically stated in each problem or in the next theorem.

▶ **Theorem 1**
If $F'(x) = f(x)$, then every antiderivative of f is of the form $F(x) + C$, where C is a real number.

A hint for the proof of Theorem 1 is given in Exercise 26. ◀

Example 1
Suppose that the slope of the tangent line to a curve at (x, y) is given by $y' = F'(x) = x^2$. The equation of the curve must be of the form $y = F(x) = \frac{1}{3}x^3 + C$. After one antiderivative F such as the one given by $F(x) = \frac{1}{3}x^3$ is found, we know by Theorem 1 that no further search for antiderivatives is necessary because *every* possible antiderivative is given by an expression of the form $\frac{1}{3}x^3 + C$. If marginal revenue is given by $R'(x) = x^2$, then the revenue function is given by an expression of the form $R(x) = \frac{1}{3}x^3 + C$; if more information is known, such as $R(0) = 0$, then $0 = 0 + C$ and C can be determined. (Discussion of the determination of C in this manner will be postponed until Section 12.3.) ■

The notation most commonly used to indicate that *every* antiderivative of f is an expression of the form $F(x) + C$ is

$$\int f(x)\, dx = F(x) + C.$$

The symbol $\int$ is called an **integral sign** and when placed before $f(x)\, dx$, it indicates that *the most general antiderivative of f is to be found;* the process of finding the most general antiderivative of f is often called **integration**. In the notation above, $f(x)$ is called the **integrand** and C is called the **constant of integration**. If $\int x^2\, dx$ is written, we say that we are required to **integrate** the given function, and we complete this operation by writing

$$\int x^2\, dx = \tfrac{1}{3}x^3 + C;$$

Chap. 12] Antiderivatives and Mathematical Models

the result of integration can be checked by recalling that this process is the inverse of differentiation so that either

$$D_x(\tfrac{1}{3}x^3 + C) = x^2$$

or

$$d(\tfrac{1}{3}x^3 + C) = x^2 \, dx$$

is a foolproof check. This type of check will aid in proving the various parts of Theorem 2, in which we state some formulas that can be used for the integration of certain functions.

▶ **Theorem 2**

(i) $\int dx = x + C.$

(ii) $\int kf(x) \, dx = k \int f(x) \, dx,$ where k is a real number.

(iii) $\int [f(x) + g(x)] \, dx = \int f(x) \, dx + \int g(x) \, dx.$

(iv) $\int x^n \, dx = \dfrac{x^{n+1}}{n+1} + C,$ where n is a real number and $n \neq -1.$

The proof of Theorem 2 is left as Exercises 27 and 28. ◀

Example 2

$$\int \frac{dx}{x^2} = \int x^{-2} \, dx = \frac{x^{-1}}{-1} + C = -\frac{1}{x} + C.$$

Notice that the integrand $1/x^2$ was first rewritten in the form x^n (with $n = -2$) so that Theorem 2(iv) could be applied. The answer is checked as follows:

$$D_x\left(-\frac{1}{x} + C\right) = -D_x x^{-1} + 0 = -(-1)x^{-2} = \frac{1}{x^2},$$

which is the original integrand. A check of the answer in this manner is optional. ∎

Example 3

$$\int (2x^3 + x^5) \, dx = 2 \int x^3 \, dx + \int x^5 \, dx \quad \text{by Theorem 2(ii) and (iii)}$$

$$= 2\left(\frac{x^4}{4} + C_1\right) + \frac{x^6}{6} + C_2 \quad \text{by Theorem 2(iv)}$$

$$= \frac{x^4}{2} + \frac{x^6}{6} + (2C_1 + C_2)$$

$$= \frac{x^4}{2} + \frac{x^6}{6} + C.$$

With practice, the first step above can be omitted; also the use of constants C_1 and C_2 is unnecessary as they can be combined, so the addition of one constant C at the end is all that is required. In the future we shall write

$$\int (2x^3 + x^5)\, dx = \frac{2x^4}{4} + \frac{x^6}{6} + C$$

$$= \frac{x^4}{2} + \frac{x^6}{6} + C. \blacksquare$$

Example 4

$$\int (2x + 1)(x - 2)\, dx = \int (2x^2 - 3x - 2)\, dx$$

$$= \frac{2x^3}{3} - \frac{3x^2}{2} - 2x + C. \blacksquare$$

Theorem 2(*iii*) is easily extended to handle addition (or subtraction) of three or more terms, as we assumed above. Observe that part (*i*) was used in integrating the third term in Example 4. It was necessary to multiply $(2x + 1)$ and $(x - 2)$ together before integrating because in general it can be shown that

$$\int f(x)g(x)\, dx \neq \left(\int f(x)\, dx \right)\left(\int g(x)\, dx \right).$$

Example 5

$$\int \frac{\sqrt[3]{t} + \sqrt{t^5} + t^{-2/3}}{t}\, dt = \int \left(\frac{t^{1/3}}{t} + \frac{t^{5/2}}{t} + \frac{t^{-2/3}}{t} \right) dt$$

$$= \int (t^{-2/3} + t^{3/2} + t^{-5/3})\, dt$$

$$= \frac{t^{1/3}}{\frac{1}{3}} + \frac{t^{5/2}}{\frac{5}{2}} + \frac{t^{-2/3}}{-\frac{2}{3}} + C$$

$$= 3t^{1/3} + \frac{2}{5} t^{5/2} - \frac{3}{2} t^{-2/3} + C. \blacksquare$$

APPLICATIONS

The material in Sections 12.1 and 12.2 will be applied in Section 12.3 and in Chapters 13 and 14.

EXERCISES

Suggested minimum assignment: Exercises 3, 8, 9, 11, 16, 18, 21, and 23.

In Exercises 1–10 perform the indicated integration.

1. $\int x^3\, dx.$

2. $\int x^{-3}\, dx.$

3. $\int (x^2 - 3x)\, dx.$ 4. $\int (1 + 2x^4)\, dx.$

5. $\int \frac{1}{t^2}\, dt.$ 6. $\int \sqrt[3]{t}\, dt.$

7. $\int (x^\pi - x^2 + 3x + 8)\, dx.$ 8. $\int \frac{1 + x^{\sqrt{3}}}{2x^2}\, dx.$

9. $\int (x - 1)(2x - 9)\, dx.$ 10. $\int x(1 + \sqrt{x})\, dx.$

In Exercises 11–14 perform the indicated integration and check your answer.

11. $\int (5x^6 - 2x^2)\, dx.$ 12. $\int \sqrt{x}(1 + 2x^2)\, dx.$

13. $\int \frac{(x^2 + 1)^2}{x^2}\, dx.$ 14. $\int (5x - 1)(x)(7x - 3)\, dx.$

15. If $g(x) = (x^5/15) + 3$ and $f(x) = x^4/3$, is g an antiderivative of f? Why?
16. If $F(x) = x^3$ and $f(x) = 3x^2 + 1$, is F an antiderivative of f? Why?
17. If $F(x) = 4x^3$ and $f(x) = x^4$, is F an antiderivative of f? Why?

In Exercises 18–25 perform the indicated integration.

18. $\int x\sqrt[3]{x}\, dx.$ 19. $\int \frac{\sqrt{x} + \sqrt[3]{x}}{x^2}\, dx.$

20. $\int (2 + \sqrt{x})^2\, dx.$ 21. $\int \left(2 + \frac{1}{\sqrt{x^3}}\right)^2 dx.$

22. $\int \frac{dx}{x\sqrt{3x}}.$ 23. $\int x^{-5}(2x - 1)^2\, dx.$

24. $\int \sqrt[7]{y^6}\, dy.$ 25. $\int (2y - 3)^3\, dy.$

26. Prove Theorem 1. (*Hint:* Let G be any antiderivative of f. Use Theorem 10 of Section 11.7.)
27. (a) Prove Theorem 2(*i*). (b) Prove Theorem 2(*ii*).
    (*Hint:* Show that the derivative of the right side of each formula is equal to the integrand of the left side.)
28. (a) Prove Theorem 2(*iii*). (b) Prove Theorem 2(*iv*).

12.2 An Important Integration Formula

In Section 10.3 we presented the important power rule for differentiation, and in this section we study the following corresponding **power rule for integration:**

$$\int [f(x)]^n f'(x)\, dx = \frac{[f(x)]^{n+1}}{n + 1} + C, \qquad (12.1)$$

provided that n is a real number and $n \neq -1$. The validity of (12.1) for $n \neq -1$ can be checked by differentiation:

$$D_x \left(\frac{[f(x)]^{n+1}}{n+1} + C \right) = \frac{1}{n+1}(n+1)[f(x)]^{(n+1)-1}f'(x) + 0$$
$$= [f(x)]^n f'(x).$$

Notice that the power rule for differentiation was used in making the above check, and of course we assumed that f was differentiable. If we substitute

$$u = f(x),$$

then

$$du = f'(x)\, dx,$$

and we can state Theorem 3.

▶ **Theorem 3**
If f is differentiable and $u = f(x)$, then

$$\int u^n\, du = \frac{u^{n+1}}{n+1} + C,$$

provided that n is a real number and that $n \neq -1$. ◀

Theorem 3 is the form of the important **power rule for integration** to which we shall refer frequently, but it is a somewhat difficult formula to apply correctly at first, and therefore we illustrate with some examples.

Example 1

$$\int (2x^2 + 1)^8 (4x)\, dx.$$

With $u = 2x^2 + 1$ and $n = 8$, it follows that $du = 4x\, dx$ and that the given problem fits exactly the left side of the formula in Theorem 3. Therefore,

$$\int (2x^2 + 1)^8 (4x\, dx) = \int u^8\, du = \frac{u^9}{9} + C$$
$$= \frac{1}{9}(2x^2 + 1)^9 + C. \blacksquare$$

Example 2

$$\int (2x^2 + 1)^8 x\, dx.$$

This example differs from Example 1 in that the number 4 is missing from the integrand. Since the number 4 is needed as part of the du (see Example

1), we simply insert the needed 4 and counteract this operation by also multiplying by $\frac{1}{4}$ so as not to change the integrand. Hence we have

$$\int (2x^2 + 1)^8 (\tfrac{1}{4})(4x\, dx).$$

By Theorem 2(*ii*) of Section 12.1, the constant $\frac{1}{4}$ can be brought across the integral sign to yield $\frac{1}{4}\int (2x^2 + 1)^8 (4x\, dx)$. Except for the extra $\frac{1}{4}$ the problem is the same as Example 1, so that

$$\int (2x^2 + 1)^8 x\, dx = \frac{1}{4}\int (2x^2 + 1)^8 (4x\, dx)$$

$$= \frac{1}{4}\int u^8\, du = \frac{1}{4}\frac{u^9}{9} + C = \frac{1}{36}(2x^2 + 1)^9 + C. \blacksquare$$

Examples 1 and 2 can be checked by differentiation. Example 2 does not check unless the extra $\frac{1}{4}$ is obtained in front of the integral sign as shown above. The insertion of a nonzero constant to make the exact required du appear, together with the corresponding insertion of the reciprocal of that constant in front of the integral sign, is a legitimate procedure that is frequently needed. The insertion of a *variable* in the integrand and the reciprocal of that *variable* in front of the integral sign, however, is *not* a legitimate procedure—only constants can be brought across the integral sign. For the problem

$$\int (2x^2 + 1)^8\, dx,$$

it might be tempting to put $4x$ in the integrand and $1/4x$ in front of the integral sign, but this leads to the wrong answer—as can be checked by differentiation. In order to solve the integration problem

$$\int (2x^2 + 1)^8\, dx,$$

it is necessary to expand $(2x^2 + 1)^8$ and then integrate each term.

Example 3
Suppose that it is desired to find the equation $y = F(x)$ of a demand curve (see Example 1 of Section 11.2). Assume that it is known that the rate of change of price y with respect to demand x is given by

$$\frac{dy}{dx} = F'(x) = -\frac{1}{(\tfrac{1}{5}x + 1)^2},$$

provided that $0 \le x \le 100$. (Observe that since $dy/dx < 0$ throughout the interval $[0, 100]$, it follows that the price decreases as the demand increases.) Then $F(x)$ can be found by using the power rule for integration as follows:

Sec. 12.2] An Important Integration Formula

$$F(x) = \int F'(x)\,dx = \int \frac{(-1)\,dx}{(\tfrac{1}{5}x+1)^2}$$

$$= (-1)\int \left(\tfrac{1}{5}x+1\right)^{-2} dx$$

$$= (-1)(5)\int \left(\tfrac{1}{5}x+1\right)^{-2}\left(\tfrac{1}{5}dx\right)$$

$$= -5\int u^{-2}\,du$$

$$= -5\,\frac{u^{-1}}{-1} + C$$

$$= \frac{5}{u} + C$$

$$= \frac{5}{\tfrac{1}{5}x+1} + C.$$

Hence the equation of the demand curve (for $0 \le x \le 100$) is

$$y = \frac{5}{\tfrac{1}{5}x+1} + C.$$

More information is needed in order to find a specific value for the constant C; for example, if it were known that $y = 2$ when $x = 10$, then it could be determined that $C = \tfrac{1}{3}$. ∎

Example 4

$$\int \frac{(4x+80)\,dx}{\sqrt{x^2+40x+1}} = \int (x^2+40x+1)^{-1/2}(4x+80)\,dx$$

$$= 2\int (x^2+40x+1)^{-1/2}(2x+40)\,dx$$

$$= 2\int u^{-1/2}\,du = 2\,\frac{u^{1/2}}{\tfrac{1}{2}} + C$$

$$= 4\sqrt{u} + C = 4\sqrt{x^2+40x+1} + C. \quad \blacksquare$$

Notice that if $u = x$, then $du = dx$, and the power rule for integration,

$$\int u^n\,du = \frac{u^{n+1}}{n+1} + C, \quad \text{provided that } n \ne -1,$$

reduces to

$$\int x^n\,dx = \frac{x^{n+1}}{n+1} + C,$$

which is Theorem 2(iv) of Section 12.1.

EXERCISES

Suggested minimum assignment: Exercises 1, 5, 11, 12, 14, 18, 20, and 21.

Perform the indicated integration in Exercises 1–16.

1. $\int (3x^2 + 1)^7 (6x)\, dx.$

2. $\int x(3x^2 + 1)^7\, dx.$

3. $\int \sqrt{4x + 3}\, dx.$

4. $\int \dfrac{dx}{3x^5}.$

5. $\int x\sqrt{x^2 + 3}\, dx.$

6. $\int \dfrac{t\, dt}{(5t^2 + 1)^2}.$

7. $\int \dfrac{x\, dx}{\sqrt{2x^2 + 5}}.$

8. $\int \dfrac{(x + 1)\, dx}{\sqrt{100x^2 + 200x + 77}}.$

9. $\int \dfrac{(1 + \sqrt{x})^5}{\sqrt{x}}\, dx.$

10. $\int \sqrt[3]{8y + 27}\, dy.$

11. $\int (x^2 + \sqrt{3x})\, dx.$

12. $\int (2x + 1)^{10}\, dx.$

13. $\int x\sqrt{x}\, dx.$

14. $\int (7x + 3)^{8/3}\, dx.$

15. $\int \dfrac{dx}{(3x + 1)^4}.$

16. $\int \sqrt{x^3 + 7x - 3}\,(3x^2 + 7)\, dx.$

17. Check your answer to Exercise 7.
18. Check your answer to Exercise 14.

In Exercises 19 and 20 select the only one of the three parts that can be done by use of Theorem 3. Then work out the one that you select.

19. (a) $\int x^2 \sqrt{9 + x^2}\, dx.$ (b) $\int x\sqrt{9 + x^2}\, dx.$

    (c) $\int \sqrt{9 + x^2}\, dx.$

20. (a) $\int \dfrac{dx}{\sqrt{x^2 + 4x + 1}}.$ (b) $\int \dfrac{(x + 1)\, dx}{\sqrt{x^2 + 4x + 1}}.$

    (c) $\int \dfrac{(x + 2)\, dx}{\sqrt{x^2 + 4x + 1}}.$

21. Find the equation $y = F(x)$ of a demand curve if it is known that the rate of change of price y with respect to demand x is given by $dy/dx = F'(x) = -100/(3x + 1)^3$, provided that $0 \le x \le 5$.

22. Find the equation $y = C(x)$ of a cost function if the marginal cost $C'(x)$ is given by $C'(x) = (1 + x/3)^3$, provided that $0 \le x \le 6$.

Sec. 12.2] An Important Integration Formula

12.3 Differential Equations†

In this section we shall give the reader additional practice in using the integration techniques studied in Sections 12.1 and 12.2. Determination of the constant of integration, provided that certain **initial conditions** are given, will be stressed. **Differential equations,** a subject of importance in many applications of calculus, will be introduced next.

A **differential equation** is an equation that involves derivatives of an unknown function. For example,

$$\frac{d^2y}{dx^2} - \frac{dy}{dx} + 2y = 8x^2 \tag{12.2}$$

is a differential equation in which y is an unknown function of x. Usually, the object is to find a function f, as given by $y = f(x)$, which satisfies the equation; such a function is called a **solution** of the differential equation.

Example 1
Let us demonstrate that the function f given by

$$y = f(x) = 4x^2 + 4x - 2$$

is a solution of (12.2).

$$\frac{dy}{dx} = 8x + 4,$$

$$\frac{d^2y}{dx^2} = 8.$$

Substitution in (12.2) yields

$$8 - (8x + 4) + 2(4x^2 + 4x - 2) = 8x^2,$$

which simplifies to the identity

$$8x^2 = 8x^2. \blacksquare$$

We have verified in Example 1 that f is a solution of the differential equation (12.2), although we do not claim to have explained how f was found in the first place. Instead of saying that "the function f given by $y = f(x) = 4x^2 + 4x - 2$ is a solution of (12.2)," we shall find it convenient to say "$y = 4x^2 + 4x - 2$ is a solution of (12.2)."

The **order** of a differential equation is the order of the highest-ordered derivative appearing in the equation. The order of equation (12.2) is 2

---

† This section is a prerequisite to the discussion of exponential growth and decay in Section 14.5 and the discussion of mathematical modeling in Sections 12.4 and 12.5; if these sections are to be omitted, this section may be considered optional.

because d^2y/dx^2 is the highest-ordered derivative that appears. The equation

$$\frac{dy}{dx} = -\frac{x}{y} \qquad (12.3)$$

has order 1. Equation (12.3) can be written as

$$y\,dy = -x\,dx,$$

in which the x's and y's are placed on opposite sides of the equation; when this can be done, we say that it is possible to **separate the variables.** In the equation $dy/dx = x + y$, it is *not* possible to separate the variables. The only type of first-order differential equation that we shall learn to solve is the type for which it is possible to separate the variables. After separating the variables, the next step is to integrate each side of the equation; we demonstrate this procedure in the next example by solving (12.3).

Example 2
Solve $dy/dx = -x/y$.
Separate the variables to obtain

$$y\,dy = -x\,dx,$$

and integrate as follows:

$$\int y\,dy = \int -x\,dx,$$

$$\frac{y^2}{2} + C_1 = -\frac{x^2}{2} + C_2,$$

$$\frac{x^2}{2} + \frac{y^2}{2} = C_2 - C_1.$$

The last equation can be written

$$x^2 + y^2 = C,$$

where $C = 2(C_2 - C_1)$. Either

$$y = \sqrt{C - x^2}$$

or

$$y = -\sqrt{C - x^2}$$

can be shown to satisfy the original equation, provided that $C > 0$ and $-\sqrt{C} < x < \sqrt{C}$. ∎

In Example 3 enough information is given to determine a specific value for C.

Example 3
Find the solution of
$$\frac{dy}{dx} = x^2$$
which also satisfies the condition that $y = \frac{14}{3}$ when $x = 2$.

We separate the variables to obtain
$$dy = x^2\, dx$$
and integrate to get
$$y = \tfrac{1}{3}x^3 + C.$$
By substituting $y = \frac{14}{3}$ and $x = 2$, C can be determined as follows:
$$\tfrac{14}{3} = \tfrac{1}{3}(2)^3 + C,$$
$$C = 2.$$
The desired solution is
$$y = \tfrac{1}{3}x^3 + 2. \quad \blacksquare$$

A condition such as $y = \frac{14}{3}$ when $x = 2$, in Example 3, is called an **initial condition**. Without the initial condition we stop with the **general solution** $y = \tfrac{1}{3}x^3 + C$; an appropriate initial condition makes it possible to determine a special case $y = \tfrac{1}{3}x^3 + 2$ of the general solution, and such a special case is called a **particular solution**. Geometrically, the general solution

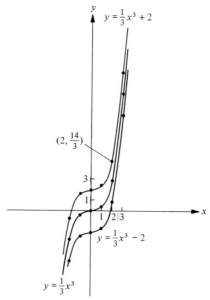

Figure 12.1. *Family of curves* $y = \tfrac{1}{3}x^3 + C$.

$y = \frac{1}{3}x^3 + C$ represents a family of curves (one for each real number C). See Figure 12.1. The particular solution is the one curve (shown in color) that passes through the point $(2, \frac{14}{3})$.

Example 4 illustrates again the use of an initial condition to determine a constant of integration.

Example 4

Suppose that for the production of a certain item, a firm has determined that its marginal cost is given by

$$C'(x) = 50 + 4x - \frac{x^2}{20}, \quad \text{provided that } 0 \le x \le 60.$$

Also it is known that the fixed cost is 100 [this means that $C(0) = 100$]. Find the cost of producing 30 items.

$$C'(x) = \frac{dC}{dx} = 50 + 4x - \frac{x^2}{20},$$

$$dC = \left(50 + 4x - \frac{x^2}{20}\right) dx,$$

$$C(x) = \int \left(50 + 4x - \frac{x^2}{20}\right) dx$$

$$= 50x + 4\frac{x^2}{2} - \frac{1}{20}\frac{x^3}{3} + B,$$

where B is the constant of integration. Since $C(0) = 100$, we can find B as follows:

$$100 = 0 + 0 - 0 + B$$

and

$$B = 100.$$

Hence the cost function is given by

$$C(x) = 50x + 2x^2 - \frac{x^3}{60} + 100.$$

To compute the cost of producing 30 items, let $x = 30$ to obtain

$$C(30) = 50(30) + 2(30)^2 - \frac{(30)^3}{60} + 100 = 2950. \blacksquare$$

The only other type of differential equation that we shall learn to solve is a second-order equation of the form $d^2y/dx^2 = f(x)$. Consider, as an example,

$$\frac{d^2y}{dx^2} = x^5.$$

Sec. 12.3] Differential Equations

This can be written

$$\frac{dy'}{dx} = x^5$$

or

$$dy' = x^5\, dx.$$

Then

$$y' = \frac{x^6}{6} + C_1$$

so that

$$dy = \left(\frac{x^6}{6} + C_1\right) dx.$$

Therefore,

$$y = \frac{x^7}{42} + C_1 x + C_2.$$

The method of solution consists of separating the variables and integrating, and then repeating the process.

The next example involves the same ideas as above, but initial conditions will be given that make possible the determination of both C_1 and C_2. Also, the reader should be alert in Example 5 for the use of the power rule for integration (see Section 12.2). Recall from Chapter 10 that if the location of a particle moving along a line is given by the equation of motion $s = f(t)$, then the velocity v is given by $v = ds/dt$ and the acceleration a is given by $a = dv/dt$. We shall start with the acceleration and work back to the equation of motion.

Example 5

Suppose that the acceleration a of a particle moving along a line is given by

$$a = \sqrt{2t + 1}.$$

Find the equation of motion for the particle if it is known that when $t = 4$ the particle has position $s = \frac{81}{5}$ and velocity $v = 15$.

$$\frac{dv}{dt} = \sqrt{2t + 1},$$

$$dv = \sqrt{2t + 1}\, dt,$$

$$v = \int \sqrt{2t + 1}\, dt = \frac{1}{2}\int (2t + 1)^{1/2} 2\, dt$$

$$= \frac{1}{2}\int u^{1/2}\, du = \frac{1}{2}\frac{u^{3/2}}{\frac{3}{2}} + C_1 = \frac{1}{3}(2t + 1)^{3/2} + C_1.$$

Since $v = 15$ when $t = 4$,

$$15 = \frac{1}{3}(9)^{3/2} + C_1,$$

$$= \frac{1}{3}(27) + C_1,$$

and hence
$$C_1 = 6.$$
Therefore,
$$v = \frac{ds}{dt} = \frac{1}{3}(2t+1)^{3/2} + 6,$$
$$ds = \left[\frac{1}{3}(2t+1)^{3/2} + 6\right]dt,$$
$$s = \frac{1}{3}\frac{1}{2}\int (2t+1)^{3/2}(2\,dt) + \int 6\,dt$$
$$= \frac{1}{6}\frac{(2t+1)^{5/2}}{\frac{5}{2}} + 6t + C_2.$$

Next, substitute $s = \frac{81}{5}$ and $t = 4$ to determine C_2 as follows:
$$\frac{81}{5} = \frac{1}{15}(9)^{5/2} + 24 + C_2,$$
$$\frac{81}{5} = \frac{1}{15}(243) + 24 + C_2,$$
$$C_2 = -24.$$

Therefore, the required equation of motion is
$$s = \frac{1}{15}(2t+1)^{5/2} + 6t - 24,$$
and the position s of the particle can be calculated for a given value of t. ∎

APPLICATIONS

Example 6
A certain publishing company would like to price a new book so as to maximize its annual profit from sales of the book. The company considers it reasonable to charge any price y between $8 and $25. From experience with a similar book it is believed that the annual demand x would be 3900 if $y = \$10$. Also, from a study of the demand curve of the similar book, it is expected that the rate of change dy/dx of price with respect to demand can be approximated by $dy/dx = -y^2/40{,}000$. The revenue from sales of the book will be the price y per book times the demand x. The annual cost is given by $(8100 + x)$. Find the price y that will maximize the annual profit P from sales of the new book.

In the differential equation
$$\frac{dy}{dx} = -\frac{y^2}{40{,}000},$$
the variables can be separated to obtain
$$dx = -40{,}000 y^{-2}\,dy.$$

Integrating both sides, we get

$$\int dx = -40{,}000 \int y^{-2}\, dy,$$

$$x = -40{,}000 \, \frac{y^{-1}}{-1} + C$$

$$= \frac{40{,}000}{y} + C.$$

Since $x = 3900$ if $y = 10$, it follows that $C = -100$ and that the equation of the demand curve is given by

$$x = \frac{40{,}000}{y} - 100.$$

The profit equals the revenue minus the cost, and hence

$$P = yx - (8100 + x)$$

$$= y\left(\frac{40{,}000}{y} - 100\right) - \left[8100 + \left(\frac{40{,}000}{y} - 100\right)\right]$$

$$= 32{,}000 - 100y - \frac{40{,}000}{y}.$$

By the methods of Chapter 11 it can be shown that maximum profit occurs at $y = \$20$. ∎

EXERCISES

Suggested minimum assignment: Exercises 1, 2(b), 3(a), 4, 5, 7, and 9.

1. State the order of each of the following differential equations.

   (a) $\dfrac{dy}{dx} = x^3 + 8x + 3.$ (b) $x\dfrac{d^2y}{dx^2} + \dfrac{dy}{dx} - 8y = 0.$

   (c) $\dfrac{d^6y}{dx^6} + \left(\dfrac{dy}{dx}\right)^7 = 3.$ (d) $\dfrac{d^2y}{dx^2} = x + y.$

2. Solve each of the following differential equations.

   (a) $\dfrac{dy}{dx} = 2x.$ (b) $\dfrac{dy}{dx} = x^2 + 8x + 3.$

   (c) $\dfrac{d^2y}{dx^2} = x^2 + 2x.$ (d) $(y+1)\dfrac{dy}{dx} = \sqrt{x}.$

   (e) $\dfrac{dy}{dx} = \sqrt{xy}.$ (f) $\dfrac{d^2y}{dx^2} = 4x^3.$

3. Find the solution of each of the following differential equations that passes through the point (1, 3).

   (a) $\dfrac{dy}{dx} = 4x.$ (b) $\dfrac{dy}{dx} = 3x^2 + 4.$

4. Sketch at least three members of the family of curves that arises in solving Exercise 3(a), including the one that passes through (1, 3).
5. Find the equation of the curve that satisfies $d^2y/dx^2 = 2x$ if it is known that the curve has slope 2 at the point $(1, 4)$.
6. Find the equation of the curve that satisfies $d^2y/dx^2 = \sqrt{2x + 3}$ if it is known that the curve has slope 4 at the point $(3, 0)$.
7. The velocity v of a particle moving along a line is given by

$$v = \frac{ds}{dt} = t\sqrt[3]{t^2 + 4}.$$

Find the equation of motion for the particle if it is known that $s = 8$ when $t = 2$.
8. The acceleration a of a particle moving along a line is given by

$$a = 6t + 3.$$

Find the equation of motion for the particle if it is known that when $t = 3$, the particle has position $s = 8$ and velocity $v = 20$. Also, find s when $t = 2$.
9. Suppose that marginal cost is given by

$$C'(x) = 1000 - 2x.$$

The fixed cost is 2000. Find $C(x)$. Find $C(10)$.
10. Suppose that marginal revenue is given by

$$R'(x) = \frac{x^2}{10} + \frac{x}{5} + 100,$$

and that $R(0) = 0$. Find $R(x)$.
11. An object is shot vertically upward from the ground. Let s be the height in feet above the ground after t seconds. The acceleration due to gravity is given by $a = -32$ feet per second per second. If $s = 0$ when $t = 0$ and velocity $v = v_0$ when $t = 0$, find a formula that expresses s in terms of t.
12. Suppose that the rate of change of price with respect to weekly demand for an item is given by

$$\frac{dy}{dx} = -\frac{y^3}{1000},$$

provided that $1.5 \leq y \leq 3$. If the weekly demand is $x = 100$ when the price is $y = 2$, find the equation of the demand curve.
13. In Example 6 show that maximum profit occurs at $y = \$20$.

12.4 Mathematical Models (Optional)

In the minds of many readers the word *model* is associated with model airplanes, model cars, dress models, human fashion models, wax models, model buildings, and so on. We can probably agree that *a model is a*

perceived substitute representation of reality. As a verb, *to model* means to construct idealized representations of a real situation. Of course, more than one *substitute representation* of a single real-world situation often can be constructed. Furthermore, each substitute representation is likely to be idealized to the extent that some aspects of the real situation are approximated with considerable accuracy while other aspects are not approximated at all. The next example illustrates these ideas by means of models that are nonmathematical.

Example 1
An architectural firm plans to bid for a contract to design a new civic center in a certain city. The firm prepares two representations of the center; one is a preliminary blueprint and the other is a miniature balsa-wood replica of the projected center. Both are models of the proposed civic center, but the nature of these models is quite different. To the consulting engineer who must advise the city council on the costs of the center, the blueprint model is a sufficient approximation while the balsa-wood model is almost useless. On the other hand, to a group of citizens who will advise on the aesthetic aspects of the projected center, the opposite is true. ∎

We shall be concerned with models that are mathematical in nature. At the very outset it should be emphasized that in mathematical models, as in nonmathematical models, complete realism is *not* expected or required. The degree of realism desired from a mathematical model depends upon several considerations, as illustrated in the next example.

Example 2
The purpose of this example is to illustrate the construction of several mathematical models of a real-world situation. Consider an object falling toward the earth after being dropped from a point above the earth. It is natural to raise certain questions about the motion of such an object; specifically, can we predict the location and velocity of the object after a certain time interval, and can we predict the time it will take to hit the ground? The modeling process is begun by *assuming* that the only force acting on the object is that due to the gravity of the earth. This means that the acceleration $a = -g$ of the object is due only to the earth's gravity. Thus a mathematical model of the vertical motion of the object is simply the differential equation

$$\frac{dv}{dt} = -g,$$

where the positive direction is upward. By using the techniques of this chapter we can find from this equation that the vertical velocity v is given by

$$v = -gt + v_0,$$

where v_0 represents the upward velocity of the object at the instant ($t = 0$) of release. Further integration produces an equation for the vertical position s of the object relative to a point on the earth's surface directly below the point of release,

$$s = s_0 + v_0 t - \tfrac{1}{2} g t^2,$$

where s_0 represents the height of the point of release. Approximate answers to the questions raised earlier in the example can be found from these derived equations. The suitability of these approximate answers depends upon several factors; among these factors is the adequacy of the assumption that the only force acting on the object was that due to gravity. This may be a realistic assumption for a small steel ball dropped from 10 feet, where the effect of air friction is small, but the assumption would be quite unrealistic for a parachute drop from 10,000 feet. In the latter case the basic assumption of the model could be modified to assume no wind and that the only forces acting on the object will be those due to gravity *and* air friction. This, of course, will lead to a new and more complex model, namely,

$$\frac{dv}{dt} = -g - kv,$$

where k is a constant depending upon the nature of the falling object. Again the techniques of integration (which are more complex in this model and will be explained in Chapter 14) produce equations for velocity and position, namely,

$$v = -\frac{g}{k}(1 - e^{-kt}) \quad \text{and} \quad s = s_0 + \frac{g}{k^2}(-kt - e^{-kt} + 1),$$

under the further assumption that the initial velocity is zero, and where e is a number † that is approximately 2.718. Again, approximate answers to the questions raised at the outset of the example can be obtained from these two equations. For objects released near the surface of the earth, the results obtained by the second model are likely to be superior to those obtained from the first model, but then the second model was also more complex than the first. Next, consider an object dropped from a spacecraft in space partway to the moon; neither of the first two models would be satisfactory and a third model would have to be constructed. The third model would allow for the gravitational force of other objects in space, the rotation of the earth, variable atmospheric resistance, and other factors. Obviously, such a model would be much more complex than either of the first two models. ∎

As illustrated in Example 2, an increase in the reality of the model often adds to the model's complexity. In other words, to achieve a greater degree of realism, one ordinarily pays a price in the complexity of the model; the question often becomes one of whether the greater realism is worth the

† The number e will be discussed in Section 14.1.

Sec. 12.4] Mathematical Models

price. Ordinarily, the extra realism gained from the second model of Example 2, when applied to a steel ball dropped from 10 feet, is not worth the added complexity. On the other hand, the use of the simple first model of Example 2, when applied to a parachutist dropped from 10,000 feet, would be quite unrealistic. Thus in mathematical modeling, one frequently engages in a trade-off between the reality of the model, on the one hand, and the complexity of the model, on the other hand.

EXERCISES

1. Use the techniques of this chapter to verify that
$$v = -gt + v_0$$
and
$$s = -\tfrac{1}{2}gt^2 + v_0 t + s_0$$
follow from the model presented in Example 2, namely,
$$\frac{dv}{dt} = -g.$$

2. In Exercise 1 let $g = 32$ feet per second per second, $v_0 = 128$ feet per second, and $s_0 = 768$ feet. Use the equations in Exercise 1 to predict the velocity v and the height s above the ground when $t = 6$ seconds. Also predict the number of seconds from the time the object is released ($t = 0$) until it hits the ground.

3. Consider the following nonmathematical modeling problem. A small planned community has lost many of its beautiful old trees. A replanting plan is needed to arrange for adding some new trees to the community. The landscape architect must consider different types of trees and various possible locations for the trees. He also has to consider the amount of money available for purchasing trees, which is not known exactly. He also has to allow for planting the trees over either a 1- or a 2-year period. Discuss some types of models that the landscape architect might present to the people of the community.

4. Read the applications of previous chapters and identify at least four mathematical models that were used.

12.5 Mathematical Models (Continued) (Optional)

A technique of constructing mathematical models is diagramed in Figure 12.2. It begins with the identification and statement of the real situation that is to be modeled. Then certain idealizing assumptions are made that permit the construction of a model. The model is subjected to appropriate mathematical techniques, and the resulting theoretical results are compared with the experimental or observed results of the real-world situation. If the comparison is not satisfactory, then the model should be refined via more

realistic assumptions, or the experimental or observed results should be examined for error.

Many early attempts at mathematical modeling are thwarted because of an immature insistence on complete realism from the initial model. Real-life situations are often so complex that it is impossible to formulate a satisfactory model on the first try; more often it is necessary to construct a succession of models (as indicated in Figure 12.2) before satisfactory theo-

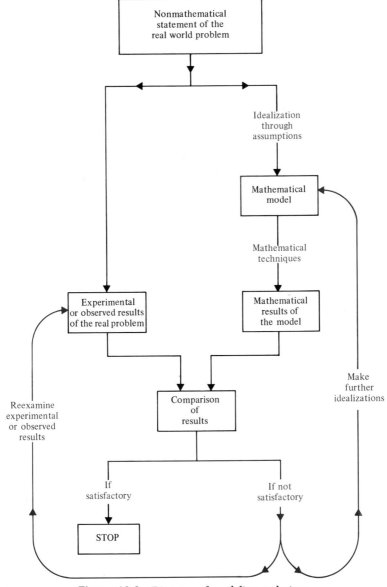

Figure 12.2. *Diagram of modeling technique.*

Sec. 12.5] Mathematical Models

retical results are obtained. Moreover, the final model, although satisfactory, is likely to be less than completely realistic.

The attempts made in obtaining a successful model may initiate new theories and understandings of the real situation that otherwise might not have been obvious. In some cases the deductions obtained from the model lead to contradictions of previously held theories. These contradictions often lead to improved theories about reality.

The importance of the idealizing assumptions should be emphasized in any serious use of models. Far too often, in an attempt to "sell" one's model, some limitations imposed by assumptions are minimized or ignored; on the other hand, critics of a particular model are often so blinded by the idealizations that they fail to recognize the model's usefulness, however limited.

As one might expect, there are many types of models; for example, some are deterministic in nature, while some are probabilistic in nature; some are simply descriptive models, while others are decision models. Models also vary in generality. Some are quite specific and are designed for a particular real situation. Others are general enough to apply to a variety of real problems. Mathematical techniques used in the construction of models also vary considerably. A breadth of knowledge of mathematics is an obvious advantage in this regard.

We now provide an illustration of the use of the model construction technique diagramed in Figure 12.2. In Exercises 1 and 2 the reader is asked to construct two other common mathematical models.

Example 1

One of the best-known mathematical models is the Leontief input–output model, which was described in Example 4 of Section 5.5 and illustrated in Example 6 of Section 5.6.

Nonmathematical statement of problem: See Example 4 of Section 5.5.
The assumptions:
1. Each activity produces exactly one commodity (that is, no joint production and no by-product).
2. Each commodity is produced by exactly one activity (that is, no alternative ways of producing a commodity).
3. The internal consumption of commodity i is a linear combination of the production of all the commodities, where the coefficients of combination are the nonnegative constants $c_{i1}, c_{i2}, \ldots, c_{in}$. This means that,

$$\text{(the input of commodity } i \text{ to the system)} = c_{i1}x_1 + \cdots + c_{in}x_n$$

where $c_{ij} \geq 0$. (This is the linearity assumption.)
4. The production of, and the demand for, each commodity is non-negative.
5. There is no surplus or deficiency in the production (the stability assumption).

The mathematical model: Because of assumption 5, the production of activity i is divided between the internal consumption of the system and the external demands. Hence, using assumption 3 we have the single linear equation

$$x_i = (c_{i1}x_1 + \cdots + c_{in}x_n) + d_i.$$

Because of assumptions 1 and 2 we have a model consisting of exactly n such equations, namely the system given at the end of Example 4 of Section 5.5.

Mathematical techniques: The Gauss–Jordan method or the use of an inverse matrix is illustrated in Example 6 of Section 5.6 can be used to solve the system of linear equations provided that a solution exists which will satisfy assumption 4.

Mathematical results: A solution $(x_1, \ldots, x_n)$ of the model (system of linear equations), subject to assumption 4, reveals a level of production that theoretically will maintain the stability of the system (assumption 5).

Comparison of results: The validity of these theoretical results will depend upon the degree to which the assumptions of the model fit reality. Two areas of nonfit may be

1. The model does not allow for commodities (such as labor) that are necessary to the system but not products of the activities. These commodities, which are necessary in production, have a constraining influence on the production.
2. There may be alternative ways of producing a given commodity which cannot be reflected in the model because of assumption 2.

Of course, the assumptions and the model can be changed to incorporate these items at the expense of a more complicated model. These matters are discussed on pp. 302–303 of Gale [23] and pp. 170–171 of Boot [10].

Example 2

The cost accounting problem as discussed on p. 150 of Kemeny et al. [38] also uses the mathematical model of Example 1 within an accounting context. Briefly, the problem is one of determining the total costs that should be assigned to various departments of an organization, where these costs occur from external services and also from the internal services performed for the other departments.

Let x_j represent the total cost assigned to department j.

Let c_{ij} represent the fraction of department j's cost that is charged to department i. In other words, if department j performs a service for department i, then department i must be charged for this service.

Let d_i represent the costs generated by department i in excess of those generated internally.

If the assumptions of Example 1 are reworded to apply within the accounting context and if these assumptions are imposed, then the same mathematical model results. Thus the mathematical theory and techniques that were developed for the model of the original problem can also be brought to

bear on this second problem. A variation of the cost accounting problem with associated model and theory is presented on pp. 262–272 of Johnston et al. [34]. ∎

EXERCISES

1. The price of a stock (or the level of exports, or the consumer price index, or various other items) may be classified as being in one of three states at sequential points in time—an increasing state, a decreasing state, or a constant state. A constant state is defined to mean that the deviation from the previous state is less than some predetermined amount. Suppose that the probability of movement of the stock price from state to state after a specified time interval is known or approximated, as shown in Figure 12.3. We can see from the figure that if the stock price is in an increased state initially, then after the first time interval (step) the probability of being in the increased state is 60 per cent, in the constant state is 20 per cent, and in the decreased state is 20 per cent. After k time intervals (steps), the problem is to find the probability that the item is in any one of the three states if the item is in an increased state initially.

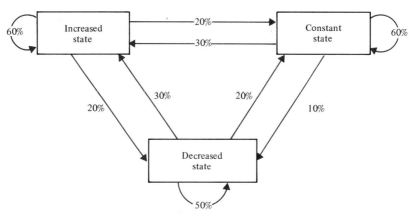

Figure 12.3

(a) Construct a matrix equation as a model of the problem just described provided the following assumptions are made. (*Hint:* See Example 7 of Section 5.3.)
  (1) There is a finite number of states and the system must be in exactly one of these states at every step.
  (2) Either the initial state is known, or the probability for obtaining each state initially is known.
  (3) For all i and j, the probability of the transition from state i to state j is the same for all steps.
(b) Solve the matrix equation found in part (a) after two steps.
2. Consider a conflict situation such as that described in Example 2 of Section 7.3.

(a) Construct a linear programming problem associated with a two-person zero-sum matrix game as a model of the conflict situation provided the following assumptions are made.
   (1) There are exactly two adversaries (the two-person assumption).
   (2) One adversary's gain is the other's loss (the zero-sum assumption).
   (3) The goal of each adversary is conservative in that each will adopt a strategy that provides maximum defense against an all-knowing opponent.
   (4) The concept of expected value governs the choice of strategy.
(b) Solve the linear programming problem found in part (a). (*Note:* This part is lengthy.)

3. In Example 2 of Section 12.4 we used the equation $dv/dt = -g$ as a mathematical model. The constant g is known by experimentation to be about 32 feet per second per second (the minus sign just indicates that gravity is causing the object to come *down*). Suppose that a person drops a pencil from a tall building and uses the equation $s = s_0 + v_0 t - \frac{1}{2}gt^2$ to predict the time until the pencil hits the ground. Because of air friction the experimental results differ slightly from the prediction. Instead of complicating the model to something like $dv/dt = -g - kv$, as discussed in Section 12.4, the person considers using a model such as $dv/dt = -30$. Discuss the advantages and disadvantages of using $dv/dt = -c$, where c is a constant somewhat less than g.

NEW VOCABULARY

antiderivative 12.1
integral sign 12.1
integration 12.1
integrand 12.1
constant of integration 12.1
power rule for integration 12.2
differential equation 12.3
solution of a differential equation 12.3

order of a differential equation 12.3
separate the variables 12.3
initial condition 12.3
general solution 12.3
particular solution 12.3
model 12.4

13 The Definite Integral

Integral calculus includes two basic operations—integration, and evaluation of definite integrals. The purpose of this chapter is to define the definite integral and to explore the very important relationship between the antiderivative and the definite integral. The vehicle for showing this relationship is the Fundamental Theorem of Integral Calculus.

Actually, the basic ideas leading to the definite integral were known for centuries before the development of the differential calculus in the 17th century. Eudoxus (408–355 B.C.) and Archimedes (287–212 B.C.) contributed methods whereby formulas for areas and volumes could be constructed. Little was done to develop the subject further, however, until the 17th century.

An indication of the importance of the definite integral in probability and statistics will be given in the applications of Sections 13.2, 13.3, 13.5, and 13.6.

Prerequisites: 1.1, 1.2, 3.1–3.9, Chapter 9, 10.1–10.5, 12.1, 12.2; the first part of 11.6 is needed for 13.4.
Suggested Sections for Moderate Emphasis: 13.1–13.4.
Suggested Sections for Minimum Emphasis: 13.1–13.3.

13.1 Definition

An interesting problem that can often be solved by methods of calculus is the problem of finding the area of a region surrounded by curves whose equations are known. In statistics such an area might represent the probability of an event, as we shall learn in Example 4 of Section 13.2. In economics such an area might represent a quantity called *consumers' surplus* or a quantity called *producers' surplus*, as we shall illustrate in Example 4 of Section 13.4. In this section we first learn how to approximate the area of a region in order to motivate the definition of the **definite integral**.

Consider the problem of finding the area of the region bounded by $y = f(x) = 4x - \frac{1}{2}x^2$, the x-axis, and the vertical lines $x = 1$ and $x = 6$ (see Figure 13.1). Using the fact that the area of a rectangle is the product of its length and width, we shall approximate the desired area by adding the areas of the four rectangles shown in Figure 13.1. We begin by deciding arbitrarily to divide the closed interval $[1, 6]$ into four subintervals by means of the

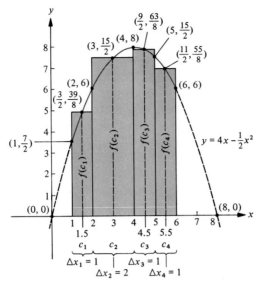

Figure 13.1. *Approximation of area with rectangles.*

division points 1, 2, 4, 5, and 6. This set $\{1, 2, 4, 5, 6\}$ of division points is called a **partition** of $[1, 6]$. The widths of the approximating rectangles shown in the figure are $\Delta x_1 = 2 - 1 = 1$, $\Delta x_2 = 4 - 2 = 2$, $\Delta x_3 = 5 - 4 = 1$, and $\Delta x_4 = 6 - 5 = 1$. We next select point c_1 in, or at an end of, the first subinterval and draw a vertical line segment from this point to the curve $y = f(x)$; we let the length $f(c_1)$ of this segment be the length (height) of the first approximating rectangle. Arbitrarily choose c_1 to be the x-coordinate of the midpoint (the point equidistant from the endpoints) of the first subinterval; hence $c_1 = \frac{3}{2}$. The area of the first rectangle is

$$[f(c_1)](\Delta x_1) = [f(\tfrac{3}{2})](1) = f(\tfrac{3}{2}) = 4(\tfrac{3}{2}) - \tfrac{1}{2}(\tfrac{3}{2})^2 = \tfrac{39}{8}.$$

Similarly, suppose that we decide to let c_2, c_3, and c_4 be the x-coordinates of the midpoints of the other subintervals, respectively; hence $c_2 = 3$, $c_3 = \frac{9}{2}$, $c_4 = \frac{11}{2}$, and the lengths of the other rectangles are $f(3)$, $f(\tfrac{9}{2})$, and $f(\tfrac{11}{2})$, respectively. The sum S_4 of the areas of the four rectangles is

$$\begin{aligned}
S_4 &= f(c_1)\,\Delta x_1 + f(c_2)\,\Delta x_2 + f(c_3)\,\Delta x_3 + f(c_4)\,\Delta x_4 \\
&= [f(\tfrac{3}{2})](1) + [f(3)](2) + [f(\tfrac{9}{2})](1) + [f(\tfrac{11}{2})](1) \\
&= (\tfrac{39}{8})(1) + (\tfrac{15}{2})(2) + (\tfrac{63}{8})(1) + (\tfrac{55}{8})(1) \\
&= \tfrac{277}{8} = 34.625 \text{ square units.}
\end{aligned}$$

By methods we shall learn in the next section it can be shown that the exact area under the curve and above the x-axis between $x = 1$ and $x = 6$ is $34\tfrac{1}{6}$ square units. Notice that the actual area and the approximated area differ by less than $\tfrac{1}{2}$ square unit. Ordinarily, a closer approximation can be obtained by using more subintervals. A very wide subinterval tends to yield more

Sec. 13.1] Definition

error in the approximation; the width of the largest subinterval is called the **norm** of the partition. In the example above the norm of the partition was 2, since $\Delta x_2 = 2$ was larger than Δx_1, Δx_3, or Δx_4. Good approximations ordinarily are obtained by choosing division points so that the number n of subintervals is large and in such a way that the norm of the partition is small; the choice of location of points such as c_1, c_2, c_3, and c_4 above also has an effect on the accuracy of an approximation.

Next we prepare for one of the most important definitions in calculus—the definition of the definite integral. The discussion that follows presents a generalization of the ideas given in the illustration above. Start with a function f defined on a closed interval $[a, b]$. The reader may refer to Figure 13.2 to visualize one possible graph of f, although what follows can be done without reference to a graph. Divide $[a, b]$ into a number n of subintervals by means of the division points $x_0, x_1, x_2, \ldots, x_{n-1}, x_n$, where

$$a = x_0 < x_1 < x_2 < \cdots < x_{n-1} < x_n = b.$$

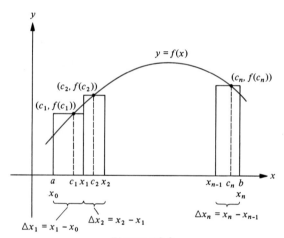

Figure 13.2

In other words, choose a partition $\{x_0, x_1, x_2, \ldots, x_{n-1}, x_n\}$ of $[a,b]$. Let

$$\Delta x_1 = x_1 - x_0,$$
$$\Delta x_2 = x_2 - x_1,$$
$$\vdots$$
$$\Delta x_n = x_n - x_{n-1}.$$

Let δ_n, the norm of the partition, be the largest of the numbers Δx_1, $\Delta x_2, \ldots, \Delta x_n$. Choose numbers $c_1, c_2, \ldots, c_n$ to satisfy

$$x_0 \leq c_1 \leq x_1,$$
$$x_1 \leq c_2 \leq x_2,$$
$$\vdots$$
$$x_{n-1} \leq c_n \leq x_n.$$

Chap. 13] The Definite Integral

Form the sum
$$S_n = f(c_1)\,\Delta x_1 + f(c_2)\,\Delta x_2 + \cdots + f(c_n)\,\Delta x_n.$$
Next we think of repeating the above process, letting n get larger and larger (we require that $\delta_n \to 0$ as $n \to \infty$ in order to make sure that $\Delta x_1, \Delta x_2, \ldots, \Delta x_n$ all become very small as n gets large). This suggests the idea of a limit and leads to the conjecture that S_n may approach some finite number as $n \to \infty$ and $\delta_n \to 0$. In other words, we are interested in finding
$$\lim_{\substack{n \to \infty \\ \delta_n \to 0}} S_n$$
if this limit exists.

▶ **Definition 1**
Let a function f be defined on $[a, b]$. If
$$\lim_{\substack{n \to \infty \\ \delta_n \to 0}} S_n$$
exists, then it is called the **definite integral** of f from a to b and will be denoted temporarily by $I_a^b(f)$.

Several comments about Definition 1 need to be made. First, in Figure 13.2, f is continuous on $[a, b]$ and $f(x)$ is *nonnegative* on $[a, b]$; in this case S_n is the sum of the areas of the n rectangles and represents an approximation of the area above the x-axis and below the curve $y = f(x)$ between $x = a$ and $x = b$. The exact area is defined to be the limit of S_n, namely, $I_a^b(f)$. Second, in Definition 1 when we say that
$$\lim_{\substack{n \to \infty \\ \delta_n \to 0}} S_n$$
exists, we mean that this limit does not depend upon the choice of the points $x_1, x_2, \ldots, x_{n-1}$ or $c_1, c_2, \ldots, c_n$; at any stage of the limiting process S_n does depend upon the selection of these points, but when we say that $I_a^b(f)$ exists, the *limit* of S_n does *not* depend upon their selection. Third, it is *not* correct in general to write $I_a^b(f) = \lim_{n \to \infty} S_n$ because n could become large without δ_n becoming small; this would happen if the first subinterval is never made smaller as $n \to \infty$, because δ_n would always be as large as Δx_1. If, however, all subintervals are *equal* in length at every stage of the limiting process, then δ_n does approach 0 as $n \to \infty$; this idea is used in optional Section 13.5, where definite integrals are evaluated using Definition 1.

We began this section with an example in which we computed a sum S_4 as an approximation of an area—that sum was an approximation of $I_1^6(f)$, where $f(x) = 4x - \frac{1}{2}x^2$. Then we considered the sum S_n, and finally we defined the definite integral of a function f from a to b (where $a < b$) by

$$I_a^b(f) = \lim_{\substack{n \to \infty \\ \delta_n \to 0}} S_n,$$

provided that this limit exists. Whenever this limit exists, f is said to be **integrable** on $[a, b]$. One can see from the definition why people often speak of the "limit of a sum" when discussing a definite integral.

APPLICATIONS

Example 1

The method of approximating an area by adding the areas of appropriate rectangles has been learned in this section. It has perhaps been implied that an equation $y = f(x)$ is always known in this type of problem. In applications it is frequently true that data are observed in some type of experiment and points are plotted from the data, but the correct equation of the curve passing through these points is not known. In the illustration that follows, it is demonstrated how an area can be approximated without knowledge of the equation for the correct curve passing through the given points.

Each day the temperature (in degrees Fahrenheit) is recorded at various times. Suppose that during the hours of darkness (when temperature often does not vary greatly) the temperature is not recorded as often as during the daytime hours. The temperatures on a particular winter day are shown in Table 13.1. The data are transferred to a graph in the manner shown in Figure 13.3. The temperature y is a function f (that is not known exactly) of the number of hours x after midnight—the domain of f is $[0, 24]$. Suppose that we are asked to approximate the area under $y = f(x)$ and above the x-axis between $x = 0$ and $x = 24$ (in Example 4 of Section 13.3 a use for this area will be given). In other words, an approximation of $I_0^{24}(f)$ is desired. Since 18 temperature readings are given, the approximation is accomplished by adding the areas of the 18 rectangles shown in Figure 13.3. As the rectangles are constructed, our estimation of the exact area is slightly high when the temperature is decreasing and slightly low when the temperature is increasing—these errors should almost cancel out over a period of 1 day. Hence the sum S_{18} gives a rather accurate approximation of $I_0^{24}(f)$:

$$\begin{aligned} S_{18} &= f(c_1)\,\Delta x_1 + f(c_2)\,\Delta x_2 + \cdots + f(c_{18})\,\Delta x_{18} \\ &= (16)(2) + (15)(2) + (13)(2) + (12)(2) + (15)(1) + (18)(1) \\ &\quad + (21)(1) + (24)(1) + (28)(1) + (32)(1) + (31)(1) + (30)(1) \\ &\quad + (28)(1) + (24)(1) + (20)(1) + (17)(1) + (15)(2) + (13)(2) \\ &= 456. \quad \blacksquare \end{aligned}$$

Table 13.1. TEMPERATURES ON A WINTER DAY.

Time	mid-night	2	4	6	8	9	10	11	noon	1	2	3	4	5	6	7	8	10
Temperature	16	15	13	12	15	18	21	24	28	32	31	30	28	24	20	17	15	13

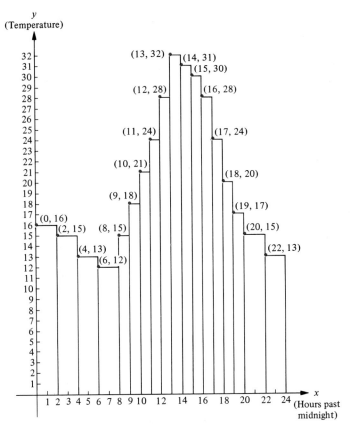

Figure 13.3. *Temperatures on a winter day.*

EXERCISES

Suggested minimum assignment: Exercises 1, 5, 7, 9, and 10.

In Exercises 1–6 sketch $y = f(x)$ between $x = a$ and $x = b$. From the given data, draw the appropriate rectangles as in Figure 13.1. Then compute S_n as an approximation of $I_a^b(f)$.

1. Given: $f(x) = x^2$, $a = 1$, $b = 3$; $n = 3$; division points are $1, 2, \frac{5}{2}, 3$; and $c_1 = \frac{3}{2}$, $c_2 = \frac{9}{4}$, $c_3 = \frac{11}{4}$.
2. Same as Exercise 1, except that $c_1 = 1$, $c_2 = \frac{5}{2}$, $c_3 = \frac{11}{4}$.
3. Given: $f(x) = x^3$, $a = 0$, $b = 2$; $n = 4$; $\Delta x_1 = \Delta x_2 = \Delta x_3 = \Delta x_4 = \frac{1}{2}$; and $c_1 = \frac{1}{4}$, $c_2 = \frac{7}{8}$, $c_3 = \frac{9}{8}$, $c_4 = \frac{7}{4}$.
4. Same as Exercise 3, except that $c_1 = 0$, $c_2 = \frac{1}{2}$, $c_3 = \frac{5}{4}$, $c_4 = 2$.
5. Given: $f(x) = \frac{1}{2}x^2 + 1$, $a = -1$, $b = 2$; $n = 5$; the partition is $\{-1, -\frac{1}{2}, 0, \frac{1}{2}, \frac{3}{2}, 2\}$; and $c_1 = -1$, $c_2 = -\frac{1}{4}$, $c_3 = \frac{1}{4}$, $c_4 = 1$, $c_5 = 2$.
6. Same as Exercise 5, except that $c_1 = -1$, $c_2 = -\frac{1}{2}$, $c_3 = 0$, $c_4 = \frac{1}{2}$, $c_5 = \frac{3}{2}$.

Sec. 13.1] Definition

7. Let $\{3, \frac{7}{2}, 4, \frac{17}{4}, \frac{19}{4}, \frac{11}{2}, 6\}$ be a partition of $[3, 6]$. Find $\Delta x_1, \Delta x_2, \Delta x_3, \Delta x_4, \Delta x_5,$ and Δx_6. Also find the norm δ_6 of this partition.

8. Let $\{1, \frac{3}{2}, 2, 3\}$ be a partition of $[1, 3]$. Find $\Delta x_1, \Delta x_2,$ and Δx_3. Also find the norm δ_3 of this partition.

9. Suppose that $[-2, 4]$ is partitioned into 12 subintervals of equal length. What is the value of each of $\Delta x_1, \ldots, \Delta x_{12}$?

10. Repeat the definition of the definite integral of a function f from a to b without reference to the text.

11. If $f(x) \geq 0$ on $[a, b]$ and f is integrable on $[a, b]$, we learned in this section that $I_a^b(f)$ gives the area under $y = f(x)$ and above the x-axis between $x = a$ and $x = b$. We also learned how to calculate a sum S_n of the areas of n rectangles as an approximation of the exact area. There are several other ways of approximating area, one of which is to calculate a sum T_n of the areas of n trapezoids (see Figure 13.4), using the fact that the area of a trapezoid is the product of one half the sum of the parallel sides and the altitude. The area of the first trapezoid in Figure 13.4 is

$$\tfrac{1}{2}\Delta x_1[f(x_0) + f(x_1)],$$

the area of the second trapezoid is

$$\tfrac{1}{2}\Delta x_2[f(x_1) + f(x_2)],$$

and the area of the last trapezoid is

$$\tfrac{1}{2}\Delta x_n[f(x_{n-1}) + f(x_n)].$$

(a) Use the data of Exercise 1 to find the sum T_3 of the appropriate three trapezoids.

(b) If $\Delta x_1, \Delta x_2, \ldots, \Delta x_n$ are all equal to the same number, Δx, find a general formula for T_n.

12. Three sides of a parcel of land are straight lines and the fourth side curves gently. A sketch of the land made to scale is bounded by $x = 0$,

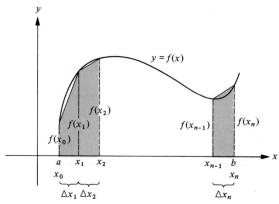

Figure 13.4. *Approximation of area with trapezoids.*

$x = 4$, $y = 0$, and $y = f(x)$. The equation $y = f(x)$ is unknown, except that it passes through $(0, 1.52)$, $(.5, 1.49)$, $(1, 1.45)$, $(1.5, 1.43)$, $(2, 1.41)$, $(2.5, 1.42)$, $(3, 1.46)$, and $(3.5, 1.50)$. The scale is fixed so that $I_0^4(f)$ equals the area in acres of the parcel of land. A buyer agrees to pay \$5000 per acre for the land. Approximate $I_0^4(f)$ by the method of Example 1, and use the result to find the total price of the land.

13. The sum
$$5 + 10 + 15 + 20 + 25 + 30 + 35 + 40$$
can be written briefly as
$$\sum_{k=1}^{8} 5k,$$
which is an example of the **summation notation** (also see optional Section 8.2). The Greek letter Σ (sigma) suggests the word *sum*. The notation $\sum_{k=1}^{8} 5k$ indicates that the values 1, 2, 3, 4, 5, 6, 7, and 8 are to be substituted successively for k in the expression $5k$ and that the results are to be added. The letter k is called the **index of summation**, but any other letter may be used; $\sum_{j=1}^{8} 5j$ represents the same sum. We did not use the summation notation when writing S_n in this section, but the instructor may wish to do so because it may be written conveniently as
$$S_n = \sum_{k=1}^{n} f(c_k)\,\Delta x_k.$$
To gain familiarity with summation notation, do each of the following:

(a) Write the sum represented by $\sum_{k=1}^{5} \dfrac{k}{k+1}$.

(b) Show that $\sum_{k=1}^{8} 5k = 5 \sum_{k=1}^{8} k$.

(c) Write $1^2 + 2^2 + 3^2 + \cdots + n^2$ using summation notation.
(d) Write $15 + 20 + 25 + 30 + 35 + 40$ using summation notation.
(e) Write $f(c_1)\,\Delta x_1 + f(c_2)\,\Delta x_2 + f(c_3)\,\Delta x_3$ using summation notation.

13.2 Fundamental Theorem of Integral Calculus

In Section 13.1 we defined the definite integral of a function f from a to b as a limit of a certain sum, provided that this limit exists. There is a condition that is sufficient to guarantee the existence of $I_a^b(f)$. The following

theorem, whose proof is beyond the scope of this book, makes it clear that many functions that we shall encounter are integrable on an interval $[a, b]$.

▶ **Theorem 1**
If f is continuous on $[a, b]$, then $I_a^b(f)$ exists. ◀

There are functions that are not continuous on $[a, b]$ which are still integrable on $[a, b]$, but Theorem 1 guarantees that every continuous function on a closed interval is integrable on that interval.

The next theorem that we state relates the definite integral (defined in Section 13.1) to the idea of antiderivative (defined in Section 12.1), and also provides a practical method of evaluating many definite integrals. Because of its importance, Theorem 2 is frequently called the **Fundamental Theorem of Integral Calculus**.†

▶ **Theorem 2**
If f is continuous on $[a, b]$ and a function F given by $F(x) = \int f(x)\, dx$ is an antiderivative of f on $[a, b]$, then

$$I_a^b(f) = F(b) - F(a).$$

A proof of Theorem 2 will be discussed in optional Section 13.6. ◀

The remarkable feature of the Fundamental Theorem is that it permits the evaluation of many definite integrals by using the seemingly unrelated concept of an antiderivative. We illustrate in Example 1.

Example 1
If $f(x) = x^2 + 1$, use Theorem 2 to evaluate the definite integral of f from $a = 2$ to $b = 3$. In other words, evaluate $I_2^3(f)$.

An antiderivative F of f on $[2, 3]$ is given by

$$F(x) = \int f(x)\, dx = \int (x^2 + 1)\, dx = \frac{x^3}{3} + x + C,$$

where C is any particular real number. Hence

$$I_2^3(f) = F(3) - F(2)$$
$$= \left(\frac{3^3}{3} + 3 + C\right) - \left(\frac{2^3}{3} + 2 + C\right)$$
$$= (12 + C) - \left(\frac{14}{3} + C\right)$$
$$= \frac{22}{3}. \blacksquare$$

† Some authors call it just the Fundamental Theorem of Calculus rather than the Fundamental Theorem of Integral Calculus.

In the type of problem illustrated in Example 1, the constant of integration C always cancels out, and hence we may as well simplify things by always selecting $C = 0$ when we use Theorem 2.

Instead of $I_a^b(f)$, a customary notation for the definite integral of f from a to b is $\int_a^b f(x)\,dx$. The notation $\int_a^b f(x)\,dx$ was not used in Section 13.1 because of the danger of confusing the definite integral with an antiderivative of f which was denoted by $\int f(x)\,dx$. The two concepts are quite different. One important thing about Theorem 2, however, is that it gives the relationship between these important concepts. In Example 1 we could write

$$\int_2^3 (x^2 + 1)\,dx = \left(\frac{x^3}{3} + x\right)\Big|_2^3$$
$$= \left(\frac{3^3}{3} + 3\right) - \left(\frac{2^3}{3} + 2\right)$$
$$= \frac{22}{3}.$$

The symbol $|_2^3$ indicates that 3 is to be substituted first; then from this result is subtracted the number obtained by substituting 2. (Notice that we selected $C = 0$.)

For the symbol $\int_a^b f(x)\,dx$, $f(x)$ is called the **integrand**, a is called the **lower limit**, b is called the **upper limit**, and x is called the **variable of integration**. Provided that the hypothesis of Theorem 2 is met, the conclusion of the Fundamental Theorem of Integral Calculus can be written

$$\int_a^b f(x)\,dx = F(b) - F(a).$$

To evaluate the definite integral of f from a to b, find an antiderivative F of f on $[a, b]$, substitute the upper limit b to obtain $F(b)$, and then subtract $F(a)$, the result of substituting the lower limit a.

Example 2

Let $f(x) = 4x - \frac{1}{2}x^2$, $a = 1$, and $b = 6$. In Section 13.1 we approximated the area under $y = f(x)$ and above the x-axis between $x = a$ and $x = b$, and we learned that the exact area is given by $I_1^6(f) = \int_1^6 (4x - \frac{1}{2}x^2)\,dx$. Calculate this area using the Fundamental Theorem of Integral Calculus.

First find the antiderivative (with $C = 0$) of the function f.

$$F(x) = \int \left(4x - \frac{1}{2}x^2\right) dx = 4\frac{x^2}{2} - \frac{1}{2}\frac{x^3}{3} = 2x^2 - \frac{1}{6}x^3.$$

Sec. 13.2] Fundamental Theorem of Integral Calculus

Then

$$\int_1^6 \left(4x - \frac{1}{2}x^2\right) dx = \left(2x^2 - \frac{1}{6}x^3\right)\Big|_1^6$$
$$= \left[2(6)^2 - \frac{1}{6}(6)^3\right] - \left[2(1)^2 - \frac{1}{6}(1)^3\right]$$
$$= (72 - 36) - \left(2 - \frac{1}{6}\right)$$
$$= 34\frac{1}{6}. \blacksquare$$

It follows from our discussion in this section and the previous section that if f is *continuous* and *nonnegative* on $[a, b]$, then the area bounded by $y = f(x)$, the x-axis, and the lines $x = a$ and $x = b$ is given by $\int_a^b f(x)\, dx$.

Example 3
Find the area bounded by $y = f(x) = 27/(2x + 3)^2$, the x-axis, and the lines $x = 0$ and $x = 3$.

The rational function f is continuous and nonnegative on $[0, 3]$. The desired area (see Figure 13.5) is given by $\int_0^3 27/(2x + 3)^2\, dx$. First find the antiderivative (with $C = 0$) of the function f.

$$F(x) = \int \frac{27}{(2x + 3)^2}\, dx = 27 \int (2x + 3)^{-2}\, dx$$
$$= \frac{27}{2} \int (2x + 3)^{-2}(2\, dx) = \frac{27}{2} \int u^{-2}\, du$$
$$= \frac{27}{2} \frac{u^{-1}}{-1} = -\frac{27}{2}(2x + 3)^{-1}$$
$$= \frac{-27}{4x + 6}.$$

Therefore,

$$\int_0^3 \frac{27}{(2x + 3)^2}\, dx = \frac{-27}{4x + 6}\Big|_0^3$$
$$= \frac{-27}{12 + 6} - \frac{-27}{0 + 6}$$
$$= -\frac{3}{2} + \frac{9}{2}$$
$$= 3. \blacksquare$$

In optional Section 13.5 we shall demonstrate how to evaluate some rather simple definite integrals by using the *definition* of a definite integral from Section 13.1 and carrying out a somewhat tedious limiting process; this demonstration will be helpful in gaining a better understanding of the definite integral. For most purposes, however, it is best to evaluate

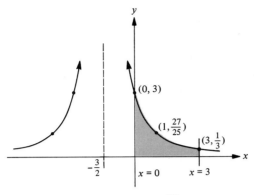

Figure 13.5. $y = \dfrac{27}{(2x+3)^2}$.

$\int_a^b f(x)\,dx$ by finding an antiderivative F of f on $[a, b]$, if possible, and then calculating $F(b) - F(a)$ according to the Fundamental Theorem of Integral Calculus. If, however, an antiderivative F cannot be found but the continuity requirement in Theorem 2 is satisfied, then $\int_a^b f(x)\,dx$ can be approximated by computing a sum S_n as explained in Section 13.1.

APPLICATIONS

*Example 4 Use of Probability Density Functions in Calculating Probability

For a discrete measurement, that is, a measurement having a finite or countably infinite number of values, the probability of an event can be determined by adding the weights of the outcomes comprising the event (see Example 6 of Section 3.9). For a continuous measurement, however, such a procedure is impossible; an event E of obtaining an outcome less than or equal to some real number b is an interval of real numbers and hence has an uncountable number of outcomes. The purpose of this example is to illustrate a method for finding probabilities of events corresponding to continuous measurements by means of a probability density function. In this section we shall limit consideration to continuous measurements whose ranges are finite closed intervals; we shall call such measurements **bounded continuous measurements**.

In Example 7 of Section 10.2 we explained how to obtain a probability density function p if the cumulative probability distribution function F were known. Below we define p in a manner that does not depend upon knowing F; the definition, of course, is consistent with our discussion in Example 7 of Section 10.2. A **probability density function** p of a *bounded* continuous measurement is a function, whose domain is the set of real numbers, that possesses the following three properties:

1. $p(x) \geq 0$.

2. $\int_a^b p(x)\,dx = 1$, where $[a, b]$ is the range of the measurement.

3. $p(x) = 0$ if $x < a$ or $x > b$.

The probability of an event $E = \{s : s \leq c\}$ that the value s of a bounded continuous measurement is less than or equal to some real number c, where $a \leq c \leq b$, is denoted by $P(E)$. The connection between the definition of a probability density function p and the discussion in Example 7 of Section 10.2 is the definition $P(E) = \int_a^c p(x)\,dx$. Since $P(E)$ also equals $F(c)$, where F is the cumulative probability distribution function, we can write

$$P(E) = F(c) = \int_a^c p(x)\,dx.$$

As an illustration of the ideas above, suppose that an insurance company is interested in knowing at what age x in years the death of a man is likely to occur if the man is presently 40 years old. As a hypothetical case (that is not claimed to agree with mortality tables), suppose that x is in the interval $[40, 100]$ and that the probability density function (see Figure 13.6) is given by

$$p(x) = \begin{cases} \frac{1}{36,000}(-x^2 + 140x - 4000) & \text{if } 40 \leq x \leq 100, \\ 0 & \text{otherwise.} \end{cases}$$

Notice from the graph that $p(x) \geq 0$ whenever $40 \leq x \leq 100$ and $p(x) = 0$ otherwise; thus properties 1 and 3 of a probability density function are satisfied. Next we verify property 2. The function p is continuous on $[40, 100]$, and hence we can use Theorem 2.

$$\int_{40}^{100} \frac{1}{36,000}(-x^2 + 140x - 4000)\,dx$$

$$= \frac{1}{36,000}\left(-\frac{x^3}{3} + 70x^2 - 4000x\right)\Big|_{40}^{100}$$

$$= \frac{1}{36,000}\left[\left(-\frac{1,000,000}{3} + 700,000 - 400,000\right)\right.$$

$$\left. - \left(-\frac{64,000}{3} + 112,000 - 160,000\right)\right]$$

$$= \frac{1}{36,000}\left(-\frac{936,000}{3} + 348,000\right)$$

$$= \frac{1}{36,000}(36,000) = 1.$$

This means that the company considers it a certainty that the man will die by the time he reaches age 100.

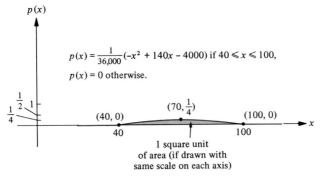

Figure 13.6. *Probability density function.*

Now determine the probability of the event that the man will die on or before his 60th birthday. Hence $E = \{s : s \leq 60\}$, and

$$P(E) = \int_{40}^{60} \frac{1}{36{,}000}(-x^2 + 140x - 4000)\, dx$$

$$= \frac{1}{36{,}000}\left(-\frac{x^3}{3} + 70x^2 - 4000x\right)\Big|_{40}^{60}$$

$$= \frac{7}{27}.$$

(The arithmetic in the last step has been omitted.)

In summary, for a bounded continuous measurement whose range is $S = [a, b]$, the probability of an event $E = \{s : s \leq c\}$, where $a \leq c \leq b$, can be found by

$$P(E) = F(c),$$

where F is the cumulative probability distribution function, or by

$$P(E) = \int_a^c p(x)\, dx,$$

where p is the probability density function. Moreover, the two functions F and p are related according to

$$p(c) = F'(c),$$

provided that $F'(c)$ exists (see Example 7 of Section 10.2). ∎

*Example 5 Mean or Expected Value

A very important concept in the study of statistics is that of the mean or expected value of a measurement (also see optional Section 8.3). It is one of three measures of central tendency discussed in this book. The others are the mode and the median.

For *discrete measurements* the **mean** (or **expected value**) is defined as

$$\mu = s_1 w(s_1) + s_2 w(s_2) + s_3 w(s_3) + \cdots,$$

where $S = \{s_1, s_2, s_3, \ldots\}$ is the sample space of all values of the discrete measurement and $w(s_1), w(s_2), w(s_3), \ldots$ are their respective weights. (The mean may not exist when S is countably infinite.)

As an illustration suppose that a man agrees to draw a single card from an ordinary well-shuffled deck with the following stipulations: If the man draws an ace he will collect $5, if he draws a king he will collect $4, and if he draws anything else he will pay $1. The probabilities of drawing an ace, king, or other card are $\frac{1}{13}, \frac{1}{13},$ and $\frac{11}{13}$, respectively. We have $S = \{s_1, s_2, s_3\} = \{5, 4, -1\}$, $w(s_1) = \frac{1}{13}$, $w(s_2) = \frac{1}{13}$, and $w(s_3) = \frac{11}{13}$. The expected value then is

$$\mu = 5(\tfrac{1}{13}) + 4(\tfrac{1}{13}) + (-1)(\tfrac{11}{13}) = -\tfrac{2}{13}.$$

In other words, the man can expect to lose $\$\frac{2}{13}$ per draw in the long run. In *one* draw, however, it is impossible to lose $\$\frac{2}{13}$; rather, the man will win $5, win $4, or lose $1.

For *bounded continuous measurements* the **mean** (or **expected value**) can be calculated by the formula

$$\mu = \int_a^b x p(x)\, dx,$$

where $[a, b]$ is the range of the measurement. A definition of μ and a derivation of this formula is given in Example 3 of Section 13.5. Using the illustration of Example 4 of this section, we calculate the expected time of death, or the mean death age, to be

$$\mu = \int_{40}^{100} \frac{1}{36{,}000} x(-x^2 + 140x - 4000)\, dx = 70. \quad \blacksquare$$

*Example 6 Variance

Once the mean of a measurement is known, it is desirable to have some indication of how scattered the values of the measurement are (also see optional Section 8.4). One such indicator, called the *variance of the measurement*, is denoted by σ^2.

For *discrete measurements* the **variance** is defined as

$$\sigma^2 = (s_1 - \mu)^2 w(s_1) + (s_2 - \mu)^2 w(s_2) + (s_3 - \mu)^2 w(s_3) + \cdots,$$

where $S = \{s_1, s_2, s_3, \ldots\}$ is the sample space of all values of the discrete measurement, $w(s_1), w(s_2), w(s_3), \ldots$ are their respective weights, and μ is the mean of the measurement. Notice that σ^2 can be viewed as the mean of the squares of the differences between the measurement values and measurement mean. The variance may not exist when S is countably infinite.

For *bounded continuous measurements* the **variance** can be calculated by the formula

$$\sigma^2 = \int_a^b (x - \mu)^2 p(x)\, dx,$$

where $[a, b]$ is the range of the measurement, μ its mean, and p the probability density function.

In both cases the positive square root σ of σ^2 is called the **standard deviation** of the measurement. ∎

EXERCISES

Suggested minimum assignment: Exercises 1, 6, 7, 9, 11, 12, and 15.

In Exercises 1–8 evaluate the given definite integral.

1. $\int_1^3 x^2\, dx.$

2. $\int_{-4}^2 (2x + 3)\, dx.$

3. $\int_{-1}^1 (x + 1)(x - 1)\, dx.$

4. $\int_1^4 \sqrt{x}\, dx.$

5. $\int_0^1 (3x^2 + 4x + 7)\, dx.$

6. $\int_1^2 \dfrac{x^3 + 2}{x^2}\, dx.$

7. $\int_{-1}^0 (2x + 1)^8\, dx.$

8. $\int_0^3 x\sqrt{x^2 + 16}\, dx.$

9. Why is it not possible to use Theorem 2 to attempt to evaluate $\int_{-1}^1 \dfrac{dx}{x^2}$?

10. Is the function f defined by $f(x) = \dfrac{1}{(x + 1)^2}$ integrable on $[0, 5]$?

11. For the symbol $\int_3^4 x^2\, dx$ state the integrand, the upper limit, the lower limit, and the variable of integration.

In Exercises 12–16 find the area bounded by $y = f(x)$, the x-axis, and the lines $x = a$ and $x = b$.

12. $f(x) = x + 3$, $a = -1$, $b = 5$.
13. $f(x) = x^2$, $a = 3$, $b = 4$.
14. $f(x) = x^3 + 2x$, $a = 1$, $b = 2$.
15. $f(x) = \sqrt{4x + 5}$, $a = 1$, $b = 5$.
16. $f(x) = \dfrac{10}{(1 - x)^3}$, $a = -2$, $b = 0$.

17. In Example 4 find the probability that the man will die on or before his 50th birthday.

18. (a) Show that $\int_{-1}^1 \tfrac{3}{4}(1 - x^2)\, dx = 1.$

    (b) Show that the function p defined by

$$p(x) = \begin{cases} \frac{3}{4}(1 - x^2) & \text{if } -1 \leq x \leq 1, \\ 0 & \text{otherwise,} \end{cases}$$

is a probability density function of a bounded continuous measurement.

(c) Show that

$$\int_{-1}^{1} (x)(\tfrac{3}{4})(1 - x^2)\, dx = 0,$$

thus proving that the mean of the measurement is zero.

(d) Calculate the variance σ^2 of the measurement.

19. A man casts a pair of dice on a flat surface and observes the total that he throws. If the total is 7 or 11 he collects $3, but for any other total he pays $1. In the long run, what is the man's expected winning (or loss) per throw in this game?

13.3 Properties of Definite Integrals

In Section 13.1 we defined the definite integral of a function f from a to b as the limit of a certain sum, provided that this limit exists. In the last section we stated the Fundamental Theorem of Integral Calculus, which gives a practical method of evaluating many definite integrals. In this section we state and illustrate some basic properties of definite integrals.

▶ **Theorem 3**

Assume the existence of each definite integral; then the following properties are valid:

(i) $\int_a^b kf(x)\, dx = k \int_a^b f(x)\, dx$, where k is a real number.

(ii) $\int_a^b f(x)\, dx = \int_a^c f(x)\, dx + \int_c^b f(x)\, dx$, where $a < c < b$.

(iii) $\int_a^b [f(x) + g(x)]\, dx = \int_a^b f(x)\, dx + \int_a^b g(x)\, dx$.

(iv) $\int_a^b f(x)\, dx = \int_a^b f(t)\, dt$.

The proof of Theorem 3 depends upon the definition of the definite integral given in Section 13.1, and is left as Exercise 18. ◀

Example 1

(a) $\int_1^4 3x^2\, dx = 3 \int_1^4 x^2\, dx$; by Theorem 3(i), the multiplier 3 may be moved across the integral sign—never move anything that involves the variable of integration in this manner, however.

(b) $\int_1^4 x^2\,dx = \int_1^3 x^2\,dx + \int_3^4 x^2\,dx$; by using Theorem 3(ii) again, we could write

$$\int_1^4 x^2\,dx = \int_1^2 x^2\,dx + \int_2^3 x^2\,dx + \int_3^4 x^2\,dx.$$

(c) $\int_1^4 (x^2 + x^3)\,dx = \int_1^4 x^2\,dx + \int_1^4 x^3\,dx$; Theorem 3(iii) also can be extended to handle sums of more than two functions.

(d) $\int_1^4 x^2\,dx = \int_1^4 t^2\,dt = \int_1^4 w^2\,dw$; by Theorem 3(iv), the particular letter used as the variable of integration is immaterial—hence it is sometimes called a "dummy variable." ∎

In Section 13.1 we defined $I_a^b(f)$ or $\int_a^b f(x)\,dx$ under the assumption that $a < b$. In order to give meaning to $\int_a^b f(x)\,dx$ in cases when $a \geq b$, it is customary to make the following definition.

▶ **Definition 2**

(i) $\int_a^a f(x)\,dx = 0$, provided that $f(a)$ exists.

(ii) If $a > b$ and $\int_b^a f(x)\,dx$ exists, then

$$\int_a^b f(x)\,dx = -\int_b^a f(x)\,dx.$$

Example 2

(a) $\int_4^4 x^2\,dx = 0.$

(b) $\int_4^1 x^2\,dx = -\int_1^4 x^2\,dx.$ ∎

Theorem 3 is still valid if $a \geq b$, and in fact the restriction in part (ii) that c be between a and b also can be removed.

From our geometric interpretation of the definite integral in Sections 13.1 and 13.2 we know that if a region above the x-axis, below $y = f(x)$, and between $x = a$ and $x = b$ has an area of 5 square units and f is continuous on $[a, b]$, then $\int_a^b f(x)\,dx = 5$. See Figure 13.7. If the graph of $y = g(x)$ were below the x-axis between $x = a$ and $x = b$ (see Figure 13.8), then the numbers $g(c_1), g(c_2), \ldots, g(c_n)$ in S_n would be negative and this would lead to a negative value of the definite integral—that is, $\int_a^b g(x)\,dx = -5$. Suppose that the graph of $y = h(x)$ crosses the x-axis at a point c between a and b

Sec. 13.3] Properties of Definite Integrals

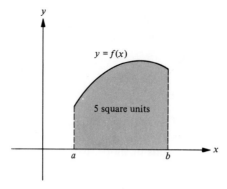

Figure 13.7. $\int_a^b f(x)\,dx = 5.$

(see Figure 13.9). Then

$$\int_a^b h(x)\,dx = \int_a^c h(x)\,dx + \int_c^b h(x)\,dx$$
$$= -6 + 4 = -2.$$

The total shaded area, however, is $|-6| + |4| = 10$ square units.

The next property, which we state without proof, is often called the **Mean Value Theorem for Integrals.**

▶ **Theorem 4**

If f is continuous on $[a, b]$, then there exists a number c, where $a \leq c \leq b$, such that

$$\int_a^b f(x)\,dx = [f(c)](b - a). \blacktriangleleft$$

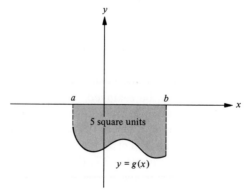

Figure 13.8. $\int_a^b g(x)\,dx = -5.$

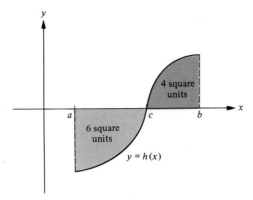

Figure 13.9. $\int_a^b h(x)\,dx = -2.$

A geometric interpretation of Theorem 4 follows. In Figure 13.10 a continuous function f on a closed interval $[a, b]$ is shown (f is chosen intentionally to be positive throughout $[a, b]$ because the geometric interpretation of Theorem 4 is more readily understood in this case). In Figure 13.10 a rectangle $KLMN$ is constructed in such a way that its area, $[f(c)](b - a)$, is equal to the area under the curve, which is given by $\int_a^b f(x)\,dx$.

There are three points in $[a, b]$ in Figure 13.10 at which c can be chosen so as to make the area under the curve, $\int_a^b f(x)\,dx$, equal the area $[f(c)](b - a)$ of rectangle $KLMN$; Theorem 4 guarantees the existence of at least one such number c in $[a, b]$.

Example 3
If $a = 0$, $b = 3$, and $f(x) = x^2$, find one value of the number c in Theorem 4.

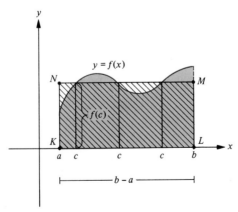

Figure 13.10. *Illustration of Theorem 4.*

Sec. 13.3] Properties of Definite Integrals

In Figure 13.11 the portion of the parabola $y = x^2$ between $x = 0$ and $x = 3$ is shown.

$$\int_a^b f(x)\,dx = \int_0^3 x^2\,dx = \frac{x^3}{3}\Big|_0^3 = \frac{27}{3} - 0 = 9.$$

Hence the area between $x = 0$ and $x = 3$ which is above the x-axis and below $y = x^2$ is 9 square units. Thus, since $f(c) = c^2$ and $b - a = 3 - 0$,

$$\int_0^3 x^2\,dx = [f(c)](b - a)$$

reduces to

$$9 = c^2(3 - 0)$$

or

$$9 = 3c^2,$$

and the only root of this equation in $[0, 3]$ is

$$c = \sqrt{3}.$$

The answer can be checked by showing that the area of the rectangle in Figure 13.11 is 9 square units; the base is $b - a = 3$ and the height is

$$f(c) = f(\sqrt{3}) = (\sqrt{3})^2 = 3,$$

and hence the area of the rectangle is 9, which equals the area under the curve. ∎

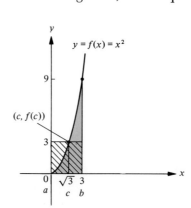

Figure 13.11

APPLICATIONS

Example 4
Let a and b be real numbers, where $a < b$, and suppose that a function f is continuous on $[a, b]$. The **average value of f on $[a, b]$** is defined to be

$$\frac{1}{b - a}\int_a^b f(x)\,dx.$$

This quantity is equal to the number $f(c)$ in Theorem 4 of this section. From either Figure 13.10 or 13.11, it can be seen why $f(c)$ is considered to be the average y value on $[a, b]$ when $y = f(x)$ is plotted on $[a, b]$.

For a specific illustration let us return to Example 1 of Section 13.1. In that example a function f with domain $[0, 24]$ was given in which temperatures for 1 day were expressed as a function of the hour of the day. The average value of this function on $[0, 24]$ is called the **average daily temperature**. Using our approximation of 456 for $I_0^{24}(f) = \int_0^{24} f(x)\,dx$ from Example 1 of Section 13.1, we find that the average daily temperature in degrees Fahrenheit is given by

$$\frac{1}{24 - 0} \int_0^{24} f(x)\,dx \approx \frac{1}{24}(456) = 19.$$

If the average daily temperature is less than 65 degrees Fahrenheit, then this average can be subtracted from 65 to obtain what is defined to be the number of **heating degree days** for the particular day in question. If the average daily temperature for a given day is equal to or greater than 65 degrees Fahrenheit, then the number of heating degree days is defined to equal 0. For our example the number of heating degree days is

$$65 - 19 = 46.$$

The number of heating degree days is usually printed in a daily newspaper each day. It can be used, for example, by fuel suppliers to estimate the amount of fuel consumption and thus to determine when to make the automatic deliveries that are promised.

Observe from the data of Example 1 of Section 13.1 that the highest recorded temperature for the day is 32 and the low is 12. These can quickly be averaged to obtain 22, which can be used in place of 19 as an estimation of the average daily temperature. This shortcut method tends to be rather inaccurate, particularly under certain weather conditions. ∎

Example 5

Definite integrals have an important role in many applications. On pp. 194–200 of his *Mathematical Techniques for Physiology and Medicine* [60], Simon finds definite integrals useful in his discussion of the theory of blood flow measurement. In the indicator dilution method of blood flow measurement, a known amount of an indicator can be injected into the bloodstream and the flow rate determined by the rate of dilution of the indicator. Sometimes the indicator is injected into the bloodstream at a constant rate for enough time so that the concentration at some point downstream is approximately constant. The quantity of the indicator substance that passes the point downstream from time t_1 to time t_2 is given by

$$\int_{t_1}^{t_2} F(t)C(t)\,dt,$$

where $F(t)$ represents the flow rate at time t and $C(t)$ represents the concentration at time t. Simon also finds the concept of the average value of a function on an interval (see Example 4) to be helpful in his discussion of the theory of blood flow measurement. ∎

*Example 6
In Example 4 of Section 13.2 we expressed the probability of the event $E = \{s : s \leq 60\}$ that a man would die on or before his 60th birthday as

$$P(E) = \int_{40}^{60} p(x)\, dx = \tfrac{7}{27}.$$

Suppose that for the same probability density function p, we wish to find the probability of the event $E_1 = \{s : 50 < s \leq 60\}$ that the man will die between ages 50 and 60. Let $E_2 = \{s : s \leq 50\}$ so that

$$P(E_1) = P(E) - P(E_2)$$
$$= \int_{40}^{60} p(x)\, dx - \int_{40}^{50} p(x)\, dx.$$

By Theorem 3(*ii*) of this section the latter expression is equal to

$$\int_{50}^{60} p(x)\, dx,$$

which can be calculated to be $\tfrac{5}{27}$. A valid geometric interpretation, which is suggested by the illustration above, is that the probability that a value of a bounded continuous measurement lies between numbers c and d is equal to the area under the graph of p and between $x = c$ and $x = d$. ∎

E X E R C I S E S

Suggested minimum assignment: Exercises 2, 5, 7, 9, 11, and 14.

In Exercises 1–8 use Figure 13.12 to evaluate the given definite integral.

1. $\displaystyle\int_a^b f(x)\, dx.$ 2. $\displaystyle\int_b^c f(x)\, dx.$ 3. $\displaystyle\int_c^d f(t)\, dt.$

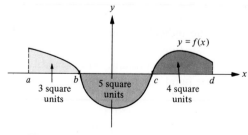

Figure 13.12. *Figure for Exercises 1–8.*

4. $\int_a^c f(x)\,dx.$ 5. $\int_b^d 3f(x)\,dx.$ 6. $\int_a^d f(x)\,dx.$

7. $\int_d^c f(x)\,dx.$ 8. $\int_b^b f(x)\,dx.$

In Exercises 9–12 find one value of the number c in Theorem 4. Give a geometric interpretation.

9. $a = 1$, $b = 3$, $f(x) = x^2$.
10. $a = 0$, $b = 2$, $f(x) = x^2$.
11. $a = -1$, $b = 2$, $f(x) = 2x$.
12. $a = 0$, $b = 5$, $f(x) = x^2 + x$.
(*Hint:* Use the quadratic formula.)

In Exercises 13–16 evaluate the given definite integral.

13. $\int_1^3 (x^3 + x^2)\,dx.$

14. $\int_{-2}^1 (3t + 2t^2)\,dt.$

15. $\int_0^4 (2x^2 - 5x + 3)\,dx.$

16. $\int_{-2}^3 |x|\,dx.$

17. Evaluate $\int_0^R \dfrac{P\pi}{2\eta l} r(R^2 - r^2)\,dr$, where R, P, η, l, and π are constants. (This definite integral occurs in the calculation of the volume of flow of blood per unit of time in an arteriole; see pp. 189–194 of Thrall et al. [63].)

18. Outline a proof of each of the following.
    (a) Theorem 3(*i*).   (b) Theorem 3(*ii*).
    (c) Theorem 3(*iii*). (d) Theorem 3(*iv*).

19. Let $f(x) = x^3 - x^2 + 2$. Find the average value (see Example 4) of f on $[1, 3]$.

20. Refer to Example 6 and write (but do not calculate) the definite integral that gives the probability of death between ages 70 and 80. Give a geometric interpretation.

13.4 Area Between Curves

In many calculus books one can read about how to use definite integrals to calculate such things as areas, volumes, centroids, arc length, and work. The desired quantity is approximated by the type of sum which, after a limiting process, leads to a definite integral. In this section we show how to use definite integrals to calculate the area between curves whose equations are known.

First we illustrate one general type of situation that will occur; in Figure 13.13 it is desired to find the area from $x = a$ to $x = b$ between an upper curve $y = f(x)$ and a lower curve $y = g(x)$. We partition $[a, b]$ into n subintervals and choose arbitrary points $c_1, \ldots, c_n$ in these subintervals just as we did in Section 13.1. The desired area can be approximated if rec-

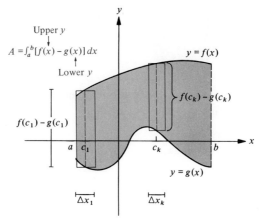

Figure 13.13. *Area using vertical strips.*

tangles are drawn in such a way (the first rectangle and a typical kth rectangle are shown in Figure 13.13) that the sum S_n of the areas of these rectangles is given by

$$S_n = [f(c_1) - g(c_1)]\,\Delta x_1 + \cdots + [f(c_n) - g(c_n)]\,\Delta x_n.$$

To get the exact area, let $n \to \infty$ and let the norm δ_n of the partition approach zero; from the definition of a definite integral this limit yields $\int_a^b [f(x) - g(x)]\,dx$, and hence the area A of the desired region is given by

$$A = \int_a^b [f(x) - g(x)]\,dx. \qquad (13.1)$$

The upper curve $y = f(x)$ is always written first in (13.1) in order that $f(x) - g(x)$ will be nonnegative, and thereby assure that A is positive when $b > a$.

Example 1
Find the area of the region bounded by $y = \tfrac{1}{2}x^2$ and $y = 4\sqrt{x}$.
    Always sketch the given curves accurately enough to be certain which area is desired and to find key points of intersection. The parabola $y = \tfrac{1}{2}x^2$ is symmetric to the y-axis and passes through $(0, 0)$, $(2, 2)$, and $(4, 8)$. The equation $y = 4\sqrt{x}$ is the equation of the upper half of the parabola $y^2 = 16x$ and passes through $(0, 0), (1, 4)$, and $(4, 8)$. See Figure 13.14. When we compare our problem with the general case illustrated in Figure 13.13, it can be seen that in this example $a = 0$, $b = 4$, $f(x) = 4\sqrt{x}$, and $g(x) = \tfrac{1}{2}x^2$. Notice that $y = 4\sqrt{x}$ is the upper curve throughout *all* the interval $[0, 4]$; hence $f(x) - g(x) = 4\sqrt{x} - \tfrac{1}{2}x^2$ is nonnegative on $[0, 4]$. From (13.1) the area A of the shaded region in Figure 13.14 is given by

$$A = \int_0^4 \left(4\sqrt{x} - \frac{1}{2}x^2\right) dx$$
$$= \left(4\frac{x^{3/2}}{\frac{3}{2}} - \frac{1}{2}\frac{x^3}{3}\right)\Big|_0^4$$
$$= \left[\left(\frac{8}{3}\right)(8) - \left(\frac{1}{2}\right)\left(\frac{64}{3}\right)\right] - [0 - 0] = \frac{32}{3}. \blacksquare$$

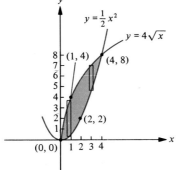

Figure 13.14

Some regions are shaped so that they do not have one curve as an upper boundary and another one as a lower boundary; instead, between $y = c$ and $y = d$, there may be one curve $x = f(y)$ which forms a right boundary and another curve $x = g(y)$ which serves as a left boundary (see Figure 13.15). It is convenient in this case to solve the equations for x in terms of y, if possible. After forming S_n from horizontal rectangles and passing to the limit, we obtain

$$A = \int_c^d [f(y) - g(y)]\, dy. \tag{13.2}$$

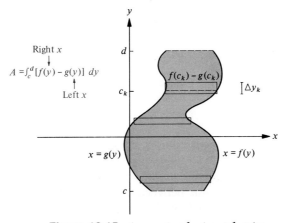

Figure 13.15. *Area using horizontal strips.*

Sec. 13.4] Area Between Curves

The curve $x = f(y)$ which serves as the right boundary is always written first in (13.2) in order that $f(y) - g(y)$ will be nonnegative, and thereby assure that A is positive when $d > c$.

Example 1 was selected so that it could also be worked using (13.2), and this is shown in Example 2.

Example 2

Rework Example 1 using (13.2). See Figure 13.16.

Notice that between $y = 0$ and $y = 8$ the curve $x = \sqrt{2y}$ is always on the right and $x = \frac{1}{16}y^2$ is always on the left so that, by (13.2), the area A is given by

$$A = \int_0^8 \left(\sqrt{2y} - \frac{1}{16}y^2\right) dy$$

$$= \sqrt{2} \int_0^8 y^{1/2}\, dy - \frac{1}{16}\int_0^8 y^2\, dy$$

$$= \left(\sqrt{2}\,\frac{y^{3/2}}{\frac{3}{2}} - \frac{1}{16}\frac{y^3}{3}\right)\Big|_0^8$$

$$= \left[\left(\frac{2\sqrt{2}}{3}\right)(8\sqrt{8}) - \frac{1}{16}\frac{(8)^3}{3}\right] - (0 - 0)$$

$$= \left(\frac{2\sqrt{2}}{3}\right)(8)(2\sqrt{2}) - \left(\frac{1}{2}\right)\left(\frac{64}{3}\right) = \frac{64}{3} - \frac{32}{3} = \frac{32}{3}. \blacksquare$$

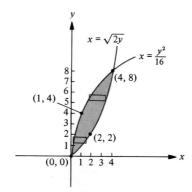

Figure 13.16

Example 3

Find the area bounded by the parabola $x = y^2 - 1$ and the line $y = x - 1$. See Figure 13.17.

In this example it is preferable to solve the equations for x in terms of y and use formula (13.2) because, between $y = -1$ and $y = 2$, $x = f(y) = y + 1$ is always the right boundary of the region and $x = g(y) = y^2 - 1$ is always the left boundary.

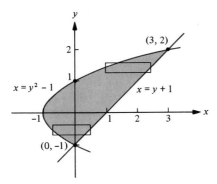

Figure 13.17

$$A = \int_{-1}^{2} [(y+1) - (y^2 - 1)]\, dy$$

$$= \int_{-1}^{2} (-y^2 + y + 2)\, dy = \left(-\frac{y^3}{3} + \frac{y^2}{2} + 2y\right)\Big|_{-1}^{2}$$

$$= \left(-\frac{8}{3} + 2 + 4\right) - \left(\frac{1}{3} + \frac{1}{2} - 2\right) = \frac{9}{2}. \blacksquare$$

The difficulty with using (13.1) in Example 3 is that between $x = -1$ and $x = 0$ the lower boundary of the region is the parabola, but between $x = 0$ and $x = 3$ the lower boundary is the line. By dividing the region into two parts (a necessary technique for some problems but not recommended in Example 3), the area can be set up using (13.1) as follows:

$$A = \int_{-1}^{0} [\sqrt{x+1} - (-\sqrt{x+1})]\, dx + \int_{0}^{3} [\sqrt{x+1} - (x-1)]\, dx.$$

Now that the reader has seen some examples, we suggest a procedure to follow for finding the area between curves:

1. Draw a figure and mark clearly the area to be found. Label key points of intersection of the given curves (find these points in the process of plotting points or by solving the equations algebraically).
2. Decide whether to use (13.1), use (13.2), or divide the region into more than one part so that one of the formulas applies to each part. Show some approximating rectangles on your figure—vertical rectangles if you use (13.1) or horizontal rectangles if you use (13.2). Use (13.1) if the region has a single upper and a single lower boundary, or apply (13.2) if the region has a single right and a single left boundary.
3. Write the necessary definite integrals for the area. [Set them up so that the answer will be positive—in (13.1) let $a < b$ so that the Δx_k in an approximating sum S_n are positive, and in order that the $f(c_k) - g(c_k)$ are positive, let $y = f(x)$ be the upper curve and $y = g(x)$ the lower curve; in (13.2) let $c < d$, and let $x = f(y)$ be the bounding curve on the right and $x = g(y)$ the one on the left.] When you use (13.1), the

Sec. 13.4] Area Between Curves

upper and lower limits should be *x* limits and the integrand a function of *x*—for (13.2) the upper and lower limits should be *y* limits and the integrand a function of *y*.
4. Evaluate the definite integrals by using the Fundamental Theorem of Integral Calculus, and check the answer for gross errors by estimating the answer from the figure. (If you cannot evaluate the definite integrals, calculate an approximating sum as done in Section 13.1.)

APPLICATIONS

Example 4

Suppose that the equation of the demand curve for a certain style of record player is given by

$$y = 250 - \tfrac{5}{6}x.$$

In this equation *y* is the selling price in dollars and *x* is the number of units demanded by consumers each week. Suppose that the supply curve for the same item has the equation

$$y = \frac{x^2}{200} + 78,$$

where *x* is the number of record players that would be supplied weekly by producers for sale at *y* dollars each. The demand and supply curves are both shown in Figure 13.18.

It can be verified that the point (120, 150) satisfies each of the given equations; this point of intersection can be found by solving simultaneously the given equations of the line and the parabola. We shall assume that we are dealing with the situation of pure competition, in which case the actual selling price is determined by the intersection of the demand and supply curves. Hence 120 of our particular style of record player are sold each week for $150 apiece.

From Figure 13.18 it can be seen that some consumers are willing to pay up to $250 for one of the record players. When they find that the cost is only

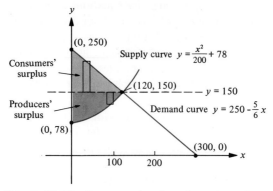

Figure 13.18. *Consumers' and producers' surpluses.*

$150, they feel that they have saved money. The total amount saved in this way each week by the 120 consumers who make purchases is called **consumers' surplus** (CS) and is represented by the area above $y = 150$ and below $y = 250 - \frac{5}{6}x$. In this problem the area is a triangle and could be found without the use of definite integrals, but the use of (13.1) of this section is a more general approach for this type of problem; the calculation of consumers' surplus using (13.1) is as follows:

$$CS = \int_0^{120} [(250 - \tfrac{5}{6}x) - (150)]\, dx$$
$$= \int_0^{120} (100 - \tfrac{5}{6}x)\, dx$$
$$= (100x - \tfrac{5}{12}x^2)\Big|_0^{120}$$
$$= (12{,}000 - 6000) - (0 - 0)$$
$$= \$6000.$$

The producers of the record player also realize a gain, called the **producers' surplus**, in the sense that Figure 13.18 shows that some of them would have built the record players even if they had known that the selling price were below $150. The producers' surplus (PS) is represented by the area below $y = 150$ and above $y = (x^2/200) + 78$, and is computed using (13.1) as follows:

$$PS = \int_0^{120} \left[150 - \left(\frac{x^2}{200} + 78\right)\right] dx$$
$$= \int_0^{120} \left(72 - \frac{x^2}{200}\right) dx$$
$$= \left(72x - \frac{x^3}{600}\right)\Big|_0^{120}$$
$$= (8640 - 2880) - (0 - 0)$$
$$= \$5760.$$

In our computation of consumers' surplus and producers' surplus in this example, we have assumed that the equations of the demand and supply curves are valid at least in the interval $[0, 120]$. ∎

EXERCISES

Suggested minimum assignment: Exercises 3, 7, 8, and 13.

In each of Exercises 1–10 find the area of the region bounded by the given curves.

1. $y = x^2$, $y = 0$, $x = 1$, $x = 2$.
2. $y = x^2$, $y = 2x$.
3. $y = x^2$, $y = x + 2$.
4. $x = y^2$, $x = 4$.

5. $x = y^2$, $y = x - 2$.
6. $y = \sqrt{x}$, $x = 2y$. Work by two methods (see Examples 1 and 2).
7. $y = x^3$, $y = 8$, $x = 0$. Work by two methods (see Examples 1 and 2).
8. $x = y^2 - 4$, $x = -y^2$.
9. $y = x^2 - 2x - 3$, $y = 0$.
10. $y = x^2 - 2x - 3$, $y = 2x - 3$.
11. Find the total area between $x = -2$ and $x = 0$ that is bounded by the x-axis and the parabola $y = x^2 - 2x - 3$. (*Hint:* The parabola crosses the x-axis between $x = -2$ and $x = 0$.)
12. (a) Find the area bounded by $y = 2$, $x = 9$, and $x = 5 - 4y - y^2$.
    (b) How much of the area in part (a) lies in the first quadrant?
    (c) How much of the area in part (a) lies in the second quadrant?
    (d) How much of the area in part (a) lies in the fourth quadrant?
13. (a) Find the area bounded by $y = \sqrt{5x - 1}$, $y = 0$, and $x = 10$.
    (b) How much of the area in part (a) is above $y = 2$?
14. (a) Calculate the consumers' surplus in Example 4 using (13.2).
    (b) Calculate the producers' surplus in Example 4 using (13.2).
15. In Example 4 suppose that the equations of the demand and supply curves are, respectively, $y = 100 - (x^2/10)$ and $y = 50 + (x^2/40)$.
    (a) Compute the consumers' surplus.
    (b) Compute the producers' surplus.

13.5 More About Calculation of Definite Integrals (Optional)

In this section we shall try to develop a better understanding of definite integrals by showing how to evaluate some simple definite integrals by use of the long limit process of Definition 1 rather than the more efficient Fundamental Theorem of Integral Calculus (Theorem 2).

We start by assuming without proof two well-known formulas that are valid for each positive integer n:

$$1 + 2 + 3 + \cdots + n = \frac{n(n+1)}{2} \tag{13.3}$$

and

$$1^2 + 2^2 + 3^2 + \cdots + n^2 = \frac{n(n+1)(2n+1)}{6}. \tag{13.4}$$

Formula (13.3) is a formula for adding the first n positive integers; it may be familiar to anyone who has studied arithmetic progressions. We verify (13.3) for $n = 8$ as follows: the left side is

$$1 + 2 + 3 + 4 + 5 + 6 + 7 + 8 = 36,$$

and the right side is

$$\frac{8(8+1)}{2} = 36.$$

Formula (13.4) gives an expression for the sum of the squares of the first n positive integers and can be verified in a similar manner for any particular positive integer n. A general verification of these formulas involves the use of mathematical induction. Formula (13.3) will be used in Example 1, and (13.4) will be used in Example 2.

The evaluation of $I_a^b(f) = \int_a^b f(x)\,dx$ by Definition 1 can be very difficult, but in the examples that follow we let f be a polynomial function of degree 1 or 2 so that the procedure will not be unreasonably complicated. Polynomial functions are continuous on any interval $[a, b]$ and therefore integrable on $[a, b]$ by Theorem 1. Hence the division points for the n subintervals of $[a, b]$ may be chosen in any manner, and we elect to divide $[a, b]$ into n equal parts each of length

$$\Delta x = \frac{b-a}{n}.$$

Therefore, our division points are

$$x_0 = a,$$
$$x_1 = a + \Delta x,$$
$$x_2 = a + 2\,\Delta x,$$
$$\vdots$$
$$x_n = b = a + n\,\Delta x.$$

The points $c_1, \ldots, c_n$ in Section 13.1 may be chosen in any convenient manner, and we shall find that it works out well to let

$$c_1 = x_1,$$
$$c_2 = x_2,$$
$$\vdots$$
$$c_n = x_n.$$

The norm δ_n of the partition $\{x_0, x_1, x_2, \ldots, x_n\}$ is $\delta_n = (b-a)/n$, since the length of each subinterval is $(b-a)/n$. Therefore,

$$\lim_{n \to \infty} \delta_n = \lim_{n \to \infty} \frac{b-a}{n} = 0.$$

This means that for our examples

$$\int_a^b f(x)\,dx = \lim_{\substack{n \to \infty \\ \delta_n \to 0}} S_n$$

can be simplified to

$$\int_a^b f(x)\,dx = \lim_{n \to \infty} S_n.$$

Example 1
Use Definition 1 to evaluate $\int_1^3 x\, dx$.
Note that $a = 1$, $b = 3$, $f(x) = x$, and
$$\Delta x = \frac{b - a}{n} = \frac{2}{n}.$$

We also have (see Figure 13.19)
$$c_1 = x_1 = 1 + \frac{2}{n},$$
$$c_2 = x_2 = 1 + 2\left(\frac{2}{n}\right),$$
$$\vdots$$
$$c_n = x_n = 1 + n\left(\frac{2}{n}\right).$$

Therefore,
$$S_n = f(c_1)\,\Delta x_1 + f(c_2)\,\Delta x_2 + \cdots + f(c_n)\,\Delta x_n$$
$$= f(c_1)\,\Delta x + f(c_2)\,\Delta x + \cdots + f(c_n)\,\Delta x$$
$$= \frac{2}{n}[f(c_1) + f(c_2) + \cdots + f(c_n)].$$

Since $f(x) = x$, then $f(c_i) = c_i$ and S_n reduces to
$$S_n = \frac{2}{n}(c_1 + c_2 + \cdots + c_n)$$
$$= \frac{2}{n}\left[\left(1 + \frac{2}{n}\right) + \left(1 + 2\left(\frac{2}{n}\right)\right) + \cdots + \left(1 + n\left(\frac{2}{n}\right)\right)\right].$$

Using (13.3) and the fact that the sum of n ones is n, we obtain
$$S_n = \frac{2}{n}\left[(1 + 1 + \cdots + 1) + \frac{2}{n}(1 + 2 + \cdots + n)\right]$$
$$= \frac{2}{n}\left[n + \frac{2}{n}\frac{n(n+1)}{2}\right]$$
$$= \frac{2}{n}[n + (n + 1)] = \frac{4n + 2}{n}.$$

Therefore,
$$I_1^3(f) = \int_1^3 x\, dx = \lim_{n\to\infty} S_n = \lim_{n\to\infty} \frac{4n + 2}{n}$$
$$= \lim_{n\to\infty} \frac{4 + 2/n}{1} = \frac{4 + 0}{1} = 4. \blacksquare$$

As a check on the answer to Example 1 we observe by the Fundamental Theorem of Integral Calculus that
$$I_1^3(f) = \int_1^3 x\, dx = \frac{x^2}{2}\bigg|_1^3 = \frac{9}{2} - \frac{1}{2} = 4.$$

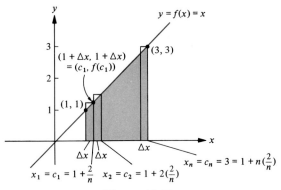

Figure 13.19

Notice how much easier the latter method is. As another check we observe that geometrically $\int_1^3 x\, dx$ represents the area of the trapezoid bounded by $x = 1$, $x = 3$, $y = 0$, and $y = x$ (see Figure 13.19), and the area of this trapezoid can easily be seen to equal 4.

Example 2

Use Definition 1 to evaluate $\int_0^3 x^2\, dx$.

In this case $a = 0$, $b = 3$, $f(x) = x^2$, and $\Delta x = 3/n$. Next observe that

$$c_1 = x_1 = 0 + 1\left(\frac{3}{n}\right) = 1\left(\frac{3}{n}\right),$$

$$c_2 = x_2 = 0 + 2\left(\frac{3}{n}\right) = 2\left(\frac{3}{n}\right),$$

$$\vdots$$

$$c_n = x_n = 0 + n\left(\frac{3}{n}\right) = n\left(\frac{3}{n}\right).$$

Some of the approximating rectangles are shown in Figure 13.20; the fact that $f(x) = x^2$ means that

$$f(c_1) = c_1^2 = 1^2\left(\frac{3}{n}\right)^2,$$

$$f(c_2) = c_2^2 = 2^2\left(\frac{3}{n}\right)^2,$$

$$\vdots$$

$$f(c_n) = c_n^2 = n^2\left(\frac{3}{n}\right)^2,$$

and the appearance of $1^2, 2^2, \ldots, n^2$ leads to a need for (13.4) in the calculation that follows.

Sec. 13.5] More About Calculation of Definite Integrals

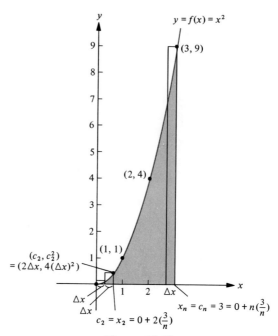

Figure 13.20

$$S_n = f(c_1)\,\Delta x_1 + f(c_2)\,\Delta x_2 + \cdots + f(c_n)\,\Delta x_n$$
$$= \frac{3}{n}[f(c_1) + f(c_2) + \cdots + f(c_n)]$$
$$= \frac{3}{n}\left[1^2\left(\frac{3}{n}\right)^2 + 2^2\left(\frac{3}{n}\right)^2 + \cdots + n^2\left(\frac{3}{n}\right)^2\right]$$
$$= \frac{3}{n}\left(\frac{3}{n}\right)^2(1^2 + 2^2 + \cdots + n^2)$$
$$= \frac{27}{n^3}\left[\frac{n(n+1)(2n+1)}{6}\right]$$
$$= \frac{9}{2}\left(\frac{2n^2 + 3n + 1}{n^2}\right).$$

Therefore,

$$I_0^3(f) = \int_0^3 x^2\,dx = \lim_{n\to\infty} \frac{9}{2}\left[\frac{2 + (3/n) + (1/n^2)}{1}\right]$$
$$= \frac{9}{2}\left(\frac{2 + 0 + 0}{1}\right)$$
$$= 9. \blacksquare$$

APPLICATIONS

*Example 3

Let us continue our discussion of the mean or expected value of a measurement that was begun in Example 5 of Section 13.2. In this example we shall define the mean of a bounded continuous measurement, where $[a, b]$ is the range of the measurement.

First, let p be the probability density function for a bounded continuous measurement with range $[a, b]$. Divide the interval $[a, b]$ into n equal subintervals, each of length Δx, as shown in Figure 13.21. From Example 6 of Section 13.3 we know that the probability of the event E_k that a value of the measurement is a number in the kth subinterval is

$$P(E_k) = \int_{x_{k-1}}^{x_k} p(x)\, dx \qquad (k = 1, 2, \ldots, n).$$

By the Mean Value Theorem for Integrals (Theorem 4 of Section 13.3), we know that if p is continuous on $[a, b]$, there is some number c_k, where $x_{k-1} \leq c_k \leq x_k$, such that

$$\begin{aligned} P(E_k) &= p(c_k)(x_k - x_{k-1}) \\ &= p(c_k)\, \Delta x \qquad (k = 1, 2, \ldots, n). \end{aligned}$$

Second, consider the mean of a discrete measurement having values $c_1, c_2, \ldots, c_n$ with respective weights $w(c_1), w(c_2), \ldots, w(c_n)$; letting $w(c_k) = P(E_k)$, we have

$$w(c_k) = P(E_k) = p(c_k)\, \Delta x \qquad (k = 1, 2, \ldots, n).$$

Since there are a finite number of values of the discrete measurement, the mean is (see Example 5 of Section 13.2)

$$\begin{aligned} c_1 w(c_1) &+ c_2 w(c_2) + \cdots + c_n w(c_n) \\ &= c_1[p(c_1)\, \Delta x] + c_2[p(c_2)\, \Delta x] + \cdots + c_n[p(c_n)\, \Delta x] \\ &= [c_1 p(c_1) + c_2 p(c_2) + \cdots + c_n p(c_n)]\, \Delta x. \end{aligned}$$

Third, the **expected value** or **mean** μ of a bounded continuous measurement with range $[a, b]$ is defined to be the limit of the last expression as the number of subintervals increases without bound. That is,

$$\mu = \lim_{n \to \infty} [c_1 p(c_1) + c_2 p(c_2) + \cdots + c_n p(c_n)]\, \Delta x,$$

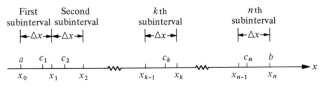

Figure 13.21

which we know is the definite integral

$$\mu = \int_a^b xp(x)\,dx.$$

An illustration of the evaluation of an expected value using the above formula was given in Example 5 of Section 13.2. ∎

EXERCISES

Suggested minimum assignment: Exercises 2, 3, 4, and 11.

1. Verify that $1 + 2 + 3 + \cdots + n = \dfrac{n(n+1)}{2}$ is true for $n = 7$.

2. Verify that $1^2 + 2^2 + 3^2 + \cdots + n^2 = \dfrac{n(n+1)(2n+1)}{6}$ is true for $n = 6$.

3. If $[3, 8]$ is divided into seven subintervals of equal length, what is the length of each subinterval? Also answer the same question for n subintervals, where n is a positive integer.

In Exercises 4–11 evaluate the given definite integrals using the method of Examples 1 and 2 of this section.

4. $\int_0^4 x\,dx.$

5. $\int_2^4 x\,dx.$

6. $\int_{-1}^2 2x\,dx.$

7. $\int_1^6 (x+1)\,dx.$

8. $\int_0^2 x^2\,dx.$

9. $\int_0^4 (x^2+2)\,dx.$

10. $\int_0^5 (x^2+x)\,dx.$

11. $\int_1^3 x^2\,dx.$

12. Define the variance of a bounded continuous measurement. (*Hint:* See Example 6 of Section 13.2 and Example 3 of this section.)

13.6 Discussion of Proof of Fundamental Theorem (Optional)

Our goal in this section is to discuss a proof of the Fundamental Theorem of Integral Calculus (Theorem 2); the proof depends upon other theorems whose proofs are beyond the scope of this text. In order to understand this discussion, it will be necessary to comprehend the concept of an **integral with a variable upper limit**, a concept that is also useful in applications related to statistics (see Examples 3 and 4).

Let f be continuous on $[a, b]$, and let x be a number such that $a \le x \le b$. In Theorem 5 of this section the **integral with a variable upper limit**

$$\int_a^x f(t)\,dt$$

will be encountered, and so an explanation is needed. To get a geometric interpretation of this expression, visualize f as a function that is positive on $[a, b]$ and hence $\int_a^x f(t)\,dt$ equals the area from a to x under $y = f(t)$ and above the t-axis (see Figure 13.22). If the function f and the number a do not change during our discussion but x is allowed to vary within $[a, b]$, then $\int_a^x f(t)\,dt$ depends only upon x and thus

$$F(x) = \int_a^x f(t)\,dt$$

defines a function F with domain $[a, b]$. If $f(t)$ is nonnegative on $[a, b]$, then $F(x)$ gives the area under the graph of f from a to x. Certainly, $F(x)$ does not depend upon the letter t that is used to label the horizontal axis; from Theorem 3(iv) we know that any letter could be used in place of t and, although x could be used, it is avoided because x is used as the variable upper limit.

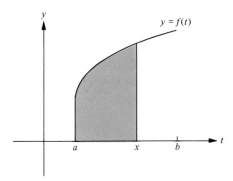

Figure 13.22. Area $= \int_a^x f(t)\,dt$.

Example 1
Consider $F(x) = \int_0^x 3t\,dt$, where $0 \leq x \leq 2$. Because $f(t) = 3t$ is nonnegative on $[0, 2]$, $F(x)$ equals the area of a triangle (see Figure 13.23) with base x and altitude $3x$. Therefore,

$$F(x) = \tfrac{1}{2}(x)(3x) = \tfrac{3}{2}x^2.$$

Next, notice that if F is differentiated, we get

$$F'(x) = \tfrac{3}{2}(2x) = 3x,$$

which is simply the integrand $3t$ evaluated at x. Theorem 5 guarantees that the latter result was no accident; that is, $F'(x) = f(x)$ on $[a, b]$ whenever f is continuous on $[a, b]$. ∎

Sec. 13.6] Discussion of Proof of Fundamental Theorem

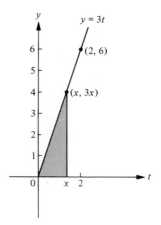

Figure 13.23

In the preceding discussion we have defined a function F in terms of a function f by means of a definite integral with a variable upper limit, namely,

$$F(x) = \int_a^x f(t)\, dt.$$

Figures 13.22 and 13.23 were used to try to give some geometric interpretation to this type of function. Theorem 5 shows that f is the derivative of F, or that F is an antiderivative of f.

▶ **Theorem 5**
Let f be continuous on $[a, b]$, and let F with domain $[a, b]$ be defined by $F(x) = \int_a^x f(t)\, dt$. Then F is differentiable on $[a, b]$ and

$$F'(x) = f(x). \blacktriangleleft$$

The proof of Theorem 5 will not be given here. The proof depends strongly on the Mean Value Theorem for Integrals (Theorem 4) and can be found in many calculus texts. Some authors call Theorem 5 the Fundamental Theorem of Integral Calculus because they feel that the development of the result that F is an antiderivative of f is of primary importance and that Theorem 2 should be considered as a corollary of Theorem 5.

Example 2
Let $F(x) = \int_2^x \sqrt{1 + t^2}\, dt$, where $2 \leq x \leq 5$. Find $F'(x)$.
In this case $a = 2$, $b = 5$, and $f(t) = \sqrt{1 + t^2}$. Since $f(t) = \sqrt{1 + t^2}$, it follows that $f(x) = \sqrt{1 + x^2}$. By Theorem 5,

$$F'(x) = f(x) = \sqrt{1 + x^2}. \blacksquare$$

We now restate Theorem 2 and give its proof, which uses Theorem 5.

▶ **Theorem 2**
If f is continuous on $[a, b]$ and F is an antiderivative of f on $[a, b]$, then
$$I_a^b(f) = \int_a^b f(x)\, dx = F(b) - F(a).$$

Proof

Statement	Reason
1. Let $G(x) = \int_a^x f(t)\, dt$, where $a \leq x \leq b$. Then G is an antiderivative of f on $[a, b]$; i.e., $G'(x) = f(x)$.	1. Theorem 5.
2. F is an antiderivative of f on $[a, b]$.	2. Hypothesis.
3. $F(x) = G(x) + C$, where C is a real number.	3. Statements 1 and 2, and Theorem 1 of Chapter 12.
4. $F(a) = G(a) + C$ $= \int_a^a f(t)\, dt + C$ $= 0 + C.$	4. Substitution of $x = a$ in statements 3 and 1; Definition 2(i).
5. $F(b) = G(b) + C$ $= \int_a^b f(t)\, dt + F(a).$	5. Substitution of $x = b$ in statements 3 and 1; by statement 4, $C = F(a)$.
6. $\int_a^b f(x)\, dx = F(b) - F(a)$.	6. Statement 5 and Theorem 3(iv). ◀

The result, $\int_a^b f(x)\, dx = F(b) - F(a)$, of Theorem 2 is also valid if $a \geq b$; this fact is a consequence of Definition 2 (see Exercises 7 and 8).

APPLICATIONS

*Example 3

In Example 4 of Section 13.2 we defined a probability density function p for bounded continuous measurements. In this example we shall reconsider the related function F, the cumulative probability distribution function, which was defined in Example 8 of Section 9.3 and discussed further in Example 7 of Section 10.2. We shall learn that for a bounded continuous measurement the function F can be expressed in terms of an integral with a variable upper limit.

Suppose that the value of a continuous measurement can be any real number s, where $a \leq s \leq b$. Suppose also that the probability density function p is continuous on $[a, b]$. Then the function F given by

$$F(x) = \begin{cases} 0 & \text{if } x < a, \\ \int_a^x p(s)\, ds & \text{if } a \leq x \leq b, \\ 1 & \text{if } x > b, \end{cases}$$

Sec. 13.6] Discussion of Proof of Fundamental Theorem

has the four properties required of a cumulative probability distribution function that are stated in Example 8 of Section 9.3. An outline of the proof of this statement follows.

Since $F(x) = 0$ if $x < a$ and $F(x) = 1$ if $x > b$, the properties

$$\lim_{x \to -\infty} F(x) = 0$$

and

$$\lim_{x \to \infty} F(x) = 1$$

are valid. Next observe that

$$F(a) = \int_a^a p(s) \, ds = 0.$$

Suppose that $a < x < b$. Then, by Theorem 5 of this section, F and p are related by the equation

$$F'(x) = p(x) \qquad \text{if } a < x < b.$$

Since $p(x) \geq 0$, $F'(x) \geq 0$ on (a, b), and hence F is a nondecreasing continuous function on (a, b). Also, observe that since p is a probability density function, $\int_a^b p(s) \, ds = 1$, and since $F(x) = \int_a^x p(s) \, ds$ if $a \leq x \leq b$, we have

$$F(b) = \int_a^b p(s) \, ds = 1.$$

From the known facts about p and F it now follows that F is a nondecreasing continuous function on $[a, b]$, and also that F has these same properties over the set of all real numbers; from this latter fact properties 3 and 4 of Example 8 of Section 9.3 are an immediate consequence. Therefore, F is a cumulative probability distribution function, and hence the probability of an event $E = \{s : s \leq c\}$, where $a \leq c \leq b$, is given by

$$P(E) = F(c) = \int_a^c p(s) \, ds.$$

For a numerical example we shall use the probability density function p from Example 4 of Section 13.2; in that example p was given by

$$p(x) = \begin{cases} \frac{1}{36,000}(-x^2 + 140x - 4000) & \text{if } 40 \leq x \leq 100, \\ 0 & \text{otherwise.} \end{cases}$$

We determine the cumulative probability distribution function F when $40 \leq x \leq 100$ as follows:

$$F(x) = \int_{40}^{x} \frac{1}{36{,}000}(-s^2 + 140s - 4000)\, ds$$

$$= \frac{1}{36{,}000}\left(-\frac{s^3}{3} + 70s^2 - 4000s\right)\bigg|_{40}^{x}$$

$$= \frac{1}{36{,}000}\left[\left(-\frac{x^3}{3} + 70x^2 - 4000x\right) - \left(-\frac{64{,}000}{3} + 112{,}000 - 160{,}000\right)\right]$$

$$= \frac{1}{36{,}000}\left(-\frac{x^3}{3} + 70x^2 - 4000x + \frac{208{,}000}{3}\right).$$

The graph of F is shown in Figure 13.24. The height of the curve at 70 is $\frac{1}{2}$; this means that the probability of death by age 70 is $\frac{1}{2}$ and agrees with the fact that the area under the graph of $y = p(x)$ between 40 and 70 is $\frac{1}{2}$ (see Figure 13.6). In general, the height of the curve at x equals the area from 40 to x under the graph of $y = p(x)$ and gives the probability of death by age x. Note that

$$F(60) - F(50) = \tfrac{7}{27} - \tfrac{2}{27} = \tfrac{5}{27},$$

which agrees with our calculation in Example 6 of Section 13.3, which showed that the probability of death between ages 50 and 60 was $\tfrac{5}{27}$.

Another graphical observation follows from the fact that $F'(x) = p(x)$, namely, that the slope of F at any number x equals the height of p at the same number x. Since $p(70) = \tfrac{1}{4}$ (see Figure 13.6), it follows that the slope of F at $(70, \tfrac{1}{2})$ is $\tfrac{1}{4}$; also since the maximum height of p is $\tfrac{1}{4}$, the maximum slope of F is $\tfrac{1}{4}$. (If the slope appears incorrect in Figure 13.24, it is because the same scale is not used on each axis.)

We showed above how to find the cumulative probability distribution function F from the probability density function p. It is even easier to find p

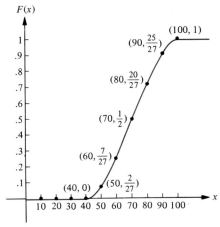

Figure 13.24. *Cumulative probability distribution function.*

Sec. 13.6] Discussion of Proof of Fundamental Theorem

if F is given because $F'(x) = p(x)$; in our example we have

$$F(x) = \frac{1}{36{,}000}\left(-\frac{x^3}{3} + 70x^2 - 4000x + \frac{208{,}000}{3}\right) \quad \text{if } 40 \leq x \leq 100,$$

and by differentiation

$$p(x) = F'(x) = \frac{1}{36{,}000}(-x^2 + 140x - 4000) \quad \text{if } 40 \leq x \leq 100. \blacksquare$$

*Example 4 Percentiles and Median of a Measurement
A cumulative probability distribution function F is particularly useful in calculating **percentiles**. For a continuous measurement the **90th percentile** is the real number c for which 90 per cent of the measurement values are less than or equal to c. In other words, if $E = \{s : s \leq c\}$, the number c will satisfy the equation

$$P(E) = .9 \quad \text{or} \quad F(c) = .9.$$

For a bounded continuous measurement with range $[a, b]$, this is equivalent to solving for c in the equation

$$\int_a^c p(x)\,dx = .9,$$

where p is the probability density function. A geometric interpretation of the latter equation is that 90 per cent of the area under the curve $y = p(x)$ will be to the left of the line $x = c$.

In general, the **kth percentile**, where $0 < k < 100$, is the solution for c of the equation

$$F(c) = \frac{k}{100}.$$

The **median** is defined to be the 50th percentile.

As an illustration, suppose that a measurement of the time required for the passage of legislation has the cumulative probability distribution function F and the probability density function p given by

$$F(x) = \begin{cases} 0 & \text{if } x < 0, \\ x^2 & \text{if } 0 \leq x \leq 1, \\ 1 & \text{if } x > 1, \end{cases} \quad \text{and} \quad p(x) = \begin{cases} 2x & \text{if } 0 \leq x \leq 1, \\ 0 & \text{otherwise.} \end{cases}$$

The 50th percentile or median time required is the (nonnegative) solution of the equation

$$F(c) = c^2 = .5,$$

or equivalently the (nonnegative) solution of

$$\int_0^c 2x\,dx = .5.$$

The result is $c = \sqrt{.5}$, which is approximately .707.

Chap. 13] The Definite Integral

For a discrete measurement, to calculate the kth percentile, we consider two cases (also see optional Section 8.3):

1. If there is a value s_i of the measurement such that $F(s_i) = k/100$, where $0 < k < 100$, then the kth percentile is $(s_i + s_{i+1})/2$, where s_{i+1} is the smallest value of the measurement that is larger than s_i. For example, consider four test scores of 60, 70, 80, and 90; then

$$F(x) = \begin{cases} 0 & \text{if } x < 60, \\ \frac{1}{4} & \text{if } 60 \le x < 70, \\ \frac{1}{2} & \text{if } 70 \le x < 80, \\ \frac{3}{4} & \text{if } 80 \le x < 90, \\ 1 & \text{if } x \ge 90. \end{cases}$$

We see that $F(70) = \frac{1}{2}$. Hence the 50th percentile (median) is $(70 + 80)/2$, or 75.

2. If there is no value s_i of the measurement such that $F(s_i) = k/100$, where $0 < k < 100$, then the kth percentile is the smallest value of the measurement such that $F(s_i) > k/100$. For example, consider three test scores of 75, 85, and 95; then

$$F(x) = \begin{cases} 0 & \text{if } x < 75, \\ \frac{1}{3} & \text{if } 75 \le x < 85, \\ \frac{2}{3} & \text{if } 85 \le x < 95, \\ 1 & \text{if } x \ge 95. \end{cases}$$

We see that there is no value s_i of the measurement such that $F(s_i) = \frac{1}{2}$; therefore, the 50th percentile (median) is 85, since $F(85) > \frac{1}{2}$ and 85 is the smallest value for which this is the case. ∎

EXERCISES

Suggested minimum assignment: Exercises 1, 2, 3, and 4.

1. If $y = \int_1^w f(z)\, dz$, where $1 \le w \le 5$, then y is a function of what variable? What is the domain of this function?

2. Let $F(x) = \int_1^x (t + 1)\, dt$, $1 \le x \le 6$. Find $F(1)$. Find $F(6)$.

In Exercises 3–6 find $F'(x)$ by the use of Theorem 5.

3. $F(x) = \int_3^x t^2\, dt$, $3 \le x \le 4$.

4. $F(x) = \int_{-1}^x \sqrt{4 + t^3}\, dt$, $-1 \le x \le 1$.

5. $F(x) = \int_1^x t\sqrt{1+t^2}\,dt$, $1 \le x \le 5$.

6. $F(x) = \int_2^x \sqrt{w}\,dw$, $2 \le x \le 4$.

7. Explain why $\int_a^b f(x)\,dx = F(b) - F(a)$ is valid if $a = b$.

8. Explain why $\int_a^b f(x)\,dx = F(b) - F(a)$ is valid if $a > b$.

9. In Example 3 calculate $F(80)$ and explain the meaning of the answer.
10. In this problem use the probability density function p in Exercise 18 of Section 13.2. Find $F(x)$ as described in Example 3 of this section. Then verify that $F(-1) = 0$, $F(1) = 1$, and $F'(x) = p(x)$.
11. In the first part of Example 4 we found the 50th percentile to be approximately .707. Find the 36th percentile.
12. A test is given to five students and the scores are 80, 85, 85, 90, and 95.
    (a) Find the 10th percentile.
    (b) Find the median.
    (c) Find the 80th percentile.

NEW VOCABULARY

division points 13.1
partition 13.1
norm of partition 13.1
definite integral 13.1
integrable 13.1
fundamental theorem of integral calculus 13.2
integrand 13.2

lower limit of definite integral 13.2
upper limit of definite integral 13.2
variable of integration 13.2
mean value theorem for integrals 13.3
integral with a variable upper limit 13.6

14 Exponential and Logarithmic Functions

Two particularly useful types of functions are the exponential and logarithmic functions. The purposes of this chapter are to define these functions, apply the techniques of differentiation and integration to them, and illustrate some of the many applications.

Historians generally credit John Napier (1550–1617) with the invention of logarithms. As a result of some of the properties of logarithmic functions, the invention was welcomed by astronomers as a significant time saver in performing various calculations; many of the great minds of the seventeenth century were thus freed from tedious calculations and were permitted more time for theoretical investigations. Today the logarithmic and exponential functions play an indispensable role in solving differential equations, which arise so often in the construction of mathematical models of real-world phenomena.

Prerequisites: For 14.1–14.3, the prerequisites are 1.1, 1.2, 3.1–3.9, Chapter 9, 10.1–10.5, 11.1–11.3, and the first part of 11.6. For 14.4, additional prerequisites from previous chapters are 10.6, 12.1, 12.2, and 13.1–13.4. Section 12.3 is needed in 14.5. The reader may skip from 14.4 directly to either 14.6 or 14.7.
Suggested Sections for Moderate Emphasis: 14.1–14.4.
Suggested Sections for Minimum Emphasis: 14.1–14.3.

14.1 Exponential Functions

Suppose that a certain amount of a quantity is present at a given starting time $(t = 0)$. Suppose also that, as time passes, the amount y that is present either always increases or always decreases. There are several practical situations in which it is reasonable to assume that the rate of change dy/dt of y with respect to time t is proportional to the amount y that is present. This relationship can be expressed as the differential equation

$$\frac{dy}{dt} = ky, \qquad (14.1)$$

where k is the constant of proportionality. If $k > 0$, then y increases as t increases; if $k < 0$, then y decreases as t increases. The differential equation (14.1) might be used in a problem about population growth of humans or animals; it is applicable in problems on decay of radioactive material (useful in determining the approximate age of certain old substances). Later in this chapter we shall show that the solution of (14.1) is the important equation

$$y = Ae^{kt}, \qquad (14.2)$$

where $e \approx 2.7$ is a number that will be introduced in this section and A is the value of y when $t = 0$. Equation (14.2) is useful for projecting the amount y expected to be present at future times. The type of function in (14.2) suggests that we begin this chapter by introducing **exponential functions**.

▶ **Definition 1**
Let b be any positive real number. The function f, whose domain is the set of all real numbers, determined by

$$f(x) = b^x,$$

is called the **exponential function with base b**.

The graph of an exponential function with base $b > 1$ has a characteristic appearance, as shown in the following example.

Example 1
Graph the exponential function f with base 2 determined by the equation

$$y = f(x) = 2^x.$$

Some corresponding values of x and y are displayed in Table 14.1, and the graph is shown in Figure 14.1. ∎

Table 14.1

x	-3	-2	-1	0	1	$\tfrac{3}{2}$	2	3
$y = 2^x$	$\tfrac{1}{8}$	$\tfrac{1}{4}$	$\tfrac{1}{2}$	1	2	$2\sqrt{2}$	4	8

In making Table 14.1, we assumed that the reader knew from algebra how to raise 2 to a rational power. Numbers with irrational exponents such as $2^{\sqrt{2}}$ can be defined by a limiting process—$2^{\sqrt{2}}$ may be thought of as the limit of the sequence

$$2^1, 2^{1.4}, 2^{1.41}, 2^{1.414}, \ldots,$$

where $1.414\ldots$ is the decimal representation of $\sqrt{2}$. As one may intuitively expect from Figure 14.1, an exponential function is continuous over the domain of all real numbers.

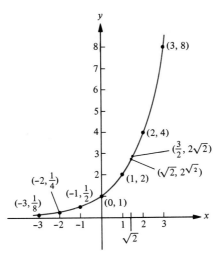

Figure 14.1. $y = f(x) = 2^x$.

As Figure 14.1 suggests, an exponential function with base $b > 1$ is an increasing function; also, if $x \to -\infty$, then $b^x \to 0$, and if $x \to \infty$, then b^x increases without bound. If $0 < b < 1$, then the graph of $y = b^x$ is the graph of a decreasing function. For example, the graph of

$$y = (\tfrac{1}{2})^x = 2^{-x}$$

is shown in Figure 14.2. Observe in this case that if $x \to \infty$, then $b^x \to 0$, and if $x \to -\infty$, then b^x increases without bound.

The graph of the exponential function with base $b = 1$ is a horizontal line because

$$f(x) = 1^x = 1$$

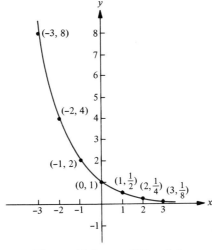

Figure 14.2. $y = (\tfrac{1}{2})^x = 2^{-x}$.

Sec. 14.1] Exponential Functions

defines a constant function (see Section 3.9). If $b \neq 1$, the range of the exponential function with base b is the set of all positive numbers—see Figures 14.1 and 14.2. Hence equations such as

$$2^x = 0 \quad \text{and} \quad (\tfrac{1}{2})^x = -1$$

have no solutions.

Properties of rational exponents were studied in Section 3.1. Some of these properties are reviewed next, and we point out that they are valid for *irrational* as well as rational exponents. If x and y are real numbers and b is a positive real number, then

$$b^x b^y = b^{x+y},$$
$$(b^x)^y = b^{xy},$$
$$\frac{b^x}{b^y} = b^x b^{-y} = b^{x-y},$$

and, if $b \neq 1$,

$$b^x = b^y \quad \text{implies that} \quad x = y.$$

We shall find that the most important exponential function is the one with base e. The number e (named after the 18th-century mathematician Euler) is used frequently in many applications of mathematics and can be defined as follows:

▶ **Definition 2**

$$e = \lim_{x \to 0} (1 + x)^{1/x}.$$

Table 14.2 suggests the existence of a limit for $(1 + x)^{1/x}$ as x approaches 0, but it is difficult to *prove* that the limit of Definition 2 exists. It does exist, however, and equals an irrational number whose value correct to five decimal places is

$$e = 2.71828.$$

In Example 2 we shall need an alternative formulation of the definition of e. Assume that $x > 0$ and let $z = 1/x$ in Definition 2. As $x \to 0$, $z \to \infty$

Table 14.2

x	$(1 + x)^{1/x}$	x	$(1 + x)^{1/x}$
.1	$(1.1)^{10} = 2.59374$	$-.1$	$(.9)^{-10} = 2.86797$
.01	$(1.01)^{100} = 2.70481$	$-.01$	$(.99)^{-100} = 2.73200$
.001	$(1.001)^{1000} = 2.71692$	$-.001$	$(.999)^{-1000} = 2.71964$
.0001	$(1.0001)^{10,000} = 2.71815$	$-.0001$	$(.9999)^{-10,000} = 2.71842$
.00001	$(1.00001)^{100,000} = 2.71827$	$-.00001$	$(.99999)^{-100,000} = 2.71830$

and we obtain

$$e = \lim_{z \to \infty} \left(1 + \frac{1}{z}\right)^z. \tag{14.3}$$

Example 2

A discussion of compound interest was begun at the outset of Section 3.9 and continued in Example 4 of Section 3.9 and Example 1 of Section 9.2. If a principal P is deposited at an annual interest rate r for t years and interest is compounded n times each year, then the amount A after t years is given by

$$A = P\left(1 + \frac{r}{n}\right)^{nt}.$$

Suppose that federal regulations limit a savings institution to a maximum annual interest rate r but that no restriction is placed on the number n of times per year that interest is compounded. To attract capital, the management decides to let $n \to \infty$; in this case we say that interest is **compounded continuously**. Let us calculate

$$\lim_{n \to \infty} P\left(1 + \frac{r}{n}\right)^{nt} = P \lim_{n \to \infty} \left[\left(1 + \frac{1}{n/r}\right)^{n/r}\right]^{rt}.$$

If $z = n/r$, then $z \to \infty$ as $n \to \infty$, and we shall assume that it is valid to write

$$P\left[\lim_{z \to \infty} \left(1 + \frac{1}{z}\right)^z\right]^{rt},$$

which equals, upon application of (14.3),

$$Pe^{rt}.$$

In Example 1 of Section 9.2 we stated that if $10 million is deposited for 1 year at 6 per cent, then the amount in the account cannot exceed $10,618,365.47, no matter how many times during the year the interest is compounded. If interest is compounded continuously at 6 per cent, the amount in the account after 1 year is

$$\begin{aligned} A = Pe^{rt} &= (10{,}000{,}000)e^{(.06)(1)} \\ &\approx (10{,}000{,}000)(1.061836547) \\ &= \$10{,}618{,}365.47. \end{aligned}$$

(From tables, $e^{.06}$ is slightly less than 1.061836547.) The reader may wish to compare the answer above with Table 9.2, in which the results of compounding daily, quarterly, and so on, are shown. ∎

APPLICATIONS

Example 3

Suppose that an amount y of some quantity has value 0 at time $t = 0$ and that, as time passes, y gradually increases but remains less than A. In this example we shall mention some situations in which this type of bounded growth might occur. An equation that has provided good experimental results for this type of growth is

$$y = A(1 - e^{-kt}). \qquad (14.4)$$

In (14.4) A and k are positive constants and time $t \geq 0$. Notice that if $t = 0$, then

$$y = A(1 - e^0) = 0.$$

The expression e^{-kt} is 1 when $t = 0$, and e^{-kt} decreases but remains positive as t increases; therefore, the expression $(1 - e^{-kt})$ is always less than 1, and hence y is always less than A. However, $y \to A$ as $t \to \infty$ because

$$\lim_{t \to \infty} A(1 - e^{-kt}) = \lim_{t \to \infty} \left(A - \frac{A}{e^{kt}} \right) = A - 0 = A.$$

In Figure 14.3, equation (14.4) is graphed for the case when $A = 10$ and $k = \frac{1}{2}$. If k were smaller, the curve would approach $y = A$ less quickly.

A curve of the type in Figure 14.3 is sometimes called a **learning curve.** It shows how learning can take place rapidly at first and then level out as it nears a maximum possible level A. Perhaps y represents the number of items produced by a worker t days after starting to learn a new job. On pp. 29–32 of *Essentials of Behavior,* Hull [30] discusses his formula for habit strength H as a function of the number N of reinforcements from the beginning of

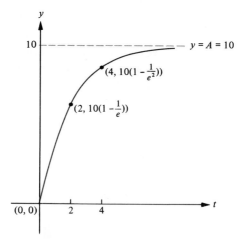

Figure 14.3. $y = 10(1 - e^{-(1/2)t})$.

learning. The formula is of the form
$$H = 1 - 10^{-lN},$$
where l is a positive constant, or
$$H = 1 - e^{-kN},$$
where $10^{-l} = e^{-k}$. The formula, therefore, is of the same type as (14.4).

Equation (14.4) might be a useful model in any situation in which a growing quantity has an upper bound A. Over extended periods of time the magnitude of a quantity is usually limited by various factors. For example, the size of a population of objects is limited by the space in which the population must live; the size of a certain individual animal is limited by hereditary factors. The growth of plants and the production of crops certainly can be enhanced by fertilizer, but there is an upper bound to this growth and production. Equation (14.4) has proved useful in studying such situations. A derivation of (14.4) will be given in Example 8 of Section 14.4. ∎

EXERCISES

Suggested minimum assignment: Exercises 1, 2, 6, 11, and 14.

1. Graph $f(x) = 3^x$. State the domain and range of f. Is f an increasing function or is f a decreasing function? State the value of $\lim_{x \to -\infty} 3^x$.

2. Graph $f(x) = (\frac{1}{3})^x = 3^{-x}$. State the domain and range of f. Is f an increasing function or is f a decreasing function? State the value of $\lim_{x \to \infty} 3^{-x}$.

3. (a) Graph the exponential function with base 4.
   (b) Graph the exponential function with base $\frac{1}{4}$.

4. Sketch $y = 2^{x+1}$. What is the relationship of this graph to the graph of $y = 2^x$?

5. Solve for x, if possible.
   (a) $3^x = 81$.   (b) $3^x = \frac{1}{81}$.   (c) $3^x = -81$.

6. What is meant by $2^{\sqrt{3}}$?

7. Simplify. (a) $4^{5/2}$.   (b) $(4^9)(4^{-7})$.
   (c) $(3^{10})^{1/5}$.   (d) $\dfrac{5^6}{5^3}$.

8. True or false: $2^{3/2} > e$. Give a reason for your answer.

9. Evaluate $(1 + x)^{1/x}$ to three decimal places for $x = \frac{1}{5}$.

10. What is the correct value of e to five decimal places?

11. Evaluate $\lim_{x \to 0} (1 + x)^{2/x}$.

12. Evaluate $\lim_{x \to \infty} 2 \left(1 + \dfrac{1}{x}\right)^x$.

Sec. 14.1] Exponential Functions

13. (a) Suppose that $P = \$100$ is invested in a savings account at an annual interest rate of $r = .04$. If interest is compounded four times a year, how much money is in the account after 1 year?
    (b) Rework part (a) if interest is compounded continuously. (Use $e^{.04} = 1.0408$.)
14. Find the amount to which \$800 will accumulate if deposited for 2 years at 5 per cent compounded continuously. (Use $e^{.1} = 1.105171$.)
15. Sketch (14.4) for the case when $A = 15$ and $k = \frac{1}{3}$.
16. If digital computer facilities are available, verify that
$$(1.001)^{1000} \approx 2.71692.$$

14.2 Logarithmic Functions

We shall find that there is a close relation between exponential functions and **logarithmic functions**. Before logarithmic functions can be understood, however, we shall have to introduce the idea of the **logarithm of a positive number a to the base b**.

▶ **Definition 3**
Let a be a positive real number and let b be any positive real number except 1. Then $\log_b a$ denotes **the logarithm of a to the base b** and
$$\log_b a = c \quad \text{means that} \quad b^c = a.$$

Example 1
$$\log_2 8 = 3, \text{ because } 2^3 = 8.$$
$$\log_{10}\left(\tfrac{1}{100}\right) = -2, \text{ because } (10)^{-2} = \tfrac{1}{100}. \blacksquare$$

In Definition 3, $\log_b a$ has a unique value for given numbers a and b because we know from Section 14.1 that there is exactly one number c such that $b^c = a$.

When the base is the number e, we shall adopt the following standard notation:
$$\log_e a = \ln a.$$
For example, when $\ln 2$ is written, the base e is understood. Logarithms with base e are called **natural logarithms**. Logarithms with base 10 are called **common logarithms**.

Example 2
$$\ln \frac{1}{e} = -1,$$
$$\ln 1 = 0,$$
$$\ln e = 1,$$
$$\ln e^2 = 2. \blacksquare$$

▶ **Definition 4**
Let b be any positive real number except 1. The function g, whose domain is the set of all positive real numbers, determined by

$$g(x) = \log_b x$$

is called the **logarithmic function with base** b.

Example 3
Graph the logarithmic function g determined by

$$y = g(x) = \log_2 x.$$

One procedure is to rewrite $y = \log_2 x$ as

$$x = 2^y,$$

and then to assign values to y and compute the corresponding values of x as shown in Table 14.3. The graph of g is shown in Figure 14.4. ∎

The domain of g in Example 3 is the set of all positive real numbers, and the range of g is the set of all real numbers. As Figure 14.4 suggests, a logarithmic function is continuous over the domain of positive real numbers.

Table 14.3

x	$\frac{1}{8}$	$\frac{1}{4}$	$\frac{1}{2}$	1	2	4	8
$y = \log_2 x$	-3	-2	-1	0	1	2	3

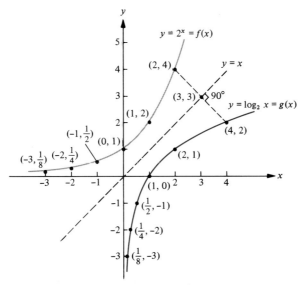

Figure 14.4. *Inverse functions.*

Sec. 14.2] Logarithmic Functions

There is a definite relationship between the functions f and g, where
$$f(x) = 2^x \quad \text{and} \quad g(x) = \log_2 x.$$
Observe that the point $(2, 4)$ satisfies the equation $y = f(x)$, whereas the point $(4, 2)$ satisfies the equation $y = g(x)$. Examination of Figure 14.4 reveals that this sort of inversion of the two coordinates of a point always occurs. It is customary to call f and g **inverse functions**—we say that g is the inverse of f and that f is the inverse of g. As Figure 14.4 indicates, the line $y = x$ is the perpendicular bisector of a segment between (x, y) and (y, x) whenever $x \neq y$; when inverse functions are plotted together, the combination is a configuration symmetric about the line $y = x$.

In Theorem 1 some of the basic properties of logarithms are listed.

▶ **Theorem 1**

Assume that a and b are positive numbers other than 1, M and N are positive numbers, and c is any real number. Then the following properties of logarithms are valid:

(i) $\log_b 1 = 0$.

(ii) $\log_b b = 1$.

(iii) $\log_b a = \dfrac{1}{\log_a b}$.

(iv) $\log_b b^c = c$.

(v) $b^{\log_b M} = M$.

(vi) $\log_b MN = \log_b M + \log_b N$.

(vii) $\log_b \dfrac{M}{N} = \log_b M - \log_b N$.

(viii) $\log_b M^c = c \log_b M$.

(ix) If $\log_b M = \log_b N$, then $M = N$.

(x) $\log_a M = \dfrac{\log_b M}{\log_b a}$.

The proofs of these properties depend primarily upon Definition 3. As an example, part (*vii*) is proved next. In words, part (*vii*) asserts that the logarithm of a quotient equals the logarithm of the numerator minus the logarithm of the denominator.

Proof of part (*vii*): Let $x = \log_b M$. Then $b^x = M$ by Definition 3. Let $y = \log_b N$. Then $b^y = N$ by Definition 3.
$$\frac{M}{N} = \frac{b^x}{b^y} = b^{x-y}.$$
By Definition 3, $M/N = b^{x-y}$ means that
$$\log_b \frac{M}{N} = x - y,$$

Chap. 14] Exponential and Logarithmic Functions

which upon substitution for x and y becomes

$$\log_b \frac{M}{N} = \log_b M - \log_b N. \blacktriangleleft$$

Example 4
(a) $\log_3 21 = \log_3 3 + \log_3 7 = 1 + \log_3 7.$
(b) $\ln \frac{7}{5} = \ln 7 - \ln 5.$
(c) $\log_{10} 3^4 = 4 \log_{10} 3.$
(d) $\ln 4 = \frac{\log_{10} 4}{\log_{10} e}.$
(e) $\ln e^x = x.$
(f) $e^{\ln x} = x.$ ∎

Example 5
Solve for x: $\log_2 x = \log_2 7 + 2 \log_2 3 - \log_2 5.$

$$\log_2 x = \log_2 \frac{(7)(3)^2}{5} = \log_2 \frac{63}{5}.$$

Therefore,

$$x = \frac{63}{5}. \blacksquare$$

Example 6
Given that $\ln 7 = 1.946$ and $\ln 11 = 2.398$, find $\ln (1/\sqrt{77})$.

$$\ln \frac{1}{\sqrt{77}} = \ln 1 - \ln (77)^{1/2}$$

$$= 0 - \frac{1}{2} \ln 77$$

$$= -\frac{1}{2} \ln [(7)(11)]$$

$$= -\frac{1}{2} (\ln 7 + \ln 11)$$

$$= -\frac{1}{2} (1.946 + 2.398)$$

$$= -2.172. \blacksquare$$

APPLICATIONS

Example 7
In Example 2 of Section 14.1 we found that if a principal P were compounded continuously at an annual interest rate r, then the amount A after t

years is given by the formula
$$A = Pe^{rt}. \tag{14.5}$$
Sometimes in a problem the unknown quantity in (14.5) is t; since natural logarithms have been introduced in this section, we are now able to solve (14.5) for t as follows:
$$e^{rt} = \frac{A}{P},$$
$$rt = \ln \frac{A}{P},$$
$$t = \frac{1}{r} \ln \frac{A}{P}. \tag{14.6}$$

Suppose, for example, that one wondered how many years it would take for a principal $P = \$700$ to accumulate to an amount $A = \$1100$ at 5 per cent compounded continuously. By (14.6)
$$t = \frac{1}{.05} \ln \frac{1100}{700}$$
$$= 20 \ln \frac{11}{7}$$
$$= 20(\ln 11 - \ln 7).$$

By use of the values of $\ln 7$ and $\ln 11$ from Example 6 of this section,
$$t = 20(2.398 - 1.946) = 9.04 \text{ years.} \blacksquare$$

Example 8
Suppose that in an experiment the data shown in Table 14.4, which relate the variables x and Y, are observed:

Table 14.4

x	1	2	4	5	7	8
Y	12	20	41	60	134	202

Apparently, Y increases rapidly as x increases. It is conjectured that x and Y are approximately related by an equation of the form
$$Y = cb^x,$$
where c is positive and the base b is greater than 1. How can one determine whether this conjecture is true?

By Theorem 1(*vi*) and (*viii*), if $Y = cb^x$, then
$$\ln Y = \ln c + x \ln b.$$

If we let $y = \ln Y$, it follows that if x is plotted against y, the graph is a line with y-intercept $\ln c$ and slope $\ln b$. Conversely, if

then
$$y = \ln c + x \ln b,$$
$$Y = e^y = e^{\ln c + x \ln b}$$
$$= e^{\ln c} e^{\ln b^x}$$
$$= cb^x.$$

Therefore the graph, if x is plotted against y, will appear to be a line if and only if the variables x and $Y = e^y$ are related by an equation of the form $Y = cb^x$.

For the data given at the beginning of this example, the values of $y = \ln Y$ can be found from a set of tables and we obtain the data given in Table 14.5. Then Figure 14.5 is constructed; the points certainly appear to lie close to the line shown in the figure. Hence we conclude that the relationship between x and Y on the interval $[1, 8]$ can be approximated by an equation of the form $Y = cb^x$. By estimating the slope $\ln b$ and the y-intercept $\ln c$ from the graph, one could obtain an estimate of b and c; more accuracy could be obtained by using a technique that we shall learn in Example 5 of Section 15.3.

In the discussion above we talked about plotting x against $\ln Y$. We could have used $\log_b Y$ instead of $\ln Y$, where b is any legitimate base. Special graph paper is available for plotting values of one variable against values of the logarithm of another variable—it is called **semilogarithmic paper.**

Applications to the biological sciences of the ideas in this example may be found on pp. 152–158 of Smith's *Biomathematics*, Vol. 1 [61]. ∎

Table 14.5

x	1	2	4	5	7	8
Y	12	20	41	60	134	202
$y = \ln Y$	2.5	3.0	3.7	4.1	4.9	5.3

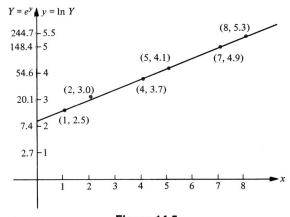

Figure 14.5

EXERCISES

Suggested minimum assignment: Exercises 1, 2, 3, 7(a) (c), and 9(c) (g) (k) (m).

1. Find each of the following:
   (a) $\log_{10} 1000$.   (b) $\ln e^3$.   (c) $\log_2 \frac{1}{4}$.
2. Solve for x, if possible:
   (a) $\log_2 16 = x$.   (b) $\log_x 16 = 2$.
   (c) $\log_5 x = -1$.   (d) $\log_5(-1) = x$.
3. Let $f(x) = e^x$ and $g(x) = \ln x$. Graph f and g on the same set of axes. Are f and g inverse functions? State the domain and range of f. State the domain and range of g.
4. Graph the exponential function with base $\frac{3}{2}$ and the logarithmic function with base $\frac{3}{2}$ on the same set of axes. What is the relationship of these functions?
5. Sketch $y = \log_{1/2} x$.
6. Sketch $y = 2\log_2(x-1)$.
7. Solve for x:
   (a) $\ln x = \ln 7 + \ln 8$.   (b) $\ln x = \ln 7 - \ln 8$.
   (c) $\ln x = 2 \ln 7$.   (d) $x = e^{\ln 5}$.
   (e) $\ln x = \dfrac{1}{\log_{10} e}$.
8. Solve for x:
   (a) $\log_3 x = (\log_3 1 + \log_3 3)(\log_3 9)$.
   (b) $\log_5 x = \log_5 3 - 2\log_5 2 + 3\log_5 5$.
9. Given that $\ln 7 = 1.946$ and $\ln 11 = 2.398$. Find each of the following:
   (a) $\ln 77$.   (b) $\ln \frac{11}{7}$.   (c) $\ln \frac{49}{11}$.
   (d) $\ln 7^2$.   (e) $(\ln 7)^2$.   (f) $\ln 7 - \ln 11$.
   (g) $\ln \sqrt{7}$.   (h) $\ln \frac{1}{11}$.   (i) $\dfrac{1}{\log_7 e}$.
   (j) $\log_{11} e$.   (k) $e^{1.946}$.   (l) $e^{2\ln 7}$.
   (m) $\log_7 11$.
10. If we use the idea of an integral with a variable upper limit from optional Section 13.6, it is possible to define $\ln x$ by
    $$\ln x = \int_1^x \frac{1}{t}\,dt, \quad x > 0.$$
    (a) Use this definition to show that $\ln 1 = 0$.
    (b) Find $D_x(\ln x)$. (*Hint:* Use Theorem 5 of Chapter 13.)
    (c) Why is the logarithmic function with base e continuous throughout its domain?
    (d) Why is the logarithmic function with base e increasing throughout its domain?
11. Prove Theorem 1(vi).

12. Sketch the function f defined by $y = f(x) = 2x + 1$. Also sketch the inverse function g of function f on the same graph and find $g(x)$.
13. (a) Why is part (iv) of Theorem 1 a special case of part $(viii)$?
    (b) Why is part (iii) of Theorem 1 a special case of part (x)?
14. Solve (14.2) on p. 578 for t.
15. The equation of a demand curve is $y = f(x) = 20 - 3x$, where $0 \le x \le 6$. Graph f and the inverse function g of f on the same axes. [*Hint:* First show that g is given by $y = g(x) = \frac{20}{3} - \frac{1}{3}x$, where $2 \le x \le 20$.] The combination is a configuration symmetric about what line?
16. Suppose for $x = 2$, $x = 4$, and $x = 5$ that Y has the values 1, 10, and 30, respectively. Plot x against $y = \ln Y$ (use $\ln 10 = 2.3$ and $\ln 30 = 3.4$). What conclusion can be drawn? (See Example 8.)

14.3 Differentiation Formulas

In this section formulas are obtained for differentiation of the logarithmic and exponential functions. These formulas, together with the fundamental knowledge about derivatives learned in Chapters 10 and 11, provide the means to solve many new applied problems. To derive the formula in Theorem 2 for differentiating a *logarithmic function*, we start with the definition of the derivative of a function (Definition 2 of Section 10.2) and proceed with the standard limit process. The formula for differentiating an *exponential function* then will be stated in Theorem 3; the derivation of the formula involves a shortcut process that makes use of Theorem 2.

▶ **Theorem 2**

If $y = \log_b u$ and $u = f(x) > 0$, where f is differentiable, then

$$D_x y = (\log_b e) \frac{1}{u} D_x u.$$

Proof: First we will find a formula for $D_x(\log_b x)$, and then the chain rule will be applied.

Statement	Reason
1. Let $g(x) = \log_b x$, and let $c > 0$ be an arbitrary element of the domain of g. Then $g'(c)$ $= \lim_{\Delta x \to 0} \dfrac{\log_b(c + \Delta x) - \log_b c}{\Delta x}$	1. Definition 2, Section 10.2.
2. $= \lim_{\Delta x \to 0} \dfrac{1}{\Delta x} \log_b \left(\dfrac{c + \Delta x}{c}\right)$	2. Theorem 1(vii), Section 14.2.
3. $= \lim_{\Delta x \to 0} \dfrac{c}{c \Delta x} \log_b \left(1 + \dfrac{\Delta x}{c}\right)$	3. Change of form.

4. $= \lim_{\Delta x \to 0} \dfrac{1}{c} \log_b \left(1 + \dfrac{\Delta x}{c}\right)^{c/\Delta x}$ 4. Theorem 1(*viii*), Section 14.2.

5. $= \lim_{z \to 0} \dfrac{1}{c} \log_b (1 + z)^{1/z}$ 5. Let $z = \dfrac{\Delta x}{c}$.

6. $= \dfrac{1}{c} \left[\lim_{z \to 0} \log_b (1 + z)^{1/z} \right]$ 6. Theorem 2(*v*), Section 9.3.

7. $= \dfrac{1}{c} \log_b \left[\lim_{z \to 0} (1 + z)^{1/z} \right]$ 7. Follows from continuity of logarithmic function.

8. $= \dfrac{1}{c} \log_b e.$ 8. Definition 2, Section 14.1.

9. $D_x(\log_b x) = (\log_b e) \dfrac{1}{x}.$ 9. Statements 1–8 and because c is an arbitrary element of the domain of g.

10. $D_x y = D_u y \, D_x u.$ 10. Theorem 2(*iii*), Section 10.4.

11. $D_x y = (\log_b e) \dfrac{1}{u} D_x u.$ 11. Statements 9 and 10. ◀

Example 1
Find $D_x y$ if $y = \log_2 (x^2 + 1)$.
    Apply Theorem 2 with $b = 2$ and $u = x^2 + 1$ to obtain

$$D_x y = (\log_2 e) \dfrac{1}{x^2 + 1} (2x)$$

$$= 2(\log_2 e) \dfrac{x}{x^2 + 1}. \blacksquare$$

The most important special case of Theorem 2 occurs when $b = e$. Fortunately, the formula is simpler in this case because $\log_b e$ equals 1 when $b = e$. Hence we have the following corollary of Theorem 2:

$$D_x \ln u = \dfrac{1}{u} D_x u. \tag{14.7}$$

Moreover, if $u = x$, we have

$$D_x \ln x = \dfrac{1}{x}.$$

Example 2
Find the slope of $f(x) = \ln (2x + 1)$ at $x = 3$.
    By (14.7)

$$f'(x) = \dfrac{1}{2x + 1} D_x(2x + 1) = \dfrac{2}{2x + 1}.$$

The desired slope, given by $f'(3)$, is found to be
$$f'(3) = \tfrac{2}{7}. \blacksquare$$

Example 3
Find y' if $y = (\ln x)/x$.

By the quotient rule
$$y' = \frac{(x)(1/x) - (\ln x)(1)}{x^2} = \frac{1 - \ln x}{x^2}. \blacksquare$$

The reader is familiar with the formula $D_x x^n = nx^{n-1}$, where n is a real number [see Theorem 1(ii) of Section 10.3]. This formula is proved below for $x > 0$ by a process called **logarithmic differentiation**.

$$y = x^n,$$
$$\ln y = n \ln x,$$
$$D_x(\ln y) = D_x(n \ln x),$$
$$\frac{1}{y} D_x y = n\left(\frac{1}{x}\right),$$
$$D_x y = \frac{ny}{x} = \frac{nx^n}{x} = nx^{n-1}.$$

The steps in the process of logarithmic differentiation illustrated above are
1. Take logarithm (preferably with base e) of both sides of the equation.
2. Differentiate both sides (note use of chain rule on left side).
3. Solve for $D_x y$ and simplify.

Logarithmic differentiation is a process to consider whenever an exponent occurs in the given function. Hence it is useful in the proof of Theorem 3 when we differentiate an exponential function. Some other situations in which the process is used will be given in the exercises.

▶ **Theorem 3**
If $y = b^u$ and $u = f(x)$, where f is differentiable, then
$$D_x y = (\ln b) b^u D_x u.$$

Outline of Proof: Use logarithmic differentiation to show that $D_x b^x = (\ln b) b^x$, and then apply the chain rule. ◀

The use of base e in Theorem 3 has the advantage (as it did in Theorem 2) of simplifying the formula. Very important special cases of Theorem 3 are
$$D_x e^u = e^u D_x u \tag{14.8}$$
and
$$D_x e^x = e^x. \tag{14.9}$$

Sec. 14.3] Differentiation Formulas

Example 4
Find $f'(x)$ if $f(x) = e^{3x} + 1/5^x$.

$$f(x) = e^{3x} + 5^{-x},$$
$$f'(x) = e^{3x}D_x(3x) + (\ln 5)5^{-x}D_x(-x)$$
$$= 3e^{3x} - (\ln 5)5^{-x}. \blacksquare$$

Example 5
In a suitable medium a full-grown bacterial cell divides by fission into two equal cells. Suppose that the number y of cells present at time t is given by

$$y = a(2)^{t/k}.$$

In this formula a and k are constants—a is the number of cells present initially and k is the average time required for a cell to double. Find the rate of change dy/dt.

By Theorem 3,

$$\frac{dy}{dt} = a(\ln 2)(2)^{t/k}D_t\left(\frac{t}{k}\right)$$
$$= \left(\frac{a}{k}\right)(\ln 2)(2)^{t/k}. \blacksquare$$

Example 6
Find all maximum and minimum points of f if $y = f(x) = xe^{-x}$.

The domain of f is the set of all real numbers. Even though f is not a rational function, we shall assume (as is the case) that the methods of Chapter 11 are applicable. By (14.8) and the product rule,

$$f'(x) = x(-e^{-x}) + e^{-x}(1) = e^{-x}(1 - x).$$

Since e^{-x} is never zero, the only critical number is $x = 1$. Therefore, $(1, 1/e)$ is the only possible maximum or minimum point. See Table 14.6 and Figure 14.6. $\blacksquare$

Table 14.6

Interval	Calculation of sign of $f'(x)$	Result	Conclusion
$(-\infty, 1)$	$(+)(+)$	$+$; f increasing	
			$(1, 1/e)$ is maximum point
$(1, \infty)$	$(+)(-)$	$-$; f decreasing	

APPLICATIONS

Example 7
In many applications data are studied over a period of time, and an attempt is made to discover an equation that approximately agrees with the data. In

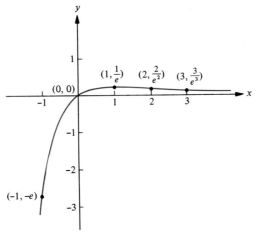

Figure 14.6. $y = xe^{-x}$.

other words, an attempt is made to find a mathematical model that agrees with the observations. Such a model can then be used to study the situation to an extent that is almost impossible from the data alone. The type of equation that is hypothesized, particularly in the biological sciences, often involves logarithmic or exponential functions.

As an example, suppose that a biologist is studying the growth of infants. He has found that the following formula predicts rather accurately the relationship between the age t of a child in months and the weight x of the child in ounces:

$$t = 2 + 2 \ln \frac{x}{2a - x}.$$

The number a is a constant and stands for the weight at age 2 months. The formula is claimed to be valid only between the ages of 2 months and 6 months.

(a) Solve for x and verify that $x = a$ when $t = 2$.

$$\ln \frac{x}{2a - x} = \frac{t}{2} - 1,$$

$$\frac{x}{2a - x} = e^{(t/2)-1},$$

$$x + xe^{(t/2)-1} = 2ae^{(t/2)-1},$$

$$x = \frac{2ae^{(t/2)-1}}{1 + e^{(t/2)-1}}.$$

When $t = 2$,

$$x = \frac{2ae^0}{1 + e^0} = \frac{2a}{2} = a.$$

Sec. 14.3] Differentiation Formulas

(b) A particular child weighs 200 ounces at age 2 months. What is its weight expected to be at age 4 months?

To answer this question, it is necessary to find x, given that $a = 200$ and $t = 4$.

$$x = \frac{2(200)e^{2-1}}{1 + e^{2-1}}$$

$$= \frac{(400)(2.71828)}{3.71828}$$

$$= 292 \text{ ounces.}$$

(c) Find the rate of growth dx/dt when $t = 4$. Assume again that $a = 200$.

Starting with

$$x = \frac{400e^{(t/2)-1}}{1 + e^{(t/2)-1}},$$

we find dx/dt by using the quotient rule and also (14.8) of this section.

$$\frac{dx}{dt} = 400 \frac{(1 + e^{(t/2)-1})e^{(t/2)-1}(\frac{1}{2}) - e^{(t/2)-1}e^{(t/2)-1}(\frac{1}{2})}{(1 + e^{(t/2)-1})^2}.$$

Rather than attempt to simplify, it is probably more efficient to substitute $t = 4$ immediately.

$$\left.\frac{dx}{dt}\right|_{t=4} = 400 \frac{(1 + e)(e)(\frac{1}{2}) - (e)(e)(\frac{1}{2})}{(1 + e)^2}$$

$$= 400 \frac{\frac{1}{2}e}{(1 + e)^2}$$

$$= 200 \frac{2.71828}{(3.71828)^2}$$

$$= 39 \text{ ounces per month.} \blacksquare$$

EXERCISES

Suggested minimum assignment: Exercises 1(a) (d) (g) (k) (l), 2, 4(a), 5(a), and 6(c).

1. Find $D_x y$ in each of the following:
   (a) $y = \log_3 (2x + 1)$.
   (b) $y = \log_{10} (x^3 + 2)$.
   (c) $y = \ln (2x + 5)$.
   (d) $y = \ln \sqrt{x}$.
   (e) $y = (1 + \ln x)^8$.
   (f) $y = x^4 \ln x$.
   (g) $y = 3^{2x}$.
   (h) $y = x2^x$.
   (i) $y = \dfrac{e^x - e^{-x}}{2}$.
   (j) $y = e^{\sqrt{2x+1}}$.
   (k) $y = \dfrac{e^{3x}}{9 + e^{3x}}$.
   (l) $y = \sqrt{1 + e^{2x}}$.

2. Prove Theorem 3.
3. Use logarithmic differentiation to show that if $u > 0$ is a differentiable function of x and n is any real number, then
$$D_x u^n = n u^{n-1} D_x u.$$
4. Use logarithmic differentiation to find $D_x y$ in each of the following. [Assume that $x > 0$ in parts (a) and (b) and that $(1 - 3x) > 0$ in part (c).]

   (a) $y = x^x$.   (b) $y = (x)^{e^x}$.   (c) $y = \dfrac{\sqrt{x^2 + 1}}{(1 - 3x)^7}$.

5. Find all maximum and minimum points of each of the given functions. Assume that the methods of Chapter 11 are applicable.
   (a) $y = f(x) = x^2 \ln x$.   (b) $y = f(x) = x \ln x$.
   (c) $y = f(x) = e^{-x^2}$.   (d) $y = f(x) = 2xe^{-x^2}$.
6. Find the slope of each of the following at the given point.

   (a) $y = \log_2 x$, $(8, \dot{3})$.   (b) $y = e^{2x} + \dfrac{1}{e^{2x}}$, $(0, 2)$.

   (c) $y = e^{2/x}$, $(2, e)$.
7. (a) In Example 7 find x if $a = 200$ and $t = 5$. (Use $e^{1.5} = 4.4817$.)

   (b) In Example 7 it was found that $\left.\dfrac{dx}{dt}\right|_{t=4} = 39$ ounces per month, but from Example 7 and part (a) of this exercise it can be concluded that the child gained only 35 ounces during its fifth month. How can you account for the difference?
8. In Example 6 (see Figure 14.6), find all points of inflection. Assume that the methods of Section 11.5 are applicable.
9. Consider the equation $2^x + x - 4 = 0$, which has a root between $x = 1$ and $x = 2$. Use Newton's method (see Example 8 of Section 10.3) to find a second approximation x_2 of the root if the first approximation is $x_1 = 1.5$. (Use $\sqrt{2} = 1.414$ and $\ln 2 = .693$.)

14.4 Integration Formulas

The main integration formula studied so far is the power rule for integration in Section 12.2. This rule is essentially the power rule for differentiation (Section 10.3) in reverse. In this section we shall take the formulas in Section 14.3 for differentiation of the exponential and logarithmic functions and reverse them in the proper manner to obtain two new integration formulas. With these new formulas the basic ideas of Chapters 12 and 13 can be applied to many more problems.

▶ **Theorem 4**

Let $u = f(x)$, where f is differentiable. Then
$$\int b^{f(x)} f'(x)\, dx = \int b^u\, du = \frac{b^u}{\ln b} + C.$$

Proof: It is understood above that the base b is a positive real number other than 1. In the following calculation Theorem 4 is verified; notice the use of Theorem 3 of Section 14.3.

$$D_x\left(\frac{b^u}{\ln b} + C\right) = D_x\left(\frac{b^{f(x)}}{\ln b} + C\right)$$

$$= \frac{1}{\ln b}(\ln b)b^{f(x)}f'(x)$$

$$= b^{f(x)}f'(x). \blacktriangleleft$$

Important special cases of Theorem 4 are

$$\int e^u \, du = e^u + C \tag{14.10}$$

and

$$\int e^x \, dx = e^x + C. \tag{14.11}$$

The difficulties in handling the du correctly in Theorem 4 are similar to those encountered in Section 12.2—it is suggested that the reader review the examples in Section 12.2 before proceeding.

Example 1

$$\int 3^{2x} \, dx = \frac{1}{2}\int 3^{2x}(2 \, dx)$$

$$= \frac{1}{2}\int 3^u \, du = \frac{1}{2}\frac{3^u}{\ln 3} + C$$

$$= \frac{3^{2x}}{2 \ln 3} + C. \blacksquare$$

Example 2

$$\int \frac{e^{1/x}}{x^2} \, dx = -\int e^{1/x}\left(-\frac{1}{x^2} \, dx\right)$$

$$= -\int e^u \, du = -e^u + C$$

$$= -e^{1/x} + C. \blacksquare$$

The answer to Example 2 can be checked by differentiating the answer to obtain the original integrand:

$$D_x(-e^{1/x} + C) = -e^{1/x}\left(-\frac{1}{x^2}\right) = \frac{e^{1/x}}{x^2}.$$

Example 3

$$\int [e^x + x(1 + e^{x^2})]\, dx = \int e^x\, dx + \int x\, dx + \int xe^{x^2}\, dx$$
$$= e^x + \frac{x^2}{2} + \frac{1}{2}\int e^{x^2}(2x\, dx)$$
$$= e^x + \frac{x^2}{2} + \frac{1}{2}\int e^u\, du$$
$$= e^x + \frac{x^2}{2} + \frac{1}{2}e^u + C$$
$$= e^x + \frac{x^2}{2} + \frac{1}{2}e^{x^2} + C. \blacksquare$$

Example 4

Find the area bounded by $y = e^{-x}$, $y = x + 1$, and $x = 1$. (See Figure 14.7.)

$$A = \int_0^1 [(x + 1) - e^{-x}]\, dx$$
$$= \int_0^1 x\, dx + \int_0^1 1\, dx + \int_0^1 e^{-x}(-dx)$$
$$= \left(\frac{x^2}{2} + x + e^{-x}\right)\Big|_0^1$$
$$= \left(\frac{1}{2} + 1 + e^{-1}\right) - (0 + 0 + 1)$$
$$= \frac{1}{2} + \frac{1}{e}. \blacksquare$$

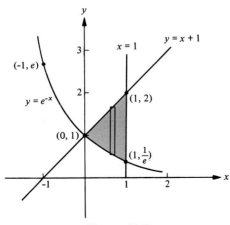

Figure 14.7

Sec. 14.4] Integration Formulas

In the last section we learned that $D_x \ln x = 1/x$; this formula is valid only when $x > 0$ because $\ln x$ is not defined if $x \leq 0$. To prepare for the next theorem, it is necessary to investigate the possibility of finding $D_x \ln |x|$. Since $\ln |x|$ has meaning except when $x = 0$, we split the problem into two cases:

1. $x > 0$.

$$D_x \ln |x| = D_x \ln x = \frac{1}{x}.$$

2. $x < 0$.

$$D_x \ln |x| = D_x \ln(-x) = \frac{1}{-x} D_x(-x) = \frac{1}{-x}(-1) = \frac{1}{x}.$$

Since the result is the same in either case, it can be concluded that

$$D_x \ln |x| = \frac{1}{x}, \qquad x \neq 0.$$

As an immediate consequence the following formula can be written:

$$\int \frac{1}{x} dx = \ln |x| + C, \qquad x \neq 0.$$

Theorem 5 is a generalization of the above result; it can be verified in a manner similar to that used in the verification of Theorem 4 of this section.

▶ **Theorem 5**
Let $u = f(x)$, where f is differentiable. Then

$$\int \frac{du}{u} = \ln |u| + C, \qquad u \neq 0.$$

The proof is left as Exercise 6. ◀

Example 5

$$\int \frac{x \, dx}{16 - x^2} = -\frac{1}{2} \int \frac{-2x \, dx}{16 - x^2}$$

$$= -\frac{1}{2} \int \frac{du}{u} = -\frac{1}{2} \ln |u| + C$$

$$= -\frac{1}{2} \ln |16 - x^2| + C. \ \blacksquare$$

It was pointed out in Section 12.2 that the power rule for integration,

$$\int u^n \, du = \frac{u^{n+1}}{n+1} + C,$$

is valid for any real number n except $n = -1$. Theorem 5 takes care of the exceptional case when $n = -1$.

At this stage the three basic tools at our disposal for doing integration problems are Theorem 4, Theorem 5, and the power rule for integration. Example 6 is designed to demonstrate the use of all three of these formulas.

Example 6

$$\int e^x \left[1 + \frac{1}{e^x + 1} + (e^x + 1)^2\right] dx$$

$$= \int e^x \, dx + \int \frac{e^x \, dx}{e^x + 1} + \int (e^x + 1)^2 e^x \, dx$$

$$= \int e^x \, dx + \int \frac{du}{u} + \int u^2 \, du$$

$$= e^x + \ln|u| + \frac{u^3}{3} + C$$

$$= e^x + \ln|e^x + 1| + \frac{(e^x + 1)^3}{3} + C.$$

In this answer it is acceptable to write $\ln(e^x + 1)$ instead of $\ln|e^x + 1|$ because the quantity $e^x + 1$ is always positive. ∎

Example 7
Find the area bounded by $y = 4/(2x + 1)$, $y = 0$, $x = -2$, and $x = -1$. (See Figure 14.8.)

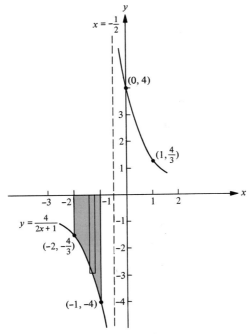

Figure 14.8

$$A = \int_{-2}^{-1} \left(0 - \frac{4}{2x+1}\right) dx$$

$$= -2 \int_{-2}^{-1} \frac{2\, dx}{2x+1}$$

$$= -2 \ln|2x+1| \Big|_{-2}^{-1}$$

$$= -2(\ln 1 - \ln 3)$$

$$= 2 \ln 3. \blacksquare$$

In using Theorem 5, it is important to avoid numbers that make the denominator zero. In Example 7 the denominator $2x + 1$ equals zero when $x = -\frac{1}{2}$ (this corresponds to the discontinuity in Figure 14.8 at $x = -\frac{1}{2}$). Observe that the interval $[-2, -1]$ of integration does not include $-\frac{1}{2}$.

APPLICATIONS

Example 8

Before reading this example, the reader should review the equation

$$y = A(1 - e^{-kt})$$

for bounded growth that was presented in Example 3 of Section 14.1. A reasonable assumption to make when y cannot exceed the value A is that the rate of growth dy/dt is proportional to the difference $A - y$. In this example we shall show that the solution of the differential equation

$$\frac{dy}{dt} = k(A - y),$$

subject to the condition that $y = 0$ when $t = 0$, is

$$y = A(1 - e^{-kt}).$$

Recall from Section 12.3 that our procedure for solving a differential equation is to separate the variables and then integrate.

$$\frac{dy}{A - y} = k\, dt,$$

$$-\int \frac{-dy}{A - y} = \int k\, dt,$$

$$-\ln|A - y| = kt + C.$$

Since $y = 0$ when $t = 0$,

$$-\ln|A| = 0 + C$$

and

$$C = -\ln|A|.$$

Chap. 14] Exponential and Logarithmic Functions

Because A and $A - y$ are positive, the absolute-value signs may be omitted and the solution may be written

$$-\ln(A - y) = kt - \ln A.$$

This result can be solved for y as follows:

$$\ln(A - y) = \ln A - kt,$$
$$A - y = e^{\ln A - kt}$$
$$= e^{\ln A} e^{-kt}$$
$$= A e^{-kt},$$
$$y = A - A e^{-kt} = A(1 - e^{-kt}). \blacksquare$$

Example 9

Often a single mathematical model will apply to a variety of applied problems. In fact, to the observant researcher, the application of a model in one discipline may suggest the application of the same model in other disciplines. The differential equation

$$\frac{dy}{dt} = k(A - y)$$

of Example 8 may be thought of as a special case of a differential equation of the form

$$\frac{dy}{dt} = ay + b. \tag{14.12}$$

In this example we shall solve (14.12) and mention some other applications in which this type of equation might arise.

Rewrite (14.12) as

$$\frac{dy}{y + (b/a)} = a\,dt.$$

This separates the variables, so we proceed to integrate each side of the equation and then to solve for y.

$$\ln\left|y + \frac{b}{a}\right| = at + C,$$
$$\left|y + \frac{b}{a}\right| = e^{at+C},$$
$$y = \pm e^{at+C} - \frac{b}{a}$$
$$= c e^{at} - \frac{b}{a}, \tag{14.13}$$

where $c = \pm e^C$. (The reader may wish to verify that the result agrees with the solution in Example 8 with $a = -k$, $b = kA$, and $c = -A$.)

Suppose that the size y of a population of objects will be affected by the

birth rate, death rate, immigration (entrance) rate, and emmigration (exit) rate. Under the assumptions that the birth rate r_B and the death rate r_D are proportional to the size of the existing population, and that the immigration and emmigration rates (r_i and r_e, respectively) are constant, we have

$$\frac{dy}{dt} = r_B - r_D + r_i - r_e$$
$$= k_1 y - k_2 y + r_i - r_e$$
$$= (k_1 - k_2)y + (r_i - r_e)$$
$$= ay + b,$$

where $a = k_1 - k_2$ and $b = r_i - r_e$. We have found that the solution of this differential equation is

$$y = ce^{at} - \frac{b}{a},$$

and thus the size of the population at time t is given by

$$y = ce^{(k_1 - k_2)t} - \frac{r_i - r_e}{k_1 - k_2}.$$

On pp. 293–298 of Batschelet [6] the applications mentioned above, which lead to equations of the form of (14.12), are discussed along with applications concerning nerve excitation, diffusion, and cooling. ∎

Example 10

Extensive use of exponential functions, logarithmic functions, and definite integrals is made in *The Growth and Structure of Human Populations, A Mathematical Investigation*, by Coale [17]. He defines a stable population as a population that is attained if fertility and mortality schedules remain unchanged over a long period of time, and he obtains formulas for the calculation of many different quantities associated with a stable population. For example, let x be the variable representing age, and let the value h of x be the highest age in the stable population; also, suppose that the rate r of increase in the stable population is known and the proportion $p(x)$ surviving to age x is known. On p. 17 Coale shows that the birth rate b is given by

$$b = \frac{1}{\int_0^h e^{-rx} p(x)\, dx}. \quad \blacksquare$$

EXERCISES

Suggested minimum assignment: Exercises 1(a) (c) (e), 2(c) (d), 3(b), and 7(b) (e).

1. Perform the indicated integration.

(a) $\int 7^{4x}\, dx.$ (b) $\int 2^{1+x^2} x\, dx.$

(c) $\int \dfrac{e^x - e^{-x}}{2}\, dx.$ (d) $\int \dfrac{dx}{2x+1}.$

(e) $\int \dfrac{x^2\, dx}{x^3+2}.$ (f) $\int \dfrac{\ln x}{x}\, dx.$

2. Perform the indicated integration.

(a) $\int \dfrac{dx}{2^x}.$ (b) $\int 3xe^{x^2}\, dx.$

(c) $\int \dfrac{e^{2x}}{e^{2x}+1}\, dx.$ (d) $\int \dfrac{e^{2x}}{(e^{2x}+1)^2}\, dx.$

(e) $\int \dfrac{2t+3}{t^2+3t+1}\, dt.$ (f) $\int \dfrac{1/x}{\ln x}\, dx.$

3. Evaluate the following definite integrals.

(a) $\int_0^3 e^{x/3}\, dx.$ (b) $\int_0^{e^2-1} \dfrac{dt}{t+1}.$ (c) $\int_0^{\ln 2}(e^x - 1)^{10} e^x\, dx.$

4. Sketch the graph of $y = \ln|x|$.

5. By long division verify that $\dfrac{x}{x+1} = 1 - \dfrac{1}{x+1}$. Use this identity to find

$\int \dfrac{x}{x+1}\, dx.$

6. Prove Theorem 5.

7. Draw a graph and find the area bounded by the given curves.
    (a) $y = e^x,\ y = 0,\ x = 0,\ x = 2.$
    (b) $y = \ln x,\ x = 0,\ y = 0,\ y = 1.$
    (c) $y = 3^x,\ x = 0,\ y = 3x.$
    (d) $y = \dfrac{1}{2x},\ y = 0,\ x = 1,\ x = 2.$
    (e) $y = \dfrac{1}{1-x},\ x = -4,\ x = 0,\ y = 0.$

8. Suppose that y represents the ratio of voters who would vote for candidate Q if an election were held today, and that x represents the amount of money that Q spends on television advertisements. Assume that dy/dx is directly proportional to the amount of money spent and to the ratio of those who would not vote for Q today. In other words, assume that

$$\dfrac{dy}{dx} = kx(1-y),$$

where k is the constant of proportionality. Solve this equation for y as a function of x, using methods similar to those demonstrated in Examples 8 and 9. Assume that $y = .3$ when $x = 0$ in order to determine the constant of integration.

14.5 Exponential Growth and Decay

At the outset of this chapter we stated that in many applications the assumption is made that the rate of change dy/dt of the amount y of a quantity with respect to time t is proportional to the amount y that is present. In other words, we assumed the differential equation

$$\frac{dy}{dt} = ky, \qquad (14.14)$$

where k is the constant of proportionality. We also stated that the solution of (14.14) is

$$y = Ae^{kt}, \qquad (14.15)$$

where $A > 0$ is the value of y at a given starting time $t = 0$. If $k > 0$, (14.15) is called the **exponential growth law**; if $k < 0$, (14.15) is called the **exponential decay law**.

Example 1
In this example we obtain (14.15) from (14.14) using methods of Section 12.3. Recall from Section 12.3 that (14.14) is called a first-order differential equation and that this type of equation can often be solved by separating the variables and integrating.

$$\frac{dy}{y} = k\, dt,$$

$$\int \frac{dy}{y} = k \int dt.$$

Applying Theorem 5 of Section 14.4, we obtain

$$\ln |y| = kt + C. \qquad (14.16)$$

Next find C. Since $y = A$ when $t = 0$,

$$\ln |A| = C,$$

and hence (14.16) becomes

$$\ln |y| = kt + \ln |A|.$$

Because the amount y of the quantity present is assumed to always be positive, the absolute-value signs may be deleted to yield

$$\ln y = kt + \ln A.$$

Next we solve for y as follows:

$$\ln y - \ln A = kt,$$
$$\ln \frac{y}{A} = kt,$$
$$\frac{y}{A} = e^{kt},$$
$$y = Ae^{kt}. \blacksquare$$

W. F. Libby [43] received a Nobel prize for his efforts in showing that by determination of the radioactivity of carbon it is possible to establish accurately the age of organic matter. When a plant or animal dies, the intake of radioactive carbon ceases and the radioactive decay of carbon 14 takes over. Suppose that carbon 14 is the only radioactive substance in a given specimen. It is known that the rate at which carbon 14 disintegrates is proportional to the amount that remains, and hence (14.14) applies. It is also known that it takes about 5760 years for one half of a quantity of carbon 14 to decay—in other words, the **half-life** of carbon 14 is approximately 5760 years.

Example 2

Suppose that a piece of wood found in an archaeological dig is compared with a piece of new wood and it is determined that the specimen contains 85 per cent of its original amount A of carbon 14. Estimate the age of the specimen. (See Table 14.7.)

Table 14.7

y	t
A	0
$\frac{1}{2}A$	5760
$.85A$	?

The amount y of carbon 14 remaining after t years by (14.15) is
$$y = Ae^{kt}.$$
The constant of proportionality k can be determined by using the fact that the half-life of carbon 14 is 5760 years.
$$\tfrac{1}{2}A = Ae^{k(5760)},$$
$$\tfrac{1}{2} = e^{5760k},$$
$$5760k = \ln \tfrac{1}{2},$$
$$k = \tfrac{1}{5760}(\ln 1 - \ln 2) = \tfrac{1}{5760}(-\ln 2).$$

Therefore, the equation
$$y = Ae^{-(t \ln 2)/5760}$$

Sec. 14.5] Exponential Growth and Decay

is valid for any problem of this type in which the radioactive substance is carbon 14. Since 85 per cent of A remains in the given specimen,

$$.85A = Ae^{-(t \ln 2)/5760},$$

$$\ln .85 = \frac{-t \ln 2}{5760},$$

$$t = \frac{-5760 \ln .85}{\ln 2}$$

$$= \frac{-5760(-.16252)}{.69315}$$

$$= 1351 \text{ years.} \blacksquare$$

Example 3 involves exponential growth rather than exponential decay. We shall not assume (14.15), but rather we shall reemphasize the entire process by starting with (14.14).

Example 3

Suppose that the number of bacteria in a culture increases at a rate proportional to the number present. If the initial count of bacteria is 500 and after 2 hours the count is 800, how many are expected to be present after 3 hours? Also find the number of hours elapsed until the count reaches 1500.

We assume the equation

$$\frac{dy}{dt} = ky,$$

where y is the number of bacteria present after t hours and k is the constant of proportionality. The known and required data are summarized in Table 14.8.

Table 14.8

y	t
500	0
800	2
?	3
1500	?

The procedure in this type of problem is to find a solution of the given differential equation that relates y and t and that agrees with Table 14.8. Then if either y or t is given, the other one can be found.

$$\frac{dy}{y} = k\,dt,$$

$$\int \frac{dy}{y} = k \int dt,$$

$$\ln|y| = kt + C. \tag{14.17}$$

Since $y = 500$ when $t = 0$,

$$\ln 500 = k(0) + C$$

and

$$C = \ln 500.$$

When we substitute $C = \ln 500$ in (14.17) and note that $|y| = y$ because $y > 0$, we obtain

$$\ln y = kt + \ln 500. \tag{14.18}$$

The constant k can be determined by substituting $y = 800$ and $t = 2$:

$$\ln 800 = 2k + \ln 500,$$
$$2k = \ln 800 - \ln 500 = \ln \tfrac{800}{500},$$
$$k = \tfrac{1}{2} \ln \tfrac{8}{5}.$$

Equation (14.18) becomes

$$\ln y = (\tfrac{1}{2} \ln \tfrac{8}{5})t + \ln 500. \tag{14.19}$$

Equation (14.19) can be solved for y as follows:

$$y = e^{[(1/2)\ln(8/5)]t + \ln 500}$$
$$= e^{[(1/2)\ln(8/5)]t} e^{\ln 500}$$
$$= 500 e^{[(1/2)\ln(8/5)]t}, \tag{14.20}$$

which is of the form of (14.15).

If $t = 3$, y can be found from (14.20),

$$y = 500 e^{(3/2)\ln(8/5)} = 500 e^{\ln(8/5)^{3/2}}$$
$$= 500(\tfrac{8}{5})^{3/2} = 500(\tfrac{8}{5})\sqrt{8/5}$$
$$= 800(1.2649) = 1012 \text{ bacteria}.$$

If $y = 1500$, t can be found from (14.20),

$$1500 = 500 e^{[(1/2)\ln(8/5)]t},$$
$$e^{\ln(8/5)^{t/2}} = 3,$$
$$\left(\frac{8}{5}\right)^{t/2} = 3,$$
$$\frac{t}{2} \ln \frac{8}{5} = \ln 3,$$
$$t = \frac{2 \ln 3}{\ln 1.6} = \frac{2(1.0986)}{.4700} = 4.675 \text{ hours}. \blacksquare$$

APPLICATIONS

The mathematical techniques necessary for solving problems involving (14.14) or (14.15) have been illustrated in the preceding examples. In Example 4 we mention some other settings in which the type of analysis learned in this section might be applicable.

Example 4

(a) Straight-line depreciation was mentioned briefly at the outset of Section 3.8. Sometimes an object, however, such as a new car, seems to depreciate more rapidly at first than later on. In this case it might be logical to assume that the rate of change of value of the object is proportional to its value. In other words, an equation of the type (14.14) might be assumed under some circumstances when computing depreciation.

(b) The equation $A = Pe^{rt}$ was used when dealing with problems in which interest was compounded continuously (see Example 2 of Section 14.1 and Example 7 of Section 14.2). This equation has the same form as (14.15). In fact, (14.15) is sometimes called the **compound interest law**.

(c) In studying population growth, the assumption that the rate of change of population is proportional to the population is often valid over limited periods of time. In other words, equation (14.14) might be assumed in this type of problem for some interval of time. Example 3 can be considered to be in this category. Some discussion of the underlying assumptions made when using (14.14) may be found on pp. 7–8 of An Introduction to Mathematical Ecology by Pielou [51].

(d) On pp. 285–292 Batschelet [6] discusses several applications of (14.14), including the following one adapted from p. 305 of Ackerman [1]. Let x be the number of ionizing particles that bombard a given unit of tissue during radiation treatment. Let A be the original number of undamaged molecules of a chemical compound present in the given unit of tissue, and let y be the number of undamaged molecules after exposure to radiation. Note that if $x = 0$, then $y = A$, and as x increases, y decreases. It may be assumed that

$$\frac{dy}{dx} = ky,$$

where k is a negative constant of proportionality; this equation is of the form (14.14), although the independent variable is not time, as it has been in our other applications.

(e) In a chemical reaction (14.14) may be applicable in determining the amount of a substance present after a certain amount of time. Suppose that 2 pounds of a substance are present initially in a chemical reaction, but after 30 minutes only 1.2 pounds remain. If (14.14) is assumed, it can be shown (Exercise 10) that after 3 hours only slightly less than .1 pound of the substance is present. ∎

Example 5

In an article [15] entitled "Turnover and Tenure in the Canadian House of Commons 1867–1968," Casstevens and Denham develop a measure c of the turnover rate of the deputies. The measure is obtained by starting with the differential equation

$$\frac{dM}{dt} = -cM, \qquad (14.21)$$

where M represents the number of deputies still holding office after t units of time have elapsed from some initial point in time; the constant c is the constant of proportionality and represents the turnover rate per Canadian deputy. The method requires the reasonable assumption (14.21) that the rate of change of M with respect to t will be directly proportional to M. The solution of the differential equation (14.21) is

$$M = M_0 e^{-ct}, \qquad (14.22)$$

where M_0 is the number of deputies in the house at the beginning of a particular session (265 in 1953).

Taking the natural logarithm of both sides of (14.22), one gets a linear equation in t and $\ln M$ (see Example 8 of Section 14.2). Historical data such as those given in Table 14.9 (where t measures the number of general elections) can be used in the linear equation to obtain the slope $(-c)$ of the line of best fit for the 1953 house. (Use the method of least squares, to be explained in Example 5 of Section 15.3.) In this manner one can develop a turnover rate c for each house following a general election. A table of such rates for the Canadian House of Commons shows that the rates increased slightly over the years 1867–1958. The half-life and expectation can be calculated from the turnover rates.

Table 14.9. 1953 House of Commons.

t	0	1	2	3	4	5	6
M	265	146	73	48	32	21	9

Several interesting possibilities arise from this discussion: (1) Turnover rates c_i for different legislative bodies (or different subsets of the same body) can be calculated and compared [reasons for any differences (or similarities) can be investigated]; (2) institutions other than legislatures, such as universities, can be compared with respect to their faculty turnover rates c_i; (3) the effect of a proposed change on the turnover rate for a certain institution can be ascertained in advance by comparing the turnover rates c_i of those similar institutions that have already effected the change with those that have not effected the change; and (4) in those instances in which the turnover rates have very little change, a degree of predictive value can be obtained. ∎

EXERCISES

Suggested minimum assignment: Exercises 1, 3, and 7.

1. Solve (14.14) for y in terms of t if it is known that $y = 900$ when $t = 0$ and that $y = 1500$ when $t = 5$.
2. Solve (14.14) for y in terms of t if it is known that $y = 20$ when $t = 0$ and that $y = 5$ when $t = 10$.

Sec. 14.5] Exponential Growth and Decay

3. The rate of decay of a radioactive element is proportional to the amount present. Suppose that a sample weighs 200 milligrams. At the end of 2 years 150 milligrams remain. How much will remain after 3 years? (Use $\sqrt{3} = 1.73205$.)
4. Rework Example 2 if the specimen contains 91 per cent (instead of 85 per cent) of its original amount of carbon 14. (Use $\ln .91 = -.09431$.)
5. In this exercise use the values of $\sqrt{\frac{8}{5}}$ and $\ln 1.6$ given in Example 3.
   (a) In Example 3 find y when $t = 1$.
   (b) In Example 3 find t if $y = 1000$. (Use $\ln 2 = .69315$.)
6. In 18 years a quantity of a radioactive element decays to 65 per cent of the original amount present. Find the half-life of the element. (Use $\ln 2 = .69315$ and $\ln .65 = -.43078$.)
7. A new car is purchased for $4000. During the next 10 years the rate of change in the value of the car is expected to be proportional to its value, and after 10 years the car is expected to be worth $100 scrap value. What will the value of the car be 5 years from now? (Use $\sqrt{10} = 3.162$.)
8. The population of a certain state has increased from 3.2 million to 4.8 million during the last 20 years. Assume the exponential growth law; what is the population expected to be 5 years from now? (Use $\sqrt[4]{1.5} = 1.107$.)
9. A radioactive element present in freshly cut hay exists to an extent estimated to be five times the safe level for feeding to dairy cows. The element is iodine 131, which has a half-life of 8 days. Assume the exponential decay law; how many days must pass before it is safe to feed the hay to the cows? (Use $\ln 2 = .69315$ and $\ln 5 = 1.60944$.)
10. Verify the statement made in Example 4(e).

14.6 Integration Using Table of Integration Formulas

In practice, integration problems may be encountered that require formulas that have not been studied in this text. To help remedy this situation, a list of 25 selected integration formulas is given on p. 656; only the first seven of these formulas should look familiar. These 25 formulas should suffice for many applications as well as for the examples and exercises of this section. Much longer tables of integration formulas are given in handbooks of mathematical tables.† The use of a table of integration formulas is not as easy as one might expect. Some basic knowledge of integration is needed in order to select the right formula and to use it correctly.

Example 1

$$\int \frac{dx}{9 - 4x^2}.$$

† A total of 596 formulas are given in *Standard Mathematical Tables*, 19th ed., Chemical Rubber Company, Cleveland, Ohio, 1971, pp. 395–444.

A study of the table of integration formulas on p. 656 should reveal that formula 9 is the one to use. First we rewrite

$$\int \frac{dx}{9 - 4x^2} = \frac{1}{2} \int \frac{2\, dx}{(3)^2 - (2x)^2}.$$

Then formula 9 can be applied with $a = 3$, $u = 2x$, and $du = 2\, dx$ to obtain

$$\int \frac{dx}{9 - 4x^2} = \frac{1}{2} \left[\frac{1}{2(3)} \ln \left| \frac{3 + 2x}{3 - 2x} \right| \right] + C$$

$$= \frac{1}{12} \ln \left| \frac{3 + 2x}{3 - 2x} \right| + C.$$

An alternative procedure in this example is to write

$$\int \frac{dx}{9 - 4x^2} = \frac{1}{4} \int \frac{dx}{(\frac{3}{2})^2 - x^2}$$

and to apply formula 9 with $a = \frac{3}{2}$, $u = x$, and $du = dx$. (In Exercise 1 the reader is asked to carry out the problem by this method.) ∎

Example 2

$$\int \frac{dx}{8 + 4x - 4x^2}.$$

It may appear that this problem does not fit any of the 25 formulas in the table on p. 656 and that a more comprehensive table of integration formulas is needed. However, by completing the square the problem may be rewritten as

$$\int \frac{dx}{9 - (4x^2 - 4x + 1)} = \int \frac{dx}{(3)^2 - (2x - 1)^2},$$

and formula 9 can be used again exactly as in Example 1 except that $u = 2x - 1$ instead of $u = 2x$. The answer is

$$\int \frac{dx}{8 + 4x - 4x^2} = \frac{1}{2} \left[\frac{1}{2(3)} \ln \left| \frac{3 + (2x - 1)}{3 - (2x - 1)} \right| \right] + C$$

$$= \frac{1}{12} \ln \left| \frac{1 + x}{2 - x} \right| + C. \ \blacksquare$$

Example 3

$$\int x^2 e^{5x}\, dx.$$

Prepare to use formula 18 with $u = 5x$, $du = 5\, dx$, and $n = 2$ by observing that

$$\int x^2 e^{5x}\, dx = \tfrac{1}{125} \int (5x)^2 e^{5x}(5\, dx).$$

Hence, by formula 18,

$$\int x^2 e^{5x}\, dx = \tfrac{1}{125}[(5x)^2 e^{5x} - 2\int (5x)^1 e^{5x}(5\, dx)].$$

The problem is not completed, but formula 18, known as a **reduction formula**, has reduced the power of x from the second power to the first power. Another application of the formula will reduce the exponent from 1 to 0. Note that $n = 1$ this time, whereas $u = 5x$ and $du = 5\, dx$ are the same.

$$\int x^2 e^{5x}\, dx = \tfrac{1}{125}(5x)^2 e^{5x} - \tfrac{2}{125}\left(5x e^{5x} - \int e^{5x} 5\, dx\right).$$

The only remaining integration can be done by formula 6, which was studied in Section 14.4.

$$\int x^2 e^{5x}\, dx = \tfrac{1}{5} x^2 e^{5x} - \tfrac{2}{25} x e^{5x} + \tfrac{2}{125} e^{5x} + C. \;\blacksquare$$

APPLICATIONS

Example 4
The purpose of this example is to illustrate the use of the table of integration formulas in developing a mathematical model for the study of epidemics. A pertinent reference is p. 20 of Bailey [5].

Let x represent the number of animals susceptible to a certain disease at time t, and let the total population be $n + 1$. In the interval of time from t to $t + \Delta t$, the change in the number of susceptibles is

$$x(t + \Delta t) - x(t),$$

and the average rate of change is

$$\frac{x(t + \Delta t) - x(t)}{\Delta t}.$$

As $\Delta t \to 0$ we obtain the instantaneous rate of change

$$\lim_{\Delta t \to 0} \frac{x(t + \Delta t) - x(t)}{\Delta t} = \frac{dx}{dt}.$$

If one assumes that this rate is proportional to both the number of susceptibles and the number of nonsusceptibles (infectors), we obtain the differential equation

$$\frac{dx}{dt} = -\beta x(n + 1 - x),$$

where $-\beta$ is the constant of proportionality. (The minus sign is used because, in an epidemic, the number of susceptibles is decreasing and it is desired to have $\beta > 0$.) Rearranging the latter equation to separate the variables, we get

$$\frac{dx}{x(n + 1 - x)} = -\beta\, dt$$

or
$$-\beta t = \int \frac{dx}{x(n+1-x)}.$$

The integration can be done by formula 11 on p. 656 with $u = x$, $du = dx$, $b = -1$, and $a = n+1$. This yields
$$-\beta t = -\frac{1}{n+1} \ln \left| \frac{n+1-x}{x} \right| + C.$$

We assume that at $t = 0$ there are n susceptibles and one infectious animal. Since x is positive and not more than n, the absolute-value sign can be dropped; also we can substitute $x = n$ and $t = 0$ in order to find C.
$$0 = -\frac{1}{n+1} \ln \frac{n+1-n}{n} + C,$$
$$C = \frac{1}{n+1} \ln \frac{1}{n} = \frac{-1}{n+1} \ln n.$$

Hence
$$-\beta t = -\frac{1}{n+1} \ln \left(\frac{n+1-x}{x} \right) - \frac{1}{n+1} \ln n$$

and
$$\beta t = \frac{1}{n+1} \left[\ln \left(\frac{n+1-x}{x} \right) + \ln n \right].$$

We proceed to solve for x in terms of t as follows:
$$\ln \frac{n(n+1) - nx}{x} = (n+1)\beta t,$$
$$\frac{n(n+1) - nx}{x} = e^{(n+1)\beta t},$$
$$nx + xe^{(n+1)\beta t} = n(n+1),$$
$$x(n + e^{(n+1)\beta t}) = n(n+1),$$
$$x = \frac{n(n+1)}{n + e^{(n+1)\beta t}}.$$

From the last result it can be shown that as $t \to \infty$, $x \to 0$. This means that, under our assumptions, the disease will eventually spread through the entire population; the rapidity of spreading, or diffusion, through the population is determined by β. ∎

Example 5
Variations of the model presented in Example 4 have been adapted in other ways. On pp. 147–174 Richardson [54] uses a variation to study the diffusion of "war fever" among a population. On pp. 392–401 Dodd [20] uses a variation to study the diffusion of rumors or messages among a population. ∎

Sec. 14.6] Integration Using Table of Integration Formulas

EXERCISES

Suggested minimum assignment: Exercises 1, 4, 5, 8, and 11.

1. Find
$$\frac{1}{4}\int \frac{dx}{(\frac{3}{2})^2 - x^2}$$
by using formula 9 of the table of integration formulas with $a = \frac{3}{2}$, $u = x$, and $du = dx$. (See Example 1.)

In Exercises 2–9 perform the indicated integration, using the table of integration formulas on p. 656. State which formula or formulas are used.

2. $\int \sqrt{9 + 4x^2}\, dx.$

3. $\int \sqrt{4x^2 - 4x + 10}\, dx.$ [Hint: $4x^2 - 4x + 10 = (2x - 1)^2 + 9$; see Exercise 2.]

4. $\int \dfrac{dx}{16x^2 - 9}.$

5. $\int \dfrac{x\, dx}{(3 + 2x)^2}.$

6. $\int \dfrac{2x\, dx}{5 + 4x}.$

7. $\int xe^{2x}\, dx.$

8. $\int x^3 e^x\, dx.$ Also check the answer by differentiation.

9. $\int x^2 \ln 2x\, dx.$

10. Use the result of Example 3 to evaluate $\displaystyle\int_0^1 x^2 e^{5x}\, dx.$

11. Evaluate $\displaystyle\int_{1/2}^{e^2/2} \ln 2x\, dx.$

12. An interesting discussion of population models is given in *The Encyclopedia of the Biological Sciences*, 2nd ed., Van Nostrand Reinhold Company, New York, 1970, pp. 755–757. One model is the equation
$$\frac{dN}{dt} = rN\left(1 - \frac{N}{K}\right).$$
In this equation N is the number of individuals in the population, t is time, r is the intrinsic rate of natural increase, and K is the environmental capacity in terms of the number of animals. Show that
$$\frac{N}{1 - N/K} = Ce^{rt}.$$

14.7 Integrals with Infinite Limits

In our previous discussion of a definite integral, $\int_a^b f(x)\, dx$, the numbers a and b were assumed to be real numbers. In this section we shall give meaning to certain integrals of the types

$$\int_a^\infty f(x)\, dx, \quad \int_{-\infty}^b f(x)\, dx, \quad \int_{-\infty}^\infty f(x)\, dx,$$

for which the upper limit, the lower limit, or both the upper and lower limits are infinite. These types of integrals are known as **improper integrals**; there are also other types of improper integrals that will not be discussed in this book.

Example 1
Consider the improper integral

$$\int_1^\infty \frac{4}{x^2}\, dx,$$

with an infinite upper limit.

First, we imagine definite integrals having upper limits that get larger and larger, such as $\int_1^{100} 4/x^2\, dx$, $\int_1^{1000} 4/x^2\, dx$, and, in general, $\int_1^t 4/x^2\, dx$, where t increases without bound. (The particular letter t used here has no special significance—a different letter could be used.) If $\int_1^\infty 4/x^2\, dx$ is to be assigned any meaning, it seems logical that it should be defined by

$$\int_1^\infty \frac{4}{x^2}\, dx = \lim_{t \to \infty} \int_1^t \frac{4}{x^2}\, dx,$$

provided that the limit exists.

We calculate as follows:

$$\lim_{t \to \infty} \int_1^t \frac{4}{x^2}\, dx = \lim_{t \to \infty} \int_1^t 4x^{-2}\, dx$$

$$= \lim_{t \to \infty} \frac{4x^{-1}}{-1} \bigg|_1^t$$

$$= \lim_{t \to \infty} \left(-\frac{4}{t} + 4 \right)$$

$$= 0 + 4$$

$$= 4.$$

Therefore, we shall agree to write

$$\int_1^\infty \frac{4}{x^2}\, dx = 4.$$

Geometrically, the answer may be interpreted as the area to the right of $x = 1$ between the curve $y = 4/x^2$ and the x-axis. See Figure 14.9. As $t \to \infty$ the area given by $\int_1^t 4/x^2 \, dx$ keeps increasing and approaches the limit of 4. ∎

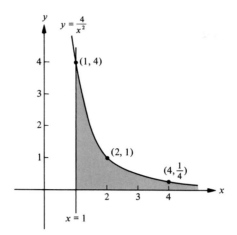

Figure 14.9

In Example 1 the function f given by $f(x) = 4/x^2$ is continuous on $[1, \infty)$. By this statement we mean that f is continuous on any interval $[1, t]$, where $t > 1$. By Theorem 1 of Section 13.2, $\int_1^t 4/x^2 \, dx$ exists for any real number $t > 1$. As we study improper integrals of the type $\int_a^\infty f(x) \, dx$, we shall insist that f be continuous on $[a, \infty)$. Therefore, in this book we shall not consider improper integrals such as $\int_0^\infty 4/x^2 \, dx$ or $\int_{-1}^\infty 4/x^2 \, dx$, because the rational function f given by $f(x) = 4/x^2$ is discontinuous at $x = 0$ (there is a vertical asymptote at $x = 0$); a discussion of such improper integrals may be found in more advanced texts.

▶ **Definition 5**
If f is continuous on $[a, \infty)$, then

$$\int_a^\infty f(x) \, dx = \lim_{t \to \infty} \int_a^t f(x) \, dx,$$

provided that the latter limit exists. If $\lim_{t \to \infty} \int_a^t f(x) \, dx$ exists, $\int_a^\infty f(x) \, dx$ is said to **converge**; otherwise, it is said to **diverge**.

The improper integral $\int_1^\infty 4/x^2 \, dx$ of Example 1 converges. Next we give an example of an improper integral that diverges.

Example 2
Evaluate $\int_0^\infty 4/(x + 1) \, dx$, if possible.

The function f given by $f(x) = 4/(x + 1)$ is continuous on $[0, \infty)$. Therefore,

$$\int_0^\infty \frac{4}{x + 1}\, dx = \lim_{t \to \infty} \int_0^t \frac{4}{x + 1}\, dx,$$

provided that this latter limit exists.

$$\lim_{t \to \infty} \int_0^t \frac{4}{x + 1}\, dx = \lim_{t \to \infty} 4 \ln |x + 1| \Big|_0^t$$
$$= \lim_{t \to \infty} (4 \ln |t + 1| - 4 \ln 1)$$
$$= 4 \lim_{t \to \infty} \ln |t + 1|.$$

However, $\lim_{t \to \infty} \ln |t + 1|$ does not exist because $\ln |t + 1|$ can be made as large as desired by selecting a large enough value of t. We conclude that $\int_0^\infty 4/(x + 1)\, dx$ does not exist and say that it diverges.

Geometrically, the area under $y = 4/(x + 1)$ to the right of $x = 0$ becomes larger without bound as x increases (see Figure 14.10). One might say that there is an infinite amount of area under the curve in Figure 14.10 but only a finite amount of area under the curve in Figure 14.9; this interesting fact is difficult to believe at first, and it certainly is not obvious from the figures. ■

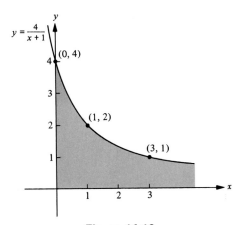

Figure 14.10

The next definition is similar to Definition 5.

▶ **Definition 6**
If f is continuous on $(-\infty, b]$, then

$$\int_{-\infty}^b f(x)\, dx = \lim_{v \to -\infty} \int_v^b f(x)\, dx,$$

provided that the latter limit exists. If $\lim_{v \to -\infty} \int_v^b f(x)\,dx$ exists, $\int_{-\infty}^b f(x)\,dx$ is said to **converge**; otherwise, it is said to **diverge**.

Example 3
Determine whether $\int_{-\infty}^1 e^x\,dx$ converges or diverges; if it converges, find its value.

$$\int_{-\infty}^1 e^x\,dx = \lim_{v \to -\infty} \int_v^1 e^x\,dx,$$

provided that this latter limit exists. Recall that an exponential function is continuous for all real numbers, and hence Definition 6 applies for this example.

$$\lim_{v \to -\infty} \int_v^1 e^x\,dx = \lim_{v \to -\infty} e^x \Big|_v^1$$

$$= \lim_{v \to -\infty} (e - e^v)$$

$$= e - 0$$

$$= e.$$

Therefore, $\int_{-\infty}^1 e^x\,dx$ converges and its value is e. The region whose area is represented by $\int_{-\infty}^1 e^x\,dx$ is indicated in Figure 14.11. ∎

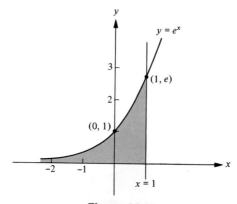

Figure 14.11

In Definition 7 we consider the type of improper integral for which both the upper and lower limits are infinite.

▶ **Definition 7**
Suppose that f is continuous for all real numbers. Then

$$\int_{-\infty}^\infty f(x)\,dx = \int_{-\infty}^0 f(x)\,dx + \int_0^\infty f(x)\,dx, \qquad (14.23)$$

provided that the latter two improper integrals each converge, in which

case $\int_{-\infty}^{\infty} f(x)\, dx$ is said to **converge**. If either (or both) of the improper integrals on the right of (14.23) diverges, then $\int_{-\infty}^{\infty} f(x)\, dx$ is said to **diverge**.

In (14.23) of Definition 7 any real number c can be used instead of 0. One should attempt to evaluate *separately* each of the improper integrals on the right side of (14.23), using Definitions 6 and 5, respectively; $\int_{-\infty}^{\infty} f(x)\, dx$ converges if *both* of the improper integrals on the right converge, and in this case equals the sum of their values.

Example 4
Consider $\int_{-\infty}^{\infty} e^{-x^2/2}\, dx$.

The function f given by $f(x) = e^{-x^2/2}$ is continuous for all real numbers. Therefore,

$$\int_{-\infty}^{\infty} e^{-x^2/2}\, dx = \int_{-\infty}^{0} e^{-x^2/2}\, dx + \int_{0}^{\infty} e^{-x^2/2}\, dx, \qquad (14.24)$$

provided that the latter two improper integrals converge. By Definitions 6 and 5 the improper integrals on the right side of (14.24) are equal to

$$\lim_{v \to -\infty} \int_{v}^{0} e^{-x^2/2}\, dx$$

and

$$\lim_{t \to \infty} \int_{0}^{t} e^{-x^2/2}\, dx,$$

respectively, provided that these limits exist. At this stage we can go no further with (14.24) because

$$\int e^{-x^2/2}\, dx$$

cannot be found by any of the integration formulas we have studied. By methods beyond the scope of this book it can be shown that both of the improper integrals on the right side of (14.24) converge and that each of them has the value $\frac{1}{2}\sqrt{2\pi}$. Therefore, $\int_{-\infty}^{\infty} e^{-x^2/2}\, dx$ converges and

$$\int_{-\infty}^{\infty} e^{-x^2/2}\, dx = \tfrac{1}{2}\sqrt{2\pi} + \tfrac{1}{2}\sqrt{2\pi} = \sqrt{2\pi}.$$

The answer can be interpreted geometrically as the area of the region beneath the curve $y = e^{-x^2/2}$ and above the x-axis. See Figure 14.12. ∎

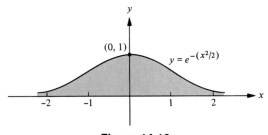

Figure 14.12

In statistics the result of Example 4 turns out to be very important; it is related to a study of probability. We shall pursue this idea further in Example 5.

APPLICATIONS

*Example 5 Normal Distribution

Suppose that the set of outcomes for an experiment is the sample space U, and suppose that a continuous measurement X with domain U and range S has been defined. Assume that S is the set of all real numbers, so that the given continuous measurement is *not* bounded. In Example 4 of Section 13.2 a probability density function p for a *bounded* continuous measurement was defined; if the bounded requirement is removed, then a probability density function p must be a continuous function such that

1. $p(x) > 0$ for any real number x.
2. $\int_{-\infty}^{\infty} p(x)\, dx = 1.$

There is no change in the four properties required for the corresponding cumulative probability distribution function F when the continuous measurement is not bounded (see Example 8 of Section 9.3). Also, for the unbounded case, the relationships between the two functions F and p are given by the two equations

$$F(x) = \int_{-\infty}^{x} p(s)\, ds$$

and

$$F'(x) = p(x),$$

where x is *any* real number (compare with Example 3 of Section 13.6). The probability of an event E, where $E = \{s : s \leq c\}$ is given by

$$P(E) = F(c) = \int_{-\infty}^{c} p(x)\, dx.$$

If $E_1 = \{s : c \leq s < d\}$, then

$$P(E_1) = F(d) - F(c) = \int_{c}^{d} p(x)\, dx.$$

If $E_2 = \{s : s > c\}$, then

$$P(E_2) = \int_{c}^{\infty} p(x)\, dx = 1 - F(c).$$

(In the inequalities used in expressing E, E_1, and E_2, either $<$ or $\leq$ can be written and also either $>$ or $\geq$.) The mean of the measurement (see Example 5 of Section 13.2) is found by

$$\mu = \int_{-\infty}^{\infty} xp(x)\, dx.$$

The variance σ^2 and the standard deviation σ (see Example 6 of Section 13.2) are determined from

$$\sigma^2 = \int_{-\infty}^{\infty} (x - \mu)^2 p(x)\, dx.$$

In statistics many different probability density functions for unbounded continuous measurements are studied. The most important of these, because it represents quite accurately the distribution of a large number of outcomes for many practical experiments, is called a **normal probability density function** and is given by

$$p(x) = \frac{1}{\sqrt{2\pi}\,\sigma} e^{-(x-\mu)^2/2\sigma^2}, \tag{14.25}$$

where μ is the mean of the measurement and $\sigma > 0$ is the standard deviation of the measurement. In this case the measurement is said to have the **normal distribution**, which is sometimes also called the **Gaussian distribution**. If one measures the height in inches of a large number of adult males, a certain "normal" amount of deviation of the results from the mean is expected; the normal probability density function, with appropriate values of μ and σ, usually will reflect the situation quite accurately. On p. 95 of *Statistics in Psychology and Education*, Garrett [24] lists several instances for which there is evidence of the accuracy of the normal distribution; his five major headings are biological statistics, anthropometrical data, social and economic data, psychological measurements, and errors of observation.

We shall focus our attention on the normal probability density function with $\mu = 0$ and $\sigma = 1$; ordinarily, an understanding of this case will enable one to handle all the other cases also. If $\mu = 0$ and $\sigma = 1$, (14.25) reduces to

$$p(x) = \frac{1}{\sqrt{2\pi}} e^{-x^2/2};$$

in this case p is called the **standardized normal probability density function**. The function p is continuous for all real numbers, and $p(x) > 0$ for any real number x. From Example 4

$$\int_{-\infty}^{\infty} e^{-x^2/2}\, dx = \sqrt{2\pi},$$

and hence

$$\int_{-\infty}^{\infty} \frac{1}{\sqrt{2\pi}} e^{-x^2/2}\, dx = \int_{-\infty}^{\infty} p(x)\, dx = 1.$$

Therefore, p does meet the requirements of a probability density function; its graph is shown in Figure 14.13. The corresponding cumulative proba-

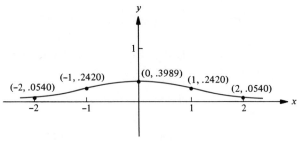

Figure 14.13. $y = p(x) = \dfrac{1}{\sqrt{2\pi}} e^{-x^2/2}$.

bility distribution function F is given by

$$F(x) = \int_{-\infty}^{x} p(s)\, ds = \int_{-\infty}^{x} \frac{1}{\sqrt{2\pi}} e^{-s^2/2}\, ds.$$

The graph of F is shown in Figure 14.14. Tables of function values $p(x)$ and $F(x)$ have been carefully computed and may be found in many statistics books; a few of these values are given in Table 14.10.

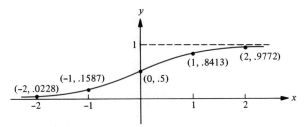

Figure 14.14. $y = F(x) = \displaystyle\int_{-\infty}^{x} \frac{1}{\sqrt{2\pi}} e^{-s^2/2}\, ds.$

Table 14.10. Values of $p(x)$ and $F(x)$ for the standard normal distribution.

x	$p(x)$	$F(x)$
0	.3989	.5000
.25	.3867	.5987
.50	.3521	.6915
.75	.3011	.7734
1.00	.2420	.8413
1.25	.1826	.8944
1.50	.1295	.9332
1.75	.0863	.9599
2.00	.0540	.9772
2.50	.0175	.9938
3.00	.0044	.9987
3.50	.0009	.9998

Suppose that a measurement has the **standard normal distribution**—in other words, $\mu = 0$ and $\sigma = 1$. What is the probability that an outcome is between -1 and 1? The answer is

$$\int_{-1}^{1} p(x)\, dx = F(1) - F(-1).$$

Table 14.10 does not include negative values of x (such as -1). Because of symmetry (see Figures 14.13 and 14.14) we have

$$p(-x) = p(x)$$

and

$$F(-x) = 1 - F(x).$$

Therefore,

$$F(1) - F(-1) = .8413 - (1 - .8413)$$
$$= .6826.$$

The result above means that slightly over 68 per cent of the outcomes are within one standard deviation of the mean (this is also true for normal distributions with different values of μ and σ). Similarly, it can be shown for a normal distribution that about 95.4 per cent of the outcomes are within two standard deviations of the mean.

Suppose that a company produces a certain type of light bulb. The life in hours of bulbs of this type is assumed to have a normal distribution with $\mu = 600$ and $\sigma = 30$. What is the probability that a given bulb will last longer than 645 hours? In other words, what is the probability that the outcome will be more than $\frac{45}{30}$ or $\frac{3}{2}$ standard deviations above the mean? If one converts to the standard normal distribution, what is the probability of the event $E = \{s : s > 1.5\}$?

$$P(E) = \int_{1.5}^{\infty} p(x)\, dx = 1 - F(1.5)$$
$$= 1 - .9332$$
$$= .0668.$$

In other words, only 6.68 per cent of the bulbs are expected to last longer than 645 hours. ∎

EXERCISES

Suggested minimum assignment: Exercises 1, 3, 5, 10, 11, and 13.

In Exercises 1–6 determine whether the given improper integral converges or diverges. If it converges, find its value and draw a sketch showing the region whose area is given by the improper integral.

1. $\int_{0}^{\infty} e^{-x}\, dx.$

2. $\int_{1}^{\infty} \frac{dx}{x}.$

Sec. 14.7] Integrals with Infinite Limits

3. $\displaystyle\int_{-\infty}^{0} \frac{dx}{(x-2)^2}.$
4. $\displaystyle\int_{-\infty}^{3} 2^x\, dx.$
5. $\displaystyle\int_{-\infty}^{\infty} \frac{x\, dx}{x^2+1}.$
6. $\displaystyle\int_{-\infty}^{\infty} e^x\, dx.$

In Exercises 7–14 determine whether the given improper integral converges or diverges. If it converges, find its value.

7. $\displaystyle\int_{3}^{\infty} \frac{dx}{x^3}.$
8. $\displaystyle\int_{1}^{\infty} \frac{4\, dx}{x^{3/2}}.$
9. $\displaystyle\int_{1}^{\infty} \frac{4\, dx}{x^{2/3}}.$
10. $\displaystyle\int_{-\infty}^{-1} \frac{dx}{2x-1}.$
11. $\displaystyle\int_{0}^{\infty} xe^{-x^2}\, dx.$
12. $\displaystyle\int_{-\infty}^{\infty} \frac{x\, dx}{\sqrt{16+x^2}}.$
13. $\displaystyle\int_{-\infty}^{\infty} \frac{x\, dx}{(x^2+4)^2}.$
14. $\displaystyle\int_{-\infty}^{-5} \frac{2x^3\, dx}{x^4+1}.$

15. Evaluate $\displaystyle\int_{-\infty}^{\infty} x(1/\sqrt{2\pi})e^{-x^2/2}\, dx.$ With respect to Example 5, what is represented by the given improper integral?
16. In Figure 14.14 what is the slope of the curve at $(0, .5)$?
17. Suppose that the heights in inches of adult women are normally distributed with $\mu = 64$ and $\sigma = 3$.
    (a) What is the probability that a woman selected at random is less than $62\tfrac{1}{2}$ inches tall?
    (b) What percentage of women are 5 feet 10 inches or taller?
    (c) What percentage of women are between 5 feet 1 inch and 5 feet 7 inches?

NEW VOCABULARY

exponential function with base b 14.1
compounded continuously 14.1
logarithm of a positive number to the base b 14.2
natural logarithms 14.2
common logarithms 14.2
logarithmic function with base b 14.2

inverse functions 14.2
logarithmic differentiation 14.3
exponential growth law 14.5
exponential decay law 14.5
half-life 14.5
reduction formula 14.6
improper integral 14.7
improper integral converges 14.7
improper integral diverges 14.7

15 Functions of More Than One Variable

In previous chapters we have given many illustrations of quantities that were dependent upon only one variable. In many practical situations, however, a quantity often depends upon several variables. For example, the cost of an item may be thought of as depending upon labor, materials, overhead expenses, and transportation expenses. The sales of an item may be dependent upon the price of the item and the amount spent on advertising. Productivity is often thought of as dependent upon the capital invested as well as the labor force. The demand for an item is related to its price and also to the income of the consumer; the demand might also depend upon the price of a similar competing item. Such situations are necessarily more complicated than those in which there is only one independent variable. Our purpose in this chapter, therefore, is to present a brief introduction to the geometry and calculus of functions of more than one variable and to present techniques for the solution of certain optimization problems that cannot be solved by the methods of Chapter 11. Such optimization problems will include those in which we seek to maximize or minimize certain functions that are subject to specified restrictions; the solution of these problems will be accomplished by the Lagrange multiplier method, the discovery of which is credited to Joseph Louis Lagrange (1736–1813).

Prerequisites: 1.1, 1.2, Chapter 3, Chapter 9, 10.1–10.5, 11.1–11.5, 14.1–14.3.
Suggested Sections for Moderate Emphasis: 15.1–15.3.
Suggested Sections for Minimum Emphasis: 15.1, 15.2.

15.1 Introduction

We have frequently expressed a function f of one variable by an equation

$$y = f(x).$$

The graph was drawn in two dimensions and was called a curve. The variable x was called the independent variable, and y was known as the dependent variable. In this chapter a function f of two variables can be expressed by an equation

$$z = f(x, y).$$

The graph of such a function will be a **surface** lying in three-dimensional space (see Section 3.10). If $z = f(x, y)$ expresses the dependence of z upon x and y, then z is called the **dependent variable,** whereas both x and y are **independent variables.**

Example 1
Let $z = f(x, y) = 9 - x^2 - y^2$. Then
$$f(1, 2) = 9 - 1 - 4 = 4$$
and
$$f(0, 0) = 9 - 0 - 0 = 9. \blacksquare$$

The proper functional notation is shown in Example 1. Given values of the independent variables, one can determine (as above) a *unique* value of the dependent variable; in this basic respect the idea of a function of more than one independent variable is similar to that of a function of one independent variable. The graph of the function in Example 1 will be discussed in Example 2.

Example 2
Sketch $z = 9 - x^2 - y^2$.

To obtain the curve of intersection of the surface $z = 9 - x^2 - y^2$ in the xy-plane, we solve simultaneously the equations
$$\begin{cases} z = 9 - x^2 - y^2, \\ z = 0. \end{cases}$$

The curve of intersection is the circle
$$x^2 + y^2 = 9,$$
with center at the origin and radius 3; this circle is shown in Figure 15.1. Similarly, the equation of the surface is solved simultaneously with $x = 0$, and then with $y = 0$, to obtain the parabolas shown in the yz-plane and xz-plane, respectively. In general, surfaces can be drawn by sketching the curves of intersection with the coordinate planes or with planes parallel to the coordinate planes. These curves of intersection give a clear outline of the surface that opens downward and has $(0, 0, 9)$ as a point on the surface that is higher than all surrounding points; we say that $(0, 0, 9)$ is a (*relative*) *maximum point.* The surface extends below the xy-plane, although this portion is not shown in Figure 15.1. $\blacksquare$

Suppose that a function f is defined by an equation
$$z = f(x, y).$$
Unless otherwise stated, it is implied that the domain of f is the set of ordered pairs (x, y) of *real* numbers that have a unique image $f(x, y)$ in the

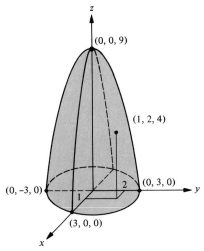

Figure 15.1. $z = 9 - x^2 - y^2$.

range of f. For example, the implied domain of the function f defined by

$$z = f(x, y) = 9 - x^2 - y^2$$

is the set of *all* ordered pairs (x, y) of real numbers. In other words, if R is the set of real numbers, the domain of f is the Cartesian product $R \times R$. The range of f is $\{z : z \leq 9\}$.

Example 3
In Figure 15.2 a sketch of the surface

$$z = y^2 - x^2$$

is shown. The graph is relatively difficult to draw, but the interesting thing is the general shape of the surface, particularly the appearance near the origin. The surface is sometimes called a **saddle surface,** and the origin may

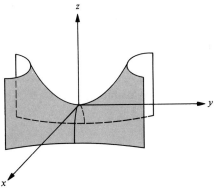

Figure 15.2. $z = y^2 - x^2$.

Sec. 15.1] Introduction

be called a **saddle point**. The origin is the minimum point on the parabola $z = y^2$ shown in the yz-plane, but it is the maximum point on the parabola $z = -x^2$ shown in the xz-plane. There are points near the origin that are higher and points that are lower; therefore, the origin is neither a (relative) maximum point nor a (relative) minimum point. ∎

We shall not attempt a lengthy discussion of limits and continuity for functions of more than one variable; we shall make a few comments of an intuitive nature, however, concerning the example

$$z = f(x, y) = 9 - x^2 - y^2.$$

For any point in the xy-plane near $(1, 2)$, the z-coordinate of the surface is close to 4 (see Figure 15.1); in a manner that is analogous with our presentation of limits of values of functions of a single variable, this concept can be expressed by saying that the *limit* of $f(x, y)$, as (x, y) approaches $(1, 2)$, is 4. That is,

$$\lim_{(x,y) \to (1,2)} (9 - x^2 - y^2) = 4.$$

This statement would be correct even if there were a hole in the surface at $(1, 2, 4)$, in which case we would say that the surface had a discontinuity at this point. In the given problem, however, the limit of $f(x, y)$ as $(x, y) \to (1, 2)$ is equal to the value of f at $(1, 2)$; that is,

$$\lim_{(x,y) \to (1,2)} f(x, y) = f(1, 2) = 4,$$

and thus f is said to be *continuous* at $(1, 2)$. In fact, f is continuous at each point of the xy-plane.

Example 4
As an example of a function of *three* independent variables, suppose that the profit P made by a builder of homes for building one house is thought of as depending upon the revenue R received for the house, the labor cost L charged to the house, and the cost M of materials used in building the house. Since profit equals revenue minus cost, we have

$$P = f(R, L, M) = R - L - M.$$

Because there are three independent variables, a sketch will not be given. It is still possible to define the concepts of limit, continuity, and partial derivative (see Section 15.2) for the function even though a geometric interpretation is not readily available. ∎

EXERCISES

Suggested minimum assignment: Exercises 1(a), 3(a), 4, 5, 8, and 9.

1. Suppose that $f(x, y) = 2x^2 + y^2 + 2xy - 3$.
   (a) Find $f(2, 1)$.   (b) Find $f(-1, 1)$.

2. Suppose that $f(x, y) = x + 2^{xy}$.
   (a) Find $f(3, 1)$.   (b) Find $f(0, 2)$.
3. Suppose that $f(w, x, y) = w^2 + wxy + \log_2 x$.
   (a) Find $f(1, 2, 3)$.   (b) Find $f(2, 1, 3)$.
4. Given that $z = f(x, y) = 4 - x^2 - y^2$.
   (a) Sketch the graph of f.
   (b) What is the implied domain of f?
   (c) What is the range of f?
   (d) Find $\lim_{(x,y)\to(2,0)} f(x, y)$.
   (e) What point on the surface is a relative maximum point?
5. If $f(x, y) = (x + 1)/y$, which ordered pairs (x, y) of real numbers are *not* in the domain of f?
6. Suppose that a function of n independent variables is given. What type of implied domain does such a function have? State your answer in terms of a Cartesian product and also in terms of vectors (see Section 5.2).
7. Sketch the plane $x + y + z = 4$ by drawing the lines of intersection of this plane with each of the coordinate planes.
8. (a) Sketch the surface $z = x^2 + y^2$ by drawing the curves of intersection of this surface with each of the planes $x = 0$, $y = 0$, and $z = 4$.
   (b) What point on the surface is a relative minimum point?
   (c) Does the surface have any points of discontinuity?
9. Suppose that the cost C of an item depends upon w, x, and y according to the formula

$$C = f(w, x, y) = w^2 + x + y.$$

   Name the dependent and independent variables.
10. Your choice of a new car is a function of many variables. Name some of these variables.

15.2 Partial Derivatives

For functions of more than one variable, the important concept of a partial derivative plays a role analogous to that of the ordinary derivative discussed in previous chapters. We now define a partial derivative of a function of two variables, and later in this section we shall give a geometric interpretation.

▶ **Definition 1**

Let $z = f(x, y)$. The functions determined by the following limits, if the limits exist, are called the **partial derivative of f with respect to x** and the **partial derivative of f with respect to y**, respectively:

(i) $\dfrac{\partial f}{\partial x} = \lim_{\Delta x \to 0} \dfrac{f(x + \Delta x, y) - f(x, y)}{\Delta x}.$

(ii) $\dfrac{\partial f}{\partial y} = \lim\limits_{\Delta y \to 0} \dfrac{f(x, y + \Delta y) - f(x, y)}{\Delta y}$.†

The key thing to observe in (i) is that the definition is similar to the definition of the ordinary derivative in Section 10.2, except that we have a second independent variable y that is kept fixed. Thus $\partial f / \partial x$ can be found by using ordinary differentiation rules provided that y is treated as a constant. Similarly, $\partial f / \partial y$ can be calculated using previous techniques by remembering to treat x as a constant.

Example 1
Find $\partial f / \partial x$ and $\partial f / \partial y$ if
$$z = f(x, y) = x^4 y^3 + 3x.$$
$$\dfrac{\partial f}{\partial x} = y^3(4x^3) + 3 = 4x^3 y^3 + 3.$$
$$\dfrac{\partial f}{\partial y} = x^4(3y^2) + 0 = 3x^4 y^2.\ \blacksquare$$

Some comments about notation need to be made. In a problem like Example 1 it is customary to use the notation $\partial z/\partial x$ or $f_x(x, y)$ as well as $\partial f/\partial x$. Hence in Example 1

$$\dfrac{\partial z}{\partial x} = \dfrac{\partial f}{\partial x} = f_x(x, y) = 4x^3 y^3 + 3.$$

Also,
$$\dfrac{\partial z}{\partial y} = \dfrac{\partial f}{\partial y} = f_y(x, y) = 3x^4 y^2.$$

If these partial derivatives are to be evaluated for $x = -1$ and $y = 2$, then the notation that we shall use is

$$\left.\dfrac{\partial z}{\partial x}\right|_{(-1,2)} = \left.\dfrac{\partial f}{\partial x}\right|_{(-1,2)} = f_x(-1, 2) = 4(-1)(8) + 3 = -29.$$

Also,
$$\left.\dfrac{\partial z}{\partial y}\right|_{(-1,2)} = \left.\dfrac{\partial f}{\partial y}\right|_{(-1,2)} = f_y(-1, 2) = 3(1)(4) = 12.$$

The reader should think of a symbol such as $\partial f/\partial x$ as one quantity—no separate meanings will be assigned to ∂f and ∂x.

For a function of more than two variables it is possible to calculate a partial derivative with respect to each of the independent variables. In differentiating with respect to any one variable, all other independent variables should be treated as constants.

† In this book $\dfrac{\partial f}{\partial x}$ and $\dfrac{\partial f}{\partial y}$ are abbreviated notations for $\dfrac{\partial}{\partial x}[f(x, y)]$ and $\dfrac{\partial}{\partial y}[f(x, y)]$, respectively.

Example 2
A **utility function** of several commodities gives a quantitative measure of the desirability for a particular consumer to possess various quantities of these products. Suppose that a utility function of three commodities is given by

$$u = f(w, x, y) = 5w + 6x + 8y - \frac{w^2}{4} - \frac{x^2}{2} - y^2 + xy + 2wy.$$

Find $\partial u/\partial w$, $\partial u/\partial x$, and $\partial u/\partial y$, which are called the **marginal utilities** of the three products.

Holding x and y fixed and differentiating with respect to w, we calculate

$$\frac{\partial u}{\partial w} = 5 - \frac{1}{2}w + 2y.$$

Similarly,

$$\frac{\partial u}{\partial x} = 6 - x + y$$

and

$$\frac{\partial u}{\partial y} = 8 - 2y + x + 2w. \blacksquare$$

Let us return to the case of a function of two variables given by

$$z = f(x, y)$$

and give a geometric interpretation of partial derivatives. In particular, we shall use as an illustration

$$z = f(x, y) = 9 - x^2 - y^2$$

from Examples 1 and 2 of Section 15.1; attention will be focused on the point $(1, 2, 4)$ of this surface.

$$f_x(x, y) = -2x,$$
$$f_x(1, 2) = -2.$$

In Figure 15.3 a portion of the surface $z = 9 - x^2 - y^2$ is shown. With y held constant ($y = 2$), we can interpret $f_x(1, 2)$ in a manner similar to the interpretation given for ordinary derivatives. That is, $f_x(1, 2) = -2$ represents the rate of change of z with respect to x at $(1, 2, 4)$ with y held constant. In the plane $y = 2$, in other words, the slope of the tangent line (see Figure 15.3) to the curve of intersection of $z = 9 - x^2 - y^2$ and $y = 2$ at $(1, 2, 4)$ is -2. Similarly, for a geometric interpretation of $\partial f/\partial y$, consider

$$f_y(x, y) = -2y$$

and

$$f_y(1, 2) = -4.$$

In the plane $x = 1$, the slope of the tangent line (see Figure 15.4) to the curve of intersection of $z = 9 - x^2 - y^2$ and $x = 1$ at $(1, 2, 4)$ is -4.

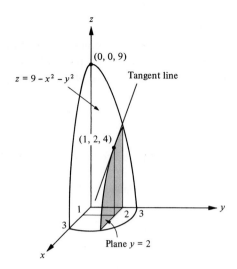

Figure 15.3

For a function of more than two variables no similar geometric interpretation is available. A partial derivative with respect to one of the variables, evaluated at a point, may be thought of as a rate of change with respect to that variable with all other independent variables being held constant.

A partial derivative of a function of several variables is also a function of the same variables and may be differentiated again to obtain a second partial derivative. This type of calculation will be important in the next section; we illustrate the technique and notation in Examples 3 and 4.

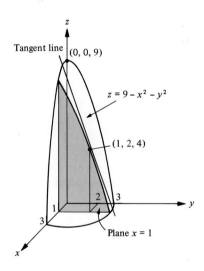

Figure 15.4

Example 3
Given $z = f(x, y) = ye^{2x} + x^2$.

$$\frac{\partial z}{\partial x} = \frac{\partial f}{\partial x} = f_x(x, y) = 2ye^{2x} + 2x.$$

$$\frac{\partial^2 z}{\partial x^2} = \frac{\partial^2 f}{\partial x^2} = f_{xx}(x, y) = 4ye^{2x} + 2.$$

$$\left.\frac{\partial^2 z}{\partial x^2}\right|_{(0,1)} = \left.\frac{\partial^2 f}{\partial x^2}\right|_{(0,1)} = f_{xx}(0, 1) = 4(1)e^0 + 2 = 6.$$

$$\frac{\partial}{\partial y}\left(\frac{\partial z}{\partial x}\right) = \frac{\partial^2 z}{\partial y\, \partial x} = \frac{\partial^2 f}{\partial y\, \partial x} = f_{xy}(x, y) = 2e^{2x}.$$

$$\frac{\partial z}{\partial y} = \frac{\partial f}{\partial y} = f_y(x, y) = e^{2x}.$$

$$\frac{\partial}{\partial x}\left(\frac{\partial z}{\partial y}\right) = \frac{\partial^2 z}{\partial x\, \partial y} = \frac{\partial^2 f}{\partial x\, \partial y} = f_{yx}(x, y) = 2e^{2x}.$$

$$\frac{\partial^2 z}{\partial y^2} = \frac{\partial^2 f}{\partial y^2} = f_{yy}(x, y) = 0. \blacksquare$$

Note in Example 3 that $\partial^2 f/\partial y\, \partial x$ and $f_{xy}(x, y)$ each technically mean that f is to be differentiated partially *first with respect to x and then with respect to y*. In Example 3 the answers for $f_{xy}(x, y)$ and $f_{yx}(x, y)$ were the same; therefore, if the differentiation had been done in the wrong order, it would have made no difference in this particular problem. In general, whenever f and its partial derivatives are continuous, $f_{xy}(x, y) = f_{yx}(x, y)$.

Example 4
If $z = f(w, x, y) = (wx/y) + w \ln w$, find $\partial z/\partial w$ and $\partial^2 z/\partial y\, \partial w$.

$$\frac{\partial z}{\partial w} = \frac{x}{y} + w\left(\frac{1}{w}\right) + \ln w$$

$$= \frac{x}{y} + 1 + \ln w.$$

Notice that x and y were treated as constants, and the product rule was used to differentiate $w \ln w$.

$$\frac{\partial^2 z}{\partial y\, \partial w} = \frac{\partial}{\partial y}\left(\frac{x}{y} + 1 + \ln w\right)$$

$$= x\left(-\frac{1}{y^2}\right) + 0 + 0$$

$$= -\frac{x}{y^2}. \blacksquare$$

Sec. 15.2] Partial Derivatives

APPLICATIONS

Example 5

The productivity or output of any company may usually be thought of as a function of several inputs. As a simple example, suppose that the number P of ironing boards produced by a firm in one day depends upon the number x of workers in the labor force on that day and the number y of machines of a certain type that are used. Suppose that from many days of experience it has been found that

$$P = f(x, y) = x^2 + 5xy + 2y^2$$

gives an accurate estimate of P when x and y are known. Assume that $5 \leq x \leq 50$ and $2 \leq y \leq 15$. If 20 workers are present on a given day and 5 machines are operating, then the number of ironing boards that are expected to be completed is given by

$$f(20, 5) = (20)^2 + 5(20)(5) + 2(5)^2 = 950.$$

We define $\partial P/\partial x$ to be the **marginal productivity of labor** and $\partial P/\partial y$ to be the **marginal productivity of machines**. Since $P = x^2 + 5xy + 2y^2$,

$$\frac{\partial P}{\partial x} = 2x + 5y$$

and

$$\frac{\partial P}{\partial y} = 5x + 4y.$$

The interpretation below of these partial derivatives in this example is analogous to the interpretation of marginal cost in Example 2 of Section 10.6 (that example should be reviewed before proceeding). If $x = 20$ and $y = 5$, then $\partial P/\partial x|_{(20, 5)}$ gives the approximate number of additional ironing boards that can be produced in 1 day if one worker is added to the labor force, and the number of machines is kept fixed at 5; the computation yields

$$\left.\frac{\partial P}{\partial x}\right|_{(20,5)} = 2(20) + 5(5) = 65.$$

On the other hand, if x is kept fixed at 20 and the number of machines is increased from 5 to 6, then the approximate number of extra ironing boards produced in 1 day is

$$\left.\frac{\partial P}{\partial y}\right|_{(20,5)} = 5(20) + 4(5) = 120. \blacksquare$$

EXERCISES

Suggested minimum assignment: Exercises 1, 7, 9, 12, 13, and 15(b).

1. If $z = f(x, y) = x^3 + 3xy^2 + y^3$, find $\partial f/\partial x$ and $\partial f/\partial y$.
2. If $z = f(w, x, y)$, define $\partial f/\partial w$ in a manner similar to Definition 1.

3. If $z = f(x, y) = x^2 + e^{xy}$, find $\partial z/\partial x$ and $\partial z/\partial y$.

4. If $z = f(x, y) = \dfrac{x}{y} + \dfrac{y}{x}$, find $f_x(x, y)$ and $f_y(x, y)$.

5. If $z = f(x, y) = 2x^3y^2 - 7xy + 3$, find $f_x(2, 1)$ and $f_y(2, 1)$.

6. If $z = f(w, x, y) = wxy + \ln wxy$, find $\partial f/\partial w$.

7. If $z = f(w, x, y) = e^{xyw^2} + \dfrac{x}{w+y}$, find $\dfrac{\partial z}{\partial x}$ and $\dfrac{\partial z}{\partial y}$.

8. Let a utility function be given by
$$u = f(w, x, y) = 5wxy - w^2 - x^2 - y^2.$$
Find the marginal utility $\partial u/\partial x$.

9. If $z = f(x, y) = 9 - x^2 - y^2$, give a geometric interpretation of $f_y(2, 1)$. Draw a sketch.

10. In the plane $y = 3$, find the slope of the tangent line to the curve of intersection of $z = x^2 + y^2$ and the plane $y = 3$ at $(1, 3, 10)$.

11. Let $z = x^3y^3 + 2x^2y$. Show that $\dfrac{\partial^2 z}{\partial y\, \partial x} = \dfrac{\partial^2 z}{\partial x\, \partial y}$.

12. Let $z = f(x, y) = ye^x + 2xe^y$. Show that $f_{xy}(x, y) = f_{yx}(x, y)$.

13. Let $z = xy - y \ln x$. Find $\dfrac{\partial^2 z}{\partial x^2}\bigg|_{(1,4)}$ and $\dfrac{\partial^2 z}{\partial y^2}\bigg|_{(1,4)}$.

14. Let $z = f(x, y) = xe^x + x(1 + 2y)^4$. Find $f_{xx}(x, y)$ and $f_{yy}(x, y)$.

15. If $z = f(w, x, y) = (w + 2x + 3y)^8$, calculate each of the following:

(a) $\dfrac{\partial^2 f}{\partial x^2}$. (b) $\dfrac{\partial^2 z}{\partial y\, \partial w}$. (c) $\dfrac{\partial^2 z}{\partial w\, \partial x}$.

16. In Example 5 find $\dfrac{\partial P}{\partial x}\bigg|_{(10,3)}$ and interpret the answer.

17. Suppose that the output of a company is given in terms of three inputs by
$$P = f(w, x, y) = 2w + 3x + 4y + wx.$$
Find the marginal productivity of each of the inputs.

15.3 Maximum and Minimum Points

In Chapter 11 we defined a relative maximum point and a relative minimum point for functions of one variable. For functions of two variables, rather than make formal definitions of relative maxima and relative minima, we shall use examples to illustrate the analogous concepts. As in Chapter 11 the word "relative" will often be omitted by agreement. In this section we seek only maxima and minima that occur at *interior* points of the domain. Referring back to Figure 15.1, we say that the function f defined by

$$z = f(x, y) = 9 - x^2 - y^2$$

has the maximum point $(0, 0, 9)$ and has no minimum points. The point $(0, 0, 9)$ is called a maximum point because all nearby points on the surface are *less* than 9 units above the xy-plane. If a surface has several "humps" in it, there might be many maximum and minimum points—for example, a putting green on a golf course often has many maximum and minimum points.

Observe that for $z = f(x, y) = 9 - x^2 - y^2$ it follows that

$$\left.\frac{\partial f}{\partial x}\right|_{(0,0)} = 0$$

and

$$\left.\frac{\partial f}{\partial y}\right|_{(0,0)} = 0.$$

These results are consistent with our geometric interpretation in Section 15.2 and with Figure 15.1; notice that the parabolas shown in Figure 15.1 are the curves of intersection of the surface with the planes $y = 0$ and $x = 0$, and in these planes the tangent lines to these curves of intersection have slope 0 at $(0, 0, 9)$. According to the next definition, the point $(0, 0, 9)$ is a **critical point** of f.

▶ **Definition 2**
A point on a surface $z = f(x, y)$ at which $\partial f/\partial x$ and $\partial f/\partial y$ are both zero is called a **critical point** of f.

In Definition 3 of Section 11.3 we required that the value of the derivative equal 0 at a critical point, whereas in Definition 2 we require that *both* first partial derivatives of f equal 0 at a critical point. In Chapter 11 most of the discussion was restricted to rational functions in order that the derivatives would exist and be continuous at each interior point of the domain. *In this chapter we shall assume that the necessary partial derivatives exist and are continuous* in order to avoid theoretical complications; with this understanding, Theorem 1, which is stated without proof, is valid.

▶ **Theorem 1**
Let $z = f(x, y)$. Each maximum or minimum point of f is a critical point of f. ◀

Theorem 1 is analogous to Theorem 2 of Chapter 11; as before, the converse of Theorem 1 is *not* true. In other words, a critical point may be neither a maximum nor minimum point—the critical points are just the only possible *candidates* for being maximum or minimum points. Example 3 of Section 15.1 furnishes an example of this last statement. The saddle point at the origin is a critical point because $\partial z/\partial x|_{(0,0)} = 0$ and $\partial z/\partial y|_{(0,0)} = 0$, but it is neither a maximum nor a minimum point as is suggested by the graph. Fortunately, there is a theorem from advanced calculus that usually enables

one to determine whether a critical point is a maximum point, a minimum point, or neither. This theorem is stated below; it involves second partial derivatives and is analogous to the second derivative test stated in Theorem 6 of Section 11.5.

▶ **Theorem 2**
Suppose that $z = f(x, y)$ and $(x_1, y_1, z_1) = (x_1, y_1, f(x_1, y_1))$ is a critical point of f. Let constants A, B, C, and D be defined by

$$A = f_{xx}(x_1, y_1),$$
$$B = f_{xy}(x_1, y_1),$$
$$C = f_{yy}(x_1, y_1),$$

and

$$D = AC - B^2.$$

(i) If $D > 0$ and $A < 0$, then (x_1, y_1, z_1) is a maximum point.
(ii) If $D > 0$ and $A > 0$, then (x_1, y_1, z_1) is a minimum point.
(iii) If $D < 0$, then (x_1, y_1, z_1) is neither a maximum nor minimum point. (It is a saddle point.)
(iv) If $D = 0$, no conclusion may be made using this theorem. ◀

Example 1
Let us return to $z = f(x, y) = 9 - x^2 - y^2$ and verify by Theorem 2 that $(0, 0, 9)$ is a maximum point.

$$f_x(x, y) = -2x,$$
$$f_y(x, y) = -2y.$$

Setting these partial derivatives equal to 0, we find that there is a critical point at $x = 0$, $y = 0$. The critical point is $(0, 0, 9)$.

$$f_{xx}(x, y) = -2,$$
$$f_{xy}(x, y) = 0,$$
$$f_{yy}(x, y) = -2,$$
$$A = f_{xx}(0, 0) = -2,$$
$$B = f_{xy}(0, 0) = 0,$$
$$C = f_{yy}(0, 0) = -2,$$
$$D = AC - B^2 = (-2)(-2) - (0)^2 = 4.$$

It follows by Theorem 2(i) that $(0, 0, 9)$ is a maximum point. ∎

Example 2
Let us return to the function in Example 3 of Section 15.1 given by

$$z = f(x, y) = y^2 - x^2$$

and verify by Theorem 2 that $(0, 0, 0)$ is neither a maximum nor a minimum point.

$$f_x(x, y) = -2x,$$
$$f_y(x, y) = 2y.$$

If we equate these partial derivatives to 0, it follows that $(0, 0, 0)$ is the only critical point. By using the pattern of Example 1, it can easily be shown that $A = -2$, $B = 0$, $C = 2$, and $D = -4$. Since $D < 0$, Theorem 2(*iii*) guarantees that $(0, 0, 0)$ is neither a maximum nor a minimum point (see Figure 15.2). ∎

Example 3
Suppose that the cost z in dollars of producing x units of gadget 1 and y units of gadget 2 per day is given by

$$z = f(x, y) = x^3 - xy + \frac{y^2}{2} - 10y + 1000.$$

Find nonnegative values of x and y, if possible, so that the cost is minimized.

$$f_x(x, y) = 3x^2 - y,$$
$$f_y(x, y) = -x + y - 10.$$

Setting these partial derivatives equal to 0 in order to find critical points, we are led to the system

$$\begin{cases} 3x^2 - y = 0, \\ y - x - 10 = 0. \end{cases}$$

Substituting from the first equation into the second, we obtain

$$3x^2 - x - 10 = 0,$$
$$(3x + 5)(x - 2) = 0,$$
$$x = -\tfrac{5}{3}, \quad x = 2.$$

Since x must be nonnegative, $x = -\tfrac{5}{3}$ may be discarded. If $x = 2$, then

$$y = 3x^2 = 12,$$

and there is a critical point at $x = 2$, $y = 12$.

$$f_{xx}(x, y) = 6x,$$
$$f_{xy}(x, y) = -1,$$
$$f_{yy}(x, y) = 1,$$
$$A = f_{xx}(2, 12) = 12,$$
$$B = f_{xy}(2, 12) = -1,$$
$$C = f_{yy}(2, 12) = 1,$$
$$D = AC - B^2 = 11.$$

By Theorem 2(*ii*) there is a minimum at $x = 2$, $y = 12$, since $D > 0$ and $A > 0$. The minimum cost is

$$f(2, 12) = 8 - (2)(12) + \tfrac{144}{2} - 120 + 1000 = \$936. \ \blacksquare$$

For the reader who has studied the characteristic value problem in matrix algebra (see Section 5.10), a procedure is stated in Exercise 12 that can be used in place of Theorem 2 in Examples 1–3.

Detection of maximum and minimum points for functions of three or more variables is generally a more difficult problem. Definition 2 can be generalized; that is, all first partial derivatives can be equated to zero to obtain critical points as before. Moreover, the critical points obtained in this manner are the *only* candidates for maximum or minimum points—this is because Theorem 1 can be generalized to cover functions of three or more variables. Formal tests for maxima and minima comparable to Theorem 2, however, are too complicated for this introductory text. Fortunately, in many applied problems, a critical point can be determined to be a maximum point or a minimum point by other means. We shall examine one example in which the existence of a minimum point is almost obvious.

Example 4
Find all maximum and minimum points of f if

$$z = f(w, x, y) = (w - 1)^2 + (x - 2)^2 + (y - 3)^2 + 4.$$

We calculate

$$\frac{\partial f}{\partial w} = 2(w - 1), \quad \frac{\partial f}{\partial x} = 2(x - 2), \quad \frac{\partial f}{\partial y} = 2(y - 3).$$

Setting these partial derivatives equal to 0, we find that the only critical point is at $w = 1$, $x = 2$, $y = 3$. Since

$$f(1, 2, 3) = 0 + 0 + 0 + 4 = 4,$$

the critical point is

$$(w, x, y, z) = (1, 2, 3, 4).$$

Because of the squared terms $(w - 1)^2$, $(x - 2)^2$, and $(y - 3)^2$, which must be nonnegative, it is clear that the value of f is never less than 4 and hence $(1, 2, 3, 4)$ is a minimum point. ∎

APPLICATIONS

Example 5
Suppose that points have been plotted on a two-dimensional graph from some given data. The art of finding the equation of a curve which in some sense comes closest to passing through the plotted points is known as **curve fitting**. In this example we shall begin with some points that nearly lie on a straight line and try to find a line $y = mx + b$ that best fits the data according to the **method of least squares**. The line $y = mx + b$ is called a **line of regression**. (Also see Example 5 of Section 5.5.)

Let the given points be

$$(4, 5), \quad (2, 3), \quad (-1, 1), \quad \text{and} \quad (-4, -2).$$

In Figure 15.5 we require the sum of the squares of the vertical distances between the given points and $y = mx + b$ to be a minimum. For example,

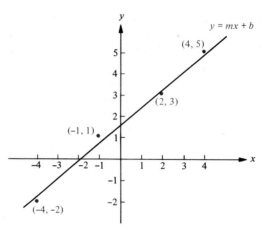

Figure 15.5. *Line of regression.*

the point (4, 5) has a y-coordinate of 5 and at $x = 4$ the line $y = mx + b$ has a y-coordinate of $4m + b$. The square of the difference can be written either $(4m + b - 5)^2$ or $(5 - 4m - b)^2$. This type of expression is found for each point, and the sum is the following function of m and b which is to be minimized:

$$f(m, b) = (4m + b - 5)^2 + (2m + b - 3)^2 + (-m + b - 1)^2 + (-4m + b + 2)^2,$$

$$\frac{\partial f}{\partial m} = 2(4m + b - 5)(4) + 2(2m + b - 3)(2) + 2(-m + b - 1)(-1) + 2(-4m + b + 2)(-4),$$

$$\frac{\partial f}{\partial b} = 2(4m + b - 5) + 2(2m + b - 3) + 2(-m + b - 1) + 2(-4m + b + 2).$$

Setting the partial derivatives equal to 0 yields the system

$$\begin{cases} 74m + 2b = 66, \\ 2m + 8b = 14. \end{cases}$$

The solution of this system is

$$m = \tfrac{125}{147}, \qquad b = \tfrac{226}{147},$$

at which the only critical point occurs. In Exercise 10 the reader is asked to use Theorem 2 of this section to verify that the critical point is a *minimum* point. Hence the line of regression is

$$y = \tfrac{125}{147}x + \tfrac{226}{147}. \blacksquare$$

EXERCISES

Suggested minimum assignment: Exercises 1, 4, 5, 8, and 9.

In Exercises 1-6 find all critical points and then use Theorem 2 to determine, if possible, whether each critical point is a maximum point, a minimum point, or neither.

1. $z = f(x, y) = 15x^2 - 40xy + 25y^2 + 20x + 500$.
2. $z = f(x, y) = (x - 1)^2 - 2(y - 2)^2 + 5$.
3. $z = f(x, y) = x^2 + 2y^2 - 2x - 8y + 14$.
4. $z = f(x, y) = 4x - 6y - x^2 - y^2 + 7$.
5. $z = f(x, y) = x^3 + x^2 + y^2 - xy + \frac{31}{16}$.
6. $z = f(x, y) = x^3 + y^3 + 9xy$.
7. A cost function is given by

$$C(x, y) = x^2 + xy + y^2 - 9x - 6y + 50.$$

Find x and y so as to minimize cost.

8. In one day a manufacturing process produces a number x of one item and a number y of a second item. The profit made from the sale of these items is given by

$$P(x, y) = xy + 6x + 11y - x^2 - 2y^2.$$

Find x and y so as to maximize profit.

9. Let $z = f(w, x, y) = 50 - (w + 2)^2 - (x - 4)^2 - (y - 6)^2$. Find all critical points after setting $\partial f/\partial w$, $\partial f/\partial x$, and $\partial f/\partial y$ equal to zero. By inspection tell whether each critical point is a maximum or minimum point.

10. In Example 5 verify by Theorem 2 that the point at which $m = \frac{125}{147}$ and $b = \frac{226}{147}$ is a minimum point.

11. Find the equation of the line of regression (see Example 5) for the points $(3, -5)$, $(0, -1)$, and $(-4, 3)$.

12. On p. 331 of Campbell [13] a procedure is discussed that can be used instead of Theorem 2 to determine, in certain cases, whether a critical point is a maximum point, a minimum point, or neither. Form the symmetric matrix

$$\begin{bmatrix} A & B \\ B & C \end{bmatrix},$$

where A, B, and C are the same as in Theorem 2. Then find the characteristic values of this matrix. If both characteristic values are positive, the critical point is a minimum point; if both values are negative, the point is a maximum point; if one value is positive and one is negative, the point is neither a maximum nor a minimum point; and otherwise no conclusion can be made. (This method can also be generalized for functions of more than two independent variables.)

In Example 3 test the critical point at $x = 2$, $y = 12$ by the matrix method.

15.4 Lagrange Multipliers

In many applied problems involving maximum and minimum points, there are one or more restrictions on the independent variables. Assume that these restrictions, or **constraints** as they are often called, can be given by means of equations. A sample constraint might be

$$w + x + y = 100{,}000.$$

This equation might express the fact that the sum of radio, television, and magazine advertising expenses for 1 month is $100,000.

In this section we shall solve more maximum and minimum problems; the problems differ from those of Section 15.3, however, in that one or more constraints will appear. Suppose that a function of n variables $(n \geq 2)$ expressed by

$$f(x_1, x_2, \ldots, x_n)$$

is to be maximized or minimized subject to just one constraint given by

$$g(x_1, x_2, \ldots, x_n) = 0.$$

(The case of more than one constraint will be considered later in this section.) The calculation of critical points for this type of problem will be accomplished by a method known as the **method of Lagrange multipliers.** The method is named after the French mathematician Joseph L. Lagrange (1736–1813). The method consists of the following steps:

1. Let a new function F be defined by

$$F(x_1, x_2, \ldots, x_n, \lambda) = f(x_1, x_2, \ldots, x_n) - \lambda g(x_1, x_2, \ldots, x_n),$$

where λ is called a **Lagrange multiplier.**

2. Form the system of $n + 1$ equations

$$\begin{cases} \dfrac{\partial F}{\partial x_1} = 0, \\[4pt] \dfrac{\partial F}{\partial x_2} = 0, \\[2pt] \vdots \\[2pt] \dfrac{\partial F}{\partial x_n} = 0, \\[4pt] \dfrac{\partial F}{\partial \lambda} = 0. \end{cases}$$

3. Find all values of $x_1, x_2, \ldots, x_n, \lambda$ that satisfy the above system. Among the sets of values of $x_1, x_2, \ldots, x_n$ that satisfy the system can be found coordinates of all the maximum and minimum points if any such points exist. (The method of Lagrange multipliers does not distinguish maximum points from minimum points or from points that are neither.)

Example 1
We shall repeat Example 3 of Section 11.4, for which the critical points can be found by the method of Lagrange multipliers. In that problem a fenced area given by

$$f(x, y) = xy$$

was to be maximized subject to the restriction that the cost of the fence be $1200; that is,

$$2x + 2x + 3y = 1200.$$

Let

$$F(x, y, \lambda) = xy - \lambda(4x + 3y - 1200).$$

(Notice that all terms of the constraint equation were transposed to one side to obtain the quantity multiplied by λ.) Next we form the system

$$\begin{cases} \dfrac{\partial F}{\partial x} = y - 4\lambda = 0, \\[4pt] \dfrac{\partial F}{\partial y} = x - 3\lambda = 0, \\[4pt] \dfrac{\partial F}{\partial \lambda} = -(4x + 3y - 1200) = 0. \end{cases}$$

From the first two equations

$$\lambda = \frac{y}{4} = \frac{x}{3},$$

and hence

$$y = \frac{4}{3}x.$$

This result can be substituted in the third equation to obtain

$$4x + 3\left(\frac{4}{3}x\right) - 1200 = 0,$$

$$8x = 1200,$$
$$x = 150.$$

Therefore,

$$\lambda = \frac{x}{3} = 50,$$

and

$$y = \frac{4}{3}x = 200.$$

The point $(x, y, f(x, y)) = (150, 200, 30{,}000)$ is a critical point. From Example 3 of Section 11.4 we know that it is a *maximum* point; this means that the height z of the surface $z = xy$ is greater when z is evaluated at the point $(x, y) = (150, 200)$ than at any other nearby point on the line $4x + 3y = 1200$. ∎

Observe in Example 1 that, in finding $\partial F/\partial x$, we hold λ as well as y fixed in doing the differentiation—F is a function of three variables, and λ is treated just like x and y in doing the routine partial differentiation necessary in forming the system of equations. The value 50 found for λ in solving the system was only a means to an end and was not a part of the final answer; however, there are situations in more advanced work in which the value of λ is significant.

In this brief discussion we do *not* present a method for testing a critical point to determine whether it is a maximum point, a minimum point, or neither. In many applied problems this is *not* a severe handicap, however, because the existence of a specific optimum can often be demonstrated from the nature of the problem; Example 1 is an illustration, since between the two extremes corresponding to very small x or very small y (in the domain), it follows that there must be a maximum area. It then can be reasoned that a specified optimum point, which is known to exist, must be one of the critical points, and it becomes a matter of identifying which one. In Example 1, (150, 200, 30,000) is the only candidate and hence the maximum area of 30,000 must occur when $x = 150$ and $y = 200$.

Example 2
Find positive values of w, x, and y that minimize

$$f(w, x, y) = x^3 - wx^2 + \frac{y^2}{2} - 10y + 1000$$

subject to the constraint

$$wx - y = 0.$$

$$F(w, x, y, \lambda) = x^3 - wx^2 + \frac{y^2}{2} - 10y + 1000 - \lambda(wx - y).$$

$$\begin{cases} \dfrac{\partial F}{\partial w} = -x^2 - \lambda x = 0, & (15.1) \\[6pt] \dfrac{\partial F}{\partial x} = 3x^2 - 2wx - \lambda w = 0, & (15.2) \\[6pt] \dfrac{\partial F}{\partial y} = y - 10 + \lambda = 0, & (15.3) \\[6pt] \dfrac{\partial F}{\partial \lambda} = -(wx - y) = 0. & (15.4) \end{cases}$$

From (15.1), $\lambda x = -x^2$, and since $x \neq 0$ it follows that $\lambda = -x$. Substituting $\lambda = -x$ in (15.2), we obtain

$$3x^2 - 2wx + wx = 0,$$
$$x(3x - w) = 0,$$

and since $x \neq 0$,

$$w = 3x.$$

Substituting $w = 3x$ in (15.4), we find that
$$y = wx = 3x^2.$$
In (15.3) replace y by $3x^2$ and λ by $-x$ to obtain
$$3x^2 - x - 10 = 0,$$
or
$$(3x + 5)(x - 2) = 0.$$
Since $x > 0$, it must follow that
$$x = 2.$$
Hence
$$\lambda = -x = -2,$$
$$w = 3x = 6,$$
and
$$y = 10 - \lambda = 12.$$

Therefore, if f has a minimum value, that value will occur when $w = 6$, $x = 2$, and $y = 12$. Example 3 of Section 15.3 is the same problem; in this section, however, w was kept and the Lagrange multiplier method used, whereas in Section 15.3 $w = y/x$ had already been substituted in $f(w, x, y)$. In Section 15.3 we found that *minimum* cost occurs at $w = 6$, $x = 2$, and $y = 12$. ∎

Suppose that $f(x_1, x_2, \ldots, x_n)$ is to be maximized or minimized subject to the m constraints
$$\begin{cases} g_1(x_1, x_2, \ldots, x_n) = 0, \\ g_2(x_1, x_2, \ldots, x_n) = 0, \\ \vdots \\ g_m(x_1, x_2, \ldots, x_n) = 0. \end{cases}$$
In this case the method of Lagrange multipliers is generalized in the following manner: Let
$$F(x_1, x_2, \ldots, x_n, \lambda_1, \lambda_2, \ldots, \lambda_m)$$
$$= f(x_1, x_2, \ldots, x_n) - \lambda_1 g_1 - \lambda_2 g_2 - \cdots - \lambda_m g_m,$$
where g_i is an abbreviation for $g_i(x_1, x_2, \ldots, x_n)$. Then solve the system of $n + m$ equations
$$\begin{cases} \dfrac{\partial F}{\partial x_1} = 0, \\ \vdots \\ \dfrac{\partial F}{\partial x_n} = 0, \\ \dfrac{\partial F}{\partial \lambda_1} = 0, \\ \vdots \\ \dfrac{\partial F}{\partial \lambda_m} = 0. \end{cases}$$

Example 3
A cost function C of three nonnegative variables w, x, and y is given by
$$C(w, x, y) = wy - \tfrac{1}{2}x^2y + 2w + xy + 4y^2 - 9x + 35.$$
Minimize C subject to the two constraints
$$x^2 - 6y - 2w = 0$$
and
$$w + x + y = 10.$$
Let
$$F(w, x, y, \lambda_1, \lambda_2) = C(w, x, y) - \lambda_1(x^2 - 6y - 2w) - \lambda_2(w + x + y - 10).$$
Write the system of equations
$$\begin{cases} \dfrac{\partial F}{\partial w} = y + 2 + 2\lambda_1 - \lambda_2 = 0, \\[4pt] \dfrac{\partial F}{\partial x} = -xy + y - 9 - 2\lambda_1 x - \lambda_2 = 0, \\[4pt] \dfrac{\partial F}{\partial y} = w - \dfrac{1}{2}x^2 + x + 8y + 6\lambda_1 - \lambda_2 = 0, \\[4pt] \dfrac{\partial F}{\partial \lambda_1} = -(x^2 - 6y - 2w) = 0, \\[4pt] \dfrac{\partial F}{\partial \lambda_2} = -(w + x + y - 10) = 0. \end{cases}$$

It can be shown that the only solution of this system (for which w, x, and y are nonnegative) is given by $w = 5$, $x = 4$, $y = 1$, $\lambda_1 = -\tfrac{3}{2}$, and $\lambda_2 = 0$. The cost, when $w = 5$, $x = 4$, and $y = 1$, is
$$C(5, 4, 1) = 14.$$
Therefore, if there is a minimum value of C, then that value is 14. Values of w, x, and y that satisfy the constraints and that are close to $(5, 4, 1)$ can be shown to yield a larger cost than 14, which suggests (but does not prove) that there is a *minimum* point at $w = 5$, $x = 4$, and $y = 1$. ∎

For the reader who has studied linear programming (see Chapter 6), the following comments may be beneficial. In linear programming the function to be optimized (maximized or minimized) has to be a linear function, whereas the function to be maximized or minimized in the method of Lagrange multipliers may be a linear function but very often is not. In linear programming the constraints must also be linear. When using Lagrange multipliers, the constraints do *not* have to be linear. Furthermore, in setting up a linear programming problem, one must arrange that the independent variables are nonnegative, whereas there does not have to be any such requirement in Lagrange multiplier problems. The methods of Lagrange multipliers and linear programming, along with the techniques of Chapter 11, provide useful tools for solving optimization problems.

APPLICATIONS

Example 4

Suppose that a utility function (see Example 2 of Section 15.2) of three products is given by

$$u(w, x, y) = 4wxy.$$

The costs of the three goods are $1, $2, and $3, respectively. A consumer can spend $54 and wishes to maximize the utility. The budget restriction is given by the equation

$$w + 2x + 3y = 54,$$

where the quantities w, x, and y of the products must be nonnegative. Use the method of Lagrange multipliers to maximize the utility function subject to the budget constraint.

Let

$$F(w, x, y, \lambda) = 4wxy - \lambda(w + 2x + 3y - 54).$$

$$\begin{cases} \dfrac{\partial F}{\partial w} = 4xy - \lambda = 0, \\[4pt] \dfrac{\partial F}{\partial x} = 4wy - 2\lambda = 0, \\[4pt] \dfrac{\partial F}{\partial y} = 4wx - 3\lambda = 0, \\[4pt] \dfrac{\partial F}{\partial \lambda} = -(w + 2x + 3y - 54) = 0. \end{cases}$$

From the first three equations

$$\lambda = 4xy = 2wy = \tfrac{4}{3}wx.$$

Since $y \neq 0$ (if $y = 0$, the value of the utility function is 0, which clearly is not a maximum value of the utility function), we can divide $4xy = 2wy$ by y to obtain

$$w = 2x.$$

Also $2wy = \tfrac{4}{3}wx$ implies that

$$y = \tfrac{2}{3}x.$$

Substituting in the fourth equation of the system above, we obtain

$$-[2x + 2x + 3(\tfrac{2}{3}x) - 54] = 0,$$
$$6x = 54,$$
$$x = 9.$$

Hence

$$y = \tfrac{2}{3}x = 6,$$
$$\lambda = 4xy = 216,$$

and

$$w = 2x = 18.$$

If $w = 18$, $x = 9$, and $y = 6$ the budget constraint is satisfied, and if there is a maximum value of u, then such a value occurs at $(18, 9, 6)$. Nearby points, which satisfy the constraint, make the value of the utility function smaller than its value at $w = 18$, $x = 9$, and $y = 6$; this suggests (but does not prove) that there is a *maximum* point at $w = 18$, $x = 9$, and $y = 6$. ∎

Example 5
Several books give applications of the method of Lagrange multipliers to problems in business and economics. For the reader who wishes to study the method further or who wishes to see more applications in business and economics, we recommend the books by Kim [40], McAdams [46], and Theodore [62]. ∎

EXERCISES

Suggested minimum assignment: Exercises 1, 5, 9, 10, and 11.

In Exercises 1–10 use the method of Lagrange multipliers under the assumption that the desired optimum exists.

1. Minimize $f(x, y) = x^2 + y^2$ subject to $x + 2y = 10$.
2. Minimize $f(w, x, y) = 2w^2 + x^2 + y^2$ subject to $w + x + y = 50$.
3. Find three positive numbers whose sum is 36 and whose product is maximum.
4. Minimize $f(w, x, y) = w + x^2 + (y^2/2)$ subject to $w + 4x + y = 15$.
5. Find positive numbers w, x, and y so that $w + x + 2y$ is maximized subject to the restriction that $w^2 + x^2 + y^2 = 54$.
6. Find positive numbers w, x, and y so that $w^2 + x^3 + y^2$ is minimized subject to the constraint $w + 3x + y = 28$.
7. Maximize the utility function $f(x, y) = x\sqrt{y}$ subject to the budget constraint $2x + y = 3000$.
8. Rework Example 4 of Section 11.4.
9. Three sides of a rectangular plot of ground are to be fenced so that the total area of the plot is 5000 square feet. Find the length and width of the plot so that the number of feet of fencing is minimum.
10. The daily production P of an item is a function of three inputs as given by
$$P = wx + 6y - x^2 - y^2.$$
Find w, x, and y so as to maximize the daily production subject to the restriction
$$w + x + y = 10.$$
11. Rework Exercise 10 by replacing w with $10 - x - y$ (thus expressing P as a function of x and y) and then using methods of Section 15.3.
12. Maximize the volume lwh of a closed rectangular box subject to the two constraints

and
$$w = \tfrac{1}{2}l$$
$$2wl + 2wh + 2lh = 300.$$

(The last equation requires that the total surface area of the box equal 300 square units.)

NEW VOCABULARY

surface 15.1
saddle surface 15.1
saddle point 15.1
partial derivative 15.2
utility function 15.2
marginal utilities 15.2

critical point 15.3
constraint 15.4
method of Lagrange multipliers 15.4
Lagrange multiplier 15.4

References

Many of the following books and articles have been referred to in the applied examples of this book. Others have been referred to for proofs of certain theorems. The references marked with an asterisk have many applications of topics in finite mathematics or calculus, and it is recommended that those readers who are interested in applications should at least scan the relevant parts of some of these books.

[1] ACKERMAN, E., *Biophysical Science*, Prentice-Hall, Inc., Englewood Cliffs, N.J., 1962.

[2] ALLEN, L. E., and CALDWELL, M. C., "Modern Logic and Judicial Decision Making: A Sketch of One View," *Law and Contemporary Problems*, 28, 1963, pp. 213–270.

[3] ALLING, D. W., "The After History of Pulmonary TB: A Stochastic Model," *Biometrics*, 14, 1958, pp. 527–547.

[4] ARNOLD, B. J., *Logic and Boolean Algebra*, Prentice-Hall, Inc., Englewood Cliffs, N.J., 1962.

[5] BAILEY, N. T. J., *The Mathematical Theory of Epidemics*, Charles Griffin & Company Ltd., London, 1957.

*[6] BATSCHELET, EDWARD, *Introduction to Mathematics for Life Scientists*, Biomathematics, Vol. 2, Springer-Verlag, New York, 1971.

[7] BELLMAN, R., and BLACKWELL, D., "Red Dog, Blackjack Poker," *Scientific American*, 184, 1951, pp. 44–47.

[8] BERNADELLI, H., "Population Waves," *J. Burma Res. Soc.*, 31, 1941, pp. 1–18.

*[9] BISHIR, J. W., and DREWES, D. W., *Mathematics in the Behavioral and Social Sciences*, Harcourt Brace Jovanovich, Inc., New York, 1970.

[10] BOOT, J. C. G., *Mathematical Reasoning in Economics and Management Science*, Prentice-Hall, Inc., Englewood Cliffs, N.J., 1967.

*[11] BREMS, HANS, *Quantitative Economic Theory*, John Wiley & Sons, Inc., New York, 1968.

[12] CAMPBELL, H. G., *An Introduction to Matrices, Vectors, and Linear Programming*, 2nd. ed., Prentice-Hall, Inc., Englewood Cliffs, N.J., 1977.

*[13] CAMPBELL, H. G., *Linear Algebra with Applications: Including Linear Programming*, Prentice-Hall, Inc., Englewood Cliffs, N.J., 1971.

*[14] CAMPBELL, H. G., *Matrices with Applications*, Prentice-Hall, Inc., Englewood Cliffs, N.J., 1968.

[15] CASSTEVENS, T. W., and DENHAM, W. A., III, "Turnover and Tenure in the Canadian House of Commons 1867–1968," *Canadian J. Political Sci.*, 3, No. 4, 1970.

[16] CHARNES, A., COOPER, W. W., and MILLIER, M. H., "Application of Linear Programming to Financial Budgeting and the Costing of Funds," *J. Business Univ. Chicago*, 32, 1959, pp. 20–46.

*[17] COALE, A. J., *The Growth and Structure of Human Populations, A Mathematical Investigation*, Princeton University Press, Princeton, N.J., 1972.

[18] DANTZIG, GEORGE, *Linear Programming and Extensions*, Princeton University Press, Princeton, N.J., 1963.

[19] DEUTSCH, K. W., and MADOW, W. G., "A Note on the Appearance of Wisdom in Large Bureaucratic Organizations," *Behavioral Sci.*, 6, 1961, pp. 72–78.

[20] DODD, S. C., "Diffusion Is Predictable: Testing Probability Models for Laws of Interaction," *Amer. Sociolog. Rev.*, 20, 1955.

[21] DORFMAN, R., *Application of Linear Programming to the Theory of the Firm, Including an Analysis of Monopolistic Firms by Non-linear Programming*, University of California Press, Berkeley, Calif., 1951.

[22] FREUND, J. E., *College Mathematics with Business Applications*, Prentice-Hall, Inc., Englewood Cliffs, N.J., 1969.

[23] GALE, DAVID, *The Theory of Linear Economic Models*, McGraw-Hill Book Company, New York, 1960.

[24] GARRETT, H. F., *Statistics in Psychology and Education*, 5th ed., David McKay Company, Inc., New York, 1958.

[25] GASS, S. I., *Linear Programming*, 3rd. ed., McGraw-Hill Book Company, New York, 1969.

[26] GOREN, C. H., *Goren's New Contract Bridge Complete*, Doubleday & Company, Inc., Garden City, N.Y., 1957.

[27] GRAY, VIRGINIA, "Innovation in the States: A Diffusion Study," *Amer. Political Sci. Rev.*, 67, No. 4, 1973.

[28] HAYS, W. L., *Quantification in Psychology*, Brooks/Cole Publishing Company, Monterey, Calif., 1967.

*[29] HOHN, F. E., *Applied Boolean Algebra*, 2nd ed., Macmillan Publishing Co., Inc., New York, 1966.

[30] HULL, C. L., *Essentials of Behavior*, Yale University Press, New Haven, Conn., 1951.

*[31] IJIRI, YUJI, *Management Goals and Accounting for Control*, Rand McNally & Company, Skokie, Ill., 1965.

[32] IJIRI, Y., LEVY, F. K., and LYON, R. C., "A Linear Programming Model for Budgeting and Financial Planning," *J. Accounting Res.*, 1, 1963, pp. 198–212.

[33] ITT, Federal Electric Corporation, *Boolean Algebra*, Prentice-Hall, Inc., Englewood Cliffs, N.J., 1966.

*[34] JOHNSTON, J. B., PRICE, G. B., and VAN VLECK, F. S., *Linear Equations and Matrices*, Addison-Wesley Publishing Company, Inc., Reading, Mass., 1966.

[35] KATTSOFF, L. O., and SIMONE, A. J., "A Logical Model for a Problem in Production Management," *APICS* (American Production and Inventory Control Society) *Quart. Bull.*, 6, No. 2, 1965.

References

[36] KAY, PAUL, ed., *Explorations in Mathematical Anthropology*, The MIT Press, Cambridge, Mass., 1971.

[37] KEMENY, J. G., MORGENSTERN, OSKAR, and THOMPSON, G. L., "A Generalization of the von Neumann Model of an Expanding Economy," *Econometrica*, 24, 1950, pp. 115–135.

*[38] KEMENY, J. G., SCHLEIFER, A., SNELL, J. L., and THOMPSON, G. L., *Finite Mathematics with Business Applications*, 2nd ed., Prentice-Hall, Inc., Englewood Cliffs, N.J., 1972.

*[39] KEMENY, J. G., SNELL, J. L., and THOMPSON, G. L., *Introduction to Finite Mathematics*, 2nd ed., Prentice-Hall, Inc., Englewood Cliffs, N.J., 1966.

[40] KIM, T. K., *Introductory Mathematics for Economic Analysis*, Scott, Foresman and Company, Glenview, Ill., 1971.

[41] KLINE, MORRIS, *Mathematics—A Cultural Approach*, Addison-Wesley Publishing Company, Inc., Reading, Mass., 1962.

[42] KORT, FRED, "Simultaneous Equations and Boolean Algebra in the Analysis of Judicial Decisions," in Hans Baade, ed., *Jurimetrics*, Basic Books, Inc., New York, 1963.

[43] LIBBY, W. F., *Radiocarbon Dating*, University of Chicago Press, Chicago, 1952.

[44] LUCE, R. D., and RAIFFA, HOWARD, *Games and Decisions: Introduction and Critical Survey*, John Wiley & Sons, Inc., New York, 1957.

[45] MACDONALD, J., *Strategy in Poker, Business, and War*, W. W. Norton & Company, New York, 1950.

[46] MCADAMS, A. K., *Mathematical Analysis for Management Decisions*, Macmillan Publishing Co., Inc., New York, 1970.

[47] MCGILL, W. J., "Stochastic Latency Mechanisms," in Luce, R. D., Bush, R. R., and Galantes, E. E., eds., *Handbook of Mathematical Psychology*, John Wiley & Sons, Inc., New York, 1963.

[48] MCKINSEY, J. C. C., *Introduction to the Theory of Games*, McGraw-Hill Book Company, New York, 1952.

[49] MILLER, IRWIN, and FREUND, J. E., *Probability and Statistics for Engineers*, Prentice-Hall, Inc., Englewood Cliffs, N.J., 1965.

[50] NEWMAN, J. R., *The World of Mathematics*, Vol. 1, Simon and Schuster, New York, 1956.

[51] PIELOU, E. C., *An Introduction to Mathematical Ecology*, John Wiley & Sons, Inc., New York, 1969.

[52] RAPOPORT, ANATOL, *Fights, Games, and Debates*, University of Michigan Press, Ann Arbor, Mich., 1960.

[53] RAPOPORT, ANATOL, *Two-Person Game Theory: The Essential Ideas*, University of Michigan Press, Ann Arbor, Mich., 1966.

[54] RICHARDSON, L. F., "War Moods," *Psychometrika*, Vol. 13, 1948.

[55] RIKER, W. H., *The Theory of Coalitions*, Yale University Press, New Haven, Conn., 1963.

- *[56] SEARLE, S. R., *Matrix Algebra for the Biological Sciences*, John Wiley & Sons, Inc., New York, 1966.
- *[57] SEARLE, S. R., and HAUSMAN, W. H., *Matrix Algebra for Business and Economics*, John Wiley & Sons, Inc. (Interscience Division), New York, 1970.
- [58] SHAPLEY, L. S., and SHUBIK, M., "A Method for Evaluating the Distribution of Power in a Committee System," *Amer. Political Sci. Rev.*, 48, 1954, pp. 787–792.
- [59] SHUBIK, MARTIN, *Strategy and Market Structure*, John Wiley & Sons, Inc., New York, 1959.
- [60] SIMON, WILLIAM, *Mathematical Techniques for Physiology and Medicine*, Academic Press, Inc., New York, 1972.
- [61] SMITH, C. A. B., *Biomathematics*, Vol. 1, Hafner Press, New York, 1966.
- [62] THEODORE, C. A., *Applied Mathematics: An Introduction*, rev. ed., Richard D. Irwin, Inc., Homewood, Ill., 1971.
- *[63] THRALL, R. M., MORTIMER, J. A., REBMAN, K. R., and BAUM, R. F., *Some Mathematical Models in Biology*, Report No. 40241-R-7, rev. ed., University of Michigan Press, Ann Arbor, Mich., 1967.
- [64] ULMER, SIDNEY, "Leadership in the Michigan Supreme Court," in G. Shubert, ed., *Judicial Decision Making*, The Free Press, New York, 1963.
- [65] *United States Life Tables by Causes of Death 1959–1961*, Vol. 1, No. 6, National Health Center for Health Statistics, Public Health Service, Washington, D.C., 1967.
- [66] VON NEUMANN, J., and MORGANSTERN, OSKAR, *Theory of Games and Economic Behavior*, 3rd. ed., Princeton University Press, Princeton, N.J., 1953.
- *[67] WILLIAMS, J. D., *The Compleat Strategyst*, rev. ed., McGraw-Hill Book Company, New York, 1966.

Table of Integration Formulas

1. $\int du = u + C.$

2. $\int kf(x)\,dx = k\int f(x)\,dx.$

3. $\int [f(x) + g(x)]\,dx = \int f(x)\,dx + \int g(x)\,dx.$

4. $\int u^n\,du = \dfrac{u^{n+1}}{n+1} + C,\ n \neq -1.$

5. $\int b^u\,du = \dfrac{b^u}{\ln b} + C,\ b > 0,\ b \neq 1.$

6. $\int e^u\,du = e^u + C.$

7. $\int \dfrac{du}{u} = \ln|u| + C,\ u \neq 0.$

8. $\int u\,dv = uv - \int v\,du.$

9. $\int \dfrac{du}{a^2 - u^2} = \dfrac{1}{2a}\ln\left|\dfrac{a+u}{a-u}\right| + C.$

10. $\int \dfrac{du}{u^2 - a^2} = \dfrac{1}{2a}\ln\left|\dfrac{u-a}{u+a}\right| + C.$

11. $\int \dfrac{du}{u(a+bu)} = -\dfrac{1}{a}\ln\left|\dfrac{a+bu}{u}\right| + C.$

12. $\int \dfrac{du}{u^2(a+bu)} = -\dfrac{1}{au} + \dfrac{b}{a^2}\ln\left|\dfrac{a+bu}{u}\right| + C.$

13. $\int \dfrac{du}{u(a+bu)^2} = \dfrac{1}{a(a+bu)} - \dfrac{1}{a^2}\ln\left|\dfrac{a+bu}{u}\right| + C.$

14. $\int \dfrac{u\,du}{a+bu} = \dfrac{1}{b^2}(a+bu - a\ln|a+bu|) + C.$

15. $\int \dfrac{u^2\,du}{a+bu} = \dfrac{1}{b^3}\left[\tfrac{1}{2}(a+bu)^2 - 2a(a+bu) + a^2\ln|a+bu|\right] + C.$

16. $\int \dfrac{u\,du}{(a+bu)^2} = \dfrac{1}{b^2}\left(\dfrac{a}{a+bu} + \ln|a+bu|\right) + C.$

17. $\int \dfrac{u^2\,du}{(a+bu)^2} = \dfrac{1}{b^3}\left(a+bu - \dfrac{a^2}{a+bu} - 2a\ln|a+bu|\right) + C.$

18. $\int u^n e^u \, du = u^n e^u - n \int u^{n-1} e^u \, du, n > 0.$

19. $\int \dfrac{e^u}{u^n} \, du = -\dfrac{e^u}{(n-1)u^{n-1}} + \dfrac{1}{n-1} \int \dfrac{e^u \, du}{u^{n-1}}, n > 0, n \neq 1.$

20. $\int \ln u \, du = u \ln u - u + C, u > 0.$

21. $\int u^n \ln u \, du = u^{n+1} \left[\dfrac{\ln u}{n+1} - \dfrac{1}{(n+1)^2} \right] + C, u > 0, n \neq -1.$

22. $\int u \sqrt{a + bu} \, du = \dfrac{6bu - 4a}{15b^2}(a + bu)^{3/2} + C, a + bu > 0.$

23. $\int \dfrac{\sqrt{a^2 - u^2}}{u} \, du = \sqrt{a^2 - u^2} - a \ln \left| \dfrac{a + \sqrt{a^2 - u^2}}{u} \right| + C, a^2 > u^2.$

24. $\int \sqrt{a^2 + u^2} \, du = \dfrac{u}{2} \sqrt{a^2 + u^2} + \dfrac{a^2}{2} \ln |u + \sqrt{a^2 + u^2}| + C.$

25. $\int \sqrt{u^2 - a^2} \, du = \dfrac{u}{2} \sqrt{u^2 - a^2} - \dfrac{a^2}{2} \ln |u + \sqrt{u^2 - a^2}| + C, u^2 > a^2.$

Table of Integration Formulas

Answers to Odd-numbered Exercises

1.1 Page 5

1. $A = \{$Vt., N.H., Me., Mass., Conn., R.I.$\}$, $A = \{x : x$ is a state in New England$\}$.
3. $C = \{-1, 1\}$, $C = \{x : x^2 - 1 = 0\}$.
5. $E = \{3\}$, E is the set of all positive integers that satisfy the equation $(x - 3)(x + 2) = 0$.
7. One answer is: G is the set of all integers between 1 and 3 inclusive, or $G = \{x : 1 \leq x \leq 3, x$ is an integer$\}$.
9. One answer is: $\{3, 4\} \subset A$, $3 \in A$.
13. $\{$Chicago, Cleveland, Pittsburgh$\}$, $\{$Chicago, Cleveland$\}$, $\{$Chicago, Pittsburgh$\}$, $\{$Pittsburgh, Cleveland$\}$, $\{$Chicago$\}$, $\{$Cleveland$\}$, $\{$Pittsburgh$\}$, $\emptyset$.   15. True.   17. False, $\{2, 3\} \subset U$.   19. False, $\emptyset \subset U$.   21. True.
23.

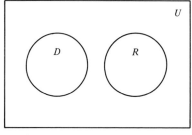 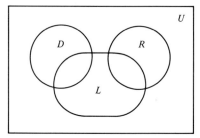

25. $\frac{2}{3}$, 0, 5, .44444$\cdots$, and $-\frac{9}{8}$ are rational.   27. (a), (b), (e), and (f) are true.

1.2 Page 13

1. (a) $A \cup B = \{1, 2, 3, 4, 5, 6\}$. (b) $A \cap B = \{4\}$. (c) $A' = \{5, 6\}$.
(d) $A \cup (A' \cap B) = \{1, 2, 3, 4, 5, 6\}$. (e) $A \cup \emptyset = \{1, 2, 3, 4\}$.
(f) $B \cap B' = \emptyset$. (g) $U \cup A = \{1, 2, 3, 4, 5, 6\}$. (h) $\emptyset \cap B = \emptyset$.
3. (a) $P \cup R = \{x : (x \in P)$ or $(x \in R)\}$. (b) $P \cap R = \{x : (x \in P)$ and $(x \in R)\}$.
(c) $R' = \{x : (x \notin R)\}$.
5. (a) $\emptyset$. (b) U. (c) U. (d) A.   7. $L \cap D$.   9. $(L \cap D) \subseteq D$.   11. C'.
13. (a) Z. (b) K. (c) I. (d) N. (e) U. (f) $\emptyset$. (g) K. (h) Z.
15. For Exercise 1, $A \triangle B = \{1, 2, 3, 5, 6\}$; for Exercise 2, $A \triangle B = \{a, d\}$.
17. $B \cap (A \cap B \cap C)'$ is one answer.

Answers to Odd-numbered Exercises

658

1.3 Page 19

1.

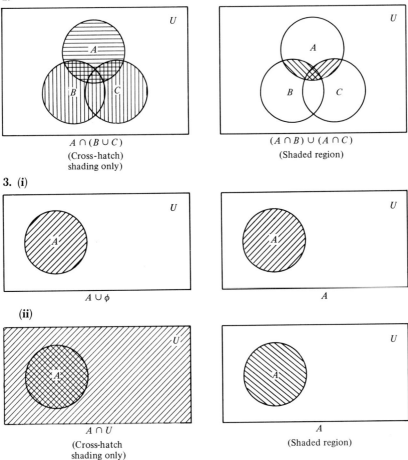

$A \cap (B \cup C)$
(Cross-hatch shading only)

$(A \cap B) \cup (A \cap C)$
(Shaded region)

3. (i)

$A \cup \phi$

A

(ii)

$A \cap U$
(Cross-hatch shading only)

A
(Shaded region)

5. Statements:

$$\begin{aligned} A \cap \emptyset &= (A \cap \emptyset) \cup \emptyset \\ &= \emptyset \cup (A \cap \emptyset) \\ &= (A \cap A') \cup (A \cap \emptyset) \\ &= A \cap (A' \cup \emptyset) \\ &= A \cap A' \\ &= \emptyset. \end{aligned}$$

Table

$x \in A$	$x \in \emptyset$	$x \in A \cap \emptyset$
True	False	False
False	False	False

Because the second and third columns in the table are identical, the definition of set equality is satisfied and the conclusion follows.

Answers to Odd-numbered Exercises

7. Statements:
$$A = A \cap U$$
$$= A \cap (A \cup A')$$
$$= (A \cap A) \cup (A \cap A')$$
$$= (A \cap A) \cup \emptyset$$
$$= A \cap A.$$

Table

$x \in A$	$x \in A \cap A$
True	True
False	False

Because the two columns in the table are identical, the definition of set equality is satisfied and the conclusion follows.

9. Given $A \cap X = A$ for every A. Hence if $A = U$, we have $U \cap X = U$. Also, $U \cap X = X$ by Theorem 3. Therefore, $U = U \cap X = X$.

11. Statements:
$$A \cap (A \cup B) = (A \cup \emptyset) \cap (A \cup B)$$
$$= A \cup (\emptyset \cap B)$$
$$= A \cup \emptyset$$
$$= A.$$

Table

(1) $x \in A$	(2) $x \in B$	(3) $x \in A \cup B$	(4) $x \in A \cap (A \cup B)$
True	True	True	True
True	False	True	True
False	True	True	False
False	False	False	False

Because the first and fourth columns in the table are identical, the definition of set equality is satisfied and the conclusion follows.

13. Use table.   **15.** Use table.   **17.** Use table.   **21.** Use table.
23. Use table.

1.4 Page 25

1. (a) 800; (b) 100.   **3.** 22.   **5.** 41.   **7.** 14.   **9.** 44.   **11.** 34.
13. 66.   **15.** 4.   **17.** 19.   **19.** 75.
21. Statements:
$$U = A \cup A';$$
$$n(U) = n(A \cup A');$$
$$n(U) = n(A) + n(A') - n(A \cap A');$$
$$n(U) = n(A) + n(A') - n(\emptyset);$$
$$n(U) - n(A) = n(A').$$

Answers to Odd-numbered Exercises

1.5 Page 29

1. $A \times B = \{(a, x), (a, y), (a, z), (b, x), (b, y), (b, z), (c, x), (c, y), (c, z)\}$.
3.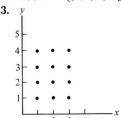
5. $n(F \times M) = n(F) \cdot n(M)$, where F is the set of females and M is the set of males. $n(M \times F)$ is also an answer; 63.

7. 24; $n(D \times R \times I)$ is one answer, where D is the set of Democratic candidates, R is the set of Republican candidates, and I is the set of independent candidates.

2.1 Page 35

1. (b) and (c) are statements. 3. (a) and (b) are statements.
5. (a) $p \rightarrow q$, "if ... then," conditional. (b) $p \rightarrow (q \wedge r)$, "if ... then," conditional. (c) $\sim(p \wedge q)$, "not," negation. 7. (a) $p \wedge \sim q$, "and," conjunction. (b) $p \rightarrow q$, "if ... then," conditional. (c) $p \leftrightarrow (q \wedge r)$, "if and only if," biconditional. 9. (a) "or," disjunction. (b) "if ... then," conditional. (c) "not," negation. 11. Either the computer is available or I will not get my project finished (but not both).

2.2 Page 40

1.

p	q	$[p \rightarrow (\sim q)] \wedge p$
T	T	F
T	F	T
F	T	F
F	F	F

3.

p	q	$[(\sim p) \vee (\sim q)] \leftrightarrow \sim(p \wedge q)$
T	T	T
T	F	T
F	T	T
F	F	T

5.

p	q	r	$[(\sim r) \vee q)] \vee (p \wedge r)$
T	T	T	T
T	T	F	T
T	F	T	T
T	F	F	T
F	T	T	T
F	T	F	T
F	F	T	F
F	F	F	T

Answers to Odd-numbered Exercises

7.

p	q	r	$[(p \vee r) \vee q] \rightarrow (p \wedge q)$
T	T	T	T
T	T	F	T
T	F	T	F
T	F	F	F
F	T	T	F
F	T	F	F
F	F	T	F
F	F	F	T

9. $q \rightarrow p$ is one answer; $p \vee (\sim q)$ is another answer.   11. 16.
13. (a) p: Excessive bail shall be required.
   q: Excessive fines shall be imposed.
   r: Cruel punishment shall be inflicted.
   s: Unusual punishment shall be inflicted.
   (b) $[(\sim p) \wedge (\sim q)] \wedge \sim (r \wedge s)$.

2.3 Page 45

1. Neither.   3. Tautology; equivalence.   5. Neither.   7. Neither.
9. Show that the evaluations of both are the same.   11. Show that the evaluations of both are the same.   13. Yes; the conditional $p \veebar (p \wedge q) \rightarrow (\sim q)$ is a tautology.
15. (a) p: Person shall be convicted of treason.
   q: Testimony is given by two witnesses to the same overt act.
   r: Confession is given in open court.
   (b) $\sim(q \vee r) \rightarrow (\sim p)$.
   (c) $(q \vee r) \vee (\sim p)$.

2.4 Page 49

1. $(\sim p) \wedge q$.   3. p.   5. $q \vee p$.
7. Statements: $p \equiv p \wedge U$
   $\equiv p \wedge [p \vee (\sim p)]$
   $\equiv (p \wedge p) \vee [p \wedge (\sim p)]$
   $\equiv (p \wedge p) \vee C$
   $\equiv p \wedge p$.
9. Given $p \wedge x \equiv p$ for every p. Hence if $p \equiv U$, we have $U \wedge x \equiv U$, but $U \wedge x \equiv x$ by Theorem 3. Therefore, $U \equiv U \wedge x \equiv x$.
11. Statements: $p \wedge (p \vee q) \equiv (p \vee C) \wedge (p \vee q)$
   $\equiv p \vee (C \wedge q)$
   $\equiv p \vee C$
   $\equiv p$.

29. (a) p: Equality of rights under the law shall be *denied* by *the United States* on account of sex.
   q: Equality of rights under the law shall be *abridged* by *the United States* on account of sex.
   r: Equality of rights under the law shall be *denied* by *any state* on account of sex.

Answers to Odd-numbered Exercises

s: Equality of rights under the law shall be *abridged* by *any state* on account of sex.
(b) $\sim[(p \vee q) \vee (r \vee s)]$. (c) $[(\sim p) \wedge (\sim q)] \wedge [(\sim r) \wedge (\sim s)]$.

2.5 Page 54

1. Valid argument form because the given conditional is a tautology, as can be proved by an evaluation table or by Theorem 23. Premises are

(1) $p \vee q$,
(2) $\sim p$.

Conclusion is q.

3. Fallacious argument form because the given conditional is *not* a tautology, as can be shown in an evaluation table. Premises are

(1) $p \vee q$,
(2) q.

Conclusion is p.

5. Nonargument form because the given statement form is not a conditional.

7. Valid argument form because the given conditional is a tautology, as can be proved by an evaluation table, or because the conclusion is implied by premises through use of the theorems of this chapter. Premises are

(1) $p \vee (\sim q)$.
(2) $(p \vee q) \wedge q$.

Conclusion is $p \wedge q$.

9. Valid. **11.** Fallacious. **13.** Valid.

23. Argument form is fallacious in Exercise 11, hence argument is unsound. The first premise is false in Exercise 13, hence argument is unsound, even though form is valid.

2.6 Page 60

1.

3.
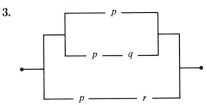

5. $\{[(p \vee r) \wedge (q \vee p)] \wedge r\} \vee r$ is one answer.
7. $\{[p \wedge (q \vee r)] \vee (q \wedge p)\} \vee r$ is one answer.
9. $(p \vee q) \wedge (s \vee t)$.

Answers to Odd-numbered Exercises

2.7 Page 67

1.

p	q	p ∧ (p ∨ q)
1	1	1
1	0	1
0	1	0
0	0	0

3.

p	q	[p ∨ (~q)] ∨ [q ∨ (~p)]
1	1	1
1	0	1
0	1	1
0	0	1

5.

p	q	r	(p ∨ r) ∧ (~q)
1	1	1	0
1	1	0	0
1	0	1	1
1	0	0	1
0	1	1	0
0	1	0	0
0	0	1	1
0	0	0	0

7.

p	q	r	(p ∧ q) ∨ [r ∧ (~q)]
1	1	1	1
1	1	0	1
1	0	1	1
1	0	0	0
0	1	1	0
0	1	0	0
0	0	1	1
0	0	0	0

9. Equivalent.   **11.** Not equivalent.
13. (a) $p \wedge q$. (b) $(\sim p) \wedge q$. (c) $(p \wedge q) \vee [(\sim p) \wedge q]$. (d) Equivalent.
15. (a) p: Excessive bail shall be required.
    q: Excessive fines shall be imposed.
    r: Cruel punishment shall be inflicted.
    s: Unusual punishment shall be inflicted.
(b) $[(\sim p) \wedge (\sim q)] \wedge \sim(r \wedge s)$.
(c) $[(\sim p) \wedge (\sim q)] \wedge [(\sim r) \vee (\sim s)]$.

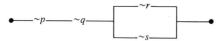

(d) $(p \vee q) \vee (r \wedge s)$.

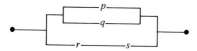

Answers to Odd-numbered Exercises

(e) A punishment must be both cruel and unusual to be a violation.
17. (a) *p*: Person shall be convicted of treason.
   q: Testimony is given by two witnesses to the same overt act.
   r: Confession is given in open court.
   (b) $\sim(q \vee r) \rightarrow (\sim p)$.   (c) $(q \vee r) \vee (\sim p)$.
   (d)

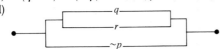

   (e) $[(\sim q) \wedge (\sim r)] \wedge p$; •————~q————~r————p————•

3.1 Page 76

1. (a) 81. (b) $\frac{1}{9}$. (c) 1. (d) 4. (e) 5. (f) 2. (g) 9. (h) $\frac{1}{27}$. (i) -4.
3. (a) $\frac{1}{25}$. (b) $\frac{81}{4}$. (c) 6. (d) 324. (e) $a^6 b^5$. (f) a^5, provided that $a \neq 0$. (g) 5.
(h) $a^7 b^9$, provided that $a \neq 0$, $b \neq 0$.
5. (a) $x^{-1}y^{-1/3}$, provided that $x \neq 0$, $y \neq 0$. (b) $x^{-1/6}y^{7/2}$, provided that $x > 0$, $y > 0$. (c) $a^{-1/6}b^{-3}$, provided that $a > 0$, $b > 0$.
7. (a) If r is a negative even integer, then a and b must be positive real numbers. If $r = 0$ or if r is a negative odd integer, then a and b can be any real numbers except 0. If r is a positive integer, then a and b can be any real numbers except that $b \neq 0$.
(b) $(a/b)^r = (ab^{-1})^r = a^r(b^{-1})^r = a^r b^{-r} = a^r/b^r$.

3.2 Page 81

1. (a) $x(x + y)$. (b) $(x - 4)(x + 3)$. (c) $(x - 3)(x + 3)$. (d) $y(2 - x)(2 + x)$.
3. (a) $(x - 3)(x + 4)$. (b) $x(x - 1)(x - 2)$. (c) $y(2x - 1)(x + 4)$.
(d) $(x - 2)(x^2 + 2x + 4)$.
5. (a) $x(x - 1)(x^2 + x + 1)$. (b) $(x - \sqrt{3})(x + \sqrt{3})(x^2 + 3)$.
(c) $(2a + 3b)(2a + 3b)$. (d) $(x - 2)(x^2 + 2x + 4)(x + 2)(x^2 - 2x + 4)$.
7. (a) $x^2 - 4y^2$. (b) $x^3 - 5x^2 + 6x$. (c) $14x^2 - 5x - 24$.
(d) $16a^2 + 40ab + 25b^2$.
9. $x^4 + 8x^3 y + 24x^2 y^2 + 32xy^3 + 16y^4$.   11. $27x^3 - 27x^2 y + 9xy^2 - y^3$.
13. (a) $\sqrt{3}/3$. (b) $5\sqrt{2}$.   15. (a) $\dfrac{3 + x}{x^2 + 2}$. (b) $\dfrac{7x + 8}{x^2}$. (c) $\dfrac{13}{24}$. (d) $\dfrac{11x - 4}{x^2 - 16}$.
17. $\dfrac{30}{(x - 5)(x^2 - 1)}$.   19. $\dfrac{ab}{a + b}$.   21. $\dfrac{(2x - y)(4x + y)}{(2x + y)(2x - 5y)}$.
23. None is correct.

3.3 Page 88

1. (a) 4. (b) $-\frac{7}{2}$. (c) $\dfrac{11 + a}{2a}$.   3. $\dfrac{3y^2 - y}{2y^3 - x}$.   5. 15.
7. (a) 3, 4, 5. (b) 4, 5. (c) 6, $-\frac{3}{2}$. (d) 1, 6.
9. (a) $-\dfrac{7}{2} \pm \dfrac{\sqrt{37}}{2}$. (b) $\dfrac{1}{2} \pm \dfrac{\sqrt{7}}{2}$.   11. $\left\{ \dfrac{9 + \sqrt{69}}{2}, \dfrac{9 - \sqrt{69}}{2} \right\}$.
13. (a) 2, 3. (b) $-\frac{1}{2}$.   15. (a) $x = 2$, $y = 8$; $x = -1$, $y = 5$.
(b) $x = 1$, $y = 1$; $x = \frac{25}{4}$, $y = -\frac{5}{2}$.
17. (a) Discriminant is -24, solutions not real.
   (b) Discriminant is 2, solutions are real.
   (c) Discriminant is 0, solutions are real.

Answers to Odd-numbered Exercises

19. (a) $2 + 6i$. (b) $4 - 7i$.

3.4 Page 93

1. $A = 25\pi$ square inches, $C = 10\pi$ inches. **3.** $A = 72$ square units.
5. 12. **7.** 15. **9.** $(16 + 2\pi)$ square feet. **11.** (a) $\dfrac{256\pi}{3}$ cubic inches.
(b) 64π square inches. **13.** (a) 54π cubic inches. (b) 45π square inches.
15. $r = 2$ feet, $h = 4$ feet. **17.** (a) 576 cubic inches. (b) \$43.20.

3.5 Page 99

1. $\frac{2}{3}$, 0, and π. **3.** (a), (c), (d) are true; (b) is false.
5. (a) $(-1, 6)$, open;

(b) $[-2, 1)$, half-open;

(c) $[-8, -6]$, closed;

(d) $(4, 7]$, half-open.

7. (a) $\{x: -3 < x < 8\}$. (b) $\{x: \pi \leq x \leq 5\}$. (c) $\{x: -19 \leq x \leq -5\}$.
(d) $\{x: -3 < x \leq 3\}$. **9.** (a) 4. (b) 0. (c) 2. (d) $\frac{4}{5}$. (e) 7. (f) 8. (g) 2.
11. (a) 7. (b) 7. (c) 7. (d) 16.
13. Let x be the number of men who sign the petition, and let y be the number of women who sign the petition. Then $x + y \geq 100$ and also $y \geq 30$ in order for the petition to be considered.
15. (a) $x > \frac{11}{3}$. (b) $x = -\frac{3}{2}, \frac{13}{2}$. (c) $\frac{10}{3} \geq x \geq -\frac{8}{3}$. (d) $x > 2$, $x < -18$.

3.6 Page 106

1. $(-3, 5)$, 2nd quadrant;
$(4, -1)$, 4th quadrant;
$(2, 7)$, 1st quadrant;
$(-5, -4)$, 3rd quadrant.

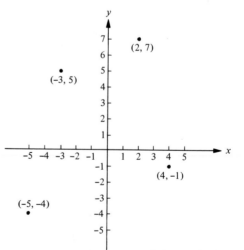

3. (a) 5. (b) $\sqrt{53}$. (c) 9. **5.** $\sqrt{113}$. **7.** (a) $\Delta x = 8$, $\Delta y = -7$.
(b) $\Delta x = -8$, $\Delta y = 7$. **9.** Yes, because $1^4 + 2^3 - 8(1) - 1 = 0$.
11. $(x + 1)^2 + (y + 2)^2 = 9$. **13.** (a) Center $(7, -10)$, radius 2. (b) Center

Answers to Odd-numbered Exercises

(0, 0), radius $\sqrt{29}$. (c) Center $(-1, 4)$, radius 5. (d) Center $(\frac{5}{2}, -\frac{1}{2})$, radius 3. **15.** Yes, because the person is only 13 miles from the sender.

3.7 Page 116

1. $f(1) = 10$, $f(3) = 30$, $f(5) = 50$, $f(7) = 70$, $R = \{10, 30, 50, 70\}$.
3. $f(3) = 13$; $R = \{y : y \geq 4\}$;

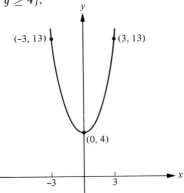

5. $D = \{1, 2, 3\}$; $R = \{4, 5, 6\}$. The number 2 has assigned to it both 5 and 7—each element in the domain of a *function* has only *one* image.
7. y is the dependent variable and x is the independent variable; $g(5) = \frac{1}{5}$.
9. (a) $\{x : x \neq -2\}$. (b) set of all real numbers. (c) $\{x : x \neq \pm 1\}$.
(d) $\{x : x \geq 0\}$.   **11.** $f(-3) = -\frac{1}{3}$; $f(a + b) = \dfrac{1}{2a + 2b + 3}$.   **13.** (a) 9.
(b) 5. (c) 1. (d) $x^2 + 6x + 9$. (e) $6x + 9$. (f) $2x + h$.
15. $D = \{x : 0 \leq x \leq 10\}$; $C(5) = 44$.   **17.** $P(x) = 2x - 40$; $P(50) = 60$; fixed cost 40; 20.   **19.** (a) Yes, because the range of w is a subset of the set of real numbers. (b) No, because the range of X might not satisfy the two defining properties of a weighting function (weights must be nonnegative and total to 1).

3.8 Page 124

1. (a) 7. (b) -5. (c) 0. (d) No slope.   **3.** $\frac{7}{11}$.   **5.** (a) $y - 3 = 2(x - 5)$ or $y - 5 = 2(x - 6)$. (b) $y + 3 = \frac{5}{3}(x + 2)$ or $y - 2 = \frac{5}{3}(x - 1)$.
(c) $y - 1 = -\frac{7}{5}(x - 3)$ or $y - 8 = -\frac{7}{5}(x + 2)$.   **7.** (a) $y + 1 = \frac{2}{3}(x - 6)$.
(b) $y - 2 = -5(x + 4)$.
9. (a) $y = 3x + 2$;

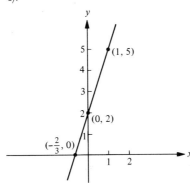

Answers to Odd-numbered Exercises

667

(b) $y = \frac{2}{3}x - 1$. (c) $y = -7x$.
11. (a) $y + 5 = \frac{3}{2}(x + 7)$ or $y - 1 = \frac{3}{2}(x + 3)$. (b) $y = \frac{3}{2}x + \frac{11}{2}$.
(c) $3x - 2y + 11 = 0$. (d) $\dfrac{x}{-\frac{11}{3}} + \dfrac{y}{\frac{11}{2}} = 1$. **13.** $5x + 2y - 24 = 0$.
15. $-\frac{3}{2}$. **17.** (a) 60 degrees Celsius. (b) 86 degrees Fahrenheit. (c) -40 degrees Fahrenheit.

3.9 Page 132

1. $8405.00. **3.** (a) $10n$. (b) $\dfrac{1}{3n - 1}$. **5.** (a) Degree 5, set of all real numbers. (b) Degree 8, set of all real numbers. **9.** $8.50.
11.

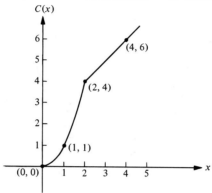

13. $1849.11. **15.** $y = 1.3x = f(x)$, where x is the cost and y is the selling price.

3.10 Page 138

1.

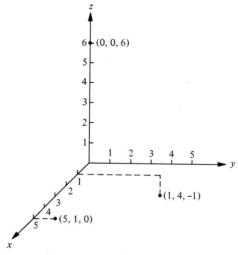

3. (a) $z = 0$. (b) $y = 0$. (c) $x = 0$. (d) $z = 3$. **5.** (a) $\sqrt{29}$. (b) 7.
7. $2x + 10y + 2z - 7 = 0$. **9.** (a) $d = \sqrt{x^2 + y^2 + z^2}$. (b) 7.

Answers to Odd-numbered Exercises

668

11. $(x - h)^2 + (y - k)^2 + (z - l)^2 = r^2$.   13. $\sqrt{41}$.
15. $5x + 4z - 20 = 0$.

4.1 Page 149

1. $\frac{3}{10}$.   3. $\frac{1}{2}$.   5. $\frac{4}{9}$.   7. $U = \{mmm, mmf, mfm, mff, fmm, fmf, ffm, fff\}$; $E = \{mmf, mfm, fmm\}$. $P(E) = \frac{3}{8}$.   9. $\frac{5}{12}$.   11. $U = \{w, y, o\}$; $E = \{y, o\}$; $w(w) = \frac{6}{11}$; $w(y) = \frac{3}{11}$; $w(o) = \frac{2}{11}$; $P(E) = \frac{5}{11}$.   13. $\frac{1}{4}$.

4.2 Page 153

1. (a) $U = \{1, 2, 3, 4, 5, 6\}$. (b) $E_1 = \{2, 4, 6\}$, $E_2 = \{3, 4, 5, 6\}$, $E_1 \cap E_2 = \{4, 6\}$, $E_1 \cup E_2 = \{2, 3, 4, 5, 6\}$, $E_2' = \{1, 2\}$. (c) $P(E_1) = \frac{1}{2}$, $P(E_2) = \frac{2}{3}$. (d) $P(E_1 \cap E_2) = \frac{1}{3}$. (e) $P(E_1 \cup E_2) = \frac{5}{6}$. (f) $P(E_2') = \frac{1}{3}$.
3. (a) $U = \{mmm, mmf, mfm, mff, fmm, fmf, ffm, fff\}$.
(b) $E_1 = \{mmm, mmf, mfm, mff\}$, $E_2 = \{mmm, fff\}$, $E_1 \cap E_2 = \{mmm\}$, $E_1 \cup E_2 = \{mmm, mmf, mfm, mff, fff\}$, $E_2' = \{mmf, mfm, mff, fmm, fmf, ffm\}$.
(c) $P(E_1) = \frac{1}{2}$, $P(E_2) = \frac{1}{4}$. (d) $P(E_1 \cap E_2) = \frac{1}{8}$. (e) $P(E_1 \cup E_2) = \frac{5}{8}$. (f) $P(E_2') = \frac{3}{4}$.
5. (a) $U = \{O_1, O_2, O_3\}$. (b) $E_1 = \{O_1, O_2\}$, $E_2 = \{O_2, O_3\}$, $E_1 \cap E_2 = \{O_2\}$, $E_1 \cup E_2 = \{O_1, O_2, O_3\}$, $E_2' = \{O_1\}$. (c) $P(E_1) = \frac{7}{12}$; $P(E_2) = \frac{3}{4}$.
(d) $P(E_1 \cap E_2) = \frac{1}{3}$. (e) $P(E_1 \cup E_2) = 1$. (f) $P(E_2') = \frac{1}{4}$.
7. $P(E_1 \cap E_2) = 0$; $P(E_1 \cup E_2) = \frac{13}{18}$; $P(E_2') = \frac{7}{9}$.
9. (a) .87. (b) .08.
11. (a)   (b) .77. (c) .25. (d) .07.

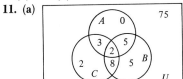

4.3 Page 160

1. $-\$\frac{1}{19}$.   3. $\$\frac{1}{2}$.   5. $-\$\frac{1}{2}$.   7. $-\$.40$; odds are 9989 to 11 against winning a prize.   9. 2.2 hours.   11. Odds are 5 to 1 against winning; $-\$\frac{1}{3}$.   13. City B because the expected value (crowd) is 6400 compared to 4400 for city A.   15. \$6.30.

4.4 Page 168

1. (a) $U = \{JC, JD, JH, JS, QC, QD, QH, QS, KC, KD, KH, KS\}$.
(b) $E_1 = \{KC, KD, KH, KS\}$; $E_2 = \{JD, QD, KD\}$, $E_1 \cap E_2 = \{KD\}$.
(c) $P(E_1) = \frac{1}{3}$, $P(E_2) = \frac{1}{4}$, $P(E_1 \cap E_2) = \frac{1}{12}$. (d) $P(E_1|E_2) = \frac{1}{3}$, $P(E_2|E_1) = \frac{1}{4}$.
3. $\frac{1}{2}$.   5. $\frac{1}{3}$.   7. $\frac{2}{3}$.   9. (a) $\frac{4}{7}$. (b) $\frac{2}{3}$. (c) $\frac{1}{2}$.   11. $\frac{1}{20}$.   13. (a) \$.125.
(b) \$0.   15. $\frac{2}{11}$.   17. $\frac{15}{22}$.   19. $\frac{6}{7}$.

4.5 Page 175

1. Not independent.   3. $\frac{3}{100}$.   5. $\frac{72}{100}$.   7. $\frac{1}{12}$.   9. (a) No. (b) Yes.
(c) Yes.   11. $\frac{19}{50}$.
13. Statements:
  (1) E_2 is independent of E_1 means $P(E_2|E_1) = P(E_2)$.
  (2) $\dfrac{P(E_1 \cap E_2)}{P(E_1)} = P(E_2|E_1) = P(E_2)$.

Answers to Odd-numbered Exercises

(3) $\dfrac{P(E_2 \cap E_1)}{P(E_1)} = P(E_2)$.

(4) $\dfrac{P(E_2 \cap E_1)}{P(E_2)} = P(E_1)$.

(5) $P(E_1|E_2) = P(E_1)$.

(6) E_1 is independent of E_2.

17. (a) $\frac{3}{4}$. (b) $\frac{7}{8}$. (c) $\frac{15}{16}$.

4.6 Page 181

1. $E = \{(ND_2D_3)\}$; $P(E) = \frac{3}{4} \cdot \frac{13}{51} \cdot \frac{12}{50} = \frac{39}{850}$.
3. $E = \{WWL, WLW, LWW\}$; $P(E) = \frac{1}{4} + \frac{1}{24} + \frac{1}{8} = \frac{5}{12}$.
5. $P(E) = \frac{3}{16} + \frac{1}{24} + \frac{1}{8} = \frac{17}{48}$. 7. $E = \{RW, DW, IW\}$, $P(E) = \frac{39}{100}$.
9. $\frac{1}{12} \cdot \frac{1}{3} + \frac{1}{9} \cdot \frac{4}{10} + \frac{5}{36} \cdot \frac{5}{11} + \frac{1}{6} + \frac{5}{36} \cdot \frac{5}{11} + \frac{1}{9} \cdot \frac{4}{10} + \frac{1}{12} \cdot \frac{1}{3} + \frac{1}{18} = \frac{244}{495}$.
11. (a) $\frac{3}{10}$. (b) $\frac{1}{10}$. (c) 7.

Supplementary Exercises
Page 183

1. (a) $\frac{13}{16}$. (b) $\$\frac{33}{16}$.
3. (a) $\frac{11}{16}$. (b) 0 cents.
5. (a) $\frac{6}{11}$. (b) $\frac{3}{5}$.
7. (a) $\frac{1}{6}$. (b) $\frac{1}{9}$.
9. (a) .0012. (b) .0688. (c) .9312.
11. No; $P(B|A) = \frac{4}{17}$ whereas $P(B) = \frac{1}{4}$.

4.7 Page 193

1. (a) 20. (b) $\frac{3}{10}$.   3. $P_2^4 = 12$; $P_5^7 = 2520$; $P_3^{22} = 9240$; $P_8^8 = 40{,}320$.
5. 120.   7. 5.   9. 9999.   11. (a) 360. (b) $\frac{1}{15}$. (c) $\frac{14}{15}$.   13. $\frac{551}{1111}$.
15. $\dfrac{P_2^8 + P_2^8 + P_2^8}{P_4^{10}} = \dfrac{1}{30}$.   17. (a) 120. (b) 20. (c) 20. (d) $\frac{1}{6}$. (e) $\$5$ million.

4.8 Page 201

1. $C_2^6 = 15$; $C_3^5 = 10$; $C_{18}^{20} = 190$.   3. 1140.   5. 16.   7. $\dfrac{C_5^{26}}{C_5^{52}}$.
9. $\dfrac{C_2^{40} C_3^{60}}{C_5^{100}}$.   11. (a) 60. (b) 6. (c) 0. (d) 252. (e) $\dfrac{11}{42}$.
13. $\dfrac{C_4^4 C_2^8 + C_1^1 C_5^{11} - C_4^4 C_1^1 C_1^7}{C_6^{12}}$.   15. $\dfrac{1}{2}$ if nine vote; $\dfrac{93}{256}$ if eight vote.

4.9 Page 208

1. The 6 ordered partitions are $\{\{a,b\}, \{c,d\}\}$, $\{\{a,c\}, \{b,d\}\}$, $\{\{a,d\}, \{b,c\}\}$, $\{\{b,c\}, \{a,d\}\}$, $\{\{b,d\}, \{a,c\}\}$, and $\{\{c,d\}, \{a,b\}\}$.
3. 2520.   5. $\dfrac{52!}{5!5!5!5!32!}$.   7. (a) $\dfrac{48!5!}{52!}$.   (b) $\dfrac{4 \cdot 48!5!}{52!}$.   9. $\dfrac{5}{22}$.
11. (a) $\dfrac{52!}{13!13!13!13!}$.   (b) $\dfrac{48!13!}{52!9!}$.   (c) $\dfrac{4 \cdot 48!13!}{52!9!}$.

Answers to Odd-numbered Exercises

4.10 Page 216

1. $\frac{1}{4}$; yes; because the two trial outcomes are equally likely. 3. $\frac{432}{2401}$.
5. $\frac{513}{2401}$. 7. (a) .354294. (b) .885735. 9. .31744 (3, 4, or 5 people constitute a majority of the sample). 11. $\frac{27}{32}$. 13. (a) .032. (b) $2.81.
15. .1372.

5.1 Page 225

1. $\begin{bmatrix} 1 & 5 \\ 6 & 4 \end{bmatrix}$. 3. [9 12 5]. 5. Impossible. 7. (a) $\begin{bmatrix} 2 & -5 & 12 \\ 2 & 21 & 5 \end{bmatrix}$.
(b) 2 by 3. (c) 5. 9. $[a_{ij}]_{(2,3)}$. 11. No.
13. $A + B = \begin{bmatrix} 1 & 1 \\ 7 & 6 \end{bmatrix}$; $B + A = \begin{bmatrix} 1 & 1 \\ 7 & 6 \end{bmatrix}$.

15. Reasons: (1) Change in notation.
    (2) Definition of addition (Definition 3).
    (3) Addition of real numbers is commutative.
    (4) Definition of addition (Definition 3).
    (5) Change in notation.
17. $A + (B + C) = (A + B) + C$ by Associative Property (Theorem 2);
        $= (B + A) + C$ by Commutative Property (Theorem 1).

5.2 Page 234

1. (7, 4, 0). 3. (−3, 12, −8). 5. 2. 7. $\sqrt{21}$.
9.

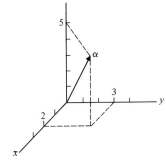

$||\alpha||$ represents the length of the arrow shown in the graph.

11. (6, 3, 4); 0.
13. $\left\| \begin{bmatrix} \frac{1}{\sqrt{35}} & \frac{3}{\sqrt{35}} & \frac{5}{\sqrt{35}} \end{bmatrix} \right\| = 1$.
15. $\alpha + \beta = (a_1, \cdots, a_n) + (b_1, \cdots, b_n)$ by change of notation;
   $= (a_1 + b_1, \cdots, a_n + b_n)$ by definition of addition;
   $= (b_1 + a_1, \cdots, b_n + a_n)$ addition of real numbers is commutative;
   $= (b_1, \cdots, b_n) + (a_1, \cdots, a_n)$ by definition of addition;
   $= \beta + \alpha$ by change of notation.
17. No, because $\alpha - \beta = \alpha + (-1)\beta$, whereas $\beta - \alpha = \beta + (-1)\alpha$, and these two results may not be the same as can be demonstrated by an example.
19. Clothing industry, $15 profit; construction industry, $26 profit.

Answers to Odd-numbered Exercises

5.3 Page 243

1. $AB = \begin{bmatrix} 3 & -2 \\ 0 & 3 \end{bmatrix}$; $BA = \begin{bmatrix} 3 & 0 \\ -6 & 3 \end{bmatrix}$.

3. AB does not exist; because the number of columns of A is not the same as the number of rows of B; $BA = \begin{bmatrix} 2 & 9 & 7 \\ 2 & 21 & 13 \end{bmatrix}$.

5. $AB = [17]$; $BA = \begin{bmatrix} 8 & 24 \\ 3 & 9 \end{bmatrix}$.

7. $AB = \begin{bmatrix} (2x + 3y) \\ (4x + 5y) \end{bmatrix}$; BA does not exist because the number of columns of B is not the same as the number of rows of A.

9. $A^2 = \begin{bmatrix} 16 & 24 \\ 32 & 48 \end{bmatrix}$.

11. $A(BC) = \begin{bmatrix} 2 & 3 \\ 4 & 5 \end{bmatrix} \begin{bmatrix} 0 & 1 \\ 8 & 7 \end{bmatrix} = \begin{bmatrix} 24 & 23 \\ 40 & 39 \end{bmatrix}$;

$(AB)C = \begin{bmatrix} 6 & 11 \\ 10 & 19 \end{bmatrix} \begin{bmatrix} 4 & 2 \\ 0 & 1 \end{bmatrix} = \begin{bmatrix} 24 & 23 \\ 40 & 39 \end{bmatrix}$.

13. (a) $n = p$. (b) m by q. (c) $m = q$. (d) p by n. (e) $m = n$. (f) m by m.

15. Statements:

$A(B + C)$

$= \begin{bmatrix} a_{11} & a_{12} \\ a_{21} & a_{22} \end{bmatrix} \begin{bmatrix} (b_{11} + c_{11}) & (b_{12} + c_{12}) \\ (b_{21} + c_{21}) & (b_{22} + c_{22}) \end{bmatrix}$

$= \begin{bmatrix} a_{11}(b_{11} + c_{11}) + a_{12}(b_{21} + c_{21}) & a_{11}(b_{12} + c_{12}) + a_{12}(b_{22} + c_{22}) \\ a_{21}(b_{11} + c_{11}) + a_{22}(b_{21} + c_{21}) & a_{21}(b_{12} + c_{12}) + a_{22}(b_{22} + c_{22}) \end{bmatrix}$

$= \begin{bmatrix} (a_{11}b_{11} + a_{12}b_{21}) + (a_{11}c_{11} + a_{12}c_{21}) & (a_{11}b_{12} + a_{12}b_{22}) + (a_{11}c_{12} + a_{12}c_{22}) \\ (a_{21}b_{11} + a_{22}b_{21}) + (a_{21}c_{11} + a_{22}c_{21}) & (a_{21}b_{12} + a_{22}b_{22}) + (a_{21}c_{12} + a_{22}c_{22}) \end{bmatrix}$

$= \begin{bmatrix} (a_{11}b_{11} + a_{12}b_{21}) & (a_{11}b_{12} + a_{12}b_{22}) \\ (a_{21}b_{11} + a_{22}b_{21}) & (a_{21}b_{12} + a_{22}b_{22}) \end{bmatrix} + \begin{bmatrix} (a_{11}c_{11} + a_{12}c_{21}) & (a_{11}c_{12} + a_{12}c_{22}) \\ (a_{21}c_{11} + a_{22}c_{21}) & (a_{21}c_{12} + a_{22}c_{22}) \end{bmatrix}$

$= \begin{bmatrix} a_{11} & a_{12} \\ a_{21} & a_{22} \end{bmatrix} \begin{bmatrix} b_{11} & b_{12} \\ b_{21} & b_{22} \end{bmatrix} + \begin{bmatrix} a_{11} & a_{12} \\ a_{21} & a_{22} \end{bmatrix} \begin{bmatrix} c_{11} & c_{12} \\ c_{21} & c_{22} \end{bmatrix}$

$= AB + AC$.

17. $\begin{bmatrix} 0 & 0 & 0 & 0 & 0 \\ 0 & 0 & 0 & 0 & 0 \\ 0 & 0 & 0 & 0 & 0 \\ 1 & 0 & 1 & 0 & 0 \\ 1 & 1 & 2 & 0 & 0 \end{bmatrix}$.

19. $\begin{array}{c} A \quad B \\ \begin{bmatrix} 2 & 6 \\ 4 & 10 \\ 2 & 7 \end{bmatrix} \begin{array}{l} \text{Electricity} \\ \text{Oil} \\ \text{Gas} \end{array} \end{array}$

Answers to Odd-numbered Exercises

5.4 Page 254

1. (a) E. (b) B. (c) A, B. (d) B. (e) A, B, D. (f) B, D, F. (g) A, B, C. (h) none.

5. $\begin{bmatrix} 3 & 5 & 1 \\ \hline 4 & 0 & 6 \\ 9 & 2 & 1 \end{bmatrix}$.

7. (ii) $(A + B)^T = \begin{bmatrix} 7 & 5 \\ 4 & 8 \end{bmatrix}$; $A^T + B^T = \begin{bmatrix} 7 & 5 \\ 4 & 8 \end{bmatrix}$.

   (iii) $(AB)^T = \begin{bmatrix} 14 & 28 \\ 12 & 15 \end{bmatrix}$; $B^T A^T = \begin{bmatrix} 14 & 28 \\ 12 & 15 \end{bmatrix}$.

   (iv) $(kA)^T = \begin{bmatrix} 3 & 9 \\ 12 & 15 \end{bmatrix}$; $kA^T = \begin{bmatrix} 3 & 9 \\ 12 & 15 \end{bmatrix}$.

9. Statements:
   - (1) Let $C = A + B$
   - (2) ijth entry of $C^T = ji$th entry of C
   - (3) $= a_{ji} + b_{ji}$
   - (4) $= (ji$th entry of $A) + (ji$th entry of $B)$
   - (5) $= (ij$th entry of $A^T) + (ij$th entry of $B^T)$
   - (6) $= ij$th entry of $A^T + B^T$.

11. Statements:
    - (1) ijth entry of $(kA)^T = ji$th entry of kA
    - (2) $= k(ji$th entry of $A)$
    - (3) $= k(ij$th entry of $A^T)$
    - (4) $= ij$th entry of kA^T.

13. A, B, C, D.

15. (a) Statements: (1) $A + B = A^T + B^T$
                    (2) $= (A + B)^T$.
    (b) Statements: (1) $kA = kA^T$
                    (2) $= (kA)^T$.
    (c) Statements: (1) $A^2 = AA$
                    (2) $= A^T A^T$
                    (3) $= (AA)^T$
                    (4) $= (A^2)^T$.
    (d) Statements: (1) $AA^T = (A^T)^T A^T$
                    (2) $= (AA^T)^T$.
    Also $AA^T = A^T A$ by direct substitution of $A = A^T$.

17. Proofs are similar to those of Exercise 15.

5.5 Page 263

1. $A = \begin{bmatrix} 1 & -1 \\ 2 & 4 \end{bmatrix}$; $[A \mid B] = \begin{bmatrix} 1 & -1 & 2 \\ 2 & 4 & 1 \end{bmatrix}$.

3. $A = \begin{bmatrix} 1 & 3 & 1 \\ 0 & -1 & -2 \end{bmatrix}$; $[A \mid B] = \begin{bmatrix} 1 & 3 & 1 & 2 \\ 0 & -1 & -2 & -3 \end{bmatrix}$.

5. $(3, 0)$.   7. $(3, -2, 1)$.   9. $(1, 2)$.   11. $(1, 3, -2)$.

Answers to Odd-numbered Exercises

13. A single point which is the intersection of two lines.

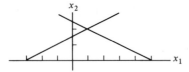

15. The graphs of the first part of the exercise will depend upon the choices of elementary operations and the order in which they are applied. A geometric interpretation of the sequence of algebraic steps is a sequence of rotations of the lines through the solution (1, 2).

5.6 Page 270

1. $\frac{1}{7}, \frac{5}{2}, \frac{1}{\sqrt{2}}$.

3. $\begin{bmatrix} -1 & \frac{1}{2} \\ 3 & -1 \end{bmatrix}$.   **5.** No inverse because matrix is not row equivalent to I_2.

7. No inverse because matrix is not square.

9. $\begin{bmatrix} -9 & -11 & 5 \\ -4 & -3 & 2 \\ 2 & 2 & -1 \end{bmatrix}$.   **11.** (2, 3).   **13.** (12, 9, 2).

15. (a) (ii) $(AB)^{-1} = \begin{bmatrix} 1 & 6 \\ 2 & 13 \end{bmatrix}^{-1} = \begin{bmatrix} 13 & -6 \\ -2 & 1 \end{bmatrix}$;

$B^{-1}A^{-1} = \begin{bmatrix} 1 & -3 \\ 0 & 1 \end{bmatrix}\begin{bmatrix} 7 & -3 \\ -2 & 1 \end{bmatrix} = \begin{bmatrix} 13 & -6 \\ -2 & 1 \end{bmatrix}$.

(b) No, because $A^{-1}B^{-1} \neq B^{-1}A^{-1}$.

17. Statements: (1) $(AB)(B^{-1}A^{-1}) = A[(BB^{-1})A^{-1}]$
(2) $\qquad\qquad\qquad = A(I_n A^{-1})$
(3) $\qquad\qquad\qquad = AA^{-1}$
(4) $\qquad\qquad\qquad = I_n$.
(5) Likewise, $(B^{-1}A^{-1})(AB) = I_n$.
(6) $B^{-1}A^{-1} = (AB)^{-1}$.

19. Statements: (1) $B = I_n B$
(2) $\qquad\quad = (A^{-1}A)B$
(3) $\qquad\quad = A^{-1}(AB)$
(4) $\qquad\quad = A^{-1}I_n$
(5) $\qquad\quad = A^{-1}$.
(6) $AB = AA^{-1} = A^{-1}A = BA$.
(7) Therefore, $B^{-1} = A$.

21. Statements: $AX = B \Rightarrow A^{-1}(AX) = A^{-1}B$
$\Rightarrow (A^{-1}A)X = A^{-1}B$
$\Rightarrow I_n X = A^{-1}B$
$\Rightarrow X = A^{-1}B$,
which is unique because A^{-1} is unique.

Answers to Odd-numbered Exercises

5.7 Page 277

1. $x_1 = 2 - \frac{1}{2}x_3,$
   $x_2 = 1 - \frac{3}{4}x_3;$ $(2, 1, 0).$
   An interpretation of the complete solution is the set of all points on the line of intersection of two planes; an interpretation of a particular solution is a point on that line.

3. $x_1 = 2 - \frac{11}{3}x_3,$
   $x_2 = 1 + \frac{4}{3}x_3;$ $\left(-\frac{5}{3}, \frac{7}{3}, 1\right).$

5. $x_1 = 3 - \frac{3}{2}x_3,$
   $x_2 = -1 - \frac{1}{2}x_3;$ $\left(\frac{3}{2}, -\frac{3}{2}, 1\right).$

7. No solution.

9. $x_1 = \frac{14}{3} - x_3 - \frac{1}{3}x_4,$
   $x_2 = \frac{-7}{3} + \frac{1}{3}x_4;$ $\left(\frac{10}{3}, \frac{-8}{3}, 1, 1\right).$

11. If x_1 is the number of barrels of P, x_2 is the number of barrels of Q, and x_3 is the number of barrels of R, then a complete solution is
$$x_1 = 4 + 2x_3,$$
$$x_2 = 8 - 5x_3.$$

5.8 Page 281

1. (b) is reduced. (a) is not reduced because $I_{(1)}$ is *not* among its columns. (c) is not reduced because $I_{(3)}$ is *not* among its columns.
3. $(1, 3, 2)$.
5. $x_1 = x_3,$
   $x_2 = -x_3;$ $(1, -1, 1).$
7. $x_1 = 7 - x_3 - 2x_4,$
   $x_2 = 5 - x_3 - x_4;$ $(4, 3, 1, 1).$
9. Exercise 5.

5.9 Page 286

1. (a) 9. (b) 18. (c) 0.

3. (a) $(1) \det \begin{bmatrix} 1 & 1 \\ 4 & 3 \end{bmatrix} + (-1)(3) \det \begin{bmatrix} 0 & 1 \\ 2 & 3 \end{bmatrix} + (4) \det \begin{bmatrix} 0 & 1 \\ 2 & 4 \end{bmatrix} = -3.$

   (b) $(-1)(3) \det \begin{bmatrix} 0 & 1 \\ 2 & 3 \end{bmatrix} + (1) \det \begin{bmatrix} 1 & 4 \\ 2 & 3 \end{bmatrix} + (-1)(4) \det \begin{bmatrix} 1 & 4 \\ 0 & 1 \end{bmatrix} = -3.$

   (c) $0 + (1) \det \begin{bmatrix} 1 & 4 \\ 2 & 3 \end{bmatrix} + (-1)(1) \det \begin{bmatrix} 1 & 3 \\ 2 & 4 \end{bmatrix} = -3.$

   (d) $(1) \det \begin{bmatrix} 1 & 1 \\ 4 & 3 \end{bmatrix} + 0 + (2) \det \begin{bmatrix} 3 & 4 \\ 1 & 1 \end{bmatrix} = -3.$

Answers to Odd-numbered Exercises

5. -54.   7. 3.   9. -120.

11. $\det(AB) = -6$; $(\det A)(\det B) = -6$.

13. (a) If the two identical rows are interchanged, then, by Theorem 8, $\det A = -\det A$. Thus $2 \det A = 0$ and hence $\det A = 0$.

(b) If each entry of the ith row of A is k times the corresponding entry of the jth row of A, then, by Theorem 9,

$$\det A = k \det B,$$

where B is a matrix having identical ith and jth rows. By part (a), $\det B = 0$; hence $\det A = 0$.

17. $\begin{bmatrix} 5 & -2 \\ -1 & \frac{1}{2} \end{bmatrix}$.   19. $\begin{bmatrix} \frac{3}{2} & 1 & -\frac{3}{2} \\ -\frac{1}{2} & 0 & \frac{1}{2} \\ 1 & 0 & 0 \end{bmatrix}$.

5.10 Page 292

1. $\lambda^2 - 5\lambda + 6 = 0$, $\lambda_1 = 3$, $\lambda_2 = 2$; two of many answers for X_1 and X_2 are

$$X_1 = \begin{bmatrix} 1 \\ 1 \end{bmatrix}, X_2 = \begin{bmatrix} 2 \\ 1 \end{bmatrix}.$$

3. $\lambda^2 - 9 = 0$, $\lambda_1 = 3$, $\lambda_2 = -3$; two of many answers for X_1 and X_2 are

$$X_1 = \begin{bmatrix} 1 \\ 1 \end{bmatrix}, X_2 = \begin{bmatrix} -1 \\ 1 \end{bmatrix}.$$

5. $\lambda_1 = 4$, $\lambda_2 = 2$; two of many answers for X_1 and X_2 are $X_1 = \begin{bmatrix} 1 \\ -3 \end{bmatrix}$, $X_2 = \begin{bmatrix} -1 \\ 1 \end{bmatrix}$.

7. $\lambda_1 = 2$, $\lambda_2 = 1$, $\lambda_3 = -2$; three of many answers for X_1, X_2, and X_3 are

$$X_1 = \begin{bmatrix} 1 \\ 0 \\ 1 \end{bmatrix}, X_2 = \begin{bmatrix} 0 \\ 1 \\ 0 \end{bmatrix}, \text{ and } X_3 = \begin{bmatrix} -1 \\ 0 \\ 1 \end{bmatrix}.$$

11. The characteristic equation is $\det(A - \lambda I_2) = 0$, or

$$\det \begin{bmatrix} a - \lambda & b \\ b & c - \lambda \end{bmatrix} = 0,$$

or

$$\lambda^2 - (a + c)\lambda + (ac - b^2) = 0.$$

If this equation is solved by the quadratic formula, we get

$$\lambda = \frac{a + c \pm \sqrt{(a + c)^2 - 4(ac - b^2)}}{2}.$$

Rearranging terms under the radical symbol, we get

$$\lambda = \frac{a + c \pm \sqrt{(a - c)^2 + 4b^2}}{2},$$

which is a real number regardless of the real values of a, b, and c.

Answers to Odd-numbered Exercises

6.1 Page 298

1.

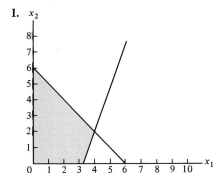

3.

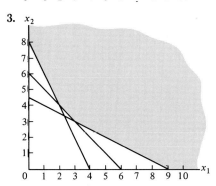

5. Structural constraints of Exercise 1 are $\begin{array}{r} x_1 + x_2 \le 6, \\ 3x_1 - x_2 \le 10. \end{array}$

Structural constraints of Exercise 3 are $\begin{array}{r} 2x_1 + x_2 \ge 8, \\ x_1 + x_2 \ge 6, \\ x_1 + 2x_2 \ge 9. \end{array}$

Nonnegativity constraints of both exercises are $\begin{array}{r} x_1 \ge 0, \\ x_2 \ge 0. \end{array}$

7.

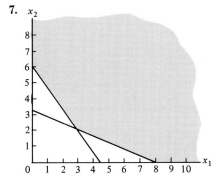

6.2 Page 304

1. Maximum $(x_1 + 2x_2) = 12$ at $(0, 6)$.
3. Minimum $(2x_1 + 3x_2) = 15$ at $(3, 3)$.

Answers to Odd-numbered Exercises

5. Minimum $(x_1 + x_2) = 2$ at $(2, 0)$.
7. Minimum $(2x_1 + 5x_2) = 16$ at $(8, 0)$ and at $(3, 2)$ and at all points on the line segment between $(8, 0)$ and $(3, 2)$.
9. Show that the family of lines representing the objective function intersects the feasible set no matter how large values of the objective function are.
11. Factory 1 should operate 3 days, and factory 2 should operate 2 days.
13. M_1 should be employed three times and M_2 should be employed one time.

6.3 Page 308

1.
$$\begin{array}{l} x_1 + 2x_2 + x_4 = 7, \quad x_1 \geq 0, \quad x_4 \geq 0, \\ x_1 - 3x_2 + x_3 - x_5 = 2, \quad x_2 \geq 0, \quad x_5 \geq 0. \\ 2x_1 + x_2 + x_3 = 4. \quad x_3 \geq 0, \end{array}$$

3. Two answers are
$$\begin{array}{l} x_3 = 14 - x_1 - 2x_2, \\ x_4 = 12 - 3x_1 - x_2; \end{array} \quad (0, 0, 14, 12); \text{ yes.}$$

$$\begin{array}{l} x_1 = 14 - 2x_2 - x_3, \\ x_4 = -30 + 5x_2 + 3x_3; \end{array} \quad (14, 0, 0, -30); \text{ no.}$$

5. Two answers are
$$\begin{array}{l} x_3 = 7 - x_1 - 2x_2, \\ x_4 = -2 + x_1 - 3x_2; \end{array} \quad (0, 0, 7, -2); \text{ no.}$$

$$\begin{array}{l} x_3 = 5 - 5x_2 - x_4, \\ x_1 = 2 + 3x_2 + x_4; \end{array} \quad (2, 0, 5, 0); \text{ yes.}$$

7. $x_1 = 4 - \frac{4}{7}x_3 + \frac{1}{7}x_4, \qquad x_1 = 12 - 4x_2 - x_4,$

$x_2 = 2 + \frac{1}{7}x_3 - \frac{2}{7}x_4; \qquad x_3 = -14 + 7x_2 + 2x_4;$

$x_1 = 5 - \frac{1}{2}x_2 - \frac{1}{2}x_3, \qquad x_2 = 10 - 2x_1 - x_3,$

$x_4 = 7 - \frac{7}{2}x_2 + \frac{1}{2}x_3; \qquad x_4 = -28 + 7x_1 + 4x_3.$

9. (a) $(0, 0, 10, 16)$, $(5, 0, 0, -14)$, $(0, 6, -14, 0)$, $(0, \frac{5}{2}, 0, \frac{7}{2})$, $(\frac{3}{2}, 0, 7, 0)$, $(1, 2, 0, 0)$.
(b) $(0, 0, 10, 16)$, $(0, \frac{5}{2}, 0, \frac{7}{2})$, $(\frac{3}{2}, 0, 7, 0)$, $(1, 2, 0, 0)$.
(c) $(0, \frac{5}{2}, 0, \frac{7}{2})$.

6.4 Page 311

1. (a)

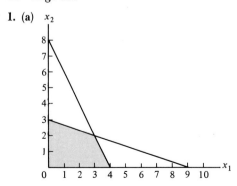

(b) $(0, 0)$, $(4, 0)$, $(9, 0)$, $(3, 2)$, $(0, 8)$, $(0, 3)$.
(c) $x_1 + 3x_2 + x_3 = 9$,
$ 2x_1 + x_2 + x_4 = 8$;
$(0, 0, 9, 8)$, $(4, 0, 5, 0)$, $(9, 0, 0, -10)$, $(3, 2, 0, 0)$, $(0, 8, -15, 0)$, $(0, 3, 0, 5)$.
(d) $(0, 0, 9, 8)$, $(4, 0, 5, 0)$, $(3, 2, 0, 0)$, $(0, 3, 0, 5)$; $(0, 0)$, $(4, 0)$, $(3, 2)$, $(0, 3)$.
(e) $(3, 2, 0, 0)$. By Theorem 1 and because the objective function evaluated at $(3, 2, 0, 0)$ is larger than the objective function evaluated at the other basic feasible solutions.
5. Minimum $(3x_1 + x_2) = 4$ at $(0, 4)$.
7. 5 thousand of P_1 and 17.5 thousand of P_2.

6.5 Page 314

1. Maximize CX subject to $AX \leq P_0$ and $X \geq 0$, where
$$A = \begin{bmatrix} 2 & 1 \\ -1 & 1 \end{bmatrix}, \quad X = \begin{bmatrix} x_1 \\ x_2 \end{bmatrix}, \quad P_0 = \begin{bmatrix} 4 \\ 5 \end{bmatrix}, \quad \text{and} \quad C = [1 \; 1].$$

3. Minimize CX subject to $AX = P_0$ and $X \geq 0$, where
$$A = \begin{bmatrix} 2 & 1 & 0 & 0 \\ 1 & 1 & 1 & 0 \\ 1 & -1 & 0 & -1 \end{bmatrix}, \quad X = \begin{bmatrix} x_1 \\ x_2 \\ x_3 \\ x_4 \end{bmatrix}, \quad P_0 = \begin{bmatrix} 3 \\ 4 \\ 2 \end{bmatrix}, \quad \text{and} \quad C = [5 \; 1 \; 0 \; 0].$$

5. Maximum $CX = 8$ at $(4, 0)$.
7. Minimum $CX = 0$ at $(0, 0)$.

6.6 Page 318

1. (a) $x_3 = 1 + x_1 - x_2$,
$x_4 = 5 - x_1 - 2x_2$; $(0, 0, 1, 5)$.
(b) $f = 1 + 3x_1 - x_2$. (c) Yes; x_1.
(d) $M_1 = \left[\begin{array}{ccc|c} -1 & 1 & 1 & 0 & 1 \\ 1 & 2 & 0 & 1 & 5 \\ \hline 3 & -1 & 0 & 0 & f-1 \end{array}\right]$.
(e) $x_3 = 6 - 3x_2 - x_4$,
$x_1 = 5 - 2x_2 - x_4$.
(f) $f = 16 - 7x_2 - 3x_4$. (g) No.
(h) Maximum $f = 16$ at $(5, 0, 6, 0)$.

6.7 Page 325

1. Maximum $(x_1 + 2x_2) = 5$ at $(3, 1, 0, 0)$.
3. Maximum $(2x_1 + x_2 + 6x_3 + x_4) = 16$ at $(2, 0, 2, 0)$.
5. Maximum $(2x_1 + x_4) = 4$ at $(2, 0, 0, 0)$.
7. Minimum $(2x_1 + x_2) = 1$ at $(0, 1)$.
9. Show that eventually in one of the applications of step D, there are no positive entries in the pivot column above the last row.
11. Show that eventually in one of the applications of step D, there are no positive entries in the pivot column above the last row.

Answers to Odd-numbered Exercises

13. (b) The intersection of the four sets is empty.

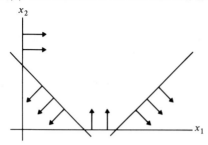

7.1 Page 333

1. (a) C plays (b) 1 cent to C. (c) $\frac{3}{2}$ and $-\frac{3}{2}$.

$$\left\{\begin{bmatrix} 4 & -1 \\ -1 & -2 \end{bmatrix}\begin{matrix} H \\ T \end{matrix}\right\} R \text{ plays.}$$

      H   T

(d) 4 and -1. (e) -1 and -2. (f) $(1, 0)$.
3. (a) $\frac{3}{2}$ and $-\frac{3}{2}$ (loss). (b) 4 and -1 (loss). (c) -1 and -2 (loss). (d) $(0, 1)$.
5. (a) 50 and 55. (b) 70 and 40. (c) 30 and 70. (d) $\frac{370}{7}$ and $\frac{370}{7}$. 7. (a) $\frac{95}{2}$ and 60. (b) 70 and 30. (c) 40 and 70. (d) $\frac{370}{7}$ and $\frac{370}{7}$.

7.2 Page 341

1. Strictly determined; $(1, 0, 0)$ for R and $(0, 1, 0)$ for C; payoff is 3.
3. Strictly determined; $(1, 0, 0)$ for R and $(0, 1, 0)$ for C; payoff is 2.
5. Not strictly determined.

7. (a)
	Goes bad	Good part
Cheap	-16	-6
Expensive	-10	-8

(b) Yes. (c) $(0, 1)$.

7.3 Page 346

1. Minimize $(x_1 + x_2)$ subject to $\begin{matrix} 2x_1 + 3x_2 \geq 1, \\ x_1 + 2x_2 \geq 1, \end{matrix}$ and $\begin{matrix} x_1 \geq 0, \\ x_2 \geq 0, \end{matrix}$

where $x_1 = p_1/v$ and $x_2 = p_2/v$.
3. Minimize $(x_1 + x_2 + x_3)$ subject to

$$\begin{matrix} 3x_1 + x_3 \geq 1, \\ 4x_2 + 2x_3 \geq 1, \\ x_1 + x_2 + 6x_3 \geq 1, \end{matrix} \text{ and } \begin{matrix} x_1 \geq 0, \\ x_2 \geq 0, \\ x_3 \geq 0, \end{matrix}$$

where $x_1 = p_1/v$, $x_2 = p_2/v$, and $x_3 = p_3/v$.
5. $v = \frac{3}{2}$ at $(\frac{1}{2}, \frac{1}{2})$. 7. $v = 0$ at $(\frac{1}{2}, \frac{1}{2})$.
13. Start with the constraints on the middle of p. 344 (but with k added to each entry of the payoff matrix) and v_1 the new value for v.

$$\begin{matrix} (a_{11} + k)p_1 + \cdots + (a_{m1} + k)p_m - v_1 \geq 0, \\ \vdots \qquad \vdots \qquad \vdots \qquad \vdots \\ (a_{1n} + k)p_1 + \cdots + (a_{mn} + k)p_m - v_1 \geq 0, \\ p_1 + \cdots + p_m = 1. \end{matrix}$$

Answers to Odd-numbered Exercises

680

Since $p_1 + \cdots + p_m = 1$, these constraints can be arranged to be

$$a_{11}p_1 + \cdots + a_{m1}p_m - (v_1 - k) \geq 0,$$
$$\vdots \qquad \vdots \qquad \vdots \qquad \vdots$$
$$a_{1n}p_1 + \cdots + a_{mn}p_m - (v_1 - k) \geq 0,$$
$$p_1 + \cdots + p_m = 1.$$

Now let $v_1 - k = v$. The solution of this problem is the same as that shown following the constraints on the middle of p. 344, where v^* is maximum v. Since $v_1 = k + v$, max $v_1 = k + v^*$ and so new row value is old row value plus k.

7.4 Page 353

1. Maximize $(y_1 + y_2)$ subject to

$$\begin{array}{c} 2y_1 + y_2 \leq 1, \\ 3y_1 + 2y_2 \leq 1, \end{array} \quad \text{and} \quad \begin{array}{c} y_1 \geq 0, \\ y_2 \geq 0, \end{array}$$

where $y_1 = q_1/w$ and $y_2 = q_2/w$.

3. Maximize $(y_1 + y_2 + y_3)$ subject to

$$\begin{array}{c} 3y_1 + y_3 \leq 1, \\ 4y_2 + y_3 \leq 1, \\ y_1 + 2y_2 + 6y_3 < 1, \end{array} \quad \text{and} \quad \begin{array}{c} y_1 \geq 0, \\ y_2 \geq 0, \\ y_3 \geq 0, \end{array}$$

where $y_1 = q_1/w$, $y_2 = q_2/w$, and $y_3 = q_3/w$.

5. $w = \frac{3}{2}$ at $(\frac{1}{2}, \frac{1}{2})$. **7.** $w = 0$ at $(\frac{2}{3}, \frac{1}{3})$.

8.1 Page 357

1.

Height	Tally marks	Frequency	Cumulative frequency
61	|	1	1
62		0	1
63	|	1	2
64	||	2	4
65	||	2	6
66	||	2	8
67	||||	4	12
68	⦀|	6	18
69	||||	4	22
70	|||	3	25
71	|||	3	28
72	||	2	30
73	||	2	32
74	|	1	33
75	|	1	34
76	|	1	35
77		0	35
78		0	35
79		0	35
80	|	1	36

Answers to Odd-numbered Exercises

Line chart omitted here. Cumulative frequency distribution function shown in the figure.

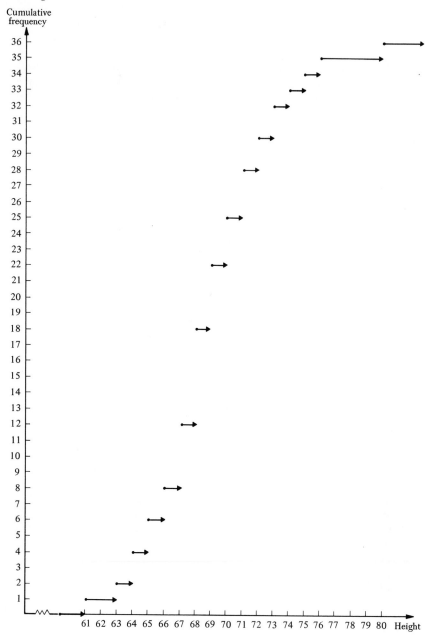

5. In a given city he might select names at random from a telephone directory. He might obtain a list of registered voters and devise a method to select names at random from the list.

Answers to Odd-numbered Exercises

8.2 Page 364

1.

Interval	Midpoint x_i of interval	Tally marks	Frequency f_i	Cumulative frequency
60.5–62.5	61.5	|	1	1
62.5–64.5	63.5	|||	3	4
64.5–66.5	65.5	||||	4	8
66.5–68.5	67.5	卌 卌	10	18
68.5–70.5	69.5	卌 ||	7	25
70.5–72.5	71.5	卌	5	30
72.5–74.5	73.5	|||	3	33
74.5–76.5	75.5	||	2	35
76.5–78.5	77.5		0	35
78.5–80.5	79.5	|	1	36

Histogram omitted here but cumulative frequency polygon shown in the figure.

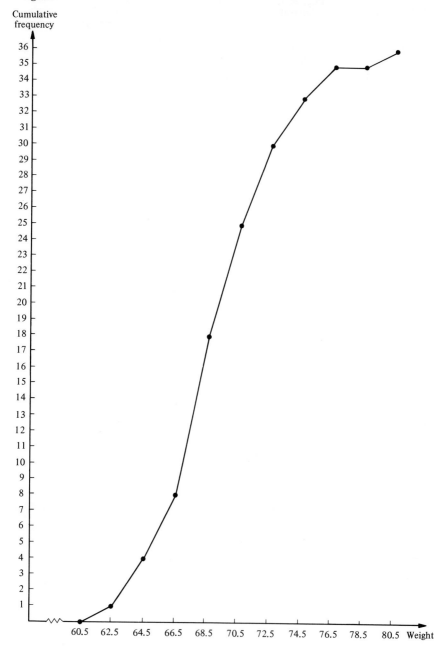

Answers to Odd-numbered Exercises

3. Answers depend upon how the intervals are selected.
5. (a) Yes. (b) No.
7. (a) $(1^2 + 3) + (2^2 + 6) + (3^2 + 9) + (4^2 + 12)$.
(b) $f(c_1)\Delta x_1 + f(c_2)\Delta x_2 + f(c_3)\Delta x_3 + f(c_4)\Delta x_4 + f(c_5)\Delta x_5$.
9. (a) $\sum_{i=1}^{5} (2i + 1)$. (b) $\sum_{i=1}^{k} x_k f_k$.

11. $\sum_{i=1}^{4} 2i^3 = 2(1^3) + 2(2^3) + 2(3^3) + 2(4^3)$
$= 2(1^3 + 2^3 + 3^3 + 4^3)$
$= 2 \sum_{i=1}^{4} i^3$.

13. $\sum_{i=1}^{3} (4 - x_i)(5 - x_i) = \sum_{i=1}^{3} (20 - 9x_i + x_i^2)$ by multiplication

$= \sum_{i=1}^{3} 20 - \sum_{i=1}^{3} 9x_i + \sum_{i=1}^{3} x_i^2$ by Theorem 1(ii)

$= 60 - \sum_{i=1}^{3} 9x_i + \sum_{i=1}^{3} x_i^2$ by Theorem 1(iii)

$= 60 - 9 \sum_{i=1}^{3} x_i + \sum_{i=1}^{3} x_i^2$.   by Theorem 1(i).

15. $\sum_{i=-2}^{2} i^2 = (-2)^2 + (-1)^2 + 0^2 + 1^2 + 2^2$
$= 2^2 + 1^2 + 1^2 + 2^2$
$= 2(1^2 + 2^2)$
$= 2 \sum_{i=1}^{2} i^2$.

8.3 Page 369

1. Mode 201; median 201; mean 167.2; mean seems best.
3. Two modes—80 and 49; median 55; mean $61\frac{1}{3}$; mean seems best.
5. (a) $\dfrac{2484}{36} = 69$.   (b) $\dfrac{275.5}{40} \approx 6.9$.
(c) $\dfrac{10.02}{20} = .501$.   (d) $\dfrac{2475}{30} = 82.5$.

7. (a) $\sum_{i=1}^{n} (x_i - \bar{x}) = \sum_{i=1}^{n} x_i - \sum_{i=1}^{n} \bar{x}$ by Theorem 1(ii)

$= n\bar{x} - \sum_{i=1}^{n} \bar{x}$ by Definition 2

$= n\bar{x} - n\bar{x}$ by Theorem 1(iii)

$= 0$.

(b) $\bar{y} = \dfrac{\sum_{i=1}^{n} y_i}{n} = \dfrac{\sum_{i=1}^{n}(x_i + c)}{n} = \dfrac{\sum_{i=1}^{n} x_i + \sum_{i=1}^{n} c}{n}$

$= \dfrac{n\bar{x} + nc}{n} = \bar{x} + c.$

(c) A real number c (either positive or negative) can be added to each number in the sample in order to make the numbers easier to work with (probably nearer to 0). The mean $\bar{y}$ of the new numbers can be found and then c can be subtracted to get the mean of the original sample.

9. (a) 25th percentile—67, median—68.5, 75th percentile—71.
   (b) 25th percentile—66.7, median—68.5, 75th percentile—71.3.
   (c) mode—67.5, mean—

$\dfrac{\sum_{i=1}^{10} x_i f_i}{36} = \dfrac{2484}{36} = 69.$

11. $16.5 + \dfrac{24 - 21}{25 - 21}(19.5 - 16.5) = 16.5 + \dfrac{3}{4}(3) = 18.75.$

8.4 Page 377

1. 19; 4. 3. 2.2. 5. 34.16; $s \approx 5.8$. 7. $s^2 \approx 14.4$; $s \approx 3.8$; 24.
9. $s^2 = \tfrac{1}{36}(171{,}903) - (69)^2 \approx 14.08$; $s \approx 3.75$.
11. $s^2 \approx \tfrac{1}{32}(1813.48) \approx 56.67.$ 13. (a) 63 inches and 67 inches. (b) .954.

9.1 Page 381

1. $n = 100 + 50t$. 3. $\Delta R = \tfrac{2}{50}$ is greater than $\Delta R = \tfrac{2}{100}$.

9.2 Page 389

1. 3;

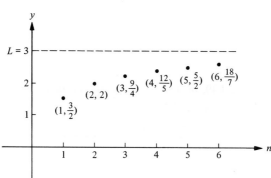

3. 1. 5. (a) No limit. (b) Limit is 1. (c) Limit is 0. (d) Limit is 0. (e) No limit.

Answers to Odd-numbered Exercises

7. $\lim_{x\to\infty}[f(x) - g(x)] = \lim_{x\to\infty}[f(x) + (-1)g(x)]$

$= \lim_{x\to\infty} f(x) + \lim_{x\to\infty}(-1)g(x)$ by Theorem 1(i)

$= \lim_{x\to\infty} f(x) - \lim_{x\to\infty} g(x)$ by Theorem 1(v).

9. (a) 0. (b) 1. (c) 2. (d) 1. (e) 0.   11. 7.   13. (a) 12. (b) $\frac{1}{3}$.

9.3 Page 397

1. (a) 2. (b) 4.   3. (a) 2. (b) 4. (c) 5. (d) Does not exist. (e) 0. (f) 2. (g) 6.
5. (a) $\frac{19}{22}$. (b) 4.   7. 8.   9. (a) 2. (b) 5. (c) 0. (d) Does not exist. (e) 2.
11. $F(x) = 0$ if $x < 2$,
$F(x) = \frac{1}{36}$ if $2 \le x < 3$,
$F(x) = \frac{3}{36}$ if $3 \le x < 4$,
$F(x) = \frac{6}{36}$ if $4 \le x < 5$,
$F(x) = \frac{10}{36}$ if $5 \le x < 6$,
$F(x) = \frac{15}{36}$ if $6 \le x < 7$,
$F(x) = \frac{21}{36}$ if $7 \le x < 8$,
$F(x) = \frac{26}{36}$ if $8 \le x < 9$,
$F(x) = \frac{30}{36}$ if $9 \le x < 10$,
$F(x) = \frac{33}{36}$ if $10 \le x < 11$,
$F(x) = \frac{35}{36}$ if $11 \le x < 12$,
$F(x) = 1$ if $x \ge 12$.

9.4 Page 404

1. No, because $\lim_{x\to 5} C(x)$ does not exist; hence $\lim_{x\to 5} C(x)$ does not equal $C(5)$.
3. Yes, because $\lim_{x\to 25} P(x) = P(25) = 2$. (Also, P is a polynomial function.)
5. (a) No. (b) Yes.   7. (a) $\frac{1}{2}$. (b) Because $\lim_{x\to 1} f(x) \ne f(1)$. (c) Yes, $x = -1$.
9. (a) $x = 1$. (b) $x = 2$, $x = 3$. (c) $x = 0$, $x = -3$.
11. $\lim_{x\to c}[f(x) + g(x)] = \lim_{x\to c} f(x) + \lim_{x\to c} g(x)$ by Theorem 2(i)

$= f(c) + g(c)$ by Definition 3.

13. (a)

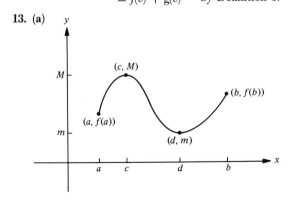

Answers to Odd-numbered Exercises

(b)

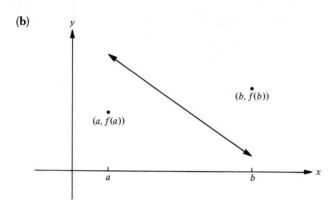

10.1 Page 415

1. (a) 1.25. (b) $-.19$.
3. Slope of secant line is $-\frac{5}{2}$.

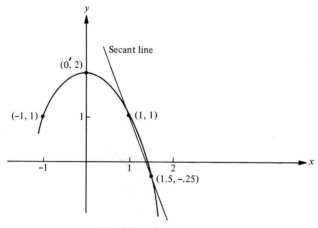

5. (a) $\Delta y = .728$; slope of secant line is 3.64. (b) $\Delta y = -1$; slope of secant line is 1. (c) 3.31.   7. (a) .120601. (b) $-.119401$. (c) 12.   9. -5; -7.

11. $f'(2) = \lim\limits_{\Delta x \to 0} \dfrac{(2 + \Delta x)^3 - 8}{\Delta x} = \lim\limits_{\Delta x \to 0} [12 + 6\,\Delta x + (\Delta x)^2] = 12$.

13. $3x + y + 4 = 0$.

15. There is no tangent line because $f'(1) = \lim\limits_{\Delta x \to 0} (\Delta y / \Delta x)$ does not exist. Also, $\lim\limits_{\Delta x \to 0} (\Delta y / \Delta x)$ does not exist because $\Delta y / \Delta x = 1$ whenever $\Delta x > 0$, but $\Delta y / \Delta x = -1$ whenever $\Delta x < 0$.

Answers to Odd-numbered Exercises

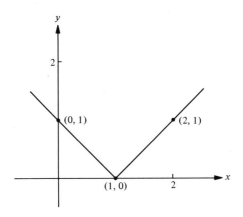

17. $x_1 = 5$.

10.2 Page 424

1. (a) -4. (b) -4.   **3.** $f'(x) = 6x - 5$; domain of f' is the set of all real numbers.   **5.** Yes; $f'(4) = -\frac{1}{9}$; no.   **7.** $-1 - 2x$; -1.   **9.** $\dfrac{7}{(2x+1)^2}$.   **11.** -12.   **13.** 2.   **15.** (a) 3 feet per second. (b) 12 feet per second. (c) 7 feet per second.   **17.** 2 seconds.   **19.** .13.

10.3 Page 431

1. 0.   **3.** $4x^3$.   **5.** $\dfrac{1}{3x^{2/3}} + 1$.   **7.** $15x^2 - 16x + 3$.

9. $5(x^2 + x)^4(2x + 1)$.   **11.** 28.   **13.** $-\frac{1}{9}$.   **15.** $\frac{3}{8}$.   **17.** $20 - \dfrac{x}{5}$.

19. $48 - \frac{3}{256}$.   **21.** 168.

23. $D_x(kf(x)) = \lim\limits_{\Delta x \to 0} \dfrac{kf(x + \Delta x) - kf(x)}{\Delta x}$

$= k \lim\limits_{\Delta x \to 0} \dfrac{f(x + \Delta x) - f(x)}{\Delta x} = kD_x f(x)$.

25. $x_2 = \frac{5}{4}$; $x_3 = \frac{239}{198} \approx 1.207$.

10.4 Page 437

1. $10x^4 + 6x$.   **3.** $\dfrac{3x + 1}{\sqrt{2x + 1}}$.   **5.** $\dfrac{2x}{(x^2 + 2)^2}$.   **7.** $-6x^2 + 8x$.

9. -20.   **11.** $-\frac{1}{2}$.   **13.** $\dfrac{5x^2 + 10x - 2}{(x + 1)^2}$.   **15.** (a) 129. (b) $\frac{5}{27}$.

17. $g(f(x)) = g(3x + 7) = (3x + 7)^2 + 1 = 9x^2 + 42x + 50$ and hence $D_x g(f(x)) = 18x + 42$; by the chain rule,

$$\dfrac{dy}{dx} = \dfrac{dy}{du}\left(\dfrac{du}{dx}\right) = (2u)(3) = 6u = 6(3x + 7) = 18x + 42.$$

Answers to Odd-numbered Exercises

19. Let n be a negative integer.

$D_x x^n = D_x \dfrac{1}{x^{-n}}$, where $-n$ is a positive integer.

$D_x \dfrac{1}{x^{-n}} = \dfrac{(x^{-n})(0) - (1)(-n)x^{-n-1}}{(x^{-n})^2} = \dfrac{nx^{-n-1}}{x^{-2n}} = nx^{-n-1+2n} = nx^{n-1}$.

21. $5t^4 + 4t - (1/t^2)$.

23. Yes; no.

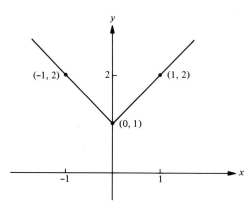

25. $\dfrac{x^2 - 1}{x^2}$; $E(x) = \dfrac{2x^2}{x^2 + 1}$.

10.5 Page 442

1. $28x^3$; $84x^2$; $168x$. **3.** $\tfrac{1}{2}x^{-1/2}$; $-\tfrac{1}{4}x^{-3/2}$; $\tfrac{3}{8}x^{-5/2}$; $-\tfrac{15}{16}x^{-7/2}$. **5.** $\dfrac{-3}{(x-4)^2}$; $\dfrac{6}{(x-4)^3}$. **7.** $210x^4$. **9.** $-\tfrac{49}{864}$. **11.** $-\dfrac{1}{2x^2}$. **13.** 24.

15. (a) Velocity is 1 meter per second; acceleration is 12 meters per second per second. (b) Velocity is 192 meters per second; acceleration is 1792 meters per second per second.   **17.** 14; speeding up.

10.6 Page 447

1. (a) $2x\,dx$. (b) $-\dfrac{3}{x^2}\,dx$. (c) dx. (d) $6x(1 + x^2)^2\,dx$. (e) 0. (f) $(3x^2 - 8x)\,dx$.

3. 1.261; 1.2.   **5.** $\dfrac{\pi}{3}(2.912)$; $.8\pi$.   **7.** $20x^4 - 24x^2 + 14x - 8$; $80x^3 - 48x + 14$; $240x^2 - 48$.   **9.** $C'(10) = 35$; $C(11) - C(10) = 34.75$; $P'(10) = 16$.   **11.** $\dfrac{dy}{dx} = 27(3x + 4)^8$; $\dfrac{d^2y}{dx^2} = 648(3x + 4)^7$.

13. $9 + \tfrac{1}{18} \approx 9.056$.

15. Yes. Assume that x is the independent variable; then
$$dy = \left[\dfrac{d}{dx}(k)\right]dx = 0\,dx = 0.$$

Answers to Odd-numbered Exercises

10.7 Page 453

1. $-\dfrac{x}{y}$.   3. $-\dfrac{y}{x}$.   5. $\dfrac{y+8}{2y-x}$.   7. $-\dfrac{4x^3+y^3}{3xy^2+5y^4}$.   9. $-\dfrac{\sqrt{y}}{\sqrt{x}}$.

11. $-\dfrac{x}{y}$; $-\dfrac{4}{y^3}$.   13. $-\dfrac{y}{x+2y}$; $\dfrac{24}{(x+2y)^3}$.   15. $\dfrac{x}{y}$; $\dfrac{16}{y^3}$; $-\dfrac{48x}{y^5}$.

17. $\tfrac{2}{9}$.   19. $-\dfrac{s+2t}{2s+t}$.   21. $s = \tfrac{1}{8}$; $\tfrac{1}{64}$.   23. $\dfrac{dh}{dr} = \dfrac{-2r-h}{r}$; $\dfrac{dr}{dh} = \dfrac{-r}{2r+h}$; $\dfrac{dr}{dh} = \dfrac{1}{dh/dr}$.

11.1 Page 460

1. $245.   3. (a) $[6, \infty)$. (b) $(-1, \infty)$. (c) $(-\infty, 20)$. (d) $(-\infty, 15]$.

5. (a)

Interval	Calculation of sign of $f(x)$	Result	Conclusion
$(0, 3)$	$(-)(-)$	+	$f(x)$ positive
$(3, 5)$	$(+)(-)$	−	$f(x)$ negative
$(5, \infty)$	$(+)(+)$	+	$f(x)$ positive

(b)

Interval	Calculation of sign of $f(x)$	Result	Conclusion
$(0, 1)$	$(-)(-)(-)$	−	$f(x)$ negative
$(1, 2)$	$(+)(-)(-)$	+	$f(x)$ positive
$(2, 3)$	$(+)(+)(-)$	−	$f(x)$ negative
$(3, 5)$	$(+)(+)(+)$	+	$f(x)$ positive

(c)

Interval	Calculation of sign of $f(x)$	Result	Conclusion
$(-\infty, -8)$	$\dfrac{(-)(+)}{(-)}$	+	$f(x)$ positive
$(-8, 1)$	$\dfrac{(-)(+)}{(+)}$	−	$f(x)$ negative
$(1, 4)$	$\dfrac{(-)(-)}{(+)}$	+	$f(x)$ positive
$(4, \infty)$	$\dfrac{(+)(-)}{(+)}$	−	$f(x)$ negative

7. (a) 0, 1. (b) $\dfrac{-2}{(x-1)^2}$; 1. (c) $\dfrac{4}{(x-1)^3}$; 1. (d) 1. (e) No.   9. $f(p)$ is negative for $2 < p < 7$.

11.2 Page 466

1. Increasing: $(4, \infty)$; decreasing: $(0, 4)$.   3. Increasing: $(-\infty, 4)$ and $(5, \infty)$; decreasing: $(4, 5)$.   5. Decreasing: $(-\infty, 0)$ and $(0, \infty)$.   7. Increasing: $(-2, 0)$ and $(3, \infty)$; decreasing: $(-\infty, -2)$ and $(0, 3)$.   9. Increasing: $(-\infty, 0)$ and $(0, \infty)$.   11. Increasing: $(-\infty, -2 - \sqrt{3})$ and $(-2 + \sqrt{3}, \infty)$;

Answers to Odd-numbered Exercises

decreasing: $(-2 - \sqrt{3}, -2 + \sqrt{3})$.   **13.** Increasing: $(0, 4)$ and $(4, \infty)$; decreasing: $(-\infty, 0)$.   **15.** Increasing: $(-\infty, 0)$; decreasing: $(0, \infty)$.   **17.** Increasing: $(-\infty, -1)$, $(-\frac{1}{5}, 1)$, and $(1, \infty)$; decreasing: $(-1, -\frac{1}{5})$.   **19.** $y' = -\dfrac{100}{(x+1)^2}$, which is negative on $(0, 20)$.

11.3 Page 475

1. $-2, 2$.   **3.** $0, 2$.
5. $(-2, 16)$ maximum point; $(2, -16)$ minimum point.

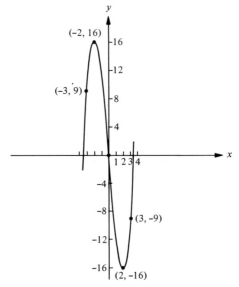

7. $(0, 0)$ maximum point; $(2, 4)$ minimum point.   **9.** $(-1, 1)$ endpoint maximum; $(0, 0)$ minimum point; $(1, 1)$ endpoint maximum.   **11.** $(0, -32)$ endpoint minimum; $(3, 1)$ endpoint maximum.   **13.** $(0, 0)$ endpoint minimum; $(2, 10)$ maximum point; $(4, 8)$ endpoint minimum.   **15.** $(-2, 30)$ maximum point; $(3, -\frac{65}{2})$ minimum point.   **17.** $(5, -\frac{625}{3})$ minimum point.   **19.** The function has no maximum or minimum points; change $>$ to $\geq$ and $<$ to $\leq$.

11.4 Page 483

1. 15.   **3.** 95.   **5.** $x = 5$.   **7.** $x = 4$.   **9.** $x = 225$ feet, $y = 300$ feet.   **11.** $r = 3$ inches, $h = 6$ inches.   **13.** $r = 2$ feet, $h = 2$ feet.
15. 5 feet high, 10 feet wide.   **17.** $\dfrac{c}{2}$.
19. $x = 10$.   **21.** (a) $s = 1$. (b) $x = 1$.

11.5 Page 490

1. $f''(x) = (x + 1)(x)(x - 1)$;

Answers to Odd-numbered Exercises

Interval	Calculation of sign of $f''(x)$	Result	Conclusion
$(-\infty, -1)$	$(-)(-)(-)$	$-$	f concave downward
$(-1, 0)$	$(+)(-)(-)$	$+$	f concave upward
$(0, 1)$	$(+)(+)(-)$	$-$	f concave downward
$(1, \infty)$	$(+)(+)(+)$	$+$	f concave upward

3. Concave upward: $(-3, 0)$ and $(3, \infty)$; concave downward: $(-\infty, -3)$ and $(0, 3)$.   **5.** $(-2, 2)$.   **7.** $(-2, -80)$ and $(2, -80)$.   **9.** Concave upward on $(0, \infty)$; concave downward on $(-\infty, 0)$; $x = 0$ is not in the domain of the rational function f.
11. Theorem 6 can be used to show that $(0, 0)$ is a minimum point and $(\frac{2}{3}, \frac{4}{27})$ is a maximum point.

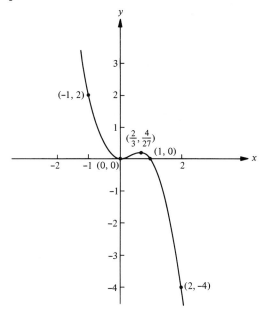

13. (a) $x = 3$. (b) $x = 12$.   **15.** 12; $P''(12) = -20 < 0$; $R'(12) = C'(12) = 134$.   **17.** $x = a/2$.
19. The function f'' is continuous at x_1, since f is a rational function. Since $f''(x_1) < 0$, there is an open interval containing x_1 throughout which $f''(x) < 0$ by Exercise 14 of Section 9.4. This interval can be chosen so that x_1 is the only critical number in the interval. By Theorem 1, f' is decreasing on this open interval. Since $f'(x_1) = 0$, $f'(x)$ changes from positive to zero to negative on this interval. By Theorem 3(i), $(x_1, f(x_1))$ is a maximum point.
21. (a) $x = 1$. (b) $x = 2$.

11.6 Page 499

1. (a) y-axis. (b) x-axis, y-axis, and origin. (c) x-axis, y-axis, and origin. (d) y-axis. (e) Origin. (f) x-axis, y-axis, and origin. (g) y-axis. (h) Origin. (i) x-axis. (j) None.   **3.** (a) $y = \frac{2}{3}$. (b) $y = \frac{1}{5}$. (c) $y = 0$. (d) None.

Answers to Odd-numbered Exercises

5.

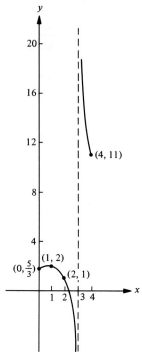

7.

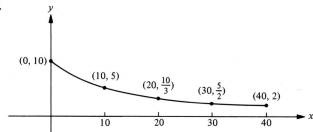

9.

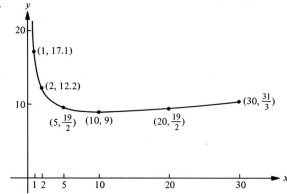

Answers to Odd-numbered Exercises

11.7 Page 505

1. (a) 0. (b) $-\frac{2}{3}$. (c) $-2, 0, 2$. 3. (a) 1. (b) 1.
5. Assume that $f'(x) < 0$ at each point of (a, b). Let x_1 and x_2 be numbers such that $a < x_1 < x_2 < b$ and show that $f(x_1) > f(x_2)$. As in the proof of Theorem 1(i),

$$\frac{f(x_2) - f(x_1)}{x_2 - x_1} = f'(c), \qquad x_1 < c < x_2.$$

Since $f'(c) < 0$ and $x_2 - x_1 > 0$, $f(x_2) - f(x_1) < 0$ and hence $f(x_1) > f(x_2)$.
7. By Theorem 10, $G(x) - F(x) = C$, where C is a constant, since two functions with equal derivatives on (a, b) differ by a constant on (a, b). Therefore, $G(x) = F(x) + C$. If $F(x) = x^2 + 3$ and $G(x) = x^2 + 7$, then $C = 4$.

12.1 Page 510

1. $\frac{x^4}{4} + C$. 3. $\frac{x^3}{3} - \frac{3x^2}{2} + C$. 5. $-\frac{1}{t} + C$.
7. $\frac{x^{\pi+1}}{\pi + 1} - \frac{x^3}{3} + \frac{3x^2}{2} + 8x + C$. 9. $\frac{2x^3}{3} - \frac{11x^2}{2} + 9x + C$.
11. $\frac{5x^7}{7} - \frac{2x^3}{3} + C$; to check, show that $D_x\left(\frac{5x^7}{7} - \frac{2x^3}{3} + C\right) = 5x^6 - 2x^2$.
13. $\frac{x^3}{3} + 2x - \frac{1}{x} + C$; to check, show that

$D_x\left(\frac{x^3}{3} + 2x - \frac{1}{x} + C\right) = \frac{(x^2 + 1)^2}{x^2}$. 15. Yes, because $g'(x) = f(x)$.

17. No, because $F'(x) \neq f(x)$. 19. $-\frac{2}{\sqrt{x}} - \frac{3}{2x^{2/3}} + C$.
21. $4x - \frac{8}{\sqrt{x}} - \frac{1}{2x^2} + C$. 23. $-\frac{2}{x^2} + \frac{4}{3x^3} - \frac{1}{4x^4} + C$.
25. $2y^4 - 12y^3 + 27y^2 - 27y + C$.
27. (a) $D_x(x + C) = 1 + 0 = 1$. Therefore, $\int 1 \, dx = \int dx = x + C$.
(b) $D_x[k \int f(x) \, dx] = k D_x[\int f(x) \, dx] = kf(x)$, from which Theorem 2(ii) follows.

12.2 Page 515

1. $\frac{1}{8}(3x^2 + 1)^8 + C$. 3. $\frac{1}{6}(4x + 3)^{3/2} + C$. 5. $\frac{1}{3}(x^2 + 3)^{3/2} + C$.
7. $\frac{1}{2}\sqrt{2x^2 + 5} + C$. 9. $\frac{1}{3}(1 + \sqrt{x})^6 + C$. 11. $\frac{1}{3}x^3 + \frac{2}{9}(3x)^{3/2} + C$.
13. $\frac{2}{5}x^{5/2} + C$. 15. $-\frac{1}{9(3x + 1)^3} + C$.
17. $D_x[\frac{1}{2}(2x^2 + 5)^{1/2} + C] = (\frac{1}{2})(\frac{1}{2})(2x^2 + 5)^{-1/2}(4x) = \frac{x}{\sqrt{2x^2 + 5}}$. 19. (b); $\frac{1}{3}(9 + x^2)^{3/2} + C$. 21. $y = \frac{50}{3(3x + 1)^2} + C$, provided that $0 \leq x \leq 5$.

12.3 Page 522

1. (a) 1. (b) 2. (c) 6. (d) 2. 3. (a) $y = 2x^2 + 1$. (b) $y = x^3 + 4x - 2$.
5. $y = \frac{1}{3}x^3 + x + \frac{8}{3}$. 7. $s = \frac{3}{8}(t^2 + 4)^{4/3} + 2$. 9. $C(x) = 1000x - x^2 + 2000$; $C(10) = 11{,}900$. 11. $s = -16t^2 + v_0 t$.

Answers to Odd-numbered Exercises

695

13. $\dfrac{dP}{dy} = \dfrac{(200 - 10y)(200 + 10y)}{y^2}$; $\dfrac{dP}{dy} > 0$ on $(8, 20)$ and $\dfrac{dP}{dy} < 0$
on $(20, 25)$. Therefore, maximum profit occurs at $y = \$20$.

12.5 Page 530

1. (a) $B_k = B_0 A^k$, where B_k is the state vector after k steps, $B_0 = [1\ 0\ 0]$, and $A = \begin{bmatrix} .6 & .2 & .2 \\ .3 & .6 & .1 \\ .3 & .2 & .5 \end{bmatrix}$. (b) $B_2 = [.48\ .28\ .24]$.

13.1 Page 537

1. $S_3 = \frac{137}{16}$.   3. $S_4 = \frac{239}{64}$.   5. $S_5 = \frac{153}{32}$.   7. $\frac{1}{2}, \frac{1}{2}, \frac{1}{4}, \frac{1}{2}, \frac{3}{4}, \frac{1}{2}; \frac{3}{4}$.   9. $\frac{1}{2}$.
11. (a) $\frac{71}{8}$. (b) $T_n = \frac{1}{2} \Delta x \{ f(x_0) + 2[f(x_1) + f(x_2) + \cdots + f(x_{n-1})] + f(x_n) \}$.
13. (a) $\frac{1}{2} + \frac{2}{3} + \frac{3}{4} + \frac{4}{5} + \frac{5}{6}$. (b) $\Sigma_{k=1}^{8} 5k = 5 + 10 + 15 + 20 + 25 + 30 + 35 + 40 = 5(1 + 2 + 3 + 4 + 5 + 6 + 7 + 8) = 5 \Sigma_{k=1}^{8} k$. (c) $\Sigma_{k=1}^{n} k^2$.
(d) Two possible answers are $\Sigma_{k=3}^{8} 5k$ and $\Sigma_{k=1}^{6} (10 + 5k)$. (e) $\Sigma_{k=1}^{3} f(c_k) \Delta x_k$.

13.2 Page 547

1. $\frac{26}{3}$.   3. $-\frac{4}{3}$.   5. 10.   7. $\frac{1}{9}$.
9. The hypothesis of Theorem 2 is not satisfied. The function f given by $f(x) = 1/x^2$ is not continuous at $x = 0$; therefore, it is not continuous on $[-1, 1]$.
11. Integrand: x^2; upper limit: 4; lower limit: 3; variable of integration: x.
13. $\frac{37}{3}$.   15. $\frac{49}{3}$.   17. $\frac{2}{27}$.   19. $\$(-\frac{1}{9})$.

13.3 Page 554

1. 3.   3. 4.   5. -3.   7. -4.   9. $\dfrac{\sqrt{39}}{3}$.   11. $\frac{1}{2}$.   13. $\frac{86}{3}$.
15. $\frac{44}{3}$.   17. $\dfrac{P\pi R^4}{8\eta l}$.   19. $\frac{23}{3}$.

13.4 Page 561

1. $\frac{7}{3}$.   3. $\frac{9}{2}$.   5. $\frac{9}{2}$.   7. 12.   9. $\frac{32}{3}$.   11. 4.   13. (a) $\frac{686}{15}$. (b) $\frac{80}{3}$.
15. (a) $\$533.33$. (b) $\$133.33$.

13.5 Page 568

1. Each side of the equation equals 28.   3. $\dfrac{5}{7}; \dfrac{5}{n}$.   5. 6.   7. $\frac{45}{2}$.
9. $\frac{88}{3}$.   11. $\frac{26}{3}$.

13.6 Page 575

1. w; $[1, 5]$.   3. x^2.   5. $x\sqrt{1 + x^2}$.   7. Left side is zero by Definition 2(i); right side is $F(b) - F(b) = 0$.   9. $\frac{20}{27}$; probability of death by age 80 is $\frac{20}{27}$.   11. .6.

Answers to Odd-numbered Exercises

696

14.1 Page 583

1. Domain: set of all real numbers; range: set of all positive numbers; increasing; 0.

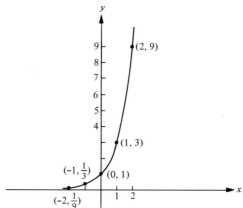

3.

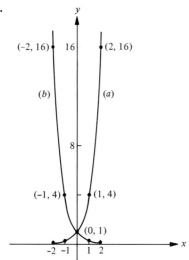

5. (a) 4. (b) −4. (c) Not possible. **7.** (a) 32. (b) 16. (c) 9. (d) 125.
9. 2.488. **11.** e^2. **13.** (a) $104.06. (b) $104.08.

15.

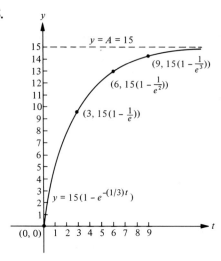

14.2 Page 590

1. (a) 3. (b) 3. (c) -2.
3. Yes, f and g are inverse functions; domain of f: set of all real numbers; range of f: set of all positive real numbers; domain of g: set of all positive real numbers; range of g: set of all real numbers.

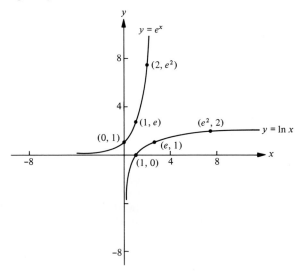

Answers to Odd-numbered Exercises

698

5.

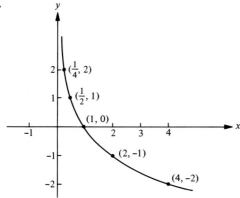

7. (a) 56. (b) $\tfrac{7}{8}$. (c) 49. (d) 5. (e) 10.   **9.** (a) 4.344. (b) .452. (c) 1.494.
(d) 3.892. (e) 3.787. (f) $-.452$. (g) .973. (h) -2.398. (i) 1.946. (j) .417. (k) 7.
(l) 49. (m) 1.232.

11. Let $x = \log_b M$; then $b^x = M$. Let $y = \log_b N$; then $b^y = N$.

$$MN = b^x b^y = b^{x+y},$$
$$\log_b MN = x + y = \log_b M + \log_b N.$$

13. (a) [*Hint:* In part (*viii*) let $M = b$; then use part (*ii*).] (b) [*Hint:* In part (*x*) let $M = b$; then use part (*ii*).]   **15.** The combination is symmetric about $y = x$.

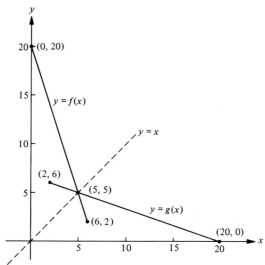

14.3 Page 596

1. (a) $\dfrac{2 \log_3 e}{2x+1}$. (b) $\dfrac{3(\log_{10} e)x^2}{x^3+2}$. (c) $\dfrac{2}{2x+5}$. (d) $\dfrac{1}{2x}$. (e) $\dfrac{8(1+\ln x)^7}{x}$.
(f) $x^3(1+4\ln x)$. (g) $2(\ln 3)3^{2x}$. (h) $2^x[1+(\ln 2)x]$. (i) $\dfrac{e^x+e^{-x}}{2}$. (j) $\dfrac{e^{\sqrt{2x+1}}}{\sqrt{2x+1}}$.

Answers to Odd-numbered Exercises

(k) $\dfrac{27e^{3x}}{(9+e^{3x})^2}$. (l) $\dfrac{e^{2x}}{\sqrt{1+e^{2x}}}$.

3. Let $y = u^n$, where $u > 0$. Then $\ln y = n \ln u$ and $\dfrac{1}{y} D_x y = \dfrac{n}{u} D_x u$. Therefore,

$$D_x y = \dfrac{ny}{u} D_x u = \dfrac{nu^n}{u} D_x u = nu^{n-1} D_x u.$$

5. (a) $\left(\dfrac{1}{\sqrt{e}}, -\dfrac{1}{2e}\right)$ minimum point. (b) $\left(\dfrac{1}{e}, -\dfrac{1}{e}\right)$ minimum point. (c) $(0, 1)$ maximum point. (d) $\left(-\dfrac{1}{\sqrt{2}}, -\sqrt{\dfrac{2}{e}}\right)$ minimum point, $\left(\dfrac{1}{\sqrt{2}}, \sqrt{\dfrac{2}{e}}\right)$ maximum point.

7. (a) 327 ounces. (b) The rate of growth is 39 ounces per month at the instant the child is 4 months old, but this rate is always changing. In fact, it can be shown that the rate of growth gradually decreases throughout the fifth month. 9. 1.39.

14.4 Page 604

1. (a) $\dfrac{1}{4 \ln 7} 7^{4x} + C$. (b) $\dfrac{1}{2 \ln 2} 2^{1+x^2} + C$ = $\dfrac{(2)^{x^2}}{\ln 2} + C$. (c) $\dfrac{e^x + e^{-x}}{2} + C$.
(d) $\dfrac{1}{2} \ln |2x + 1| + C$. (e) $\dfrac{1}{3} \ln |x^3 + 2| + C$. (f) $\dfrac{(\ln x)^2}{2} + C$.

3. (a) $3(e - 1)$. (b) 2. (c) $\tfrac{1}{11}$.   5. $x - \ln |x + 1| + C$.   7. (a) $e^2 - 1$.
(b) $e - 1$. (c) $\dfrac{2}{\ln 3} - \dfrac{3}{2}$. (d) $\tfrac{1}{2} \ln 2$. (e) $\ln 5$.

14.5 Page 611

1. $y = 900(\tfrac{5}{3})^{t/5}$.   3. 130 milligrams.   5. (a) 632 bacteria. (b) 2.95 hours.   7. $633.   9. 19 days.

14.6 Page 616

1. $\dfrac{1}{12} \ln \left|\dfrac{3 + 2x}{3 - 2x}\right| + C$.   3. Formula 24; $\tfrac{1}{4}(2x - 1) \sqrt{4x^2 - 4x + 10}$ + $\tfrac{9}{4} \ln |2x - 1 + \sqrt{4x^2 - 4x + 10}| + C$.

5. Formula 16; $\dfrac{1}{4}\left(\dfrac{3}{3 + 2x} + \ln |3 + 2x|\right) + C$.

7. Formulas 18 and 6; $\tfrac{1}{2} x e^{2x} - \tfrac{1}{4} e^{2x} + C$.
9. Formula 21; $\tfrac{1}{3} x^3 \ln 2x - \tfrac{1}{9} x^3 + C$.   11. $\tfrac{1}{2}(e^2 + 1)$.

Answers to Odd-numbered Exercises

14.7 Page 625

1. Converges; 1.

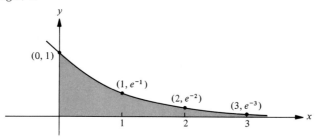

3. Converges; $\frac{1}{2}$.

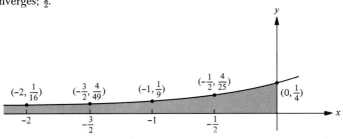

5. Diverges. **7.** Converges; $\frac{1}{18}$. **9.** Diverges. **11.** Converges; $\frac{1}{2}$.
13. Converges; 0. **15.** 0; the mean of the standard normal distribution.
17. (a) .3085. (b) 2.28 per cent. (c) 68.26 per cent.

15.1 Page 630

1. (a) 10. (b) -2.   **3.** (a) 8. (b) 10.   **5.** Those for which $y = 0$.
7.

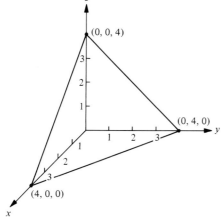

9. w, x, and y are independent variables; C is the dependent variable.

15.2 Page 636

1. $\dfrac{\partial f}{\partial x} = 3x^2 + 3y^2$; $\dfrac{\partial f}{\partial y} = 6xy + 3y^2$.   **3.** $\dfrac{\partial z}{\partial x} = 2x + ye^{xy}$; $\dfrac{\partial z}{\partial y} = xe^{xy}$.

Answers to Odd-numbered Exercises

5. $f_x(2, 1) = 17$; $f_y(2, 1) = 18$.

7. $\dfrac{\partial z}{\partial x} = yw^2 e^{xyw^2} + \dfrac{1}{w+y}$; $\dfrac{\partial z}{\partial y} = xw^2 e^{xyw^2} - \dfrac{x}{(w+y)^2}$.

9. In the plane $x = 2$, the slope of the tangent line to the curve of intersection of $z = 9 - x^2 - y^2$ and the plane $x = 2$ at $(2, 1, 4)$ is $f_y(2, 1) = -2$.
11. Each side of the equation equals $9x^2y^2 + 4x$.   **13.** 4; 0.
15. (a) $224(w + 2x + 3y)^6$. (b) $168(w + 2x + 3y)^6$. (c) $112(w + 2x + 3y)^6$.
17. $\dfrac{\partial P}{\partial w} = 2 + x$; $\dfrac{\partial P}{\partial x} = 3 + w$; $\dfrac{\partial P}{\partial y} = 4$.

15.3 Page 643

1. $(10, 8, 600)$ neither.   **3.** $(1, 2, 5)$ minimum point.   **5.** $(0, 0, \frac{31}{16})$ minimum point; $(-\frac{1}{2}, -\frac{1}{4}, 2)$ neither.   **7.** $x = 4$, $y = 1$.
9. $(w, x, y, z) = (-2, 4, 6, 50)$ is the only critical point and it is a maximum point.
11. $y = -\frac{42}{37}x - \frac{51}{37}$.

15.4 Page 650

1. Minimum $f(x, y)$ is 20 at $x = 2$, $y = 4$.   **3.** Each number is 12.
5. $w = 3$, $x = 3$, $y = 6$.   **7.** Maximum $f(x, y)$ is $10{,}000 \sqrt{10}$ at $x = 1000$, $y = 1000$.   **9.** Length: 100 feet; width: 50 feet.   **11.** $w = 6$, $x = 2$, $y = 2$.

Index

A

Abscissa, 102
Absolute value, 95–96
Absorption property, 19, 48
Acceleration, 439
 in revenue, 441
Accounting application, 249–251, 311
Addition of matrices, 221
Addition of vectors, 229
Algebra,
 of components, 34, 45
 of matrices, 219–257
 of sets, 15
Alice in Wonderland, 18
Allele, 146n., 224
Altitude of a triangle, 90
Analytic geometry, 73, 101
"And" connective, 32–34
Animal locomotion, 233
Animal navigation, 232
Antiderivative, 508
A posteriori probability, 146
A priori probability, 146
Archimedes, 532
Area between curves, 555
Area using trapezoids, 538
Argument, 50–54
Argument form, 50–54
Arithmetic mean, 365
Associative property for,
 matrix operations, 222, 237
 set operations, 20
 statement form operations, 48
Asymptotes, 492, 494
Augmented matrix, 257
Autocatalytic reaction, 491–492
Average; see Mean
Average cost, 436, 446–447, 505
Average cost function, 436, 498
Average rate of change, 380, 408, 416
Average value of a function on an
 interval, 552
Average velocity, 421

B

Bar graph, 360
Barrow, Isaac, 507
Basic feasible solution, 308
Basic solution, 307–308
Basic variables, 274, 308
Bayes theorem, 168
Biconditional statement form, 34
Bimodal, 365
Binomial expansion, 79, 212
Binomial experiment, 181, 209–216
Binomial theorem, 79
Block matrix, 247
Blocking coalition, 12
Blocks of a matrix, 247
Blood classification, 10–11
Bounded continuous measurement,
 543
Bounded growth, 582, 602
Break-even chart, 112
Break-even point, 112
Bridge hand probabilities, 200, 207

C

Calculus, 379
Carroll, Lewis, 31
Cartesian coordinate system, 101, 134
Cartesian product of sets, 26–29, 187
 definition of, 27, 28
Cauchy, Augustin, 379
Central tendency, measures of, 365,
 482
Chain property, 51
Chain rule, 432
Change of sign of factor, 457
Characteristic,
 equation, 291
 value problem, 288–292, 640
 values, 289, 291, 643
 vectors, 290, 291
Circle, 90, 105
Closed interval, 95

Closed right circular cylinder, 93
Closed switch, 56
Coalitions of voting,
  blocking, 12
  losing, 12
  winning, 12
Coefficient matrix, 257
Cofactor of an entry, 282
Cofactor matrix, 286
Column value of game, 348
Column vector, 227
Combinations, 195–201
  definition of, 195
Common logarithms, 584
Common ratio, 388
Communication models, 66, 223
Commutative property for,
  matrix operations, 222
  set operations, 15
  statement form operations, 46
Complement of a set, 9
Complement property for,
  set operations, 17
  statement form operations, 47
Complementation of a set, 9
Complete solution of a system of linear equations, 274, 307
Completing the square, 84
Complex number, 88
Component of a statement form, 33
Components of a system, 56, 57
Components of a vector, 227
Composite function, 433
Compound component, 57
Compound interest, 126, 130, 382, 581, 587–588
Compound interest law, 610
Compound statement, 33
Compound statement form, 33
Compounded continuously, 581, 587–588
Computers, electronic, 69
Concave downward, 486
Concave upward, 486
Concavity, 486
Conclusion of an argument form, 50
Conditional probability, 162–168
  definition of, 164
Conditional statement form, 34

Conformable matrices for,
  addition, 221
  multiplication, 237
Conjunction, 34
Conjunction property, 51
Connective,
  "and," 32
  "if and only if," 32, 34
  "if . . . then," 32, 34
  "not," 32, 34
  "or," 32, 33
  parallel, 57
  series, 57
Constant function, 128
Constant of integration, 508
Constraints, 296, 644
  nonnegativity, 298
  structural, 298
Consumers' surplus, 532, 561
Continuity, 398, 630
Continuous at a point, 400
Continuous measurement, 131
Continuous on an interval, 402
Continuous variable, 375, 383
Continuously compounding interest, 383
Contradiction, 43
Contrapositive property, 51
Contrapositive statement form, 41
Converge, 618–621
Converse statement form, 52
Coordinate, 5
Coordinate axes, 101, 134
Coordinate planes, 134, 628
Coordinate system, 101, 134
Coordinates, 102, 134
Cost function, 112
Countable set, 131
Countably infinite set, 131
Counting techniques,
  combinations, 195–201
  for elements of a set, 20–25
  fundamental counting principal, 187
  partitions, 203
  permutations, 185–193
Crigler, John R., 376n.
Critical number, 469
Critical point, 469, 638
Cube root, 74

Cumulative frequency, 357
Cumulative frequency distribution function, 357
Cumulative frequency polygon, 360
Cumulative probability distribution function, 394–396, 422, 543, 571–573, 622
Curve, 104
Curve fitting, 641

D

Death probabilities, 148
Decimal representation, 5
Decreasing function, 461, 463
Definite integral, 532, 535
Degeneracy, $320n$.
Degree of a polynomial function, 127
Degrees of freedom, $372n$.
Demand curve, 461, 560
DeMéré, Chevalier, 146
DeMorgan's property, 20, 48
Density of water, 474
Dependent variable, 109, 628
Derivative, 380, 417
Derived function, 417, 438
Descartes, René, 73, 101
Detachment property, 51
Determinant, 282–286
  definition of, 282–284
  effect of elementary row operations on, 284
  expansion of, 284
  use in calculation of inverse matrix, 285–286
Diagonal, main, of a matrix, 220
Diagonal block matrix, 255
Diagonal matrix, 246, 249
Difference equation, 414
Differentiable function, 418
Differential, 442
Differential equation, 516, 577, 606
Differentiation, 426
Dimension of a vector, 227
Direct functional relationship, 448
Discontinuous at a point, 400
Discrete measurement, 131
Discrete variable, 375, 383
Discriminant, 88
Disjoint sets, 8, 151

Disjunction, 34
Disjunction detachment property, 51
Disjunction property, 51
Distance, 96, 103
Distance formula, 103, 136
Distributive property for,
  matrix operations, 237
  set operations, 16
  statement form operations, 46
Diverge, 618–621
Dividing point, 458
Division points, 533
Dodgson, Charles, 31
Domain, 108, 628
Dominant column (row) of a game matrix, 351
Dot product of vectors, 229
Double negation property, 51

E

e, 580
Eigenvalues, 289
Eighth Amendment analysis, 68
Elasticity of cost, 437, 446–447
Electronic computers, 69
Elementary row operations, 259
Ellipse, 449
Empty set, 3
Endpoint, 95
  maximum, 467, 472
  minimum, 467, 472
Entry of a matrix, 220
Equal matrices, 221
Equal Rights Amendment (proposed) analysis, 68
Equal sets, 7
Equal vectors, 228
Equally likely outcomes, 142
Equation of a line, 120–123
Equation of motion, 421, 520
Equation of a plane, 136
Equations, linear system of, *see* Linear equations, systems of
Equivalent,
  components, 63
  matrices, 259
  statement forms, 41
  switches, 63
Eudoxus, 532

Euler, 580
Evaluation,
   of components, 61–66
   of statement forms, 37–40
   of switches, 62
   tables, 37
Event, 113, 141
Event space, 141
Event(s),
   definition of, 113, 141
   independent, 171–175
   mutually exclusive, 151
   probability of, 114, 143
Exclusive "or," 33, 36
Existence of limit, 384, 392
Expansion of a determinant, 284
Expected value, 156–160, 545–546, 567
   definition of, 157
   of a game to a player, 330
Experiment, 113
   binomial, 181, 209–216
   stochastic, 178–181
Experiment outcome, 113, 141
Explicit functional relationship, 448
Exponential decay law, 606
Exponential function with base b, 578
Exponential growth law, 606
Exponents, 73
Extreme point, 467

F

Factorial notation, 189
Factoring, 77
Fair game, 350
Fallacy, 50
False statement, 31
Family of curves, 518, 519
Feasible set, 297
Feasible solution, 297
Fermat, Pierre, 73
Finite interval, 456
Finite set, 5
First Amendment analysis, 39
First derivative, 438
First derivative test, 469–470
First quadrant, 101
First quartile, 367

Fixed cost, 112
Fourth quadrant, 101
Fractions, 79
Frequency, 356
Frequency distribution, 357
Function, 107
   composite, 433
   constant, 128
   continuous, 400
   cost, 112
   decreasing, 461, 463
   derivative of, 380
   derived, 417, 438
   differentiable, 418
   discontinuous, 400
   domain of, 108
   exponential, 578
   graph of, 111
   increasing, 461, 463
   inverse, 586
   linear, 128
   logarithmic, 585
   objective, 300
   of more than one variable, 627
   polynomial, 127
   probability, 113, 144, 150
   profit, 112
   quadratic, 128
   range of, 108
   rational, 128
   revenue, 111
   utility, 633, 649
   value of, 108
   weighting, 113
Fundamental counting principal, 187
Fundamental theorem of integral calculus, 540

G

Galileo, Galilei, 147
Game matrix, 328
Game theory; *see* Matrix game
Gauss–Jordan method, 278–281,
   definition of, 279
Gaussian distribution, 623
General form of equation of line, 122
General solution of a differential equation, 518
General term of a sequence, 127

Genetics, 146, 224, 241, 276
Genotypes, 146, 224, 241
Geometric representation of a vector, 228
Geometric series, 388
Geometric solution of a linear programming problem, 295–305
Geometry, analytic, 73, 101
Grand Duke of Tuscany, 147
Graph of an equation, 104
Graph of a function, 111
Graunt, John, 354
Grouped data, 360
Grouped frequency distribution, 360

H

Half-life, 607
Half-open interval, 95
Heating degree days, 553
Higher derivatives, 438
Histogram, 360
Homogeneous system of linear equations, 279, 289, 291
Horizontal asymptote, 495
Horizontal line, 101
Hypotenuse, 90

I

Identity element for matrix addition, 245
Identity element for matrix multiplication, 245
Identity matrix, 245, 249
Identity property for,
  matrix operations, 245
  set operations, 16
  statement form operations, 46
Image, 107–108
Imaginary number, 88
Implication, 42
Implicit differentiation, 449
Implicit functional relationship, 448
Impossible event, 113n.
Improper integral, 617
  converges, 618–621
  diverges, 618–621
Increasing function, 461, 463

Increment, 102
Independent events, 171–175
  definition of, 171
Independent trials, 209
Independent variable, 109, 628
Index of summation, 361, 539
Inequality, 94
  symbol(s), 94
Inequality of matrices, 313
Infinite interval, 456
Infinite series, 387
Infinite set, 5
Inflection point, 487
Initial conditions, 516, 518
Instance of a statement form, 32
Instantaneous rate of change, 380, 410, 416
Instantaneous velocity, 421
Integer programming, 303
Integers,
  negative, 5
  positive, 5
  zero, 5
Integrable, 536
Integral sign, 508
Integral with a variable upper limit, 568, 590
Integrand, 508, 541
Integration, 507, 508
Integration formulas, table of, 612, 656–657
Intercept form, 122
Intercepts, 105
Interchange of rows of a matrix, 259
International Morse code, 193
Interpolation, 369
Interquartile range, 371
Interior point, 95, 637
Intersection of sets, 7
Interval,
  closed, 95
  endpoints of, 95
  finite, 456
  half-open, 95
  infinite, 456
  interior point of, 95
  open, 95
Inverse element, 265
Inverse functions, 586
Inverse matrix, 265

Invertible matrix, 267
Irrational number, 5
Isosceles triangle, 90

L

Lagrange, Joseph Louis, 627, 644
Lagrange multiplier, 644
  method, 627, 644, 647
Latency period, 215
Learning curve, 582
Least squares method, 262, 641
Legal analysis, 11, 34, 39, 59, 68–69
Legs of a right triangle, 90
Leibniz, Gottfried, 381, 407
Length of finite interval, 456
Length of segment, 102
Length of a vector, 230
Leontief input–output model, 261, 269, 528
Leontief, Wassily W., 261
Limb action, 232
Limb traction, 232
Limit, 382, 384, 392, 630
  exists, 384, 392
Line, 118
Line chart, 357
Line of regression, 641
Linear equation, 83
Linear equations, systems of, 256–281
  homogeneous, 279, 289, 291
  solution of, 256
Linear function, 128
Linear inequality, 97, 297, 306
Linear programming, 295–327, 648
Linear programming problem,
  constraints of, 296
    nonnegativity, 298
    structural, 298
  feasible set of, 297
  matrix representation of, 313–314
  methods of solution of,
    algebraic, 305–327
    geometric, 295–305
    simplex algorithm of, 315–327
  objective equation of, 300
  objective function of, 300
  solutions of,
    basic, 307, 308
    basic feasible, 308

Linear programming problem [*cont.*]
  feasible, 297
  optimal, 308
  variables of,
    basic, 274, 308
    nonbasic, 308
    slack, 306
Logarithm of a to base b, 584
Logarithmic differentiation, 593
Logarithmic function with base b, 585
Logic, 31–72
Logical puzzle, 44
Losing coalition, 12
Lower limit of a definite integral, 541
Lower quartile, 367

M

Magnitude of a vector, 230
Main diagonal of a matrix, 220
Marginal cost, 420, 444
Marginal productivity,
  of labor, 636
  of machines, 636
Marginal profit, 420, 445
Marginal revenue, 420, 445
Marginal utilities, 633
Markov chain, 181, 240
  states of, 240
Mathematical model, 55, 65, 223, 261, 276, 301, 325, 328, 379, 523–531, 595, 603, 614
Matrix,
  augmented, 257
  block, 247
  characteristic value of, 289, 291
  characteristic vector of, 290, 291
  coefficient, 257
  cofactor, 286
  definition of, 220
  determinant of, 282–286
  diagonal, 246, 249
  diagonal block, 255
  entry of, 220
  equality, 221
  game, 328
  identity, 245, 249
  inverse, 265
  invertible, 267

Matrix [cont.]
  noninvertible, 267
  order of, 220
  partitioned, 247
  payoff, 328
  reduced, 278
  scalar, 246, 249
  skew-symmetric, 248, 249
  square, 220
  stochastic, 239
  symmetric, 248, 249
  transition, 224, 240, 253, 280
  transpose of, 247
  triangular, 246, 249
  triangular block, 256
  zero, 245, 249
Matrix game, 328–353
  C's best pure strategy of, 337
  column value of, 348
  expected value for players of, 330
  fair, 350
  mixed strategy for, 335
  optimal strategy for, 331
  payoffs of, 329
  play of, 329
  pure strategy for, 335
  R's best pure strategy for, 336
  row value of, 343
  strategy for, 329
  strictly determined, 337
  two-person, 328
  value of, 349
  zero-sum, 329
Matrix product, 235
Matrix representation of,
  a game, 329
  influence diagram, 222
  linear programming problem, 313–314
  Markov chain, 240
  stochastic experiment, 239
  transition diagram, 224
Matrix sum, 221
Maxima, 467, 637
Maximum point, 467, 638
Mean, 365–366, 545–546, 567, 622
Mean value theorem,
  for derivatives, 501–502
  for integrals, 550

Measurement, 115, 158, 355
  continuous, 131
  discrete, 131
Measures of central tendency, 365, 482
Median, 366–367, 574
Median of a triangle, 90
Method of Lagrange multipliers, 627, 644, 647
Method of least squares, 262, 641
Midpoint of an interval, 360
Minima, 467, 637
Minimum point, 467, 638
Minor of an entry, 282
Mixed strategy, 335
Mode, 365, 482
Models, see Mathematical models
Multiplication of,
  matrices, 235
  matrix by real number, 222
  sets (Cartesian product), 26–28
  vector by real number, 229
  vectors (dot product), 229
Multiplicative inverse,
  of a matrix, 265
  of a real number, 265
Mutually exclusive events, 151

N

Napier, John, 577
Natural logarithms, 584
Natural numbers, 5
Negation of a component, 57
Negation of a statement form, 34
Negative integers, 5
Negative number, 94
Newton, Isaac, 381, 407, 507
Newton's method, 428–431
Nonbasic variables, 308
Noninvertible matrix, 267
Nonnegative number, 94
Nonnegativity constraints, 298
Nonpositive number, 94
Nonvertical line, 118
Norm of a partition, 534
Normal curve, 375
Normal distribution, 375, 623
Normal probability density function, 623

"Not" connective, 34
Null set, 3
Number of elements in a set, 20–25
Numbers,
 complex, 88
 imaginary, 88
 irrational, 5
 natural, 5
 negative, 94
 nonnegative, 94
 nonpositive, 94
 positive, 94
 rational, 5
 real, 4, 94

O

Objective equation, 300
Objective function, 300
Observation, 355
Odds, 158
Open interval, 95
Open right circular cylinder, 93
Open switch, 57
Operations, elementary row, 259
Optimal lot-size formula, 481
Optimal solution, 308
Optimal strategy, 331
Optimal useful life, 490
Optimization problems, 455, 627, 648
Optimum, 455
"Or" connective,
 exclusive, 33, 36
 inclusive, 33
Order of a differential equation, 516
Order of a matrix, 220
Ordered pair, 26–27, 108n.
Ordered partition, 203
Ordinate, 102
Origin, 4, 101
Outcomes of an experiment, 113, 141
 equally likely, 142
 weights of, 114

P

Pair, ordered, 26, 108n.
Parabola, 128, 492
Parallel connection of switches, 57
Parallel system, 55, 174

Parallelepiped, 93
Parameter, 274, 308
Partial derivative, 631
Particular solution, 274
Particular solution of a differential
 equation, 518
Partition of an interval, 533
Partition of a set, 203–207
 ordered, 203
 unordered, 203
Partitioned matrix, 247
Pascal, Blaise, 146
Path of a tree diagram, 163
Payoff matrix, 328
Penny-matching game, 328
Percentile, 367, 574
Perimeter, 90–91
Permutations, 185–193
 definition of, 189
Petty, Sir William, 354
Pivot column, 320
Pivot entry, 320
Pivot row, 320
Play of game, 329
Point, 4, 102
 of inflection, 487
Point-slope form, 121
Poker, 198–200, 208
 probabilities of certain hands of,
  199
Polynomial function of degree n, 127
Population, 355
Positive integers, 5
Positive number, 94
Power rule,
 for differentiation, 426
 for integration, 511–512
Powers of a matrix, 238
Premise of an argument form, 50
Present value, 130
Probability, 113, 141–218
 a posteriori, 146
 a priori, 146
 conditional, 162–168
 of an event, 114, 143, 144
Probability density function, 424, 482,
 543–545, 571–573, 622
Probability of an event, 114, 143, 144
Probability vector, 239
Producers' surplus, 532, 561

Product,
    Cartesian, 26–28
    dot, 229
    matrix, 235
Product rule, 432
Profit function, 112
Proper subset, 3
Pure competition, 560
Pure strategy, 335
Pythagorean theorem, 90, 103, 135

Q

Quadrant, 101
Quadratic equation, 84
Quadratic formula, 85
Quadratic function, 128
Quartile, 367
Quetelet, L. A. J., 354
Quotient rule, 432

R

Random sampling, 355
Random variable, 115, 158, 355
Range, 108, 629
Range of a sample, 371
Rate of change, 416
    of slope, 439
Rational function, 128
Rational number, 5
Real line, 4, 94
Real numbers, 4, 94
Recessive column (row) of a game matrix, 351
Rectangle, 90
Rectangular (or Cartesian) coordinate system, 27, 101, 134
Reduced matrix, 278
Reduction formula, 614
Relative complement of B in A, 14
Relative maximum point, 468, 628, 638
Relative minimum point, 468, 638
Reliability of a system, 58, 174
Repeated trials, 178, 209, 240, 280
Residuals, 262
Revenue function, 111
Reverse the sense of an inequality, 97
Right circular cone, 93

Right circular cylinder, 92
Right triangle, 90
Rolle's theorem, 501–502
Root of an equation, 85
Root of a number, 74
Roulette, 156
Row equivalent matrices, 259
Row of a matrix, 220
Row value of a game, 343
Row vector, 227

S

Saddle point, 630
Saddle surface, 629
Sample, 213–214, 355
Sample data, 355
Sample space, 113, 141
    of binomial experiment, 210–211
    of equally likely outcomes, 142
    of nonequally likely outcomes, 143
    of stochastic experiment, 178–180
Scalar matrix, 246, 249
Secant line, 408, 502
Second derivative, 438
    test, 489
Second quadrant, 101
Semicircle, 90
Semilogarithmic paper, 589
Separate the variables, 517
Sequence, 126
    general term of, 127
    nth term of, 127
    of partial sums, 388
    terms of, 126
Series; *see* Infinite series
Series connection, 57
Series system, 55, 174
Set(s), 1–30
    complement of, 9, 153
    disjoint, 8, 151
    element of, 1
    empty, 3
    equality, 7
    feasible, 297
    finite, 5
    infinite, 5
    intersection of, 7
    null, 3
    proper subset of, 3

Index

Set(s) [cont.]
  subset of, 2
  subtraction of, 14
  unbounded feasible, 301
  union of, 8
  universal, 3
  Venn diagram of, 3
Sign of factor, 457
Similar triangles, 90
Simple random sampling, 355n.
Simple statement, 33
Simple statement form, 33
Simplex method, 315–327
Skew-symmetric matrix, 248, 249
Slack variables, 306
Slope of a line, 118
Slope-intercept form, 122
Solution,
  of an algebraic equation, 83
  basic, 307–308
  basic feasible, 308
  complete, 274, 307
  of a differential equation, 516
  feasible, 297
  of an inequality, 97
  optimal, 308
  particular, 274
  of a system of linear equations, 256
Solution set of an equation, 83
Sound argument, 53
Space,
  event, 141
  sample, 113, 141
Special product, 79
Specific heat, 422
Specific weight of water, 474
Sphere, 92
Square matrix, 220
Square root, 74
Standard deviation, 372, 547, 623
Standard normal distribution, 625
Standardized normal probability
  density function, 623
Straight line, 118
State vector, 240, 280
Statement, 31
Statement form(s), 31–35
  biconditional, 34
  compound, 33
  conditional, 34

Statement form(s) [cont.]
  conjunction, 34
  connectives, 31–35
  contradiction, 43
  contrapositive, 41
  converse, 52
  definition of, 32
  disjunction, 34
  equivalent, 41
  evaluation of, 37–40
  implication, 42
  instance of, 32
  negation of, 34
  simple, 33
  tautology, 42
  value of, 37
State vector, 240
States of a Markov chain, 240
Statistics, 354
Steady state vector, 280
Stimulus-object-response model,
  251–253
Stochastic experiment, 178–181
Stochastic matrix, 239
Strategy,
  column player's best pure, 337
  definition of, 329
  mixed, 335
  optimal, 331
  pure, 335
  row player's best pure, 336
Strictly determined games, 337
Structural constraints, 298
Submatrix, 247, 249
Subset, 2
Subtraction,
  of matrices, 222
  of sets, 14
  of vectors, 229
Sum of an infinite series, 388
Summation notation, 360, 539
Supply curve, 462, 560
Surface, 628
Switch(es),
  closed, 56
  connectives of, 57
  equivalent, 63
  evaluation of, 62
  negation of, 57
  off, 56

Switch(es) [cont.]
  on, 56
  open, 57
  parallel connection of, 57
  series connection of, 57
Symbols, table of, 34
Symmetric difference, 14
Symmetric matrix, 248, 249
Symmetric to origin, 493
Symmetric to x-axis, 492
Symmetric to y-axis, 492
Symmetry, 492
System(s),
  of components, 55–72
  definition of, 55
  parallel, 55
  reliability of, 58, 174
  series, 55
  of switches, 56
System of linear equations; *see* Linear equations, system of

T
Table(s),
  evaluation, 37, 62
  life expectancy, 148
  of operation symbols, 34
  poker hand probabilities, 199
Table of integration formulas, 612, 656–657
Tangent line, 411
Tautology, 42
Tenth Amendment analysis, 11
Terms of a sequence, 126
Third quadrant, 101
Third quartile, 367
Three-dimensional space, 134
Transition diagram, 224
Transition matrix, 224, 240, 253, 280
Transpose of a matrix, 247
Trapezoid, 90
Trapezoid method of approximating area, 538
Tree diagrams, 163
Trial failure, 209
Trial outcome, 178, 209
Trial success, 209
Trials, repeated,
  in binomial experiments, 209

Trials, repeated [cont.]
  in stochastic experiments, 178, 240, 280
Triangle, 90
Triangular block matrix, 256
Triangular matrix, 246, 249
  determinant of, 285
True statement, 31, 37
2 by 2 matrix game, 328
Two-person game, 328

U
Unbounded feasible set, 301
Ungrouped data, 360
Union of sets, 8
Unique solution, 256, 272, 278
Universal set, 3
Unordered partitions, 203
Unsound argument, 53
Upper limit of a definite integral, 541
Upper quartile, 367
Utility function, 633, 649

V
Valid argument form, 50
Value of,
  component, 61–62
  game, 349
  statement form, 37
Variable, 109
  basic, 274, 308
  nonbasic, 308
  of integration, 541
  slack, 306
Variance, 372, 546, 623
Vector(s), 227–234
  addition of, 229
  column, 227
  components of, 227
  dimension, 227
  dot product of, 229
  equality of, 228
  geometric representation of, 228
  length of, 230
  magnitude of, 230
  multiplication of, 229
  multiplication by a real number, 229

Index

Vector(s) [cont.]
   nonzero, 227
   probability, 239
   row, 227
   state, 240
   subtraction of, 229
   sum of, 229
   zero, 227
Vector of cofactors, 283
Velocity, 421
   in revenue, 441
Venn diagram, 3
Vertical asymptote, 494
Vertical line, 101
Von Neumann, John, 332
Voting coalitions; *see* Coalitions, of voting
Voting device, 63
Voting power, 192

W

Weight of an outcome, 114, 144
Weighting function, 114
Winning coalition, 12

X

x-axis, 101, 134
x-coordinate, 102
x-intercept, 105
xy-plane, 134
xz-plane, 134

Y

y-axis, 101, 134
y-coordinate, 102
y-intercept, 105
yz-plane, 134

Z

z-axis, 134
Zero, 5
Zero matrix, 245, 249
Zero-sum game, 329
Zero vector, 227